Enterohemorrhagic
Escherichia coli
and Other Shiga Toxin-Producing *E. coli*

Enterohemorrhagic
Escherichia coli
and Other Shiga Toxin-Producing *E. coli*

EDITED BY

Vanessa Sperandio
Department of Microbiology
University of Texas
Dallas, TX 75390

and

Carolyn J. Hovde
School of Food Science
Idaho INBRE Program
Moscow, ID 83844-3025

ASM
PRESS

Washington, DC

Library of Congress Cataloging-in-Publication Data

Enterohemorrhagic Escherichia coli and other shiga toxin-producing E. coli / edited by Vanessa Sperandio, Department of Microbiology, University of Texas, Dallas, TX, [and] Carolyn J. Hovde, School of Food Science, Idaho INBRE Program, Moscow, ID.
pages cm
Includes index.
ISBN 978-1-55581-878-4 (print) -- ISBN 978-1-55581-879-1 (electronic) 1. Escherichia coli infections. 2. Escherichia coli O157:H7. 3. Verocytotoxins. 4. Hemolytic-uremic syndrome. I. Sperandio, Vanessa, editor. II. Hovde, Carolyn J., editor.
QR201.E82E58 2015
616.9'26--dc23
2015003716
doi:10.1128/9781555818791

Printed in the United States of America

10 9 8 7 6 5 4 3 2 1
Address editorial correspondence to: ASM Press, 1752 N St., N.W., Washington, DC 20036-2904, USA.
Send orders to: ASM Press, P.O. Box 605, Herndon, VA 20172, USA.
Phone: 800-546-2416; 703-661-1593. Fax: 703-661-1501.
E-mail: books@asmusa.org
Online: http://www.asmscience.org

Contents

Contributors ix
Preface xv

OVERVIEW

1 **Overview and Historical Perspectives** 3
   James B. Kaper and Alison D. O'Brien

MICROBIOLOGY

2 **Taxonomy Meets Public Health: The Case of Shiga Toxin-Producing**
   Escherichia coli 17
   Flemming Scheutz

3 **Shiga Toxin (Stx) Classification, Structure, and Function** 37
   Angela R. Melton-Celsa

4 **Enterohemorrhagic *Escherichia coli* Genomics: Past, Present,**
   and Future 55
   Shah M. Sadiq, Tracy H. Hazen, David A. Rasko, and Mark Eppinger

PATHOGENESIS

5 **Role of Shiga/Vero Toxins in Pathogenesis** 75
   Fumiko Obata and Thomas Obrig

6 **The Locus of Enterocyte Effacement and Associated Virulence Factors**
   of Enterohemorrhagic *Escherichia coli* 97
   Mark P. Stevens and Gad M. Frankel

7 **Enterohemorrhagic *Escherichia coli* Adhesins** 131
   Brian D. McWilliams and Alfredo G. Torres

8 **Animal Models of Enterohemorrhagic *Escherichia coli* Infection** 157
   Jennifer M. Ritchie

9 **Enterohemorrhagic *Escherichia coli* Virulence Gene Regulation** 175
   Jay L. Mellies and Emily Lorenzen

INCIDENCE, EPIDEMIOLOGY, AND ECOLOGY

10 **Shiga Toxin (Verotoxin)-Producing *Escherichia coli* in Japan** 199
    Jun Terajima, Sunao Iyoda, Makoto Ohnishi, and Haruo Watanabe

11   **Animal Reservoirs of Shiga Toxin-Producing** *Escherichia coli*   **211**
Anil K. Persad and Jeffrey T. LeJeune

12   **Shiga Toxin-Producing** *Escherichia coli* **in Fresh Produce: A Food Safety Dilemma**   **231**
Peter Feng

13   **Public Health Microbiology of Shiga Toxin-Producing** *Escherichia coli*   **245**
Alfredo Caprioli, Gaia Scavia, and Stefano Morabito

DIAGNOSIS, DETECTION, AND STRAIN CHARACTERIZATION

14   **Detection of Shiga Toxin-Producing** *Escherichia coli* **from Nonhuman Sources and Strain Typing**   **263**
Lothar Beutin and Patrick Fach

CLINICAL, PATHOLOGICAL, AND PATHOPHYSIOLOGICAL ASPECTS

15   **Shiga Toxin/Verocytotoxin-Producing** *Escherichia coli* **Infections: Practical Clinical Perspectives**   **299**
T. Keefe Davis, Nicole C. A. J. van de Kar, and Phillip I. Tarr

16   **The Inflammatory Response during Enterohemorrhagic** *Escherichia coli* **Infection**   **321**
Jaclyn S. Pearson and Elizabeth L. Hartland

17   **New Therapeutic Developments against Shiga Toxin-Producing** *Escherichia coli*   **341**
Angela R. Melton-Celsa and Alison D. O'Brien

HOST DETERMINANTS OF DISEASE AND HOST RESPONSE

18   **Risk Factors for Shiga Toxin-Producing** *Escherichia coli*-**Associated Human Diseases**   **361**
Marta Rivas, Isabel Chinen, Elizabeth Miliwebsky, and Marcelo Masana

19   **Enterohemorrhagic** *Escherichia coli* **Pathogenesis and the Host Response**   **381**
Diana Karpman and Anne-lie Ståhl

20   **The Interplay between the Microbiota and Enterohemorrhagic** *Escherichia coli*   **403**
Reed Pifer and Vanessa Sperandio

PREVENTION AND CONTROL STRATEGIES

21.   **"Preharvest" Food Safety for** *Escherichia coli* **O157 and Other Pathogenic Shiga Toxin-Producing Strains**   **421**
Thomas E. Besser, Carrie E. Schmidt, Devendra H. Shah, and Smriti Shringi

22 Peri- and Postharvest Factors in the Control of Shiga Toxin-Producing
    Escherichia coli in Beef   437
    Rodney A. Moxley and Gary R. Acuff

23 Veterinary Public Health Approach to Managing Pathogenic
    Verocytotoxigenic *Escherichia coli* in the Agri-Food Chain   457
    Geraldine Duffy and Evonne McCabe

24 Clinical Studies of *Escherichia coli* O157:H7 Conjugate Vaccines in
    Adults and Young Children   477
    Shousun Chen Szu and Amina Ahmed

25 Vaccination of Cattle against *Escherichia coli* O157:H7   487
    David R. Smith

ESCHERICHIA COLI O104:H4

26 *Escherichia coli* O104:H4 Pathogenesis: An Enteroaggregative
    E. coli/Shiga Toxin-Producing *E. coli* Explosive Cocktail of
    High Virulence   505
    Fernando Navarro-Garcia

THE WAY FORWARD

27 The Way Forward   533
    Vanessa Sperandio

Index   541
About the Editors   553

Contributors

Gary R. Acuff
Department of Animal Science, 2471 TAMU, Texas A&M University,
College Station, TX 77843-2471

Amina Ahmed
Levine Children's Specialty Center - Pediatric Infectious Disease,
Carolina Medical Centers, Charlotte, NC 28203

Thomas E. Besser
Veterinary Microbiology and Pathology, Washington State University, Pullman,
WA 99164

Lothar Beutin
National Reference Laboratory for Escherichia coli, Department of Biological
Safety, Federal Institute for Risk Assessment (BfR), Diedersdorfer Weg 1,
D-12277 Berlin, Germany

Alfredo Caprioli
EU Reference Laboratory for E. coli, Dipartimento di Sanità Pubblica
Veterinaria e Sicurezza Alimentare, Istituto Superiore di Sanità,
Viale Regina Elena 299, 00161 Rome, Italy

Isabel Chinen
Servicio Fisiopatogenia, Instituto Nacional de Enfermedades Infecciosas –
ANLIS "Dr. C. G. Malbrán," (1281) Buenos Aires, Argentina

T. Keefe Davis
Division of Nephrology, Department of Pediatrics, Washington University
School of Medicine, St. Louis, MO 63110

Geraldine Duffy
Teagasc Food Research Centre, Ashtown, Dublin 15, Ireland

Mark Eppinger
Department of Biology and South Texas Center for Emerging Infectious
Diseases, University of Texas at San Antonio, San Antonio, TX 78249

Patrick Fach
Food Safety Laboratory, ANSES (French Agency for Food, Environmental
and Occupational Health and Safety), Fr-94706 Maisons-Alfort, France

Peter Feng
Division of Microbiology, U.S. Food and Drug Administration,
College Park, MD 20740-3835

Gad M. Frankel
MRC Centre for Molecular Bacteriology & Infection, Department of Life
Sciences, Imperial College London, London, SW7 2AZ, United Kingdom

Elizabeth L. Hartland
Department of Microbiology and Immunology, University of Melbourne,
Victoria 3010, and Murdoch Children's Research Institute, Royal Children's
Hospital, Parkville, Victoria 3052, Australia

Tracy H. Hazen
Institute for Genome Sciences, Department of Microbiology and Immunology,
University of Maryland School of Medicine, Baltimore, MD 21201

Sunao Iyoda
Department of Bacteriology, National Institute of Infectious Diseases, 1-23-1
Toyama, Shinjuku-ku, Tokyo 162-8640, Japan

James B. Kaper
Department of Microbiology & Immunology, University of Maryland School of
Medicine, Baltimore, MD 21122

Diana Karpman
Department of Pediatrics, Clinical Sciences, Lund University, 22185 Lund,
Sweden

Jeffrey T. LeJeune
Food Animal Health Research Program, Ohio Agricultural Research and
Development Center, The Ohio State University, Wooster, OH 4491

Emily Lorenzen
Laboratory of Chemical Biology and Signal Transduction, The Rockefeller
University, 1230 York Avenue, New York, NY 10065

Marcelo Masana
Instituto Tecnología de Alimentos, Centro de Investigación de Agroindustria,
Instituto Nacional de Tecnología Agropecuaria, INTA, (B1708WAB) Morón,
Pcia. De Buenos Aires, Argentina

Evonne McCabe
Teagasc Food Research Centre, Ashtown, Dublin 15, Ireland

Brian D. McWilliams
Department of Microbiology and Immunology, University of Texas
Medical Branch, Galveston, TX 77555

Jay L. Mellies
Department of Biology, Reed College, 3203 SE Woodstock Blvd.,
Portland, OR 97202

Angela R. Melton-Celsa
Department of Microbiology & Immunology, Uniformed Services University of the Health Sciences, 4301 Jones Bridge Road, Bethesda, MD 20814

Elizabeth Miliwebsky
Servicio Fisiopatogenia, Instituto Nacional de Enfermedades Infecciosas – ANLIS "Dr. C. G. Malbrán," (1281) Buenos Aires, Argentina

Stefano Morabito
EU Reference Laboratory for E. coli, Dipartimento di Sanità Pubblica Veterinaria e Sicurezza Alimentare, Istituto Superiore di Sanità, Viale Regina Elena 299, 00161 Rome, Italy

Rodney A. Moxley
School of Veterinary Medicine and Biomedical Sciences, University of Nebraska-Lincoln, Lincoln, NE 68685-0905

Fernando Navarro-Garcia
Department of Cell Biology, Centro de Investigación y de Estudios Avanzados del IPN (CINVESTAV-IPN), México DF, Mexico

Fumiko Obata
University of Maryland School of Medicine, 685 W. Baltimore St., HSF-1 Suite 380, Baltimore, MD 21201

Alison D. O'Brien
Department of Microbiology & Immunology, Uniformed Services University of the Health Sciences, Bethesda, MD 20814

Thomas Obrig
University of Maryland School of Medicine, 685 W. Baltimore St., HSF-1 Suite 380, Baltimore, MD 21201

Makoto Ohnishi
Department of Bacteriology, National Institute of Infectious Diseases, 1-23-1 Toyama, Shinjuku-ku, Tokyo 162-8640, Japan

Jaclyn S. Pearson
Department of Microbiology and Immunology, University of Melbourne, Victoria 3010, Australia

Anil K. Persad
Food Animal Health Research Program, Ohio Agricultural Research and Development Center, The Ohio State University, Wooster, OH 4491

Reed Pifer
Department of Microbiology and Department of Biochemistry, University of Texas Southwestern Medical Center, Dallas, TX 75390

David A. Rasko
Institute for Genome Sciences, Department of Microbiology and Immunology, University of Maryland School of Medicine, Baltimore, MD 21201

Jennifer M. Ritchie
School of Biosciences and Medicine, University of Surrey, Guildford GU27XH, United Kingdom

Marta Rivas
Servicio Fisiopatogenia, Instituto Nacional de Enfermedades Infecciosas – ANLIS "Dr. C. G. Malbrán," (1281) Buenos Aires, Argentina

Shah M. Sadiq
Department of Biology and South Texas Center for Emerging Infectious Diseases, University of Texas at San Antonio, San Antonio, TX 78249

Gaia Scavia
EU Reference Laboratory for E. coli, Dipartimento di Sanità Pubblica Veterinaria e Sicurezza Alimentare, Istituto Superiore di Sanità, Viale Regina Elena 299, 00161 Rome, Italy

Flemming Scheutz
WHO Collaborating Centre for Reference and Research on Escherichia and Klebsiella, Department of Microbiology and Infection Control, Statens Serum Institut, DK-2300 Copenhagen S, Denmark

Carrie E. Schmidt
Veterinary Microbiology and Pathology, Washington State University, Pullman, WA 99164

Devendra H. Shah
Veterinary Microbiology and Pathology, Washington State University, Pullman, WA 99164

Smriti Shringi
Veterinary Microbiology and Pathology, Washington State University, Pullman, WA 99164

David R. Smith
College of Veterinary Medicine, Mississippi State University, Mississippi State, MS 39762-6100

Vanessa Sperandio
Department of Microbiology and Department of Biochemistry, University of Texas Southwestern Medical Center, Dallas, TX 75390

Anne-lie Ståhl
Department of Pediatrics, Clinical Sciences, Lund University, 22185 Lund, Sweden

Mark P. Stevens
The Roslin Institute & Royal (Dick) School of Veterinary Studies, University of Edinburgh, Midlothian, EH25 9RG, United Kingdom

Shousun Chen Szu
Eunice Kennedy Shriver National Institute of Child Health & Human Development, National Institutes of Health, 9000 Rockville Pike, Bethesda, MD 20892

Phillip I. Tarr
Division of Gastroenterology, Hepatology, and Nutrition, Department of Pediatrics, and Department of Molecular Microbiology, Washington University School of Medicine, St. Louis, MO 63110

Jun Terajima
Department of Microbiology, National Institute of Health Sciences, Kamiyoga 1-18-1, Setagaya-ku, Tokyo 158-8501, Japan

Alfredo G. Torres
Department of Pathology and Sealy Center for Vaccine Development, University of Texas Medical Branch, Galveston, TX 77555

Nicole C. A. J. van de Kar
Division of Nephrology, Department of Pediatrics, Radboud University Medical Centre, Nijmegen, The Netherlands

Haruo Watanabe
Director, National Institute of Infectious Diseases, 1-23-1 Toyama, Shinjuku-ku, Tokyo 162-8640, Japan

Preface

This book is an exceptional compilation of our current worldwide understanding of the enterohemorrhagic *E. coli* (EHEC) and other Shiga toxin-producing *E. coli*. It spans diverse topics including microbial pathogenesis, pathophysiology of the disease, food safety, genetic analysis, veterinary microbiology, epidemiology, and environmental microbiology. It was compiled as an introduction, review, and critical overview of the pertinent areas of knowledge and brings the previous edition (Kaper JB, O'Brien AD [ed], Escherichia *coli O157:H7 and Other Shiga Toxin-Producing* E. coli *Strains*, ASM Press, 1998) up to date with the current literature.

The style and content are intended to make this volume of interest and value as a resource for research scientists, clinicians, students, health professionals, policy makers, and those in industry. In addition, we believe the text could be used for advanced courses in microbiology, food safety, infectious disease, or microbial pathogenesis.

The book contributors come from many and diverse research disciplines. Its breadth demonstrates the complexity of the problem of EHEC and Shiga toxin-producing *E. coli*. For rapid and timely dissemination, each chapter previously appeared in *Microbiology Spectrum* and is available online, but they are assembled here as a convenient hardbound reference volume. The book begins with a broad overview and historical perspective, followed by eight sections that organize information into subtopics. The text concludes with the traditional "look to the future" in chapter 27, titled "The Way Forward," which was not previously published.

We gratefully acknowledge the outstanding skill and organizational work of all those at ASM Press who made this book possible. It was a pleasure working with Greg Payne, Ellie Tupper, Kenneth April, Courtenay Brown, and Cathy Balogh.

Both of us have devoted our professional careers to understanding the EHEC in hopes of contributing to effective interventions to improve human health. As editors, we are humbled by the exceptional work done by our colleagues, the chapter authors. They brought their full and thoughtful expertise and knowledge to their writing and fueled our excitement for the book. We hope that you, as a reader, will find the topics covered to be relevant and that their depth will bring new insights to your own work.

Vanessa Sperandio
Carolyn J. Hovde
December 2014

OVERVIEW

Overview and Historical Perspectives

JAMES B. KAPER[1] and ALISON D. O'BRIEN[2]

The scope of topics covered in this book reflects the broad areas of research required for the comprehensive study of Shiga toxin-producing *Escherichia coli* (STEC) infections. Substantial progress has been made in all of these areas since the first edition of this book (1). Although this second edition brings the field up to date in all major areas of research, these pathogens have a long and complicated history, and understanding this history is valuable for a full understanding of this field. The purpose of this chapter is to set the stage for this book by examining the seminal discoveries about STEC biology, epidemiology, and pathogenesis. In this article, we refer to the cytotoxins of *E. coli* O157:H7, *E. coli* O104:H4, and other *E. coli* as Shiga toxins (Stxs; formerly called Shiga-like toxins), hence the nomenclature STEC. However, for reasons described below, a number of investigators prefer the term verotoxin (VT). We refer the reader to past discussions of nomenclature (2, 3) for a better understanding of the historical basis for the dichotomy in nomenclature. Additionally, a recently published multicenter study by Scheutz and colleagues (4) provides clear guidance on nomenclature for Stx subtypes. Scheutz reviews that typing scheme in chapter 2 of this volume.

[1]Department of Microbiology and Immunology, University of Maryland School of Medicine, Baltimore, MD 21122; [2]Department of Microbiology and Immunology, Uniformed Services University of the Health Sciences, Bethesda, MD 20814.

Enterohemorrhagic Escherichia coli *and Other Shiga Toxin-Producing* E. coli
Edited by Vanessa Sperandio and Carolyn J. Hovde
© 2015 American Society for Microbiology, Washington, DC
doi:10.1128/microbiolspec.EHEC-0028-2014.

Our understanding of what constitutes a virulent STEC isolate for humans has evolved, as have the organisms themselves. The history of STEC as an emerging pathogen can be divided into two phases. Phase one of STEC chronology is a history of the convergence of two independent laboratory-based research tracks and two independent epidemiology-based areas of investigation. One laboratory-based track focused on Stx, its discovery, characterization, and relationship to *E. coli* cytotoxins, and the other concentrated on enteropathogenic *E. coli* (EPEC) adherence characteristics. These studies led to the realization that a subset of STEC strains, including *E. coli* O157:H7, shares with EPEC both the pathogenic trait of producing attaching and effacing (A/E) intestinal lesions and the genes to provoke that lesion. The epidemiology-based areas include the search for the cause of the well-described but idiopathic hemolytic-uremic syndrome (HUS) and the search for the agent responsible for a newly described clinical syndrome called hemorrhagic colitis.

Phase two of the STEC story started with the understanding that the backbone of the *E. coli* strain that produces Stx does not necessarily have to have adherence traits similar to those of EPEC and *E. coli* O157:H7 to cause a large-scale outbreak of human disease. Witness the large 2011 STEC outbreak in Germany in which 3,816 cases were reported (including 54 deaths); 845 of those cases led to HUS (5). The etiologic agent in that outbreak was an enteroaggregative *E. coli* O104:H4 strain that had become transduced with the Stx type 2a (Stx2a)-encoding phage (6). Enteroaggregative *E. coli* (EAEC) strains that do not make Stxs are well-established etiologic agents of diarrhea. The fact that an EAEC isolate can be converted from an exclusively diarrheagenic agent to one that also causes hemorrhagic colitis and HUS in so many patients affirms the preeminent role that Stx plays in these syndromes and illustrates how readily an *E. coli* strain can evolve into a life-threatening pathogen by the acquisition, through horizontal genetic exchange, of the Stx2a-converting phage. Recognizing the essential role Stx plays in the development of severe disease caused by STEC infection, and for chronological reasons, we begin this tale with the discovery and characterization of Stxs.

HISTORY

Stxs: History of the Field

In 1898, Kioshi Shiga (7) provided the definitive description of the agent of epidemic bacterial dysentery, *Shigella dysenteriae* type 1 (Shiga's bacillus). Five years following that discovery, Conradi (8) reported that extracts of Shiga's bacillus paralyzed and killed rabbits. Similar findings were published independently by Neisser and Shiga (9). The next nearly 70 years of Stx research led to the clear separation of the endotoxic activity associated with Shiga's bacillus from the activity of the protein Stx; the partial purification of Stx (10); the discovery that high iron concentrations inhibit Stx synthesis (11); the seminal observation by Bridgwater et al. (12) and Howard (13) that Stx appears to target vascular endothelium in the brain; and the discovery by Vicari et al. that Stx is lethal for certain epithelial cells in culture (14). Although these findings were of interest to toxicologists, none of the results proved a direct role for Stx in the pathogenesis of shigellosis. Only decades later, with the evaluation of data obtained from infection of volunteers (15) and subsequently of monkeys (16) with *S. dysenteriae* type 1, was it clear that production of Stx by the organism exacerbates the severity of the intestinal and systemic lesions in human subjects and increases the intestinal pathology in primate hosts. The ultimate proof of a role for Stx in shigellosis due to Shiga's bacillus was the establishment of a connection between production of this and related toxins with the subsequent development of HUS (see below or chapters 5 and 19).

In 1972, Keusch and colleagues made the significant finding that Stx alone caused fluid accumulation and enteritis in ligated rabbit intestinal segments (17). This observation revealed that Stx can contribute to the intestinal phase of bacillary dysentery, i.e., bloody diarrhea. That this enterotoxic activity of Stx is a result of the same molecule responsible for its cytotoxic and lethal activities was convincingly demonstrated by the purification of Stx to homogeneity (first by Olsnes and Eiklid [18], followed shortly by reports from O'Brien et al. [19], Brown and colleagues [20], and Donohue-Rolfe and coworkers [21]) and the subsequent testing of that material for all three bioactivities (19).

Stxs and Verotoxins Are Different Names for the Same Family of Toxins

With the availability of purified Stx came the capacity to produce monospecific, cytotoxin-neutralizing rabbit anti-Stx antibodies. O'Brien and colleagues used such sera to ascertain that certain strains of *E. coli* produce a cytotoxin that can be neutralized by anti-Stx (22, 23), an observation that explains the original Shiga-like toxin nomenclature. The preliminary report of that discovery (23) occurred in 1977, the same year that Konowalchuk and colleagues found that certain diarrheagenic *E. coli* strains make a cytotoxin that can kill Vero cells (24), hence the name verotoxin. In 1983, O'Brien and colleagues (25) reported that a Shiga-like toxin was produced by the *E. coli* O157:H7 strain that had caused an outbreak of hemorrhagic colitis in the United States (see below) and that this toxin was the same as the verotoxin shown by Johnson et al. (26) to be produced by *E. coli* O157:H7. Thus, 1983 became the year when the paths of research on Stxs and verotoxins merged. Subsequent genetic studies showed that Stx1 (VT1) differs by none or only a single amino acid from Shiga toxin (27, 28). These studies on the Stx/VT of *E. coli* culminated in a pivotal report published by Karmali et al. (29) in that same year. In that paper Karmali and colleagues proposed that

the verotoxin produced by these organisms was linked epidemiologically to the development of HUS (see below).

The mid to late 1980s heralded the era of the molecular characterization of the genes encoding the Stx family members (reviewed in reference 30). In the mid-1980s, it was also discovered that Stx1 and Stx2 are usually encoded on lambdoid prophages in *E. coli* (31–35). In contrast, Stx of Shiga's bacillus and Stx2e (edema disease toxin) of animal STEC were later shown to be chromosomally encoded (27, 36). Subsequent genomic analysis revealed that the genome of *E. coli* O157:H7 is riddled with prophage regions that not only encode Stx but also other potential virulence factors (see chapter 4 of this volume). The toxin genes themselves show considerable variation that can correlate with epidemiological significance, with more than 100 Stx variants so far described (see chapter 2).

Intimate Adherence to Mucosal Epithelium: The Connection between EPEC and EHEC

With the genetics and biology of Stxs fairly well elucidated by the mid to late 1980s, the focus of research on STEC broadened to address the question of how *E. coli* O157:H7 adheres to epithelial cells. The primary finding that initiated a series of discoveries about *E. coli* O157:H7 adherence mechanisms was the observation by Tzipori et al. (37) that *E. coli* O157:H7 causes intestinal A/E lesions in gnotobiotic piglets and that these lesions resemble those produced by Stx-negative EPEC, albeit at different sites in the bowel of the animals. The A/E lesion is characterized by intimate adherence of the bacteria to the enterocyte membrane and effacement of the microvilli. This observation suggested to Levine (38) that these lesions might be a hallmark of *E. coli* O157:H7 and related bacteria. He proposed that the capacity of *E. coli* O157:H7 and related organisms to evoke A/E lesions, together with the production of Stxs by these microbes and the presence of a characteristic large plasmid, was sufficient to define a new category of virulent "enterohemorrhagic" *E. coli* (EHEC). With the identification of pathogenic and genetic markers for this newly recognized group of *E. coli* came the realization that the *E. coli* O26:H11 serotype, which had long been considered a classic EPEC serotype, should be reclassified as an EHEC. This reclassification was supported by the findings that strains of the O26:H11 serotype also produced Stx and possessed the large plasmid found in O157:H7 (39). Thus, the O26:H11 serotype is one example of an STEC serotype that had been associated with diarrheal disease (reviewed in reference 40) long before the discovery of *E. coli* O157:H7.

The pathognomonic A/E histopathology of EPEC, *E. coli* O157:H7, and a few additional serotypes of STEC was subsequently shown by Knutton et al. (41) to correspond *in vitro* to a lesion characterized by bacterial microcolonies intimately adherent to the surface of tissue culture cells, with accumulation of cytoskeletal actin under the bacteria. The actin accumulation was visualized in that study by a fluorescent actin stain (FAS) test in which fluoresceinated phalloidin was used as a probe. The FAS-positive phenotype of EPEC bound to HEp-2 laryngeal epithelial cells was used by Jerse et al. (42) to screen EPEC for genes required for this intimate adherence. These investigators identified the gene *eae* (for *E. coli* attaching and effacing), which was also present in O157:H7. The *eae* gene product, appropriately named intimin, was subsequently shown to be required for EPEC to cause A/E lesions in gnotobiotic pigs (43). Shortly after the discovery of EPEC *eae*, the homologous gene was cloned and sequenced from two strains of *E. coli* O157:H7 (44, 45). The intimin of *E. coli* O157:H7 was also shown to be necessary but not sufficient to induce A/E lesions *in vitro* and *in vivo* (46, 47).

An additional twist to the similarities in pathogenic mechanisms between EPEC and EHEC was the provocative finding of McDaniel et al. (48) that *eae* lies within a pathogenicity island of approximately 35 kb and that this island encodes genes for attachment, FAS reactivity, a type III secretion apparatus, and secreted proteins (48). The island was named locus of enterocyte effacement (LEE) (48). In chapter 6 of this volume, Stevens and Frankel review the LEE pathogenicity island, intimin, the type III secretion system, and other associated virulence factors of EHEC.

The discovery and characterization of LEE prompted the realization that the term EHEC represents a subset of STEC since not all STEC strains contain LEE and the large EHEC plasmid (see chapter 2). Although the majority of STEC strains associated with human disease possess LEE and the large plasmid, some strains lacking these factors, most notably O104:H4, have been implicated in human disease, thereby leading to the use of the more general term STEC rather than EHEC.

Hemolytic-Uremic Syndrome

HUS, first described in 1955 by Gasser et al. (49) in Switzerland, is defined by a triad of clinical features that include acute renal failure, thrombocytopenia, and microangiopathic hemolytic anemia. HUS is a leading cause of acute renal failure in children, and in some studies it is the most common cause of renal failure in this age group. A variety of agents, including drugs, chemicals, toxins, and various microbes, had been proposed as the cause of HUS; indeed, before 1983, most nephrologists thought that HUS was a multifactorial disease that could result from a number of initiating events (50). Because HUS occasionally occurred in outbreaks, an infectious cause was sought. The strongest documented linkage between HUS and a microorganism was the association with *S. dysenteriae* type 1, but numerous microorganisms, including *Salmonella typhi, Campylobacter jejuni, Yersinia pseudotuberculosis, Streptococcus pneumoniae,* rickettsia-like organisms, coxsackievirus, echovirus, and Epstein-Barr virus, were proposed as the causative agent (reviewed in reference 50). Several studies noted that many, if not the majority, of HUS cases were preceded by diarrhea. Interestingly, a survey of HUS in South

Africa led Kibel and Barnard in 1968 (51) to speculate that HUS is caused by an enteropathogenic strain of *E. coli* that had acquired a bacteriophage. These authors also raised concerns that treatment with antibiotics might lead to excessive bacterial destruction and enhanced absorption of toxin, with an adverse clinical outcome (see chapter 15 of this volume). Additional information on the history of HUS and various proposed pathogenic mechanisms can be found in several reviews (50, 52, 53; chapters 5 and 19 of this volume).

The key event in the linkage of HUS and STEC was the 1983 report in *The Lancet* by Karmali et al. (29) that sporadic cases of HUS were linked to the presence of verotoxin, which O'Brien et al. (25) reported was equivalent to Stx (see above), and/or *E. coli* that produced Stx in patients' stools. The toxigenic *E. coli* strains characterized by Karmali and colleagues belonged to different serogroups, thus ruling out a single strain as the cause of this disease, and serum collected from several patients contained rising titers of neutralizing antibody activity against verotoxin. This initial report was confirmed by a prospective case-control study that linked cases of HUS with isolation from stool of STEC belonging to at least six O serogroups (O26, O111, O113, O121, O145, and O157) (54). The 1985 article in *Journal of Infectious Diseases* describing the case-control study was reprinted by that journal in 2004 along with a commentary (55) describing it as one of the landmark papers published in that journal over the first centennial of its history. The connection between Stx (VT) production by an *E. coli* strain and the development of HUS after infection with that organism was most recently substantiated by the 2011 O104:H4 outbreak, showing that diarrheagenic enteroaggregative *E. coli* could cause HUS after acquiring the capacity to produce Stx2 (see below).

Hemorrhagic Colitis

In 1982, two outbreaks of a severe bloody diarrheal syndrome in Oregon and Michigan were linked to the consumption of hamburgers from a specific restaurant chain (56). This syndrome, called hemorrhagic colitis, was characterized by severe abdominal cramps, grossly bloody stools, little or no fever, and evidence of colonic mucosal edema, erosion, or hemorrhage (57). *E. coli* strains of a previously rare serotype, O157:H7, were isolated from the stools of about half the cases but from none of the healthy controls. Strains of this serotype were subsequently shown to produce Stx (see above). Numerous studies have since confirmed that O157:H7 is an important cause of hemorrhagic colitis, nonbloody diarrhea, and HUS in the United States, Canada, the United Kingdom, and Japan, as reviewed in other chapters in this volume.

The abrupt appearance of *E. coli* O157:H7 in 1982 raised questions as to whether this organism had recently emerged as a pathogen or had always been present and had simply been unrecognized. To address this issue, investigators at national laboratories in the United States, Canada, and the United Kingdom reviewed their records and *E. coli* collections and found archived *E. coli* O157:H7 strains recovered before 1982 from the stool of one patient in the United States, one patient in the United Kingdom, and six patients in Canada, some of whom had bloody diarrhea (reviewed in reference 52). The clinical syndrome of hemorrhagic colitis is so distinctive that outbreaks are unlikely to have been overlooked, although occasional cases of a hemorrhagic colitis-like syndrome of unknown etiology were reported in the 1960s and 1970s (reviewed in references 50 and 52). Thus, the available evidence indicates that the incidence of infections with O157:H7 and other STEC strains increased in the 1980s and 1990s. However, this conclusion is confounded by the increase in the number of laboratories seeking this pathogen (50) and by the 25 to 75% of patients with O157:H7 infection who present with nonbloody diarrhea (50, 58), a clinical manifestation that may go unrecognized as one of the manifestations of O157:H7 disease. Studies by Whittam and colleagues using multilocus enzyme electrophoresis demonstrated the stepwise evolution of STEC O157:H7 from an O55:H7 ancestor (59, 60), a Stx-negative serotype of EPEC that had previously been only associated with nonbloody diarrhea.

The comprehensive study of the Oregon and Michigan O157:H7 outbreaks of hemorrhagic colitis (in which no cases of HUS were noted) was published in the March 24, 1983, issue of *New England Journal of Medicine* (61). Karmali's study linking verotoxin-producing *E. coli* strains of different serogroups to HUS was published in the March 19, 1983, issue of *Lancet* (29), and 1 week later, in the March 26, 1983, issue of *Lancet*, O'Brien and colleagues (25) reported that a Shiga-like toxin was produced by the *E. coli* O157:H7 strains from the Oregon and Michigan outbreaks of hemorrhagic colitis and that this toxin was the same as the verotoxin previously shown to be produced by *E. coli* O157:H7 (26). Thus, March 1983 was a momentous month in which numerous laboratory, clinical, and epidemiological studies on HUS, bloody diarrhea, verotoxins, Shiga toxin, and *E. coli* came together to establish the field of STEC infections.

O104:H4

In May 2011, a large outbreak of gastroenteritis and HUS began in Germany, one of the largest outbreaks of STEC yet reported. In 3 months, 3,816 cases (including 54 deaths) were reported, of which 845 (22%) were HUS (5). The etiologic agent was identified as Stx-producing *E. coli* O104:H4, and sprouts were identified as the outbreak vehicle (62). Sprouts had previously been identified as the vehicle in STEC outbreaks, most notably the 1996 outbreak in Sakai City, Japan, where 12,680 cases were reported (63). The most striking clinical and epidemiological finding in the 2011 outbreak was the very high number of HUS cases (22% of all cases), with 88% of the HUS cases occurring in adults rather than in children. In contrast, the incidence of HUS in the 1996 Japan outbreak was 1% and all cases were in children. The dramatic features of the 2011 outbreak suggested that the causative agent might be a novel pathogenic variant of STEC.

Investigators quickly established that the O104:H4 strain lacked the LEE pathogenicity island present in O157:H7 and other common EHEC strains. Instead, the strain possessed an unusual combination of virulence factors that were typical of EAEC in addition to Stx (64). EAEC strains that do not make Stxs are well-established etiologic agents of nonbloody diarrhea that can be either acute or persistent in duration. Disease is seen in both children and adults, in travelers, and in people infected with human immunodeficiency virus in both the developed and developing world (reviewed in reference 65). The term "enteroaggregative" is derived from the "stacked-brick" appearance of EAEC on intestinal epithelial cells, in which large numbers of bacteria closely adhere to enterocytes in a biofilm. The pathogenesis of EAEC is poorly understood, but a variety of virulence factors have been described, including aggregative adherence fimbriae (AAF) that mediate intestinal adherence and induce inflammation, several serine protease autotransporters of *Enterobacteriaceae* (SPATEs) implicated in mucosal damage and colonization, and several other putative adhesins and toxins (reviewed in reference 66).

Analysis of the genome sequence of the O104:H4 strain from the Germany outbreak revealed that it closely resembled other EAEC strains, but that it had become transduced with the Stx2a-encoding phage (6). The genome sequence also revealed the presence of an unusual combination of SPATEs and several antibiotic-resistance factors. Other *E. coli* O104:H4 strains unrelated to the Germany outbreak did not possess Stx. Further discussion of the pathogenesis of O104:H4 is presented in chapter 26 of this volume.

Although more than 100 different serotypes of *E. coli* have been shown to produce Stx, the majority of such STEC strains are not considered to be pathogens. Several serotypes such as O26:H11, O111:NM, and O121:H19 contain the LEE and other pathogenicity islands found in O157:H7 and are clearly pathogens.

A few other STEC serotypes, such as O91:H21 and O113:H21, lack the LEE but have additional virulence factors and have been epidemiologically implicated as pathogens (67). The different STEC serotypes and their epidemiological significance are reviewed in chapter 2. In the case of the German Shiga toxin-producing enteroaggregative *E. coli* O104:H4, the addition of the Stx2a phage to a pathotype of *E. coli* that was already capable of avidly adhering to and damaging the intestinal epithelium produced a novel pathogen with profound clinical and epidemiological consequences.

THE PRESENT

The current themes and directions of STEC research span an astonishing range of topics. Multiple disciplines encompass epidemiology, animal ecology, food safety, clinical microbiology, gastroenterology, nephrology, infectious disease, toxicology, bacterial pathogenesis, cell biology, and immunology. Topics in this area of research range from farm management of livestock and manure to clinical management of end-stage renal disease. This book, edited and written by internationally recognized experts in this area, reflects this breadth of topics.

The first section of the volume describes the microbiology of STEC. In chapter 2, Flemming Scheutz describes the taxonomy of STEC and Stx toxins and relates this information to the public health significance of the different serotypes and toxin subtypes. The discussion of Stx toxin continues in chapter 3 where Angela Melton-Celsa reviews the structure and function of these toxins. Sadiq and colleagues take a genomic perspective in chapter 4 to review the history of typing and genetic analysis from distinguishing STEC strains using pregenomic methodologies to the current technology, where the genome sequences of multiple strains can be determined in a single day.

The pathogenesis of STEC infections is covered in section two. In chapter 5, Obata and Obrig discuss the role of Stx toxins in pathogenesis, with a particular emphasis on effects in the renal system. In chapter 6, Stevens and Frankel review the LEE pathogenicity island and virulence factors encoded therein, as well as other virulence factors encoded outside the LEE. Colonization of the intestinal tract is an essential first step in STEC pathogenesis, and a variety of potential adherence factors have been described, as reviewed by McWilliams and Torres in chapter 7. Unfortunately, there is no single animal model that reproduces all aspects of STEC disease, but Ritchie reviews the various models available and their advantages and disadvantages in chapter 8. The range of environments where STEC can be found—from the farm environment to the human intestine—requires numerous regulatory genetic elements to optimize expression of virulence factors and survival factors. Mellies and Lorenzen describe the complex regulation of STEC virulence in chapter 9.

The incidence, epidemiology, and ecology of STEC are reviewed in the third section. In chapter 10, Terajima et al. review the incidence and epidemiology of STEC in Japan, the site of the largest STEC outbreak reported. Animals, particularly cattle, serve as the reservoir of STEC infections, and Persad and LeJeune review this critical reservoir in chapter 11. Transmission to humans most often involves consumption of contaminated food items. Initial outbreaks of STEC disease involved improperly cooked hamburgers, an issue that was relatively easy to address by increasing cooking temperatures. However, as reviewed by Feng in chapter 12, most outbreaks in recent years involved produce that is consumed raw. Caprioli et al. review in chapter 13 the epidemiology and other public health aspects of STEC infection, with a particular emphasis on Europe. Methods for detecting STEC from nonhuman sources and strain typing are reviewed by Beutin and Fach in chapter 14.

Clinical, pathological, and pathophysiological aspects of human disease are reviewed in chapters 15 through 17. Tarr and coauthors review in chapter 15 the clinical features of STEC infections in humans, including outcomes and prognosis, and provide insights from both gastroenterological and nephrological perspectives. The inflammatory response to STEC infection and the virulence factors these pathogens have evolved to thwart this response are discussed by Pearson and Hartland in chapter 16. Unfortunately, no ideal therapy is available for STEC infections, and the use of antimicrobials is contraindicated, at least for typical EHEC infections, although investigators of the German O104:H4 outbreak reported the benefit of azithromycin treatment to reduce fecal shedding of the organism. This issue, along with novel therapeutic interventions under study, is reviewed by Melton-Celsa and O'Brien in chapter 17.

Host determinants of disease and host responses encompass factors ranging from cultural and dietary practices to host genetics and immune status to intestinal microbiota, all of which can play important roles in STEC infection and outcome. Risk factors for STEC infections are discussed by Rivas et al. in chapter 18. The host response and other aspects of STEC pathogenesis are reviewed by Karpman and Stahl in chapter 19. With the recognition that regulation of EHEC virulence factors can be influenced by commensal intestinal bacteria, the interplay between the microbiota and EHEC can be important, as reviewed by Pifer and Sperandio in chapter 20.

Prevention and control strategies to reduce or eliminate the risk of STEC infections are particularly important in the control of STEC infections. Preharvest and peri- and postharvest food safety factors are reviewed by Besser and colleagues in chapter 21 and by Moxley and Acuff in chapter 22. A veterinary public health approach to managing pathogenic STEC in the agri-food chain is discussed by Duffy and McCabe in chapter 23. Vaccines have been critical in reducing the disease burden in humans for many infectious diseases, but for STEC infections, vaccines to reduce carriage in the bovine reservoir to reduce transmission to humans may hold more promise than direct immunization of humans. In chapter 24, Szu and Ahmed present data showing that parenteral O157 lipopolysaccharide conjugate vaccines are safe and immunogenic in children and adults. However, there are multiple potential problems in vaccinating humans against STEC disease, including finding a population with a high enough incidence in which to determine vaccine efficacy and identifying an appropriate target population to vaccinate once vaccine efficacy is established (reviewed in reference 68). Success in vaccinating cattle to reduce fecal shedding of STEC has been achieved, and David Smith reviews these studies in chapter 25.

The emergence of STEC-EAEC O104:H4 in 2011 was a landmark development in the history of STEC infections. The virulence factors that combined to produce this highly virulent strain are discussed by Navarro-Garcia in chapter 26. Such a development makes one wonder what the future holds for the STEC field, and Vanessa Sperandio offers in the final chapter some speculations on current questions and future directions for investigating this fascinating and ever-changing pathogen.

ACKNOWLEDGMENTS

The opinions or assertions contained herein are the private ones of the authors and are not to be construed as official or reflecting the views of the Department of Defense, the Uniformed Services University of the Health Sciences, or the National Institutes of Health.

This work was supported by National Institutes of Health grants R01 DK58957 (JBK), R37 AI21657 (JBK), and R37 AI020148 (ADO).

We declare no conflicts of interest with regard to the manuscript.

CITATION

Kaper JB, O'Brien AD. 2014. Overview and historical perspectives. Microbiol Spectrum 2(6):EHEC-0028-2014.

REFERENCES

1. **Kaper JB, O'Brien AD (ed).** 1998. *Escherichia coli O157:H7 and Other Shiga Toxin-Producing* E. coli *Strains.* ASM Press, Washington, DC.

2. **Calderwood SB, Acheson DWK, Keusch GT, Barrett TJ, Griffin PM, Strockbine NA, Swaminathan B, Kaper JB, Levine MM, Kaplan BS, Karch H, O'Brien AD, Obrig TG, Takeda Y, Tarr PI, Wachsmuth IK.** 1996. Proposed new nomenclature for SLT (VT) family. *ASM News* **62:**118–119.

3. **Karmali MA, Lingwood CA, Petrie M, Brunton J, Gyles C.** 1996. Maintaining the existing phenotype nomenclatures for *E. coli* cytotoxins. *ASM News* **62:**167–169.

4. **Scheutz F, Teel LD, Beutin L, Pierard D, Buvens G, Karch H, Mellmann A, Caprioli A, Tozzoli R, Morabito S, Strockbine NA, Melton-Celsa AR, Sanchez M, Persson S, O'Brien AD.** 2012. Multicenter evaluation of a sequence-based protocol for subtyping Shiga toxins and standardizing Stx nomenclature. *J Clin Microbiol* **50:**2951–2963.

5. **Frank C, Werber D, Cramer JP, Askar M, Faber M, an der Heiden M, Bernard H, Fruth A, Prager R, Spode A, Wadl M, Zoufaly A, Jordan S, Kemper MJ, Follin P, Muller L, King LA, Rosner B, Buchholz U, Stark K, Krause G.** 2011. Epidemic profile of Shiga-toxin-producing *Escherichia coli* O104:H4 outbreak in Germany. *N Engl J Med* **365:**1771–1780.

6. **Rasko DA, Webster DR, Sahl JW, Bashir A, Boisen N, Scheutz F, Paxinos EE, Sebra R, Chin CS, Iliopoulos D, Klammer A, Peluso P, Lee L, Kislyuk AO, Bullard J, Kasarskis A, Wang S, Eid J, Rank D, Redman JC, Steyert SR, Frimodt-Moller J, Struve C, Petersen AM, Krogfelt KA, Nataro JP, Schadt EE, Waldor MK.** 2011. Origins of the *E. coli* strain causing an outbreak of hemolytic-uremic syndrome in Germany. *N Engl J Med* **365:**709–717.

7. **Shiga K.** 1898. Ueber den Dysenterie-bacillus (*Bacillus dysenteriae*). *Zentralbl Baktriol Orig* **24:**913–918.

8. **Conradi H.** 1903. Uber Iosliche, durch asptische Autolyse erhaltene Giftstoffe vonRuhr- und Typhus-Bazillen. *Dtsch Med Wochenschr* **29:**26–28.

9. **Neisser M, Shiga K.** 1903. Ueber freie Receptoren von Typhus- und Dysenterie-Bazillen und ueber das Dysenterie Toxin. *Dtsch Med Wochenschr* **29:**61–62.

10. **Van Heyningen WE, Gladstone GP.** 1953. The neurotoxin of *Shigella shigae*. III. The effect of iron on production of the toxin. *Br J Exp Pathol* **34:**221–229.

11. **Dubos RJ, Geiger JW.** 1946. Preparation and properties of Shiga toxin and toxoid. *J Exp Med* **84:**143–156.

12. **Bridgwater FA, Morgan RS, Rowson KE, Wright GP.** 1955. The neurotoxin of *Shigella shigae*: morphological and functional lesions produced in the central nervous system of rabbits. *Br J Exp Pathol* **36:**447–453.

13. **Howard JG.** 1955. Observations on the intoxication produced in mice and rabbits by the neurotoxin of *Shigella shigae*. *Br J Exp Pathol* **36:**439–446.

14. **Vicari G, Olitzki AL, Olitzki Z.** 1960. The action of the thermolabile toxin of *Shigella dysenteriae* on cells cultivated in vitro. *Br J Exp Pathol* **41:**179–189.

15. **Levine MM, DuPont HL, Formal SB, Hornick RB, Takeuchi A, Gangarosa EJ, Snyder MJ, Libonati JP.** 1973. Pathogenesis of *Shigella dysenteriae* 1 (Shiga) dysentery. *J Infect Dis* **127:**261–270.

16. **Fontaine A, Arondel J, Sansonetti PJ.** 1988. Role of Shiga toxin in the pathogenesis of bacillary dysentery, studied by using a Tox- mutant of *Shigella dysenteriae* 1. *Infect Immun* **56:**3099–3109.

17. **Keusch GT, Grady GF, Mata LJ, McIver J.** 1972. The pathogenesis of *Shigella* diarrhea. I. Enterotoxin production by *Shigella dysenteriae* I. *J Clin Invest* **51:**1212–1218.

18. **Olsnes S, Eiklid K.** 1980. Isolation and characterization of *Shigella shigae* cytotoxin. *J Biol Chem* **255:**284–289.

19. **O'Brien AD, LaVeck GD, Griffin DE, Thompson MR.** 1980. Characterization of *Shigella dysenteriae* 1 (Shiga) toxin purified by anti-Shiga toxin affinity chromatography. *Infect Immun* **30:**170–179.

20. **Brown JE, Griffin DE, Rothman SW, Doctor BP.** 1982. Purification and biological characterization of Shiga toxin from *Shigella dysenteriae* 1. *Infect Immun* **36:**996–1005.

21. **Donohue-Rolfe A, Keusch GT, Edson C, Thorley-Lawson D, Jacewicz M.** 1984. Pathogenesis of *Shigella* diarrhea. IX. Simplified high yield purification of Shigella toxin and characterization of subunit composition and function by the use of subunit-specific monoclonal and polyclonal antibodies. *J Exp Med* **160:**1767–1781.

22. **O'Brien AD, LaVeck GD.** 1983. Purification and characterization of a *Shigella dysenteriae* 1-like toxin produced by *Escherichia coli*. *Infect Immun* **40:**675–683.

23. **O'Brien AD, Thompson MR, Cantey JR, Formal SB.** 1977. *Production of a Shigella dysenteriae-like toxin by pathogenic Escherichia coli*, abstr. B-103. Abstr 77th Annu Meet Am Soc Microbiol. American Society for Microbiology, Washington, DC.

24. Konowalchuk J, Speirs JI, Stavric S. 1977. Vero response to a cytotoxin of *Escherichia coli*. *Infect Immun* **18**:775–779.

25. O'Brien AO, Lively TA, Chen ME, Rothman SW, Formal SB. 1983. *Escherichia coli* O157:H7 strains associated with haemorrhagic colitis in the United States produce a *Shigella dysenteriae* 1 (SHIGA) like cytotoxin. *Lancet* **i**:702.

26. Johnson WM, Lior H, Bezanson GS. 1983. Cytotoxic *Escherichia coli* O157:H7 associated with haemorrhagic colitis in Canada. *Lancet* **i**:76.

27. Strockbine NA, Jackson MP, Sung LM, Holmes RK, O'Brien AD. 1988. Cloning and sequencing of the genes for Shiga toxin from *Shigella dysenteriae* type 1. *J Bacteriol* **170**:1116–1122.

28. Takao T, Tanabe T, Hong YM, Shimonishi Y, Kurazono H, Yutsudo T, Sasakawa C, Yoshikawa M, Takeda Y. 1988. Identity of molecular structure of Shiga-like toxin I (VT1) from *Escherichia coli* O157:H7 with that of Shiga toxin. *Microb Pathog* **5**:57–69.

29. Karmali MA, Steele BT, Petric M, Lim C. 1983. Sporadic cases of haemolytic-uraemic syndrome associated with faecal cytotoxin and cytotoxin-producing *Escherichia coli* in stools. *Lancet* **i**:619–620.

30. O'Brien AD, Tesh VL, Donohue-Rolfe A, Jackson MP, Olsnes S, Sandvig K, Lindberg AA, Keusch GT. 1992. Shiga toxin: biochemistry, genetics, mode of action, and role in pathogenesis. *Curr Top Microbiol Immunol* **180**:65–94.

31. O'Brien AD, Marques LR, Kerry CF, Newland JW, Holmes RK. 1989. Shiga-like toxin converting phage of enterohemorrhagic *Escherichia coli* strain 933. *Microb Pathog* **6**:381–390.

32. O'Brien AD, Newland JW, Miller SF, Holmes RK, Smith HW, Formal SB. 1984. Shiga-like toxin-converting phages from *Escherichia coli* strains that cause hemorrhagic colitis or infantile diarrhea. *Science* **226**:694–696.

33. Scotland SM, Smith HR, Willshaw GA, Rowe B. 1983. Vero cytotoxin production in strain of *Escherichia coli* is determined by genes carried on bacteriophage. *Lancet* **ii**:216.

34. Smith HW, Green P, Parsell Z. 1983. Vero cell toxins in *Escherichia coli* and related bacteria: transfer by phage and conjugation and toxic action in laboratory animals, chickens and pigs. *J Gen Microbiol* **129**:3121–3137.

35. Strockbine NA, Marques LR, Newland JW, Smith HW, Holmes RK, O'Brien AD. 1986. Two toxin-converting phages from *Escherichia coli* O157:H7 strain 933 encode antigenically distinct toxins with similar biologic activities. *Infect Immun* **53**:135–140.

36. Weinstein DL, Jackson MP, Samuel JE, Holmes RK, O'Brien AD. 1988. Cloning and sequencing of a Shiga-like toxin type II variant from *Escherichia coli* strain responsible for edema disease of swine. *J Bacteriol* **170**:4223–4230.

37. Tzipori S, Wachsmuth IK, Chapman C, Birden R, Brittingham J, Jackson C, Hogg J. 1986. The pathogenesis of hemorrhagic colitis caused by *Escherichia coli* O157:H7 in gnotobiotic piglets. *J Infect Dis* **154**:712–716.

38. Levine MM. 1987. *Escherichia coli* that cause diarrhea: enterotoxigenic, enteropathogenic, enteroinvasive, enterohemorrhagic, and enteroadherent. *J Infect Dis* **155**:377–389.

39. Levine MM, Xu JG, Kaper JB, Lior H, Prado V, Tall B, Nataro J, Karch H, Wachsmuth K. 1987. A DNA probe to identify enterohemorrhagic *Escherichia coli* of O157:H7 and other serotypes that cause hemorrhagic colitis and hemolytic uremic syndrome. *J Infect Dis* **156**:175–182.

40. Robins-Browne RM. 1987. Traditional enteropathogenic *Escherichia coli* of infantile diarrhea. *Rev Infect Dis* **9**:28–53.

41. Knutton S, Baldwin T, Williams PH, McNeish AS. 1989. Actin accumulation at sites of bacterial adhesion to tissue culture cells: basis of a new diagnostic test for enteropathogenic and enterohemorrhagic *Escherichia coli*. *Infect Immun* **57**:1290–1298.

42. Jerse AE, Yu J, Tall BD, Kaper JB. 1990. A genetic locus of enteropathogenic *Escherichia coli* necessary for the production of attaching and effacing lesions on tissue culture cells. *Proc Natl Acad Sci USA* **87**:7839–7843.

43. Tzipori S, Gunzer F, Donnenberg MS, de Montigny L, Kaper JB, Donohue-Rolfe A. 1995. The role of the *eaeA* gene in diarrhea and neurological complications in a gnotobiotic piglet model of enterohemorrhagic *Escherichia coli* infection. *Infect Immun* **63**:3621–3627.

44. Beebakhee G, Louie M, De Azavedo J, Brunton J. 1992. Cloning and nucleotide sequence of the *eae* gene homologue from enterohemorrhagic *Escherichia coli* serotype O157:H7. *FEMS Microbiol Lett* **70**:63–68.

45. Yu J, Kaper JB. 1992. Cloning and characterization of the *eae* gene of enterohaemorrhagic *Escherichia coli* O157:H7. *Mol Microbiol* **6**:411–417.

46. Donnenberg MS, Tzipori S, McKee ML, O'Brien AD, Alroy J, Kaper JB. 1993. The role of the *eae* gene of enterohemorrhagic *Escherichia coli* in intimate attachment in vitro and in a porcine model. *J Clin Invest* **92**:1418–1424.

47. McKee ML, Melton-Celsa AR, Moxley RA, Francis DH, O'Brien AD. 1995. Enterohemorrhagic *Escherichia coli* O157:H7 requires intimin to colonize the gnotobiotic pig intestine and to adhere to HEp-2 cells. *Infect Immun* **63**:3739–3744.

48. **McDaniel TK, Jarvis KG, Donnenberg MS, Kaper JB.** 1995. A genetic locus of enterocyte effacement conserved among diverse enterobacterial pathogens. *Proc Natl Acad Sci USA* **92:**1664–1668.

49. **Gasser C, Gautier E, Steck A, Siebenmann RE, Oechslin R.** 1955. [Hemolytic-uremic syndrome: bilateral necrosis of the renal cortex in acute acquired hemolytic anemia]. *Schweiz Med Wochenschr* **85:**905–909 (In German).

50. **Karmali MA.** 1989. Infection by verocytotoxin-producing *Escherichia coli*. *Clin Microbiol Rev* **2:**15–38.

51. **Kibel MA, Barnard PJ.** 1968. The haemolytic-uraemic syndrome: a survey in Southern Africa. *S Afr Med J* **42:**692–698.

52. **Griffin PM, Tauxe RV.** 1991. The epidemiology of infections caused by *Escherichia coli* O157:H7, other enterohemorrhagic *E. coli*, and the associated hemolytic uremic syndrome. *Epidemiol Rev* **13:**60–98.

53. **Kaplan BS, Trompeter RS, Moake JL.** 1992. Introduction, p xvii–xxvii. *In* Kaplan BS, Trompeter RS, Moake JL (ed), *Hemolytic Uremic Syndrome and Thrombotic Thrombocytopenic Purpura*. Marcel Dekker, Inc, New York, NY.

54. **Karmali MA, Petric M, Lim C, Fleming PC, Arbus GS, Lior H.** 1985. The association between idiopathic hemolytic uremic syndrome and infection by verotoxin-producing *Escherichia coli*. *J Infect Dis* **151:**775–782.

55. **Blaser MJ.** 2004. Bacteria and diseases of unknown cause: hemolytic-uremic syndrome. *J Infect Dis* **189:**552–555.

56. **Centers for Disease Control and Prevention.** 1982. Isolation of *E. coli* O157:H7 from sporadic cases of hemorrhagic colitis—United States. *MMWR Morb Mortal Wkly Rep* **31:**580, 585.

57. **Riley LW.** 1987. The epidemiologic, clinical, and microbiologic features of hemorrhagic colitis. *Annu Rev Microbiol* **41:**383–407.

58. **Griffin PM.** 1995. *Escherichia coli* O157:H7 and other enterohemorrhagic *Escherichia coli*, p 739–761. *In* Blaser MJ, Smith PD, Ravdin JI, Greenberg HB, Guerrant RL (ed), *Infections of the Gastrointestinal Tract*. Raven Press, New York, NY.

59. **Feng P, Lampel KA, Karch H, Whittam TS.** 1998. Genotypic and phenotypic changes in the emergence of *Escherichia coli* O157:H7. *J Infect Dis* **177:**1750–1753.

60. **Whittam TS.** 1998. Evolution of *Escherichia coli* O157:H7 and other Shiga toxin-producing *E. coli* strains, p 195–209. *In* Kaper JB, O'Brien AD (ed), *Escherichia coli O157:H7 and Other Shiga Toxin-Producing* E. coli *Strains*. ASM Press, Washington, DC.

61. **Riley LW, Remis RS, Helgerson SD, McGee HB, Wells JG, Davis BR, Hebert RJ, Olcott ES, Johnson LM, Hargrett NT, Blake PA, Cohen ML.** 1983. Hemorrhagic colitis associated with a rare *Escherichia coli* serotype. *N Engl J Med* **308:**681–685.

62. **Buchholz U, Bernard H, Werber D, Bohmer MM, Remschmidt C, Wilking H, Delere Y, an der Heiden M, Adlhoch C, Dreesman J, Ehlers J, Ethelberg S, Faber M, Frank C, Fricke G, Greiner M, Hohle M, Ivarsson S, Jark U, Kirchner M, Koch J, Krause G, Luber P, Rosner B, Stark K, Kuhne M.** 2011. German outbreak of *Escherichia coli* O104:H4 associated with sprouts. *N Engl J Med* **365:**1763–1770.

63. **Fukushima H, Hashizume T, Morita Y, Tanaka J, Azuma K, Mizumoto Y, Kaneno M, Matsuura M, Konma K, Kitani T.** 1999. Clinical experiences in Sakai City Hospital during the massive outbreak of enterohemorrhagic *Escherichia coli* O157 infections in Sakai City, 1996. *Pediatr Int* **41:**213–217.

64. **Scheutz F, Nielsen EM, Frimodt-Moller J, Boisen N, Morabito S, Tozzoli R, Nataro JP, Caprioli A.** 2011. Characteristics of the enteroaggregative Shiga toxin/verotoxin-producing *Escherichia coli* O104:H4 strain causing the outbreak of haemolytic uraemic syndrome in Germany, May to June 2011. *Euro Surveill* **16**(24):pii=19889.

65. **Huang DB, Mohanty A, DuPont HL, Okhuysen PC, Chiang T.** 2006. A review of an emerging enteric pathogen: enteroaggregative *Escherichia coli*. *J Med Microbiol* **55:**1303–1311.

66. **Estrada-Garcia T, Navarro-Garcia F.** 2012. Enteroaggregative *Escherichia coli* pathotype: a genetically heterogeneous emerging foodborne enteropathogen. *FEMS Immunol Med Microbiol* **66:**281–298.

67. **Karmali MA, Mascarenhas M, Shen S, Ziebell K, Johnson S, Reid-Smith R, Isaac-Renton J, Clark C, Rahn K, Kaper JB.** 2003. Association of genomic O island 122 of *Escherichia coli* EDL 933 with verocytotoxin-producing *Escherichia coli* seropathotypes that are linked to epidemic and/or serious disease. *J Clin Microbiol* **41:**4930–4940.

68. **Tauxe RV.** 1998. Public health perspective on immunoprophylactic strategies for *Escherichia coli* O157:H7: who or what would we immunize, p 445–452. *In* Kaper JB, O'Brien AD (ed), *Escherichia coli O157:H7 and Other Shiga Toxin-Producing* E. coli *Strains*. ASM Press, Washington, DC.

MICROBIOLOGY

Taxonomy Meets Public Health: The Case of Shiga Toxin-Producing *Escherichia coli*

2

FLEMMING SCHEUTZ[1]

PATHOTYPES AND TAXONOMY

The term enteropathogenic *Escherichia coli* was originally used to refer to strains belonging to a limited number of O groups epidemiologically associated with infantile diarrhea (1). Subsequently, *E. coli* strains isolated from intestinal diseases have been grouped into at least six main categories on the basis of epidemiological evidence, phenotypic traits, clinical features of the disease they produce, and specific virulence factors. The well-described intestinal pathotypes or categories of diarrheagenic *E. coli* groups are enteropathogenic *E. coli* (EPEC), Shiga toxin-producing *E. coli* (STEC) or verocytotoxin-producing *E. coli* (VTEC) (including enterohemorrhagic *E. coli* [EHEC]), enterotoxigenic *E. coli* (ETEC), enteroaggregative *E. coli* (EAEC), enteroinvasive *E. coli*, and diffusely adherent *E. coli*. The general definition of an *E. coli* pathotype as "a group of strains of a single species that cause a common disease using a common set of virulence factors" (2) has been further refined for STEC to help assess the clinical and public health risks associated with different STEC strains (3). An empirical classification scheme was used to classify STEC serotypes into five "seropathotypes" (A through E) according to the reported association of serotypes with human intestinal disease, outbreaks, and hemolytic-uremic

[1]WHO Collaborating Centre for Reference and Research on *Escherichia* and *Klebsiella*, Department of Microbiology and Infection Control, Statens Serum Institut, DK-2300 Copenhagen S, Denmark.

Enterohemorrhagic Escherichia coli *and Other Shiga Toxin–Producing* E. coli
Edited by Vanessa Sperandio and Carolyn J. Hovde
© 2015 American Society for Microbiology, Washington, DC
doi:10.1128/microbiolspec.EHEC-0019-2013

syndrome (HUS) (3). This classification system uses a gradient ranging from seropathotype A (high risk) to seropathotypes D and E (minimal risk). This definition has been of considerable value in cases of human infection but is also problematic because the majority of isolates from STEC infections are not fully characterized and coupled to reliable clinical information. Although the definition of HUS is distinct, the spectrum of diarrheal disease varies considerably and may include a range of symptoms from nonbloody to scanty blood to true hemorrhagic colitis. Additionally, the use of A through E adds confusion because Shiga toxin subtypes are also named alphabetically. Most importantly, the concept of (sero)-pathotypes collides with the requirements of a good taxonomy, which separates elements of each group into subgroups that are mutually exclusive, unambiguous, and, together, include all possibilities. In practice, a good taxonomy should be simple to apply, easy to remember, and easy to use. The need to define human pathogenic STEC and to identify factors of STEC that absolutely predict the potential to cause human disease is obvious in terms of clinical management, supportive or antibiotic treatment, quarantine measurements, risk assessment, surveillance, and outbreak investigations and management. This chapter presents a brief history of the concept of pathotypes and describes the possible alternatives for categorizing STEC based on phenotypic or molecular typing.

PATHOTYPES

First discovered in 1977 (4), verocytotoxin was found to be biologically and structurally similar to Shiga toxin produced by *Shigella dysenteriae* Type 1 (5, 6). It was soon realized that antigenically distinct cytotoxins could be found in different *E. coli* serotypes (7, 8). STEC or VTEC strains are characterized by their ability to produce either one or both of these cytotoxins, referred to as Stx1 or VT1 (first described as Shiga-like toxin I, SLTI)

and Stx2 or VT2 (first described as Shiga-like toxin II, SLTII). The cytotoxin production is usually bacteriophage-mediated (9–12), and the diversity of this toxin family has since become clear.

The public health significance of STEC was first recognized in 1982, when two outbreaks in the United States affected at least 47 people in Oregon and Michigan. Nine of 12 stool cultures yielded a rare *E. coli* serotype, O157:H7, that was also isolated from a beef patty from a suspected lot of meat in Michigan (13). This strain was designated EDL933 and has since been used as the prototype STEC strain by researchers worldwide. Distinct clinical features of hemorrhagic colitis (HC) included abdominal cramps, copious bloody diarrhea described as "all blood and no stool," unaccompanied by fecal leukocytes, and no fever. Duration of illness was 2 to 9 days, and there were no deaths, complications, or sequelae in any of the cases.

In an outbreak of HC in November 1982 at a Canadian institution for elderly patients, sorbitol-negative *E. coli* O157:H7 was shown to produce verocytotoxin. Two of six sporadic cytotoxic O157:H7 strains were associated with HC, and 70% of 78 cytotoxic serotypes isolated from sporadic cases of diarrhea during 1978 to 1982 were *E. coli* O26:H11 (14). The cytotoxicity of an O26 strain (H30, described as a verocytotoxin producer by Konowalchuk et al. [4]), two of the *E. coli* O157:H7 strains from the U.S. cases of HC, and the beef patty isolate EDL933 from this outbreak could be neutralized by rabbit antiserum to purified Shiga toxin from *S. dysenteriae* Type 1, substantiating the premise that these cytotoxins were the same and that they played an important role in *E. coli* diarrheal diseases (15). In 1983, 11 of 15 sporadic cases of enteropathic HUS were shown to have evidence of infection with VTEC, indicating an association between sporadic cases of HUS and cytotoxin-producing *E. coli* strains (16). Verocytotoxin was proposed as having direct etiological importance in the pathogenesis of both HUS and HC (17).

The term EHEC, coined to refer to strains such as O157:H7 that manifest the above-mentioned clinical, epidemiological, and pathogenic features (18), further defined a pathogenic subgroup of STEC strains based on their association with disease in humans and their ability to hybridize to a DNA probe (CVD419) derived from a large plasmid present in most O157:H7 strains and the majority of other STEC strains isolated from cases of HC (19). Many of the other STEC strains belonged to classical EPEC O groups or serotypes such as O26, O111, O114, O125, O126, and O128 (4, 5, 12, 19–21), but non-EPEC O groups were also found to produce Stx: O1, O2, O4, O5, O6, O18, O45, O50, O68, O91, O103, O113, O121, and O145 (4, 5, 19). Furthermore, Stx production was also described in O groups O138, O139, and O141 isolated from weaned pigs with edema disease (12). Among the more than 472 STEC/VTEC serotypes, and apart from O157:H– and O157:H7, those in O groups O26, O103, O111, and O145 are most commonly isolated from humans worldwide (22). They, along with strains that have caused outbreaks, are clearly recognized as pathogens.

Consequently, serotypes and their association with diseases of varying severity in humans and with sporadic disease or outbreaks have been used to classify STEC into five seropathotypes (A through E) according to the reported occurrence of serotypes in human disease, in outbreaks, and in HUS (3). Seropathotype A consists of O157:H7 and O157:NM, considered to be the most virulent. Seropathotype B originally consisted of five serotypes that were similar to seropathotype A in causing severe disease and outbreaks but occurred at lower frequency, but in the United States, seropathotype B has been extended to include 13 STEC serotypes: O26:H11 and NM; O45:H2 and NM; O103:H2, H11, H25, and NM; O111:H8 and NM; O121:H19 and H7; and O145:NM (23). Seropathotype C includes serotypes infrequently implicated in sporadic HUS but not typically with outbreaks and includes O5:NM, O91:H21, O104:H21, O113:H21, O121:NM, and O165:H25. Seropathotype D is composed of 12 serotypes that have been implicated in sporadic cases of diarrhea but not with outbreaks or HUS, and seropathotype E is composed of at least 14 animal serotypes that have not been implicated in disease in humans (3).

This approach has been of considerable value in defining pathogenic STEC serotypes of importance in cases of human infection and also for STEC isolates from ruminants. However, classification of strains based on the criteria above is also problematic because the majority of isolates from STEC infections are not fully serotyped nor characterized for the presence of virulence factors. A recent Belgian study of STEC added 14 serotypes to seropathotype C (including four serotypes associated with HUS) and 54 serotypes to seropathotype D, demonstrating how versatile the definition of seropathotypes can be (24). Outbreaks with emerging or new hybrid strains with hitherto unknown virulence factors may also continuously challenge our understanding and appreciation of virulence potential. The limitation to "relevant" serotypes may therefore result in the omission or incorrect classification of specific pathotypes. Strain O104:H4 is such an example, because until the May-June 2011 outbreak in Germany this highly virulent strain would have been classified as seropathotype D on the basis of its sporadic occurrence, its lack of association with outbreaks, and its limited association with HUS (25). In fact, other STEC serotypes such as O104:H21 and O113:H21 strains lacking the *eae* gene were responsible for an outbreak and a cluster of three HUS cases in the United States and Australia, respectively (26, 27).

It has been suggested that these unusual and emerging types should have their own seropathotype designation (28). Using data from the European Surveillance System (TESSy data) as provided by the European Centre for Disease Prevention and Control (ECDC) and data available in the European Union Summary Report on Trends and Sources of Zoonoses, Zoonotic Agents and Food-borne Outbreaks in 2011 (29), the European Food

Safety Authority Panel on Biological Hazards (BIOHAZ Panel) concluded that the seropathotype classification by Karmali et al. (3) does not define pathogenic VTEC nor does it provide an exhaustive list of pathogenic serotypes. Eighty-five percent of 13,545 confirmed cases of human VTEC infection from 2007 to 2010 were not fully serotyped and could not be classified by using the seropathotype concept, and about 27% of the cases could not be assigned to a seropathotype group as they were not listed by Karmali et al. (25). However, about half of the isolates with missing H type were from cases of O157 infection (5,610 cases) and would most likely have been typed as O157:H7 or O157:H– and therefore assigned to seropathotype A. This would have expanded this group to 6,657 cases, or 87% of cases, 78% of fatal cases, 91% of the hospitalizations, 91% of the HUS cases, and 95% of the cases with bloody diarrhea (25).

Reporting of detailed clinical data is often incomplete. Of the reported confirmed VTEC cases in the European Union between 2007 and 2010, the health outcome was reported for 53% of diarrheal cases and 59% of HUS cases (25). Most patients (ca. 64%) presented with only diarrhea. Clinical information is not obtained according to standardized guidelines and definitions, and detailed information on the individual clinical course of disease is generally absent. Although the definition of HUS is distinct, the spectrum of diarrheal disease varies considerably and may include symptoms ranging from nonbloody to scanty blood to true hemorrhagic colitis. A specific STEC pathotype may even be associated with clinical presentations ranging from asymptomatic carriage to life-threatening HUS and death, as observed during outbreaks and in person-to-person transmission, whereby an index patient may experience only mild symptoms whereas secondary cases may develop into severe complications (30).

The BIOHAZ Panel concluded that "pathogenicity can neither be excluded nor confirmed for a given STEC serogroup or serotype based on the seropathotype concept

or analysis of the public health surveillance data" (25). In addition, even though the clinical manifestations of non-O157 STEC infection may differ considerably from those of O157:H7 (31), STEC O157:H7 has also been isolated from stools of healthy individuals. Many studies have tried to correlate the presence of specific virulence factors with disease or severity of disease (see discussion in reference 32). The combination of the locus of enterocyte effacement (LEE)-encoded *eae* gene for intimin and *stx2* is significantly more frequent in isolates from serotypes found in humans and is most strongly associated with disease in humans, particularly with severe disease (32–34). The reverse is true for *stx1*, which is found more frequently in serotypes not found in humans (32). Enterohemolysin, a plasmid-encoded toxin expressed by the *ehxA* gene that readily causes the hemolysis of washed sheep erythrocytes and liberates hemoglobin from the red blood cells during infection, has been linked to severe disease symptoms (35, 36).

The taxonomy of EPEC and VTEC is intimately intertwined in that many VTEC types share specific virulence factors such as pathogenicity islands, e.g., LEE (including the *eae* gene), O island (OI) 122, and plasmids with EPEC. Non-LEE (*nle*)-encoded genes have also been found in both EPEC and STEC isolates, such as the EspI/NleA effector protein encoded on a prophage (37). Several studies have indicated that the major difference between certain EPEC and STEC isolates is the absence or presence of the bacteriophage encoding Stx. However, the presence of LEE does not seem to be essential for full virulence, as a wide number of LEE-negative STEC strains have been associated with sporadic cases and small outbreaks of HC and HUS (38).

How are environmental, food, and veterinary STEC types, which by nature cannot be isolated from human cases of either HC or HUS, classified? Animals often carry types that are referred to as EHEC or could be classified in the A through E pathotype scheme without any clinical symptoms. The classification based on the clinical course of disease in humans is clearly host associated, and the term EHEC or the A through E classification of nonhuman isolates could be misleading.

PHYLOGENY AND DISEASE

An *E. coli* genome contains between 4,200 and 5,500 genes, with fewer than 2,000 genes conserved among all strains of the species (the core genome). Comparison of 61 *E. coli* and *Shigella* spp. sequenced genomes has shown that the genetic repertoire or pangenome comprises 15,741 gene families and that only 993 (6%) of the families are represented in every genome (39). The variable or "accessory" genes thus make up more than 90% of the pan-genome and about 80% of a typical genome; some of these variable genes tend to be collocated on genomic islands. Continuous gene flux has occurred during *E. coli* divergence, mainly as a result of horizontal gene transfers and deletions. This genetic plasticity accelerates the adaptation of *E. coli* to varied environments and lifestyles, as it allows multiple gene combinations that result in phenotypic diversification and the emergence of new hypervirulent strains such as the hybrid STEC and EAggEC O104:H4-B1-ST678 strain that combine resistance and virulence genes, which in classical pathogenic *E. coli* strains traditionally have been mutually exclusive (40). More than describing STEC stains as a separate group of *E. coli*, STEC represents many, if not the majority, of phylogenetic lineages of *E. coli* in general. The sequence-based method targeting housekeeping genes (using distinct sets of genes), multilocus-sequence typing (MLST), generally has less discriminatory power than pulsed-field gel electrophoresis (PFGE) but has been considered the most reliable method to determine the genetic relatedness of epidemiologically unrelated isolates. However, when in silico MLST is performed on whole-genome sequences, many of the strains appear jumbled

and less well resolved (39). *E. coli* strains are assigned by MLST to different sequence types (STs), and within each ST diverse clusters can be observed by PFGE. Three distinct sequence-based methods targeting housekeeping genes exist for *E. coli*. As of October 2013, Institut Pasteur's MLST scheme using *dinB, icdA, pabB, polB, putP, trpA, trpB,* and *uidA* genes lists 599 unique STs (http://www.pasteur.fr/recherche/genopole/PF8/mlst/EColi.html). Mark Achtman's MLST scheme using *adk, fumC, gyrB, icd, mdh, purA,* and *recA* lists 5,873 isolates belonging to 3,874 STs, of which 1,485 are found in 54 ST complexes and 4,388 isolates belonging to 3,562 unique STs have not been assigned an ST complex (mlst.warwick.ac.uk). T. Whittam's MLST scheme, the *Ec*MLST, lists allele sequences and allele profiles for 679 *E. coli* strains and uses different combinations of 15 housekeeping genes, i.e., internal fragments of the seven *aspC, clpX, fadD, icdA, lysP, mdh,* and *uidA* genes where the associated sequence type is defined as st7 or further characterized by using internal fragments of the eight additional *arcA, aroE, cyaA, dnaG, grpE, mtlD, mutS,* and *rpoS* genes where the associated sequence type is defined as st15. Some isolates are only characterized by using internal fragments of two of the seven housekeeping genes, *mdh* and *uidA*, where the associated sequence type is defined as st2 (www.shigatox.net). A comparison study is ongoing to establish correspondence between the two former MLST schemes.

There is increasing evidence that within serotype O157:[H7] there are differences in the clinical outcome and association with HUS. Molecular methods have identified different genetic lineages of *E. coli* O157:H7. Octamer-based genome scanning and microarray comparative genomic hybridization were first to identify three lineages designated I, II, and I/II (41–43). Nucleotide polymorphism-derived genotyping and phylogenetic analyses identified eight major STEC O157 lineages (44). Seven lineages are typically found in cattle, including one that does not associate with human disease and may be evolving away from human virulence and two other lineages accounting for a majority of human disease (44). The 30 sorbitol-fermenting O157 strains belong to a lineage VIII exclusively isolated from humans (44). In a study of 528 O157 strains primarily from Michigan patients but also including strains from Argentina, Australia, Canada, Germany, Japan, and the United Kingdom from 1982 to 2006, Manning et al. identified 39 single nucleotide polymorphism (SNP) genotypes that differed at 20% of SNP loci and separated them into nine distinct clades (45). The outbreak strain TW14359, implicated in a multistate outbreak associated with the consumption of bagged spinach in North America in 2006 (46, 47), was shown to be a member of clade 8, which was significantly associated with younger age (0 to 18 years) and patients with HUS, who were seven times more likely to be infected with clade 8 strains than patients with strains from clades 1 to 7 combined (45). The study revealed substantial genomic differences between clades, suggesting that an emergent subpopulation of the clade 8 lineage has acquired critical factors that contribute to more severe disease. Comparison of the phylogenetically divergent O157:H7 outbreak strains TW14359 and RIMD0509952, which caused the largest O157:H7 outbreak to date in Sakai, Japan, showed that these two strains vary in their ability to colonize or initiate the disease process (48). Interestingly, most LEE genes, the *stx2* genes, and several pO157-encoded genes that promote adherence, including type II secretion genes and their effectors *stcE* and *adfO*, are upregulated in the spinach outbreak strain, whereas flagellar and chemotaxis genes are primarily upregulated in the Sakai strain (48).

Evolutionary analyses of STEC by multilocus enzyme electrophoresis (49) and partial sequencing of 13 housekeeping genes (50) have identified two distantly related clonal groups classified as EHEC 1, including serotype O157:H7 and its inferred ancestor O55:H7, and EHEC 2, represented by several O groups (O26, O111, O118, etc.). These two clonal groups differ in their virulence and global distribution. Although several fully annotated genomic sequences exist for strains of serotype O157:H7, much less is known about the genomic composition of EHEC 2. Analysis of 24 clinical EHEC 2 strains representing serotypes O26:H11, O111:H8/H11, O118:H16, O153:H11, and O15:H11 from humans and animals by comparative genomic hybridization supports the hypothesis that extensive modular shuffling of mobile DNA elements has occurred among STEC strains, and the gene content variation of phage-related genes in EHEC 2 seems to indicate that EHEC 2 is a multiform pathogenic clonal complex, characterized by substantial intraserotype genetic variation. The heterogeneous distribution of mobile elements is especially seen in O26:H11 more than in other EHEC 2 serotypes (51). Comparative analysis of whole-genome phylogeny and of type III secretion system effectors of 114 LEE+ *E. coli* isolates shows that attaching and effacing *E. coli* is divided into five distinct genomic lineages and that the LEE+/*stx*+/*bfp*– genomes are primarily divided into two genomic lineages, the O157/O55 EHEC1 and non-O157 EHEC2 (52). Most importantly, phylogenic relatedness was independent of the presence or absence of *stx*-encoding phages, highlighting the close relation between LEE+ EPEC and STEC lineages (52). In this study of 138 whole genomes of which 114 were LEE+, *stx* genes were only found in phylogroups B1, including EHEC2, EPEC2, and unclassified attaching and effacing *E. coli*, and in E, represented by O157 EHEC1 (52). Even less is known about non-LEE STEC, but whole-genome comparative analysis of nine non-LEE genomes revealed that phage-encoded genes, including non-LEE-encoded effectors, were absent from all nine STEC genomes. Several plasmid-encoded virulence factors reportedly identified in LEE-negative STEC isolates were identified in only a subset of the nine LEE-negative isolates, further confirming the diversity of this group. Characterization of the lambdoid *stx*-encoding phages showed that although the integrase gene sequence corresponded with genomic location, it was not correlated with *stx* subtype, highlighting the mosaic nature of these phages. A wide range of basal and induced expression of the Shiga toxin genes, *stx1* and *stx2*, and the Q genes was observed (53).

TOXIN AND DISEASE

The Shiga toxin family is divided into two branches, Stx1 (almost identical to Stx from *S. dysenteriae* Type 1) and Stx2. Subtypes, denoted by Arabic letters that follow the main type name, may exhibit significant differences in biologic activity, including serologic reactivity, receptor binding, and the capacity to be activated by elastase in intestinal mucus. Variants have been defined by relatedness of sequence within a subtype that differs by one or more AAs from the prototype (54). The phylogenetic relationship of variants has been analyzed (54–60), but not all variants have been examined for all classical phenotypic differences, biologic activity, and hybridization properties. The variants are designated by toxin subtype, O group if the host strain is *E. coli* and generic name of the host bacterium if the host strain is not *E. coli*, followed by the strain name or number from which that toxin was described (54). At present, 107 variants have been identified: subtypes of Stx1 include 9 variants of Stx1a (including Shiga toxin from *S. dysenteriae*), 4 of Stx1c, and 1 of Stx1d, and subtypes of Stx2 include 21 variants of Stx2a, 16 of Stx2b, 18 of Stx2c, 18 of Stx2d, 14 of Stx2e, 2 of Stx2f, and 4 of Stx2g (Fig. 1).

Stx subtypes, and maybe specific variants, are clinically relevant. Stx2a (with or without Stx2c) seems to be highly associated with HUS (30, 61–64). The combination of *eae* and *stx2* especially has been associated with the development of HUS and bloody diarrhea (34, 62, 63, 65, 66), and an unspecified synergism between the adhesin intimin encoded by *eae* and Stx2 has been suggested (32). A German

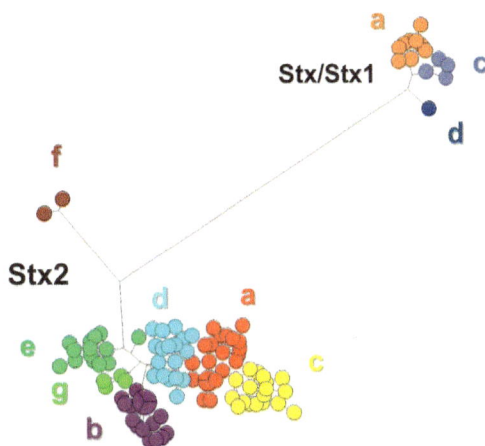

FIGURE 1 Stx subtypes and variants. Parsimony tree of 107 variants: nine variants of Stx1a (including Shiga toxin from *S. dysenteriae*), four variants of Stx1c, one variant of Stx1d, and subtypes of Stx2, including 21 Stx2a, 16 Stx2b, 18 Stx2c, 18 Stx2d, 14 Stx2e, two Stx2f, and four Stx2g variants. Data from reference 54 and updated by the author. doi:10.1128/microbiolspec.EHEC-0019-2013.f1

study of 922 patients with STEC infection found that 81 of 107 patients (76%) with sorbitol-fermenting O157 had HUS (54). This particular STEC strain is typically positive for both *stx2a* and *eae*. Sporadic adult cases of HUS in France were seen in patients infected with six different serotypes, all *vtx2* (*vtx1* and *eae* negative) (67). In Australia, a VTEC O113: H21 strain lacking *eae* was responsible for a cluster of three cases of HUS (27). In Finland, O174:H21 (68) and O:rough:K-:H49 (69), both *eae* negative and *vtx2* positive, were isolated from two separate cases of HUS. In Germany, a large outbreak in 2011 of a hybrid STEC-EAggEC strain O104:H4, *stx2a* and *eae* negative but positive for many EAggEC-associated genes, affected 3,167 patients without HUS (16 deaths) and 908 with HUS (34 deaths) (70).

In Germany, an association between the activatable subtype *vtx2d* in *eae*-negative STEC strains and HUS has been found (61). The median age of 21 years in patients with *stx2d* was considerably higher than in other patients with STEC. Stx2 subtypes Stx2a and Stx2d studied in vitro with Vero monkey kidney cells and primary human renal proximal tubule epithelial cells were at least 25 times more potent than Stx2b and Stx2c. The in vivo potency of Stx2b and Stx2c in mice was similar to that of Stx1, whereas Stx2a and Stx2d were 40 to 400 times more potent than Stx1 (71). It has been suggested that disease outbreaks select for producers of high levels of Stx2a among *E. coli* O157:H7 strains shed by animals and that Stx1 expression is unlikely to be significant in human outbreaks (72). Nearly all lineage I strains carry *stx2a*, whereas all lineage II strains carry *stx2c*, and 4 of 14 lineage I/II strains have copies of both *stx2a* and *stx2c* (73). Real-time PCR and enzyme-linked immunosorbent assay have demonstrated that lineage I and I/II strains produce significantly more *stx2a* mRNA and Stx2a than lineage II strains. However, among lineage I strains significantly more Stx2a is also produced by strains from humans than from cattle. Therefore, lineage-associated differences among *E. coli* O157:H7 strains, such as prophage content, toxin type, and toxin expression, may contribute to host isolation bias. However, the level of Stx2 production alone may also play an important role in the within-lineage association of O157:H7 strains with human clinical disease. Indeed, clade 8 strains associated with HUS overexpress Stx2a when compared to strains from clades 1 to 3, and SNPs, which may affect Stx2a expression and could be useful in the genetic differentiation of highly virulent strains, have been described (74).

Analyses of the 2006 outbreak of O157:H7 clade 8 strains in spinach suggested the presence of *stx2a* and an *stx2c* variant (75), but not all clade 8 strains have both *stx2a* and *stx2c*, and none of the strains has only *stx2c*. The presence and presumable production of the Stx2c variant alone cannot be solely responsible for the enhanced virulence attributed to the clade 8 lineage. This also is true for the production of Stx2a, because it was detected in nearly every strain representing all nine clades (45).

In a Danish outbreak with an O157:H7 strain with a toxin gene subtype profile consisting of *eae*, *stx1a*, and *stx2a*, HUS developed in 62% of the patients. HUS had been previously found in two of six (33%) patients infected with STEC. Infections with STEC O157 containing the *vtx2a* toxin profile were associated with a higher number of HUS cases (24 to 30%), whereas HUS developed in only 1 of 31 (3%) patients infected with O157:H7 *eae* + *stx2c*-positive strains and did not develop in 93 patients with *eae* + *stx1a* with (85 patients) or without (8 patients) *stx2c* (76). The association between *stx2a* and severity of disease has also been demonstrated in two animal models for clinical genotype (CG) strains (carrying *stx2a* with or without Stx2c), which induced more severe clinical symptoms, earlier and higher mortality, and more severe histopathologic lesions compared to bovine-biased genotype (BBG) strains (carrying *stx2c* only) (77). Purified Stx2a has also been shown to be more potent than Stx2c against primary human kidney cell lines and in mouse models (71). It is possible that carriage and expression of Stx2a alone are sufficient to confer increased virulence in animal models and increased expression of human disease, but alternatively, these phenotypes may result in whole or in part in other genetic factors that are correlated with Stx2a.

Stx2c has occasionally been associated with HUS but with a significantly lower risk (62, 64). The majority of 210 patients in Japan infected with Stx2c strains presented no or mild symptoms, except for 3 patients with bloody diarrhea (78). Also in Japan, 169 strains carrying only *stx2vha* (now *stx2c*) (54) were probably less virulent and caused bloody diarrhea less frequently (79). In an Austrian study of 201 STEC strains collected from patients and environmental sources, the *stx2a* and *stx2c* alleles were associated with high virulence and the ability to cause HUS, whereas *stx2d*, *stx2e*, *stx1a*, and *stx1c* occurred in milder or asymptomatic infections (80). However, caution is warranted for infections by Stx2c O157 strains, in addition to Stx2a O157 strains.

Other subtypes or variants of Stx1 and Stx2 are primarily associated with a milder course of disease (61–63). Except for the few individual, unusual cases mentioned below, HUS did not develop in a total of 825 Danish patients as follows: 426 with *vtx1a* STEC (313 *eae* positive and 113 *eae* negative), 65 with *stx1c* (5 *eae* positive and 60 *eae* negative), 13 *stx1d* (all *eae* negative), 87 with *stx2b* (1 *eae* positive and 86 *eae* negative), 12 with *stx2e* (1 *eae* positive and 11 *eae* negative), 37 with *stx2f* (all *eae* positive), 5 with *stx2g* (*eae* negative), and 180 with seven different combinations of *stx1* and *stx2* subtypes, excluding *stx2a* and *stx2d* (70 *eae* positive and 110 *eae* negative) (unpublished Danish surveillance data). In France, sporadic cases of STEC infection complicated with HUS have been described in patients infected with a clone of O103:H2, virulence type *vtx1* and *eae* (81). Among infections with VTEC strains with only the *vtx1* gene, one case of HUS has been described; a strain with the serotype O115:H18 (82) and 8 of 169 non-O157 HUS-associated strains representing 42 different ST types have been subtyped as *vtx1*-only strains (54). Data from the European Surveillance System (TESSy data), as provided by the ECDC on virulence characteristics of reported confirmed VTEC cases in 2007–2010, including all cases, hospitalized cases only, and HUS cases only, show that most HUS cases (89.2%), for which information was reported on virulence factors, were either *eae*, *vtx2* positive or *eae*, *vtx1*, *vtx2* positive and that an additional 5.9% were either *vtx2* positive or *vtx1*, *vtx2* positive but without the *eae* gene. Only 2.3% were positive for *stx1* (1.6% *eae*, *stx1* and 0.7% *stx1*) (25). None of the 124 reported infections due to O103:H2, of which all but one were *eae*, *vtx1* positive, caused HUS (25). Surveillance data rarely include detailed clinical descriptions of individual HUS cases. Follow-up on clinical data on Danish patients with unusual or rare association between HUS and the virulence profile *eae*, *vtx1* revealed possible complicating factors such as antibiotics during the acute infection, underlying nephrotic syndrome, association with outbreak cases infected with an HUS-associated outbreak O157:H7 strain,

or a double infection with O157:H7 (*eae* + *vtx1* + *vtx2*). HUS also developed in one patient infected with O55:H12 *vtx1* only who had received six different antibiotics during the acute phase and in one patient with O13,O73: K1:H18 *vtx2d* (unpublished data). The association of specific virulence factors with HUS should be carefully examined for underlying disease, epidemiologic relation, and antibiotic treatment during the acute phase of illness.

STEC producing Stx2e is closely associated with edema disease in swine, and Stx2e-producing STEC strains are probably not pathogens for humans (83). However, Stx2e-producing *E. coli* strains belonging to O groups O101 and O9 have sporadically been isolated from patients with diarrhea and HUS. These O groups in STEC strains are not usually associated with edema disease (84). Thus, STEC without both *stx2* and *eae* is, with a few exceptions, only sporadically associated with HUS, and toxin subtyping can be useful in identifying high-risk or HUSEC strains.

Stx-ASSOCIATED BACTERIOPHAGE INSERTION

In *E. coli*, the capacity to produce Shiga toxin is encoded by genes on bacteriophages, and it has been suggested that rather than describing the taxonomy of STEC bacteria, the biology and taxonomy of the bacteriophages that are hosted by certain lineages of *E. coli* could serve as the basis for refining this pathogenic group.

Several common *stx* phage insertion sites have been reported in LEE-positive STEC genomes. These insertion sites include *wrbA*, *yecE*, *torS/T*, *sbcB*, *yehV*, *argW*, *ssrA*, and *prfC* (85, 86). In LEE-negative strains, additional insertion sites are often different and include *patC*, *yciD*, *ynfH*, *serU*, *mlrA*, and *yjbM* (53) (Fig. 2). However, a number of sites have not been determined for some of the most common serotypes like O157, O26, O111, and O103

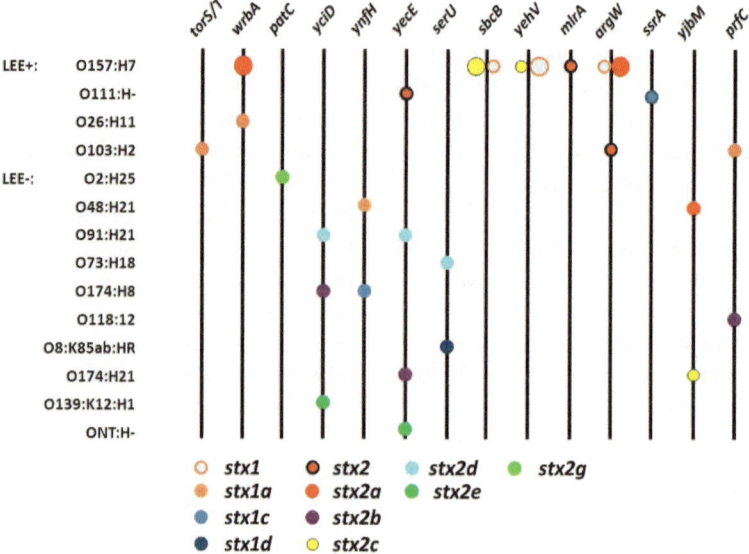

FIGURE 2 Stx bacteriophage insertion sites in LEE-positive STEC include *wrbA, yecE, torS/T, sbcB, yehV, argW, ssrA,* and *prfC*. Data from references 85, 86, and 118. In LEE-negative STEC genomes additional insertion sites are often different and include *patC, yciD, ynfH, serU, mlrA,* and *yjbM*. Data from references 53, 87, and 119. Big circles indicate the preferred bacteriophage integration site.
doi:10.1128/microbiolspec.EHEC-0019-2013.f2

(85). Insertion site occupancy by *stx* phages depends on the host strain and on the availability of the preferred locus in the host strain. For the most part, *yehV* is occupied by *stx1* encoding phages in LEE-positive O157 strains whereas *wrbA* or *argW* is preferentially selected by the *stx2a* phages and *sbcB* by *stx2c* phages (87). Phages preferentially use one insertion site, but if this primary insertion site is unavailable, then a secondary insertion site is selected (88, 89). Molecular epidemiological studies using Stx-associated bacteriophage insertion sites for strain differentiation have shown that specific bacteriophage-associated genetic factors underlie the differential virulence of CG and BBG, where genotypes that are significantly overrepresented among human isolates, as compared to cattle isolates, are referred as CG and genotypes among cattle isolates that show significantly overrepresentation or similar representation, as compared to human isolates, are referred as BBG. STEC O157 can be further classified into several genotypes (90). In a study of a panel of

419 STEC O157 strains from geographically and temporally diverse cattle- and human-origin isolates, the presence of Stx2a-associated bacteriophage sequences detected adjacent to either *wrbA* or *argW* and the detection of stx2a were significantly associated with CG compared to BBG strains, of which many (42.9%) had Stx2a-associated bacteriophage sequences adjacent to *wrbA* or *argW* but lacked detectable *stx2a*. All 281 CG strains had Stx2a-associated bacteriophage sequences inserted in *wrbA* or *argW* and carried *stx2a*. In contrast, the presence of Stx2c-associated bacteriophage sequences inserted in *sbcB* and the presence of *stx2c* were significantly more common in BBG compared to CG isolates. All 107 BBG isolates had both traits, whereas only 11.8% of CG isolates belonged to genotypes with Stx2c-associated bacteriophage sequences inserted in *sbcB*, and only 7.1% of CG isolates carried *stx2c* (87). Deep sequencing of pooled STEC O157 DNAs from human clinical cases (n = 91) and cattle (n = 102) identified 42 genotypes that could be tagged by a minimal set of 32

polymorphisms. Phylogenetic trees of these genotypes are also divided into clades that represent strains of cattle origin, or cattle and human origin (91), thus confirming that certain O157 lineages are more associated with the bovine reservoir and do not turn up in human disease.

The lineage-associated differences among STEC O157:H7 strains such as prophage content, toxin type, and toxin expression may contribute to the observed host isolation bias. However, the level of Stx2 production alone may also play an important role in the within-lineage association of STEC O157:H7 strains with human clinical disease (73).

TOXIN EXPRESSION AND Q GENES

The biology of the Stx-encoding phages contributes greatly to the production of Stx, and many research results during the past decade have contradicted the prevailing assumption that phages serve merely as agents for virulence gene transfer.

In a majority of STEC strains, expression of *stx* genes within lambdoid phages is believed to be largely under the control of the late phage promoter, pR', and the Q antiterminator protein (92). In STEC lysogenic for Stx-converting bacteriophages, expression of the *stx* genes is usually repressed and production of Stx is preceded by prophage induction (93, 94). Variations in the Q gene have been proposed to influence the quantitative expression of Stx (95). The Q gene transcription is increased under inducing conditions allowing for increased transcription of the *stx* genes that are downstream of the Q-binding site (92). Inducing factors include nutritional stress, oxidative stress, UV radiation, antibiotics (mitomycin C, quinolones, β-lactams, etc.), EDTA and other chelating agents, hydrogen peroxide, heat shock, and quorum sensing (96), but Stx1 production can also be induced through low-iron induction of p_{Stx1} (97). Different Q variants have been described. Q_{933} was found in the Stx2a-producing strain EDL933 (98). Q_{21} was found in phage 21, which does not contain any *stx* genes (99), but phage 2851 encoding vtx2c contains a Q gene with 96.9% identity to Q_{21} and has frequently been detected in Stx2c-producing O157 strains, indicating that phages related to 2851 are associated with Stx2c production in O157 strains from different locations and time periods (100), although it can be found in other Stx subtypes (101). The $Q_{O111:H-}$ was found in a Japanese O111:H– isolate, 11128, in a study of O157:H– clinical isolates related to HUS (86). Other Q genes found in Stx1- and Stx2a-producing phages show high identity to the Q_{933}, i.e., the Stx1 phage in H19-B, which shows 96.5% identity to Q_{933} (98); the Stx2a-producing phage, VT2-Sakai, which is identical to Q_{933} (102); the Stx1-producing phage, VT1-Sakai, which shows 97.4% identity with Q_{933} (103); and the Stx1-producing phage in strain Morioka, which shows 97% identity to Q_{933} (104). Using the above-mentioned Q gene sequences as references, Jacobsen (105) recently queried 287 *E. coli* genome sequences of varying quality: Two genome sequences were from Los Alamos National Laboratory and 285 genome sequences were publicly available online (106 from three sequencing projects: *Escherichia* group, *E. coli* O104:H4, and *E. coli* antibiotic resistance) from the Broad Institute of Harvard and Massachusetts Institute of Technology (www.broadinstitute.org/) and 179 from National Center for Biotechnology Information (www.ncbi.nlm.nih.gov/). Information about serotype was available for 131 (45.6%) strains and resulted in data on 57 different serotypes with a bias toward O157:H7 and O104:H4. Jacobsen identified 474 antiterminator Q genes in 235 of 287 genome sequences, of which 80 Q genes were found upstream of the *stx* operon in 62 genomes of at least 14 different serotypes. On the basis of their amino acid sequence, these 80 Q genes could be classified into five main groups (I to V) and further into 10 variants whereby variants within the main groups are distinguished by one amino acid difference. Thirteen Q genes are classified as variant Q_I, which is homolog to Q_{2851} and Q_{21} and, in accordance with previous findings, primarily found in *stx2c* phages although one *stx2a* phage was also identified. Main group Q_{II}, represented by six Q genes, is homolog to $Q_{O111:H–}$, where one is characterized as Q_{IIb} in an *stx2a* phage and five are characterized as Q_{IIa} found in *stx2a* (one) and *stx2d* (four) phages, respectively. Q_{III} is the smallest group with only two Q genes in *stx2e* phages, whereas Q_{IV} has as many as 59 Q genes, with homologs to Q_{933W}, Q_{H-19B}, $Q_{Morioka}$, and $Q_{VT1-Sakai}$. Q_{IV} can be divided into six variants, none of which is found in *stx2c*, *stx2d*, or *stx2e* phages, but Q_{IVa} is found in *stx1a* and *stx2a*, Q_{IVb} in *stx1a*, Q_{IVc} in *stx1c*, Q_{IVd} in *stx2a*, Q_{IVe} in *stx1a*, and Q_{IVf} in *stx2a* phages (105).Group V contained only two sequences that were not used in Jacobsen's study. Among all Q variants, 23 of 169 sites were conserved sites, and 32 were functionally conserved. There were 13 different Q-*vtx* combinations based on Q variants and *vtx* subtypes, but some strains carried two VT phages, resulting in 17 different Q-*vtx* genotypes. These phylogenetic analyses of the associated *stx* genes thus revealed six different *vtx* subtypes, *vtx*1a, *vtx*1c, *vtx*2a, *vtx*2c, *vtx*2d, and *vtx*2e. Q genes associated with the *stx2b* and *stx2g* phages in isolates EH250 and 7V, respectively, are apparently associated with another phage in the isolate, indicating that Q expression might have a contribution from additional Q genes in non-*stx* phages (53).

The genetic architecture does not appear to be the only factor affecting Stx expression. The inferred phylogeny based on the alignment of *stx* bacteriophages indicates a broad phylogenetic diversity also observed with whole-genome phylogeny (53). However, taken together, the present data seem to suggest that the primary sequence of Q may play a major role in the regulation and released amount of Shiga toxin. In O157, DNA sequencing of the genes flanking the *stx2c* gene from a lineage II strain, EC970520, has revealed that the Q gene is replaced by a *pphA* (serine/threonine phosphatase) homolog in this and all other lineage II strains tested with very low similarity to the antiterminator Q of lineage I strains, which appears to be a useful

molecular marker to distinguish among Stx2a- and Stx2c-encoding phages in O157 strains (73).

The great variation in Q genes compared to *stx* genes indicates that the *stx* genes have been taken up by lambdoid phages on several occasions and that there is a high genetic selection for this gene combination and that all the Q genes, functionally expressing their lysis genes, can take up the *stx* genes under certain favorable conditions.

GENOMIC ISLANDS

Sequencing of STEC O157:H7 EDL933 revealed O157:H7-specific DNA organized in genomic O islands (106), some of which are referred to as pathogenicity islands because they can contain a flexible pool of virulence-associated genes. One such pathogenicity island is LEE, which encodes a type III secretion system involved in the formation attaching and effacing lesions. The most important protein is intimin encoded by the gene *eae*, which is responsible for both the intimate adhesion of bacteria to the intestinal epithelium and the attaching and effacing lesion. Analysis of the variable C-terminal encoding sequence of *eae* defines at least 29 distinct intimin types (α1, α2, β1, β2, β3, γ1, θ1, κ, δ, ε1, ε2, ε3, ε4, ε5, ζ1, ζ3, η1, η2, ι1-A, ι1-B, ι1-C, ι2, λ, μ, υ, ο, π, ρ, σ) (107–109) that have been associated with tissue tropism. Intimin binds to the cell receptor *Tir*, which is translocated by bacteria to the enterocyte through a type III secretion system (110). One SNP, 255 T>A in *tir*, has been shown to predict the propensity of STEC O157 isolates to cause human clinical disease (111). The overrepresentation of the *tir* 255 T>A T allele in human-derived isolates versus the *tir* 255 T>A A allele suggests that these isolates have a higher propensity to cause disease. The high frequency of bovine isolates with the A allele suggests a possible bovine ecological niche for this STEC O157 subset. A Norwegian study of 167 human and nonhuman *E. coli* O157:H7/NM (nonmotile) isolates with respect to the 255 T A/T SNP in the *tir* gene was able to differentiate STEC O157 into distinct virulence clades (1 to 3 and 8) and found an overrepresentation of the T allele among human strains compared to nonhuman strains, including five of six HUS cases (112).

OI-122 is another 23-kb pathogenicity island with at least six genes with significant homology to known virulence genes: *pagC* of *Salmonella enterica* serovar Typhimurium; *sen* of *Shigella flexneri*; two non-LEE effector (*nle*) genes *nleB* and *nleE* of *Citrobacter rodentium*; and the EHEC factor for adherence gene cluster *efa1* and *efa2* found in STEC O157:H7. A modular arrangement of OI-122 genes based on their association with each other across HUS-associated non-O157 VTEC strains was proposed: module 1 contains Z4318, *pagC*, and Z4322; module 2 contains Z4323, *sen*, *nleB*, and *nleE*; and module 3 contains the *efa* gene cluster, also referred to as *lifA* (lymphocyte activation inhibitor) (113, 114). A progressive decrease in the prevalence of OI-122 genes in non-O157 VTEC strains belonging to seropathotype B through E has been suggested (3), and a Belgian study of 265 STEC isolates also observed a progressive decrease in the frequency of complete OI-122 in seropathotypes A through D, with a concomitant increase in the frequencies of incomplete and absent OI-122. The variable virulence profiles of the non-O157 serotypes indicated that complete OI-122 was more frequently present in isolates associated with HUS and that the individual genes *stx2*, *eae*, *espP*, as well as the OI-122-associated genes *sen*, *nleB*, *nleE*, and the *efa/lifA* gene cluster were significantly more often present in non-O157 STEC associated with HUS, and that the combined virulence profile *vtx2-nleE-efa/lifA* showed the strongest association with HUS (24). The presence of putative transposases in OI-122 has led to the hypothesis that its elements are acquired or lost in a modular manner. Although *pagC*, Z4322, *sen*, *nleB*, *nleE*, and *efa1/lifA* individually are more prevalent in non-O157 VTEC associated with HUS, the simultaneous presence of all of these genes strengthens the association with serious disease. Thus, after acquiring OI-122 modules from STEC O157:H7 through horizontal transfer, less pathogenic non-O157 strains could cross a virulence threshold, resulting in sufficient pathogenicity to cause HUS (115).

TAXONOMY AND PUBLIC HEALTH

More than 472 O:H serotypes of STEC have been described (28). Some seropathotypes seem to be more closely related to specific reservoirs and specific diseases, and their number and characteristics are ever changing. However, there seems to be a much closer relationship between the virulence profile, i.e., *stx* subtype, additional "cocktail" genes, and the clinical course of disease than to the serotype itself—even for O157 (76). Many of these genes are acquired by horizontal gene transfer, and it is therefore possible (and plausible) that certain types of *E. coli* that have never been associated with disease or outbreaks could acquire mobile genetic elements such as the Stx encoding bacteriophages through parallel evolution, which in turn could convert them into virulent strains capable of causing human infection and outbreaks. Understanding this complex biology of host-pathogen interaction is evidently a high-priority research subject, and a better understanding of the real genetic basis behind our classification of seropathotypes is needed. The presence of LEE and OI-122 in many of the A and B seropathotypes has been very useful but has also been challenged by the serious O104:H4 outbreak in Germany in 2011 as well as by other non-LEE STEC outbreaks and HUS cases. Furthermore, the genome sequences for some of the common non-O157 O groups, O26, O103, and O111, have revealed that these pathogenic groups have arisen multiple times by acquisition of mobile genetic elements harboring virulence genes. Lambdoid prophages and other integrative elements have played a major role in the stepwise evolution of pathogenic lineages of STEC, especially the acquisition of type 3 secretion effectors, i.e., non-LEE-encoding *nle* genes, and some of these otherwise unlinked effectors can work in concert with one another to produce a desired effect on the host cell (116, 117). More whole-genome sequence information on non-O157 STEC and STEC-related *E. coli* strains is needed to better understand and further categorize the clinically relevant pathotypes for the necessary public health response and action.

In public health the primary goal is to (i) prevent onset of disease, (ii) minimize the risk of progression of the disease in individuals or transmission of illness, and (iii) provide rehabilitation to prevent the worsening of an individual's health. The methods and techniques to categorize an STEC strain isolated from a patient at the present time require extensive analyses that are usually not available in the primary diagnostic line where public health action and management of STEC-infected patients begins. In the first-line diagnosis, three primary issues need immediate attention:

- Risk of severe disease and prognosis
- Quarantine of infected individuals
- Treatment with antibiotics, especially with long-term otherwise healthy carriers

In Denmark, the observed and well-documented association between Stx2 and *eae* or ability to colonize and persist in the gut, such as the EAEC hybrid strains, has resulted in a basic and primary definition of HUSEC for first-line public health action: *stx2* in a background of *eae*- or *aggR*-positive *E. coli* is HUS-associated and is kept "on hold" until *stx* is subtyped and further characterized. All other STEC strains with the virulence profiles of *vtx1*, *vtx1* and *eae*, *vtx2*, *vtx1* and *vtx2*, regardless of serotype, are considered "low-risk" STEC. If this HUSEC definition had been applied, 71% (1,454 of 2,046) of Danish patients would have been informed that they had a "low-risk" STEC infection during the years 1983 to 2012. Twenty-nine percent (552 of 2,046) would have been informed that they

might have an HUSEC infection. If *vtx2f* is diagnosed as unspecified *vtx2*, an additional 55 (3%) patients would have been added, but as *stx2f* requires a different and specific set of primers, this would not have been relevant in most cases. Further refining of the definition of HUSEC by *stx* subtyping to include only *stx2a* and *stx2d* as HUSEC would reduce the number of patients with HUSEC to 11% (224 of 2,062). For specific, mainly socioeconomic reasons, antibiotic treatment of patients infected with VTEC is also considered according to specific criteria, which include the isolation of identical "low-risk" types of STEC found in two separate stool specimens, no isolation of HUSEC, and a detailed characterization of the STEC strain's virulence profile and serotype. In the first line this includes the detection of the *eae, stx1, stx2, aggR,* and *aaiC* genes followed by a more detailed second-line subtyping of the *stx* genes. This simplified approach has been generally accepted by the primary diagnostic clinical microbiology laboratories and public health officers in Denmark. Though not complete in the characterization of each STEC isolate, this approach is practical and easy to use in an operational environment, which is expected to quickly (i) evaluate the risk of progression of the disease in individuals, (ii) minimize transmission of HUSEC, (iii) rehabilitate individuals to prevent the worsening of an individual's health, and (iv) reduce the socioeconomic impact on their families. However, such a simple approach should be applied with prudence because each individual case is unique, and each patient should be carefully evaluated with regard to predisposing factors, general clinical condition, contact with other STEC-infected individuals, possible link to an outbreak, and so forth. The identification of "low-risk" STEC does not per se exclude the risk of progression to severe disease, dehydration, or HUS, and patients with bloody stools and/or affected kidney function must be carefully monitored.

The versatile integration of mobile virulence factors in the stepwise and parallel evolution of pathogenic lineages of STEC collides with the requirements of a good taxonomy and complicates how public health goals are met in a timely manner. The intertwinement of the traditional taxonomy concept and the association of (sero)-pathotypes to epidemiological evidence, phenotypic traits, clinical features of the disease, and specific virulence factors is therefore an ongoing challenge, and the need to identify factors of STEC that absolutely predict the potential to cause human disease is obvious in terms of clinical management, supportive or antibiotic treatment, quarantine measurements, risk assessment, surveillance, and outbreak investigations and management.

However, recent sequence information has provided advanced analytical tools that will reduce the clinical divide between the STEC types that are associated with severe disease from the seemingly benign STEC types.

ACKNOWLEDGMENTS

I declare no conflicts of interest with regard to the manuscript.

CITATION

Scheutz F. 2014. Taxonomy meets public health: the case of Shiga toxin-producing *Escherichia coli*. Microbiol Spectrum 2(3):EHEC-0019-2013.

REFERENCES

1. **Neter E, Westphal O, Lüderitz O, Gino RM, Gorzynski EA.** 1955. Demonstration of antibodies against enteropathogenic *Escherichia coli* in sera of children of various ages. *Pediatrics* **16**:801–807.

2. **Kaper JB, Nataro JP, Mobley HL.** 2004. Pathogenic *Escherichia coli*. *Nat Rev Microbiol* **2**:123–140.

3. **Karmali MA, Mascarenhas M, Shen S, Ziebell KA, Johnson S, Reid-Smith R, Isaac-Renton J, Clark C, Rahn K, Kaper JB.** 2003. Association of genomic O island 122 of *Escherichia coli* EDL 933 with verocytotoxin-producing *Escherichia coli* seropathotypes that are linked to epidemic and/or serious disease. *J Clin Microbiol* **41**:4930–4940.

4. **Konowalchuk J, Speirs JI, Stavric S.** 1977. Vero response to a cytotoxin of *Escherichia coli*. *Infect Immun* **18**:775–779.

5. **O'Brien AD, LaVeck GD, Thompson MR, Formal SB.** 1982. Production of *Shigella dysenteria* type 1-like cytotoxin by *Escherichia coli*. *J Infect Dis* **146**:763–769.

6. **O'Brien AD, La Veck GF.** 1983. Purification and characterization of a *Shigella dysenteria* 1-like toxin produced by *Escherichia coli*. *Infect Immun* **40**:675–683.

7. **Karmali MA, Petric M, Louie S, Cheung R.** 1986. Antigenic heterogeneity of *Escherichia coli* verotoxins. *Lancet* **i**:164–165.

8. **Scotland SM, Smith HR, Rowe B.** 1985. Two distinct toxins active on Vero cells from *Escherichia coli* O157. *Lancet* **ii**:885–886.

9. **O'Brien AD, Newland JW, Miller SF, Holmes RK, Smith HW, Formal SB.** 1984. Shiga-like toxin-converting phages from *Escherichia coli* strains that cause hemorrhagic colitis or infantile diarrhea. *Science* **226**:694–696.

10. **Scotland SM, Smith HR, Willshaw GA, Rowe B.** 1983. Vero cytotoxin production in strain of *Escherichia coli* is determined by genes carried on bacteriophage. *Lancet* **ii**:216. (Letter).

11. **Smith HR, Day NP, Scotland SM, Gross RJ, Rowe B.** 1984. Phage-determined production of Vero cytotoxin in strains of *Escherichia coli* serogroup O157. *Lancet* **i**:1242–1243. (Letter).

12. **Smith HW, Green P, Parsell Z.** 1983. Vero cell toxins in *Escherichia coli* and related bacteria: transfer by phage and conjugation and toxic action in laboratory animals, chickens and pigs. *J Gen Microbiol* **129**:3121–3137.

13. **Riley LW, Remis RS, Helgerson SD, McGee HB, Wells JG, Davis BR, Herbert RJ, Olcott ES, Johnson LM, Hargret NT, Blake PA, Cohen ML.** 1983. Hemorrhagic colitis associated with a rare *Escherichia coli* serotype. *N Engl J Med* **308**:681–685.

14. **Johnson WM, Lior H, Bezanson GS.** 1983. Cytotoxic *Escherichia coli* O157:H7 associated with hemorrhagic colitis in Canada. *Lancet* **i**:76.

15. **O'Brien AD, Lively TA, Chen ME, Rothman SW, Formal SB.** 1983. *Escherichia coli* O157:H7 strains associated with haemorrhagic colitis in the United States produce a *Shigella dysenteriae* 1 (Shiga) like cytotoxin. *Lancet* **i**:702.

16. **Karmali MA, Steele BT, Petric M, Lim CB.** 1983. Sporadic cases of haemolytic-uraemic syndrome associated with faecal cytotoxin and cytotoxin-producing *Escherichia coli* in stools. *Lancet* **i**:619–620.

17. **Karmali MA, Petric M, Lim C, Fleming PC, Steele BT.** 1983. *Escherichia coli* cytotoxin

haemolytic-uraemic syndrome, and haemorrhagic colitis. *Lancet* **ii:**1299–1300.

18. **Levine MM, Edelman R.** 1984. Enteropathogenic *Escherichia coli* of classic serotypes associated with infant diarrhea: epidemiology and pathogenesis. *Epidemiol Rev* **6:**31–51.

19. **Levine MM, Xu J-G, Kaper JB, Lior H, Prado V, Nataro JP, Karch H, Wachsmuth IK.** 1987. A DNA probe to identify enterohemorrhagic *Escherichia coli* of O157:H7 and other serotypes that cause hemorrhagic colitis and hemolytic uremic syndrome. *J Infect Dis* **156:**175–182.

20. **Wade WG, Thom BT, Evans N.** 1979. Cytotoxic enteropathogenic *Escherichia coli. Lancet* **ii:**1235–1236.

21. **Scotland SM, Day NP, Willshaw GA, Rowe B.** 1980. Cytotoxic enteropathogenic *Escherichia coli. Lancet* **1:**90.

22. **Scheutz F, Strockbine NA.** 2005. *Escherichia,* p 607–624. *In* Garrity GM, Brenner DJ, Krieg NR, Staley JT (ed), Bergey's Manual of Systematic Bacteriology, **vol 2**, part B, *The Gammaproteobacteria.* Springer, New York, NY.

23. **Bosilevac JM, Koohmaraie M.** 2011. Prevalence and characterization of non-O157 Shiga toxin-producing *Escherichia coli* isolates from commercial ground beef in the United States. *Appl Environ Microbiol* **77:**2103–2112.

24. **Buvens G, Pierard D.** 2012. Virulence profiling and disease association of verocytotoxin-producing *Escherichia coli* O157 and non-O157 isolates in Belgium. *Foodborne Pathog Dis* **9:**530–535.

25. **EFSA Panel on Biological Hazards (BIOHAZ).** 2013. Scientific opinion on VTEC-seropathotype and scientific criteria regarding pathogenicity assessment. *EFSA J* **11:**3138. 1–106. 7-3-2013. European Food Safety Authority, Parma, Italy.

26. **Centers for Disease Control and Prevention.** 1995. Outbreak of acute gastroenteritis attributable to *Escherichia coli* serotype O104:H21—Helena, Montana. *Morb Mortal Wkly Rep* **44:**501–503.

27. **Paton AW, Woodrow MC, Doyle RM, Lanser JA, Paton JC.** 1999. Molecular characterization of a Shiga toxigenic *Escherichia coli* O113:H21 strain lacking *eae* responsible for a cluster of cases of hemolytic-uremic syndrome. *J Clin Microbiol* **37:**3357–3361.

28. **Mora A, Herrera A, Lopez C, Dahbi G, Mamani R, Pita JM, Alonso MP, Llovo J, Bernardez MI, Blanco JE, Blanco M, Blanco J.** 2011. Characteristics of the Shiga-toxin-producing enteroaggregative *Escherichia coli* O104:H4 German outbreak strain and of STEC strains isolated in Spain. *Int Microbiol* **14:**121–141.

29. **European Food Safety Authority and European Centre for Disease Prevention and Control.** 2013. The European Union summary report on trends and sources of zoonoses, zoonotic agents and food-borne outbreaks in 2011. *EFSA J* **11:**3129. 1–250. 4-9-2013. European Food Safety Authority, Parma, Italy.

30. **Kawano K, Okada M, Haga T, Maeda K, Goto Y.** 2008. Relationship between pathogenicity for humans and *stx* genotype in Shiga toxin-producing *Escherichia coli* serotype O157. *Eur J Clin Microbiol Infect Dis* **27:**227–232.

31. **Pai CH, Ahmed N, Lior H, Johnson WM, Sims HV, Woods DE.** 1988. Epidemiology of sporadic diarrhea due to verocytotoxin-producing *Escherichia coli*: a two-year prospective study. *J Infect Dis* **157:**1054–1057.

32. **Boerlin P, Mcewen SA, Boerlin-Petzold F, Wilson JB, Johnson RP, Gyles CL.** 1999. Associations between virulence factors of Shiga toxin-producing *Escherichia coli* and disease in humans. *J Clin Microbiol* **37:**497–503.

33. **Gyles CL.** 2007. Shiga toxin-producing *Escherichia coli*: an overview. *J Anim Sci* **85:**E45–E62.

34. **Ethelberg S, Olsen KEP, Scheutz F, Jensen C, Schiellerup P, Engberg J, Petersen AM, Olesen B, Gerner-Smidt P, Mølbak K.** 2004. Virulence factors for hemolytic uremic syndrome, Denmark. *Emerg Infect Dis* **10:**842–847.

35. **Cookson AL, Bennett J, Thomson-Carter F, Attwood GT.** 2007. Molecular subtyping and genetic analysis of the enterohemolysin gene (*ehxA*) from Shiga toxin-producing *Escherichia coli* and atypical enteropathogenic *E. coli. Appl Environ Microbiol* **73:**6360–6369.

36. **Schmidt H, Karch H.** 1996. Enterohemolytic phenotypes and genotypes of Shiga toxin-producing *Escherichia coli* O111 strains from patients with diarrhea and hemolytic-uremic syndrome. *J Clin Microbiol* **34:**2364–2367.

37. **Mundy R, Jenkins C, Yu J, Smith H, Frankel G.** 2004. Distribution of espI among clinical enterohaemorrhagic and enteropathogenic *Escherichia coli* isolates. *J Med Microbiol* **53:**1145–1149.

38. **Bettelheim KA.** 2007. The non-O157 shiga-toxigenic (verocytotoxigenic) *Escherichia coli*; under-rated pathogens. *Crit Rev Microbiol* **33:**67–87.

39. **Lukjancenko O, Wassenaar TM, Ussery DW.** 2010. Comparison of 61 sequenced *Escherichia coli* genomes. *Microb Ecol* **60:**708–720.

40. **Mellmann A, Harmsen D, Cummings CA, Zentz EB, Leopold SR, Rico A, Prior K, Szczepanowski R, Ji Y, Zhang W, McLaughlin SF, Henkhaus JK, Leopold B, Bielaszewska M, Prager R, Brzoska PM, Moore RL, Guenther S, Rothberg JM, Karch H.** 2011. Prospective genomic characterization of the German enterohemorrhagic *Escherichia coli* O104:H4 outbreak by rapid next generation sequencing technology. *PloS One* **6:**e22751.

41. Kim J, Nietfeldt J, Benson AK. 1999. Octamer-based genome scanning distinguishes a unique subpopulation of *Escherichia coli* O157:H7 strains in cattle. *Proc Natl Acad Sci USA* **96:**13288–13293.

42. Yang Z, Kovar J, Kim J, Nietfeldt J, Smith DR, Moxley RA, Olson ME, Fey PD, Benson AK. 2004. Identification of common subpopulations of non-sorbitol-fermenting, beta-glucuronidase-negative *Escherichia coli* O157:H7 from bovine production environments and human clinical samples. *Appl Environ Microbiol* **70:**6846–6854.

43. Zhang Y, Laing C, Steele M, Ziebell K, Johnson R, Benson AK, Taboada E, Gannon VP. 2007. Genome evolution in major *Escherichia coli* O157:H7 lineages. *BMC Genomics* **8:**121.

44. Bono JL, Smith TP, Keen JE, Harhay GP, McDaneld TG, Mandrell RE, Jung WK, Besser TE, Gerner-Smidt P, Bielaszewska M, Karch H, Clawson ML. 2012. Phylogeny of Shiga toxin-producing *Escherichia coli* O157 isolated from cattle and clinically ill humans. *Mol Biol Evol* **29:**2047–2062.

45. Manning SD, Motiwala AS, Springman AC, Qi W, Lacher DW, Ouellette LM, Mladonicky JM, Somsel P, Rudrik JT, Dietrich SE, Zhang W, Swaminathan B, Alland D, Whittam TS. 2008. Variation in virulence among clades of *Escherichia coli* O157:H7 associated with disease outbreaks. *Proc Natl Acad Sci USA* **105:**4868–4873.

46. Centers for Disease Control and Prevention. 2006. Ongoing multistate outbreak of *Escherichia coli* serotype O157:H7 infections associated with consumption of fresh spinach—United States, September 2006. *Morb Mortal Wkly Rep* **55:**1045–1046.

47. Grant J, Wendelboe AM, Wendel A, Jepson B, Torres P, Smelser C, Rolfs RT. 2008. Spinach-associated *Escherichia coli* O157:H7 outbreak, Utah and New Mexico, 2006. *Emerg Infect Dis* **14:**1633–1636.

48. Abu-Ali GS, Ouellette LM, Henderson ST, Whittam TS, Manning SD. 2010. Differences in adherence and virulence gene expression between two outbreak strains of enterohaemorrhagic *Escherichia coli* O157:H7. *Microbiology* **156:**408–419.

49. Whittam TS. 1998. Evolution of *Eschericia coli* O157:H7 and other Siga toxin-producing *E. coli* strains, p 195–209. *In* Kaper JB, O'Brien AD (ed), *Escherichia coli* O157:H7 and Other Shiga Toxin-Producing *E. coli*. Strains, 1st ed. ASM Press, Washington, DC.

50. Tarr CL, Large TM, Moeller CL, Lacher DW, Tarr PI, Acheson DW, Whittam TS. 2002. Molecular characterization of a serotype O121:H19 clone, a distinct Shiga toxin-producing clone of pathogenic *Escherichia coli*. *Infect Immun* **70:** 6853–6859.

51. Abu-Ali GS, Lacher DW, Wick LM, Qi W, Whittam TS. 2009. Genomic diversity of pathogenic *Escherichia coli* of the EHEC 2 clonal complex. *BMC Genomics* **10:**296.

52. Hazen TH, Sahl JW, Fraser CM, Donnenberg MS, Scheutz F, Rasko DA. 2013. Refining the pathovar paradigm via phylogenomics of the attaching and effacing *Escherichia coli*. *Proc Natl Acad Sci USA* **110:**12810–12815.

53. Steyert SR, Sahl JW, Fraser CM, Teel LD, Scheutz F, Rasko DA. 2012. Comparative genomics and *stx* phage characterization of LEE-negative Shiga toxin-producing *Escherichia coli*. *Front Cell Infect Microbiol* **2:**133.

54. Scheutz F, Teel LD, Beutin L, Pierard D, Buvens G, Karch H, Mellmann A, Caprioli A, Tozzoli R, Morabito S, Strockbine NA, Melton-Celsa AR, Sanchez M, Persson S, O'Brien AD. 2012. Multicenter evaluation of a sequence-based protocol for subtyping shiga toxins and standardizing Stx nomenclature. *J Clin Microbiol* **50:** 2951–2963.

55. Asakura H, Makino SI, Kobori H, Watarai M, Shirahata T, Ikeda T, Takeshi K. 2001. Phylogenetic diversity and similarity of active sites of Shiga toxin (Stx) in Shiga toxin-producing *Escherichia coli* (STEC) isolates from humans and animals. *Epidemiol Infect* **127:**27–36.

56. Bastian SN, Carle I, Grimont F. 1998. Comparison of 14 PCR systems for the detection and subtyping of stx genes in Shiga-toxin-producing *Escherichia coli*. *Res Microbiol* **149:**457–472.

57. De Baets L, Van der Taelen I, De Filette M, Piérard D, Allison L, De Greve H, Hernalsteens JP, Imberechts H. 2004. Genetic typing of shiga toxin 2 variants of *Escherichia coli* by PCR-restriction fragment length polymorphism analysis. *Appl Environ Microbiol* **70:**6309–6314.

58. Iwasa M, Makino S, Asakura H, Kobori H, Morimoto Y. 1999. Detection of *Escherichia coli* O157:H7 from *Musca domestica* (Diptera: Muscidae) at a cattle farm in Japan. *J Med Entomol* **36:**108–112.

59. Lee JE, Reed J, Shields MS, Spiegel KM, Farrell LD, Sheridan PP. 2007. Phylogenetic analysis of Shiga toxin 1 and Shiga toxin 2 genes associated with disease outbreaks. *BMC Microbiol* **7:**109.

60. Piérard D, Muyldermans G, Moriau L, Stevens D, Lauwers S. 1998. Identification of new vero-cytotoxin type 2 variant B-subunit genes in human and animal *Escherichia coli* isolates. *J Clin Microbiol* **36:**3317–3322.

61. Bielaszewská M, Friedrich AW, Aldick T, Schurk-Bulgrin R, Karch H. 2006. Shiga toxin activatable by intestinal mucus in *Escherichia coli* isolated from humans: predictor for a severe clinical outcome. *Clin Infect Dis* **43:**1160–1167.

62. Friedrich AW, Bielaszewská M, Zhang WL, Pulz M, Kuczius T, Ammon A, Karch H. 2002. *Escherichia coli* harboring Shiga toxin 2 gene variants: frequency and association with clinical symptoms. *J Infect Dis* **185**:74–84.

63. Persson S, Olsen KEP, Ethelberg S, Scheutz F. 2007. Subtyping method for *Escherichia coli* Shiga toxin (verocytotoxin) 2 variants and correlations to clinical manifestations. *J Clin Microbiol* **45:** 2020–2024.

64. Eklund M, Leino K, Siitonen A. 2002. Clinical *Escherichia coli* strains carrying *stx* genes: *stx* variants and *stx*-positive virulence profiles. *J Clin Microbiol* **40:**4585–4593.

65. Pedersen MG, Hansen C, Riise E, Persson S, Olsen KE. 2008. Subtype-specific suppression of Shiga toxin 2 released from *Escherichia coli* upon exposure to protein synthesis inhibitors. *J Clin Microbiol* **46:**2987–2991.

66. Rivas M, Miliwebsky E, Chinen I, Roldan CD, Balbi L, Garcia B, Fiorilli G, Sosa-Estani S, Kincaid J, Rangel J, Griffin PM. 2006. Characterization and epidemiologic subtyping of Shiga toxin-producing *Escherichia coli* strains isolated from hemolytic uremic syndrome and diarrhea cases in Argentina. *Foodborne Pathog Dis* **3:**88–96.

67. Bonnet R, Souweine B, Gauthier G, Rich C, Livrelli V, Sirot J, Joly B, Forestier C. 1998. Non-O157:H7 Stx2-producing *Escherichia coli* strains associated with sporadic cases of hemolytic-uremic syndrome in adults. *J Clin Microbiol* **36:**1777–1780.

68. Keskimäki M, Ikaheimo R, Karkkainen P, Scheutz F, Ratiner Y, Puohiniemi R, Siitonen A. 1997. Shiga toxin-producing *Escherichia coli* serotype OX3:H21 as a cause of hemolytic-uremic syndrome. *Clin Infect Dis* **24:**1278–1279.

69. Keskimäki M, Ratiner Y, Oinonen S, Leijala E, Nurminen M, Saari M, Siitonen A. 1999. Haemolytic-uraemic syndrome caused by vero toxin-producing *Escherichia coli* serotype Rough: K-:H49. *Scand J Infect Dis* **31:**141–144.

70. Rasko DA, Webster DR, Sahl JW, Bashir A, Boisen N, Scheutz F, Paxinos EE, Sebra R, Chin CS, Iliopoulos D, Klammer A, Peluso P, Lee L, Kislyuk AO, Bullard J, Kasarskis A, Wang S, Eid J, Rank D, Redman JC, Steyert SR, Frimodt-Moller J, Struve C, Petersen AM, Krogfelt KA, Nataro JP, Schadt EE, Waldor MK. 2011. Origins of the *E. coli* strain causing an outbreak of hemolytic-uremic syndrome in Germany. *N Engl J Med* **365:**709–717.

71. Fuller CA, Pellino CA, Flagler MJ, Strasser JE, Weiss AA. 2011. Shiga toxin subtypes display dramatic differences in potency. *Infect Immun* **79:**1329–1337.

72. Baker DR, Moxley RA, Steele MB, LeJeune JT, Christopher-Hennings J, Chen DG, Hardwidge PR, Francis DH. 2007. Differences in virulence among *Escherichia coli* O157:H7 strains isolated from humans during disease outbreaks and from healthy cattle. *Appl Environ Microbiol* **73:**7338–7346.

73. Zhang Y, Laing C, Zhang Z, Hallewell J, You C, Ziebell K, Johnson RP, Kropinski AM, Thomas JE, Karmali M, Gannon VP. 2010. Lineage and host source are both correlated with levels of Shiga toxin 2 production by *Escherichia coli* O157:H7 strains. *Appl Environ Microbiol* **76:**474–482.

74. Neupane M, Abu-Ali GS, Mitra A, Lacher DW, Manning SD, Riordan JT. 2011. Shiga toxin 2 overexpression in *Escherichia coli* O157:H7 strains associated with severe human disease. *Microb Pathog* **51:**466–470.

75. Uhlich GA, Sinclair JR, Warren NG, Chmielecki WA, Fratamico P. 2008. Characterization of Shiga toxin-producing *Escherichia coli* isolates associated with two multistate food-borne outbreaks that occurred in 2006. *Appl Environ Microbiol* **74:**1268–1272.

76. Søborg B, Lassen SG, Muller L, Jensen T, Ethelberg S, Molbak K, Scheutz F. 2013. A verocytotoxin-producing *E. coli* outbreak with a surprisingly high risk of haemolytic uraemic syndrome, Denmark, September–October 2012. *Euro Surveill* **18:**ii, 20350.

77. Shringi S, Garcia A, Lahmers KK, Potter KA, Muthupalani S, Swennes AG, Hovde CJ, Call DR, Fox JG, Besser TE. 2012. Differential virulence of clinical and bovine-biased enterohemorrhagic *Escherichia coli* O157:H7 genotypes in piglet and Dutch belted rabbit models. *Infect Immun* **80:**369–380.

78. Kawano K, Ono H, Iwashita O, Kurogi M, Haga T, Maeda K, Goto Y. 2012. *stx* genotype and molecular epidemiological analyses of Shiga toxin-producing *Escherichia coli* O157:H7/H- in human and cattle isolates. *Eur J Clin Microbiol Infect Dis* **31:**119–127.

79. Nishikawa Y, Zhou Z, Hase A, Ogasawara J, Cheasty T, Haruki K. 2000. Relationship of genetic type of shiga toxin to manifestation of bloody diarrhea due to enterohemorrhagic *Escherichia coli* serogroup O157 isolates in Osaka City, Japan. *J Clin Microbiol* **38:**2440–2442.

80. Orth D, Grif K, Khan AB, Naim A, Dierich MP, Wurzner R. 2007. The Shiga toxin genotype rather than the amount of Shiga toxin or the cytotoxicity of Shiga toxin in vitro correlates with the appearance of the hemolytic uremic syndrome. *Diagn Microbiol Infect Dis* **59:**235–242.

81. **Mariani-Kurkdjian P, Denamur E, Milon A, Picard B, Cave H, Lambert-Zechovsky N, Loirat C, Goullet P, Sansonetti PJ, Elion J.** 1993. Identification of a clone of *Escherichia coli* O103:H2 as a potential agent of hemolytic-uremic syndrome in France. *J Clin Microbiol* **31:**296–301.

82. **Beutin L, Krause G, Zimmermann S, Kaulfuss S, Gleier K.** 2004. Characterization of Shiga toxin-producing *Escherichia coli* strains isolated from human patients in Germany over a 3-year period. *J Clin Microbiol* **42:**1099–1108.

83. **Scheutz F, Ethelberg S.** 2008. *Nordic meeting on detection and surveillance of VTEC infections in humans. 1-30. 2008.* http://www.ssi.dk/English/HealthdataandICT/National%20Reference%20Laboratories/Bacteria/-/media/Indhold/EN%20-%20engelsk/Public%20Health/National%20Reference%20Laboratories/Nordic%20VTEC%20Report.ashx.

84. **Franke S, Harmsen D, Caprioli A, Piérard D, Wieler LH, Karch H.** 1995. Clonal relatedness of Shiga-like toxin-producing *Escherichia coli* O101 strains of human and porcine origin. *J Clin Microbiol* **33:**3174–3178.

85. **Ogura Y, Ooka T, Asadulghani, Terajima J, Nougayrede JP, Kurokawa K, Tashiro K, Tobe T, Nakayama K, Kuhara S, Oswald E, Watanabe H, Hayashi T.** 2007. Extensive genomic diversity and selective conservation of virulence-determinants in enterohemorrhagic *Escherichia coli* strains of O157 and non-O157 serotypes. *Genome Biol* **8:**R138.

86. **Ogura Y, Ooka T, Iguchi A, Toh H, Asadulghani M, Oshima K, Kodama T, Abe H, Nakayama K, Kurokawa K, Tobe T, Hattori M, Hayashi T.** 2009. Comparative genomics reveal the mechanism of the parallel evolution of O157 and non-O157 enterohemorrhagic *Escherichia coli. Proc Natl Acad Sci USA* **106:**17939–17944.

87. **Shringi S, Schmidt C, Katherine K, Brayton KA, Hancock DD, Besser TE.** 2012. Carriage of *stx2a* differentiates clinical and bovine-biased strains of *Escherichia coli* O157. *PLoS One* **7:**e51572. doi:10.1371/journal.pone.0051572.

88. **Garcia-Aljaro C, Muniesa M, Jofre J, Blanch AR..** 2009. Genotypic and phenotypic diversity among induced, stx2-carrying bacteriophages from environmental *Escherichia coli* strains. *Appl Environ Microbiol* **75:**329–336.

89. **Serra-Moreno R, Jofre J, Muniesa M.** 2007. Insertion site occupancy by stx_2 bacteriophages depends on the locus availability of the host strain chromosome. *J Bacteriol* **189:**6645–6654.

90. **Shringi S, Schmidt C, Katherine K, Brayton KA, Hancock DD, Besser TE.** 2012. Carriage of *stx2a* differentiates clinical and bovine-biased strains of *Escherichia coli* O157. *PLoS One* **7:**e51572. doi:10.1371/journal.pone.0051572.

91. **Clawson ML, Keen JE, Smith TP, Durso LM, McDaneld TG, Mandrell RE, Davis MA, Bono JL..** 2009. Phylogenetic classification of *Escherichia coli* O157:H7 strains of human and bovine origin using a novel set of nucleotide polymorphisms. *Genome Biol* **10:**R56.

92. **Brüssow H, Canchaya C, Hardt WD.** 2004. Phages and the evolution of bacterial pathogens: from genomic rearrangements to lysogenic conversion. *Microbiol Mol Biol Rev* **68:**560–602.

93. **Herold S, Karch H, Schmidt H.** 2004. Shiga toxin-encoding bacteriophages—genomes in motion. *Int J Med Microbiol* **294:**115–121.

94. **Waldor MK, Friedman DI.** 2005. Phage regulatory circuits and virulence gene expression. *Curr Opin Microbiol* **8:**459–465.

95. **Wagner PL, Neely MN, Zhang XP, Acheson DWK, Waldor MK, Friedman DI.** 2001. Role for a phage promoter in Shiga toxin 2 expression from a pathogenic *Escherichia coli* strain. *J Bacteriol* **183:**2081–2085.

96. **Fortier LC, Sekulovic O.** 2013. Importance of prophages to evolution and virulence of bacterial pathogens. *Virulence* **4:**354–365.

97. **Wagner PL, Livny J, Neely MN, Acheson DW, Friedman DI, Waldor MK.** 2002. Bacteriophage control of Shiga toxin 1 production and release by *Escherichia coli. Mol Microbiol* **44:**957–970.

98. **Plunkett G, Rose DJ, Durfee TJ, Blattner FR.** 1999. Sequence of Shiga toxin 2 phage 933W from *Escherichia coli* O157:H7: Shiga toxin as a phage late-gene product. *J Bacteriol* **181:**1767–1778.

99. **Guo HC, Kainz M, Roberts JW.** 1991. Characterization of the late-gene regulatory region of phage 21. *J Bacteriol* **173:**1554–1560.

100. **Strauch E, Schaudinn C, Beutin L.** 2004. First-time isolation and characterization of a bacteriophage encoding the Shiga toxin 2c variant, which is globally spread in strains of *Escherichia coli* O157. *Infect Immun* **72:**7030–7039.

101. **Strauch E, Hammerl JA, Konietzny A, Schneiker-Bekel S, Arnold W, Goesmann A, Puhler A, Beutin L.** 2008. Bacteriophage 2851 is a prototype phage for dissemination of the Shiga toxin variant gene 2c in *Escherichia coli* O157:H7. *Infect Immun* **76:**5466–5477.

102. **Makino K, Yokoyama K, Kubota Y, Yutsudo CH, Kimura S, Kurokawa K, Ishii K, Hattori M, Tatsuno I, Abe H, Iida T, Yamamoto K, Onishi M, Hayashi T, Yasunaga T, Honda T, Sasakawa C, Shinagawa H.** 1999. Complete nucleotide sequence of the prophage VT2-Sakai carrying the verotoxin 2 genes of the enterohemorrhagic *Escherichia coli* O157:H7 derived from the Sakai outbreak. *Genes Genet Syst* **74:**227–239.

103. **Yokoyama K, Makino K, Kubota Y, Watanabe M, Kimura S, Yutsudo CH, Kurokawa K, Ishii**

K, Hattori M, Tatsuno I, Abe H, Yoh M, Iida T, Ohnishi M, Hayashi T, Yasunaga T, Honda T, Sasakawa C, Shinagawa H. 2000. Complete nucleotide sequence of the prophage VT1-Sakai carrying the Shiga toxin 1 genes of the enterohemorrhagic *Escherichia coli* O157:H7 strain derived from the Sakai outbreak. *Gene* **258:**127–139.

104. Sato T, Shimizu T, Watarai M, Kobayashi M, Kano S, Hamabata T, Takeda Y, Yamasaki S. 2003. Genome analysis of a novel Shiga toxin 1 (Stx1)-converting phage which is closely related to Stx2-converting phages but not to other Stx1-converting phages. *J Bacteriol* **185:**3966–3971.

105. Jacobsen A. 2012. *M.Sc. thesis.* Statens Serum Institut, Technical University of Denmark, Copenhagen, Denmark.

106. Perna NT, Plunkett G 3rd, Burland V, Mau B, Glasner JD, Rose DJ, Mayhew GF, Evans PS, Gregor J, Kirkpatrick HA, Posfai G, Hackett J, Klink S, Boutin A, Shao Y, Miller L, Grotbeck EJ, Davis NW, Lim A, Dimalanta ET, Potamousis KD, Apodaca J, Anantharaman TS, Lin J, Yen G, Schwartz DC, Welch RA, Blattner FR. 2001. Genome sequence of enterohaemorrhagic *Escherichia coli* O157:H7. *Nature* **409:**529–533.

107. Blanco M, Blanco JE, Dahbi G, Alonso MP, Mora A, Coira MA, Madrid C, Juarez A, Bernardez MI, Gonzalez EA, Blanco J. 2006. Identification of two new intimin types in atypical enteropathogenic *Escherichia coli. Int Microbiol* **9:**103–110.

108. Garrido P, Blanco M, Moreno-Paz M, Briones C, Dahbi G, Blanco J, Blanco J, Parro V. 2006. STEC-EPEC oligonucleotide microarray: a new tool for typing genetic variants of the LEE pathogenicity island of human and animal Shiga toxin-producing *Escherichia coli* (STEC) and enteropathogenic *E. coli* (EPEC) strains. *Clin Chem* **52:**192–201.

109. Mora A, Blanco M, Yamamoto D, Dahbi G, Blanco JE, Lopez C, Alonso MP, Vieira MA, Hernandes RT, Abe CM, Piazza RM, Lacher DW, Elias WP, Gomes TA, Blanco J. 2009. HeLa-cell adherence patterns and actin aggregation of enteropathogenic *Escherichia coli* (EPEC) and Shiga-toxin-producing *E. coli* (STEC) strains carrying different *eae* and *tir* alleles. *Int Microbiol* **12:**243–251.

110. Garmendia J, Frankel G, Crepin VF. 2005. Enteropathogenic and enterohemorrhagic *Escherichia*

coli infections: translocation, translocation, translocation. *Infect Immun* **73:**2573–2585.

111. Bono JL, Keen JE, Clawson ML, Durso LM, Heaton MP, Laegreid WW. 2007. Association of *Escherichia coli* O157:H7 tir polymorphisms with human infection. *BMC Infect Dis* **7:**98.

112. Haugum K, Brandal LT, Lobersli I, Kapperud G, Lindstedt BA. 2011. Detection of virulent *Escherichia coli* O157 strains using multiplex PCR and single base sequencing for SNP characterization. *J Appl Microbiol* **110:**1592–1600.

113. Badea L, Doughty S, Nicholls L, Sloan J, Robins-Browne RM, Hartland EL. 2003. Contribution of Efa1/LifA to the adherence of enteropathogenic *Escherichia coli* to epithelial cells. *Microb Pathog* **34:**205–215.

114. Klapproth JA, Scaletsky ICA, McNamara BP, Lai LC, Malstrom C, James SP, Donnenberg MS. 2000. A large toxin from pathogenic *Escherichia coli* strains that inhibits lymphocyte activation. *Infect Immun* **68:**2148–2155.

115. Wickham ME, Lupp C, Mascarenhas M, Vazquez A, Coombes BK, Brown NF, Coburn BA, Deng W, Puente JL, Karmali MA, Finlay BB. 2006. Bacterial genetic determinants of non-O157 STEC outbreaks and hemolytic-uremic syndrome after infection. *J Infect Dis* **194:**819–827.

116. Newton HJ, Pearson JS, Badea L, Kelly M, Lucas M, Holloway G, Wagstaff KM, Dunstone MA, Sloan J, Whisstock JC, Kaper JB, Robins-Browne RM, Jans DA, Frankel G, Phillips AD, Coulson BS, Hartland EL. 2010. The type III effectors NleE and NleB from enteropathogenic *E. coli* and OspZ from Shigella block nuclear translocation of NF-kappaB p65. *PLoS Pathog* **6:**e1000898. doi:10.1371/journal.ppat.1000898 [doi].

117. Tree JJ, Wolfson EB, Wang D, Roe AJ, Gally DL. 2009. Controlling injection: regulation of type III secretion in enterohaemorrhagic *Escherichia coli. Trends Microbiol* **17:**361–370.

118. Shaikh N, Tarr PI. 2003. *Escherichia coli* O157:H7 Shiga toxin-encoding bacteriophages: integrations, excisions, truncations, and evolutionary implications. *J Bacteriol* **185:**3596–3605.

119. Recktenwald J, Schmidt H. 2002. The nucleotide sequence of Shiga toxin (Stx) 2e-encoding phage phiP27 is not related to other Stx phage genomes, but the modular genetic structure is conserved. *Infect Immun* **70:**1896–1908.

Shiga Toxin (Stx) Classification, Structure, and Function

ANGELA R. MELTON-CELSA[1]

Shiga toxin (Stx) is one of the most potent biological poisons known. Stx causes fluid accumulation in rabbit ileal loops; causes renal damage in mice, rabbits, greyhounds, and baboons; and is lethal to animals upon injection. However, humans encounter Stx as a consequence of infection with *Shigella dysenteriae* type 1 or certain serogroups of *Escherichia coli* such as O157:H7. There are two immunologically distinct groups of Stxs, and this review discusses toxin classification, structure, and function and the virulence associated with Stx-producing *E. coli* (STEC).

OVERVIEW

S. dysenteriae and Stx were identified in the 19th century by Neisser and Shiga (1) and Conradi (2). Approximately 80 years later, the same toxin (now called Stx1 to distinguish it from the toxin produced by *S. dysenteriae*) was found in a group of *E. coli* isolates. These bacteria caused bloody diarrhea and a serious sequela, hemolytic-uremic syndrome (HUS), a condition characterized by thrombocytopenia, hemolytic anemia, and kidney failure (3, 4). Some *E. coli* strains were later shown to produce a highly related toxin, Stx2, that has the

[1]Department of Microbiology and Immunology, Uniformed Services University of the Health Sciences, Bethesda, MD 20814.

Enterohemorrhagic Escherichia coli *and Other Shiga Toxin-Producing* E. coli
Edited by Vanessa Sperandio and Carolyn J. Hovde
© 2015 American Society for Microbiology, Washington, DC
doi:10.1128/microbiolspec.EHEC-0024-2013

same mode of action as Stx/Stx1 but is immunologically distinct. The Stxs (also known as verotoxins and previously as Shiga-like toxins) are a group of bacterial AB5 protein toxins of about 70 kDa that inhibit protein synthesis in sensitive eukaryotic cells. Protein synthesis is blocked by the Stxs through the removal of an adenine residue from the 28S rRNA of the 60S ribosome. This N-glycosidase activity of the toxin resides in the A subunit. The pentamer of identical B subunits mediates toxin binding to the cellular receptor globotriaosylceramide (Gb3). Additional commonalities between the Stx groups are that the subunit genes are encoded in an operon with the A subunit gene 5′ to that of the B subunit, that the stx operon is usually found within the sequence for an inducible, lysogenic, lambda-like bacteriophage, and that the toxins use a retrograde pathway to reach the cytoplasm. Differences between the two toxin groups are that the genes for stx/stx_{1a} are repressed by Fur when high levels of iron are present (5–7) and that $E.$ $coli$ strains that encode stx_2 are epidemiologically linked to more severe disease than those that carry stx_1 (8, 9).

TYPING AND NOMENCLATURE

Although the prototype $E.$ $coli$ Stxs from each main group, Stx1 and Stx2 (now called Stx1a and Stx2a for distinction in the nomenclature from other toxin subtypes [10]), are the most common types found in association with disease from their respective groups, subtypes of each toxin exist, as listed in Table 1. Toxin subtypes were originally only recognized when differences in biological activity and/or immunoreactivity could be demonstrated. However, as many new STEC strains were isolated and the toxin genes from those strains were sequenced, it became difficult to know if any differences found between the newly isolated gene and the prototype stx_{2a} should result in the designation of another toxin subtype. Therefore, a phylogenetic analysis of stx sequences was undertaken and a PCR typing scheme developed that enables the assignment of a toxin to a particular subtype (10).

Stx/Stx1 Subtypes

To date, no variants of Stx as produced by *Shigella* have been described, but Stx is

TABLE 1 Prototype toxins and strains that produce those toxins

Toxin type(s)	Prototype strain used for determination of stx subtype[a]	Linked with serious human disease; difference(s) from prototype toxin[b]	Reference(s)
Stx	3818T	Yes	133
Stx1a	EDL933 (makes Stx1a and Stx2a)	Yes	113
Stx1c	DG131/3	No; immunologically distinct	134, 135
Stx1d	MHI813	No; immunologically distinct, less potent	136
Stx2a	EDL933 (makes Stx1a and Stx2a)	Yes	113
Stx2b (originally named VT-2d or Stx2d)	EH250	No; the B subunit gene was not detected by methods used to detect other sx_2 B subunit genes	24
Stx2c	031	Yes, less toxic to Vero cells and mice	137
Stx2d (Stx2dact)	C165-02	Yes; more toxic after incubation with elastase, less toxic to Vero cells	138
Stx2e	S1191	No; binds to Gb4, associated with disease in pigs	139
Stx2f	T4/97	No; originally isolated in STEC from pigeons; immunologically distinct	140
Stx2g	7v	No; the stx_{2g} gene is not amplified by primers specific for stx_{2a}	141

[a]More information about the prototypes strains may be found in reference 10.
[b]Prototype toxin indicated in bold.

occasionally found in *Shigella sonnei* and type 4 *S. dystenteriae* (11, 12). Only two variants of Stx1a have been identified: Stx1c and Stx1d. Both Stx1c and Stx1a can be distinguished immunologically from Stx1 (13, 14). Stx1c and Stx1d are rarely found in human disease, and when associated with STEC isolated from patients, are linked with a mild disease course (15, 16).

Stx2a Subtypes

The first Stx2a toxin variant identified as important for human disease, Stx2c, exhibits reduced cytotoxicity on Vero cells and reacts differently than Stx2a to some monoclonal antibodies (17). Another Stx2a variant, Stx2d (Stx2d activatable), was identified because incubation with elastase from intestinal mucus increases the Vero cell cytotoxicity of the toxin (18, 19). The activatable Stx2d is associated with the most serious manifestation of STEC infection, HUS (20). Both Stx2c and Stx2d show reduced cytotoxicity for Vero cells due to 2 amino acid differences in the B subunit, but Stx2d is as toxic as Stx2a when injected into animals (21). Moreover, strains that produce Stx2d are highly virulent in a streptomycin-treated mouse model of infection (18, 22). In contrast, Stx2c is reported to show reduced toxicity compared to Stx2

when injected into mice (23). A different Stx2a subtype that was originally also named Stx2d, now known as Stx2b, is not activatable and is associated with mild disease (24, 25). Stx2e, Stx2f, and Stx2g are associated with animal STEC infection. Of the latter toxins, only Stx2e is associated with disease in the animal host; this toxin causes edema disease of swine, a rare but serious neurological disorder that is frequently fatal (26).

SHIGA TOXIN STRUCTURE

The mature A subunit of Stx/Stx1a consists of 293 aa whereas the Stx2a A chain is 4 aa longer at the C terminus. The active-site amino acid of the toxin is the glutamic acid at position 167 (27). A trypsin-sensitive region (aa 248–251) allows the A subunit to be cleaved asymmetrically into an A_1 subunit and A_2 peptide held together by a disulfide bridge (Fig. 1). The enzymatic activity of the toxin resides within A_1 while the A_2 peptide tethers A_1 to the binding moiety, and further, threads through the B pentamer. For Stx2a, the A_2 peptide, in addition to maintaining holotoxin structure, appears to partially block the site 3 Gb3-binding site (discussed further below) (28), and for Stx2d, is crucial for the activation phenotype, as the final 2 aa of the A_2 are

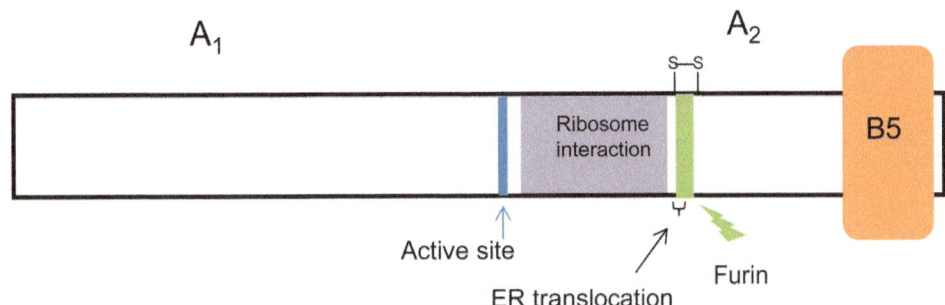

FIGURE 1 Cartoon representation of the Stx structure. The active-site glutamic acid is indicated as a vertical blue line, the ribosome interaction region is shown in purple, the protease (furin)- sensitive site is depicted in green, and the B pentamer as an orange block. The disulfide bridge that connects the A_1 subunit and the A_2 peptide is shown above the protease-sensitive site. A region important for translocation from the ER to the cytosol is indicated by a bracket. Not to scale. doi:10.1128/microbiolspec.EHEC-0024-2013.f1

cleaved when the toxin is treated with elastase (29). The pentamer of identical B subunits (each subunit equals 69 aa for Stx/Stx1a, 71 aa for Stx2a) enables the toxin to find target cells that express surface Gb3. The crystal structures of the Stx B pentamer (30), Stx (31), Stx2a (28), Stx2a complexed with adenine (32), and the solution structures of the Stx1a B pentamer alone (33) or with the trisaccharide moiety from Gb3 (34) or a Gb3 analog (35) have been reported. A ribbon diagram of the crystal structure of Stx is shown in Fig. 2.

Genetic Analyses of the Stx A Subunit

Besides the active-site glutamic acid, colored red in Fig. 2, other A subunit amino acid residues that contribute to the full enzymatic function of Stx, colored pale blue in Fig. 2,

include N75, Y77, Y114, R170, and W203 (W202 in Stx2) (27, 36–38). An analysis of truncated A_1 fragments of Stx1a in yeast confirmed that residues within aa 1239 are required for full enzymatic activity of the toxin, whereas the amino acids from 240 to 245 (and perhaps up to 251) are necessary for translocation of the A_1 from the endoplasmic reticulum (ER) to the cytosol (39). The authors of the latter study hypothesized that it is the general structure of the region between amino acid residues 240 and 251 that is recognized by an ER mechanism that directs proteins from the ER into the cytosol. To define regions of Stx1a and the ribosome that interact, Gariepy's group used yeast two-hybrid and pull-down experiments and identified three ribosomal proteins that interact with the A_1 subunit of the toxin and further showed that a

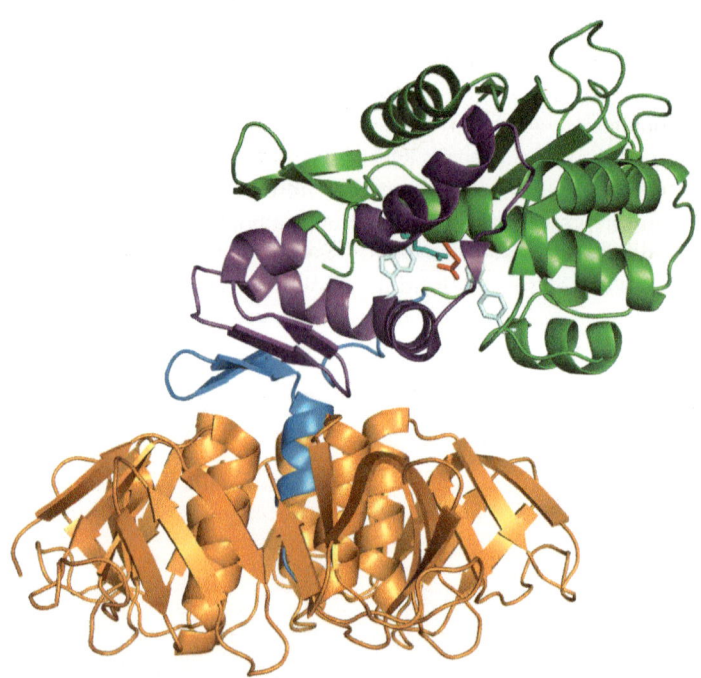

FIGURE 2 Ribbon diagram of the Stx1 crystal structure. The B pentamer is shown in orange and the A_2 in blue. The majority of the A_1 is depicted in green except for the region that interacts with the ribosome, which is shown in purple. The active residue 167 is red, and other active-site side chains are pale blue. The A_2 chain is medium blue, and the B subunits are orange. The structure (1R4Q) was drawn with PyMOL Molecular Graphics System, Version 1.5.0, Schrödinger, LLC. Figure kindly provided by Dr. James Vergis.
doi:10.1128/microbiolspec.EHEC-0024-2013.f2

conserved peptide from two of the ribosomal proteins inhibited Stx1a activity in vitro (40). They also found that the region from R170 to L233 (purple in Fig. 2) in Stx1a is important for ribosome interaction (41).

Stx AND CELL BINDING

As noted previously, the Stxs bind to the gly-cosphingolipid Gb3, a molecule composed of a lipid or ceramide component and a trisac-charide of [α-gal(1→4)-β-gal(1→4)-β-glc] (42–44). Cell lines and mice deficient in Gb3 are insensitive to the Stxs (45–47). The normal cellular function of Gb3 is not known; however, individuals with excess Gb3, a condition called Fabry's disease (48), exhibit kidney disease among other symptoms. When challenged with Stx2a, a mouse model of Fabry's disease exhibits an elevated LD_{50} (49), perhaps due to mistargeting of the toxin to multiple sites of the body rather than the kidney or to altered cellular trafficking due to changes in Gb3 subspecies populations. Stx1a and Stx2a interact with globotetraosylceramide (Gb4) in addition to Gb3, though weakly (50), whereas Stx2e binds Gb4 in preference to Gb3 (51, 52). An early report of a possible protein receptor for the Stxs (53) has not been substantiated in the literature.

Gb3 clusters within detergent-insoluble portions of membranes called "lipid rafts," which also contain cholesterol and the cholera toxin receptor monosialotetrahexosylganglioside, or GM1 (54, 55). The interaction of the Stx1a B subunit with HeLa cells causes a 2.5-fold increase in Gb3 present in the lipid raft domains (56), a result that suggests that cell binding by the B pentamer may promote stronger or additional toxin-cell interaction by recruiting more receptors to the rafts. Several reports indicate that the fatty acid chain length of the lipid component of Gb3 and the saturation state of the lipid also influence toxin binding (57, 58). The presence of cholesterol in the lipid rafts has been reported to increase toxin binding (50). Conversely, another group found no role

for cholesterol in cell binding by the B subunit of Stx1a, but they did find that reductions in the cholesterol content of cells resulted in a decrease in uptake of the B pentamer (55). More recently, Lingwood's group found that extraction of cholesterol from adult renal tissue sections enhanced Stx binding to glomeruli (59), a result that confirms the colocalization of Gb3 and cholesterol molecules. However, it may be difficult to dissect a possible role for cholesterol because Gb3 and cholesterol are both present in the lipid rafts, and alteration of either component may disrupt the integrity of the rafts. Together, these studies paint a picture of a complex interaction of the toxin with the host cell receptor and suggest that the Stxs may interact differently with cells based on the nature of the receptor environment in the cell membrane.

The high affinity of the Stxs for Gb3 is likely due to the presence of at least two and up to three Gb3-binding sites per B monomer (two of the binding sites are present between adjacent monomers), as demonstrated by modeling studies and the crystal structure of the Stx B pentamer complexed with a Gb3 trisaccharide analog (35, 60). However, precise measurements of the binding affinity of the Stxs for Gb3 are difficult because Gb3 is not soluble. Therefore, soluble forms of the tri-saccharide or trisaccharide analogs, or immobilized versions of the trisaccharide, Gb3, or Gb3 analogs are used to measure toxin-receptor interaction. When the interaction between the Stx1a B pentamer and the Gb3 trisaccharide was assessed in solution, binding site 2 (present within each monomer) was found to be the most highly occupied; site 1 was less engaged, and no interaction with site 3 was detected (34). Another study that also examined the interaction of the B pentamer with the Gb3 oligosaccharide component found that, at low concentrations of ligand, only site 2 was occupied though the authors of that study acknowledged that Stx binding is likely to be "polyvalent" at the cell surface (61). A mutational analysis of the three putative Gb3-binding sites in Stx1a confirmed the role of sites 1 and 2 for toxin-receptor interaction (62). A mutation within site 3 (W34A) reduced binding of the holotoxin to Gb3, although the mutant toxin had the same overall affinity for Gb3 as did Stx1a. Another group found that Gb3-binding site 2 alone is not sufficient to confer high avidity binding of the toxin to Gb3 and that optimal binding to the receptor required all three sites (63). Although these latter studies on the number of functional Gb3-binding sites per toxin molecule were performed with Stx/Stx1a, the presence and primary importance of site 2 were confirmed for Stx2a (64). In addition, that site 1 contributes to the Vero cell toxicity phenotype of the Stx2a group is demonstrated by the fact that a single amino acid change in the Stx2d B subunit in the site 1 binding region renders that toxin as potent for Vero cells as Stx2a (21). Finally, a point in favor of the polyvalent nature of Stx binding is that Gb3 mimics with higher densities and clustering of the Gb3 trisaccharide (up to 18 copies) demonstrate higher affinities for both Stx1a and Stx2a (65, 66).

The question of how Stx enters the host from the site of pathogen colonization in the intestine remains, due to reports that colonic tissue lacks Gb3 (67). Macropinocytosis is a possible alternate mechanism for the uptake of Stxs into cells that do not express Gb3 (68). Polarized colonic T84 cells (which lack Gb3) nonetheless take up the Stx1a B pentamer in a way that is partially blocked by inhibitors of clathrin-dependent endocytic processes (68). However, such inhibition of the macropinocytic pathway does not reduce the amount of B subunit that is transcytosed through the monolayer, a finding that suggests that no matter which entry mechanism the Stx uses to enter these cells, the toxin can reach the basolateral side. That both Stx1a and Stx2a cross polarized T84 cells without disrupting the monolayer (46, 69, 70) or inhibiting protein synthesis has also been demonstrated (71). Therefore, it may be that some Gb3-negative intestinal cells allow systemic delivery of toxin in the absence of receptor. However, we and others have shown that incubation of intestinal cells with butyrate increases the sensitivity of intestinal cell lines to the Stxs (72, 73), and we further found that expression of Gb3 on intestinal cells is exquisitely sensitive to the removal of that gut metabolite (74). This latter finding suggests that previous measurements of Gb3 levels on intestinal tissues may be underestimates if butyrate was removed before Gb3 detection. In support of the possible presence of functional Gb3 receptors on intestinal tissue is the finding that incubation of pediatric intestinal explants with Stx2a showed cell extrusion from tissues taken from both ileum and colon (71). The damage observed in the latter study was specific as pre-incubation of the toxin with antiserum to Stx2a ameliorated the tissue injury. Furthermore, both Stx1a and Stx2a bind to Paneth cells when overlaid onto normal or inflamed pediatric duodenal tissues collected during endoscopy (75). We also found that toxin overlaid into normal adult tissue binds colonic tissue (72). Finally, Malyukova et al. identified both Stx1a and Stx2a in intestinal tissues (epithelial cells and lamina propria) from O157-infected patients (68), although Gb3 could not be found in those intestinal epithelial cells (68). However, it may be that the Gb3 is masked in such tissues by cholesterol, as described earlier, or that the expression of Gb3 was reduced by removal of butyrate when those samples were processed.

Stx-Gb3 interaction leads to uptake of the toxin-receptor complex through clathrin-dependent or -independent mechanisms. However, the clathrin-dependent process of entry appears to be the most common mode of uptake of the toxin-receptor complex (76). One perhaps surprising finding is that the A subunit of the Stx is involved in toxin uptake: more holotoxin is taken up via the endocytic pathway when the holotoxin binds to a cell as compared to B pentamer alone (77). The latter study illustrates the importance of confirming findings based on the B pentamer alone with the holotoxin. In clathrin-independent uptake, the Stx B subunit induces tubular-shaped invaginations within the HeLa cell plasma membrane, the function of which is not clear (78). Curiously, a mutation in Gb3-binding site 3 of the B subunit prevented formation of the invaginations (78), although the pentamer still bound to the cells.

RETROGRADE TRAFFICKING OF THE Stxs

Stx was first shown to use a retrograde pathway to reach the cytosol more than 20 years ago in the human epidermal carcinoma cell line A431, which had been sensitized to the toxin by butyric acid treatment (79). An outline of the retrograde pathway used by the Stxs is shown in Fig. 3 and summarized as follows: after binding to Gb3, the toxin-receptor complex enters early endosomes, then traffics to the Golgi, and finally the ER (for a review see reference 80). The protease-sensitive site of the toxin A subunit is nicked either within the intestinal mucus (18) or within the Golgi (81); however, the nicked toxin retains the AB5 structure because of a disulfide bond between the A_1 and A_2 chains. The disulfide bond is reduced once the toxin gets into the ER, and only the toxin A_1 subunit enters the cytosol. The Stx receptor, Gb3, is required for the toxin-receptor complex to move through the retrograde pathway, as toxin taken up by a nonreceptor-mediated mechanism or in cells depleted of glycosphingolipids does not reach the Golgi, though the complex does reach endosomes and possibly the ER (56, 70, 82). Many cellular proteins have now been identified that are involved in the Stx retrograde pathway (reviewed in references 80, 83). An unfolded toxin A_1 chain exits the ER, apparently by subverting the ER-associated protein degradation pathway, to reach the target ribosomes (84, 85). The A_1 then removes an adenine from the alpha-sarcin loop in the 28S ribosomal subunit. The injured ribosome no longer associates with elongation factor 1 (86, 87), and protein synthesis is halted. Although much of the work describing the retrograde pathway was done in epithelial cells, the primary target cell for the Stxs is the endothelial

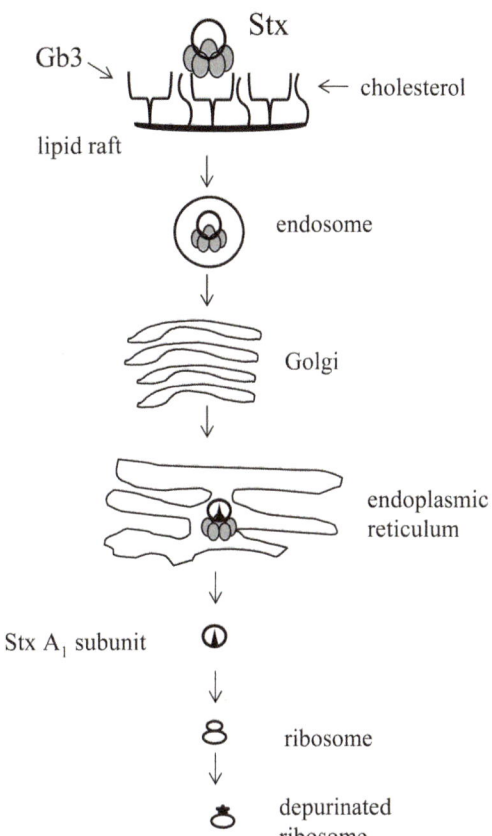

FIGURE 3 An illustration of the retrograde pathway for Stxs. The toxin binds to Gb3 within lipid rafts that contain cholesterol and that complex is internalized within an endosome. From the endosome the toxin traffics to the Golgi where it is nicked by furin if that nicking did not occur in the intestine. The nicked toxin moves to the ER where the disulfide bridge that keeps the A_1 tethered to A_2B5 is reduced. The A_1 chain then enters the cytosol and removes an adenine residue from the 28S ribosome. doi:10.1128/microbiolspec.EHEC-0024-2013.f3

cell, discussed further below, and the Stx B pentamer has been shown to enter the retrograde pathway in mesangial and glomerular vascular endothelial cells as well (88).

ACTIONS OF Stx IN TARGET CELLS

Numerous studies show that Stx inhibits protein synthesis in target cells and that active-site mutants of Stx are no longer cytotoxic.

Furthermore, a single molecule of Stx may be sufficient to kill a cell (85). Recently, however, the effects that the toxin has on cell signal transduction and immune modulation have begun to be explored (reviewed in references 89, 90). Stx-mediated damage to the ribosome induces a response in cells called a "ribotoxic stress response," which is both proinflammatory and proapoptotic (see review in reference 89). The Stxs are also associated with an unfolded protein response that comes about through stress on the ER, and the result of which may also be apoptotic (reviewed in references 89, 91). Evidence that Stx and STEC are associated with apoptosis came from rabbit studies in the mid-1980s (92, 93). Since that time, the molecular mechanisms of apoptosis due to the Stxs have been elucidated further (see review in reference 91).

Stx1a COMPARED TO Stx2a

Sx1a is about 10-fold more cytotoxic to Vero cells than Stx2a; however, the reverse is true in mice: Stx1a is 100- to 400-fold less lethal to mice than Stx2a, even though the toxins exhibit equivalent enzymatic activities/ng of protein in vitro (94, 95). Epidemiological data from human disease indicate a stronger association with severe disease for STEC strains that produce Stx2a than Stx1a alone (96–98). So an important question within the STEC field is how to explain the paradox of the differential toxicity of the Stxs in vitro as compared to in vivo. Russo et al. recently found that Stx2a is more toxic than Stx1a by the oral route in mice, data that support the hypothesis that Stx2a is more potent from the gut and not just when injected intraperitoneally (99). One possible explanation for the differential toxicity of the toxins is that in contrast to epithelial cells such as Vero, the endothelial cells targeted in HUS are more sensitive to Stx2a than to Stx1a. Obrig's laboratory did find that renal microvascular endothelial cells obtained from human glomeruli are about 1,000-fold more sensitive to Stx2a than to Stx1a, although

umbilical vein endothelial cells were equivalently sensitive (100). Since the Stxs are equally enzymatically active and they both bind Gb3, the question remains why different cell types exhibit differential sensitivity to the two Stx types. However, Gb3 does not exist in a homogeneous population within the cell, and it may be that the toxins do not bind equivalently to the Gb3 in target cells. An in vitro study with soluble forms of the trisaccharide moiety of Gb3 showed that Stx1a and Stx2a bind with similar affinity for that component of the receptor (64). Furthermore, Stx1a and Stx2a demonstrate similar affinities for renally derived Gb3, as measured by enzyme-linked immunosorbent assay, though by thin-layer chromatography Stx2a exhibits less binding than Stx1a does (101). Examinations of toxin interaction with immobilized trisaccharides suggest that Stx1a has a higher affinity for the carbohydrate than does Stx2a (50, 102). Other experiments that examined the capacity of the toxins to bind Gb3 analogs showed that Stx1a and Stx2a exhibit differential binding to those analogs in vitro (103), but whether similar analogs exist or act functionally as receptors in vivo has not been demonstrated.

Membrane cholesterol may also mask a portion of the potential Stx1a-binding Gb3 pool in Vero cell membranes (104) and adult renal glomeruli sections, as mentioned previously (59), and incubation of Stx1a but not Stx2a with Gb3, cholesterol, and phosphatidylcholine partially neutralizes the cytotoxicity of Stx1a but not Stx2a (103). Stx2a, in contrast, is neutralized by human serum amyloid component P and lipopolysaccharide (LPS) from O107 or O117 E. coli (105, 106). That the above compounds neutralize either Stx1a or Stx2a specifically, and presumably by preventing the toxins from binding to the cell, supports the hypothesis that the B pentamers of these toxins bind differently to cells. The Stx1a B pentamer is more stable than the comparable binding moiety of Stx2a (64), and that difference was shown to be at least partially due to a single amino acid residue within the B subunit (107). A study that examined the interaction of Stx1a and Stx2a with lipid rafts and subsequent intracellular trafficking in Vero cells showed that although both toxins associated with lipid rafts, Stx2a was also found in detergent-soluble (non-lipid raft) fractions of the membrane (108). In addition, some of the internalized Stx2a, but not Stx1a, colocalized with the cell surface marker transferrin, a finding that suggests that Stx2a is endocytosed from different areas of the membrane than Stx1a. However, both toxins localized to the Golgi after 1 h of incubation with the Vero cells, although there was a suggestion that Stx2a exited from the Golgi at a slower rate than Stx1a.

At the level of the immune response to the toxins, Stx1a and Stx2a elicit differential chemokine responses from human umbilical vein endothelial cells (109) and a renal tubular epithelial cell line, HK-2 (110). The HK-2 cells are more sensitive to Stx1a than to Stx2a, and Stx2a treatment induces expression of the macrophage chemoattractants macrophage inflammatory protein-1α (MIP-1α) and MIP-1β (110). However, in a baboon intoxication model, there was no change in MIP-1β in response to either Stx1a or Stx2a (111). Nonetheless, in that baboon model many characteristics of HUS are observed, and Stx2a is lethal at lower doses than Stx1a is, although the kidney injury takes longer to develop in the Stx2a-treated animals.

Taken together these studies on the differences between Stx1a and Stx2a indicate that the interaction between each toxin and the receptor is complex and influenced by several factors. Four distinctions between the Stx and Stx2a crystal structures that may contribute to the known biological differences exhibited by these toxins, as discussed above, are: (i) the Stx active site is partially blocked by the N-terminal region of the A$_2$ peptide; (ii) the Stx2a A$_2$ C terminus has a more ordered structure than that of Stx; (iii) Gb3-binding site 2 has a different conformation in each of the toxins; and (iv) Gb3-binding site 3 is partially blocked by the C-terminal 2 aa of the Stx2a A$_2$ peptide (28).

Stx ASSOCIATION WITH DISEASE

HUS cases were reported in the literature as early as the 1950s, though the cause was unknown. An understanding of the origins of infection-associated HUS was also complicated by the fact that some cases of HUS are of genetic origin (see review in reference 112). However, in 1983, bloody diarrhea and HUS were linked to certain serogroups of *E. coli*, and O'Brien et al. found that those strains produced a toxin related to the Stx of *S. dysenteriae* (3, 4, 113). The strong association between HUS and Stx2a was underscored in the largest outbreak of HUS in Germany in 2011 in which more than 800 cases of the disease were identified (114). The surprise from German outbreak was that although the implicated strain was an *E. coli*, unlike typical STEC, the epidemic isolate also encoded virulence factors associated with enteroaggregative *E. coli*. Previously, enteroaggregative *E. coli* had only rarely been associated with HUS, and in each of those cases, the implicated isolate was found to carry an Stx gene.

In animal models, purified Stx in the absence of detectable LPS is linked to kidney damage and death (99, 115). *S. dysenteriae* mutants that lack *stx* caused a milder dysentery in monkeys and produced only a trace of blood in stool compared to the wild-type isolate (116). Furthermore, Stx can be detected in the kidneys of some patients (117–119). Animal models further show that antibodies to the Stxs are protective, and humanized versions of those antibodies have completed phase I trials (120, 121).

STEC Pathogenesis

STEC pathogenesis requires ingestion of the bacterium in contaminated food or water, though the organism may occasionally be passed person to person. To cause disease, the organism colonizes the intestine, a process that likely includes attachment and proliferation. That STEC strains that carry the locus of enterocyte effacement (LEE) use the adhesin intimin (encoded by the *eae* gene) to adhere within the intestine has been shown in mouse and pig models (122, 123). In addition, STEC strains that are *eae* positive are more likely to cause disease than intimin-negative strains, even in non-O157 strains (8, 124). Zumbrun et al. recently demonstrated in mice that a high-fiber diet may influence the development of disease after O157 infection (74, 125). The influence of the higher fiber diet was 2-fold and was associated with the production of butyrate within the intestine: mice on a high-fiber diet were colonized to a greater degree than mice on a low-fiber diet, perhaps as a consequence of reduced overall levels of competitive native *Escherichia* species, and increased levels of Gb3 were observed in both the intestine and kidney.

After colonization with STEC, the innate proinflammatory response may initially be suppressed in those infected with strains that are LEE+ (see review in reference 90), a conclusion supported by the fact that STEC infection is generally not an inflammatory diarrhea. In the intestine, the STEC strains elaborate Stxs, and the patient may develop hemorrhagic colitis, the pathogenesis of which is not clearly understood. For systemic consequences, the Stxs enter the host from the intestine and reach target endothelium cells in the kidney and, in some cases, the central nervous system. The Stxs damage cells directly and cause apoptosis, as described above. The Stxs and likely LPS also elicit cytokines and other immune mediators within the kidney and from monocytes (for a review see reference 90). Endothelial cells within the glomeruli become damaged and may die and detach from the membrane. Extrusion of the cell exposes collagen, normally not present within the blood vessel, and platelets become activated. The presence of activated platelets encourages fibrin deposition and leads to the development of a thrombus and promotes a coagulation cascade. Recent studies suggest that Stx treatment of endothelial cells can induce the expression of factors that may stabilize clot formation, even in the absence of cell death or extrusion (see reviews in references 126, 127). Such damaged endothelial cells also become a site for the adhesion of leukocytes, the consequence of which is additional damage to the kidney (128). The localized prothrombotic environment in the glomerulus leads to the reduction of platelets in the serum (thrombocytopenia), hemolytic anemia, and renal damage, the triad of HUS (for reviews see references 126, 129, 130). Altered complement levels have been found in patients with STEC HUS (131, 132), but whether those changes are part of the etiology of the disease or occur as a consequence is not clear (see reference 127 for a review of the possible role for complement in STEC HUS).

SUMMARY

The Stxs are potent poisons associated with bloody diarrhea and the potentially severe disease, HUS. The Stxs act as ribotoxins that halt protein synthesis within the cell and induce apoptosis, but they can also prompt altered gene or protein expression in epithelial cells, endothelial cells, monocytes, and mesangial cells. The consequence of exposure to *E. coli* strains that elaborate one or more Stx can range from asymptomatic infection to bloody diarrhea and, in some patients, HUS. In the research arena, the Stxs have been exploited beautifully to interrogate mechanisms of retrograde transport, to localize sites of Gb3 expression, and, more recently, to determine the possible function of tubular invaginations in membranes.

ACKNOWLEDGMENTS

A special thank-you to Alison D. O'Brien for review of this chapter. This work was supported by R37 AI020148 and 2U54 AI057168.

CITATION

Melton-Celsa AR. 2014. Shiga toxin (Stx) classification, structure, and function. Microbiol Spectrum 2(4):EHEC-0024-2013.

REFERENCES

1. **Trofa AF, Ueno-Olsen H, Oiwa R, Yoshikawa M.** 1999. Dr. Kiyoshi Shiga: discoverer of the dysentery bacillus. *Clin Infect Dis* **29:**1303–1306.

2. **Conradi H.** 1903. Uber Iosliche, durch asptische Autolyse erhaltene Giftstoffe vonRuhr- und Typhus-Bazillen. *Dtsch Med Wochenschr* **29:**26–28.

3. **Karmali MA, Steele BT, Petric M, Lim C.** 1983. Sporadic cases of haemolytic-uraemic syndrome associated with faecal cytotoxin and cytotoxin-producing *Escherichia coli* in stools. *Lancet* **i:**619–620.

4. **O'Brien AO, Lively TA, Chen ME, Rothman SW, Formal SB.** 1983. *Escherichia coli* O157:H7 strains associated with haemorrhagic colitis in the United States produce a *Shigella dysenteriae* 1 (SHIGA) like cytotoxin. *Lancet* **i:**702.

5. **Calderwood SB, Mekalanos JJ.** 1987. Iron regulation of Shiga-like toxin expression in *Escherichia coli* is mediated by the fur locus. *J Bacteriol* **169:**4759–4764.

6. **Dubos RJ, Geiger JW.** 1946. Preparation and properties of Shiga toxin and toxoid. *J Exp Med* **84:**143–156.

7. **O'Brien AD, LaVeck GD, Thompson MR, Formal SB.** 1982. Production of *Shigella dysenteriae* type 1-like cytotoxin by *Escherichia coli*. *J Infect Dis* **146:**763–769.

8. **Luna-Gierke RE, Griffin PM, Gould LH, Herman K, Bopp CA, Strockbine N, Mody RK.** 2014. Outbreaks of non-O157 Shiga toxin-producing *Escherichia coli* infection: USA. *Epidemiol Infect* **7:**1–11.

9. **Friedrich AW, Bielaszewska M, Zhang WL, Pulz M, Kuczius T, Ammon A, Karch H.** 2002. *Escherichia coli* harboring Shiga toxin 2 gene variants: frequency and association with clinical symptoms. *J Infect Dis* **185:**74–84.

10. **Scheutz F, Teel LD, Beutin L, Pierard D, Buvens G, Karch H, Mellmann A, Caprioli A, Tozzoli R, Morabito S, Strockbine NA, Melton-Celsa AR, Sanchez M, Persson S, O'Brien AD.** 2012. Multicenter evaluation of a sequence-based protocol for subtyping Shiga toxins and standardizing Stx nomenclature. *J Clin Microbiol* **50:**2951–2963.

11. **Gupta SK, Strockbine N, Omondi M, Hise K, Fair MA, Mintz E.** 2007. Emergence of Shiga toxin 1 genes within *Shigella dysenteriae* type 4 isolates from travelers returning from the Island of Hispanola. *Am J Trop Med Hyg* **76:** 1163–1165.

12. **Beutin L, Strauch E, Fischer I.** 1999. Isolation of *Shigella sonnei* lysogenic for a bacteriophage encoding gene for production of Shiga toxin. *Lancet* **353:**1498.

13. **Zhang W, Bielaszewska M, Kuczius T, Karch H.** 2002. Identification, characterization, and distribution of a Shiga toxin 1 gene variant (stx(1c)) in *Escherichia coli* strains isolated from humans. *J Clin Microbiol* **40:**1441–1446.

14. **Ohmura-Hoshino M, Ho ST, Kurazono H, Igarashi K, Yamasaki S, Takeda Y.** 2003. Genetic and immunological analysis of a novel variant of Shiga toxin 1 from bovine *Escherichia coli* strains and development of bead-ELISA to detect the variant toxin. *Microbiol Immunol* **47:**717–725.

15. **Friedrich AW, Borell J, Bielaszewska M, Fruth A, Tschape H, Karch H.** 2003. Shiga toxin 1c-producing *Escherichia coli* strains: phenotypic and genetic characterization and association with human disease. *J Clin Microbiol* **41:**2448–2453.

16. **Kumar A, Taneja N, Kumar Y, Sharma M.** 2012. Detection of Shiga toxin variants among Shiga toxin-forming *Escherichia coli* isolates from animal stool, meat and human stool samples in India. *J Appl Microbiol* **113:**1208–1216.

17. **Schmitt CK, McKee ML, O'Brien AD.** 1991. Two copies of Shiga-like toxin II-related genes common in enterohemorrhagic *Escherichia coli* strains are responsible for the antigenic heterogeneity of the O157:H- strain E32511. *Infect Immun* **59:**1065–1073.

18. **Melton-Celsa AR, Darnell SC, O'Brien AD.** 1996. Activation of Shiga-like toxins by mouse and human intestinal mucus correlates with virulence of enterohemorrhagic *Escherichia coli* O91:H21 isolates in orally infected, streptomycin-treated mice. *Infect Immun* **64:**1569–1576.

19. **Kokai-Kun JF, Melton-Celsa AR, O'Brien AD.** 2000. Elastase in intestinal mucus enhances the cytotoxicity of Shiga toxin type 2d. *J Biol Chem* **275:**3713–3721.

20. **Bielaszewska M, Friedrich AW, Aldick T, Schurk-Bulgrin R, Karch H.** 2006. Shiga toxin activatable by intestinal mucus in *Escherichia coli* isolated from humans: predictor for a severe clinical outcome. *Clin Infect Dis* **43:**1160–1167.

21. **Lindgren SW, Samuel JE, Schmitt CK, O'Brien AD.** 1994. The specific activities of Shiga-like toxin type II (SLT-II) and SLT-II-related toxins of enterohemorrhagic *Escherichia coli* differ when measured by Vero cell cytotoxicity but not by mouse lethality. *Infect Immun* **62:**623–631.

22. **Lindgren SW, Melton AR, O'Brien AD.** 1993. Virulence of enterohemorrhagic *Escherichia coli* O91:H21 clinical isolates in an orally infected mouse model. *Infect Immun* **61:**3832–3842.

23. **Fuller CA, Pellino CA, Flagler MJ, Strasser JE, Weiss AA.** 2011. Shiga toxin subtypes display dramatic differences in potency. *Infect Immun* **79:**1329–1337.

24. Pierard D, Muyldermans G, Moriau L, Stevens D, Lauwers S. 1998. Identification of new vero-cytotoxin type 2 variant B-subunit genes in human and animal *Escherichia coli* isolates. *J Clin Microbiol* **36**:3317–3322.

25. Stephan R, Hoelzle LE. 2000. Characterization of Shiga toxin type 2 variant B-subunit in *Escherichia coli* strains from asymptomatic human carriers by PCR-RFLP. *Lett Appl Microbiol* **31**:139–142.

26. Moxley RA. 2000. Edema disease. *Vet Clin North Am Food Anim Pract* **16**:175–185.

27. Hovde CJ, Calderwood SB, Mekalanos JJ, Collier RJ. 1988. Evidence that glutamic acid 167 is an active-site residue of Shiga-like toxin I. *Proc Natl Acad Sci USA* **85**:2568–2572.

28. Fraser ME, Fujinaga M, Cherney MM, Melton-Celsa AR, Twiddy EM, O'Brien AD, James MN. 2004. Structure of Shiga toxin type 2 (Stx2) from *Escherichia coli* O157:H7. *J Biol Chem* **279**:27511–27517.

29. Melton-Celsa AR, Kokai-Kun JF, O'Brien AD. 2002. Activation of Shiga toxin type 2d (Stx2d) by elastase involves cleavage of the C-terminal two amino acids of the A2 peptide in the context of the appropriate B pentamer. *Mol Microbiol* **43**:207–215.

30. Stein PE, Boodhoo A, Tyrrell GJ, Brunton JL, Read RJ. 1992. Crystal structure of the cell-binding B oligomer of verotoxin-1 from *E. coli*. *Nature* **355**:748–750.

31. Fraser ME, Chernaia MM, Kozlov YV, James MN. 1994. Crystal structure of the holotoxin from *Shigella dysenteriae* at 2.5 A resolution. *Nat Struct Biol* **1**:59–64.

32. Fraser ME, Cherney MM, Marcato P, Mulvey GL, Armstrong GD, James MN. 2006. Binding of adenine to Stx2, the protein toxin from *Escherichia coli* O157:H7. *Acta Crystallogr Sect F Struct Biol Cryst Commun* **62**:627–630.

33. Richardson JM, Evans PD, Homans SW, Donohue-Rolfe A. 1997. Solution structure of the carbohydrate-binding B-subunit homopentamer of verotoxin VT-1 from *E. coli*. *Nat Struct Biol* **4**:190–193.

34. Shimizu H, Field RA, Homans SW, Donohue-Rolfe A. 1998. Solution structure of the complex between the B-subunit homopentamer of verotoxin VT-1 from *Escherichia coli* and the trisaccharide moiety of globotriaosylceramide. *Biochemistry* **37**:11078–11082.

35. Ling H, Boodhoo A, Hazes B, Cummings MD, Armstrong GD, Brunton JL, Read RJ. 1998. Structure of the Shiga-like toxin I B-pentamer complexed with an analogue of its receptor Gb3. *Biochemistry* **37**:1777–1788.

36. Aletrari MO, McKibbin C, Williams H, Pawar V, Pietroni P, Lord JM, Flitsch SL, Whitehead R, Swanton E, High S, Spooner RA. 2011. Eeyarestatin 1 interferes with both retrograde and anterograde intracellular trafficking pathways. *PLoS One* **6**:e22713.

37. Deresiewicz RL, Calderwood SB, Robertus JD, Collier RJ. 1992. Mutations affecting the activity of the Shiga-like toxin I A-chain. *Biochemistry* **31**:3272–3280.

38. Di R, Kyu E, Shete V, Saidasan H, Kahn PC, Tumer NE. 2011. Identification of amino acids critical for the cytotoxicity of Shiga toxin 1 and 2 in *Saccharomyces cerevisiae*. *Toxicon* **57**:525–539.

39. LaPointe P, Wei X, Gariepy J. 2005. A role for the protease-sensitive loop region of Shiga-like toxin 1 in the retrotranslocation of its A1 domain from the endoplasmic reticulum lumen. *J Biol Chem* **280**:23310–23318.

40. McCluskey AJ, Poon GM, Bolewska-Pedyczak E, Srikumar T, Jeram SM, Raught B, Gariepy J. 2008. The catalytic subunit of Shiga-like toxin 1 interacts with ribosomal stalk proteins and is inhibited by their conserved C-terminal domain. *J Mol Biol* **378**:375–386.

41. McCluskey AJ, Bolewska-Pedyczak E, Jarvik N, Chen G, Sidhu SS, Gariepy J. 2012. Charged and hydrophobic surfaces on the a chain of Shiga-like toxin 1 recognize the C-terminal domain of ribosomal stalk proteins. *PLoS One* **7**:e31191.

42. Lindberg AA, Brown JE, Stromberg N, Westling-Ryd M, Schultz JE, Karlsson KA. 1987. Identification of the carbohydrate receptor for Shiga toxin produced by *Shigella dysenteriae* type 1. *J Biol Chem* **262**:1779–1785.

43. Lingwood CA, Law H, Richardson S, Petric M, Brunton JL, De Grandis S, Karmali M. 1987. Glycolipid binding of purified and recombinant *Escherichia coli* produced verotoxin in vitro. *J Biol Chem* **262**:8834–8839.

44. Waddell T, Head S, Petric M, Cohen A, Lingwood C. 1988. Globotriosyl ceramide is specifically recognized by the *Escherichia coli* verocytotoxin 2. *Biochem Biophys Res Commun* **152**:674–679.

45. Jacewicz MS, Mobassaleh M, Gross SK, Balasubramanian KA, Daniel PF, Raghavan S, McCluer RH, Keusch GT. 1994. Pathogenesis of *Shigella* diarrhea: XVII. A mammalian cell membrane glycolipid, Gb3, is required but not sufficient to confer sensitivity to Shiga toxin. *J Infect Dis* **169**:538–546.

46. Acheson DW, Moore R, De Breucker S, Lincicome L, Jacewicz M, Skutelsky E, Keusch GT. 1996. Translocation of Shiga toxin across polarized intestinal cells in tissue culture. *Infect Immun* **64**:3294–3300.

47. **Okuda T, Tokuda N, Numata S, Ito M, Ohta M, Kawamura K, Wiels J, Urano T, Tajima O, Furukawa K.** 2006. Targeted disruption of Gb3/CD77 synthase gene resulted in the complete deletion of globo-series glycosphingolipids and loss of sensitivity to verotoxins. *J Biol Chem* **281:**10230–10235.

48. **Tarabuso AL.** 2011. Fabry disease. *Skinmed* **9:**173–177.

49. **Cilmi SA, Karalius BJ, Choy W, Smith RN, Butterton JR.** 2006. Fabry disease in mice protects against lethal disease caused by Shiga toxin-expressing enterohemorrhagic *Escherichia coli*. *J Infect Dis* **194:**1135–1140.

50. **Nakajima H, Kiyokawa N, Katagiri YU, Taguchi T, Suzuki T, Sekino T, Mimori K, Ebata T, Saito M, Nakao H, Takeda T, Fujimoto J.** 2001. Kinetic analysis of binding between Shiga toxin and receptor glycolipid Gb3Cer by surface plasmon resonance. *J Biol Chem* **276:**42915–42922.

51. **DeGrandis S, Law H, Brunton J, Gyles C, Lingwood CA.** 1989. Globotetraosylceramide is recognized by the pig edema disease toxin. *J Biol Chem* **264:**12520–12525.

52. **Samuel JE, Perera LP, Ward S, O'Brien AD, Ginsburg V, Krivan HC.** 1990. Comparison of the glycolipid receptor specificities of Shiga-like toxin type II and Shiga-like toxin type II variants. *Infect Immun* **58:**611–618.

53. **Devenish J, Gyles C, LaMarre J.** 1998. Binding of *Escherichia coli* verotoxins to cell surface protein on wild-type and globotriaosylceramide-deficient Vero cells. *Can J Microbiol* **44:**28–34.

54. **Katagiri YU, Mori T, Nakajima H, Katagiri C, Taguchi T, Takeda T, Kiyokawa N, Fujimoto J.** 1999. Activation of Src family kinase yes induced by Shiga toxin binding to globotriaosyl ceramide (Gb3/CD77) in low density, detergent-insoluble microdomains. *J Biol Chem* **274:**35278–35282.

55. **Kovbasnjuk O, Edidin M, Donowitz M.** 2001. Role of lipid rafts in Shiga toxin 1 interaction with the apical surface of Caco-2 cells. *J Cell Sci* **114:**4025–4031.

56. **Falguieres T, Romer W, Amessou M, Afonso C, Wolf C, Tabet JC, Lamaze C, Johannes L.** 2006. Functionally different pools of Shiga toxin receptor, globotriaosyl ceramide, in HeLa cells. *FEBS J* **273:**5205–5218.

57. **Kiarash A, Boyd B, Lingwood CA.** 1994. Glycosphingolipid receptor function is modified by fatty acid content. Verotoxin 1 and verotoxin 2c preferentially recognize different globotriaosyl ceramide fatty acid homologues. *J Biol Chem* **269:**11138–11146.

58. **Pellizzari A, Pang H, Lingwood CA.** 1992. Binding of verocytotoxin 1 to its receptor is influenced by differences in receptor fatty acid content. *Biochemistry* **31:**1363–1370.

59. **Khan F, Proulx F, Lingwood CA.** 2009. Detergent-resistant globotriaosyl ceramide may define verotoxin/glomeruli-restricted hemolytic uremic syndrome pathology. *Kidney Int* **75:**1209–1216.

60. **Nyholm PG, Magnusson G, Zheng Z, Norel R, Binnington-Boyd B, Lingwood CA.** 1996. Two distinct binding sites for globotriaosyl ceramide on verotoxins: identification by molecular modelling and confirmation using deoxy analogues and a new glycolipid receptor for all verotoxins. *Chem Biol* **3:**263–275.

61. **Thompson GS, Shimizu H, Homans SW, Donohue-Rolfe A.** 2000. Localization of the binding site for the oligosaccharide moiety of Gb3 on verotoxin 1 using NMR residual dipolar coupling measurements. *Biochemistry* **39:**13153–13156.

62. **Bast DJ, Banerjee L, Clark C, Read RJ, Brunton JL.** 1999. The identification of three biologically relevant globotriaosyl ceramide receptor binding sites on the Verotoxin 1 B subunit. *Mol Microbiol* **32:**953–960.

63. **Soltyk AM, MacKenzie CR, Wolski VM, Hirama T, Kitov PI, Bundle DR, Brunton JL.** 2002. A mutational analysis of the globotriaosylceramide-binding sites of verotoxin VT1. *J Biol Chem* **277:**5351–5359.

64. **Kitova EN, Kitov PI, Paszkiewicz E, Kim J, Mulvey GL, Armstrong GD, Bundle DR, Klassen JS.** 2007. Affinities of Shiga toxins 1 and 2 for univalent and oligovalent Pk-trisaccharide analogs measured by electrospray ionization mass spectrometry. *Glycobiology* **17:**1127–1137.

65. **Nishikawa K, Matsuoka K, Watanabe M, Igai K, Hino K, Hatano K, Yamada A, Abe N, Terunuma D, Kuzuhara H, Natori Y.** 2005. Identification of the optimal structure required for a Shiga toxin neutralizer with oriented carbohydrates to function in the circulation. *J Infect Dis* **191:**2097–2105.

66. **Watanabe M, Igai K, Matsuoka K, Miyagawa A, Watanabe T, Yanoshita R, Samejima Y, Terunuma D, Natori Y, Nishikawa K.** 2006. Structural analysis of the interaction between Shiga toxin B subunits and linear polymers bearing clustered globotriose residues. *Infect Immun* **74:**1984–1988.

67. **Holgersson J, Jovall PA, Breimer ME.** 1991. Glycosphingolipids of human large intestine: detailed structural characterization with special reference to blood group compounds and bacterial receptor structures. *J Biochem* **110:**120–131.

68. **Malyukova I, Murray KF, Zhu C, Boedeker E, Kane A, Patterson K, Peterson JR, Donowitz M, Kovbasnjuk O.** 2009. Macropinocytosis in Shiga toxin 1 uptake by human intestinal epithelial cells and transcellular transcytosis. *Am J Physiol Gastrointest Liver Physiol* **296:**G78–92.

69. **Hurley BP, Thorpe CM, Acheson DW.** 2001. Shiga toxin translocation across intestinal epithelial cells is enhanced by neutrophil transmigration. *Infect Immun* **69:**6148–6155.

70. **Philpott DJ, Ackerley CA, Kiliaan AJ, Karmali MA, Perdue MH, Sherman PM.** 1997. Translocation of verotoxin-1 across T84 monolayers: mechanism of bacterial toxin penetration of epithelium. *Am J Physiol* **273:**G1349–1358.

71. **Schuller S, Frankel G, Phillips AD.** 2004. Interaction of Shiga toxin from *Escherichia coli* with human intestinal epithelial cell lines and explants: Stx2 induces epithelial damage in organ culture. *Cell Microbiol* **6:**289–301.

72. **Zumbrun SD, Hanson L, Sinclair JF, Freedy J, Melton-Celsa AR, Rodriguez-Canales J, Hanson JC, O'Brien AD.** 2010. Human intestinal tissue and cultured colonic cells contain globotriaosylceramide synthase mRNA and the alternate Shiga toxin receptor globotetraosylceramide. *Infect Immun* **78:**4488–4499.

73. **Jacewicz MS, Acheson DW, Mobassaleh M, Donohue-Rolfe A, Balasubramanian KA, Keusch GT.** 1995. Maturational regulation of globotriaosylceramide, the Shiga-like toxin 1 receptor, in cultured human gut epithelial cells. *J Clin Invest* **96:**1328–1335.

74. **Zumbrun SD, Melton-Celsa AR, Smith MA, Gilbreath JJ, Merrell DS, O'Brien AD.** 2013. Dietary choice affects Shiga toxin-producing *Escherichia coli* (STEC) O157:H7 colonization and disease. *Proc Natl Acad Sci USA* **110:**E2126–2133.

75. **Schuller S, Heuschkel R, Torrente F, Kaper JB, Phillips AD.** 2007. Shiga toxin binding in normal and inflamed human intestinal mucosa. *Microbes Infect* **9:**35–39.

76. **Bergan J, Dyve Lingelem AB, Simm R, Skotland T, Sandvig K.** 2012. Shiga toxins. *Toxicon* **60:**1085–1107.

77. **Torgersen ML, Lauvrak SU, Sandvig K.** 2005. The A-subunit of surface-bound Shiga toxin stimulates clathrin-dependent uptake of the toxin. *FEBS J* **272:**4103–4113.

78. **Romer W, Berland L, Chambon V, Gaus K, Windschiegl B, Tenza D, Aly MR, Fraisier V, Florent JC, Perrais D, Lamaze C, Raposo G, Steinem C, Sens P, Bassereau P, Johannes L.** 2007. Shiga toxin induces tubular membrane invaginations for its uptake into cells. *Nature* **450:**670–675.

79. **Sandvig K, Garred O, Prydz K, Kozlov JV, Hansen SH, van Deurs B.** 1992. Retrograde transport of endocytosed Shiga toxin to the endoplasmic reticulum. *Nature* **358:**510–512.

80. **Sandvig K, Bergan J, Dyve AB, Skotland T, Torgersen ML.** 2010. Endocytosis and retrograde transport of Shiga toxin. *Toxicon* **56:**1181–1185.

81. **Garred O, van Deurs B, Sandvig K.** 1995. Furin-induced cleavage and activation of Shiga toxin. *J Biol Chem* **270:**10817–10821.

82. **Raa H, Grimmer S, Schwudke D, Bergan J, Walchli S, Skotland T, Shevchenko A, Sandvig K.** 2009. Glycosphingolipid requirements for endosome-to-Golgi transport of Shiga toxin. *Traffic* **10:**868–882.

83. **Sandvig K, Skotland T, van Deurs B, Klokk TI.** 2013. Retrograde transport of protein toxins through the Golgi apparatus. *Histochem Cell Biol* **140:**317–326.

84. **Spooner RA, Lord JM.** 2012. How ricin and Shiga toxin reach the cytosol of target cells: retro-translocation from the endoplasmic reticulum. *Curr Top Microbiol Immunol* **357:**19–40.

85. **Tam PJ, Lingwood CA.** 2007. Membrane cytosolic translocation of verotoxin A1 subunit in target cells. *Microbiology* **153:**2700–2710.

86. **Obrig TG, Moran TP, Brown JE.** 1987. The mode of action of Shiga toxin on peptide elongation of eukaryotic protein synthesis. *Biochem J* **244:**287–294.

87. **Furutani M, Kashiwagi K, Ito K, Endo Y, Igarashi K.** 1992. Comparison of the modes of action of a Vero toxin (a Shiga-like toxin) from *Escherichia coli*, of ricin, and of alpha-sarcin. *Arch Biochem Biophys* **293:**140–146.

88. **Warnier M, Romer W, Geelen J, Lesieur J, Amessou M, van den Heuvel L, Monnens L, Johannes L.** 2006. Trafficking of Shiga toxin/Shiga-like toxin-1 in human glomerular microvascular endothelial cells and human mesangial cells. *Kidney Int* **70:**2085–2091.

89. **Jandhyala DM, Thorpe CM, Magun B.** 2012. Ricin and Shiga toxins: effects on host cell signal transduction. *Curr Top Microbiol Immunol* **357:**41–65.

90. **Lee MS, Kim MH, Tesh VL.** 2013. Shiga toxins expressed by human pathogenic bacteria induce immune responses in host cells. *J Microbiol* **51:**724–730.

91. **Tesh VL.** 2012. The induction of apoptosis by Shiga toxins and ricin. *Curr Top Microbiol Immunol* **357:**137–178.

92. **Pai CH, Kelly JK, Meyers GL.** 1986. Experimental infection of infant rabbits with verotoxin-producing *Escherichia coli*. *Infect Immun* **51:**16–23.

93. **Keenan KP, Sharpnack DD, Collins H, Formal SB, O'Brien AD.** 1986. Morphologic evaluation of the effects of Shiga toxin and *E. coli* Shiga-like toxin on the rabbit intestine. *Am J Pathol* **125:**69–80.

94. **Tesh VL, Burris JA, Owens JW, Gordon VM, Wadolkowski EA, O'Brien AD, Samuel JE.** 1993. Comparison of the relative toxicities of Shiga-like toxins type I and type II for mice. *Infect Immun* **61:**3392–3402.

95. **Smith MJ, Teel LD, Carvalho HM, Melton-Celsa AR, O'Brien AD.** 2006. Development of a hybrid Shiga holotoxin vaccine to elicit heterologous protection against Shiga toxins types 1 and 2. *Vaccine* 24:4122–4129.

96. **Soborg B, Lassen SG, Muller L, Jensen T, Ethelberg S, Molbak K, Scheutz F.** 2013. A verocytotoxin-producing *E. coli* outbreak with a surprisingly high risk of haemolytic uraemic syndrome, Denmark, September–October 2012. *Euro Surveill* 18:ii, 20350.

97. **Boerlin P, McEwen SA, Boerlin-Petzold F, Wilson JB, Johnson RP, Gyles CL.** 1999. Associations between virulence factors of Shiga toxin-producing *Escherichia coli* and disease in humans. *J Clin Microbiol* 37:497–503.

98. **Ostroff SM, Tarr PI, Neill MA, Lewis JH, Hargrett-Bean N, Kobayashi JM.** 1989. Toxin genotypes and plasmid profiles as determinants of systemic sequelae in *Escherichia coli* O157:H7 infections. *J Infect Dis* 160:994–998.

99. **Russo LM, Melton-Celsa AR, Smith MA, Smith MJ, O'Brien DA.** 2013. Oral intoxication of mice with Shiga toxin type 2a (Stx2a) and protection by anti-Stx2a monoclonal antibody 11E10. *Infect Immun* 82:1213–1221.

100. **Louise CB, Obrig TG.** 1995. Specific interaction of *Escherichia coli* O157:H7-derived Shiga-like toxin II with human renal endothelial cells. *J Infect Dis* 172:1397–1401.

101. **Chark D, Nutikka A, Trusevych N, Kuzmina J, Lingwood C.** 2004. Differential carbohydrate epitope recognition of globotriaosyl ceramide by verotoxins and a monoclonal antibody. *Eur J Biochem* 271:405–417.

102. **Head SC, Karmali MA, Lingwood CA.** 1991. Preparation of VT1 and VT2 hybrid toxins from their purified dissociated subunits. Evidence for B subunit modulation of a subunit function. *J Biol Chem* 266:3617–3621.

103. **Gallegos KM, Conrady DG, Karve SS, Gunasekera TS, Herr AB, Weiss AA.** 2012. Shiga toxin binding to glycolipids and glycans. *PLoS One* 7:e30368.

104. **Mahfoud R, Manis A, Binnington B, Ackerley C, Lingwood CA.** 2010. A major fraction of glycosphingolipids in model and cellular cholesterol-containing membranes is undetectable by their binding proteins. *J Biol Chem* 285:36049–36059.

105. **Gamage SD, McGannon CM, Weiss AA.** 2004. *Escherichia coli* serogroup O107/O117 lipopolysaccharide binds and neutralizes Shiga toxin 2. *J Bacteriol* 186:5506–5512.

106. **Kimura T, Tani S, Matsumoto Yi Y, Takeda T.** 2001. Serum amyloid P component is the Shiga toxin 2-neutralizing factor in human blood. *J Biol Chem* 276:41576–41579.

107. **Conrady DG, Flagler MJ, Friedmann DR, Vander Wielen BD, Kovall RA, Weiss AA, Herr AB.** 2010. Molecular basis of differential B-pentamer stability of Shiga toxins 1 and 2. *PLoS One* 5:e15153.

108. **Tam P, Mahfoud R, Nutikka A, Khine AA, Binnington B, Paroutis P, Lingwood C.** 2008. Differential intracellular transport and binding of verotoxin 1 and verotoxin 2 to globotriaosylceramide-containing lipid assemblies. *J Cell Physiol* 216:750–763.

109. **Matussek A, Lauber J, Bergau A, Hansen W, Rohde M, Dittmar KE, Gunzer M, Mengel M, Gatzlaff P, Hartmann M, Buer J, Gunzer F.** 2003. Molecular and functional analysis of Shiga toxin-induced response patterns in human vascular endothelial cells. *Blood* 102:1323–1332.

110. **Lentz EK, Leyva-Illades D, Lee MS, Cherla RP, Tesh VL.** 2011. Differential response of the human renal proximal tubular epithelial cell line HK-2 to Shiga toxin types 1 and 2. *Infect Immun* 79:3527–3540.

111. **Stearns-Kurosawa DJ, Collins V, Freeman S, Tesh VL, Kurosawa S.** 2010. Distinct physiologic and inflammatory responses elicited in baboons after challenge with Shiga toxin type 1 or 2 from enterohemorrhagic *Escherichia coli*. *Infect Immun* 78:2497–2504.

112. **Bu F, Borsa N, Gianluigi A, Smith RJ.** 2012. Familial atypical hemolytic uremic syndrome: a review of its genetic and clinical aspects. *Clin Dev Immunol* 2012:370426.

113. **Riley LW, Remis RS, Helgerson SD, McGee HB, Wells JG, Davis BR, Hebert RJ, Olcott ES, Johnson LM, Hargrett NT, Blake PA, Cohen ML.** 1983. Hemorrhagic colitis associated with a rare *Escherichia coli* serotype. *N Engl J Med* 308:681–685.

114. **Frank C, Werber D, Cramer JP, Askar M, Faber M, an der Heiden M, Bernard H, Fruth A, Prager R, Spode A, Wadl M, Zoufaly A, Jordan S, Kemper MJ, Follin P, Muller L, King LA, Rosner B, Buchholz U, Stark K, Krause G.** 2011. Epidemic profile of Shiga-toxin-producing *Escherichia coli* O104:H4 outbreak in Germany. *N Engl J Med* 365:1771–1780.

115. **Sauter KA, Melton-Celsa AR, Larkin K, Troxell ML, O'Brien AD, Magun BE.** 2008. Mouse model of hemolytic-uremic syndrome caused by endotoxin-free Shiga toxin 2 (Stx2) and protection from lethal outcome by anti-Stx2 antibody. *Infect Immun* 76:4469–4478.

116. **Fontaine A, Arondel J, Sansonetti PJ.** 1988. Role of Shiga toxin in the pathogenesis of bacillary dysentery, studied by using a Tox- mutant of *Shigella dysenteriae* 1. *Infect Immun* 56:3099–3109.

117. Chaisri U, Nagata M, Kurazono H, Horie H, Tongtawe P, Hayashi H, Watanabe T, Tapchaisri P, Chongsa-nguan M, Chaicumpa W. 2001. Localization of Shiga toxins of enterohaemorrhagic *Escherichia coli* in kidneys of paediatric and geriatric patients with fatal haemolytic uraemic syndrome. *Microb Pathog* **31**:59–67.

118. Uchida H, Kiyokawa N, Horie H, Fujimoto J, Takeda T. 1999. The detection of Shiga toxins in the kidney of a patient with hemolytic uremic syndrome. *Pediatr Res* **45**:133–137.

119. Tazzari PL, Ricci F, Carnicelli D, Caprioli A, Tozzi AE, Rizzoni G, Conte R, Brigotti M. 2004. Flow cytometry detection of Shiga toxins in the blood from children with hemolytic uremic syndrome. *Cytometry B Clin Cytom* **61**:40–44.

120. Bitzan M, Poole R, Mehran M, Sicard E, Brockus C, Thuning-Roberson C, Riviere M. 2009. Safety and pharmacokinetics of chimeric anti-Shiga toxin 1 and anti-Shiga toxin 2 monoclonal antibodies in healthy volunteers. *Antimicrob Agents Chemother* **53**:3081–3087.

121. Dowling TC, Chavaillaz PA, Young DG, Melton-Celsa A, O'Brien A, Thuning-Roberson C, Edelman R, Tacket CO. 2005. Phase 1 safety and pharmacokinetic study of chimeric murine-human monoclonal antibody c alpha Stx2 administered intravenously to healthy adult volunteers. *Antimicrob Agents Chemother* **49**:1808–1812.

122. McKee ML, Melton-Celsa AR, Moxley RA, Francis DH, O'Brien AD. 1995. Enterohemorrhagic *Escherichia coli* O157:H7 requires intimin to colonize the gnotobiotic pig intestine and to adhere to HEp-2 cells. *Infect Immun* **63**:3739–3744.

123. Judge NA, Mason HS, O'Brien AD. 2004. Plant cell-based intimin vaccine given orally to mice primed with intimin reduces time of *Escherichia coli* O157:H7 shedding in feces. *Infect Immun* **72**:168–175.

124. Girardeau JP, Dalmasso A, Bertin Y, Ducrot C, Bord S, Livrelli V, Vernozy-Rozand C, Martin C. 2005. Association of virulence genotype with phylogenetic background in comparison to different seropathotypes of Shiga toxin-producing *Escherichia coli* isolates. *J Clin Microbiol* **43**:6098–6107.

125. Zumbrun SD, Melton-Celsa AR, O'Brien AD. 2014. When a healthy diet turns deadly. *Gut Microbes* **5**(1):40–43.

126. Petruzziello-Pellegrini TN, Moslemi-Naeini M, Marsden PA. 2013. New insights into Shiga toxin-mediated endothelial dysfunction in hemolytic uremic syndrome. *Virulence* **4**:556–563.

127. Keir LS, Saleem MA. 2013. Current evidence for the role of complement in the pathogenesis of Shiga toxin haemolytic uraemic syndrome. *Pediatr Nephrol* Jul 11. (Epub ahead of print.)

128. Zoja C, Buelli S, Morigi M. 2010. Shiga toxin-associated hemolytic uremic syndrome: pathophysiology of endothelial dysfunction. *Pediatr Nephrol* **25**:2231–2240.

129. Andreoli SP, Trachtman H, Acheson DW, Siegler RL, Obrig TG. 2002. Hemolytic uremic syndrome: epidemiology, pathophysiology, and therapy. *Pediatr Nephrol* **17**:293–298.

130. Maye CL, Leibowitz CS, Kurosawa S, Stearns-Kurosawa DJ. 2012. Shiga toxins and the pathophysiology of hemolytic uremic syndrome in humans and animals. *Toxins (Basel)* **4**:1261–1287.

131. Stahl AL, Sartz L, Karpman D. 2011. Complement activation on platelet-leukocyte complexes and microparticles in enterohemorrhagic *Escherichia coli*-induced hemolytic uremic syndrome. *Blood* **117**:5503–5513.

132. Thurman JM, Marians R, Emlen W, Wood S, Smith C, Akana H, Holers VM, Lesser M, Kline M, Hoffman C, Christen E, Trachtman H. 2009. Alternative pathway of complement in children with diarrhea-associated hemolytic uremic syndrome. *Clin J Am Soc Nephrol* **4**:1920–1924.

133. Hale TL, Formal SB. 1980. Cytotoxicity of *Shigella dysenteriae* 1 for cultured mammalian cells. *Am J Clin Nutr* **33**:2485–2490.

134. Koch C, Hertwig S, Lurz R, Appel B, Beutin L. 2001. Isolation of a lysogenic bacteriophage carrying the stx(1(OX3)) gene, which is closely associated with Shiga toxin-producing *Escherichia coli* strains from sheep and humans. *J Clin Microbiol* **39**:3992–3998.

135. Paton AW, Beutin L, Paton JC. 1995. Heterogeneity of the amino-acid sequences of *Escherichia coli* Shiga-like toxin type-I operons. *Gene* **153**:71–74.

136. Burk C, Dietrich R, Acar G, Moravek M, Bulte M, Martlbauer E. 2003. Identification and characterization of a new variant of Shiga toxin 1 in *Escherichia coli* ONT:H19 of bovine origin. *J Clin Microbiol* **41**:2106–2112.

137. Paton AW, Paton JC, Manning PA. 1993. Polymerase chain reaction amplification, cloning and sequencing of variant *Escherichia coli* Shiga-like toxin type II operons. *Microb Pathog* **15**:77–82.

138. Persson S, Olsen KE, Ethelberg S, Scheutz F. 2007. Subtyping method for *Escherichia coli* Shiga toxin (verocytotoxin) 2 variants and correlations to clinical manifestations. *J Clin Microbiol* **45**:2020–2024.

139. Weinstein DL, Jackson MP, Samuel JE, Holmes RK, O'Brien AD. 1988. Cloning and sequencing of a Shiga-like toxin type II variant from *Escherichia coli* strain responsible for edema disease of swine. *J Bacteriol* **170**:4223–4230.

140. **Schmidt H, Scheef J, Morabito S, Caprioli A, Wieler LH, Karch H.** 2000. A new Shiga toxin 2 variant (Stx2f) from *Escherichia coli* isolated from pigeons. *Appl Environ Microbiol* **66:**1205–1208.

141. **Leung PH, Peiris JS, Ng WW, Robins-Browne RM, Bettelheim KA, Yam WC.** 2003. A newly discovered verotoxin variant, VT2g, produced by bovine verocytotoxigenic *Escherichia coli*. *Appl Environ Microbiol* **69:**7549–7553.

Enterohemorrhagic *Escherichia coli* Genomics: Past, Present, and Future

4

SHAH M. SADIQ,[1] TRACY H. HAZEN,[2] DAVID A. RASKO,[2] and
MARK EPPINGER[1]

INTRODUCTION

O157:H7 is the most common enterohemorrhagic *Escherichia coli* (EHEC) serotype in North America, and it has been the principal causative agent of numerous food-poisoning outbreaks worldwide (1, 2). Initially *E. coli* O157:H7 was recognized as a human pathogen in 1982 when it was isolated from 47 persons in two states who had developed bloody diarrhea after consuming hamburgers contaminated with this organism (3). Since then, *E. coli* O157:H7 has emerged as a major enteric pathogen, capable of causing localized infections and large outbreaks of gastrointestinal disease (4). Data accumulated from 1982 to 1996 showed that approximately two-thirds of the 3,000 cases of *E. coli* infections from 139 recognized outbreaks were associated with the ingestion of contaminated food products, whereas 22% of the reported cases were from direct person-to-person transmission and 10% were from drinking water (5). Surveillance data have demonstrated a high prevalence of *E. coli* O157:H7 among cattle and their environment, but a relatively low incidence of human infection. This supports the potential hypothesis that a subset of *E. coli* O157:H7 harbored

[1]Department of Biology, and South Texas Center for Emerging Infectious Diseases, University of Texas at San Antonio, San Antonio, TX 78249; [2]Institute for Genome Sciences, Department of Microbiology and Immunology, University of Maryland School of Medicine, Baltimore, MD 21201.

Enterohemorrhagic Escherichia coli *and Other Shiga Toxin-Producing* E. coli
Edited by Vanessa Sperandio and Carolyn J. Hovde
© 2015 American Society for Microbiology, Washington, DC
doi:10.1128/microbiolspec.EHEC-0020-2013

in cattle may be responsible for the majority of human disease (6). To minimize or eradicate adverse effects on public health, the *E. coli* O157:H7 lineage has been the focus of numerous epidemiological, microbiological, genomic, forensic, and diagnostic studies. Overall, it is estimated that *E. coli* O157:H7 alone causes more than 76,000 infections and 61 deaths in humans due to severe complications annually in the United States (7). Symptoms include a range of gastrointestinal morbidities, such as severe abdominal cramping accompanied with little or no associated fever and a watery diarrhea that leads to severe bloody diarrhea (8). Although many infected individuals remain asymptomatic, approximately 15 to 20% of people infected with EHEC present severe enough symptoms to require hospitalization. In such severe cases, patients display renal dysfunction known as hemolytic-uremic syndrome (HUS), hemorrhagic colitis, and central nervous system failure with potentially lethal outcomes (9–11).

The nomenclature surrounding EHEC and other Shiga-toxin containing *E. coli* (STEC) can be confusing; hence, the molecular definitions for EHEC and STEC that we employ in this article are outlined in Fig. 1. The definitions are based on key virulence factors of the EHEC, STEC, and enteropathogenic *E. coli* (EPEC) pathogenic variants or pathovars (1, 2). We briefly introduce virulence factors as a way to define the pathogens and isolates we are examining. The three virulence factors are the *eae* gene that encodes intimin and is used as a surrogate marker for the locus of enterocyte effacement (LEE region); the *bfpA* gene, used as a marker for the bundle-forming pilus operon that has been described as essential in the binding to the epithelial cells; and the *stx* gene (in this case, this means any Shiga toxin genes). Figure 1 demonstrates how these three genes separate the EPEC, EHEC, and STEC isolates. EHEC isolates contain both the *eae* and *stx* genes, whereas STEC isolates contain only the *stx* gene. The EPEC can be separated into two groups: typical EPEC, containing both *eae* and *bfp*, and

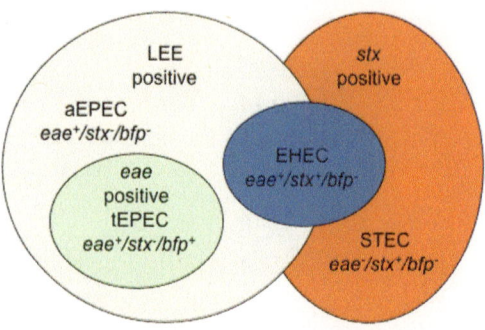

FIGURE 1 Figure depicts the molecular differences that define each of the attaching and effacing *E. coli*. The *eae* gene encodes the intimin protein on the LEE region; the *bfp* gene in this case is the presence of the bundle-forming pilus operon, and the *stx* gene encodes the Shiga toxin. These three features are classically used to define the pathotypes, including EHEC. doi.10.1128/microbiolspec.EHEC-0020-2013.f1

atypical EPEC, containing only the *eae* gene. One can quickly identify the potential flaws in these molecular definitions, especially since each of these virulence factors is encoded on a mobile element: *stx* on the phage; *eae* on the LEE genomic island, and *bfp* on the EAF plasmid. The ease with which each or many of these genes can be lost has not been defined in many strains, and thus, the lack of any of these features may prevent the proper categorization of the isolate to the appropriate pathovar. This highlights the crucial need for the development of more stable genetic biomarkers from genomic information that is addressed later in this article.

It has been noted that recent outbreaks, especially those associated with green leafy vegetables, have been associated with increased virulence, as measured by the number of individuals that present in health care facilities. Three *E. coli* O157:H7 outbreaks in 2006 from ingested fresh produce, referred to as spinach, Taco Bell, and Taco John outbreaks (http://www.cdc.gov/ecoli/), captured the attention of both the public health and lay communities. For example, the spinach outbreak caused illness in 199 people from 26 states after they ingested fresh spinach contaminated with *E. coli* O157:H7 (12). Three deaths were attributed to the spinach outbreak: two elderly women and a 2-year-old child. Among the ill, 51% were hospitalized, and in 16% (12) of the cases, infection progressed to HUS and kidney failure. The high number of patients hospitalized and the high rate of kidney failure suggest that this outbreak was due to a more virulent strain of *E. coli* O157:H7. The isolates for the most recent outbreaks have been examined by genome sequencing as a part of the epidemiological and microbiological examination of the outbreak.

The introduction of microbial sequencing has opened up the opportunity for comparative analysis of many genomes to identify regions of the genome that may be associated with the greater virulence described above. The sequencing of the *E. coli* MG1655 isolate was the beginning of *E. coli* genome sequencing (13); this was followed by the sequencing of two EHEC O157:H7 isolates, EDL933 (14) and the Sakai isolate (15). The sequencing of these two isolates and the associated comparative analysis provided significant insights into two key points of the evolution of *E. coli* in general and EHEC specifically: (i) the isolates of EHEC were closely related and the analysis could distinguish differences in each group, and (ii) a significant amount of diversity within and between the EHEC isolates existed, but there was >1 Mb of DNA that was unique in the EHEC isolates that was not present in the laboratory-adapted K-12 isolate. This represented approximately 20% of the genome that was unique in the pathogen, providing ample opportunity for functional characterization. Continued genome sequencing of *E. coli* and EHEC specifically has resulted in significant insights into the evolution of these important pathogens, but with the advent of new sequencing technologies, we no longer sequence one or two prototype isolates; rather, we sequence "collections" or "outbreaks" to define the molecular markers of these pathogens. This article highlights where we started with the molecular characterization of *E. coli* and EHEC and where we are today with high-throughput technologies.

RESERVOIRS FOR HUMAN HEALTH

Cattle are recognized as a main reservoir of STEC O157:H7. The significant differences in host prevalence, transmissibility, and virulence phenotypes among strains from bovine and human sources are of major interest to the public health community and livestock industry (16). Genomic analysis revealed divergence into three lineages: lineage I and lineage I/II strains are commonly associated with human disease, whereas lineage II strains are overrepresented in the asymptomatic bovine host reservoir (17). Growing evidence suggests that genotypic differences between these lineages, such as polymorphisms in Shiga toxin subtypes and synergistically acting virulence factors, are correlated with phenotypic differences in virulence, host ecology, and epidemiology (18). To assess the genomic plasticity on a genomewide scale, the whole genome of strain EC869, a bovine-associated *E. coli* O157:H7 isolate, was sequenced. Comparative phylogenomic analysis of this key isolate enabled placement of the bovine lineage II strains within the genetically homogeneous *E. coli* O157:H7 clade (18). Identification of polymorphic loci that are anchored both in the chromosomal backbone and horizontally acquired regions allowed association of bovine genotypes with altered virulence phenotypes and host prevalence. Polymorphic markers are valuable in the development of a robust typing system critical for forensic, diagnostic, and epidemiological studies of this emerging human pathogen.

TYPING AND GENETIC ANALYSIS OF EHEC OUTBREAKS WITH PREGENOMIC METHODOLOGIES

Unlike other *E. coli* serotypes, the O157:H7 lineage is distinguished by its genetically highly homogeneous population structure, comparable to clonal microbial species such as *Yersinia pestis* (19) or *Bacillus anthracis* (20). With potentially lethal and widespread outbreaks, a large-scale and in-depth survey of genetic and architectural polymorphisms is a crucial prerequisite to obtain insights into the natural pathogenome evolution and extent of bacterial disease virulence genotypes. Genetic heterogeneity among O157:H7 EHEC strains has been established by using a broad panel of targeted- and whole-genome-based typing assays to determine diversity and evolutionary relationships among EHEC isolates, such as multilocus sequence typing (21), octamer- and PCR-based genome scanning, (22, 23), phage typing (24–26), multiple-locus variable-number tandem repeat analysis (27, 28), microarrays (29), microarray-based comparative genome hybridization (17), nucleotide polymorphism assays (19, 30, 31), pulsed-field electrophoresis (PFGE) (32), subtractive hybridization (33, 34), and optical mapping (12, 35, 36).

Pulsed-Field Gel Electrophoresis

PFGE has been widely employed as a molecular typing method in epidemiological investigations of EHEC (32, 37). According to differences in the XbaI PFGE patterns, EHEC O157:H7 isolates are classified into different types (types I to V and ND) (38). Alterations in PFGE patterns after restriction digest are results of genomic rearrangements driven by recombination of prophages and mobile elements (39, 40). PFGE, though considered a first-line molecular screening tool, provides insight into the genome structure and changes in restriction patterns and potentially identifies spontaneous genomic rearrangements or recombination of mobile elements (39, 40). In short-term epidemiological studies, it may be that the PFGE pattern will be useful as an inclusion or exclusion criterion when *E. coli* O157:H7 isolates are examined over a defined time window in an outbreak situation. However, even now, we are seeing the greater use of whole-genome sequencing for the investigation of outbreaks of *E. coli* (41–46).

Multilocus Sequence Typing

Sequenced-based methods, such as multilocus sequence typing (MLST), have been powerful subtyping tools in molecular epidemiology. These methods have the advantage of being easily standardized and automated. MLST was first developed for *Neisseria meningitidis* in 1998 to overcome the poor reproducibility between laboratories applying older molecular typing schemes (47). The principle behind the MLST scheme is to identify internal nucleotide sequences of approximately 400 to 500 bp in multiple housekeeping genes. Unique sequences (alleles) are assigned a random integer number, and a unique combination of alleles at each locus, an "allelic profile," specifies the sequence type. In this new era of high-throughput sequencing, it may be more rational to use whole-genome sequence data for typing. Two MLST schemes exist for *E. coli*: *E. coli* scheme 1, which employs seven genes (*adk, fumC, gyrB, icd, mdh, purA, recA*) (48), and *E. coli* scheme 2, which employs eight genes (*dinB, icdA, pabB, polB, putP, trpA, trpB, uidA*) (49). Though these MLST methods are generally congruent, there are some differences for some strains.

These MLST systems first determined that EHEC isolates are genetically highly clonal but could also be separated into multiple clades, suggesting that there are multiple evolutionary paths to generate a fully virulent EHEC isolate (50).

Multiple-Locus Variable Number Tandem Repeat Analysis

As a result of the poor discriminatory ability of MLST for *E. coli* O157:H7, it was decided to target short tandem repeats, which are areas of the bacterial genome that evolve rapidly. Targeting of these elements, which often vary in number among different strains of the same species (the definition of a variable-number tandem repeat), has successfully been used to discriminate between strains of prokaryotes (51). Multiple-locus variable-number tandem repeat analysis involves determination of the number of repeats at multiple loci, thereby providing a powerful tool for assessing the genetic relationships between bacterial strains of the same species. Multiple-locus variable-number tandem repeat analysis has several advantages over PFGE because, like MLST, the output is highly objective, making the data amenable to automated computer analysis for the rapid detection of outbreaks and easy to compare across laboratories.

Whole-Genome Mapping

Whole-genome mapping has been used for detailed genome comparisons to differentiate closely related *E. coli* O157:H7 strains based on alterations in the chromosomal architectures (insertions, deletions, and rearrangements). The sizing and positioning of lateral acquired genomic regions in EHEC are crucial in assessing the Shiga toxin virulence status determined by Stx allele prevalence and respective chromosomal insertion sites of the converting prophages (12, 35, 36). Comparative map analysis reveals valid biological markers to trace evolution and also assists in genome assembly for molecular epidemiology outbreak investigations (36).

Genome Sequence-Based High-Resolution Genotyping

Octamer-Based Genome Scanning Typing Assay

The rapid accumulation of whole-genome sequence data for O157:H7 has allowed the development of high-resolution subtyping methods that enable inter- and intraspecies bacterial genome comparisons. Sequence-based phylogenetic assays have determined that EHEC strains comprise three highly related but distinct populations with a global prevalence that differs in genotype and host ecology. Polymorphisms were identified using high-density octamer-based genome scanning (OBGS) analysis by testing multiple OBGS primer combinations in independent reactions

screening a diverse strains set. Polymorphic OBGS products that were specific to lineage I or lineage II strains excluding the polymorphisms are not found within prophage, insertion sequences, or plasmid sequences. OBGS first demonstrated that the *E. coli* O157:H7 clonal complex had diverged into two highly related lineages, designated lineages I and II, that were found to be disproportionately represented among human and bovine isolates, respectively (22). Lineage II strains are found less frequently associated with human disease due either to inefficient transmission from bovine sources or to attenuated virulence in humans and the bovine host (26).

Lineage-Specific Polymorphism Typing Assay

Further analyses led to a refined classification system, termed the lineage-specific polymorphism assay (LSPA), that ultimately partitions *E. coli* O157:H7 strains into three lineages, I, I/II, and II, according to a PCR-based assay and polyacrylamide gel separation by testing the repeat length at six genic and intergenic chromosomal loci (52). In reference to the EDL933 genome, the LSPA markers comprise a 9-base insertion in gene *Z5935*, a 78-base insertion in the *yhcG* gene, a 9-base deletion in the *rbsB* gene, a 9-base insertion in the *rtcB* gene, and an 18-base insertion in the intergenic region spanning the *arp-iclR* genes. These techniques showed divergence into three different but interlinked lineages; lineage I and lineage I/II strains are commonly associated with human infections, whereas lineage II strains are overrepresented in the asymptomatic bovine host reservoir. Genetic evidence also suggests that distributions of genotypes between these lineages, such as polymorphisms in Shiga toxin subtypes and synergistically acting virulence factors, are correlated with phenotypic differences of major biological relevance in virulence, host ecology, and epidemiology. Identification of polymorphic loci that are anchored both in the chromosomal backbone and in horizontally acquired regions allowed us to associate bovine genotypes with altered virulence phenotypes and host prevalence. Numerous novel lineage II-specific genome signatures, some of which appear to be intimately associated with the altered pathogenic potential and niche adaptation within the bovine rumen and further discriminate bovine super shedders, have been cataloged (18, 26, 36).

Whole-Genome Sequence Typing/Single Nucleotide Polymorphism Typing

Despite multiple genomes of this lineage having become available in the genomics era (14, 15, 18, 36, 53, 54), the biological insight into epidemiology and disease mechanism suffers from a lack of markers for accurate typing and genotype/phenotype association. High-resolution phylogenomic approaches allow the dynamics of pathogenome evolution to be followed at a high level of phylogenetic accuracy and resolution. Whereas the current molecular markers and typing assays used by public health microbiology laboratories may be adequate for routine surveillance and identification of *E. coli* O157:H7, these approaches lack the polymorphic markers and discriminatory power to study the relatedness of strains of unknown provenance, which becomes of particular importance when investigating outbreak strains that form clonal complexes with only a few genetic polymorphisms. With the increase of next-generation sequencing technologies. whole-genome sequences typing approaches, such as the discovery of single nucleotide polymorphisms (SNPs), have gained popularity. SNP typing not only provides stable genetic markers to study evolution but also offers greater phylogenetic resolution. SNP discovery approaches have yielded thousands of high-quality SNPs as critical bases for the development of a refined phylogenomic framework (36, 55, 56). SNP typing allowed the elucidation of the evolutionary origin and emergence of the pathogenic O157:H7 lineage, and the determination of the genetic relationships to the intermediate immobile O157:H(−) and ancestral EPEC O55:H7 serotypes (56). A SNP-based clade typing assay that detects SNPs in 96 loci has been applied to more than 500 clinical *E. coli* O157 strains. This resulted in a refined LSPA-6 lineage classification and further separated EHEC isolates into nine distinct clades. The frequency and distribution of Shiga toxin-converting prophages and form of clinical disease manifestation could also be elucidated (30, 57). Recently, Manning et al. (30) used SNP analysis to identify a group (clade 8) of hypervirulent *E. coli* O157:H7 strains found in the United States. In 2006, food-borne outbreaks involving O157 contamination of fresh produce (e.g., spinach) were associated with more severe infection, causing higher rates of HUS and more frequencies of hospitalization, demonstrating that increased virulence had been acquired (30). *In silico* scoring of SNPs was successfully deployed by Eppinger et al. to investigate the degree of genetic heterogeneity in stains derived from a single outbreak of human disease (36) and to establish biologically relevant markers among strains from clinical settings and the animal reservoir (36). These enriched mutational database resources will provide a robust foundation to better associate genotypic group profiles and virulence phenotypes within *E. coli* O157:H7.

HISTORICAL VIEW OF EHEC/STEC GENOMICS

Whereas *Helicobacter pylori* was the first organism to have multiple isolates sequenced (58, 59), EHEC was the first *E. coli* pathotype with genomes of multiple isolates available for comparison (14, 15). Considering the relatedness of these isolates, it can be argued as the true beginning of comparative genomics as we know it. Irrespective of when comparative genomics began, the scope of the analyses has changed over the years. We have transitioned from sequencing prototype isolates from a group of pathogens to sequencing large numbers of isolates in an effort to understand the population structure of the pathogen associated with human health and to explore the concept of genomic epidemiology. This section briefly covers some of the important publications on genomics associated with *E. coli* O157 isolates.

First View of the Interpathotype Comparisons: EDL993 and Sakai Genomes

The first genome of *E. coli* was published in 1997 by the Blattner group and signaled a change in our understanding of model organisms (13). It could be argued that the sequencing of the MG1655 genome was the first nonpathogen genome to be decoded, as it had long been known that this isolate could no longer colonize humans or cause disease in humans. The sequencing of the EDL933 genome (14) provided the opportunity to examine two genomes from the same species. These early comparisons identified that there was >1 Mb of unique genomic material, encoding 1,387 genes, associated with the EDL933 genome that was not in the MG1655 genome. It was assumed that the majority of this DNA, present in what was termed "O-islands," would be associated with the ability to cause disease. Although it was true that the majority of genes that had been identified to be associated with human disease were in these O157-specifc regions, including Shiga-toxin phage and the LEE region, it was clear that not all of these regions were directly associated with disease and could be considered part of the central metabolism of these organisms. It was also noted that the genome of the EDL933 isolate contained an ~400-kb chromosomal inversion when compared to MG1655 genome. It is still unclear if this genomic inversion contributes to the pathogenesis of EDL933. One key point that was highlighted in this first genomic comparison was the identification of 18 identifiable prophage regions ranging in size from ~7 kb to 62 kb that were designated BP-933 [A-Q]. These phage regions include a number of potential secreted effectors (14, 60). Overall, the EDL933 genome provided a starting point for the study of potentially new genes associated with pathogenesis; however, one must keep in mind that at this point the comparisons

were only between two genomes, and thus the comparison of two points on a continuum does not provide a deep insight into the pathogen.

The Sakai genome, as mentioned above, represented the start of "intrapathotype" comparisons of *E. coli*, as well as the two most closely related bacterial genomes to be sequenced at this point in history (15). The genome of Sakai was 859 kb larger than that of the MG1655 genome and included two plasmids, a large virulence plasmid of 92,721 bp (pO157) and a cryptic plasmid of 3,306 bp (pOSAK1). Comparative genomics identified >1,600 genes that were Sakai specific when compared to MG1655, with 131 of them having a direct correlation to virulence. Interestingly, there were 24 prophage in the Sakai genome (18 in EDL933), and much of the unique DNA was associated with horizontal transfer or was suggested to have been transferred into the genome.

These seminal publications provided the spark that allowed others to start functionally characterizing these genes and the role they play in either survival or pathogenesis in the human model systems. They also allowed comparison of EHEC to other pathotypes of *E. coli*. Two studies examined the pan-genome of *E. coli* and included numerous EHEC isolates (61, 62). Each of these studies identified that there was an open genome in *E. coli*, suggesting that these organisms continue to collect more DNA. In each case, more sequencing would continue to identify new genes in all additional divergent isolates. Additionally, each study demonstrated that there were pathotype-specific adaptations that made each pathotype unique and contributed to the niche specialization that each had displayed (61, 62). At this point, a limited number of genomes were available (<20 complete and draft genomes), and comparisons were limited to mostly prototype isolates. The study by Rasko et al. highlighted that there were significant genetic differences between the EHEC isolates, but this observation could also be attributed to the fact that the annotation of different genomes could have a significant impact on the findings that were gene based (62).

Additional interesting genomes of the O157 serotype were published by Kulasekara et al. in 2009 (63); they sequenced an isolate, TW14359, that was associated with HUS. In this study they identified an additional 70 kb of genetic material that was not present in either of the reference O157 isolates. These regions encoded additional putative Type III secreted effectors and a gene for an anaerobic nitric oxide reductase, *norV*. It was suggested that the *norV* gene could be used as a marker for increased virulence leading to HUS; however, screening of large numbers of isolates suggested that the *norV* gene was associated with the loss of *stx1*, but the direct impact on pathogenesis was not confirmed. This publication is important in that it attempts to meld comparative genomics with disease presentation and gene presence or absence. This can be considered the beginning of the field we now know as genomic epidemiology.

Also in 2009, the first non-O157 EHEC genomes were published when Ogura et al. (53) published the genomes of isolates from serotypes O26, O111, and O103. These genomes were compared to the Sakai genome. This study demonstrated that the EHEC genomes were routinely larger than the MG1655 genome and contained a diverse array of phage and integrative elements, sometimes associated with the virulence gene catalog. Most importantly, it demonstrated how different serogroups and lineages of *E. coli* could evolve into EHEC isolates and demonstrated that there was a genomic core among similar isolates by comparing the gene content of 345 orthologous genes in all *E. coli* isolates that were sequenced to date. This is similar to the presentations in Fig. 2 and Fig. 3 in this article; however, we now avoid the problems caused by calling genes and curation errors by using the unannotated whole genome sequence for comparison (see below). Additionally, the study demonstrated that not all features of the genomes were the same. For example, the LEE pathogenicity island was inserted at different locations in the genome in these isolates, but it appeared to be functional as all isolates were derived from individuals who were ill.

As the cost of sequencing decreased, there started to be studies that examined large numbers of isolates. With EHEC, these large-scale sequencing projects were often linked to outbreaks of food-borne disease in the mid-2000s.

How Outbreaks Have Shaped the Sequencing of EHEC

From the early days of genomics, the majority of the focus has been on the EHEC O157:H7 isolates associated with human disease. In recent years, with the decrease in the cost of sequencing we have observed an increase in sequencing a significant number of isolates associated with disease within an outbreak setting. Some of these studies have attempted to associate genetic changes with increases in virulence or markers that laboratories could use as identifiers of outbreak strains. The study by Eppinger et al. describes the sequencing of a large number of isolates (36) to reconstruct the "anatomy of the outbreak." In this study the authors sequenced 25 isolates and through comparative genomics identified mutational and structural markers in the genome core and mobilome that allowed distinguishing these outbreak isolates among three almost simultaneous outbreaks that occurred in the United States as well as other O157 isolates that have been sequenced with other outbreaks and clinical sources as well as from animal reservoirs. Using a panel of >1,200 SNPs they were able to identify specific lineages of *E. coli* O157:H7. This type of study demonstrates how the genomics of a pathogen can distill the identification to a very fine scale. This is possible among the O157 isolates, since genomically they are all very similar when compared to the other *E. coli* strains (Fig. 2); however, there are significant differences when only the O157 isolates are compared (Fig. 3).

The *E. coli* outbreak in Europe in 2011 provided an opportunity to highlight the use of genomics to examine the bacterial isolates in near real time (41–44, 46). Additionally, the results provided insight into the evolution

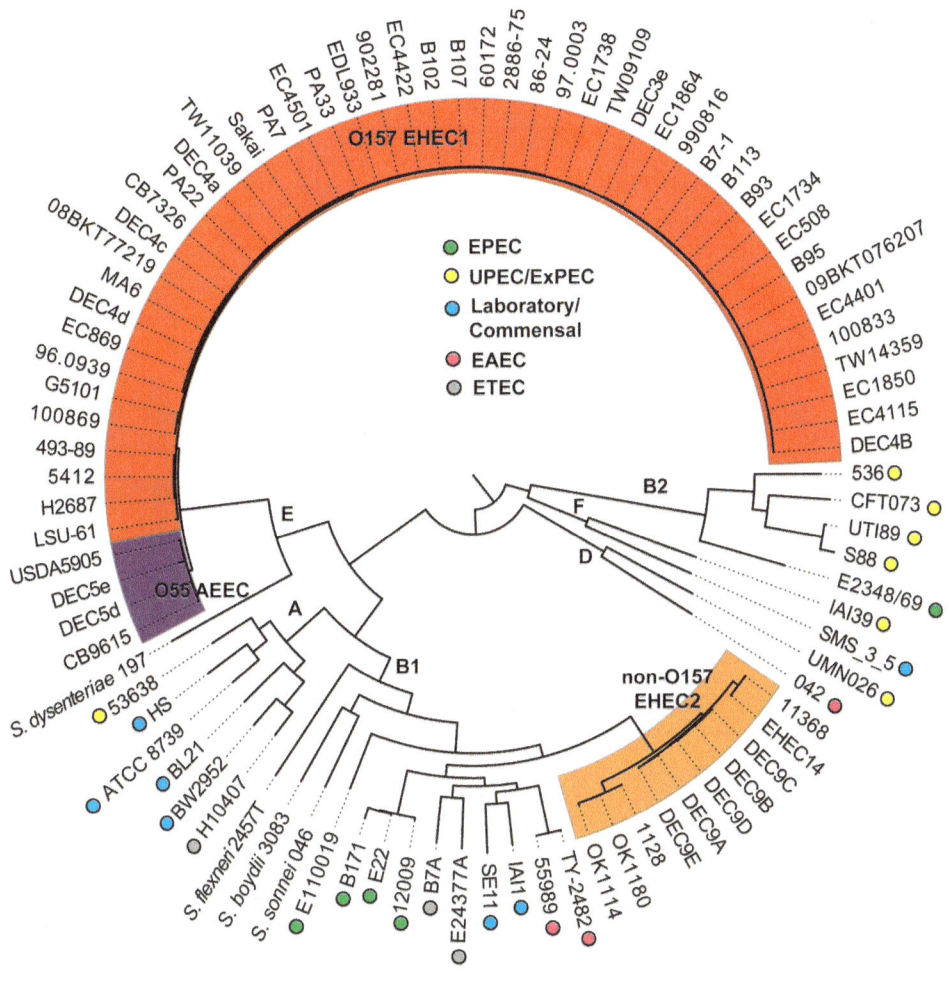

0.06

FIGURE 2 Phylogeny of *E. coli* reference isolates, including EHEC. The reference used for the SNP calling is *E. coli* HS as a true commensal isolate (62). The tree is maximum likelihood with 100 bootstraps made using RAxML, as previously described (68). This figure highlights, as does MLST, that when compared to a large and diverse collection of isolates, EHEC, and specifically the O157 and O55 (in red and purple) and other serotypes (highlighted in yellow), demonstrates a high level of similarity. This high-level of similarity extends within these groups. Only once these clades are examined in detail can one begin to identify regions that are diagnostic in these clades. doi.10.1128/microbiolspec.EHEC-0020-2013.f2

of bacterial pathogens as a species and challenged the definitions and boundaries of the STEC pathovar. The genome sequencing of the isolates associated with this outbreak and comparison to isolates that were in the database identified that the majority of the

chromosome of the outbreak-associated isolate was most similar to EAEC isolate 55989 (61). There were significant changes in the genome in that it had acquired the Shiga-toxin phage, as well as an antimicrobial resistance plasmid. Neither of these features had been

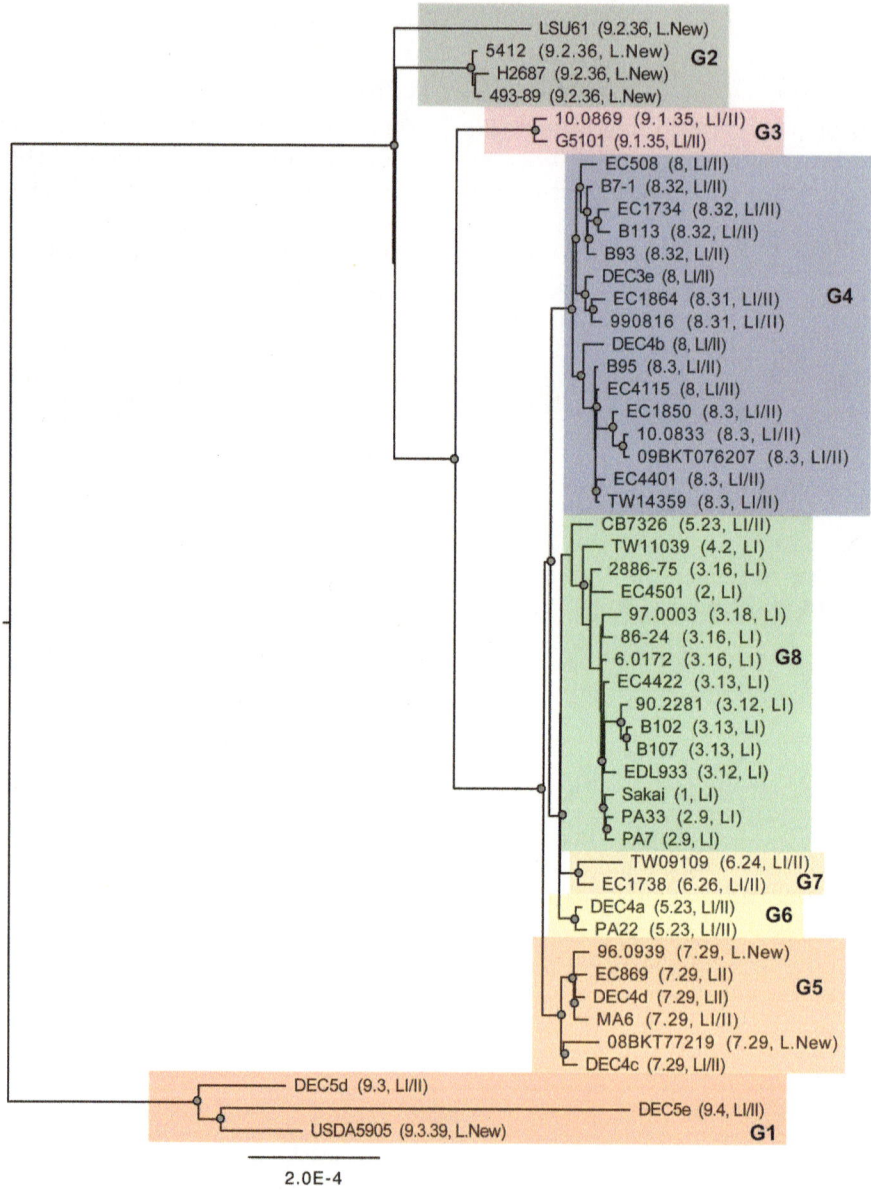

FIGURE 3 Phylogeny of O157 *E. coli* reference isolates demonstrating that there is variation within the O157 clade that is not observed in the larger genomic comparisons. This multi-whole genome alignment contains 4,093,272 bp of sequence. The tree is maximum likelihood with 100 bootstraps made using RAxML, as previously described (68). The color indicates distinguishable clades within this selection of 50 EHEC isolates. The additional numbers in parentheses are the clade and LSPA designations, as described in the text. doi:10.1128/microbiolspec.EHEC-0020-2013.f3

previously identified in EAEC isolates, but both potentially contribute to the pathogenicity of the bacteria. Thus the definition of the isolate was technically STEC, but the majority of the gene markers identified pointed toward an EAEC isolate (64). Many names were provided to this new type of isolate with this collection of virulence factors, including "entero-aggregative-haemorrhagic" *E. coli*, (46) but the most appropriate name proposed would be enteroaggregative, Shiga toxin-producing *E. coli* O104:H4. The isolates in this outbreak highlighted the issues with the current nomenclature that is largely based on the identification of virulence-associated genes that are present on mobile elements.

The genomic studies also highlighted that the determination of the presence or absence of specific genes or gene combinations could not really predict the virulence of an isolate and that further functional characterization is required. In addition to a novel collection of virulence factors in the European outbreak isolates, it was demonstrated that, like EHEC isolates, the Shiga toxin genes were activated by the exposure to subclinical doses of antibiotics (42). As time goes on, more detailed studies of the functional virulence of these isolates will be examined and the true impact of the novel combinations of virulence factors will be identified.

GENOMICS-GUIDED EXAMINATION OF PATHOGENESIS

The study of the evolution of pathogenesis in EHEC has mainly focused on the traditional virulence factors, such as LEE, Shiga toxin, a limited number of adhesins, and plasmid encoded features. Molecular evidence suggests that EHEC isolates have acquired the majority of their virulence factors through horizontal gene transfer and thereby acquisition of the LEE pathogenicity-associated island and the Stx genes were two crucial steps in the evolution of EHEC O157 from a commensal ancestor (53). The LEE pathogenicity-associated island is clearly a mosaic structure, which arose from multiple recombination events with foreign DNA, as did the large EHEC-hemolysin plasmid (65, 66). LEE can be found on chromosomes of EHEC next to tRNA genes at different locations (67), suggesting that LEE has been acquired on more than one occasion.

Additionally, other virulence factors found on the O157 genomes, such as fimbrial and non-fimbrial adhesins, iron uptake systems, and non-LEE effectors, are also thought to be required for the full virulence of EHEC, but their prevalence among non-O157 EHEC isolates, and in some cases among the O157 EHEC isolates, has not yet been systematically analyzed. Global genomic differences (or conservation) between O157 and non-O157 EHEC strains have recently been addressed in the study by Hazen et al. (68), which demonstrated that the O55 and O157 isolates formed closely related clades, but the non-O157/non-O55 EHEC (also known to some as EHEC2) were divergent. We demonstrate this point in Fig. 2, using a collection of reference *E. coli* isolates from each of the pathovars as well as a selection of the O157, O55, and non-O157 EHEC isolates that have been recently sequenced. This clearly demonstrates that although the virulence factor typing has grouped these isolates together, pathogenomic analysis suggests they are significantly different.

That there has been a parallel evolution and independent acquisition of the virulence factors in each of the serogroups (69) has been proposed. Hazen et al. (68) recently noted that there appears to be a pattern of secreted effectors that are common in EHEC strains when compared to other EAEC isolates. This suggests that there is a complement of effectors that are "tuned" to the EHEC genome content for optimal regulation, interaction, and secretion; however, this does not necessarily have to be linked to human virulence and further characterization of this orchestrated program is required.

Are There Genomic Differences in Recent Isolates That Have Come from Green Leaf Vegetables Rather Than Most of the Original Isolates That Were Beef Related in Some Way?

Recent Shiga-toxigenic *E. coli* O157:H7 outbreaks have been linked to consumption of fresh produce (12, 63). It is generally recognized that bacterial attachment to vegetal matrices would constitute the first step in contamination of fresh produce, but how and when this occurs are under some debate. Cellular appendages, such as curli fibers, and cellulose, a constituent of extracellular matrix, have been suggested to be involved in *E. coli* attachment and persistence in fresh produce. A comparative evaluation was conducted on the ability of STEC O157:H7 strains EDL933 and 86-24, linked to two independent food-borne disease outbreaks in humans, and their mutants deficient in curli and/or cellulose expression to colonize and to firmly attach to spinach leaf. Inoculated spinach leaves were incubated at 22°C, and at 0, 24, and 48 h after incubation loosely and strongly attached *E. coli* O157:H7 populations were determined. Curli-expressing *E. coli* O157:H7 strains developed stronger association with leaf surface, whereas curli-deficient mutants attached to spinach at significantly ($P<0.01$) lower numbers. Attachment of cellulose-impaire mutants to spinach leaves was not significantly different from strains that contained curli; however, the relative attachment strength of *E. coli* O157:H7 to spinach increased with incubation time for the curli-expressing strains. Laser scanning confocal microscopy analysis of inoculated leaves revealed that curli-expressing *E. coli* O157:H7 was surrounded by extracellular structures strongly immunostained with anticurli antibodies. Production of cellulose was not required to develop strong attachment to spinach leaf. These results indicate that curli fibers are essential for strong attachment of *E. coli* O157:H7 to spinach whereas cellulose is dispensable.

PUBLIC HEALTH VIEW—WHAT ARE THE NEXT STEPS?

Whole bacterial genome sequencing in medical microbiology is fast becoming a reality; however, the challenge of converting the primary sequence data into useful clinical or public health action remains unmet, because experience with such data is limited. For the identification of an isolate or an outbreak strain, comparison with genomes of very closely related organisms is required. What is clear is that a large sequence database comprising a comprehensive panel of well-chosen isolates is necessary if genome sequencing is to be useful in the field. Perhaps the most comprehensive single-species genome data available thus far are for *E. coli*. This database became extremely useful when the outbreak in Europe occurred in 2011 (41–43). This outbreak resulted in 3,816 identified STEC infections and 54 deaths (45). It was rapidly determined that the etiological agent was *E. coli*, but the isolate had features of EAEC and was Shiga toxin positive but lacked the majority of features associated with EHEC (LEE, *ehx*, etc.), suggesting that this isolate may simply be an EAEC isolate that had acquired the Shiga toxin phage. Rapid comparative genomics highlighted that this was the case and that the isolate had also acquired an antibiotic resistance plasmid. That we had large collections of isolates already sequenced allowed the accurate and rapid determination of the genomic origin of this isolate. In the future these databases of isolate collections will become invaluable in the public health setting in transmitting information from the sequencer to the physician treatment paradigm for any patient. We are still far away from this reality, as the software for these analyses are somewhat custom, but as the benchtop sequencers become more prevalent in clinical laboratories, the methodology must follow.

CONCLUSION

The use of modern techniques like whole-genome sequence typing can benefit researchers interested in surveillance (18), forensics (70), diagnostics, and epidemiological studies of microbial pathogens and pathogenome evolution, and holds the potential to discover novel *E. coli* O157:H7 virulence genotypes that are intimately associated with adverse human disease outcomes. Use of these techniques is not limited to *E. coli* O157:H7 or EHEC as this research can provide principles and valuable benchmarking data crucial in the modeling of bacterial outbreak-associated population dynamics. Once data are available, the scientific community will be key to the analysis of the gathered sequence data. Application of postgenomic tools to the generated data can help to understand variability in gene content and activity. These technologies include DNA microarray analysis or RNAseq for gene profiling and differentiation of strains; optical mapping of chromosomes for discovery of novel gene insertions and deletions; and phenotypic microarray of pathogens' physiological and metabolic capabilities for strain characterization. Virulence assays in host model systems allow validation of genetic predictions and association of isolate-specific gene content to outbreak-associated physiological and virulence capabilities. Better and more effective schemes for epidemiological studies can be developed to trace the source of the outbreaks, or ultimately link the source to the infection in outbreak investigations. The identification of subtle but defined polymorphisms would allow us to investigate the drivers of pathogenome evolution and analyze mutations that are correlated with human disease and severity. This will enable us to track and distinguish separate unrelated outbreaks and classify isolates as members of an outbreak population, and also to refine standards currently used in outbreak investigations. The potential discovery of novel polymorphisms complements current techniques used to classify strains, and is crucial in the development of rapid methods of detecting and tracking microbial outbreaks. Studies of non-O157 EHEC are currently under way by several independent research teams (68). These data sets will potentiate each other in gathering a more complete and refined picture of the *E. coli* O157:H7 pathogenome evolution and its pathogenic potential in relation to the *E. coli* species biology.

ACKNOWLEDGMENTS

D.A.R. and T.H.H. are supported by federal funds from the National Institute of Allergy and Infectious Diseases, National Institutes of Health, Department of Health and Human Services under grant number 1U19AI090873 and funds from the State of Maryland. M.E. is supported by the Department of Biology and South Texas Center for Emerging Infectious Diseases, University of Texas at San Antonio. This review is further based upon work at the University of Texas at San Antonio supported by the Army Research Office of the Department of Defense under Contract No. W911NF-11-1-0136. We thank Sara S. K. Koenig for critically reading the manuscript and all research groups who have produced data discussed in this work.

CITATION

Sadiq SM, Hazen TH, Rasko DA, Eppinger M. 2014. Enterohemorrhagic *Escherichia coli* genomics: past, present, and future. Microbiol Spectrum 2(4):EHEC-0020-2013.

REFERENCES

1. **Kaper JB, Nataro JP, Mobley HL.** 2004. Pathogenic *Escherichia coli*. *Nat Rev Microbiol* **2:**123–140.
2. **Croxen MA, Law RJ, Scholz R, Keeney KM, Wlodarska M, Finlay BB.** 2013. Recent advances in understanding enteric pathogenic *Escherichia coli*. *Clin Microbiol Rev* **26:**822–880.
3. **Riley LW, Remis RS, Helgerson SD, McGee HB, Wells JG, Davis BR, Hebert RJ, Olcott ES, Johnson LM, Hargrett NT, Blake PA, Cohen ML.** 1983. Hemorrhagic colitis associated with a rare *Escherichia coli* serotype. *N Engl J Med* **308:**681–685.
4. **Mead PS, Griffin PM.** 1998. *Escherichia coli* O157:H7. *Lancet* **352:**1207–1212.
5. **Karch H, Bielaszewska M, Bitzan M, Schmidt H.** 1999. Epidemiology and diagnosis of Shiga toxin-producing *Escherichia coli* infections. *Diagn Microbiol Infect Dis* **34:**229–243.
6. **Clawson ML, Keen JE, Smith TP, Durso LM, McDaneld TG, Mandrell RE, Davis MA, Bono JL.** 2009. Phylogenetic classification of *Escherichia coli* O157:H7 strains of human and bovine

origin using a novel set of nucleotide polymorphisms. *Genome Biol* **10**:R56.

7. **Griffin PM, Tauxe RV.** 1991. The epidemiology of infections caused by *Escherichia coli* O157:H7, other enterohemorrhagic *E. coli*, and the associated hemolytic uremic syndrome. *Epidemiol Rev* **13**:60–98.

8. **Griffin PM, Ostroff SM, Tauxe RV, Greene KD, Wells JG, Lewis JH, Blake PA.** 1988. Illnesses associated with *Escherichia coli* O157:H7 infections. A broad clinical spectrum. *Ann Intern Med* **109**:705–712.

9. **Riley DG, Gray JT, Loneragan GH, Barling KS, Chase CC Jr.** 2003. *Escherichia coli* O157:H7 prevalence in fecal samples of cattle from a southeastern beef cow-calf herd. *J Food Prot* **66**:1778–1782.

10. **Cimolai N, Morrison BJ, Carter JE.** 1992. Risk factors for the central nervous system manifestations of gastroenteritis-associated hemolytic-uremic syndrome. *Pediatrics* **90**:616–621.

11. **Besser RE, Griffin PM, Slutsker L.** 1999. *Escherichia coli* O157:H7 gastroenteritis and the hemolytic uremic syndrome: an emerging infectious disease. *Annu Rev Med* **50**:355–367.

12. **Kotewicz ML, Mammel MK, LeClerc JE, Cebula TA.** 2008. Optical mapping and 454 sequencing of *Escherichia coli* O157: H7 isolates linked to the US 2006 spinach-associated outbreak. *Microbiology* **154**:3518–3528.

13. **Blattner FR, Plunkett G 3rd, Bloch CA, Perna NT, Burland V, Riley M, Collado-Vides J, Glasner JD, Rode CK, Mayhew GF, Gregor J, Davis NW, Kirkpatrick HA, Goeden MA, Rose DJ, Mau B, Shao Y.** 1997. The complete genome sequence of *Escherichia coli* K-12. *Science* **277**:1453–1462.

14. **Perna NT, Plunkett G 3rd, Burland V, Mau B, Glasner JD, Rose DJ, Mayhew GF, Evans PS, Gregor J, Kirkpatrick HA, Posfai G, Hackett J, Klink S, Boutin A, Shao Y, Miller L, Grotbeck EJ, Davis NW, Lim A, Dimalanta ET, Potamousis KD, Apodaca J, Anantharaman TS, Lin J, Yen G, Schwartz DC, Welch RA, Blattner FR.** 2001. Genome sequence of enterohaemorrhagic *Escherichia coli* O157:H7. *Nature* **409**:529–533.

15. **Hayashi T, Makino K, Ohnishi M, Kurokawa K, Ishii K, Yokoyama K, Han CG, Ohtsubo E, Nakayama K, Murata T, Tanaka M, Tobe T, Iida T, Takami H, Honda T, Sasakawa C, Ogasawara N, Yasunaga T, Kuhara S, Shiba T, Hattori M, Shinagawa H.** 2001. Complete genome sequence of enterohemorrhagic *Escherichia coli* O157:H7 and genomic comparison with a laboratory strain K-12. *DNA Res* **8**:11–22.

16. **Ferens WA, Hovde CJ.** 2011. *Escherichia coli* O157:H7: animal reservoir and sources of human infection. *Foodborne Pathog Dis* **8**:465–487.

17. **Zhang Y, Laing C, Steele M, Ziebell K, Johnson R, Benson AK, Taboada E, Gannon VP.** 2007. Genome evolution in major *Escherichia coli* O157:H7 lineages. *BMC Genomics* **8**:121.

18. **Eppinger M, Mammel MK, Leclerc JE, Ravel J, Cebula TA.** 2011. Genome signatures of *Escherichia coli* O157:H7 isolates from the bovine host reservoir. *Appl Environ Microbiol* **77**:2916–2925.

19. **Eppinger M, Mammel MK, LeClerc JE, Ravel J, Cebula TA.** 2011. Genomic anatomy of *Escherichia coli* O157:H7 outbreaks. *Proc Natl Acad Sci USA* **108**:20142–20147.

20. **Medini D, Donati C, Tettelin H, Masignani V, Rappuoli R.** 2005. The microbial pan-genome. *Curr Opin Genet Dev* **15**:589–594.

21. **Afset JE, Anderssen E, Bruant G, Harel J, Wieler L, Bergh K.** 2008. Phylogenetic backgrounds and virulence profiles of atypical enteropathogenic *Escherichia coli* strains from a case-control study using multilocus sequence typing and DNA microarray analysis. *J Clin Microbiol* **46**:2280–2290.

22. **Kim J, Nietfeldt J, Benson AK.** 1999. Octamer-based genome scanning distinguishes a unique subpopulation of *Escherichia coli* O157:H7 strains in cattle. *Proc Natl Acad Sci USA* **96**:13288–13293.

23. **Ohnishi M, Terajima J, Kurokawa K, Nakayama K, Murata T, Tamura K, Ogura Y, Watanabe H, Hayashi T.** 2002. Genomic diversity of enterohemorrhagic *Escherichia coli* O157 revealed by whole genome PCR scanning. *Proc Natl Acad Sci USA* **99**:17043–17048.

24. **Ratnam S, March SB, Ahmed R, Bezanson GS, Kasatiya S.** 1988. Characterization of *Escherichia coli* serotype O157:H7. *J Clin Microbiol* **26**:2006–2012.

25. **Ahmed R, Bopp C, Borczyk A, Kasatiya S.** 1987. Phage-typing scheme for *Escherichia coli* O157: H7. *J Infect Dis* **155**:806–809.

26. **Bono JL, Smith TP, Keen JE, Harhay GP, McDaneld TG, Mandrell RE, Jung WK, Besser TE, Gerner-Smidt P, Bielaszewska M, Karch H, Clawson ML.** 2012. Phylogeny of Shiga toxin-producing *Escherichia coli* O157 isolated from cattle and clinically ill humans. *Mol Biol Evol* **29**:2047–2062.

27. **Keys C, Kemper S, Keim P.** 2005. Highly diverse variable number tandem repeat loci in the *E. coli* O157:H7 and O55:H7 genomes for high-resolution molecular typing. *J Appl Microbiol* **98**:928–940.

28. **Noller AC, McEllistrem MC, Pacheco AG, Boxrud DJ, Harrison LH.** 2003. Multilocus variable-number tandem repeat analysis distinguishes outbreak and sporadic *Escherichia coli* O157:H7 isolates. *J Clin Microbiol* **41**:5389–5397.

29. Jackson SA, Mammel MK, Patel IR, Mays T, Albert TJ, LeClerc JE, Cebula TA. 2007. Interrogating genomic diversity of *E. coli* O157:H7 using DNA tiling arrays. *Forensic Sci Int* **168:**183–199.

30. Manning SD, Motiwala AS, Springman AC, Qi W, Lacher DW, Ouellette LM, Mladonicky JM, Somsel P, Rudrik JT, Dietrich SE, Zhang W, Swaminathan B, Alland D, Whittam TS. 2008. Variation in virulence among clades of *Escherichia coli* O157:H7 associated with disease outbreaks. *Proc Natl Acad Sci USA* **105:**4868–4873.

31. Zhang W, Qi W, Albert TJ, Motiwala AS, Alland D, Hyytia-Trees EK, Ribot EM, Fields PI, Whittam TS, Swaminathan B. 2006. Probing genomic diversity and evolution of *Escherichia coli* O157 by single nucleotide polymorphisms. *Genome Res* **16:**757–767.

32. Gerner-Smidt P, Kincaid J, Kubota K, Hise K, Hunter SB, Fair MA, Norton D, Woo-Ming A, Kurzynski T, Sotir MJ, Head M, Holt K, Swaminathan B. 2005. Molecular surveillance of shiga toxigenic *Escherichia coli* O157 by PulseNet USA. *J Food Prot* **68:**1926–1931.

33. Steele M, Ziebell K, Zhang Y, Benson A, Johnson R, Laing C, Taboada E, Gannon V. 2009. Genomic regions conserved in lineage II *Escherichia coli* O157:H7 strains. *Appl Environ Microbiol* **75:**3271–3280.

34. Steele M, Ziebell K, Zhang Y, Benson A, Konczy P, Johnson R, Gannon V. 2007. Identification of *Escherichia coli* O157:H7 genomic regions conserved in strains with a genotype associated with human infection. *Appl Environ Microbiol* **73:**22–31.

35. Kotewicz ML, Jackson SA, LeClerc JE, Cebula TA. 2007. Optical maps distinguish individual strains of *Escherichia coli* O157:H7. *Microbiology* **153:**1720–1733.

36. Eppinger M, Mammel MK, Leclerc JE, Ravel J, Cebula TA. 2011. Genomic anatomy of *Escherichia coli* O157:H7 outbreaks. *Proc Natl Acad Sci USA* **108:**20142–20147.

37. Barrett TJ, Lior H, Green JH, Khakhria R, Wells JG, Bell BP, Greene KD, Lewis J, Griffin PM. 1994. Laboratory investigation of a multistate food-borne outbreak of *Escherichia coli* O157:H7 by using pulsed-field gel electrophoresis and phage typing. *J Clin Microbiol* **32:**3013–3017.

38. Izumiya H, Terajima J, Wada A, Inagaki Y, Itoh KI, Tamura K, Watanabe H. 1997. Molecular typing of enterohemorrhagic *Escherichia coli* O157:H7 isolates in Japan by using pulsed-field gel electrophoresis. *J Clin Microbiol* **35:**1675–1680.

39. Akiba M, Masuda T, Sameshima T, Katsuda K, Nakazawa M. 1999. Molecular typing of *Escherichia coli* O157:H7 (H-) isolates from cattle in Japan. *Epidemiol Infect* **122:**337–341.

40. Murase T, Yamai S, Watanabe H. 1999. Changes in pulsed-field gel electrophoresis patterns in clinical isolates of enterohemorrhagic *Escherichia coli* O157:H7 associated with loss of Shiga toxin genes. *Curr Microbiol* **38:**48–50.

41. Rohde H, Qin J, Cui Y, Li D, Loman NJ, Hentschke M, Chen W, Pu F, Peng Y, Li J, Xi F, Li S, Li Y, Zhang Z, Yang X, Zhao M, Wang P, Guan Y, Cen Z, Zhao X, Christner M, Kobbe R, Loos S, Oh J, Yang L, Danchin A, Gao GF, Song Y, Yang H, Wang J, Xu J, Pallen MJ, Aepfelbacher M, Yang R. 2011. Open-source genomic analysis of Shiga-toxin-producing *E. coli* O104:H4. *N Engl J Med* **365:**718–724.

42. Rasko DA, Webster DR, Sahl JW, Bashir A, Boisen N, Scheutz F, Paxinos EE, Sebra R, Chin CS, Iliopoulos D, Klammer A, Peluso P, Lee L, Kislyuk AO, Bullard J, Kasarskis A, Wang S, Eid J, Rank D, Redman JC, Steyert SR, Frimodt-Moller J, Struve C, Petersen AM, Krogfelt KA, Nataro JP, Schadt EE, Waldor MK. 2011. Origins of the *E. coli* strain causing an outbreak of hemolytic-uremic syndrome in Germany. *N Engl J Med* **365:**709–717.

43. Mellmann A, Harmsen D, Cummings CA, Zentz EB, Leopold SR, Rico A, Prior K, Szczepanowski R, Ji Y, Zhang W, McLaughlin SF, Henkhaus JK, Leopold B, Bielaszewska M, Prager R, Brzoska PM, Moore RL, Guenther S, Rothberg JM, Karch H. 2011. Prospective genomic characterization of the German enterohemorrhagic *Escherichia coli* O104:H4 outbreak by rapid next generation sequencing technology. *PLoS One* **6:**e22751.

44. Grad YH, Lipsitch M, Feldgarden M, Arachchi HM, Cerqueira GC, Fitzgerald M, Godfrey P, Haas BJ, Murphy CI, Russ C, Sykes S, Walker BJ, Wortman JR, Young S, Zeng Q, Abouelleil A, Bochicchio J, Chauvin S, Desmet T, Gujja S, McCowan C, Montmayeur A, Steelman S, Frimodt-Moller J, Petersen AM, Struve C, Krogfelt KA, Bingen E, Weill FX, Lander ES, Nusbaum C, Birren BW, Hung DT, Hanage WP. 2012. Genomic epidemiology of the *Escherichia coli* O104:H4 outbreaks in Europe, 2011. *Proc Natl Acad Sci USA* **109:**3065–3070.

45. Frank C, Werber D, Cramer JP, Askar M, Faber M, an der Heiden M, Bernard H, Fruth A, Prager R, Spode A, Wadl M, Zoufaly A, Jordan S, Kemper MJ, Follin P, Muller L, King LA, Rosner B, Buchholz U, Stark K, Krause G. 2011. Epidemic profile of Shiga-toxin-producing *Escherichia coli* O104:H4 outbreak in Germany. *N Engl J Med* **365:**1771–1780.

46. Brzuszkiewicz E, Thurmer A, Schuldes J, Leimbach A, Liesegang H, Meyer FD, Boelter J, Petersen H, Gottschalk G, Daniel R. 2011.

Genome sequence analyses of two isolates from the recent *Escherichia coli* outbreak in Germany reveal the emergence of a new pathotype: entero-aggregative-haemorrhagic *Escherichia coli* (EAHEC). *Arch Microbiol* **193**:883–891.

47. Maiden MC, Bygraves JA, Feil E, Morelli G, Russell JE, Urwin R, Zhang Q, Zhou J, Zurth K, Caugant DA, Feavers IM, Achtman M, Spratt BG. 1998. Multilocus sequence typing: a portable approach to the identification of clones within populations of pathogenic microorganisms. *Proc Natl Acad Sci USA* **95**:3140–3145.

48. Wirth T, Falush D, Lan R, Colles F, Mensa P, Wieler LH, Karch H, Reeves PR, Maiden MC, Ochman H, Achtman M. 2006. Sex and virulence in *Escherichia coli*: an evolutionary perspective. *Mol Microbiol* **60**:1136–1151.

49. Jaureguy F, Landraud L, Passet V, Diancourt L, Frapy E, Guigon G, Carbonnelle E, Lortholary O, Clermont O, Denamur E, Picard B, Nassif X, Brisse S. 2008. Phylogenetic and genomic diversity of human bacteremic *Escherichia coli* strains. *BMC Genomics* **9**:560.

50. Whittam TS, Wachsmuth IK, Wilson RA. 1988. Genetic evidence of clonal descent of *Escherichia coli* O157:H7 associated with hemorrhagic colitis and hemolytic uremic syndrome. *J Infect Dis* **157**:1124–1133.

51. van Belkum A, Scherer S, van Alphen L, Verbrugh H. 1998. Short-sequence DNA repeats in prokaryotic genomes. *Microbiol Mol Biol Rev* **62**:275–293.

52. Yang Z, Kovar J, Kim J, Nietfeldt J, Smith DR, Moxley RA, Olson ME, Fey PD, Benson AK. 2004. Identification of common subpopulations of non-sorbitol-fermenting, beta-glucuronidase-negative *Escherichia coli* O157:H7 from bovine production environments and human clinical samples. *Appl Environ Microbiol* **70**:6846–6854.

53. Ogura Y, Ooka T, Iguchi A, Toh H, Asadulghani M, Oshima K, Kodama T, Abe H, Nakayama K, Kurokawa K, Tobe T, Hattori M, Hayashi T. 2009. Comparative genomics reveal the mechanism of the parallel evolution of O157 and non-O157 enterohemorrhagic *Escherichia coli*. *Proc Natl Acad Sci USA* **106**:17939–17944.

54. Eppinger M, Daugherty S, Agrawal S, Galens K, Sengamalay N, Sadzewicz L, Tallon L, Cebula TA, Mammel MK, Feng P, Soderlund R, Tarr PI, Debroy C, Dudley EG, Fraser CM, Ravel J. 2013. Whole-genome draft sequences of 26 enterohemorrhagic *Escherichia coli* O157:H7 strains. *Genome Announc* **1**:e0013412.

55. Norman KN, Strockbine NA, Bono JL. 2012. Association of nucleotide polymorphisms within the O-antigen gene cluster of *Escherichia coli* O26, O45, O103, O111, O121, and O145 with serogroups and genetic subtypes. *Appl Environ Microbiol* **78**:6689–6703.

56. Leopold SR, Magrini V, Holt NJ, Shaikh N, Mardis ER, Cagno J, Ogura Y, Iguchi A, Hayashi T, Mellmann A, Karch H, Besser TE, Sawyer SA, Whittam TS, Tarr PI. 2009. A precise reconstruction of the emergence and constrained radiations of *Escherichia coli* O157 portrayed by backbone concatenomic analysis. *Proc Natl Acad Sci USA* **106**:8713–8718.

57. Abu-Ali GS, Ouellette LM, Henderson ST, Lacher DW, Riordan JT, Whittam TS, Manning SD. 2010. Increased adherence and expression of virulence genes in a lineage of *Escherichia coli* O157:H7 commonly associated with human infections. *PLoS One* **5**:e10167.

58. Tomb JF, White O, Kerlavage AR, Clayton RA, Sutton GG, Fleischmann RD, Ketchum KA, Klenk HP, Gill S, Dougherty BA, Nelson K, Quackenbush J, Zhou L, Kirkness EF, Peterson S, Loftus B, Richardson D, Dodson R, Khalak HG, Glodek A, McKenney K, Fitzegerald LM, Lee N, Adams MD, Hickey EK, Berg DE, Gocayne JD, Utterback TR, Peterson JD, Kelley JM, Cotton MD, Weidman JM, Fujii C, Bowman C, Watthey L, Wallin E, Hayes WS, Borodovsky M, Karp PD, Smith HO, Fraser CM, Venter JC. 1997. The complete genome sequence of the gastric pathogen *Helicobacter pylori*. *Nature* **388**:539–547.

59. Alm RA, Ling LS, Moir DT, King BL, Brown ED, Doig PC, Smith DR, Noonan B, Guild BC, deJonge BL, Carmel G, Tummino PJ, Caruso A, Uria-Nickelsen M, Mills DM, Ives C, Gibson R, Merberg D, Mills SD, Jiang Q, Taylor DE, Vovis GF, Trust TJ. 1999. Genomic-sequence comparison of two unrelated isolates of the human gastric pathogen *Helicobacter pylori*. *Nature* **397**:176–180.

60. Tobe T, Beatson SA, Taniguchi H, Abe H, Bailey CM, Fivian A, Younis R, Matthews S, Marches O, Frankel G, Hayashi T, Pallen MJ. 2006. An extensive repertoire of type III secretion effectors in *Escherichia coli* O157 and the role of lambdoid phages in their dissemination. *Proc Natl Acad Sci USA* **103**:14941–14946.

61. Touchon M, Hoede C, Tenaillon O, Barbe V, Baeriswyl S, Bidet P, Bingen E, Bonacorsi S, Bouchier C, Bouvet O, Calteau A, Chiapello H, Clermont O, Cruveiller S, Danchin A, Diard M, Dossat C, Karoui ME, Frapy E, Garry L, Ghigo JM, Gilles AM, Johnson J, Le Bouguenec C, Lescat M, Mangenot S, Martinez-Jehanne V, Matic I, Nassif X, Oztas S, Petit MA, Pichon C, Rouy Z, Ruf CS, Schneider D, Tourret J, Vacherie B, Vallenet D, Medigue C, Rocha EP, Denamur E. 2009. Organised genome dynamics

in the *Escherichia coli* species results in highly diverse adaptive paths. *PLoS Genet* **5:**e1000344.

62. **Rasko DA, Rosovitz MJ, Myers GS, Mongodin EF, Fricke WF, Gajer P, Crabtree J, Sebaihia M, Thomson NR, Chaudhuri R, Henderson IR, Sperandio V, Ravel J.** 2008. The pangenome structure of *Escherichia coli*: comparative genomic analysis of *E. coli* commensal and pathogenic isolates. *J Bacteriol* **190:**6881–6893.

63. **Kulasekara BR, Jacobs M, Zhou Y, Wu Z, Sims E, Saenphimmachak C, Rohmer L, Ritchie JM, Radey M, McKevitt M, Freeman TL, Hayden H, Haugen E, Gillett W, Fong C, Chang J, Beskhlebnaya V, Waldor MK, Samadpour M, Whittam TS, Kaul R, Brittnacher M, Miller SI.** 2009. Analysis of the genome of the *Escherichia coli* O157:H7 2006 spinach-associated outbreak isolate indicates candidate genes that may enhance virulence. *Infect Immun* **77:**3713–3721.

64. **Scheutz F, Nielsen EM, Frimodt-Moller J, Boisen N, Morabito S, Tozzoli R, Nataro JP, Caprioli A.** 2011. Characteristics of the enteroaggregative Shiga toxin/verotoxin-producing Escherichia coli O104:H4 strain causing the outbreak of haemolytic uraemic syndrome in Germany, May to June 2011. *Euro Surveill* **16**(24):ii, 19889.

65. **Boerlin P, Chen S, Colbourne JK, Johnson R, De Grandis S, Gyles C.** 1998. Evolution of enterohemorrhagic *Escherichia coli* hemolysin plasmids and the locus for enterocyte effacement in shiga toxin-producing *E. coli. Infect Immun* **66:**2553–2561.

66. **Morabito S, Tozzoli R, Oswald E, Caprioli A.** 2003. A mosaic pathogenicity island made up of the locus of enterocyte effacement and a pathogenicity island of *Escherichia coli* O157:H7 is frequently present in attaching and effacing *E. coli. Infect Immun* **71:**3343–3348.

67. **Wieler LH, McDaniel TK, Whittam TS, Kaper JB.** 1997. Insertion site of the locus of enterocyte effacement in enteropathogenic and enterohemorrhagic *Escherichia coli* differs in relation to the clonal phylogeny of the strains. *FEMS Microbiol Lett* **156:**49–53.

68. **Hazen TH, Sahl JW, Fraser CM, Donnenberg MS, Scheutz F, Rasko DA.** 2013. Refining the pathovar paradigm via phylogenomics of the attaching and effacing *Escherichia coli. Proc Natl Acad Sci USA* **110:**12810–12815.

69. **Reid SD, Herbelin CJ, Bumbaugh AC, Selander RK, Whittam TS.** 2000. Parallel evolution of virulence in pathogenic *Escherichia coli. Nature* **406:**64–67.

70. **Rasko DA, Worsham PL, Abshire TG, Stanley ST, Bannan JD, Wilson MR, Langham RJ, Decker RS, Jiang L, Read TD, Phillippy AM, Salzberg SL, Pop M, Van Ert MN, Kenefic LJ, Keim PS, Fraser-Liggett CM, Ravel J.** 2011. *Bacillus anthracis* comparative genome analysis in support of the Amerithrax investigation. *Proc Natl Acad Sci USA* **108:**5027–5032.

PATHOGENESIS

Role of Shiga/Vero Toxins in Pathogenesis

5

FUMIKO OBATA[1] and TOM OBRIG[1]

ACTIVITIES OF Stx AND LPS IN RENAL DISEASE

Shiga Toxin Actions

It is generally accepted that all actions of Shiga toxin (Stx) depend on its interaction with the receptor, globotriaosylceramide (Gb_3), on eukaryotic cells. Although alternative receptors for Stx have been postulated, no definitive data have been forthcoming in support. Stx holotoxin is internalized by receptor-mediated endocytosis, retrograde transported via the Golgi apparatus and processed through in the endoplasmic reticulum, and released into the cytoplasm where it enzymatically inactivates ribosomes and inhibits protein synthesis (Fig. 1). However, it is important to note that, in addition to Stx holotoxin, the B-subunit alone can interact with Gb_3 in a physiologically meaningful manner where it activates signal transduction pathways in target cells (Fig. 1) (1). An additional but unexplained anomaly is the interaction of Stx with eukaryotic cells in a Gb_3-independent manner that leads to induction of cytokines by these cells (2). As shown in Fig. 1, intracellular responses to Stx are diverse, including inhibition of protein synthesis, activation of cellular stress responses, and induction of cytokines and chemokines. It is likely that these different schemes take place in cell-specific activities during Shiga toxin-producing *Escherichia*

[1]University of Maryland School of Medicine, Baltimore, MD 21201.
Enterohemorrhagic Escherichia coli *and Other Shiga Toxin-Producing* E. coli
Edited by Vanessa Sperandio and Carolyn J. Hovde
© 2015 American Society for Microbiology, Washington, DC
doi:10.1128/microbiolspec.EHEC-0005-2013

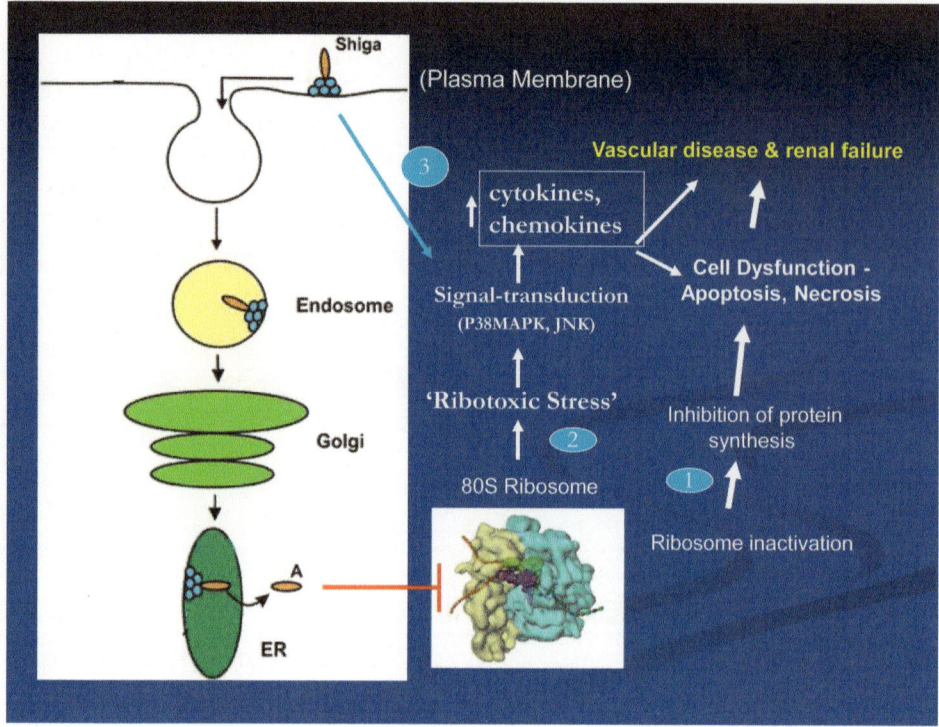

FIGURE 1 Schema: Shiga toxin interaction with eukaryotic cells. doi:10.1128/microbiolspec.EHEC-0005-2013.f1

coli (STEC) infections in humans, culminating in typical hemolytic-uremic syndrome (HUS). As depicted, it is clear that in some cases Stx can result in activation of p38 mitogen-activated protein kinase as well as apoptotic and necrotic cell death (Fig. 1). The topic of HUS renal disease has been reviewed recently (3–5).

Cell Types Responsive to Stx

The high number of Stx-sensitive cell types makes more difficult identification of more important events responsible for HUS. Renal microvascular endothelial cells are generally accepted to be the primary target of Stxs in HUS. Data in support of this concept come from many sources, most notably autopsy kidney pathology samples showing swollen and detached endothelial cells accompanied by thrombi (6). Such human renal microvascular endothelial cells were also shown to be very sensitive to Stx in vitro (7). However, other cells that make up the human renal glomerulus are also sensitive to Stx, including podocytes and mesangial cells (8, 9). In addition, extraglomerular epithelial cell types of the human kidney have been postulated to be targets of Stx, including proximal tubule and collecting duct cells (8, 10, 11). Cell types in the blood circulation that may be key to development of HUS and that are sensitive to Stx include platelets, neutrophils, and monocytes (12–16).

In summary, most, if not all, of the cell types mentioned may well have a role in STEC-related kidney disease and typical HUS. The relative importance and role of these cell types in STEC HUS remain to be determined. For example, it is not clear which of the renal cell types are actually responsible for renal failure in STEC HUS, although apoptosis of tubules appears to be a common feature (8, 17). The relative contributions in HUS disease of renal microvascular coagulation and thrombosis (i.e., endothelial cells), imbalance of fluid and electrolytes (i.e., nephron tubules), and altered filtration barrier function (i.e., endothelial and podocyte cells) have yet to be elucidated for typical HUS. If in vitro cell culture studies are pertinent to HUS in patients, the sensitivity (50% lethal dose) of human renal cells to Stx2 (endothelial, 0.1 pM > podocyte, 0.5 pM >> proximal tubule, 10 pM) suggests the renal filtration barrier is at considerable risk (8).

Inflammatory Cells, Chemokines, and Renal Thrombosis

A primary feature in the renal pathology of STEC HUS is microvascular coagulation and thrombosis. In humans and in a murine model of HUS, the interaction of Stx and lipopolysaccharide (LPS) with circulating cells and resident renal cells appears to have a causal role in microvascular thrombosis (18, 19). In a series of studies of the Stx/LPS murine model of

HUS, a pathway leading to fibrin deposition was revealed (Fig. 2). LPS activation of cells such as endothelial and renal tubule cells elicited chemokines (monocyte chemotactic protein 1 [MCP-1], macrophage inflammatory protein 1 [MIP-1] alpha, RANTES) known as chemoattractants for monocyte/macrophage cells and coactivators of platelets. In this response, Stx enhances the effects of but does not replace LPS. The response was associated with renal fibrin deposition (12, 20). In the murine model, simultaneous neutralization of these three chemokines inhibited LPS/Stx-induced monocyte accumulation and fibrin deposition in the kidneys (20). Further, administration of adenosine A2a receptor agonists to Stx/LPS mice also reduced monocyte and fibrin accumulation in the kidneys. As shown in Fig. 3, adenosine A2a receptor agonists act as anti-inflammatory agents in monocytes, platelets, and endothelial cells (21). Taken together, these studies indicate that both LPS and Stx are required for maximal renal fibrin deposition and that platelets may

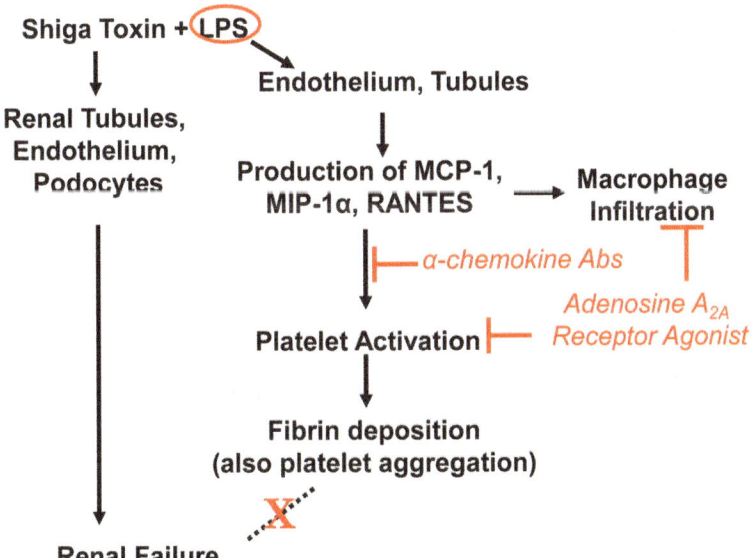

FIGURE 2 Proposed pathways of Stx and LPS actions in mice. Data derived from a Stx/LPS murine model of HUS indicate that LPS is the primary elicitor of fibrin deposition in kidneys. This pathway requires chemokines and platelets but is not responsible for renal failure. Stx is responsible for renal failure in this murine model in a process that involves nonendothelial renal cell types. doi:10.1128/microbiolspec.EHEC-0005-2013.f2

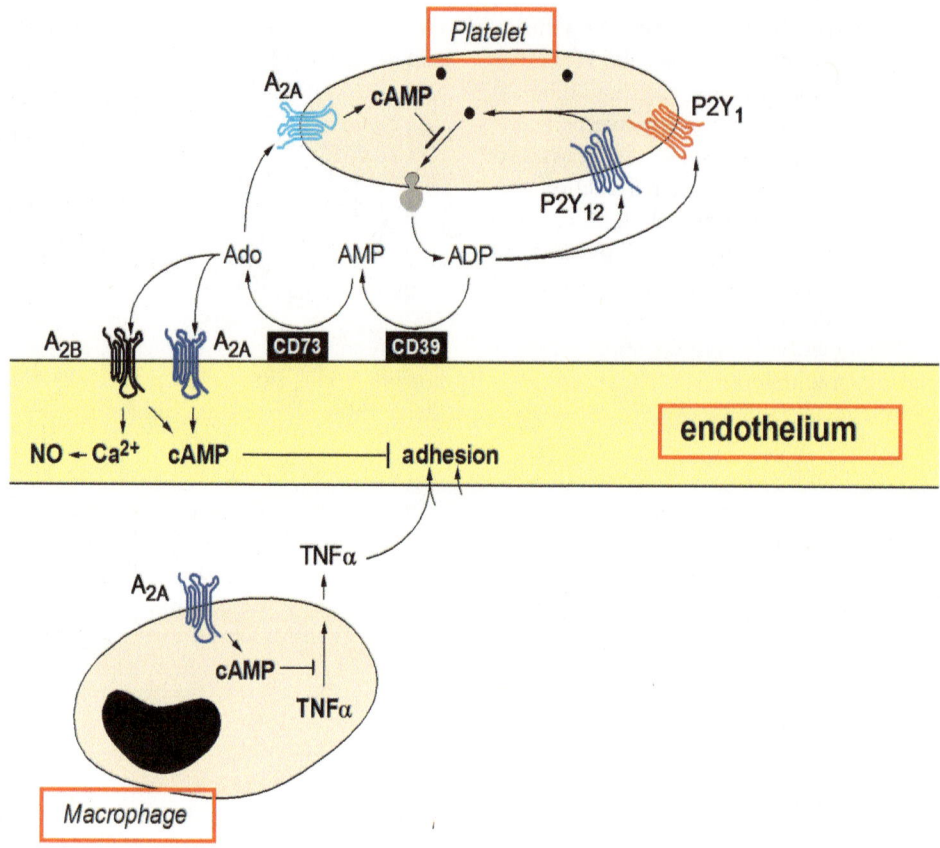

FIGURE 3 Anti-inflammatory actions of adenosine in HUS. Data derived from an Stx/LPS murine model of HUS suggest adenosine A2a receptor agonist, i.e., adenosine, effectively blocks the actions of LPS (enhanced by Stx2) at the level of different renal cell types to prevent platelet activation and coagulation. doi:10.1128/microbiolspec.EHEC-0005-2013.f3

be required. Because mice deficient in MCP-1 have sharply reduced platelet deposition after exposure to Stx/LPS, we have suggested that this chemokine serves as a coactivator of platelets in typical HUS (Keepers TR, unpublished data). The primary activators of platelet activation are thrombin or adenosine diphosphate. Our renal gene array analysis of the LPS response in mice indicated that LPS strongly elicited fibrinogen mRNA, the precursor of fibrin (Obrig T, unpublished data). In addition, it is noteworthy that selective elimination of monocytes from mice prior to the above studies had no effect on the ability of Stx/LPS to elicit renal fibrin deposition, suggesting the

chemokines are being generated from other cell types such as renal tubules (20). Important conclusions from the murine HUS model are that LPS, not Stx, is the initial primary elicitor of renal coagulation and thrombosis, but Stx, not LPS, is the lethal agent of STEC.

In the murine Stx/LPS model of HUS, monocyte migration into the kidneys was restricted to the extraglomerular space in contrast to polymorphonuclear leukocytes (PMN), which, in addition, migrated into the glomeruli. The latter may be important in humans because neutrophilia has been implicated as a primary risk factor for HUS disease and increased neutrophil migration into the kidneys was a key observation in HUS renal biopsies (22, 23). In the murine model of HUS, the neutrophil chemotactic factors chemokine ligand 1 (CXCL1) keratinocyte-derived chemokine (KC) and CXCL2 (MIP-2) were induced in the kidneys by LPS (15). The induction was at the transcriptional level and was enhanced by Stx2. Administration of neutralizing antibodies for these neutrophil chemotactic factors prevented the movement of neutrophils into the kidneys. It was also demonstrated that vascular cell adhesion molecule 1 (VCAM-1) was induced in the kidneys simultaneously with CXCL-1 and CXCL-2 in response to Stx2/LPS in mice (Fig. 4). VCAM-1 is known to assist movement of neutrophils across the endothelium and appeared to exhibit this function for neutrophils in the Stx2/LPS murine model of HUS. However, the relative importance of renal neutrophils in Stx-induced renal failure has yet to be determined in mice and humans.

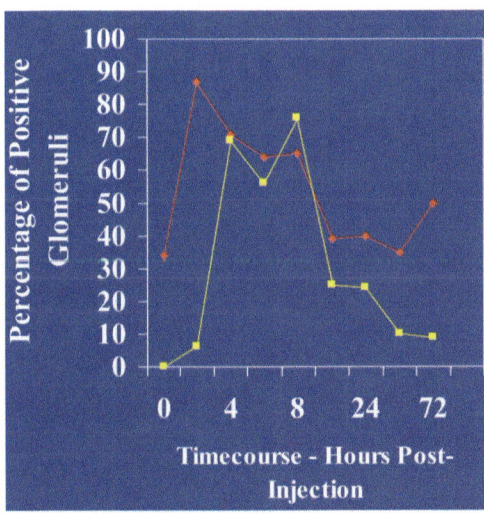

FIGURE 4 Neutrophil-endothelial cell interactions in HUS. In the Stx2/LPS murine model of HUS, analysis of renal gene activation and neutrophil infiltration into kidneys demonstrates a concomitant increase in PMNs and VCAM-1 expression, suggesting a mechanism of PMN-endothelial association. ♦, Neutrophils in the glomeruli; ■, VCAM-1 in the glomeruli. doi:10.1128/microbiolspec.EHEC-0005-2013.f4

Renal Gene Array Analysis of Murine Responses to Stx2 and LPS

Much information is now available regarding the biological effects of Stx2 and LPS on kidneys in the murine HUS model. The following is a synopsis of the more pertinent gene microarray data obtained from temporal studies of the murine renal responses to Stx2, LPS, or Stx2/LPS (19). On the basis of the total of both up- and downregulated genes, five times more renal genes responded to LPS than to Stx2 over the 72-h time course. Response to LPS was mostly early, whereas Stx2 responses occurred later in the 72-h time course. These results are more meaningful when viewed in the larger picture of HUS where renal failure occurs later in the time course in both mice and humans. It should be emphasized that Stx2, rather than LPS, is the lethal factor in the murine HUS model. The gene array data revealed different roles for LPS and Stx2 in the renal physiological responses. LPS responses were mostly inflammatory, stress related, or cell defensive in nature. In contrast, Stx2 responses were related to cell repair and involved cell proliferation and differentiation or cell cycle control genes. An interesting finding was that renal genes downregulated by Stx2 included membrane transporters, which appeared to signal a protective survival mode and slowing of cell metabolism.

The renal genes most upregulated by Stx2 or LPS are depicted in Fig. 5. As expected from the inflammatory responses described above, LPS induced a number of chemokine genes that code for chemotactic factors for monocytes and neutrophils. These tend to be "immediate" response genes, which attract monocytes and neutrophils into the kidneys and set the stage for a broad inflammatory response in the kidneys. Such LPS "immediate" response genes are mentioned in the literature in descriptions of typical HUS, i.e., MCP-1, MIP-2alpha, and the murine interleukin-8 mimic, KC. It was also observed that interferon-gamma-inducible protein-10 (IP-10) (CXCL10) was induced by LPS and by

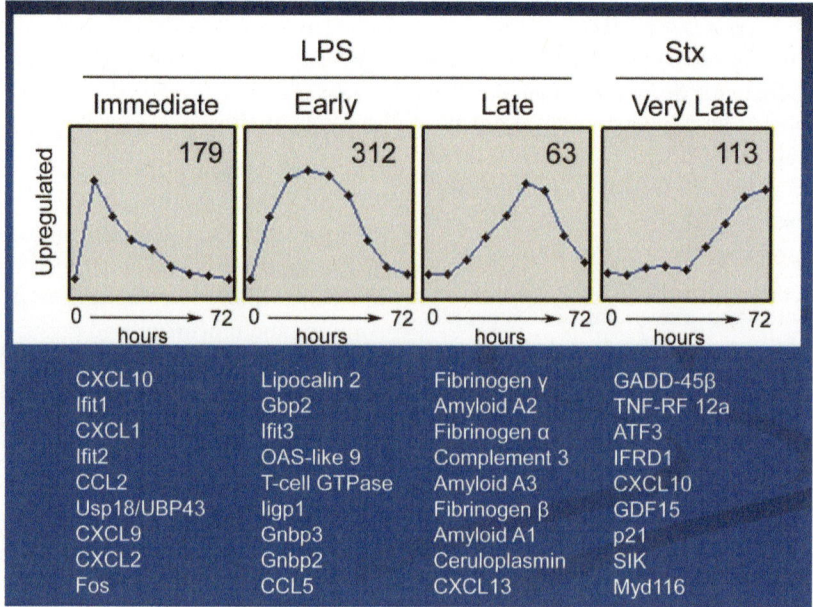

FIGURE 5 Renal gene activation in the Stx/LPS murine model. Shown are the nine most upregulated genes in the temporal response of mice to either LPS or Stx2. Gene microarrays were employed to analyze kidney gene activation over a 72-h response of C57BL/6 mice to 300 µg/kg of LPS or 100 ng/kg of Stx2.
doi:10.1128/microbiolspec.EHEC-0005-2013.f5

Stx2, albeit in early and late parts of the HUS disease time course, respectively. Related to renal coagulation and thrombosis in HUS, LPS induced a set of fibrinogen genes "late" in the time course of the murine model of HUS concomitant with the appearance of fibrin deposition and coagulation in the renal microvasculature of HUS (Fig. 5). These data agree with our observation that LPS is responsible, in part, for fibrin deposition in the Stx2/LPS murine model of HUS (19). Amyloid protein, which has been reported to be a Stx-sensitizing factor in HUS, is induced at the mRNA level by LPS in mice, as shown in Fig. 5, as a renal "late" gene product (24). More recently, complement has been identified as a factor that may contribute to renal failure in atypical HUS.

Products of some of the genes shown in Fig. 5 have been examined by investigators as potential biomarkers for diagnostic purposes. For example, IP-10 has been identified as a urine biomarker for other kidney diseases such as lupus nephritis (25, 26). Lipocalin 2 (neutrophil gelatinase-associated lipocalin), an LPS-induced "early" gene (Fig. 5), is a common urine biomarker for numerous renal diseases, including STEC-HUS (27).

How Valid Is the Murine Model of HUS for Translation to the Human Disease?

A large volume of data exists for mouse models of Stx-HUS (28). The two common experimental approaches for these murine models are either oral infection with STEC or injection with purified Stx plus or minus LPS (17, 19, 29, 30). In virtually all cases these are lethality models within 4 to 12 days after exposure to the agents and are accompanied by renal damage. Where examined, these murine models usually exhibit the three hallmarks of HUS: thrombocytopenia, hemolytic anemia, and renal failure. However, every animal model has its limitations, and for the murine models of HUS, the renal microvascular endothelial cells do not express Gb_3 and are resistant to Stx action. This is important if one believes that the primary target of Stx is the renal microvascular endothelium. Indeed, human renal endothelial cells in vitro are very sensitive to Stx, and the pathology of human kidneys in HUS describes swollen and detached glomerular endothelial cells. But it is surprising why such human glomerular endothelium is not killed by Stx in HUS kidneys. This suggests either a more indirect action of Stx in human HUS or dominant survival activities are activated within the endothelium after exposure to Stx. An alternative explanation is that the primary target of Stx in human kidneys is not the endothelium, but rather glomerular podocytes and extraglomerular tubules along the nephron. Support for this exists for HUS in mice and humans where urine specific gravity changes, chemokines are increased in the urine, and biomarkers of damaged podocytes and tubule cells are detected.

Mouse models have been helpful in separating the actions of Stx and LPS in HUS. In general, and as described above, LPS is the primary inducer of cytokines and chemokines where Stx enhances the activity of LPS. The complexity of inflammation in HUS is critical but has yet to be fully delineated in murine models and in human HUS. The murine model mirrors typical HUS of humans as resting platelets are resistant to Stx and require preactivation with LPS (19). However, it is most important to reiterate that Stx, not LPS, is responsible for the renal failure in typical HUS. In conclusion, the murine responses to Stx and LPS include most of the features of STEC-HUS in humans.

ACTIVITIES OF Stx IN CNS DISEASE

CNS Symptoms of Animal Models

In either an oral inoculation of STEC model or purified Stx injection animal model, the most common and most frequently reported central nervous system (CNS) impairment is paralysis of extremities. Most frequently, the hind legs are affected first, followed by the forelegs. Other symptoms include anorexia, lethargy, ataxic gait, recumbency (the affected animals lose strength to hold their body in an upright position), convulsions, seizure, coma, and death.

STEC oral administration animal models are summarized in Table 1. The oral inoculation models of STEC that describe CNS symptoms are limited to pig and mouse. Pigs develop "edema disease" with Stx2e-producing *E. coli* and present CNS symptoms (Table 2). Experimentally, the edema disease-like state is reproducible with Stx2-producing *E. coli* that has been isolated from human patients. CNS symptoms are only seen in Stx2 (both Stx2 and Stx2e) producers, but not in non-Stx2 producers. This indicates a strong association of Stx2 with CNS impairment.

LPS is an outer membrane component of gram-negative bacteria and a strong inflammation inducer. The involvement of LPS in STEC-associated CNS symptoms was tested by using LPS nonresponder mouse C3H/HeJ (29). C3H/HeJ did present CNS symptoms when given Stx2-producing *E. coli* but not when Stx-nonproducer was inoculated. This again suggests a strong involvement of Stx in CNS symptoms. The difference between LPS-responder mouse (C3H/HeN) and C3H/HeJ in CNS symptoms was that C3H/HeN showed a progressive time course of CNS symptoms whereas C3H/HeJ showed a "biphasic" response in that they developed milder CNS symptoms and recovered once, but then progressed to a severe form of CNS impairment. This suggests that even though Stx2 may be the central cause of CNS symptoms, addition of LPS response may contribute to the progress of the disease.

To further study the action of Stx2 in CNS disease, different animals were tested with purified Stx2. Stx2 injection animal models with CNS complications are summarized in Table 3. Also, LPS involvement or contribution to Stx2-associated CNS disease was tested in some reports. The reproducible results of

TABLE 1 STEC oral administration model with CNS descriptions

Ref.	Animal	*E. coli* strain	CNS[a]	Histopathology[b,c,d]	IHC[e]/TUNEL[b,c,f]	Model notes	Other assays
31	Pig	RCH/86 (Stx2+)	Yes	HE: CL cap, small inf, small hrrg, fib in sub and cap	ND	Gnotobiotic (cesarean section derived)	NA
32	Pig	86-24 (Stx2+)	Yes	Gross: MO and CL hrrg and necPAS: MO, CL, and Sc, cap swl nec, peri deposits	ND	Suckling (colostrum provided)	NA
32	Pig	87-23 Stx (–)	No	No lesion	ND	Suckling (colostrum provided)	NA
33	Pig	S1191 (Stx2e+) M112 (Stx2e+)	Yes	EM: myo and cap nec not apop, mono apop	TUNEL + myo in MO (5/11 pigs)	3-w-o	NA
33	Pig	Strain 123 (nonpathogenic *E. coli*)	No	No lesion	ND	3-w-o	NA
34	Pig	sakai	Yes	LFB: mye deg, hrrg, pyk and prolif cap, peri ede	ND	Neonatal	NA
35	ICR mouse	E32511/HSC (Stx2c+)[i]	ND	EM: cap ede in CR ctx, mye degHE: hrrg and ede in CR ctx only in CNS symptom (+) mice	Immuno EM-DAB[g]: Stx2+ in CR ctx pyr and deg mye	Sm, MMC[h]	Tracer (i.v.) detected in cap and deg mye
29	C3H/HeN mouse	86-24, 86BL or 134 (Stx2+)	Yes	ND	ND	Fasted	NA
29	C3H/HeN mouse	87-23, 87BL (Stx2–)	Yes	ND	ND	Fasted	NA
29	C3H/HeJ mouse	86-24, 86BL or 134 (Stx2+)	Yes (biphasic)	ND	ND	Fasted	NA
29	C3H/HeJ mouse	87-23, 87BL (Stx2–)	No	ND	ND	Fasted	NA
36	C57BL/6 mouse	N-9 (Stx1+/Stx2+)	ND	HE: infilt, hrrg, cap with fib in CR ctxLFB: No deg mye in hippo	Anti-Stx + hippo	PCM[j]	NA
37	IQI mouse	EDL931 (Stx1+/Stx2+)	Yes	HE: ede, fib in cap, neu deg, cap prolif	ND	Gnotobiotic	Brain TNFα increased
55	C57BL/6 mouse	Smr N-9 (Stx1+/Stx2+)	ND	ND	TUNEL + hippo neu during CNS symptom (+)	PCM	Serum Stx ↑, TNFα ↑, IL10 ↑TLC-anti-PkMab[k] brain +
38	IQI mouse	O157:H7 strain 6 (Stx1+/Stx2+)	Yes	HE: CR ctx and CL neu nec and slight loss of Purkinje	ND	Gnotobiotic	ND

(Continued on next page)

TABLE 1 (Continued)

Ref.	Animal	E. coli strain	CNS[a]	Histopathology[b,c,d]	IHC[e]/TUNEL[b,c,f]	Model notes	Other assays
56	ICR mouse	E32511/HSC (Stx2c+)	Yes	ND	GFAP[m] ↑, AQP4↓, casp3↑[n]neu cer Sc ventral and MO dorsal	Sm, MMC	ISH[l] Gb3 synthase

[a]Detailed CNS symptoms are summarized in Table 2.

[b]Histopathology analysis keys are Gross, gross observation in nonstained tissue; HE, hematoxylin-esosin stain that stains cytoplasm in pink and nucleus blue, light microscopic findings (LM); PAS, periodic acid-Schiff stain that detects polysaccharides, glycoproteins, and glycolipid, LM; LFB, Luxol fast blue stain that stains myelin in blue, LM; EM, electron microscopic findings; ND, not described; NA, not applicable.

[c]CNS regions and cell type abbreviations are CR, cerebrum; ctx, cortex; hippo, hippocampus; str, striatum; CL, cerebellum; MO, medulla oblongata; Sc, spinal cord; cer cervical; tho, thoracic; lum, lumbaris; sub, subarachinoid space; BS, brain stem is used where midbrain, pons, or medulla oblongata is not specified. Histopathologic feature abbreviations are cap, endothelial cells or capillaries; inf, infarction; hrrg, hemorrhage; fib, fibrin deposition; nec, necrosis; swl, swelling; peri, perivascular; myo, myocytes; apop, apoptotic; mono, monocytes; mye, myelin; deg, degeneration; pyk, pyknotic nuclei; prolif, proliferation/hyperplasia; ede, edema; pyr, pyramidal neuron; inflt, infiltration of blood cells to parenchyma; neu, neuron; Purkinje, Purkinje cells are large neurons in CL.

[d]Histopathologic feature abbreviations are cap, endothelial cells or capillaries; inf, infarction; hrrg, hemorrhage; fib, fibrin deposition; nec, necrosis; swl, swelling; peri, perivascular; myo, myocytes; apop, apoptotic; mono, monocytes; mye, myelin; deg, degeneration; pyk, pyknotic nuclei; prolif, proliferation/hyperplasia; ede, edema; pyr, pyramidal neuron; inflt, infiltration of blood cells to parenchyma; neu, neuron; Purkinje, Purkinje cells are large neurons in CL.

[e]IHC, immunohistochemistry, immunodetection of the target in the tissue sections.

[f]TUNEL, terminal deoxynucleotidyl transferase dUTP nick end labeling detects DNA fragmentation that is a hallmark of apoptosis.

[g]Immuno-EM-DAB, immunodetection of the target with 3,3′-diaminobenzidine deposition by electron microscopy.

[h]Sm, streptomycin; MMC, mitomycin C.

[i]Sm[r], MMC[r].

[j]PCM, protein calorie malnutrition.

[k]TLC-anti-PkMab (thin layer chromatography with anti-Pk monoclonal antibody detectin).

[l]ISH, in situ hybridization.

[m]GFAP, glial fibrillary acidic protein, an astrocyte marker; an increase of GFAP suggests astrogliosis.

[n]IHC for activated (cleaved) caspase-3.

hind-leg paralysis and high frequency of convulsions and seizures with purified Stx confirm the central role of the toxin in STEC-associated CNS disease. Human STEC patients present CNS symptoms that range from eye involvement (diplopia, hallucinations, and cortical blindness), behavioral changes (hyperactivity, distractibility, irritability, and altered sensorium), posturing/coordination difficulties (poor fine-motor coordination, hemiplegia, ataxia, and clumsiness), to severe symptoms such as seizures, dysregulation of breathing, and alteration in consciousness such as coma. Within these varieties of symptoms, ataxia or hemiparesis resembles Stx-associated animal CNS symptoms. Also, it is notable that in human patients, seizures are a frequent observation. This resemblance between patients and animal models of STEC/Stx suggests there is a great possibility that analyzing these animal models may give some clues to define the mechanisms of CNS impairment in Stx-associated disease.

CNS Histopathology of Animal Models

In animal models with STEC oral inoculation that describe CNS symptoms, most exhibit defective capillaries (pig [31–34], mouse [35, 36]). Those capillary lesions are mostly related to endothelial cell weakening that appears as hemorrhage, with leaked red blood cells in parenchyma. Noncapillary components in the parenchyma such as neurons and myelin defects were seen in some mouse STEC models (35, 37, 38), but not others (36). In purified Stx2 injection models, similar lesions involving capillary/endothelial cells were found in

TABLE 2 Observed CNS symptoms in animal models[a]

Ref.	Animal	Model	ANOX	LTHG	HL para	FL para	ATX	RCM[b]	CV/TR	SZR	Coma	Death	Other
46	Baboon	Stx1	ND[c]	ND	ND	ND	ND	ND	ND	ND	ND	Yes	
67	Baboon	Stx1	Yes	ND	ND	ND	ND	ND	ND	3/6 (50%)	Yes	Yes	
31	Pig	STEC	Yes	Yes	Yes	ND	Yes	Yes	Yes	Yes	Yes	Yes	
71	Pig	STEC	ND	ND	ND	ND	Yes	Yes	ND	ND	ND	Yes	Diarrhea, then CNS+
32	Pig	STEC	ND	ND	Yes	Yes	ND	Yes	Yes	ND	ND	Yes	Paddling
39	Pig	Stx2e i.v.	Yes	ND	ND	ND	Yes	ND	Yes	ND	Yes	Yes	Paddling, extensor rigidity, dyspnea
40	Pig	Sup Stx2e i.v.	ND	ND	ND	ND	Yes	Yes	Yes	ND	Yes	Yes	Paddling, extensor rigidity
33	Pig	STEC	ND	ND	ND	ND	Yes (1/11)	Yes (1/11)	ND	ND	ND	ND	
41	Rabbit	Stx1	Yes	Yes	Yes	ND	Yes	ND	ND	ND	ND	Yes	
42	Rabbit	Stx1 i.v.	Yes	Yes	Yes	Yes	ND	Yes	No	ND	ND	Yes	Ruffled fur, rapid respiration
52	Rabbit	Stx2 i.v.	ND	ND	Yes (50%)	Yes (50%)	Yes (33%)	ND	Yes (50%)	ND	ND	Yes (50%)	Opisthotonic posture
47	Rabbit	Stx2 i.v. and i.t.	Yes	Yes	Yes	Yes	ND	Yes	ND	ND	ND	Yes	
68	Rabbit	Stx2 i.v.	Yes	ND	Yes	Yes	ND	ND	ND	ND	ND	Yes	
54	Rabbit	Stx2 i.v.	Yes	ND	Yes	Yes	ND	ND	ND	ND	ND	Yes	Dyspnea
43	Rabbit	Stx2 i.v.	Yes	ND	Yes (83.3%)	ND	Yes (83.3%)	ND	ND	ND	ND	Yes	
48	Rabbit	Stx2 i.v.	ND	ND	Yes (25%)	ND	ND	ND	Yes (25%)	ND	ND	ND	
57	Rat	Stx2 i.c.v.	ND	Yes	Yes	ND	ND	ND	ND	Yes	ND	Yes	Crawling
35	Mouse	STEC	ND	Yes	Yes	Yes	ND	ND	ND	ND	ND	Death	Deformity of backbone, loss of pain
29	Mouse	STEC	ND	ND	Yes	Yes	Yes	ND	Yes	ND	Yes	Yes	Jerky rhythmic motion
36	Mouse	STEC	Yes	Yes	Yes	ND	ND	ND	(Yes)[d]	ND	ND	Yes	Ruffled fur, jerky rhythmic motion
37	Mouse	STEC	Yes	Yes	Yes	ND	ND	ND	ND	ND	ND	Yes	
44	Mouse	Stx2 i.v.	ND	ND	Yes	ND	ND	ND	ND	ND	ND	Yes	
44	Mouse	Stx2 +LPS i.v.	ND	ND	ND	ND	ND	ND	Yes	Yes	ND	Yes	
38	Mouse	STEC	Yes	Yes	Yes	ND	ND	ND	Yes	ND	ND	Yes	Ruffled fur

(Continued on next page)

TABLE 2 (Continued)

Ref.	Animal	Model	ANOX	LTHG	HL para	FL para	ATX	RCM[b]	CV/TR	SZR	Coma	Death	Other
53	Mouse	Stx2 i.p.	ND	Yes	Yes	ND	Yes	ND	Yes	Yes	ND	Yes	Retain sense (pain)
56	Mouse	STEC	ND	ND	Yes	ND	ND	ND	Yes	ND	ND	Yes	Spinal deformity

[a]Abbreviations for CNS symptoms are ANOX, anorexia; LTHG, lethargy; HL para, hind-leg paralysis; FL para, foreleg paralysis; ATX, ataxic gait; RCM, recumbency, difficulty holding body upright by itself; CV/TR, convulsions/tremors; SZR, seizure. Injection route abbreviations: i.c.v., intracerebroventricular; i.p., intraperitoneal; i.t., intrathecal; i.v., intravenous.

[b]Lateral, sternal, or dorsal recumbency; the animal is lying down with leaning on its side, abdomen, or back, having difficulty holding its body upright.

[c]ND, not described.

[d]Shivering.

pig (39, 40), rabbit (41–43), and mouse (44, 45). In contrast, other models did not have these lesions but rather lesions related to neuronal degeneration (baboon [46], rabbit [43, 47, 48], rat [49, 50], mouse [51]) or myelin degeneration (baboon [46], rabbit [52], rat [49]). Also, some reports showed normal appearance of neurons (rabbit [47], striatal neurons; mouse [53] lumbar spinal cord neurons). As all models exhibit similar CNS symptoms such as hind-leg paralysis, the difference in histopathological lesions may be due to involvement of different parts of CNS, different time points in the disease, or species-specific sensitivities. The mechanism of inducing CNS symptoms may be weakening of endothelial cells/capillary composition-caused neurotoxicity or direct effect of Stx in neuronal toxicity. The observation of lamellipodia-like processes of glial origin interrupting synaptic connections at the lumbar spinal cord interneuron to motor neuron may explain the resulting hind leg paralysis (mouse [53]). A similar observation is reported in a rat model of striatum neurons (51).

CNS Molecular Physiology of Animal Model

Molecular marker analysis in STEC or Stx animal models suggests possible mechanisms for Stx-associated CNS impairment.

The apoptotic nature of Stx-associated lesions has been described. Terminal deoxynucleotidyltransferase-mediated dUTP-biotin nick end labeling (TUNEL) stain detects fragmented DNA and therefore is often used as an apoptotic assay. Capillaries (pig [33], rabbit [43, 54]), neurons (mouse [55], rabbit [43]), and glial cells (rabbit [43]) have been detected as TUNEL positive. Activated caspase-3 targeted immunohistochemistry has been used for another marker of apoptotic cells. Neurons (mouse [56]) and capillaries (rabbit [54]) have been detected positive. Another pro-apoptotic marker, bax, was found increased in rat neurons (57). Along with electron microscopy observation (rat [49]), some neurons and capillary cells (endothelial cells and pericytes) undergo apoptosis, but some appear as necrotic (rabbit [33]). Careful and detailed information of which area of the CNS and what types of cells in that area present apoptotic features may help elucidate these conflicting results.

Aquaporin 4 (AQP4) is mostly expressed in astrocyte foot processes that have a direct contact with capillaries in the CNS. The reduction of AQP4 suggests that there is alteration in astrocytic foot process, which is important to strengthen the blood-brain barrier (BBB). AQP4 expression decreased in Stx2-injected rat (50) and STEC-infected mouse (56), while astrocytic activation marker glial fibrillary acidic protein increased. This suggests Stx-associated astrocyte activation may participate in weakening the BBB.

An increase in tumor necrosis factor alpha in STEC-inoculated mouse (37) and Stx2-injected rabbit (43) brain along with serum

TABLE 3 Shiga toxin and/or LPS administration model with CNS descriptions

Ref.	Animal[a]	Toxin	CNS[b]	Gb3/Stx binding	Histopathology[c,d,e]	Imaging	IHC[f]	Injection route[g]	Other assays
66	Baboon[h]	Stx1	ND	ND	EM: mye deg, peri ede, large neu and glia deg, cap normal	ND	ND	i.v.	
67	Baboon	Stx1	Yes	ND	ND	ND	ND	i.v.	
39	Weaned YL pigs	Stx2e	Yes	ND	Gross: Sc ede, CL ede and hrrg HE: CL hrrg but not ede or cap nec, no lesions in CR, BS, thalamus	ND	ND	i.v.	
40	Weaned YL pigs	Sup[i] Stx2e	Yes	ND	HE: CL, sub ede, hrrg, cap nec HE: MB, fib nec , peri eos	ND	ND	i.v.	
41	NZW rabbit	Stx1	Yes	TLC-Stx1 over lay of CL, BS and Sc (+) at LacCer[j] position	HE: BS, Sc, CL cap narrowed, peri ede, cap damage and fib, Purkinje decreased	ND	ND	i.v.	
42	NZW rabbit	Stx1	Yes	^{125}I-Stx1 tissue distribution high in cecum, brain, small intestine, colon, Sc	HE: cerv Sc hrrg, inf, ede, fib in cap, lum Sc fib cap, pyk cap	ND	Anti-Stx (+) in Sc cap	i.v.	
52	JW rabbit	Stx2	Yes	ND	EM: mye deg, axo normal	MRI: V3 (24 h), later BS, cc, lateral amy, CL vermis	Immuno EM-DAB[k]: anti-Stx2 (+) at luminal side of cap, deg mye	i.v.	Tracer
47	JW rabbit	Stx2	Yes	ND	CR ctx, CL ctx, Sc, pyk neu, str neu not affected, inf in MB and CL	ND	Anti-Stx2 IHC (+) cap and sub	i.v. and i.t.	Stx2 ↑ in CSF
65	JW rabbit	Stx2	ND	ND	ND	MRI: CL at 82 h	ND	i.v.	Stx2 ↑ in CSF
65	JW rabbit	Stx2	ND	ND	ND	MRI: the rear of CL (contact w CSF) at 48 h	ND	i.t.	ND
68	JW rabbit	Stx2	Yes	ND	ND	MRI: BS and cerv Sc dorsal close to death	ND	i.v.	Baroreflex function
69	JW rabbit	Stx1	ND	ND	ND	MRI: MB, BS and Sc ede, cerv Sc dorsal hematoma	ND	i.v.	Stx1 ↑ slightly in CSF

Ref	Animal	Toxin			Pathology		IHC/Other	Route	Notes
54	JW rabbit young	Stx2	Yes	ND	HE: inf CL, myo thickening, fib, pyk and fragmented, pons myo nec	ND	Anti-Stx2 IHC(+) in pons cap and myo, anti-ssDNA IHC small number cap (+), IHC caspase-3 and-9 small number (+)	i.v.	
43	JW rabbit	Stx2	Yes	Anti-Gb₃ (+) in cap lum Sc	HE: inf lum Sc with hrrg and fib cap, str neu pyk, hippos pyr neu apop	ND	TUNEL + in hippo (DG, CA1, CR ctx neu, pons glia, cap IHC: Ib4↑ (microglia activated) in lum Sc and thalamus	i.v.	qRT-PCR: TNFα↑ and IFNβ ↑
70	JW rabbit	Stx2	ND	ND	ND	MRI: Gd leak (↑permeability)	ND	i.v.	
48	DB rabbit	Stx2	Yes	ND	HE: neu deg in the BS	ND	ND	i.v.	
49	SD rat	Stx2	ND	ND	ND	ND	Anti-Stx IHC (+) peri	i.p.	
49	SD rat	Stx2	ND	ND	EM: irregular shape neu, mye deg, hypertrophic axo, astro phago mye, neu apop, deg, vacuol, peri astro ede, non-peri astro gliosis, oligo pathologic	ND	Anti-Stx2 (+) in neu, immune-EM-DAB: anti-Stx2 (+) in neu fibers and astro nucleus	i.c.v.	
58	SD rat	Stx2	ND	ND	ND	ND	Stx2 (+) in anterior hippo astro, ips hippo Stx2(+) astro and neu, cont hippo Stx2(+) neuropils	i.c.v.	Str and CR ctx neu↓, cap↓ NADPH-d/NOS activity
57	SD rat	Stx2	Yes	Anti-Gb₃↑ CA1, str neu	ND	ND	Anti-Stx2↑ MAP2 ↑ CA1 neu.Anti-bax ↑neu (inner ctx, CA1, subV dorsal str, hypothalamic peri	i.c.v.	
50	SD rat	Sup Stx2	ND	ND	Nissle: neu pyk hypertrophy axo, ede ctx, subV CL.	ND	Anti-AQP4↓ cp	i.p.	

(Continued on next page)

TABLE 3 Shiga toxin and/or LPS administration model with CNS descriptions (Continued)

Ref.	Animal[a]	Toxin	CNS[b]	Gb3/Stx binding	Histopathology[c,d,e]	Imaging	IHC[f]	Injection route[g]	Other assays
35	ICR mouse	Stx2	Yes	ND	ND	ND	Immune-EM-DAB: anti-Stx2 (+) mye deg, lyso pyr ctx	i.p.	Tracer
44	C57BL/6 mouse	Stx2	Yes	ND	HE: pyk oligo astro nuclei, hrrg sub	ND	ND	i.v.	
44	C57BL/6 mouse	Stx2 + LPS	Yes	ND	HE: hrrg BS sub severe than Stx2 alone	ND	ND	i.v.	
45	ICR mouse	Stx2	ND	ND	HE: congestion CL and hippo, hrrg CL, neu and glia normal	ND	Anti-Stx2 IHC (+) in RBC and cap CL, MB and thalamus, anti-Stx2 IHC (−) in neu, glia	i.v.	
63	C57BL/6 a4galt−/−	Stx1 and Stx2	Not susceptible to Stx1	Anti-Gb3 became negative in cap	ND	ND	ND	Not specified	
53	C57BL/6 mouse	Stx2	Yes	Anti-Gb3 (+) in neu mouse and human Sc, human cap	EM: glia lamellipodia-like foot process interrupts synapse at motor neu of lum Sc		Immuno-gold EM: anti-Gb3 and anti-Stx2 double positive in motor neu of lum Sc	i.p.	
51	NIH mouse	Stx2	ND	ND	EM: neu, astro and peri ede, synaptic disruption, oligo defect	NA	NA	i.v.	Behavioral motor test +
61	Human	NA	NA	DRGm, Stx1 binding (+) neu and cap	NA	NA	NA	NA	
61	Rabbit	NA	NA	DRG, Stx1 binding (+) neu and cap	NA	NA	NA	NA	
61	Rat	NA	NA	DRG, Stx1 binding (+) neu	NA	NA	NA	NA	
62	Human	NA	NA	DRG, anti-Gb3 and Stx1 binding (+) neu and cap	NA	NA	NA	NA	
62	Rabbit	NA	NA	DRG, anti-Gb3 and Stx1 binding (+) neu and cap	NA	NA	NA	NA	

62	Rat	NA	NA	DRG, anti-Gb$_3$ and Stx1 binding (+) neu	NA	NA	NA
62	Mouse	NA	NA	DRG, anti-Gb$_3$ and Stx1 binding (+) neu	NA	NA	NA
60	C57BL/6 mouse	NA	NA	Anti-Gb$_3$ (+) neu at olf, CR ctx, str, hippo, hypothalamus, CVOs, CL, MO, Sc V3 ependyma	NA	NA	NA

[a]Animal keys: YL, Yorkshire-Landrace; NZW, New Zealand White; JW, Japanese White; DB, Dutch Belted; SD, Sprague-Dawley.

[b]Detailed CNS symptoms are summarized in Table 2. ND, not described; NA, not applicable.

[c]Histopathology analysis keys are Gross, gross observation in non-stained tissue; HE, hematoxylin-esosin stain that stains cytoplasm in pink and nucleus blue, light microscopic findings (LM); PAS, periodic acid-Schiff stain that detects polysaccharides, glycoproteins and glycolipid, LM; LFB, Luxol fast blue stain that stains myelin in blue, LM; EM, electron microscopic findings.

[d]CNS regions and cell type abbreviations are CR, cerebrum; ctx, cortex; hippo, hippocampus; DG, dentate gyrus; str, striatum and other basal ganglia; CL, cerebellum; MO, medulla oblongata; MB, midbrain; BS, brain stem is used where midbrain, pons, or medulla oblongata are not specified; Sc, spinal cord; cerv, cervical; tho, thoracic; lum, lumbaris; sub, subarachnoid space. Histopathologic feature abbreviations are cap, endothelial cells or capillaries; inf, infarction; hrrg, hemorrhage; fib, fibrin deposition; nec, necrosis; swl, swelling; peri, perivascular; myo, myocytes; apop, apoptotic; mono, monocytes; mye, myelin; deg, degeneration; pyk, pyknotic nuclei; prolif, proliferation/hyperplasia; ede, edema; pyr, pyramidal neuron; neu, neuron; Purkinje, Purkinje cells are large neurons in CL; V3, third ventricle; cc, corpus callosum; amy, amygdala; ips, ipsilateral, injection side of brain; cont, contlateral, opposite of injection side; subV, subventricular region; cp, choroid plexus; CVO, circumventricular organs.

[e]Histopathologic feature abbreviations are cap, endothelial cells or capillaries; inf, infarction; hrrg, hemorrhage; fib, fibrin deposition; nec, necrosis; swl, swelling; peri, perivascular; myo, myocytes; apop, apoptotic; mono, monocytes; mye, myelin; degeneration; pyk, pyknotic nuclei; prolif, proliferation/hyperplasia; ede, edema; pyr, pyramidal neuron; inft, infiltration of blood cells to parenchyma; neu, neuron; Purkinje, Purkinje cells are large neurons in CL; glia, glial cells such as astrocytes, microglia, and oligodendrocytes); eos, eosinophilic globules, deposits); axo, axon, axoplasm; astro, astrocyte; oligo, oligodendrocyte; phago, phagocytosis; lyso, lysosome; RBC, red blood cells.

[f]IHC, immunohistochemistry, immunodetection of the target in the tissue sections.

[g]Injection route abbreviations: i.v., intravenous; i.t. (intrathecal, injection from cysterna magna that makes it possible to inject into cerebrospinal fluid (CSF); i.p, intraperitoneal; i.c.v., intracerebroventricular injection that injects solution directly into CNS parenchyma of cerebral cortex/ventricle.

[h]Baboon in this chart is Papio c. cynocephalus, or Papio c. Anubis

[i]Sup, E. coli culture supernatant.

[j]LacCer, lactosylceramide; adding galactose to LacCer completes Gb$_3$.

[k]Immuno-EM-DAB: immunodetection of the target with 3,3'-diaminobenzidine deposition by electron microscopy.

[l]Immuno-gold EM: immunodetection of the target with 5- to 10-nm gold particle allows precise localization as well as double labeling.

[m]DRG, dorsal root ganglion, a peripheral nervous system structure consisting of sensory neurons and other cell types.

tumor necrosis factor alpha increase in STEC-inoculated rabbit (55) suggests Stx-associated inflammation in the CNS.

Ca^{2+} imaging and electrophysiological study are useful tools to assess direct physiological action of Stx in fresh brain slices. Our group showed Stx2-associated neuronal glutamate release in mouse brain slice (cerebral cortex) indirectly by recording intracellular Ca^{2+} in astrocyte (53). Recently, it is shown that Stx2 induces depolarization of neurons in the thalamic area of female rat (58).

Receptor Gb$_3$ Expression in Animal Central and Peripheral Nervous Systems (CNS, PNS)

Shiga toxin receptor localization in the animal nervous system has been described for different species. There are three ways to localize Shiga toxin receptor. First is to perform anti-Stx immunodetection in tissues of STEC-infected or Stx-injected animals (rabbit [42, 47, 52], rat [49, 59], mouse [35, 36]). Second is to incubate a naïve tissue section with Stx followed by anti-Stx immunodetection (pig [60]). Third is to recognize Gb$_3$ as an Stx receptor with anti-Gb$_3$ immunodetection in tissues. Detecting anti-Gb$_3$ immunoreaction in the naïve tissue provides a basal expression level and cell types that would be influenced by Stx initially in the course of disease. These include neurons in the mouse spinal cord (53) and other regions of CNS (61). In the Stx-administered tissue, it may or may not indicate the spontaneous Stx receptor expression but certainly indicates cell types responsive to Stx. The cell types that are positive in either of the analyses above often include small vessel endothelial cells (rabbit [42, 43, 47, 52, 62, 63], mouse [45, 64]), neurons (rat [49, 57, 59], mouse [35, 45, 53, 61]), and glial cells (rat [49, 57, 59], mouse [45, 61]). Miyatake and colleagues compared the peripheral nervous system (dorsal root ganglion) of different species with the same method and found that human and rabbit expressed Stx receptor in endothelial cells and neurons, whereas rat and

mouse expression was restricted to neurons (62, 63). Our group reported that throughout the mouse CNS, the only nonneuronal cell type to exhibit anti-Gb$_3$ immunoreactivity was the third ventricle ependymal cell (61). Studies have suggested, in the naïve state, humans and rabbits express Stx receptor in their vessels as well as neurons, and rodents appear to express Gb$_3$ mainly in neurons. However, it was shown that Stx receptors in the rat CNS are induced by Stx administration (57). Among different species, the receptor expression patterns in different regions of CNS, the cell types, and the amount expressed may be different, but all models present with common CNS impairment such as hind-leg paralysis. This may be interpreted as expression of Stx receptor in endothelial cells is not necessary for toxin to be able to internalize into the CNS parenchyma to have an effect.

In 2006, Okuda et al. (64) reported a 4galt knockout mouse that lacks Gb$_3$ synthase (alpha 1,4-galactosyltransferase) and therefore produces no Gb$_3$. In this mouse, originally Gb$_3$-positive vessels lost their anti-Gb$_3$ immunoreactivity and became Stx resistant. Gb$_3$ synthase probe has been applied for an in situ hybridization in the mouse (56) and rat (58) CNS. While metabolic pathway enzymes such as Gb$_3$ synthase, a glycosyltransferase, add the terminal galactose to complete Gb$_3$, other glycosyltransferases in the pathway are unique in each step of glycolipid synthesis, and there are catabolic pathway enzymes as well (see Fig. 6). All these enzymes participate in determining the amount of Gb$_3$ in the cell. Measuring these Gb$_3$-associated enzymes may provide more insight into Stx receptor regulation.

Discussion about How Stx Enters CNS of Animals

Purified Stx peripheral injection (intraperitoneal [i.p.] or intravenous [i.v.]) is able to induce CNS impairment similar to that of STEC oral infection, suggesting that there is a direct effect of Stx on CNS parenchymal cells. The rat model of intraventricular purified Stx2

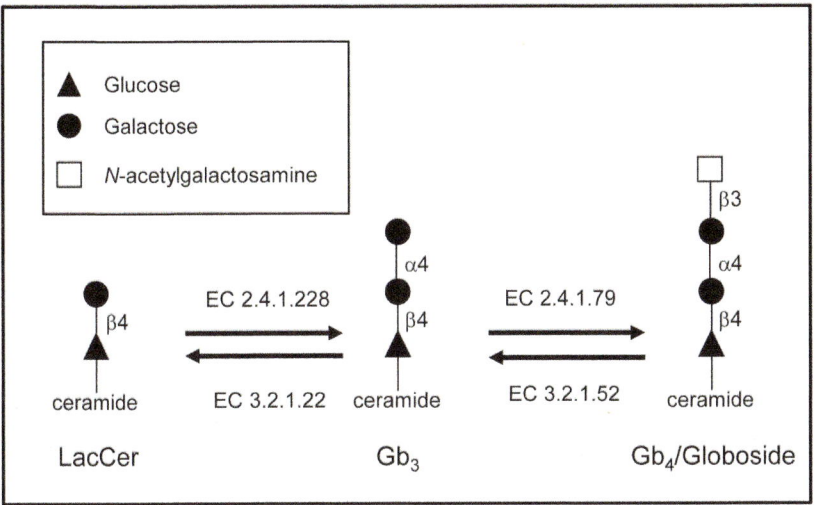

FIGURE 6 Metabolic and catabolic pathway enzymes for Gb$_3$ synthesis. A part of Gb$_3$ synthesis pathway is shown. From lactosylceramide (LacCer) to Gb$_3$, alpha 1, 4-galactosyltransferase (EC 2.4.1.228) adds a galactose to LacCer to produce Gb$_3$. Likewise, UDP-GalNAc: beta 1,3-galactosaminyltransferase (EC 2.4.1.79) works on Gb$_3$ to make Gb$_4$. In the catabolic pathway, beta-hexosaminidase (EC 3.2.1.52) degrades Gb$_4$ to Gb$_3$, and alpha-galactosidase (EC 3.2.1.22) makes LacCer from Gb$_3$. doi:10.1128/microbiolspec.EHEC-0005-2013.f6

injection in which purified Stx2 is inoculated directly into CNS parenchyma also induces similar CNS symptoms such as lethargy, hind-leg weakness, or paralysis (57). These results suggest that Stx released from STEC internalizes into the blood and then transfers to CNS parenchyma and asserts its toxicity.

The route and CNS region of Stx permeabilization are of great interest to explain which part of the CNS is most likely influenced by Stx. Stx injected by i.v. has been detected in cerebrospinal fluid (CSF) (rabbit [47, 65]). This suggests there is translocation of Stx from blood to CSF. A reduction of AQP1 in choroid plexus in rat with Stx (i.p.) suggests that there is weakening of the blood-CSF barrier in this location that may allow Stx to enter CSF from the blood. The ependymal cells lining at the third ventricle are a border between CSF and CNS parenchyma. Our group showed in mouse CNS that ependymal cells at the third ventricle are expressing Gb$_3$ in a naïve state (61). The tracer horseradish peroxidase that is injected intrathecally into CSF crossed and entered ependymal cells and

parenchyma (rabbit [52]), and magnetic resonance imaging showed the third ventricle area with a bright signal that is an indication of leakiness into the fluid in this area. Taken together, it is reasonable to think that Stx uses blood-CSF barrier penetration as one of the routes into CNS parenchyma. On the other hand, Stx injected i.p. was detected in the perivascular area in rat (49), and BBB weakening was suggested by the reduction of AQP4 (rat [50], mouse [56], and by tracer horseradish peroxidase (i.v.) detection in parenchyma (mouse [35]). These results suggest that Stx can also use the BBB crossing route to enter the CNS. An important fact to note is that purified Stx by itself, without any other bacterial component, can enter CNS and assert its toxicity regardless of differences in receptor-expressing cell types among different species.

ACKNOWLEDGMENTS

This work was supported by National Institutes of Health grants 5U01AI075778-05 and 1R56AI090144-01A1 to F.O.

CITATION

Obata F, Obrig T. 2014. Role of Shiga/vero toxins in pathogenesis. Microbiol Spectrum 2(3):EHEC-0005-2013.

REFERENCES

1. **Ohmura M, Yamamoto M, Tomiyama-Miyaji C, Yuki Y, Takeda Y, Kiyono H.** 2005. Nontoxic Shiga toxin derivatives from *Escherichia coli* possess adjuvant activity for the augmentation of antigen-specific immune responses via dendritic cell activation. *Infect Immun* **73:**4088–4097.
2. **Tesh VL, Ramegowda B, Samuel JE.** 1994. Purified Shiga-like toxins induce expression of proinflammatory cytokines from murine peritoneal macrophages. *Infect Immun* **62:**5085–5094.
3. **Obrig TG.** 2010. *Escherichia coli* Shiga toxin mechanisms of action in renal disease. *Toxins (Basel)* **2:**2769–2794.
4. **Obrig TG, Karpman D.** 2012. Shiga toxin pathogenesis: kidney complications and renal failure. *Curr Top Microbiol Immunol* **357:**105–136.
5. **Karpman D, Sartz L, Johnson S.** 2010. Pathophysiology of typical hemolytic uremic syndrome. *Semin Thromb Hemost* **36:**575–585.
6. **Habib R, Mathieu H, Royer P.** 1967. [Hemolytic-uremic syndrome of infancy: 27 clinical and anatomic observations]. *Nephron* **4:**139–172 (In French).
7. **Obrig TG, Louise CB, Lingwood CA, Boyd B, Barley-Maloney L, Daniel TO.** 1993. Endothelial heterogeneity in Shiga toxin receptors and responses. *J Biol Chem* **268:**15484–15488.
8. **Psotka MA, Obata F, Kolling GL, Gross LK, Saleem MA, Satchell SC, Mathieson PW, Obrig TG.** 2009. Shiga toxin 2 targets the murine renal collecting duct epithelium. *Infect Immun* **77:**959–969.
9. **Simon M, Cleary TG, Hernandez JD, Abboud HE.** 1998. Shiga toxin 1 elicits diverse biologic responses in mesangial cells. *Kidney Int* **54:**1117–1127.
10. **Shibolet O, Shina A, Rosen S, Cleary TG, Brezis M, Ashkenazi S.** 1997. Shiga toxin induces medullary tubular injury in isolated perfused rat kidneys. *FEMS Immunol Med Microbiol* **18:**55–60.
11. **Hughes AK, Stricklett PK, Kohan DE.** 1998. Cytotoxic effect of Shiga toxin-1 on human proximal tubule cells. *Kidney Int* **54:**426–437.
12. **Ghosh SA, Polanowska-Grabowska RK, Fujii J, Obrig T, Gear AR.** 2004. Shiga toxin binds to activated platelets. *J Thromb Haemost* **2:**499–506.
13. **Karpman D, Manea M, Vaziri-Sani F, Stahl AL, Kristoffersson AC.** 2006. Platelet activation in hemolytic uremic syndrome. *Semin Thromb Hemost* **32:**128–145.
14. **Fernandez GC, Rubel C, Dran G, Gomez S, Isturiz MA, Palermo MS.** 2000. Shiga toxin-2 induces neutrophilia and neutrophil activation in a murine model of hemolytic uremic syndrome. *Clin Immunol* **95:**227–234.
15. **Roche JK, Keepers TR, Gross LK, Seaner RM, Obrig TG.** 2007. CXCL1/KC and CXCL2/MIP-2 are critical effectors and potential targets for therapy of *Escherichia coli* O157:H7-associated renal inflammation. *Am J Pathol* **170:**526–537.
16. **Foster GH, Armstrong CS, Sakiri R, Tesh VL.** 2000. Shiga toxin-induced tumor necrosis factor alpha expression: requirement for toxin enzymatic activity and monocyte protein kinase C and protein tyrosine kinases. *Infect Immun* **68:**5183–5189.
17. **Eaton KA, Friedman DI, Francis GJ, Tyler JS, Young VB, Haeger J, Abu-Ali G, Whittam TS.** 2008. Pathogenesis of renal disease due to enterohemorrhagic *Escherichia coli* in germ-free mice. *Infect Immun* **76:**3054–3063.
18. **Stahl AL, Sartz L, Nelsson A, Bekassy ZD, Karpman D.** 2009. Shiga toxin and lipopolysaccharide induce platelet-leukocyte aggregates and tissue factor release, a thrombotic mechanism in hemolytic uremic syndrome. *PLoS One* **4:**e6990.
19. **Keepers TR, Psotka MA, Gross LK, Obrig TG.** 2006. A murine model of HUS: Shiga toxin with lipopolysaccharide mimics the renal damage and physiologic response of human disease. *J Am Soc Nephrol* **17:**3400–3414.
20. **Keepers TR, Gross LK, Obrig TG.** 2007. Monocyte chemoattractant protein 1, macrophage inflammatory protein 1 alpha, and RANTES recruit macrophages to the kidney in a mouse model of hemolytic-uremic syndrome. *Infect Immun* **75:**1229–1236.
21. **Hasko G, Linden J, Cronstein B, Pacher P.** 2008. Adenosine receptors: therapeutic aspects for inflammatory and immune diseases. *Nat Rev Drug Discov* **7:**759–770.
22. **Walters MD, Matthei IU, Kay R, Dillon MJ, Barratt TM.** 1989. The polymorphonuclear leucocyte count in childhood haemolytic uraemic syndrome. *Pediatr Nephrol* **3:**130–134.
23. **Inward CD, Howie AJ, Fitzpatrick MM, Rafaat F, Milford DV, Taylor CM.** 1997. Renal histopathology in fatal cases of diarrhoea-associated haemolytic uraemic syndrome. *Pediatr Nephrol* **11:**556–559.
24. **Griener TP, Strecker JG, Humphries RM, Mulvey GL, Fuentealba C, Hancock RE, Armstrong GD.** 2011. Lipopolysaccharide renders transgenic mice expressing human serum amyloid P component sensitive to Shiga toxin 2. *PLoS One* **6:**e21457.

25. **Kawachi H, Han GD, Miyauchi N, Hashimoto T, Suzuki K, Shimizu F.** 2009. Therapeutic targets in the podocyte: findings in anti-slit diaphragm antibody-induced nephropathy. *J Nephrol* **22:** 450–456.

26. **Das L, Brunner HI.** 2009. Biomarkers for renal disease in childhood. *Curr Rheumatol Rep* **11:**218–225.

27. **Trachtman H, Christen E, Cnaan A, Patrick J, Mai V, Mishra J, Jain A, Bullington N, Devarajan P.** 2006. Urinary neutrophil gelatinase-associated lipocalin in D+HUS: a novel marker of renal injury. *Pediatr Nephrol* **21:**989–994.

28. **Mohawk KL, O'Brien AD.** 2011. Mouse models of *Escherichia coli* O157:H7 infection and Shiga toxin injection. *J Biomed Biotechnol* **2011:**258185.

29. **Karpman D, Connell H, Svensson M, Scheutz F, Alm P, Svanborg C.** 1997. The role of lipopolysaccharide and Shiga-like toxin in a mouse model of *Escherichia coli* O157:H7 infection. *J Infect Dis* **175:**611–620.

30. **Wadolkowski EA, Sung LM, Burris JA, Samuel JE, O'Brien AD.** 1990. Acute renal tubular necrosis and death of mice orally infected with *Escherichia coli* strains that produce Shiga-like toxin type II. *Infect Immun* **58:**3959–3965.

31. **Tzipori S, Chow CW, Powell HR.** 1988. Cerebral infection with *Escherichia coli* O157:H7 in humans and gnotobiotic piglets. *J Clin Pathol* **41:**1099–1103.

32. **Dean-Nystrom EA, Pohlenz JF, Moon HW, O'Brien AD.** 2000. *Escherichia coli* O157:H7 causes more-severe systemic disease in suckling piglets than in colostrum-deprived neonatal piglets. *Infect Immun* **68:**2356–2358.

33. **Matise I, Sirinarumitr T, Bosworth BT, Moon HW.** 2000. Vascular ultrastructure and DNA fragmentation in swine infected with Shiga toxin-producing *Escherichia coli*. *Vet Pathol* **37:** 318–327.

34. **Shringi S, Garcia A, Lahmers KK, Potter KA, Muthupalani S, Swennes AG, Hovde CJ, Call DR, Fox JG, Besser TE.** 2012. Differential virulence of clinical and bovine-biased enterohemorrhagic *Escherichia coli* O157:H7 genotypes in piglet and Dutch belted rabbit models. *Infect Immun* **80:**369–380.

35. **Fujii J, Kita T, Yoshida S, Takeda T, Kobayashi H, Tanaka N, Ohsato K, Mizuguchi Y.** 1994. Direct evidence of neuron impairment by oral infection with verotoxin-producing *Escherichia coli* O157:H- in mitomycin-treated mice. *Infect Immun* **62:**3447–3453.

36. **Kurioka T, Yunou Y, Kita E.** 1998. Enhancement of susceptibility to Shiga toxin-producing *Escherichia coli* O157:H7 by protein calorie malnutrition in mice. *Infect Immun* **66:**1726–1734.

37. **Isogai E, Isogai H, Kimura K, Hayashi S, Kubota T, Fujii N, Takeshi K.** 1998. Role of tumor necrosis factor alpha in gnotobiotic mice infected with an *Escherichia coli* O157:H7 strain. *Infect Immun* **66:**197–202.

38. **Taguchi H, Takahashi M, Yamaguchi H, Osaki T, Komatsu A, Fujioka Y, Kamiya S.** 2002. Experimental infection of germ-free mice with hyper-toxigenic enterohaemorrhagic *Escherichia coli* O157:H7, strain 6. *J Med Microbiol* **51:**336–343.

39. **MacLeod DL, Gyles CL, Wilcock BP.** 1991. Reproduction of edema disease of swine with purified Shiga-like toxin-II variant. *Vet Pathol* **28:**66–73.

40. **Gannon VP, Gyles CL, Wilcock BP.** 1989. Effects of *Escherichia coli* Shiga-like toxins (verotoxins) in pigs. *Can J Vet Re.* **53:**306–312.

41. **Zoja C, Corna D, Farina C, Sacchi G, Lingwood C, Doyle MP, Padhye VV, Abbate M, Remuzzi G.** 1992. Verotoxin glycolipid receptors determine the localization of microangiopathic process in rabbits given verotoxin-1. *J Lab Clin Med* **120:** 229–238.

42. **Richardson SE, Rotman TA, Jay V, Smith CR, Becker LE, Petric M, Olivieri NF, Karmali MA.** 1992. Experimental verocytotoxemia in rabbits. *Infect Immun* **60:**4154–4167.

43. **Takahashi K, Funata N, Ikuta F, Sato S.** 2008. Neuronal apoptosis and inflammatory responses in the central nervous system of a rabbit treated with Shiga toxin-2. *J Neuroinflammation* **5:**11.

44. **Sugatani J, Igarashi T, Munakata M, Komiyama Y, Takahashi H, Komiyama N, Maeda T, Takeda T, Miwa M.** 2000. Activation of coagulation in C57BL/6 mice given verotoxin 2 (VT2) and the effect of co-administration of LPS with VT2. *Thromb Res* **100:**61–72.

45. **Nishikawa K, Matsuoka K, Kita E, Okabe N, Mizuguchi M, Hino K, Miyazawa S, Yamasaki C, Aoki J, Takashima S, Yamakawa Y, Nishijima M, Terunuma D, Kuzuhara H, Natori Y.** 2002. A therapeutic agent with oriented carbohydrates for treatment of infections by Shiga toxin-producing *Escherichia coli* O157:H7. *Proc Natl Acad Sci USA* **99:**7669–7674.

46. **Taylor CM, Williams JM, Lote CJ, Howie AJ, Thewles A, Wood JA, Milford DV, Raafat F, Chant I, Rose PE.** 1999. A laboratory model of toxin-induced hemolytic uremic syndrome. *Kidney Int* **55:**1367–1374.

47. **Mizuguchi M, Tanaka S, Fujii I, Tanizawa H, Suzuki Y, Igarashi T, Yamanaka T, Takeda T, Miwa M.** 1996. Neuronal and vascular pathology produced by verocytotoxin 2 in the rabbit central nervous system. *Acta Neuropathol (Berl)* **91:**254–262.

48. **Garcia A, Marini RP, Catalfamo JL, Knox KA, Schauer DB, Rogers AB, Fox JG.** 2008. Intravenous Shiga toxin 2 promotes enteritis and renal injury characterized by polymorphonuclear leukocyte infiltration and thrombosis in Dutch Belted rabbits. *Microbes Infect* **10:**650–656.

49. **Goldstein J, Loidl CF, Creydt VP, Boccoli J, Ibarra C.** 2007. Intracerebroventricular administration of Shiga toxin type 2 induces striatal neuronal death and glial alterations: an ultrastructural study. *Brain Res* **1161:**106–115.

50. **Lucero MS, Mirarchi F, Goldstein J, Silberstein C.** 2012. Intraperitoneal administration of Shiga toxin 2 induced neuronal alterations and reduced the expression levels of aquaporin 1 and aquaporin 4 in rat brain. *Microb Pathog* **53:**87–94.

51. **Tironi-Farinati C, Geoghegan PA, Cangelosi A, Pinto A, Loidl CF, Goldstein J.** 2013. A translational murine model of sub-lethal intoxication with Shiga toxin 2 reveals novel ultrastructural findings in the brain striatum. *PLoS One* **8:**e55812.

52. **Fujii J, Kinoshita Y, Kita T, Higure A, Takeda T, Tanaka N, Yoshida S.** 1996. Magnetic resonance imaging and histopathological study of brain lesions in rabbits given intravenous verotoxin 2. *Infect Immun* **64:**5053–5060.

53. **Obata F, Tohyama K, Bonev AD, Kolling GL, Keepers TR, Gross LK, Nelson MT, Sato S, Obrig TG.** 2008. Shiga toxin 2 affects the central nervous system through receptor globotriaosylceramide localized to neurons. *J Infect Dis* **198:** 1398–1406.

54. **Mizuguchi M, Sugatani J, Maeda T, Momoi T, Arima K, Takashima S, Takeda T, Miwa M.** 2001. Cerebrovascular damage in young rabbits after intravenous administration of Shiga toxin 2. *Acta Neuropathol (Berl)* **102:**306–312.

55. **Kita E, Yunou Y, Kurioka T, Harada H, Yoshikawa S, Mikasa K, Higashi N.** 2000. Pathogenic mechanism of mouse brain damage caused by oral infection with Shiga toxin-producing *Escherichia coli* O157:H7. *Infect Immun* **68:**1207–1214.

56. **Amran MY, Fujii J, Suzuki SO, Kolling GL, Villanueva SY, Kainuma M, Kobayashi H, Kameyama H, Yoshida S.** 2013. Investigation of encephalopathy caused by Shiga toxin 2c-producing *Escherichia coli* infection in mice. *PLoS One* **8:**e58959.

57. **Tironi-Farinati C, Loidl CF, Boccoli J, Parma Y, Fernandez-Miyakawa ME, Goldstein J.** 2010. Intracerebroventricular Shiga toxin 2 increases the expression of its receptor globotriaosylceramide and causes dendritic abnormalities. *J Neuroimmunol* **222:**48–61.

58. **Meuth SG, Gobel K, Kanyshkova T, Ehling P, Ritter MA, Schwindt W, Bielaszewska M, Lebiedz** P, **Coulon P, Herrmann AM, Storck W, Kohmann D, Muthing J, Pavenstadt H, Kuhlmann T, Karch H, Peters G, Budde T, Wiendl H, Pape HC.** 2013. Thalamic involvement in patients with neurologic impairment due to Shiga toxin 2. *Ann Neurol* **73:** 419–429.

59. **Boccoli J, Loidl CF, Lopez-Costa JJ, Creydt VP, Ibarra C, Goldstein J.** 2008. Intracerebroventricular administration of Shiga toxin type 2 altered the expression levels of neuronal nitric oxide synthase and glial fibrillary acidic protein in rat brains. *Brain Res* **1230:**320–333.

60. **Winter KR, Stoffregen WC, Dean-Nystrom EA.** 2004. Shiga toxin binding to isolated porcine tissues and peripheral blood leukocytes. *Infect Immun* **72:**6680–6684.

61. **Obata F, Obrig T.** 2010. Distribution of Gb(3) immunoreactivity in the mouse central nervous system. *Toxins (Basel)* **2:**1997–2006.

62. **Ren J, Utsunomiya I, Taguchi K, Ariga T, Tai T, Ihara Y, Miyatake T.** 1999. Localization of verotoxin receptors in nervous system. *Brain Res* **825:**183–188.

63. **Utsunomiya I, Ren J, Taguchi K, Ariga T, Tai T, Ihara Y, Miyatake T.** 2001. Immunohistochemical detection of verotoxin receptors in nervous system. *Brain Res Brain Res Protoc* **8:**99–103.

64. **Okuda T, Tokuda N, Numata S, Ito M, Ohta M, Kawamura K, Wiels J, Urano T, Tajima O, Furukawa K.** 2006. Targeted disruption of Gb3/CD77 synthase gene resulted in the complete deletion of globo-series glycosphingolipids and loss of sensitivity to verotoxins. *J Biol Chem* **281:** 10230–10235.

65. **Fujii J, Kinoshita Y, Yamada Y, Yutsudo T, Kita T, Takeda T, Yoshida S.** 1998. Neurotoxicity of intrathecal Shiga toxin 2 and protection by intrathecal injection of anti-Shiga toxin 2 antiserum in rabbits. *Microb Pathog* **25:**139–146.

66. **Taylor CM, Williams JM, Lote CJ, Howie AJ, Thewles A, Wood JA, Milford DV, Raafat F, Chant I, Rose PE.** 1999. A laboratory model of toxin-induced hemolytic uremic syndrome. *Kidney Int* **55:**1367–1374.

67. **Siegler RL, Pysher TJ, Lou R, Tesh VL, Taylor FB Jr.** 2001. Response to Shiga toxin-1, with and without lipopolysaccharide, in a primate model of hemolytic uremic syndrome. *Am J Nephrol* **21:**420–425.

68. **Yamada Y, Fujii J, Murasato Y, Nakamura T, Hayashida Y, Kinoshita Y, Yutsudo T, Matsumoto T, Yoshida S.** 1999. Brainstem mechanisms of autonomic dysfunction in encephalopathy-associated Shiga toxin 2 intoxication. *Ann Neurol* **45:**716–723.

69. **Fujii J, Kinoshita Y, Yutsudo T, Taniguchi H, Obrig T, Yoshida SI.** 2001. Toxicity of Shiga

toxin 1 in the central nervous system of rabbits. *Infect Immun* **69:**6545–6548.

70. **Fujii J, Kinoshita Y, Matsukawa A, Villanueva SY, Yutsudo T, Yoshida S.** 2009. Successful steroid pulse therapy for brain lesion caused by Shiga toxin 2 in rabbits. *Microb Pathog* **46:**179–184.

71. **Tzipori S, Gunzer F, Donnenberg MS, de Montigny L, Kaper JB, Donohue-Rolfe A.** 1995. The role of the eaeA gene in diarrhea and neurological complications in a gnotobiotic piglet model of enterohemorrhagic *Escherichia coli* infection. *Infect Immun* **63:**3621–3627.

The Locus of Enterocyte Effacement and Associated Virulence Factors of Enterohemorrhagic *Escherichia coli*

6

MARK P. STEVENS[1] and GAD M. FRANKEL[2]

INTRODUCTION

Enterohemorrhagic *Escherichia coli* (EHEC) was first recognized as a cause of human disease in 1983 and is associated with diarrhea and hemorrhagic colitis, which may be complicated by life-threatening renal and neurological sequelae (reviewed in reference 242). EHEC strains are defined by their ability to produce one or more Shiga toxins (Stx), which mediate the systemic complications of EHEC infections (reviewed in reference 243), and to induce attaching and effacing (A/E) lesions on intestinal epithelia. The ability of EHEC to induce such lesions is shared by enteropathogenic *E. coli* (EPEC), *Escherichia albertii* (formerly classified as *eae*-positive *Hafnia alvei*), and the murine pathogen *Citrobacter rodentium*. The A/E histopathology was first described in gnotobiotic piglets challenged with a strain of EHEC serotype O157:H7 (1) but has subsequently been observed in ruminant reservoirs and diverse animal models (reviewed in reference 244).

A/E lesions are characterized by intimate bacterial adherence to the apical surface of enterocytes, sometimes on raised pedestals (pseudopodia), and destruction of nearby microvilli (Fig. 1A and B). Though reports of such lesions

[1]The Roslin Institute and Royal (Dick) School of Veterinary Studies, University of Edinburgh, Midlothian, EH25 9RG, United Kingdom; [2]MRC Centre for Molecular Bacteriology and Infection, Department of Life Sciences, Imperial College London, London, SW7 2AZ, United Kingdom.

Enterohemorrhagic Escherichia coli *and Other Shiga Toxin-Producing* E. coli
Edited by Vanessa Sperandio and Carolyn J. Hovde
© 2015 American Society for Microbiology, Washington, DC
doi:10.1128/microbiolspec.EHEC-0007-2013

during human infection are lacking, Knutton et al. described that EPEC strains of many serogroups were able to form A/E lesions on human duodenal biopsies cultured ex vivo (2).

Reasoning that the electron-dense pedestals may contain polymerized F-actin, the authors subsequently reported the use of fluorescein-labeled phallotoxin to stain F-actin at sites of

(A)

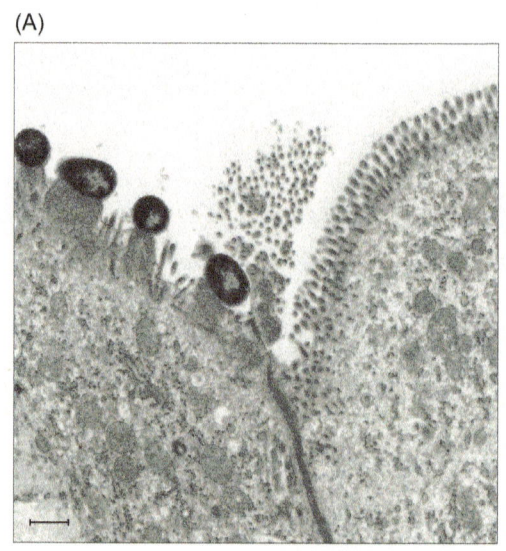

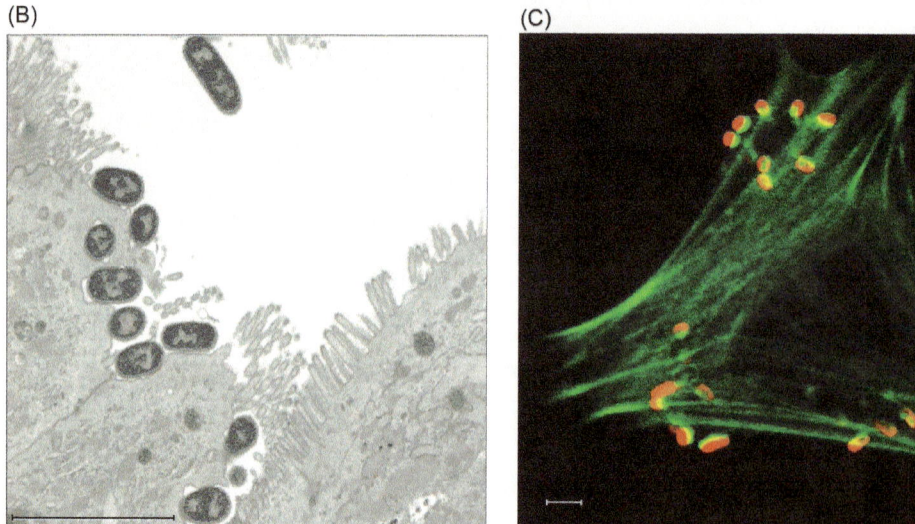

FIGURE 1 (A) Transmission electron micrograph (TEM) showing A/E lesions induced by EHEC O111:H– strain E45035N in the spiral colon of a neonatal calf (note raised electron-dense pedestals and microvillus effacement relative to proximal uninfected enterocyte). From reference 233; scale bar = 1 μm. (B) TEM of A/E lesions induced by EHEC O157:H7 strain 85-170 12 h after inoculation of a bovine ligated ileal loop (note intimate adherence but relative absence of elongated pedestals). From reference 84; scale bar = 5 μm. (C) Fluorescence micrograph showing nucleation of F-actin under EHEC O103:H3 strain PMK5 adhering to a HeLa cell (green, F-actin detected with Oregon green[514]-phalloidin; red, bacteria stained with rabbit anti-O103 typing serum detected with anti-rabbit immunoglobulin-Alexa[568]). From reference 240; scale bar = 5 μm.
doi:10.1128/microbiolspec.EHEC-0007-2013.f1

adherence, forming the basis of a fluorescent-actin staining (FAS) test for A/E lesion formation by fluorescence microscopy (3) (Fig. 1C). Many other host cell proteins are recruited to A/E lesions, sometimes in a host-, pathotype-, serotype-, and time-dependent manner, and the molecular mechanisms underlying lesion formation have been the subject of intense study. This article focuses on the factors required for A/E lesion formation, their mode of action, and role in persistence, pathogenesis, and protection during EHEC infection.

THE LOCUS OF ENTEROCYTE EFFACEMENT

The locus of enterocyte effacement (LEE) is required for A/E lesion formation and is the most intensively studied of all virulence factors of attaching and effacing *E. coli*. The role of LEE-encoded genes in adherence to epithelial cells and nucleation of F-actin was first described after screening random Tn*phoA* mutants of EPEC O127:H6 for mutants deficient in these processes. This identified the *E. coli* attaching and effacing (*eae*) gene encoding intimin (4) and other genes required for adherence and actin nucleation in a single ca. 35-kb locus that is conserved among A/E pathogens, including strains of EHEC serotypes O157:H7 and O26:H11 (5). Sequencing of a limited region of the LEE of the prototype EPEC O127:H6 strain E2348/69 identified four genes predicted to encode components of a Type III secretion system (T3SS), based on homology to *Yersinia* Lcr/Ysc and *Shigella* Mxi/Spa proteins (6). These genes were initially designated *sepA–D* (secretion of EPEC proteins) (6), but many were renamed *esc* to conform to the nomenclature of homologous *Yersinia* T3SS genes upon complete sequencing of the LEE of E2348/69 (7). Type III secretion is one of multiple pathways for export of proteins in bacteria and, consistent with a role in this process, the product of the *sepB* (*escN*) gene, which encodes a predicted inner membrane ATPase, was found to be required for secretion of several EPEC O127:H6 proteins required for A/E lesion formation, designated Esps (*E. coli* secreted proteins) (6). Secretion of Esps was also detected in EHEC O157:H7 and EHEC O26:H11 (8, 9) and found to be *escN*-dependent (9), supporting the existence of a conserved locus for secretion of proteins involved in A/E lesion formation.

The LEE of EPEC O127:H6 strain E2348/69 contains 41 open reading frames that are mostly organized in five polycistronic operons (7). Soon after the sequencing of the EPEC O127:H6 strain E2348/69 LEE, the sequence of the homologous region of the genome of the prototype EHEC O157:H7 strain EDL933 was reported (10), followed by the LEE of *C. rodentium* (11) and rabbit EPEC strains (12, 13). Many other LEE sequences have since emerged with the advent of high-throughput genome sequencing. These typically exhibit significantly lower %GC content than the flanking DNA (ca. 38% compared to the ca. 50% genomic mean), are inserted at tRNA loci, and contain remnants of mobile genetic elements, indicating that they are likely to have been acquired by horizontal transfer. The cloned LEE of EPEC O127:H6 strain E2348/69 is necessary and sufficient to confer the ability to form pedestals upon a laboratory-adapted *E. coli* K-12 strain (14); however, the cloned EHEC O157:H7 LEE cannot (15), consistent with the requirement for non-LEE-encoded proteins in pedestal formation by EHEC O157:H7, as discussed below.

The genetic organization of the EHEC O157:H7 strain EDL933 LEE is depicted in Fig. 2. The LEE region in this strain is inserted at the equivalent of 82 min on the *E. coli* K-12 chromosome, proximal to the *selC* selenocysteine tRNA locus, and shares the 41 genes found in EPEC O127:H6 strain E2348/69 in the same order, but is larger owing to the presence of a cryptic P4 family prophage at one end (10). Predicted structural components of the T3SS exhibit remarkable conservation among A/E pathogens; however, greater levels of amino acid sequence divergence exist

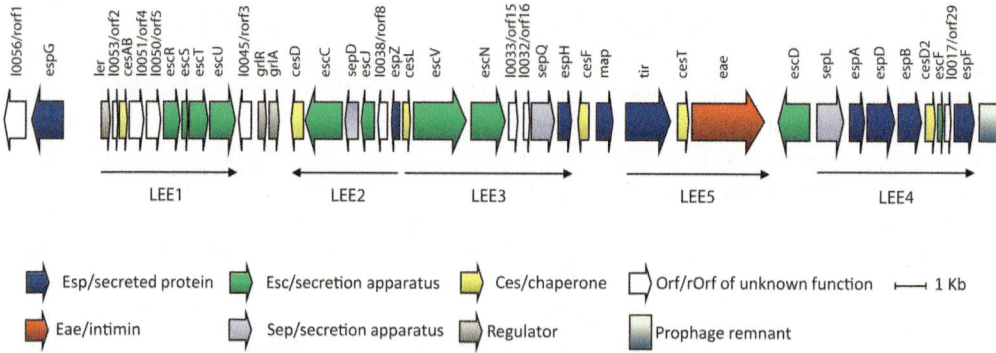

FIGURE 2 Genetic organization of LEE of *E. coli* O157:H7. Open reading frames are represented by thick arrows, and putative polycistronic operons are designated by thin arrows. Clear arrows represent open reading frames of unknown function and are designated *orf* or *rorf*, depending on the direction of transcription relative to *eae*. doi:10.1128/microbiolspec.EHEC-0007-2013.f2

among components that are predicted interact with host cells. Indeed, comparative analyses have suggested that the LEE comprises units under distinct selection pressures (16). Analysis of the insertion sites and flanking regions of the LEE in serogroup O26, O103, and O111 EHEC strains and atypical EPEC has indicated that while insertions tend to be restricted to the *selC*, *pheU*, and *pheV* tRNA loci, there is marked variation in the flanking regions (17, 18; reviewed in reference 19). For example, substrates for the LEE-encoded T3SS may be encoded in the flanking regions, such as Ibe 5′ of the LEE in atypical EPEC (20), and a conserved cassette encoding NleE, NleB, and EspL 3′ of the *LEE4* operon in EHEC O26, O103, and O111 (18). In some strains the *lifA/efa-1* gene or a truncated variant thereof may be located 3′ of *LEE4* as part of a mosaic island comprising the LEE and O-island 122 (18, 21). LifA/Efa1 and the truncated LifA/Efa1′ (Z4332) have recently been reported to be secreted by the LEE-encoded T3SS (22) and (together with other O-island 122-encoded genes) have been associated with isolates that cause epidemic and serious disease (23). In EHEC O103:H2 strain RW1374, the LEE, *lifA/efa-1*, and other virulence-associated genes form a pathogenicity-related island spanning ca. 111 kb at *pheV* (24), whereas in other cases effector genes may

be present in the genome but not physically linked to the LEE, owing to activity of bacteriophages in the capture and transfer of genes encoding Type III secreted proteins (18, 25). As more genomes are sequenced, it is evident that the nature of lateral transfer of the LEE, and genes associated with the function of the T3SS, is highly variable, and further studies are required to define the consequences of such.

Regulation of the LEE is highly complex, with inputs from numerous global regulators, bacteriophage-encoded regulators, and LEE-encoded regulators. Control occurs at transcriptional, translational, posttranslational, and bacterial population levels (reviewed in reference 245). Of particular interest, LEE expression and the efficiency of pedestal formation are enhanced by passage in the mammalian host (26), and evidence exists that LEE expression is sensitive to host stress-related catecholamine hormones that are detected by bacterial adrenergic sensor kinases (reviewed in reference 27). Such studies reinforce the challenge of understanding LEE regulation and functions in cell-based systems devoid of cues from the host. Moreover, it is evident that the repertoire of regulators is strain-dependent and that studies on LEE regulation in single strains are therefore to be interpreted with caution. For example, lysogeny with Shiga toxin 2-encoding bacteriophages was recently reported to repress Type III secretion in phage-type (PT) 21/28 EHEC O157:H7 strains that are commonly isolated from humans in the United Kingdom and to account for a distinct level of T3SS activity relative to Stx2-minus PT32 strains (28).

The LEE-Encoded Type III Secretion System

T3SSs are key virulence factors of Gram-negative enteric pathogens and the four major genera of plant pathogenic bacteria and serve to inject bacterial proteins directly into host cells (reviewed in reference 29). A needle complex spans both the inner and outer bacterial membranes, and a subset of Type III secreted proteins forms a "translocon" that interacts with the eukaryotic cell membrane and mediates the delivery of secreted "effector" proteins into target cells. Effectors of A/E *E. coli* then modulate cellular processes to the benefit of the pathogen. Predicted functions have been assigned to LEE-encoded genes on the basis of homology with components of T3SSs in other bacteria (30); a schematic representation of the organization of the apparatus is shown in Fig. 3. Predicted protein-protein interactions within the T3SS apparatus have been confirmed experimentally in some cases, for example, by yeast-2-hybrid analysis (31) and analysis of purified needle complexes (32). Transmission electron microscopy of the purified EPEC O127:H6 Type III secretion needle complex has indicated that it comprises cylindrical inner and outer rings similar to the flagellar basal body (33), consistent with the shared evolution of such systems.

Systematic mutagenesis of the LEE in *C. rodentium* has been used to assess the role of all 41 genes in regulation of LEE expression, the level and hierarchy of Type III secretion, the ability to form pedestals via the FAS test, and virulence in mice (34). Nineteen genes essential for secretion of both translocator and effector proteins were identified (*escR*, *escS*, *escT*, *escU*, *escC*, *escJ*, *escV*, *escN*, *escD*, *escF*,

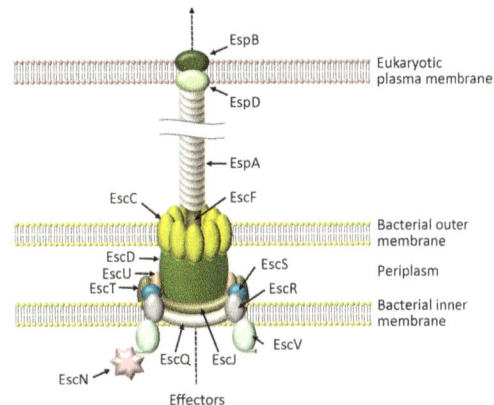

FIGURE 3 Schematic representation of the Type III secretion apparatus showing the predicted spatial organization of LEE-encoded proteins. Adapted from reference 30.
doi:10.1128/microbiolspec.EHEC-0007-2013.f3

escE, *orf4*, *escL*, *rorf3*, *escI*, *orf12*, *escA*, *sepQ*, and *escG*) with mutations in a further four genes (*orf3*, *rorf6*, *orf16*, and *sepL*) impairing secretion of translocators preferentially (34). Real-time imaging has subsequently indicated a distinct order in the efficiency of secretion of specific effectors in a manner dependent on intra-bacterial effector concentration, effector-chaperone interactions, and the efficiency of bacterial attachment to target cells (35). Although these studies indicated that EspZ is injected into cells early after infection, recent work has indicated that EspZ arrests translocation of effectors from within the host cell through an ill-defined "translocation stop" activity (36). Indeed, ectopic expression of EspZ inside eukaryotic cells renders them refractory to actin pedestal formation and prior infection of cultured cells with one A/E pathogen prevents superinfection by another (36). Mutants lacking *espZ* cause elevated cytotoxicity, probably owing to uncontrolled injection of effectors (36, 37). The importance of *sepL*, *escD*, *sepQ*, *escV*, *rorf8*, *escC*, *escR*, and *orf4* in EspA secretion by EHEC O157:H7 was confirmed by analysis of transposon mutants that exhibit reduced adherence to Caco-2 cells in vitro (38).

Mutations affecting components of the needle complex and translocon were found to impair intestinal colonization of calves by signature-tagged transposon mutagenesis of EHEC O157:H7 (39) and EHEC O26:H– (40). A retrospective transposon-directed insertion-site sequencing analysis of the library of EHEC O157:H7 strain EDL933 mutants screened in calves by Dziva et al. identified 51 attenuating mutations in the LEE, many of which affected predicted needle complex components (41), extending earlier observations on the role of the LEE genes without further use of cattle. Screening of mutants in complex pools can exaggerate fitness costs; however, EHEC mutants with defects affecting the *escN*-encoded inner membrane ATPase were also found to be highly attenuated when tested in isolation in infant rabbits (42) and calves (39, 40), confirming a key role for the LEE-encoded T3SS in EHEC persistence and pathogenesis.

Inhibitors of Type III secretion in A/E pathogens have been identified (43–45). Many of these belong to a family of salicylidene acylhydrazides and appear to act in part by inhibition of transcription of LEE genes in EHEC O157:H7 (45), rather than by interacting with basal components of the T3SS apparatus and interfering with needle complex assembly as reported for *Shigella* sp. (46). Different salicylidene acylhydrazides produce distinct patterns of LEE repression (45), and the targets of such drugs have so far been mapped to genes outside the LEE (44, 47), providing insights into how LEE expression and T3SS function are controlled. The potential prophylactic and therapeutic applications of such inhibitors have received relatively little attention.

Translocon Components

A filamentous extension of the T3SS needle complex is transiently expressed on the surface of A/E pathogens in the early stages of lesion formation and comprises mostly the *LEE4*-encoded protein EspA (48, 49) (Fig. 4). EspA filaments in EPEC O127:H6 can vary in length up to 600 nm and are ca. 12 nm in diameter (33, 50). Such filaments form a physical bridge between the bacteria and host cells that is required for the translocation of EspB and Tir into host cells (48, 49, 51, 52) (Fig. 4A). This led to the hypothesis that EspA filaments comprise a hollow channel through which effector proteins are injected into host cells, a notion supported by resolution of the three-dimensional structure, which showed that EPEC O127:H6 EspA subunits polymerized in a helical tube of 120 Å diameter with 5.6 subunits per turn enclosing a central channel of 25 Å diameter (53). Immunogold electron microscopy revealed that Tir is secreted from the tips of EspA filaments (54) (Fig. 4B and C), providing direct evidence that EspA filaments are hollow conduits through which effectors are secreted. Secretion and intracellular stability of EspA require the *LEE1*-encoded chaperone CesAB (55), and the filaments are elongated by addition of subunits at the growing tip (54). EspA binds directly to the needle complex protein EscF, which is required for EspA filament assembly and effector translocation (33, 56).

It remains unclear precisely how EspA filaments are connected to the host cell surface and the translocation pore is created. The *LEE4*-encoded Type III secreted EspB and EspD proteins are believed to mediate formation of the translocation pore on the basis of homology to the *Yersinia* YopB and YopD, their presence in the host cell plasma membrane, and ability to form pores in erythrocyte membranes (48, 57–61). EspD is required for assembly of EspA filaments (48, 58), interacts directly with EspA and EspB (62), and requires the chaperones CesD and CesD2 for secretion (63). EspB interacts with EspA and is required for the translocation of effectors, including Tir (64). However, EspA filaments can bind to host cell membranes in the absence of EspB, indicating that they may interact directly with cellular components (64). Indeed, studies using single, double, or triple mutants of EPEC O127:H6 lacking EspA filaments, intimin, and bundle-forming pili

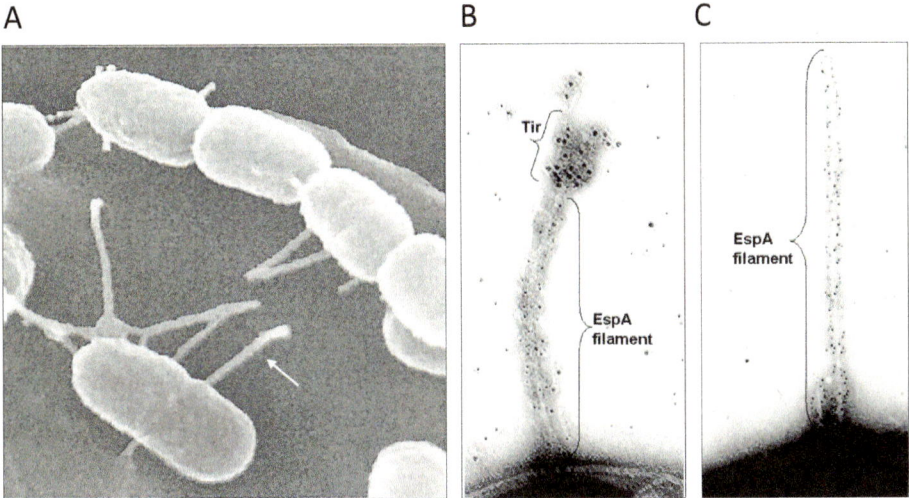

FIGURE 4 **(A) Scanning electron micrograph showing EspA filaments (arrow) of EHEC O26:H11 strain H19 attaching to the surface of an erythrocyte. From reference 241. (B) Transmission electron micrograph of an EspA filament of wild-type EPEC O127:H6 strain E2348/69 showing Tir issuing from the tip. EspA filaments were immunolabeled with anti-EspA conjugated to 5-nm diameter gold particles, and Tir was detected with anti-Tir conjugated to 10-nm diameter gold particles. (C) The specificity of Tir staining was confirmed using the same gold-labeled antibodies but an isogenic *tir* mutant. Panels B and C from reference 54.** doi:10.1128/microbiolspec.EHEC-0007-2013.f4

indicate that EspA filaments play a role in intimin-independent adherence to HEp-2 and Caco-2 cells, as an *espA bfpA eae* triple mutant was less adherent than a *bfpA eae* double mutant (65). EspA filaments appear shorter in EHEC O157:H7, and their expression is heterogeneous at the bacterial population level (66) but is coordinated with expression of *LEE5* in single cells (67). EspA filaments are typically absent from bacteria at the time of intimate adherence to host cells on pedestals, and the basis of loss or disassembly of the translocon after initial attachment remains ill-defined.

It is clear that *LEE4*-encoded translocon components play pivotal roles in intestinal colonization by EHEC, as random and refined mutants are highly attenuated in cattle (39–41, 68) and in mice (69). Moreover, EspA is considered an important constituent of vaccines for control of EHEC in cattle (70, 71). In this regard it is noteworthy that EspA filaments of EPEC and EHEC are antigenically distinct (72) and that variations in EspA primary

sequences exist between serogroup O157 and non-O157 EHEC strains. Indeed, antibodies induced by immunization of cattle with Type III secreted proteins from EHEC O26:H11, O103:H2, or O111:NM failed to cross-react with EspA from EHEC O157:H7 (73), leading the authors to conclude that cross-serogroup protection due to EspA may be limited.

Although EspB is acknowledged to be a key part of the T3SS translocon, it can also be detected in the cytoplasm of infected cells (74), where it modulates cellular processes. For example, EspB binds to the host proteins α-catenin, α1-antitrypsin, and myosin that regulate the actin network leading to alterations in cell morphology (reviewed in reference 75). The myosin-binding domain of EspB inhibits the interaction of myosins with actin leading to microvillus effacement, and mutants lacking this domain lack the ability to efface enterocytes, suppress phagocytosis, or colonize the intestines of mice (76). EspB was also implicated in the ability of EHEC to suppress NF-κB activation and synthesis of proinflammatory cytokines (77); however, the extent to which this was dependent on EspB per se, or its role in delivery of other effectors, was not elucidated.

Intimin

Intimin is a 94- to 97-kDa outer membrane adhesin produced by all EHEC strains and related A/E pathogens and is encoded by the *eae* gene. Though first identified in EPEC O127:H6 (4), a homolog in EHEC O157:H7 that exhibits 83% amino acid sequence identity was subsequently reported to mediate bacterial adherence to cultured cells and intestinal colonization in gnotobiotic piglets (78, 79). Subsequent studies established that EHEC O157:H7 intimin plays a pivotal role in persistence and pathogenesis in mice, infant rabbits, neonatal calves and lambs, and adult cattle and sheep (80–84). It has also been reported to drive mucosal inflammatory responses; for example, it induces a T-helper cell type 1 (T_H1) response characterized by mucosal thickening and infiltration of CD4+ T cells during infection of mice with *C. rodentium* (85), and can augment mitogen-stimulated proliferation of spleen CD4+ T lymphocytes and cells from organized lymphoid tissues (85, 86). Intimin is a component of chimeric and multivalent subunit vaccines that control EHEC in experimental models (reviewed in reference 246), and immunization with intimin alone has been reported to confer protection against intestinal colonization by EHEC O157:H7 when delivered by live-attenuated *Salmonella* sp. to cattle and mice (87, 88) and recombinant plants to mice (82). Moreover, neonatal piglets suckling dams vaccinated with EHEC O157:H7 intimin are passively protected (89), consistent with the ability of antibodies directed against the carboxyl-terminal domain to inhibit adherence of EHEC O157:H7 to HEp-2 cells (90).

Intimin shares significant homology with invasin proteins of *Yersinia pseudotuberculosis* and *Yersinia enterocolitica*, the carboxyl termini of which bind to β1-chain integrins to facilitate bacterial invasion of eukaryotic cells. Frankel et al. demonstrated that the carboxyl-terminal 280 amino acids of intimin (Int280) from EPEC O127:H6 and EHEC O157:H7 could directly bind to HEp-2 cells (91), and the same region also mediates binding of intimin from EHEC O26:H– to host cells (92). Furthermore, expression of the carboxyl-terminal two-thirds of intimin was sufficient to restore adherence of an EHEC O157:H7 Δ*eae* mutant (93). Separate reports indicated that binding of purified intimin to eukaryotic cells could only be detected if the cells were pre-infected with EPEC or EHEC, indicating that a bacterial factor also influences intimin-mediated adherence (94, 95).

Studies to identify the host cell receptor for intimin initially focused on a 90-kDa membrane protein (Hp90) that became tyrosine phosphorylated during EPEC infection of epithelial cells and localized under adherent bacteria (96). EPEC intimin was later shown to bind to tyrosine-phosphorylated Hp90 but not to nonphosphorylated Hp90, and it was hypothesized that the bacteria signal to host cells to induce phosphorylation of a membrane protein that is subsequently bound by intimin (95). Kenny et al. later reported Hp90 to be the tyrosine-phosphorylated version of a 78-kDa EPEC O127:H6 protein that is translocated into the eukaryotic cell plasma membrane via the LEE-encoded T3SS where it acts as the receptor for intimin (51). Hp90 was thus renamed Tir (for translocated intimin receptor). Tir was independently discovered in EHEC O26:H– as EspE, an 80-kDa protein that is translocated into host cells where it appears as a 90-kDa tyrosine-phosphorylated protein associated with adherent bacteria (97). EHEC O157:H7 Tir is also delivered into host cells where it acts as a receptor for intimin; however, it is not tyrosine phosphorylated (97, 98).

The three-dimensional structures of the carboxyl-terminal domain of intimin and the Int280-Tir complex were first resolved for the EPEC O127:H6 protein by X-ray crystallography and nuclear magnetic resonance

(99–101). These studies indicate that intimin contains four distinct domains within the carboxyl-terminal 380 residues that extend from the outer membrane. Domains D1, D2, and D3 belong to the immunoglobulin superfamily and are predicted to comprise an articulated rod connected to the membrane-anchored amino-terminal portion by a flexible linker comprising two glycine residues. The terminal D4 domain is predicted to be accessible to the target cell and shares similarity with C-type lectins, a family of proteins that recognize cell surface carbohydrates. The most distal immunogloblin-like domain (D3) and D4 C-type lectin domain form a rigid superdomain that binds Tir (99), principally through contacts in a β-hairpin motif at the tip of the extracellular domain of Tir and an analogous region of the intimin D4 domain (101). This portion of Tir is critical for intimin binding (102–104). The crystal structure of the EPEC O127:H6 Int280-Tir complex reveals the intimin-binding domain of Tir to be a dimer, with the two parallel α-helices of the Tir extracellular domains aligning to form a four-helix bundle bound by two intimin molecules (101). Analysis of the crystal structure of the Tir-binding domain of EHEC O157:H7 intimin (Int188) at 2.8 Å resolution suggests that a similar conformation and contacts are adopted (105), extending earlier predictions from a yeast-2-hybrid analysis (106). Although it is unclear if this structure is formed by the native proteins in vivo or by A/E pathogens with other intimin and Tir sequences, it predicts that the proteins bind in a plane roughly parallel to the surfaces of the bacteria and host cell, consistent with the estimated gap between the host cell plasma membrane and intimately attached bacteria of just 100 Å.

Early phylogenetic analysis of *eae* genes identified distinct subtypes, designated Int-α, -β, -γ, -δ, -ε, and -θ (107–109), though further intimin subtypes are now recognized. These differ in sequence in the carboxyl-terminal cell-binding domain and are often associated with specific clonal lineages of EPEC and EHEC. For example, Int-α is associated with EPEC clone 1 serotypes O55:H6 and O127:H6, whereas Int-γ is found in EHEC clone 1 serotypes, including O157:H7 and its progenitor EPEC serotype O55:H7. As the carboxyl-terminal domain of intimin mediates binding to Tir and host cell surfaces, it was anticipated that divergence of intimin subtypes may influence the avidity and specificity of adherence. Int-γ is required for colonization of the surface and glandular epithelium of the large intestine in gnotobiotic piglets (78). However, expression of Int-α from EPEC O127:H6 in an EHEC O157:H7 Δeae mutant was reported to cause a shift in intestinal tissue tropism, with colonization of the terminal ileum as well as the surface of the large intestine in a manner similar to wild-type EPEC (110). In subsequent studies, precise chromosomal replacement of EHEC O157:H7 *eae* with EPEC O127:H6 *eae* did not alter tissue tropism in piglets (111), and the basis of the disparity is unclear.

Support for the role of intimin in mediating intestinal tissue tropism derives from in vitro organ culture studies using human and porcine intestinal explants. Adherence of EHEC O157:H7 to human intestinal mucosa and the formation of A/E lesions are restricted to follicle-associated epithelium (FAE) overlying ileal Peyer's patches (112), whereas Int-α-expressing EPEC O127:H7 can adhere to human small intestinal explants from a variety of sites (113). Expression of EPEC O127:H6 Int-α in an EHEC O157:H7 Δeae mutant without exchanging Tir results in an EHEC strain capable of adhering to explants from various intestinal sites from human (114) and pigs (115). A similar extension of tropism for human and porcine explants was observed when EHEC O157:H7 was engineered to express Int-β from *C. rodentium* (116). In reciprocal experiments EHEC O157:H7 Int-γ conferred upon an EPEC O127:H6 Δeae mutant a tropism for FAE overlying Peyer's patches (113). FAE at the recto-anal junction has been reported to be a key site of persistence of EHEC O157:H7 in cattle (117); however, while Int-γ is known to be required for efficient colonization of this site by EHEC O157:H7

(118), it is unclear if different subtypes vary in their specificity for this region of the bovine gut. Indeed, non-O157 EHEC has been reported to colonize squamous epithelium in the bovine terminal rectum independently of intimin (119), and studies have indicated that factors other than those encoded by the LEE play a role in adherence of EHEC O157:H7 to rectal squamous epithelial cells in culture (120).

The ability of intimin to bind to eukaryotic cell surfaces in the absence of Tir and influence tissue tropism suggests the existence of a cellular coreceptor(s). Sinclair and O'Brien demonstrated that the carboxyl-terminal domain of EHEC O157:H7 Int-γ binds in a specific and saturable manner to HEp-2 cells with a dissociation constant of 84 nM, consistent with the existence of a single host cell receptor (121). By affinity purification and sequencing of peptides derived from a 110-kDa intimin-binding HEp-2 cell protein, the receptor was identified as nucleolin, a protein involved in regulation of cell growth that can be expressed at the cell surface (121). Cell surface-localized nucleolin was observed to colocalize with bound purified Int-γ and with EHEC O157:H7 adhering to HEp-2 cells (121), and with EHEC O157:H7 adhering to porcine and bovine intestinal mucosa in vivo (122). Antinucleolin antibodies partially inhibit EHEC O157:H7 adherence to cultured cells (121). Interestingly, surface expression of nucleolin is enhanced by Stx2, and it is believed that this may partially explain the ability of Shiga toxin to promote EHEC O157:H7 adherence to Hep-2 cells and intestinal colonization in mice (122). Intimin subtypes α, β, and γ bind to nucleolin with equal affinity, indicating that the distribution of nucleolin along the gastrointestinal tract is unlikely to determine tissue tropism of EHEC and EPEC (123). All three intimin subtypes bind to nucleolin with a lower avidity than to Tir. Furthermore, binding of intimin α, β, or γ to Tir in vitro blocks the interaction between intimin and nucleolin (123), suggesting that Tir and nucleolin compete for intimin binding.

Sites of adherence of EHEC O157:H7 to porcine and bovine tissue are also enriched in β1-chain integrins (122). Consistent with a role for such factors as coreceptors, EPEC O127:H6 intimin-α can bind to β1-chain integrins expressed on the surface of human lymphocytes or in enzyme-linked immunosorbent assays (124). However, β1-chain integrins were reported to be dispensable for intimin-mediated adherence as inactivation of the β1-chain integrin gene or the addition of antagonists of integrin function, such as EDTA or anti-β1 chain antibody, did not affect EPEC adherence or pedestal formation (94). Conversely, Muza-Moons et al. suggested that disruption of tight junctions by EPEC in an EspF-dependent manner leads to redistribution of β1-chain integrins from the basolateral to apical surface to promote bacterial adherence (125). Studies using isogenic single and double eae and tir mutants of EHEC O157:H7 in calves have indicated that intimin-Tir interactions are more significant than intimin-coreceptor interactions during colonization of the bovine intestines, as mutation of eae in a tir mutant did not exert further attenuation (84).

Tir

In addition to its role as the translocated receptor for intimin (above), Tir is required to activate actin assembly and recruit cytoskeletal proteins at the site of bacterial adherence. Intimin-Tir interactions are required to cluster Tir to initiate this process (126). The mechanism of insertion of Tir into the apical plasma membrane of enterocytes is ill-defined, but phosphorylated intermediates of EPEC O127:H6 Tir can be detected in the cytoplasm and a delay exists between Tir injection and intimin-Tir interaction (102, 127), implying that it may first be found in the cytoplasm prior to insertion. Phosphorylation of tyrosine residue 474 in a 12-amino-acid motif in the carboxyl-terminal domain of EPEC O127:H7 Tir is required to trigger local actin assembly through recruitment of the host cell adaptor protein Nck (102, 126, 128, 129). Such phosphorylation requires redundant cellular tyrosine kinases, including c-Fyn, Abl, Arg, and Etk, and was proposed to allow Nck to bind a proline-rich domain of the neural Wiskott-Aldrich syndrome protein (N-WASP), which in turn stimulates the actin-nucleating activity of the cellular Arp2/3 complex (102, 129, 130). WASP-interacting protein (WIP), which binds Nck and a WASP-homology 1 (WH1) domain of N-WASP, may also act as linker between Nck and N-WASP, and such interactions may act in synergy with Nck interactions with the N-WASP proline-rich domain as interference in either interaction does not abolish pedestal formation (131). Recent studies have indicated that WIP is not essential for pedestal formation by typical EPEC (132).

N-WASP is also required for pedestal formation by EHEC O157:H7 (133); however, in contrast to EPEC O127:H6, the Tir of EHEC O157:H7 lacks the Y474 residue and is not tyrosine phosphorylated upon entry into host cells (98, 134). Instead, a conserved Asn-Pro-Tyr (NPY_{458}) motif in a 12-amino-acid motif in the cytoplasmic domain of EHEC O157:H7 Tir is required for Nck-independent actin assembly (135–136) through recruitment of insulin receptor tyrosine kinase substrate (IRTKS) (137) and insulin receptor substrate p53 (IRSp53) (138). This motif is also found in EPEC O127:H7 Tir as NPY_{454} and can initiate actin assembly independently of Nck, albeit inefficiently (139). In EHEC O157:H7 the pathway of actin assembly mediated by recruitment of IRTKS/IRSp53 to Tir is amplified by another Type III secreted effector protein termed $EspF_U$ or TccP (<u>T</u>ir <u>c</u>ytoskeleton <u>c</u>oupling <u>p</u>rotein), which contains 47-amino-acid proline-rich repeats that bind to IRTKS/IRSp53 (138) and the GTPase-binding domain of N-WASP (140, 141). Structural studies have indicated that the proline-rich repeats in $EspF_U$/TccP compete with an auto-inhibitory domain of N-WASP for binding of its GTPase-binding domain, thereby relieving N-WASP auto-inhibition (142, 143). The number of such

repeats varies in natural isolates and is associated with the efficiency of actin assembly (144, 145). Remarkably, the Arp2/3 complex is recruited to actin pedestals formed in a Tir- and $EspF_U$/TccP-dependent process even in the absence of N-WASP, implying that redundant as yet unknown pathways are used to ensure actin assembly under adherent bacteria (146). Indeed, an EPEC O125:H6 strain unable to use the Nck or IRTKS/IRSp53 pathways was nevertheless able to form typical A/E lesions on human intestinal explants cultured ex vivo (147).

It is noteworthy that differences in the relative importance of Nck-dependent and Nck-independent pathways have been detected between immortalized epithelial lines in culture and human intestinal explants cultured ex vivo (148) as well as between infection of cultured cells and animals (149). During *C. rodentium* infection in mice or EHEC infection of human intestinal biopsies, IRTKS, but not IRSp53, was recruited to sites of bacterial attachment sites, and N-WASP was recruited to pedestals even in the absence of tyrosine residues required for Nck or IRTKS binding (149). Thus, despite the key role of the Tir:Nck and Tir:IRTKS/IRSp53 pathways in actin polymerization in cultured epithelial cells, they appear not to be essential for N-WASP recruitment or A/E lesion formation on explants or in mice. *C. rodentium* mutants with Y451 or Y471 substitutions were outcompeted by the wild-type strain during mixed infection of mice (149), suggesting that such pathways do contribute to persistence. Further, while an EHEC O157:H7 $espF_U$/tccP mutant exhibited no obvious defect in A/E lesion formation or early intestinal colonization of calves (84), such mutants appear impaired in the efficiency of formation and tissue distribution of A/E lesions in gnotobiotic piglets and infant rabbits (150).

The Tir:Nck and Tir:IRTKS/IRSp53 pathways appear conserved in isolates of EPEC belonging to lineage 1 and EHEC O157:H7, respectively (151). Typical isolates of EHEC O157:H7 carry $espF_U$/tccP on prophage

CP-933U/Sp14 but also harbor a related pseudogene (*z1385/ecs2715*) on prophage CP-933 M/Sp4 (*tccP2/espF$_M$*). Sorbitol-fermenting EHEC O157:H– strains contain an intact copy of *tccP2/espF$_M$*, which can restore function to an EspF$_U$/tccP mutant and encodes a protein with near-identical proline-rich repeats but a distinct 80-amino acid amino terminus, which shares 42.7% identity with the amino terminus of EspF$_U$/TccP (151). Interestingly, non-O157 EHEC isolates in general possess a Tir with a conserved Y474 residue for Nck recruitment and an intact copy of *tccP2/espF$_M$* (151), suggesting that they are able to use both the Tir:Nck and Tir:IRTKS/IRSp53 pathways. The same applies to a high proportion of EPEC lineage 2 strains (152), indicating that varied and independent solutions to form pedestals

have evolved. The repertoire of Tir, EspF$_U$/TccP, and EspF$_M$/TccP2 pathways may explain the recruitment of distinct sets of cytoskeletal proteins to the pedestals formed by different A/E pathogens as well as distinct pedestal morphologies (Fig. 1A and B). Modulation of the cytoskeleton by A/E pathogens is summarized in Fig. 5.

Tir plays a key role in persistence and pathogenesis of EHEC, with *tir* mutants of EHEC O157:H7 being attenuated in infant rabbits (83), calves (84), and colonization of the terminal rectum of steers (118). Moreover, Tir is an important component of vaccines for control of EHEC O157:H7 in cattle (70, 71). Despite intensive research into the molecular mechanisms underlying actin assembly during A/E lesion formation, the advantages to the

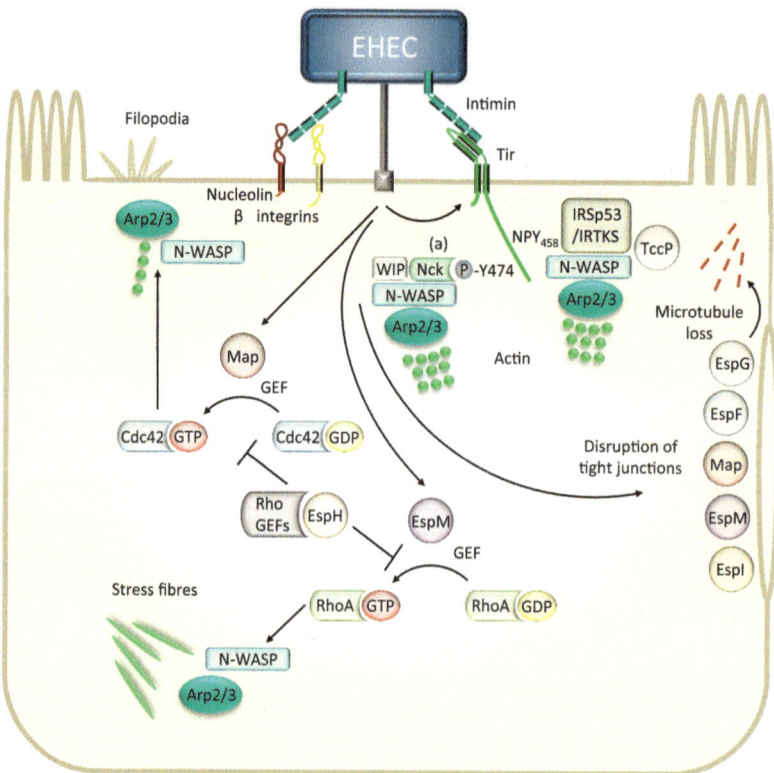

FIGURE 5 Diagram summarizing the activities of a subset of EHEC Type III secreted proteins on the cytoskeleton. Note (a), the Tir:Nck pathway dependent on phosphorylation of the residue equivalent to tyrosine 474 of EPEC O127:H6 Tir operates in some non-O157 EHEC but not prototype *E. coli* O157:H7 strains. Effectors are represented by circles. Adapted from reference 164. doi:10.1128/microbiolspec.EHEC-0007-2013.f5

pathogen of forming pedestals remain open to question. It has been reported that N-WASP is required for efficient translocation of Type III secreted effectors (146), and a role in stabilizing the translocon and/or clustering Tir for interactions with intimin is plausible. In turn, stabilization of intimin-Tir interactions may help prevent bacterial detachment from cells under flow. Pedestals have also been reported to be mobile on the cell surface (153), and a role in spread between enterocytes and evasion of phagocytosis by remodeling of the cell surface cannot be excluded.

Other Type III Secreted Effectors

In addition to EspB and Tir, five other effector proteins are encoded within the LEE of the prototype EHEC O157:H7 strains EDL933 and RIMD 0509952 (EspF, EspG, EspH, EspZ, and Map) (Fig. 2). As discussed above, the boundaries of the LEE are less defined in some EHEC strains, with additional effector proteins encoded in the flanking regions. Upon sequencing of the complete genome sequences of the above strains (154, 155), more than 60 candidate Type III secreted proteins were identified on the basis of homology with effectors in other pathogens, of which 49 were judged to be potentially functional (25). The T3SS-dependent secretion of 31 candidate effectors of EHEC O157:H7 was confirmed by analysis of the secreted proteome of an EHEC O157:H7 ΔsepL strain with an isogenic ΔsepL ΔescR double mutant incapable of Type III secretion (25). SepL regulates the hierarchy of secretion in EHEC, and mutation of sepL causes the secretion of effectors in preference to translocon components (156). Subsequent studies have indicated that the ability of the carboxyl-terminal portion of SepL to bind to Tir is sufficient to delay the export of effectors while the EspABD translocon components are secreted (157).

To confirm that candidate effectors identified by bioinformatics or proteomic analysis are translocated into eukaryotic cells in a T3SS-dependent manner, assays to detect the

activity of reporter fusions have been used. Commonly, candidate non-LEE-encoded effectors have been fused to TEM-1 β-lactamase, enabling translocation to be detected by a shift in the wavelength of fluorescence emitted by target cells loaded with a fluorescent β-lactamase substrate (CCF2/AM) (158). Fusions to epitopes (e.g., FLAG) or *Bordetella pertussis* adenylate cyclase (CyaA) have also been employed for this purpose (25).

In total, 39 Type III secreted effector proteins were experimentally validated in EHEC O157:H7 (25). It is evident that lambdoid bacteriophages have played a key role in the lateral transfer of such effectors, both in EHEC O157:H7 (25) and non-O157 EHEC (18). Prophage-encoded effector genes are frequently carried downstream of tail fiber genes and exhibit a low %GC content relative to the prophage backbone (18, 25). Analysis of the content and insertion site of effector-encoding prophages in non-O157 EHEC suggests distinct evolutionary histories (18). Although many effector-encoding prophages identified by genome sequencing were initially considered defective, it is now evident that many are inducible and can release their DNA from EHEC O157:H7, possibly owing to recombination and other interprophage interactions (159). Many effectors encoded outside the LEE are designated Nle (non-LEE-encoded) effectors. Effector proteins of EHEC O157:H7 strain RIMD 0509952 and their activities, where known or inferred from studies with homologs in other A/E pathogens, are listed in Table 1. A notable absence from the genomes of the sequenced EHEC O157:H7 strains EDL933 and RIMD 0509952 is the non-LEE-encoded effector cycle-inhibiting factor (Cif), which was first described in an EHEC O103:H2 strain (160) and is found in many non-O157 EHEC. Cif arrests the cell cycle at G1/S and G2/M phases and inhibits other cellular processes by deamidation of ubiquitin or the ubiquitin-like protein NEDD8 that regulates cullin-RING-ubiquitin ligase complexes (reviewed in reference 161).

It is clear that effectors are often multifunctional. Indeed, EspF has been described

TABLE 1 Type III secreted effector proteins of EHEC O157:H7 and their activities, where known or inferred from homologs in other A/E pathogens[a]

Effector	No. of alleles[b]	Subcellular localization	Interacting partners or substrates	Homologs	Function[c]
EspB	1	Plasma membrane, cytosol	α1-Antitrypsin, α-catenin, myosin-1c	*Yersinia* YopD	Pore-forming translocon component; microvillus effacement; antiphagocytosis; disruption of adherens junctions
EspF	1	Plasma membrane, cytosol, mitochondria, nucleus	14-3-3zeta, ABCF2, actin, Arp2, CK18, N-WASP, profiling, SNX9, ZO-1/-2	EspF$_U$ (TccP)	Disrupts mitochondrial function, nucleolus, tight junctions and intermediate filaments. Inactivates NHE3 and SGLT-1; activates SNX9 and N-WASP; inhibits PI3K-dependent phagocytosis and induces apoptosis
EspG	1	Cytosol, Golgi	Arf1/6, PAK1/2/3, tubulin, GM130, Rab1	*Shigella* VirA	Disrupts microtubules, tight junctions and paracellular permeability; sequesters ADP-ribosylating factor (Arf) to modulate GTPase signaling; stimulates p21-activated kinases (PAKs) to inhibit endomembrane trafficking; binds GM130 and inactivates Rab1 to disrupt Golgi structure and protein secretion; induces calpain protease and necrosis in absence of Tir
EspH	1	Plasma membrane, pedestal	DH-PH Rho guanine nucleotide exchange factors (GEFs)	None known	Inhibits RhoGTPase signaling and FCγR-mediated phagocytosis; causes cell detachment via disassembly of focal adhesions and remodels brush border; promotes actin nucleation and pedestal elongation by recruiting N-WASP and Arp2/3 via WIP
EspZ	1	Plasma membrane, mitochondria	CD98, translocase of inner mitochondrial membrane 17b (TIM17b)	None known	Inhibits apoptosis and cytotoxicity; regulates Type III secretion via "translocation stop" activity
Map	1	Mitochondria	Na$^+$/H$^+$ exchanger regulatory factor (NHERF)-1 and -2, Cdc42	*Salmonella* SopE/SopE2, *Shigella* IpgB	GEF for Cdc42 that induces transient filopodia formation; causes mitochondrial dysfunction; inactivates sodium-D-glucose cotransporter (SGLT-1) in a manner that may result in net fluid secretion; disrupts tight junctions, causing loss of epithelial barrier integrity

(Continued on next page)

TABLE 1 *(Continued)*

Effector	No. of alleles[b]	Subcellular localization	Interacting partners or substrates	Homologs	Function[c]
Tir	1	Plasma membrane	Intimin, 14-3-3tau, α-actinin, cortactin, CK18, IQGAP1, IRTKS, IRSp53, Nck, PI3K, SHP-1, Talin, Vinculin	None known	Receptor for intimin, actin pedestal formation, regulates activities of Map and EspG
EspI/NleA	1	Plasma membrane, Golgi	Syntrophin, Sec23/24, MALS3, PDZK11, SNX27, TCOF1, NHERF-1 and -2, MAGI-3, SAP97 and -102 PSD-95	None known	Inhibits protein export from the endoplasmic reticulum by disrupting COPII function; disrupts tight junctions
EspJ	1	Cytosol, mitochondria	Unknown	*Pseudomonas* HopF	Inhibits phagocytosis mediated by complement receptor 3- and Fcγ-receptor
EspK	1	Cytosol	Unknown	*Salmonella* GogB	Unknown, influences intestinal colonization of calves by EHEC O157:H7
EspL	4 (1)	Pedestal	Annexin 2	*Shigella* OspD	Promotes F-actin bundling activity of annexin 2
EspM	2	Cytosol	RhoA	*Salmonella* SopE/SopE2, *Shigella* IpgB	GEF for RhoA that induces actin stress fibers and modulates pedestal formation; disrupts tight junctions and monolayer integrity
EspN	1	Unknown	Unknown	*E. coli* CNF1, *Salmonella* Arizonae SARI_01330 SARI _01464	Unknown
EspO	2	Unknown	Integrin-linked kinase (ILK)	*Shigella* OspE	Regulates EspM2-mediated RhoA activity and stabilizes focal adhesions to block cell detachment
EspR	4 (1)	Unknown	Unknown	*Shigella flexneri* SF1757	Unknown
EspV	1 (1)	Cytosol	Unknown	*Pseudomonas* AvrA	Alters cell morphology
EspW	1	Unknown	Unknown	*Pseudomonas* HopPmaA, HopW1	Unknown
EspX	7 (1)	Unknown	Unknown	*Shigella sonnei* SSON_0027, *Shigella dysenteriae* Sd1012_0237	Unknown
EspY	5 (1)	Unknown	Unknown	*Salmonella* SopD	Unknown

(Continued on next page)

TABLE 1 Type III secreted effector proteins of EHEC O157:H7 and their activities, where known or inferred from homologs in other A/E pathogens[a] *(Continued)*

Effector	No. of alleles[b]	Subcellular localization	Interacting partners or substrates	Homologs	Function[c]
NleB	3 (1)	Cytosol	Glyceraldehyde 3-phosphate dehydrogenase (GAPDH)	*Salmonella* SseK	Inhibits TNFα-induced activation of NF-κB and proinflammatory responses by transferring N-acetyl-D-glucosamine to GAPDH, thereby disrupting TRAF2-GAPDH interaction to suppress TRAF2 polyubiquitination and NF-κB activation
NleC	1	Cytosol, nucleus	p65 (RelA), p50, c-Rel, IκB, acetyltransferase p300	*Photobacterium* AIP56	Zinc metalloprotease that cleaves p65 (RelA), p50, c-Rel and IκB to inhibit NF-κB activation
NleD	1	Cytosol	c-Jun N-terminal kinase (JNK), p38 mitogen-activated protein kinase (MAPK)	*Pseudomonas* HopAP1, HopH1	Zinc metalloprotease that cleaves JNK and MAPK to inhibit induction of apoptosis and proinflammatory responses
NleE	1	Cytosol	TAB2 and -3	*Shigella* OspZ	Blocks IκB degradation to inhibit activation of NF-κB, proinflammatory responses and neutrophil transepithelial migration. Uses S-adenosyl-L-methionine-dependent methyltransferase activity to modify Npl4 zinc finger domains in TAB2 and TAB3, which regulate NF-κB signaling
NleF	1	Unknown	Caspase-4, -8, and -9, Tmp21	None known	Inhibitor of caspase activation and apoptosis; binds the COPI-vesicle receptor Tmp21 involved in Golgi function and slows protein secretion
NleG/NleI	14 (5)	Cytosol	UBE2D2	*Salmonella* Typhi STY1076	E3 ubiquitin ligases analogous to eukaryotic RING finger and U-box enzymes
NleH	2	Plasma membrane, cytosol, endoplasmic reticulum	Bax-inhibitor 1, NHERF2, ribosomal protein S3 (RPS3)	*Shigella* OspG	Binds Bax-inhibitor 1 to block apoptosis; sequesters RPS3 to inhibit NF-κB signaling
TccP/ EspF_U	2 (1)	Pedestal	N-WASP, IRSp53, IRTKS, cortactin	EspF	Relieves auto-inhibition of N-WASP to stimulate the Arp2/3 complex to polymerize actin and form pedestals

[a]Adapted from reference 164.

[b]Allele number is shown for the Sakai outbreak strain of EHEC O157:H7 (RIMD 0509952). Values in parentheses denote the number of predicted pseudogenes.

[c]References in support of functions may be found in the text and reference 164.

as the "Swiss army knife" of A/E pathogens owing to its effect on epithelial barrier function, apoptosis, mitochondrial function, cytoskeletal components, microvillus effacement, and other processes (reviewed in reference 162). It is also evident that effectors may act in redundant, synergistic, and antagonistic ways. As a consequence, mutation of effector genes does not produce attenuation at the same level as defects affecting the translocon or needle complex, which result in loss of secretion all effectors. The LEE-encoded effectors EspF, EspG, EspH, and Map have been reported to be required for full colonization of the intestines of infant rabbits by EHEC O157:H7, albeit that the effect was modest and tissue-specific in some cases (42). Moreover, mutations affecting 29 or the 39 effectors of EHEC O157:H7 strain EDL933 were detected by screening pools of random transposon mutants in calves (39, 41); however, the extent of negative selection was often modest, and in the case of *map* and *nleD* mutants, could not be reproduced when the mutants were screened in isolation (39, 163). Functions of Type III secreted proteins of EHEC are considered below by functional categories and are reviewed in detail elsewhere (164).

MODULATION OF INNATE IMMUNITY

In recent years it has become clear that EHEC and other A/E pathogens inject multiple non-LEE-encoded effectors to suppress pro-inflammatory responses, converging on inhibition of the activation of host transcription factors including NF-κB and activator protein-1. NF-κB comprises a dimer of p50 and p65 subunits that are retained in the cytoplasm by a family of inhibitory proteins (IκB). Activation of NF-κB requires phosphorylation of IκB by the IκB kinase complex, which is subsequently ubiquitinated and degraded by the proteasome, allowing p50/p65 to translocate to the nucleus to bind consensus sequences upstream of genes associated with the inflammatory response. Remarkably, EHEC and

other A/E pathogens disrupt these processes at multiple levels, as summarized in Fig. 6. For example, NleB suppresses tumor necrosis factor alpha-mediated NF-κB activation (165, 166) by transferring *N*-acetyl-D-glucosamine to the glycolysis enzyme glyceraldehyde-3-phosphate dehydrogenase, disrupting its ability to interact with the tumor necrosis factor receptor-associated factor 2 (TRAF2), which is required for NF-κB activation (167). In EPEC it appears that Tir also impairs TRAF function by interacting with the TRAF2 adaptor protein and inducing its proteasome-independent degradation (168). It has also been reported that EPEC Tir interferes with signaling via Toll-like receptors by binding to the host cell tyrosine phosphatase SHP-1 in a manner dependent on phosphorylated immunoreceptor tyrosine-based inhibition motifs in Tir (169). The association of Tir with SHP-1 facilitates the recruitment of SHP-1 to the adaptor TRAF6, thereby inhibiting the ubiquitination of TRAF6 and the subsequent induction of proinflammatory cytokines and intestinal immunity (169).

At another level, NleC inactivates the NF-κB subunits p50 and p65 by means of a zinc-dependent endopeptidase activity (170–173). Another zinc metalloprotease (NleD) degrades key kinases (c-Jun N-terminal kinase [JNK] and mitogen-activated protein kinase p38) involved in activation of activator protein-1, which like NF-κB controls expression of genes involved in the inflammatory response (170). A further effector, NleE, interferes with degradation of IκB to prevent translocation of NF-κB to the nucleus (165, 166, 174) and uses *S*-adenosyl-L-methionine-dependent methyltransferase activity to modify Npl4 zinc finger domains in the cellular proteins TAB2 and TAB3, which regulate NF-κB signaling (175). NleH also acts in concert with these effectors, binding to a subunit of NF-κB transcriptional complexes (ribosomal protein S3), preventing its phosphorylation by the kinase IκB kinase complex-β and translocation to the nucleus (176, 177). NleH1 and NleH2 also suppress IκB degradation by interfering with the ubiqitination

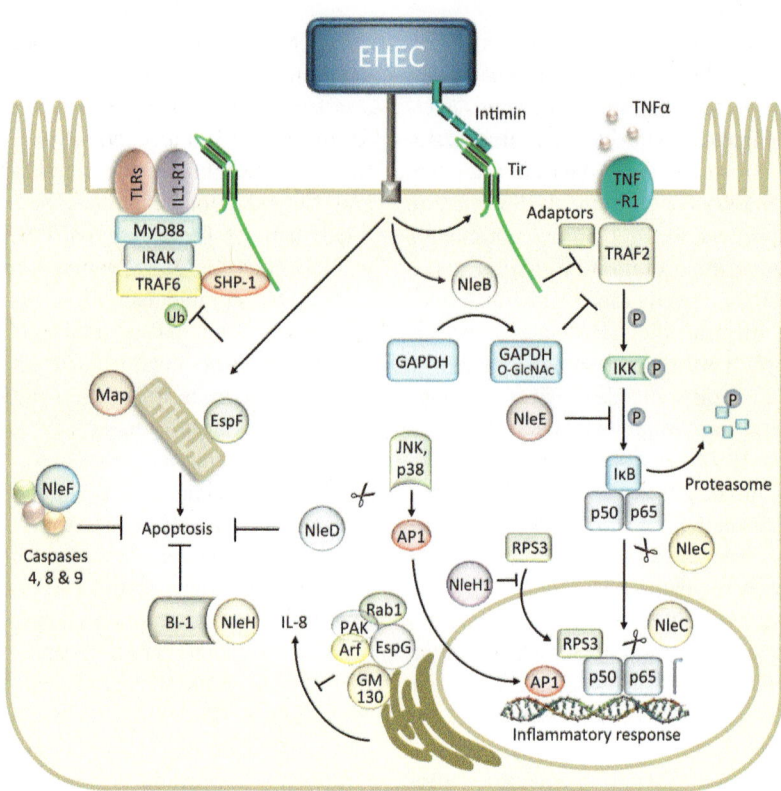

FIGURE 6 Diagram summarizing the activities of a subset of EHEC Type III secreted proteins on signaling pathways leading to inflammation and apoptosis. Effectors are represented by circles. Adapted from reference 164. doi:10.1128/microbiolspec.EHEC-0007-2013.f6

of phosphorylated IκB, thereby aiding retention of NF-κB in the cytoplasm (178). The functional significance of inhibition of pro-inflammatory responses through these pathways is evidenced by the fact that mutants lacking the effectors above typically elicit elevated proinflammatory cytokine responses and are attenuated in animal models, albeit that combination of mutations produces stronger phenotypes owing to functional redundancy. One should also caution that subtle differences may exist in the activities of homologs between EHEC and other A/E pathogens, as well as between members of the same family of effectors, as recently reported for NleH1 and NleH2 of EHEC O157:H7 (179).

Another level at which Type III secreted effectors act in concert to repel host innate defenses is the inhibition of phagocytosis.

EHEC and EPEC are able to disable phagocytosis by macrophages via the FCγ receptor (FCγR) and complement receptor (CR3), which, respectively, mediate uptake of particles opsonised with immunoglobulin G or complement fragment C3bi (180). Such inhibition is partly dependent on EspJ (180), though its mode of action is ill-defined at the time of writing. The myosin-binding effector EspB also plays a role in suppressing phagocytosis of EPEC, possibly by interfering with myosin-actin interactions required to form phagocytic cups and force phagosome closure during uptake (76). EspB may act in concert with EspH to inhibit remodeling of the cytoskeleton during phagocytosis, as the ability of EspH to inactivate RhoGTPase signaling has been associated with blocking of uptake of EPEC O127:H6 by macrophages (181).

In addition to the effectors above, EspF has been described to play a key role in preventing phagocytosis by macrophages and translocation across microfold-cell-like cells in culture (180, 182–184). In the case of EPEC O127:H6 EspF, this involves inhibition of PI3 kinase-dependent pathways (183, 185), but not FCγR- or CR3-dependent opsonophagocytosis (181), suggesting that it acts at a distinct level to EspJ. Antiphagocytic activity of EPEC O127:H6 EspF has been mapped to the amino-terminal 101 residues (183), which contain binding sites for sorting nexin 9 (SNX9), actin, and profilin (186–188), and it is possible that it acts in concert with EspB and EspH to disrupt the cytoskeletal rearrangements required to form and close phagosomes. Paradoxically, interactions between EspF and SNX9 have been reported to remodel the apical plasma membrane and facilitate EPEC invasion in epithelial cells (189), indicating the potential for cell type-specific effects. Moreover, distinct *espF* alleles may confer distinct activities as comparative studies found EspF of EHEC O157:H7 to be far less effective in inhibiting macrophage phagocytosis than the variant in EPEC O127:H6, possibly owing to weaker interactions with SNX9 (184). Conversely, EspF from EHEC O157:H7 was more effective than the EPEC O127:H6 EspF in preventing uptake by primary epithelial cells from the site of EHEC O157:H7 persistence in the bovine terminal rectum (184).

Modulation of the Cytoskeleton

The role of Tir and adaptor proteins in formation of actin pedestals is well established (above); however, it is becoming clear that other effectors are injected that act on the host cell cytoskeleton (summarized in Fig. 5). For example, mitochondrial-associated protein (Map) and EspM are WxxxE-family effectors homologous to the *Shigella* IpgB and *Salmonella* SopE proteins that act as guanine nucleotide exchange factors (GEFs) to activate GTPases that regulate the actin network. Map activates Cdc42 to trigger transient formation of filopodia at attachment sites (127, 190, 191), whereas EspM activates RhoA to form actin stress fibers (192, 193). A further WxxxE effector (EspT) has a putative GEF activity for Cdc42 and Rac1 to trigger membrane ruffling, and filopodia formation has been found in a minority of EPEC strains, but so far not in EHEC (194). The relevance of the GEF activity of Map and EspM in vivo is not fully understood, but recent research with *Salmonella* sp. has indicated that activation of RhoGTPases by SopE triggers the NOD1 signaling pathway leading to inflammation, indicating that their activity may extend beyond changes to cell architecture (195). Interestingly, A/E pathogens also inject EspH, which has an opposing activity to Map, EspM, and EspT. EspH inactivates cellular Dbl-homology and pleckstrin-homology (DH-PH) Rho family GEFs by competitively binding to tandem DH-PH domains, resulting in inhibition of activation of RhoA (181). The bacterial GEFs Map, EspM, and EspT lack sequence homology with DH-PH domains, and recent studies have shown that EspH does not directly interfere with their activity (196). Rather, EHEC and EPEC appear to use EspH to inactivate endogenous RhoGEFs while injecting their own GEFs to modulate Rho GTPase signaling to their advantage (196). The interplay between these effectors is such that phenotypes detected by expression of effectors in isolation are to be interpreted with caution. For example, expression of EspH alone causes disassembly of focal adhesions, cell detachment, caspase-3 activation, and cytotoxicity, but its activity is normally offset by the opposing activity of EspM2 and, in EPEC, EspT (196). Overexpression of EspH has been reported to result in elongation of actin pedestals, and conversely, *espH* mutation causes them to shorten (197). It is now understood that EspH plays a role in recruitment of N-WASP and the Arp2/3 complex to assemble actin at sites of bacterial attachment (198). It is possible this involves recruitment of the WASP-interacting protein WIP, as the WIP-binding domain of N-WASP and carboxyl terminus of Tir are required for the effect (198). A further effector reported to subvert actin dynamics is EspV (199). Ectopic expression of EspV causes nuclear condensation, cell rounding, and formation of dendrite-like projections; however, it is relatively rarely found in EHEC and its mode of action is unknown.

Actin is not the only component of the cytoskeleton, and evidence is growing that effectors act on other structures involved in cell archietcture. For example, EspG and EspG2 interact with tubulin, deplete microtubules at sites of attachment, and disrupt the integrity of tight junctions and epithelial paracellular permeability (200–204). In a further example of the multifunctional nature of effectors, recent work has also established that EspG also interferes with protein secretion, including release of interleukin-8, by disrupting endoplasmic reticulum to Golgi transport via interactions with ADP-ribosylation factors, p21-activated kinases, Rab1 GTPase, and Golgi matrix protein GM130 (205–208). Intermediate filaments are also targets for effectors of A/E pathogens, and evidence exists that Tir recruits the cytokeratins CK8 and CK18 to pedestals to influence actin accretion (209) in a manner associated with direct binding of Tir to CK18 (209) and 14-3-3tau (210), which acts as an adaptor protein of CK18. Depletion of either CK18 or 14-3-3tau impaired pedestal formation (209, 210), implying a significant role. Spectrin, which polymerizes under the plasma membrane to form a scaffold important for cell shape and membrane integrity, is also recruited to EPEC O127:H6 and EHEC O157:H7 pedestals, and siRNA-mediated knockdown of spectrin and spectrin-associated proteins impaired pedestal formation (211).

Modulation of Apoptosis

Synergistic and antagonist interactions of Type III secreted effectors influence host cell survival during infection by A/E pathogens, as summarized in Fig. 6. For example, EspF, Map, and Cif induce apoptosis whereas NleD, NleF, NleH1, and NleH2 appear to counteract the process. EspF triggers apoptosis by damaging mitochondria and reducing the levels or function of Abcf2, which suppresses caspase 3 activation (212–216). Map also targets mitochondria, disrupting their membrane potential (127, 217), and may act in concert with EspF to trigger cell death via mitochondrial damage. In EPEC and some EHEC, Cif induces a delayed form of apoptosis characterized by activation of caspases, accumulation of cleaved caspase-3, and exposure of the phosphatidylserine on the cell surface (218). The basis of this effect is ill-defined but may be a consequence of Cif-mediated arrest of the cell cycle and subversion of the ubiquitin-dependent degradation pathway. The relevance of such events during infection is open to question. Indeed, one study reported a decrease in the number of apoptotic cells in the ileum and ileal Peyer's patches of rabbits following infection with rabbit enteropathogenic *E. coli* (219).

NleH effectors block the induction of multiple apoptotic pathways, including those stimulated by staurosporine, brefeldin A, tunicamycin (220), and *Clostridium difficile* toxin B (221). This appears to be due to the ability of NleH1 and NleH2 to interact with the anti-apoptotic protein Bax inhibitor-1 (BI-1), as knockdown of BI-1 abolished their anti-apoptotic activity (220). NleH was also observed to inhibit cleavage of the procaspase-3 at sites of attachment of *C. rodentium* to the murine gut (220), implying that it would prevent infected cells from entering apoptosis in a caspase-3-dependent manner. NleH proteins have been reported to bind to Na(+)/H(+) exchanger regulatory factor 2 (NHERF2), and this may limit its ability to block apoptosis, as overexpression of NHERF2 appears to sequester NleH and impede its anti-apoptotic activity (222). In recent studies NleF has also been reported to prevent epithelial cell apoptosis in a manner associated with direct binding to caspase-4, -8, and -9 (223). NleD-mediated cleavage of JNK may also help reduce apoptosis as JNK stimulates this process.

Induction of Diarrhea

It is evident that the LEE-encoded T3SS and selected effectors play key roles in the induction of enteritis and fluid loss during infection of animals; however, the precise mechanisms involved in EHEC-induced diarrhea are not well understood. It is likely that a combination of factors lead to net fluid accumulation in the gut lumen, including a loss of adsorptive capacity owing to microvillus effacement and enterocyte extrusion, disruption of tight junctions, inhibition of water adsorption channels (aquaporins), and dysregulation of ion and glucose transporters (reviewed in reference 224). EspB has been reported to contribute to loss of microvilli (76), and EspF, EspG, Map, EspI/NleA, and EspM disrupt tight junctions and monolayer integrity (201, 203, 225–227). EspF has been reported to downregulate the activity of Na(+)/H(+) exchanger 3 (228), which is the main Na(+)-absorbing isoform in the mammalian small intestine. It is not yet clear how binding of the Na(+)/H(+) exchanger regulatory factor 2 by Map, EspI, and NleH affects the activity of such ion exchangers. It is evident that EPEC O127:H6 EspF, Map, Tir, and intimin act cooperatively to inactivate the sodium-D-glucose cotransporter (SGLT-1), which plays a key role in the uptake of fluid from the lumen in the normal intestine (229). However, it is not clear if the same applies during EHEC infection, and a lack of tractable models of EHEC-induced diarrhea has hindered our understanding of the role of specific effectors in the process.

LifA/Efa1, Z4332 and ToxB

Quantitative analysis of the secreted proteome of EPEC O127:H7 by stable isotope labeling with amino acids in cell culture-based mass spectrometry recently revealed that LifA is a substrate for secretion by the LEE-encoded T3SS (22). LifA (lymphostatin) is a 366-kDa protein that inhibits mitogen-activated lymphocyte proliferation and proinflammatory cytokine synthesis (230) and is nearly identical to an adhesin of EHEC O111:H– termed EHEC factor for adherence 1 (Efa1) (231). LifA/Efa1 is highly conserved in non-O157 EHEC and plays a significant role in intestinal colonization of calves by EHEC strains of serotypes O5:H–, O26:H–, and O111:H– (232, 233). In each of these strains, mutations affecting *lifA/efa1* impair adherence, but in EHEC O5:H– and O111:H– this could not be separated from a posttranscriptional reduction in the expression and secretion of EspA and Tir (233). LifA/Efa1 of an EHEC O26:H– strain influences adherence independently of effects on EspA (232), and the basis of pleotropic effects of *lifA/efa1* mutation in non-O157 EHEC is unknown.

It is assumed that lymphostatin activity does not require Type III secretion as it is detectable in cell-free lysates and can be transferred to laboratory-adapted *E. coli* K-12 lacking a T3SS (230). Moreover, its ability to promote adherence independently of effects on Type III secretion and in a manner sensitive to antibody indicates that a portion of the protein may be surface localized (232, 234). It is evident, however, that some Type III secreted proteins can be secreted from *E. coli* by alternative pathways (e.g., the EPEC serine protease EspC) (235). Predicted glycosyltransferase and cysteine protease motifs of LifA/Efa1 were dispensable for intestinal colonization of calves by EHEC O26:H– (232), and an earlier report claiming roles for such motifs during *C. rodentium* infection of mice (236) can be explained by the fact that the motif deletions caused truncation.

Most EHEC O157:H7 strains lack *lifA/efa1* but contain a truncated variant (*z4332*) and a full-length pO157-encoded homolog (*toxB*). An extremely high frequency of carriage of *lifA/efa1*, *z4332*, and/or *toxB* exists in LEE-positive strains, with physical linkage of such genes to the LEE in many cases. Deng et al. recently confirmed that Z4332 and ToxB are also effectors of the T3SS (22). Mutation of either factor has been reported to impair Type III secretion at a posttranscriptional level

(237, 238) but does not affect a lymphostatin-like activity of EHEC O157:H7 (239). Intestinal colonization of calves or lambs by a Stx-negative *E. coli* O157:H7 strain was not affected by single or double *z4332* or *toxB* mutations (237). As with other Type III secreted effectors, it is likely LifA/Efa1 and ToxB are multifunctional and their location and activities inside host cells require study.

FUTURE PERSPECTIVES

Research on the LEE-encoded T3SS and the repertoire and functions of secreted effectors has expanded enormously. Today, it represents one of the most vibrant areas of study on *E. coli* O157:H7 and other EHEC strains, yet significant challenges remain. It is already clear that analysis of the activities of effectors in isolation cannot account for synergistic or antagonist activities with other effectors, or reproduce the timing and intracellular concentration of effectors delivered in vivo. Moreover, a relatively small number of prototype strains are used, and variation in the repertoire, sequence, and regulation of effectors between strains may exert distinct phenotypes. The sources of novel effectors and regulators of LEE function, and frequency and impact of acquisition of these on the virulence of EHEC, require further investigation. Moreover, research should recognize that EHEC can infect multiple hosts and that effectors that appear unimportant in surrogate rodent- or cell-based assays may play key roles in adaptation to humans or reservoir hosts. Unraveling the activities in effectors in reservoir hosts is presently hindered by the paucity of reagents to study signaling pathways and detect proteins in ruminant cells relative to murine and human systems.

While there is much merit in the intellectual pursuit of the molecular mechanisms by which constituents of the T3SS and its effectors act, future research should also consider how we may exploit this knowledge for practical gain. Vaccines that target intimin,

Tir, and translocon components show promise in control of EHEC O157 in cattle (reference 246); however, the extent of cross-protection against other EHEC strains is ill-defined. Inhibitors of Type III secretion have proven useful for "chemical genetics" to understand LEE function and regulation, but delivery of such molecules at the required sites, concentrations, and times to ablate T3SS activity poses a formidable challenge. Given the array of effectors dedicated to disarming the innate response, future research could consider the extent to which this mediates persistence in reservoirs and prevents development of effective adaptive immunity, as this may inform the design of live vaccines.

ACKNOWLEDGMENTS

The authors gratefully acknowledge the financial support of the Biotechnology and Biological Sciences Research Council, Medical Research Council and The Wellcome Trust. We apologize to those authors whose work could not be described or cited, owing to constraints of space.

CITATION

Stevens MP, Frankel GM. 2014. The locus of enterocyte effacement and associated virulence factors of enterohemorrhagic *Escherichia coli*. Microbiol Spectrum 2(4):EHEC-0007-2013.

REFERENCES

1. **Tzipori S, Wachsmuth IK, Chapman C, Birden R, Brittingham J, Jackson C, Hogg J.** 1986. The pathogenesis of hemorrhagic colitis caused by *Escherichia coli* O157:H7 in gnotobiotic piglets. *J Infect Dis* **154:**712–716.
2. **Knutton S, Lloyd DR, McNeish AS.** 1987. Adhesion of enteropathogenic *Escherichia coli* to human intestinal enterocytes and cultured human intestinal mucosa. *Infect Immun* **55:**69–77.
3. **Knutton S, Baldwin T, Williams PH, McNeish AS.** 1989. Actin accumulation at sites of bacterial adhesion to tissue culture cells: basis of a new diagnostic test for enteropathogenic and

enterohemorrhagic *Escherichia coli. Infect Immun* **57:**1290–1298.

4. **Jerse AE, Yu J, Tall BD, Kaper JB.** 1990. A genetic locus of enteropathogenic *Escherichia coli* necessary for the production of attaching and effacing lesions on tissue culture cells. *Proc Natl Acad Sci USA* **87:**7839–7843.

5. **McDaniel TK, Jarvis KG, Donnenberg MS, Kaper JB.** 1995. A genetic locus of enterocyte effacement conserved among diverse enterobacterial pathogens. *Proc Natl Acad Sci USA* **92:**1664–1668.

6. **Jarvis KG, Girón JA, Jerse AE, McDaniel TK, Donnenberg MS, Kaper JB.** 1995. Enteropathogenic *Escherichia coli* contains a putative type III secretion system necessary for the export of proteins involved in attaching and effacing lesion formation. *Proc Natl Acad Sci USA* **92:**7996–8000.

7. **Elliott SJ, Wainwright LA, McDaniel TK, Jarvis KG, Deng YK, Lai LC, McNamara BP, Donnenberg MS, Kaper JB.** 1998. The complete sequence of the locus of enterocyte effacement (LEE) from enteropathogenic *Escherichia coli* E2348/69. *Mol Microbiol* **28:**1–4.

8. **Ebel F, Deibel C, Kresse AU, Guzman CA, Chakraborty T.** 1996. Temperature- and medium-dependent secretion of proteins by Shiga toxin-producing *Escherichia coli. Infect Immun* **64:**4472–4479.

9. **Jarvis KG, Kaper JB.** 1996. Secretion of extracellular proteins by enterohemorrhagic *Escherichia coli* via a putative type III secretion system. *Infect Immun* **64:**4826–4829.

10. **Perna NT, Mayhew GF, Posfai G, Elliott S, Donnenberg MS, Kaper JB, Blattner FR.** 1998. Molecular evolution of a pathogenicity island from enterohemorrhagic *Escherichia coli* O157:H7. *Infect Immun* **66:**3810–3817.

11. **Deng W, Li Y, Vallance BA, Finlay BB.** 2001. Locus of enterocyte effacement from *Citrobacter rodentium*: sequence analysis and evidence for horizontal transfer among attaching and effacing pathogens. *Infect Immun* **69:**6323–6335.

12. **Tauschek M, Strugnell RA, Robins-Browne RM.** 2002. Characterization and evidence of mobilization of the LEE pathogenicity island of rabbit-specific strains of enteropathogenic *Escherichia coli. Mol Microbiol* **44:**1533–1550.

13. **Zhu C, Agin TS, Elliott SJ, Johnson LA, Thate TE, Kaper JB, Boedeker EC.** 2001. Complete nucleotide sequence and analysis of the locus of enterocyte effacement from rabbit diarrheagenic *Escherichia coli* RDEC-1. *Infect Immun* **69:**2107–2115.

14. **McDaniel TK, Kaper JB.** 1997. A cloned pathogenicity island from enteropathogenic *Escherichia coli* confers the attaching and effacing phenotype on *E. coli* K-12. *Mol Microbiol* **23:**399–407.

15. **Pósfai G, Koob MD, Kirkpatrick HA, Blattner FR.** 1997. Versatile insertion plasmids for targeted genome manipulations in bacteria: isolation, deletion, and rescue of the pathogenicity island LEE of the *Escherichia coli* O157:H7 genome. *J Bacteriol* **179:**4426–4428.

16. **Castillo A, Eguiarte LE, Souza V.** 2005. A genomic population genetics analysis of the pathogenic enterocyte effacement island in *Escherichia coli*: the search for the unit of selection. *Proc Natl Acad Sci USA* **102:**1542–1547.

17. **Müller D, Benz I, Liebchen A, Gallitz I, Karch H, Schmidt MA.** 2009. Comparative analysis of the locus of enterocyte effacement and its flanking regions. *Infect Immun* **77:**3501–3513.

18. **Ogura Y, Ooka T, Iguchi A, Toh H, Asadulghani M, Oshima K, Kodama T, Abe H, Nakayama K, Kurokawa K, Tobe T, Hattori M, Hayashi T.** 2009. Comparative genomics reveal the mechanism of the parallel evolution of O157 and non-O157 enterohemorrhagic *Escherichia coli. Proc Natl Acad Sci USA* **106:**17939–17944.

19. **Schmidt MA.** 2010. LEEways: tales of EPEC, ATEC and EHEC. *Cell Microbiol* **12:**1544–1552.

20. **Buss C, Müller D, Rüter C, Heusipp G, Schmidt MA.** 2009. Identification and characterization of Ibe, a novel type III effector protein of A/E pathogens targeting human IQGAP1. *Cell Microbiol* **11:**661–677.

21. **Konczy P, Ziebell K, Mascarenhas M, Choi A, Michaud C, Kropinski AM, Whittam TS, Wickham M, Finlay B, Karmali MA.** 2008. Genomic O island 122, locus for enterocyte effacement, and the evolution of virulent verocytotoxin-producing *Escherichia coli. J Bacteriol* **190:**5832–5840.

22. **Deng W, Yu HB, de Hoog CL, Stoynov N, Li Y, Foster LJ, Finlay BB.** 2012. Quantitative proteomic analysis of type III secretome of enteropathogenic *Escherichia coli* reveals an expanded effector repertoire for attaching/effacing bacterial pathogens. *Mol Cell Proteomics* **11:**692–709.

23. **Karmali MA, Mascarenhas M, Shen S, Ziebell K, Johnson S, Reid-Smith R, Isaac-Renton J, Clark C, Rahn K, Kaper JB.** 2003. Association of genomic O island 122 of Escherichia coli EDL 933 with verocytotoxin-producing *Escherichia coli* seropathotypes that are linked to epidemic and/or serious disease. *J Clin Microbiol* **41:**4930–4940.

24. **Jores J, Wagner S, Rumer L, Eichberg J, Laturnus C, Kirsch P, Schierack P, Tschäpe H, Wieler LH.** 2005. Description of a 111-kb pathogenicity island (PAI) encoding various virulence features in the enterohemorrhagic *E. coli*

(EHEC) strain RW1374 (O103:H2) and detection of a similar PAI in other EHEC strains of serotype O103:H2. *Int J Med Microbiol* **294**:417–425.

25. Tobe T, Beatson SA, Taniguchi H, Abe H, Bailey CM, Fivian A, Younis R, Matthews S, Marches O, Frankel G, Hayashi T, Pallen MJ. 2006. An extensive repertoire of type III secretion effectors in *Escherichia coli* O157 and the role of lambdoid phages in their dissemination. *Proc Natl Acad Sci USA* **103**:14941–14946.

26. Brady MJ, Radhakrishnan P, Liu H, Magoun L, Murphy KC, Mukherjee J, Donohue-Rolfe A, Tzipori S, Leong JM. 2011. Enhanced actin pedestal formation by enterohemorrhagic *Escherichia coli* O157:H7 adapted to the mammalian host. *Front Microbiol* **2**:226.

27. Hughes DT, Sperandio V. 2008. Inter-kingdom signalling: communication between bacteria and their hosts. *Nat Rev Microbiol* **6**:111–120.

28. Xu X, McAteer SP, Tree JJ, Shaw DJ, Wolfson EB, Beatson SA, Roe AJ, Allison LJ, Chase-Topping ME, Mahajan A, Tozzoli R, Woolhouse ME, Morabito S, Gally DL. 2012. Lysogeny with Shiga toxin 2-encoding bacteriophages represses type III secretion in enterohemorrhagic *Escherichia coli*. *PLoS Pathog* **8**:e1002672.

29. Büttner D. 2012. Protein export according to schedule: architecture, assembly, and regulation of type III secretion systems from plant- and animal-pathogenic bacteria. *Microbiol Mol Biol Rev* **76**:262–310.

30. Pallen MJ, Beatson SA, Bailey CM. 2005. Bioinformatics analysis of the locus for enterocyte effacement provides novel insights into type-III secretion. *BMC Microbiol* **5**:9.

31. Creasey EA, Delahay RM, Daniell SJ, Frankel G. 2003. Yeast two-hybrid system survey of interactions between LEE-encoded proteins of enteropathogenic *Escherichia coli*. *Microbiology* **149**:2093–2106.

32. Ogino T, Ohno R, Sekiya K, Kuwae A, Matsuzawa T, Nonaka T, Fukuda H, Imajoh-Ohmi S, Abe A. 2006. Assembly of the type III secretion apparatus of enteropathogenic *Escherichia coli*. *J Bacteriol* **188**:2801–2811.

33. Sekiya K, Ohishi M, Ogino M, Tamano K, Sasakawa C, Abe A. 2001. Supermolecular structure of the enteropathogenic *Escherichia coli* type III secretion system and its direct interaction with the EspA-sheath-like structure. *Proc Natl Acad Sci USA* **98**:11638–11643.

34. Deng W, Puente JL, Gruenheid S, Li Y, Vallance BA, Vázquez A, Barba J, Ibarra JA, O'Donnell P, Metalnikov P, Ashman K, Lee S, Goode D, Pawson T, Finlay BB. 2004. Dissecting virulence: systematic and functional analyses of a

pathogenicity island. *Proc Natl Acad Sci USA* **101**:3597–3602.

35. Mills E, Baruch K, Charpentier X, Kobi S, Rosenshine I. 2008. Real-time analysis of effector translocation by the type III secretion system of enteropathogenic *Escherichia coli*. *Cell Host Microbe* **3**:104–113.

36. Berger CN, Crepin VF, Baruch K, Mousnier A, Rosenshine I, Frankel G. 2012. EspZ of enteropathogenic and enterohemorrhagic *Escherichia coli* regulates type III secretion system protein translocation. *MBio* **3**:e00317-12.

37. Shames SR, Deng W, Guttman JA, de Hoog CL, Li Y, Hardwidge PR, Sham HP, Vallance BA, Foster LJ, Finlay BB. 2010. The pathogenic *E. coli* type III effector EspZ interacts with host CD98 and facilitates host cell prosurvival signalling. *Cell Microbiol* **12**:1322–1339.

38. Tatsuno I, Kimura H, Okutani A, Kanamaru K, Abe H, Nagai S, Makino K, Shinagawa H, Yoshida M, Sato K, Nakamoto J, Tobe T, Sasakawa C. 2000. Isolation and characterization of mini-Tn5Km2 insertion mutants of enterohemorrhagic *Escherichia coli* O157:H7 deficient in adherence to Caco-2 cells. *Infect Immun* **68**:5943–5952.

39. Dziva F, van Diemen PM, Stevens MP, Smith AJ, Wallis TS. 2004. Identification of *Escherichia coli* O157:H7 genes influencing colonization of the bovine gastrointestinal tract using signature-tagged mutagenesis. *Microbiology* **150**:3631–3645.

40. van Diemen PM, Dziva F, Stevens MP, Wallis TS. 2005. Identification of enterohemorrhagic *Escherichia coli* O26:H− genes required for intestinal colonization in calves. *Infect Immun* **73**:1735–1743.

41. Eckert SE, Dziva F, Chaudhuri RR, Langridge GC, Turner DJ, Pickard DJ, Maskell DJ, Thomson NR, Stevens MP. 2011. Retrospective application of transposon-directed insertion site sequencing to a library of signature-tagged mini-Tn5Km2 mutants of *Escherichia coli* O157:H7 screened in cattle. *J Bacteriol* **193**:1771–1776.

42. Ritchie JM, Waldor MK. 2005. The locus of enterocyte effacement-encoded effector proteins all promote enterohemorrhagic *Escherichia coli* pathogenicity in infant rabbits. *Infect Immun* **73**:1466–1474.

43. Gauthier A, Robertson ML, Lowden M, Ibarra JA, Puente JL, Finlay BB. 2005. Transcriptional inhibitor of virulence factors in enteropathogenic *Escherichia coli*. *Antimicrob Agents Chemother* **49**:4101–4109.

44. Rasko DA, Moreira CG, Li de R, Reading NC, Ritchie JM, Waldor MK, Williams N, Taussig R, Wei S, Roth M, Hughes DT, Huntley JF, Fina MW, Falck JR, Sperandio V. 2008. Targeting

QseC signaling and virulence for antibiotic development. *Science* **321**:1078–1080.

45. **Tree JJ, Wang D, McInally C, Mahajan A, Layton A, Houghton I, Elofsson M, Stevens MP, Gally DL, Roe AJ.** 2009. Characterization of the effects of salicylidene acylhydrazide compounds on type III secretion in *Escherichia coli* O157:H7. *Infect Immun* **77**:4209–4220.

46. **Veenendaal AK, Sundin C, Blocker AJ.** 2009. Small-molecule type III secretion system inhibitors block assembly of the *Shigella* type III secreton. *J Bacteriol* **191**:563–570.

47. **Wang D, Zetterström CE, Gabrielsen M, Beckham KS, Tree JJ, Macdonald SE, Byron O, Mitchell TJ, Gally DL, Herzyk P, Mahajan A, Uvell H, Burchmore R, Smith BO, Elofsson M, Roe AJ.** 2011. Identification of bacterial target proteins for the salicylidene acylhydrazide class of virulence-blocking compounds. *J Biol Chem* **286**:29922–29931.

48. **Ebel F, Podzadel T, Rohde M, Kresse AU, Kramer S, Deibel C, Guzman CA, Chakraborty T.** 1998. Initial binding of Shiga toxin-producing *Escherichia coli* to host cells and subsequent induction of actin rearrangements depend on filamentous EspA-containing surface appendages. *Mol Microbiol* **30**:147–161.

49. **Knutton S, Rosenshine I, Pallen MJ, Nisan I, Neves BC, Bain C, Wolff C, Dougan G, Frankel G.** 1998. A novel EspA-associated surface organelle of enteropathogenic *Escherichia coli* involved in protein translocation into epithelial cells. *EMBO J.* **17**:2166–2176.

50. **Daniell SJ, Takahashi N, Wilson R, Friedberg D, Rosenshine I, Booy FP, Shaw RK, Knutton S, Frankel G, Aizawa S.** 2001. The filamentous type III secretion translocon of enteropathogenic *Escherichia coli. Cell Microbiol* **3**:865–871.

51. **Kenny B, DeVinney R, Stein M, Reinscheid DJ, Frey EA, Finlay BB.** 1997. Enteropathogenic *E. coli* (EPEC) transfers its receptor for intimate adherence into mammalian cells. *Cell* **91:** 511–520.

52. **Wolff C, Nisan I, Hanski E, Frankel G, Rosenshine I.** 1998. Protein translocation into host epithelial cells by infecting enteropathogenic *Escherichia coli. Mol Microbiol* **28**:143–155.

53. **Daniell SJ, Kocsis E, Morris E, Knutton S, Booy FP, Frankel G.** 2003. 3D structure of EspA filaments from enteropathogenic *Escherichia coli. Mol Microbiol* **49**:301–308.

54. **Crepin VF, Shaw R, Abe CM, Knutton S, Frankel G.** 2005. Polarity of enteropathogenic *Escherichia coli* EspA filament assembly and protein secretion. *J Bacteriol* **187**:2881–2889.

55. **Creasey EA, Friedberg D, Shaw RK, Umanski T, Knutton K, Rosenshine I, Frankel G.** 2003.

CesAB is an enteropathogenic *Escherichia coli* chaperone for the type-III translocator proteins EspA and EspB. *Microbiology* **149**:3639–3647.

56. **Wilson RK, Shaw RK, Daniell S, Knutton S, Frankel G.** 2001. Role of EscF, a putative needle complex protein, in the type III protein translocation system of enteropathogenic *Escherichia coli. Cell Microbiol* **3**:753–762.

57. **Ide T, Laarmann S, Greune L, Schillers H, Oberleithner H, Schmidt MA.** 2001. Characterization of translocation pores inserted into plasma membranes by type III-secreted Esp proteins of enteropathogenic *Escherichia coli. Cell Microbiol* **3**:669–679.

58. **Kresse AU, Rohde M, Guzman CA.** 1999. The EspD protein of enterohemorrhagic *Escherichia coli* is required for the formation of bacterial surface appendages and is incorporated in the cytoplasmic membranes of target cells. *Infect Immun* **67**:4834–4842.

59. **Wachter C, Beinke C, Mattes M, Schmidt MA.** 1999. Insertion of EspD into epithelial target cell membranes by infecting enteropathogenic *Escherichia coli. Mol Microbiol* **31**:1695–1707.

60. **Warawa J, Finlay BB, Kenny B.** 1999. Type III secretion-dependent hemolytic activity of enteropathogenic *Escherichia coli. Infect Immun* **67**:5538–5540.

61. **Shaw RK, Daniell S, Ebel F, Frankel G, Knutton S.** 2001. EspA filament-mediated protein translocation into red blood cells. *Cell Microbiol* **3**:213–222.

62. **Luo W, Donnenberg MS.** 2011. Interactions and predicted host membrane topology of the enteropathogenic *Escherichia coli* translocator protein EspB. *J Bacteriol* **193**:2972–2980.

63. **Neves BC, Mundy R, Petrovska L, Dougan G, Knutton S, Frankel G.** 2003. CesD2 of enteropathogenic *Escherichia coli* is a second chaperone for the type III secretion translocator protein EspD. *Infect Immun* **71**:2130–2141.

64. **Hartland EL, Daniell SJ, Delahay RM, Neves BC, Wallis T, Shaw RK, Hale C, Knutton S, Frankel G.** 2000. The type III protein translocation system of enteropathogenic *Escherichia coli* involves EspA-EspB protein interactions. *Mol Microbiol* **35**:1483–1492.

65. **Cleary J, Lai LC, Shaw RK, Straatman-Iwanowska A, Donnenberg MS, Frankel G, Knutton S.** 2004. Enteropathogenic *Escherichia coli* (EPEC) adhesion to intestinal epithelial cells: role of bundle-forming pili (BFP), EspA filaments and intimin. *Microbiology* **150**:527–538.

66. **Roe AJ, Yull H, Naylor SW, Woodward MJ, Smith DG, Gally DL.** 2003. Heterogeneous surface expression of EspA translocon filaments

by *Escherichia coli* O157:H7 is controlled at the posttranscriptional level. *Infect Immun* **71:**5900–5909.

67. **Roe AJ, Naylor SW, Spears KJ, Yull HM, Dransfield TA, Oxford M, McKendrick IJ, Porter M, Woodward MJ, Smith DG, Gally DL.** 2004. Co-ordinate single-cell expression of LEE4- and LEE5-encoded proteins of *Escherichia coli* O157:H7. *Mol Microbiol* **54:**337–352.

68. **Naylor SW, Roe AJ, Nart P, Spears K, Smith DG, Low JC, Gally DL.** 2005. *Escherichia coli* O157:H7 forms attaching and effacing lesions at the terminal rectum of cattle and colonization requires the LEE4 operon. *Microbiology* **151:**2773–2781.

69. **Nagano K, Taguchi K, Hara T, Yokoyama S, Kawada K, Mori H.** 2003. Adhesion and colonization of enterohemorrhagic *Escherichia coli* O157:H7 in cecum of mice. *Microbiol Immunol* **47:**125–132.

70. **McNeilly TN, Mitchell MC, Rosser T, McAteer S, Low JC, Smith DG, Huntley JF, Mahajan A, Gally DL.** 2010. Immunization of cattle with a combination of purified intimin-531, EspA and Tir significantly reduces shedding of *Escherichia coli* O157:H7 following oral challenge. *Vaccine* **28:**1422–1428.

71. **Potter AA, Klashinsky S, Li Y, Frey E, Townsend H, Rogan D, Erickson G, Hinkley S, Klopfenstein T, Moxley RA, Smith DR, Finlay BB.** 2004. Decreased shedding of *Escherichia coli* O157:H7 by cattle following vaccination with type III secreted proteins. *Vaccine* **22:**362–369.

72. **Neves BC, Shaw RK, Frankel G, Knutton S.** 2003. Polymorphisms within EspA filaments of enteropathogenic and enterohemorrhagic *Escherichia coli*. *Infect Immun* **71:**2262–2265.

73. **Asper DJ, Sekirov I, Finlay BB, Rogan D, Potter AA.** 2007. Cross reactivity of enterohemorrhagic *Escherichia coli* O157:H7-specific sera with non-O157 serotypes. *Vaccine* **25:**8262–8269.

74. **Taylor KA, O'Connell CB, Luther PW, Donnenberg MS.** 1998. The EspB protein of enteropathogenic *Escherichia coli* is targeted to the cytoplasm of infected HeLa cells. *Infect Immun* **66:**5501–5507.

75. **Hamada D, Hamaguchi M, Suzuki KN, Sakata I, Yanagihara I.** 2010. Cytoskeleton-modulating effectors of enteropathogenic and enterohemorrhagic *Escherichia coli*: a case for EspB as an intrinsically less-ordered effector. *FEBS J* **277:**2409–2415.

76. **Iizumi Y, Sagara H, Kabe Y, Azuma M, Kume K, Ogawa M, Nagai T, Gillespie PG, Sasakawa C, Handa H.** 2007. The enteropathogenic *E. coli* effector EspB facilitates microvillus effacing and antiphagocytosis by inhibiting myosin function. *Cell Host Microbe* **2:**383–392.

77. **Hauf N, Chakraborty T.** 2003. Suppression of NF-kappa B activation and proinflammatory cytokine expression by Shiga toxin-producing *Escherichia coli*. *J Immunol* **170:**2074–2082.

78. **Donnenberg MS, Tzipori S, McKee ML, O'Brien AD, Alroy J, Kaper JB.** 1993. The role of the *eae* gene of enterohemorrhagic *Escherichia coli* in intimate attachment in vitro and in a porcine model. *J Clin Investig* **92:**1418–1424.

79. **Yu J, Kaper JB.** 1992. Cloning and characterization of the *eae* gene of enterohaemorrhagic *Escherichia coli* O157:H7. *Mol Microbiol* **6:**411–417.

80. **Cornick NA, Booher SL, Moon HW.** 2002. Intimin facilitates colonization by *Escherichia coli* O157:H7 in adult ruminants. *Infect Immun* **70:**2704–2707.

81. **Dean-Nystrom EA, Bosworth BT, Moon HW, O'Brien AD.** 1998. *Escherichia coli* O157:H7 requires intimin for enteropathogenicity in calves. *Infect Immun* **66:**4560–4563.

82. **Judge NA, Mason HS, O'Brien AD.** 2004. Plant cell-based intimin vaccine given orally to mice primed with intimin reduces time of *Escherichia coli* O157:H7 shedding in feces. *Infect Immun* **72:**168–175.

83. **Ritchie JM, Thorpe CM, Rogers AB, Waldor MK.** 2003. Critical roles for *stx2*, *eae*, and *tir* in enterohemorrhagic *Escherichia coli*-induced diarrhea and intestinal inflammation in infant rabbits. *Infect Immun* **71:**7129–7139.

84. **Vlisidou I, Dziva F, La Ragione RM, Best A, Garmendia J, Hawes P, Monaghan P, Cawthraw SA, Frankel G, Woodward MJ, Stevens MP.** 2006. Role of intimin-Tir interactions and the Tir-cytoskeleton coupling protein in the colonization of calves and lambs by *Escherichia coli* O157:H7. *Infect Immun* **74:**758–764.

85. **Higgins LM, Frankel G, Connerton I, Gonçalves NS, Dougan G, MacDonald TT.** 1999. Role of bacterial intimin in colonic hyperplasia and inflammation. *Science* **285:**588–591.

86. **Gonçalves NS, Hale C, Dougan G, Frankel G, MacDonald TT.** 2003. Binding of intimin from enteropathogenic *Escherichia coli* to lymphocytes and its functional consequences. *Infect Immun* **71:**2960–2965.

87. **Khare S, Alali W, Zhang S, Hunter D, Pugh R, Fang FC, Libby SJ, Adams LG.** 2010. Vaccination with attenuated *Salmonella enterica* Dublin expressing *E. coli* O157:H7 outer membrane protein intimin induces transient reduction of fecal shedding of *E. coli* O157:H7 in cattle. *BMC Vet Res* **6:**35.

88. **Oliveira AF, Cardoso SA, Almeida FB, de Oliveira LL, Pitondo-Silva A, Soares SG, Hanna ES.** 2012. Oral immunization with attenuated

Salmonella vaccine expressing *Escherichia coli* O157:H7 intimin gamma triggers both systemic and mucosal humoral immunity in mice. *Microbiol Immunol* **56**:513–522.

89. **Dean-Nystrom EA, Gansheroff LJ, Mills M, Moon HW, O'Brien AD.** 2002. Vaccination of pregnant dams with intimin(O157) protects suckling piglets from *Escherichia coli* O157:H7 infection. *Infect Immun* **70**:2414–2418.

90. **Gansheroff LJ, Wachtel MR, O'Brien AD.** 1999. Decreased adherence of enterohemorrhagic *Escherichia coli* to HEp-2 cells in the presence of antibodies that recognize the C-terminal region of intimin. *Infect Immun* **67**:6409–6417.

91. **Frankel G, Candy DC, Everest P, Dougan G.** 1994. Characterization of the C-terminal domains of intimin-like proteins of enteropathogenic and enterohemorrhagic *Escherichia coli*, *Citrobacter freundii*, and *Hafnia alvei*. *Infect Immun* **62**:1835–1842.

92. **Deibel C, Dersch P, Ebel F.** 2001. Intimin from Shiga toxin-producing *Escherichia coli* and its isolated C-terminal domain exhibit different binding properties for Tir and a eukaryotic surface receptor. *Int J Med Microbiol* **290**:683–691.

93. **McKee ML, O'Brien AD.** 1996. Truncated enterohemorrhagic *Escherichia coli* (EHEC) O157:H7 intimin (EaeA) fusion proteins promote adherence of EHEC strains to HEp-2 cells. *Infect Immun* **64**:2225–2233.

94. **Liu H, Magoun L, Leong JM.** 1999. beta1-chain integrins are not essential for intimin-mediated host cell attachment and enteropathogenic *Escherichia coli*-induced actin condensation. *Infect Immun* **67**:2045–2049.

95. **Rosenshine I, Ruschkowski S, Stein M, Reinscheid DJ, Mills SD, Finlay BB.** 1996. A pathogenic bacterium triggers epithelial signals to form a functional bacterial receptor that mediates actin pseudopod formation. *EMBO J* **15**:2613–2624.

96. **Rosenshine I, Donnenberg MS, Kaper JB, Finlay BB.** 1992. Signal transduction between enteropathogenic *Escherichia coli* (EPEC) and epithelial cells: EPEC induces tyrosine phosphorylation of host cell proteins to initiate cytoskeletal rearrangement and bacterial uptake. *EMBO J* **11**:3551–3560.

97. **Deibel C, Kramer S, Chakraborty T, Ebel F.** 1998. EspE, a novel secreted protein of attaching and effacing bacteria, is directly translocated into infected host cells, where it appears as a tyrosine-phosphorylated 90 kDa protein. *Mol Microbiol* **28**:463–474.

98. **DeVinney R, Stein M, Reinscheid D, Abe A, Ruschkowski S, Finlay BB.** 1999. Enterohemorrhagic *Escherichia coli* O157:H7 produces Tir, which is translocated to the host cell membrane but is not tyrosine phosphorylated. *Infect Immun* **67**:2389–2398.

99. **Batchelor M, Prasannan S, Daniell S, Reece S, Connerton I, Bloomberg G, Dougan G, Frankel G, Matthews S.** 2000. Structural basis for recognition of the translocated intimin receptor (Tir) by intimin from enteropathogenic *Escherichia coli*. *EMBO J* **19**:2452–2464.

100. **Kelly G, Prasannan S, Daniell S, Fleming K, Frankel G, Dougan G, Connerton I, Matthews S.** 1999. Structure of the cell-adhesion fragment of intimin from enteropathogenic *Escherichia coli*. *Nat Struct Biol* **6**:313–318.

101. **Luo Y, Frey EA, Pfuetzner RA, Creagh AL, Knoechel DG, Haynes CA, Finlay BB, Strynadka NC.** 2000. Crystal structure of enteropathogenic *Escherichia coli* intimin-receptor complex. *Nature* **405**:1073–1077.

102. **Kenny B.** 1999. Phosphorylation of tyrosine 474 of the enteropathogenic *Escherichia coli* (EPEC) Tir receptor molecule is essential for actin nucleating activity and is preceded by additional host modifications. *Mol Microbiol* **31**:1229–1241.

103. **de Grado M, Abe A, Gauthier A, Steele-Mortimer O, DeVinney R, Finlay BB.** 1999. Identification of the intimin-binding domain of Tir of enteropathogenic *Escherichia coli*. *Cell Microbiol* **1**:7–17.

104. **Hartland EL, Batchelor M, Delahay RM, Hale C, Matthews S, Dougan G, Knutton S, Connerton I, Frankel G.** 1999. Binding of intimin from enteropathogenic *Escherichia coli* to Tir and to host cells. *Mol Microbiol* **32**:151–158.

105. **Yi Y, Ma Y, Gao F, Mao X, Peng H, Feng Y, Fan Z, Wang G, Guo G, Yan J, Zeng H, Zou Q, Gao GF.** 2010. Crystal structure of EHEC intimin: insights into the complementarity between EPEC and EHEC. *PLoS One* **5**:e15285.

106. **Liu H, Radhakrishnan P, Magoun L, Prabu M, Campellone KG, Savage P, He F, Schiffer CA, Leong JM.** 2002. Point mutants of EHEC intimin that diminish Tir recognition and actin pedestal formation highlight a putative Tir binding pocket. *Mol Microbiol* **45**:1557–1573.

107. **Adu-Bobie J, Frankel G, Bain C, Gonçalves AG, Trabulsi LR, Douce G, Knutton S, Dougan G.** 1998. Detection of intimins alpha, beta, gamma, and delta, four intimin derivatives expressed by attaching and effacing microbial pathogens. *J Clin Microbiol* **36**:662–668.

108. **Oswald E, Schmidt H, Morabito S, Karch H, Marchés O, Caprioli A.** 2000. Typing of intimin genes in human and animal enterohemorrhagic and enteropathogenic *Escherichia coli*: characterization of a new intimin variant. *Infect Immun* **68**:64–71.

109. **Zhang WL, Kohler B, Oswald E, Beutin L, Karch H, Morabito S, Caprioli A, Suerbaum S, Schmidt H.** 2002. Genetic diversity of intimin genes of attaching and effacing *Escherichia coli* strains. *J Clin Microbiol* **40:**4486–4492.

110. **Tzipori S, Gunzer F, Donnenberg MS, de Montigny L, Kaper JB, Donohue-Rolfe A.** 1995. The role of the *eaeA* gene in diarrhea and neurological complications in a gnotobiotic piglet model of enterohemorrhagic *Escherichia coli* infection. *Infect Immun* **63:**3621–3627.

111. **Mallick EM, Brady MJ, Luperchio SA, Vanguri VK, Magoun L, Liu H, Sheppard BJ, Mukherjee J, Donohue-Rolfe A, Tzipori S, Leong JM, Schauer DB.** 2012. Allele- and Tir-independent functions of intimin in diverse animal infection models. *Front Microbiol* **3:**11.

112. **Phillips AD, Navabpour S, Hicks S, Dougan G, Wallis T, Frankel G.** 2000. Enterohaemorrhagic *Escherichia coli* O157:H7 target Peyer's patches in humans and cause attaching/effacing lesions in both human and bovine intestine. *Gut* **47:**377–381.

113. **Phillips AD, Frankel G.** 2000. Intimin-mediated tissue specificity in enteropathogenic *Escherichia coli* interaction with human intestinal organ cultures. *J Infect Dis* **181:**1496–1500.

114. **Fitzhenry RJ, Pickard DJ, Hartland EL, Reece S, Dougan G, Phillips AD, Frankel G.** 2002. Intimin type influences the site of human intestinal mucosal colonisation by enterohaemorrhagic *Escherichia coli* O157:H7. *Gut* **50:**180–185.

115. **Girard F, Batisson I, Frankel GM, Harel J, Fairbrother JM.** 2005. Interaction of enteropathogenic and Shiga toxin-producing *Escherichia coli* and porcine intestinal mucosa: role of intimin and Tir in adherence. *Infect Immun* **73:**6005–6016.

116. **Mundy R, Schüller S, Girard F, Fairbrother JM, Phillips AD, Frankel G.** 2007. Functional studies of intimin in vivo and ex vivo: implications for host specificity and tissue tropism. *Microbiology* **153:**959–967.

117. **Naylor SW, Low JC, Besser TE, Mahajan A, Gunn GJ, Pearce MC, McKendrick IJ, Smith DG, Gally DL.** 2003. Lymphoid follicle-dense mucosa at the terminal rectum is the principal site of colonization of enterohemorrhagic *Escherichia coli* O157:H7 in the bovine host. *Infect Immun* **71:**1505–1512.

118. **Sheng H, Lim JY, Knecht HJ, Li J, Hovde CJ.** 2006. Role of *Escherichia coli* O157:H7 virulence factors in colonization at the bovine terminal rectal mucosa. *Infect Immun* **74:**4685–4693.

119. **Kudva IT, Hovde CJ, John M.** 2013. Adherence of non-O157 Shiga toxin-producing *Escherichia coli* to bovine recto-anal junction squamous epithelial cells appears to be mediated by mechanisms distinct from those used by O157. *Foodborne Pathog Dis* **10:**375–381.

120. **Kudva IT, Griffin RW, Krastins B, Sarracino DA, Calderwood SB, John M.** 2012. Proteins other than the locus of enterocyte effacement-encoded proteins contribute to *Escherichia coli* O157:H7 adherence to bovine rectoanal junction stratified squamous epithelial cells. *BMC Microbiol* **12:**103.

121. **Sinclair JF, O'Brien AD.** 2002. Cell surface-localized nucleolin is a eukaryotic receptor for the adhesin intimin-gamma of enterohemorrhagic *Escherichia coli* O157:H7. *J Biol Chem* **277:**2876–2885.

122. **Sinclair JF, Dean-Nystrom EA, O'Brien AD.** 2006. The established intimin receptor Tir and the putative eucaryotic intimin receptors nucleolin and beta1 integrin localize at or near the site of enterohemorrhagic *Escherichia coli* O157:H7 adherence to enterocytes in vivo. *Infect Immun* **74:**1255–1265.

123. **Sinclair JF, O'Brien AD.** 2004. Intimin types alpha, beta, and gamma bind to nucleolin with equivalent affinity but lower avidity than to the translocated intimin receptor. *J Biol Chem* **279:**33751–33758.

124. **Frankel G, Lider O, Hershkoviz R, Mould AP, Kachalsky SG, Candy DCA, Cahalon L, Humphries MJ, Dougan G.** 1996. The cell-binding domain of intimin from enteropathogenic *Escherichia coli* binds to beta1 integrins. *J Biol Chem* **271:**20359–20364.

125. **Muza-Moons MM, Koutsouris A, Hecht G.** 2003. Disruption of cell polarity by enteropathogenic *Escherichia coli* enables basolateral membrane proteins to migrate apically and to potentiate physiological consequences. *Infect Immun* **71:**7069–7078.

126. **Campellone KG, Rankin S, Pawson T, Kirschner MW, Tipper DJ, Leong JM.** 2004. Clustering of Nck by a 12-residue Tir phosphopeptide is sufficient to trigger localized actin assembly. *J Cell Biol* **164:**407–416.

127. **Kenny B, Ellis S, Leard AD, Warawa J, Mellor H, Jepson MA.** 2002. Co-ordinate regulation of distinct host cell signalling pathways by multifunctional enteropathogenic *Escherichia coli* effector molecules. *Mol Microbiol* **44:**1095–1107.

128. **Campellone KG, Giese A, Tipper DJ, Leong JM.** 2002. A tyrosine-phosphorylated 12-amino-acid sequence of enteropathogenic *Escherichia coli* Tir binds the host adaptor protein Nck and is required for Nck localization to actin pedestals. *Mol Microbiol* **43:**1227–1241.

129. **Gruenheid S, DeVinney R, Bladt F, Goosney F, Gelkop S, Gish GD, Pawson T, Finlay BB.** 2001.

Enteropathogenic *E. coli* Tir binds Nck to initiate actin pedestal formation in host cells. *Nat Cell Biol* **3:**856–859.

130. **Rohatgi R, Nollau P, Ho H, Kirschner MW, Mayer BJ.** 2001. Nck and phosphatidylinositol 4,5-bisphosphate synergistically activate actin polymerization through the N-WASP-Arp2/3 pathway. *J Biol Chem* **276:**26448–26452.

131. **Lommel S, Benesch S, Rottner K, Franz T, Wehland J, Kuhn R.** 2001. Actin pedestal formation by enteropathogenic *Escherichia coli* and intracellular motility of *Shigella flexneri* are abolished in N-WASP-defective cells. *EMBO Rep* **2:**850–857.

132. **Garber JJ, Takeshima F, Antón IM, Oyoshi MK, Lyubimova A, Kapoor A, Shibata T, Chen F, Alt FW, Geha RS, Leong JM, Snapper SB.** 2012. Enteropathogenic *Escherichia coli* and vaccinia virus do not require the family of WASP-interacting proteins for pathogen-induced actin assembly. *Infect Immun* **80:**4071–4077.

133. **Lommel S, Benesch S, Rohde M, Wehland J, Rottner K.** 2004. Enterohaemorrhagic and enteropathogenic *Escherichia coli* use different mechanisms for actin pedestal formation that converge on N-WASP. *Cell Microbiol* **6:**243–254.

134. **DeVinney R, Puente JL, Gauthier A, Goosney D, Finlay BB.** 2001. Enterohaemorrhagic and enteropathogenic *Escherichia coli* use a different Tir-based mechanism for pedestal formation. *Mol Microbiol* **41:**1445–1458.

135. **Brady MJ, Campellone KG, Ghildiyal M, Leong JM.** 2007. Enterohaemorrhagic and enteropathogenic *Escherichia coli* Tir proteins trigger a common Nck-independent actin assembly pathway. *Cell Microbiol* **9:**2242–2253.

136. **Campellone KG, Brady MJ, Alamares JG, Rowe DC, Skehan BM, Tipper DJ, Leong JM.** 2006. Enterohaemorrhagic *Escherichia coli* Tir requires a C-terminal 12-residue peptide to initiate EspF-mediated actin assembly and harbours N-terminal sequences that influence pedestal length. *Cell Microbiol* **8:**1488–1503.

137. **Vingadassalom D, Kazlauskas A, Skehan B, Cheng HC, Magoun L, Robbins D, Rosen MK, Saksela K, Leong JM.** 2009. Insulin receptor tyrosine kinase substrate links the *E. coli* O157:H7 actin assembly effectors Tir and EspF(U) during pedestal formation. *Proc Natl Acad Sci USA* **106:**6754–6759.

138. **Weiss SM, Ladwein M, Schmidt D, Ehinger J, Lommel S, Städing K, Beutling U, Disanza A, Frank R, Jänsch L, Scita G, Gunzer F, Rottner K, Stradal TE.** 2009. IRSp53 links the enterohemorrhagic *E. coli* effectors Tir and EspF_U for actin pedestal formation. *Cell Host Microbe* **5:**244–258.

139. **Campellone KG, Leong JM.** 2005. Nck-independent actin assembly is mediated by two phosphorylated tyrosines within enteropathogenic *Escherichia coli* Tir. *Mol Microbiol* **56:**416–432.

140. **Campellone KG, Robbins D, Leong JM.** 2004. EspF_U is a translocated EHEC effector that interacts with Tir and N-WASP and promotes Nck-independent actin assembly. *Dev Cell* **7:**217–228.

141. **Garmendia J, Phillips AD, Carlier MF, Chong Y, Schüller S, Marches O, Dahan S, Oswald E, Shaw RK, Knutton S, Frankel G.** 2004. TccP is an enterohaemorrhagic *Escherichia coli* O157:H7 type III effector protein that couples Tir to the actin-cytoskeleton. *Cell Microbiol* **6:**1167–1183.

142. **Cheng HC, Skehan BM, Campellone KG, Leong JM, Rosen MK.** 2008. Structural mechanism of WASP activation by the enterohaemorrhagic *E. coli* effector EspF(U). *Nature* **454:**1009–1013.

143. **Sallee NA, Rivera GM, Dueber JE, Vasilescu D, Mullins RD, Mayer BJ, Lim WA.** 2008. The pathogen protein EspF(U) hijacks actin polymerization using mimicry and multivalency. *Nature* **454:**1005–1008.

144. **Campellone KG, Cheng HC, Robbins D, Siripala AD, McGhie EJ, Hayward RD, Welch MD, Rosen MK, Koronakis V, Leong JM.** 2008. Repetitive N-WASP-binding elements of the enterohemorrhagic *Escherichia coli* effector EspF(U) synergistically activate actin assembly. *PLoS Pathog* **4:**e1000191.

145. **Garmendia J, Carlier MF, Egile C, Didry D, Frankel G.** 2006. Characterization of TccP-mediated N-WASP activation during enterohaemorrhagic *Escherichia coli* infection. *Cell Microbiol* **8:**1444–1455.

146. **Vingadassalom D, Campellone KG, Brady MJ, Skehan B, Battle SE, Robbins D, Kapoor A, Hecht G, Snapper SB, Leong JM.** 2010. Enterohemorrhagic *E. coli* requires N-WASP for efficient type III translocation but not for EspF_U-mediated actin pedestal formation. *PLoS Pathog* **6:**e1001056.

147. **Bai L, Schüller S, Whale A, Mousnier A, Marches O, Wang L, Ooka T, Heuschkel R, Torrente F, Kaper JB, Gomes TA, Xu J, Phillips AD, Frankel G.** 2008. Enteropathogenic *Escherichia coli* O125:H6 triggers attaching and effacing lesions on human intestinal biopsy specimens independently of Nck and TccP/TccP2. *Infect Immun* **76:**361–368.

148. **Schüller S, Chong Y, Lewin J, Kenny B, Frankel G, Phillips AD.** 2007. Tir phosphorylation and Nck/N-WASP recruitment by enteropathogenic and enterohaemorrhagic *Escherichia coli* during

ex vivo colonization of human intestinal mucosa is different to cell culture models. *Cell Microbiol* **9**:1352–1364.

149. **Crepin VF, Girard F, Schüller S, Phillips AD, Mousnier A, Frankel G.** 2010. Dissecting the role of the Tir:Nck and Tir:IRTKS/IRSp53 signalling pathways in vivo. *Mol Microbiol* **75**:308–323.

150. **Ritchie JM, Brady MJ, Riley KN, Ho TD, Campellone KG, Herman IM, Donohue-Rolfe A, Tzipori S, Waldor MK, Leong JM.** 2008. EspF$_U$, a type III-translocated effector of actin assembly, fosters epithelial association and late-stage intestinal colonization by *E. coli* O157:H7. *Cell Microbiol* **10**:836–847.

151. **Ogura Y, Ooka T, Whale A, Garmendia J, Beutin L, Tennant S, Krause G, Morabito S, Chinen I, Tobe T, Abe H, Tozzoli R, Caprioli A, Rivas M, Robins-Browne R, Hayashi T, Frankel G.** 2007. TccP2 of O157:H7 and non-O157 enterohemorrhagic *Escherichia coli* (EHEC): challenging the dogma of EHEC-induced actin polymerization. *Infect Immun* **75**:604–612.

152. **Whale AD, Hernandes RT, Ooka T, Beutin L, Schüller S, Garmendia J, Crowther L, Vieira MA, Ogura Y, Krause G, Phillips AD, Gomes TA, Hayashi T, Frankel G.** 2007. TccP2-mediated subversion of actin dynamics by EPEC 2—a distinct evolutionary lineage of enteropathogenic *Escherichia coli*. *Microbiology* **153**:1743–1755.

153. **Sanger JM, Chang R, Ashton F, Kaper JB, Sanger JW.** 1996. Novel form of actin-based motility transports bacteria on the surfaces of infected cells. *Cell Motil Cytoskeleton* **34**:279–287.

154. **Perna NT, Plunkett G 3rd, Burland V, Mau B, Glasner JD, Rose DJ, Mayhew GF, Evans PS, Gregor J, Kirkpatrick HA, Pósfai G, Hackett J, Klink S, Boutin A, Shao Y, Miller L, Grotbeck EJ, Davis NW, Lim A, Dimalanta ET, Potamousis KD, Apodaca J, Anantharaman TS, Lin J, Yen G, Schwartz DC, Welch RA, Blattner FR.** 2001. Genome sequence of enterohaemorrhagic *Escherichia coli* O157:H7. *Nature* **409**:529–533.

155. **Hayashi T, Makino K, Ohnishi M, Kurokawa K, Ishii K, Yokoyama K, Han CG, Ohtsubo E, Nakayama K, Murata T, Tanaka M, Tobe T, Iida T, Takami H, Honda T, Sasakawa C, Ogasawara N, Yasunaga T, Kuhara S, Shiba T, Hattori M, Shinagawa H.** 2001. Complete genome sequence of enterohemorrhagic *Escherichia coli* O157:H7 and genomic comparison with a laboratory strain K-12. *DNA Res* **8**:11–22.

156. **Deng W, Li Y, Hardwidge PR, Frey EA, Pfuetzner RA, Lee S, Gruenheid S, Strynadka NC, Puente JL, Finlay BB.** 2005. Regulation of type III secretion hierarchy of translocators and effectors in attaching and effacing bacterial pathogens. *Infect Immun* **73**:2135–2146.

157. **Wang D, Roe AJ, McAteer S, Shipston MJ, Gally DL.** 2008. Hierarchal type III secretion of translocators and effectors from *Escherichia coli* O157:H7 requires the carboxy terminus of SepL that binds to Tir. *Mol Microbiol* **69**:1499–1512.

158. **Charpentier X, Oswald E.** 2004. Identification of the secretion and translocation domain of the enteropathogenic and enterohemorrhagic *Escherichia coli* effector Cif, using TEM-1 beta-lactamase as a new fluorescence-based reporter. *J Bacteriol* **186**:5486–5495.

159. **Asadulghani M, Ogura Y, Ooka T, Itoh T, Sawaguchi A, Iguchi A, Nakayama K, Hayashi T.** 2009. The defective prophage pool of *Escherichia coli* O157:prophage-prophage interactions potentiate horizontal transfer of virulence determinants. *PLoS Pathog* **5**:e1000408.

160. **Marchès O, Ledger TN, Boury M, Ohara M, Tu X, Goffaux F, Mainil J, Rosenshine I, Sugai M, De Rycke J, Oswald E.** 2003. Enteropathogenic and enterohaemorrhagic *Escherichia coli* deliver a novel effector called Cif, which blocks cell cycle G2/M transition. *Mol Microbiol* **50**:1553–1567.

161. **Taieb F, Nougayrède JP, Oswald E.** 2011. Cycle inhibiting factors (cifs): cyclomodulins that usurp the ubiquitin-dependent degradation pathway of host cells. *Toxins (Basel)* **3**:356–368.

162. **Holmes A, Mühlen S, Roe AJ, Dean P.** 2010. The EspF effector, a bacterial pathogen's Swiss army knife. *Infect Immun* **78**:4445–4453.

163. **Marchés O, Wiles S, Dziva F, La Ragione RM, Schüller S, Best A, Phillips AD, Hartland EL, Woodward MJ, Stevens MP, Frankel G.** 2005. Characterization of two non-locus of enterocyte effacement-encoded type III-translocated effectors, NleC and NleD, in attaching and effacing pathogens. *Infect Immun* **73**:8411–8417.

164. **Wong AR, Pearson JS, Bright MD, Munera D, Robinson KS, Lee SF, Frankel G, Hartland EL.** 2011. Enteropathogenic and enterohaemorrhagic *Escherichia coli*: even more subversive elements. *Mol Microbiol* **80**:1420–1438.

165. **Nadler C, Baruch K, Kobi S, Mills E, Haviv G, Farago M, Alkalay I, Bartfeld S, Meyer TF, Ben-Neriah Y, Rosenshine I.** 2010. The type III secretion effector NleE inhibits NF-kappaB activation. *PLoS Pathog* **6**:e1000743.

166. **Newton HJ, Pearson JS, Badea L, Kelly M, Lucas M, Holloway G, Wagstaff KM, Dunstone MA, Sloan J, Whisstock JC, Kaper JB, Robins-Browne RM, Jans DA, Frankel G, Phillips AD, Coulson BS, Hartland EL.** 2010. The type III effectors NleE and NleB from enteropathogenic *E. coli* and OspZ from *Shigella* block nuclear translocation of NF-kappaB p65. *PLoS Pathog* **6**:e1000898.

167. **Gao X, Wang X, Pham TH, Feuerbacher LA, Lubos ML, Huang M, Olsen R, Mushegian A, Slawson C, Hardwidge PR.** 2013. NleB, a bacterial effector with glycosyltransferase activity, targets GAPDH function to inhibit NF-κB activation. *Cell Host Microbe* **13:**87–99.

168. **Ruchaud-Sparagano MH, Mühlen S, Dean P, Kenny B.** 2011. The enteropathogenic *E. coli* (EPEC) Tir effector inhibits NF-κB activity by targeting TNFα receptor-associated factors. *PLoS Pathog* **7:**e1002414.

169. **Yan D, Wang X, Luo L, Cao X, Ge B.** 2012. Inhibition of TLR signaling by a bacterial protein containing immunoreceptor tyrosine-based inhibitory motifs. *Nat Immunol* **13:**1063–1071.

170. **Baruch K, Gur-Arie L, Nadler C, Koby S, Yerushalmi G, Ben-Neriah Y, Yogev O, Shaulian E, Guttman C, Zarivach R, Rosenshine I.** 2011. Metalloprotease type III effectors that specifically cleave JNK and NF-κB. *EMBO J* **30:**221–231.

171. **Mühlen S, Ruchaud-Sparagano MH, Kenny B.** 2011. Proteasome-independent degradation of canonical NFkappaB complex components by the NleC protein of pathogenic *Escherichia coli*. *J Biol Chem* **286:**5100–5107.

172. **Pearson JS, Riedmaier P, Marchès O, Frankel G, Hartland EL.** 2011. A type III effector protease NleC from enteropathogenic *Escherichia coli* targets NF-κB for degradation. *Mol Microbiol* **80:**219–230.

173. **Yen H, Ooka T, Iguchi A, Hayashi T, Sugimoto N, Tobe T.** 2010. NleC, a type III secretion protease, compromises NF-κB activation by targeting p65/RelA. *PLoS Pathog* **6:**e1001231.

174. **Vossenkämper A, Marchès O, Fairclough PD, Warnes G, Stagg AJ, Lindsay JO, Evans PC, Luong le A, Croft NM, Naik S, Frankel G, MacDonald TT.** 2010. Inhibition of NF-κB signaling in human dendritic cells by the enteropathogenic *Escherichia coli* effector protein NleE. *J Immunol* **185:**4118–4127.

175. **Zhang L, Ding X, Cui J, Xu H, Chen J, Gong YN, Hu L, Zhou Y, Ge J, Lu Q, Liu L, Chen S, Shao F.** 2011. Cysteine methylation disrupts ubiquitin-chain sensing in NF-κB activation. *Nature* **481:**204–208.

176. **Gao X, Wan F, Mateo K, Callegari E, Wang D, Deng W, Puente J, Li F, Chaussee MS, Finlay BB, Lenardo MJ, Hardwidge PR.** 2009. Bacterial effector binding to ribosomal protein s3 subverts NF-kappaB function. *PLoS Pathog* **5:**e1000708.

177. **Wan F, Weaver A, Gao X, Bern M, Hardwidge PR, Lenardo MJ.** 2011. IKKβ phosphorylation regulates RPS3 nuclear translocation and NF-κB function during infection with *Escherichia coli* strain O157:H7. *Nat Immunol* **12:**335–343.

178. **Royan SV, Jones RM, Koutsouris A, Roxas JL, Falzari K, Weflen AW, Kim A, Bellmeyer A, Turner JR, Neish AS, Rhee KJ, Viswanathan VK, Hecht GA.** 2010. Enteropathogenic *E. coli* non-LEE encoded effectors NleH1 and NleH2 attenuate NF-κB activation. *Mol Microbiol* **78:**1232–1245.

179. **Pham TH, Gao X, Tsai K, Olsen R, Wan F, Hardwidge PR.** 2012. Functional differences and interactions between the *Escherichia coli* type III secretion system effectors NleH1 and NleH2. *Infect Immun* **80:**2133–2140.

180. **Marchès O, Covarelli V, Dahan S, Cougoule C, Bhatta P, Frankel G, Caron E.** 2008. EspJ of enteropathogenic and enterohaemorrhagic *Escherichia coli* inhibits opsono-phagocytosis. *Cell Microbiol* **10:**1104–1115.

181. **Dong N, Liu L, Shao F.** 2010. A bacterial effector targets host DH-PH domain RhoGEFs and antagonizes macrophage phagocytosis. *EMBO J* **29:**1363–1376.

182. **Martinez-Argudo I, Sands C, Jepson MA.** 2007. Translocation of enteropathogenic *Escherichia coli* across an in vitro M cell model is regulated by its type III secretion system. *Cell Microbiol* **9:**1538–1546.

183. **Quitard S, Dean P, Maresca M, Kenny B.** 2006. The enteropathogenic *Escherichia coli* EspF effector molecule inhibits PI-3 kinase-mediated uptake independently of mitochondrial targeting. *Cell Microbiol* **8:**972–981.

184. **Tahoun A, Siszler G, Spears K, McAteer S, Tree J, Paxton E, Gillespie TL, Martinez-Argudo I, Jepson MA, Shaw DJ, Koegl M, Haas J, Gally DL, Mahajan A.** 2011. Comparative analysis of EspF variants in inhibition of *Escherichia coli* phagocytosis by macrophages and inhibition of *E. coli* translocation through human- and bovine-derived M cells. *Infect Immun* **79:**4716–4729.

185. **Celli J, Olivier M, Finlay BB.** 2001. Enteropathogenic *Escherichia coli* mediates antiphagocytosis through the inhibition of PI 3-kinase-dependent pathways. *EMBO J* **20:**1245–1258.

186. **Alto NM, Weflen AW, Rardin MJ, Yarar D, Lazar CS, Tonikian R, Koller A, Taylor SS, Boone C, Sidhu SS, Schmid SL, Hecht GA, Dixon JE.** 2007. The type III effector EspF coordinates membrane trafficking by the spatiotemporal activation of two eukaryotic signaling pathways. *J Cell Biol* **178:**1265–1278.

187. **Marchès O, Batchelor M, Shaw RK, Patel A, Cummings N, Nagai T, Sasakawa C, Carlsson SR, Lundmark R, Cougoule C, Caron E, Knutton S, Connerton I, Frankel G.** 2006. EspF of enteropathogenic *Escherichia coli* binds sorting nexin 9. *J Bacteriol* **188:**3110–3115.

188. Peralta-Ramírez J, Hernandez JM, Manning-Cela R, Luna-Muñoz J, Garcia-Tovar C, Nougayréde JP, Oswald E, Navarro-Garcia F. 2008. EspF interacts with nucleation-promoting factors to recruit junctional proteins into pedestals for pedestal maturation and disruption of paracellular permeability. *Infect Immun* **76**:3854–3868.

189. Weflen AW, Alto NM, Viswanathan VK, Hecht G. 2010. *E. coli* secreted protein F promotes EPEC invasion of intestinal epithelial cells via an SNX9-dependent mechanism. *Cell Microbiol* **12**:919–929.

190. Berger CN, Crepin VF, Jepson MA, Arbeloa A, Frankel G. 2009. The mechanisms used by enteropathogenic *Escherichia coli* to control filopodia dynamics. *Cell Microbiol* **11**:309–322.

191. Huang Z, Sutton SE, Wallenfang AJ, Orchard RC, Wu X, Feng Y, Chai J, Alto NM. 2009. Structural insights into host GTPase isoform selection by a family of bacterial GEF mimics. *Nat Struct Mol Biol* **16**:853–860.

192. Arbeloa A, Bulgin RR, MacKenzie G, Shaw RK, Pallen MJ, Crepin VF, Berger CN, Frankel G. 2008. Subversion of actin dynamics by EspM effectors of attaching and effacing bacterial pathogens. *Cell Microbiol* **10**:1429–1441.

193. Arbeloa A, Garnett J, Lillington J, Bulgin RR, Berger CN, Lea SM, Matthews S, Frankel G. 2010. EspM2 is a RhoA guanine nucleotide exchange factor. *Cell Microbiol* **12**:654–664.

194. Bulgin R, Arbeloa A, Goulding D, Dougan G, Crepin VF, Raymond B, Frankel G. 2009. The T3SS effector EspT defines a new category of invasive enteropathogenic *E. coli* (EPEC) which form intracellular actin pedestals. *PLoS Pathog* **5**:e1000683.

195. Keestra AM, Winter MG, Auburger JJ, Frässle SP, Xavier MN, Winter SE, Kim A, Poon V, Ravesloot MM, Waldenmaier JF, Tsolis RM, Eigenheer RA, Bäumler AJ. 2013. Manipulation of small Rho GTPases is a pathogen-induced process detected by NOD1. *Nature* **496**:233–237.

196. Wong AR, Clements A, Raymond B, Crepin VF, Frankel G. 2012. The interplay between the *Escherichia coli* Rho guanine nucleotide exchange factor effectors and the mammalian RhoGEF inhibitor EspH. *MBio* **3**:e00250-11.

197. Tu X, Nisan I, Yona C, Hanski E, Rosenshine I. 2003. EspH, a new cytoskeleton-modulating effector of enterohaemorrhagic and enteropathogenic *Escherichia coli. Mol Microbiol* **47**:595–606.

198. Wong AR, Raymond B, Collins JW, Crepin VF, Frankel G. 2012. The enteropathogenic *E. coli* effector EspH promotes actin pedestal formation and elongation via WASP-interacting protein (WIP). *Cell Microbiol* **14**:1051–1070.

199. Arbeloa A, Oates CV, Marchès O, Hartland EL, Frankel G. 2011. Enteropathogenic and enterohemorrhagic *Escherichia coli* type III secretion effector EspV induces radical morphological changes in eukaryotic cells. *Infect Immun* **79**:1067–1076.

200. Hardwidge PR, Deng W, Vallance BA, Rodriguez-Escudero I, Cid VJ, Molina M, Finlay BB. 2005. Modulation of host cytoskeleton function by the enteropathogenic *Escherichia coli* and *Citrobacter rodentium* effector protein EspG. *Infect Immun* **73**:2586–2594.

201. Matsuzawa T, Kuwae A, Abe A. 2005. Enteropathogenic *Escherichia coli* type III effectors EspG and EspG2 alter epithelial paracellular permeability. *Infect Immun* **73**:6283–6289.

202. Matsuzawa T, Kuwae A, Yoshida S, Sasakawa C, Abe A. 2004. Enteropathogenic *Escherichia coli* activates the RhoA signaling pathway via the stimulation of GEF-H1. *EMBO J* **23**:3570–3582.

203. Tomson FL, Viswanathan VK, Kanack KJ, Kanteti RP, Straub KV, Menet M, Kaper JB, Hecht G. 2005. Enteropathogenic *Escherichia coli* EspG disrupts microtubules and in conjunction with Orf3 enhances perturbation of the tight junction barrier. *Mol Microbiol* **56**:447–464.

204. Shaw RK, Smollett K, Cleary J, Garmendia J, Straatman-Iwanowska A, Frankel G, Knutton S. 2005. Enteropathogenic *Escherichia coli* type III effectors EspG and EspG2 disrupt the microtubule network of intestinal epithelial cells. *Infect Immun* **73**:4385–4390.

205. Clements A, Smollett K, Lee SF, Hartland EL, Lowe M, Frankel G. 2011. EspG of enteropathogenic and enterohemorrhagic *E. coli* binds the Golgi matrix protein GM130 and disrupts the Golgi structure and function. *Cell Microbiol* **13**:1429–1439.

206. Dong N, Zhu Y, Lu Q, Hu L, Zheng Y, Shao F. 2012. Structurally distinct bacterial TBC-like GAPs link Arf GTPase to Rab1 inactivation to counteract host defenses. *Cell* **150**:1029–1041.

207. Germane KL, Spiller BW. 2011. Structural and functional studies indicate that the EPEC effector, EspG, directly binds p21-activated kinase. *Biochemistry* **50**:917–919.

208. Selyunin AS, Sutton SE, Weigele BA, Reddick LE, Orchard RC, Bresson SM, Tomchick DR, Alto NM. 2011. The assembly of a GTPase-kinase signalling complex by a bacterial catalytic scaffold. *Nature* **469**:107–111.

209. Batchelor M, Guignot J, Patel A, Cummings N, Cleary J, Knutton S, Holden DW, Connerton I, Frankel G. 2004. Involvement of the intermediate filament protein cytokeratin-18 in actin pedestal formation during EPEC infection. *EMBO Rep* **5**:104–110.

210. **Patel A, Cummings N, Batchelor M, Hill PJ, Dubois T, Mellits KH, Frankel G, Connerton I.** 2006. Host protein interactions with enteropathogenic *Escherichia coli* (EPEC): 14-3-3tau binds Tir and has a role in EPEC-induced actin polymerization. *Cell Microbiol* **8:**55–71.

211. **Ruetz TJ, Lin AE, Guttman JA.** 2012. Enterohaemorrhagic *Escherichia coli* requires the spectrin cytoskeleton for efficient attachment and pedestal formation on host cells. *Microb Pathog* **52:**149–156.

212. **Crane JK, McNamara BP, Donnenberg MS.** 2001. Role of EspF in host cell death induced by enteropathogenic *Escherichia coli*. *Cell Microbiol* **3:**197–211.

213. **Nagai T, Abe A, Sasakawa C.** 2005. Targeting of enteropathogenic *Escherichia coli* EspF to host mitochondria is essential for bacterial pathogenesis: critical role of the 16th leucine residue in EspF. *J Biol Chem* **280:**2998–3011.

214. **Nougayrède JP, Donnenberg MS.** 2004. Enteropathogenic *Escherichia coli* EspF is targeted to mitochondria and is required to initiate the mitochondrial death pathway. *Cell Microbiol* **6:**1097–1111.

215. **Nougayrède JP, Foster GH, Donnenberg MS.** 2007. Enteropathogenic *Escherichia coli* effector EspF interacts with host protein Abcf2. *Cell Microbiol* **9:**680–693.

216. **Zhao S, Zhou Y, Wang C, Yang Y, Wu X, Wei Y, Zhu L, Zhao W, Zhang Q, Wan C.** 2013. The N-terminal domain of EspF induces host cell apoptosis after infection with enterohaemorrhagic *Escherichia coli* O157:H7. *PLoS One* **8:**e55164.

217. **Kenny B, Jepson M.** 2000. Targeting of an enteropathogenic *Escherichia coli* (EPEC) effector protein to host mitochondria. *Cell Microbiol* **3:**417–426.

218. **Samba-Louaka A, Nougayrède JP, Watrin C, Oswald E, Taieb F.** 2009. The enteropathogenic *Escherichia coli* effector Cif induces delayed apoptosis in epithelial cells. *Infect Immun* **77:**5471–5477.

219. **Heczko U, Carthy CM, O'Brien BA, Finlay BB.** 2001. Decreased apoptosis in the ileum and ileal Peyer's patches: a feature after infection with rabbit enteropathogenic *Escherichia coli* O103. *Infect Immun* **69:**4580–4589.

220. **Hemrajani C, Berger CN, Robinson KS, Marchès O, Mousnier A, Frankel G.** 2010. NleH effectors interact with Bax inhibitor-1 to block apoptosis during enteropathogenic *Escherichia coli* infection. *Proc Natl Acad Sci USA* **107:**3129–3134.

221. **Robinson KS, Mousnier A, Hemrajani C, Fairweather N, Berger CN, Frankel G.** 2010. The enteropathogenic *Escherichia coli* effector NleH inhibits apoptosis induced by *Clostridium difficile* toxin B. *Microbiology* **156:**1815–1823.

222. **Martinez E, Schroeder GN, Berger CN, Lee SF, Robinson KS, Badea L, Simpson N, Hall RA, Hartland EL, Crepin VF, Frankel G.** 2010. Binding to Na(+)/H(+) exchanger regulatory factor 2 (NHERF2) affects trafficking and function of the enteropathogenic *Escherichia coli* type III secretion system effectors Map, EspI and NleH. *Cell Microbiol* **12:**1718–1731.

223. **Blasche S, Mörtl M, Steuber H, Siszler G, Nisa S, Schwarz F, Lavrik I, Gronewold TM, Maskos K, Donnenberg MS, Ullmann D, Uetz P, Kögl M.** 2013. The *E. coli* effector protein NleF is a caspase inhibitor. *PLoS One* **8:**e58937.

224. **Viswanathan VK, Hodges K, Hecht G.** 2009. Enteric infection meets intestinal function: how bacterial pathogens cause diarrhoea. *Nat Rev Microbiol* **7:**110–119.

225. **Dean P, Kenny B.** 2004. Intestinal barrier dysfunction by enteropathogenic *Escherichia coli* is mediated by two effector molecules and a bacterial surface protein. *Mol Microbiol* **54:**665–675.

226. **Thanabalasuriar A, Koutsouris A, Weflen A, Mimee M, Hecht G, Gruenheid S.** 2010. The bacterial virulence factor NleA is required for the disruption of intestinal tight junctions by enteropathogenic *Escherichia coli*. *Cell Microbiol* **12:**31–41.

227. **Simovitch M, Sason H, Cohen S, Zahavi EE, Melamed-Book N, Weiss A, Aroeti B, Rosenshine I.** 2010. EspM inhibits pedestal formation by enterohaemorrhagic *Escherichia coli* and enteropathogenic *E. coli* and disrupts the architecture of a polarized epithelial monolayer. *Cell Microbiol* **12:**489–505.

228. **Hodges K, Alto NM, Ramaswamy K, Dudeja PK, Hecht G.** 2008. The enteropathogenic *Escherichia coli* effector protein EspF decreases sodium hydrogen exchanger 3 activity. *Cell Microbiol* **10:**1735–1745.

229. **Dean P, Maresca M, Schüller S, Phillips AD, Kenny B.** 2006. Potent diarrheagenic mechanism mediated by the cooperative action of three enteropathogenic *Escherichia coli*-injected effector proteins. *Proc Natl Acad Sci USA* **103:**1876–1881.

230. **Klapproth JM, Scaletsky IC, McNamara BP, Lai LC, Malstrom C, James SP, Donnenberg MS.** 2000. A large toxin from pathogenic *Escherichia coli* strains that inhibits lymphocyte activation. *Infect Immun* **68:**2148–2155.

231. **Nicholls L, Grant TH, Robins-Browne RM.** 2000. Identification of a novel genetic locus that is required for in vitro adhesion of a clinical isolate of enterohaemorrhagic *Escherichia coli* to epithelial cells. *Mol Microbiol* **35:**275–288.

232. **Deacon V, Dziva F, van Diemen PM, Frankel G, Stevens MP.** 2010. Efa-1/LifA mediates intestinal colonization of calves by enterohaemorrhagic *Escherichia coli* O26:H- in a manner independent of glycosyltransferase and cysteine protease motifs or effects on type III secretion. *Microbiology* **156:**2527–2536.

233. **Stevens MP, van Diemen PM, Frankel G, Phillips AD, Wallis TS.** 2002. Efa1 influences colonization of the bovine intestine by Shiga toxin-producing *Escherichia coli* serotypes O5 and O111. *Infect Immun* **70:**5158–5166.

234. **Badea L, Doughty S, Nicholls L, Sloan J, Robins-Browne RM, Hartland EL.** 2003. Contribution of Efa1/LifA to the adherence of enteropathogenic *Escherichia coli* to epithelial cells. *Microb Pathog* **34:**205–215.

235. **Vidal JE, Navarro-García F.** 2008. EspC translocation into epithelial cells by enteropathogenic *Escherichia coli* requires a concerted participation of Type V and III secretion systems. *Cell Microbiol* **10:**1975–1986.

236. **Klapproth JM, Sasaki M, Sherman M, Babbin B, Donnenberg MS, Fernandes PJ, Scaletsky IC, Kalman D, Nusrat A, Williams IR.** 2005. *Citrobacter rodentium lifA/efa1* is essential for colonic colonization and crypt cell hyperplasia in vivo. *Infect Immun* **73:**1441–1451.

237. **Stevens MP, Roe AJ, Vlisidou I, van Diemen PM, La Ragione RM, Best A, Woodward MJ, Gally DL, Wallis TS.** 2004. Mutation of *toxB* and a truncated version of the *efa-1* gene in *Escherichia coli* O157:H7 influences the expression and secretion of locus of enterocyte effacement-encoded proteins but not intestinal colonization in calves or sheep. *Infect Immun* **72:**5402–5411.

238. **Tatsuno I, Horie M, Abe H, Miki T, Makino K, Shinagawa H, Taguchi H, Kamiya S, Hayashi T, Sasakawa C.** 2001. *toxB* gene on pO157 of enterohemorrhagic *Escherichia coli* O157:H7 is required for full epithelial cell adherence phenotype. *Infect Immun* **69:**6660–6669.

239. **Abu-Median AB, van Diemen PM, Dziva F, Vlisidou I, Wallis TS, Stevens MP.** 2006. Functional analysis of lymphostatin homologues in enterohaemorrhagic *Escherichia coli*. *FEMS Microbiol Lett* **258:**43–49.

240. **Stevens MP, Marchès O, Campbell J, Huter V, Frankel G, Phillips AD, Oswald E, Wallis TS.** 2002. Intimin, Tir, and Shiga toxin 1 do not influence enteropathogenic responses to Shiga toxin-producing *Escherichia coli* in bovine ligated intestinal loops. *Infect Immun* **70:**945–952.

241. **Neves BC, Shaw RK, Frankel G, Knutton S.** 2003. Polymorphisms within EspA filaments of enteropathogenic and enterohemorrhagic *Escherichia coli*. *Infect Immun* **71:**2262–2265.

242. **Davis TK, van der Kar NCAJ, Tarr PI.** Chapter 15, this volume.

243. **Obata F, Obrig T.** Chapter 5, this volume.

244. **Ritchie JM.** Chapter 8, this volume.

245. **Mellies JL, Lorenzen E.** Chapter 9, this volume.

246. **Smith DR.** Chapter 25, this volume.

Enterohemorrhagic *Escherichia coli* Adhesins

7

BRIAN D. McWILLIAMS[1] and ALFREDO G. TORRES[1,2]

INTRODUCTION

Among the thousands of bacterial species contained within the intestinal gut flora, it is accepted that each species requires the use of adhesin proteins, or some combination thereof, that bring the bacteria closer to the epithelia and allow them to colonize the intestine. In a similar way, enteric pathogens also require surface-localized adhesins for colonization of the host intestine and eventual establishment of disease. Enterohemorrhagic *Escherichia coli* (EHEC) and, in general, Shiga toxin-producing *E. coli* (STEC) strains are known to contain a large number of proteins responsible for adhesion and contribute to establishment, persistence, and tissue tropism observed during infection with these pathogens. Understanding how these adhesins work is critical to having a full picture of the pathogenic and pathophysiological process associated with EHEC. Further, because adhesins play such an important role in virulence, they are targets for therapeutic intervention. Thus, this review summarizes the current knowledge on the adhesive proteins in EHEC, emphasizing up-to-date information and discussing gaps in knowledge and future directions in the study of these virulence factors.

[1]Department of Microbiology and Immunology, University of Texas Medical Branch, Galveston, TX 77555; [2]Department of Pathology and Sealy Center for Vaccine Development, University of Texas Medical Branch, Galveston, TX 77555.

Enterohemorrhagic Escherichia coli *and Other Shiga Toxin-Producing* E. coli
Edited by Vanessa Sperandio and Carolyn J. Hovde
© 2015 American Society for Microbiology, Washington, DC
doi:10.1128/microbiolspec.EHEC-0003-2013

LOCUS OF ENTEROCYTE EFFACEMENT

The study of adhesion in *E. coli* strains that produce an intimate attachment to the epithelia dates back to 1990, when a single gene, *eae*, was discovered in enteropathogenic *E. coli* (EPEC) through a transposon-based mutagenesis system. The product of this gene was established as necessary for the formation of attaching and effacing lesions (A/E lesions) (1). A follow-up study indicated that this gene was conserved in EHEC (2), and early in vivo studies showed that this gene's protein product, intimin, was important for intimate attachment and colonization of the intestine in both piglets and humans (3, 4). Though intimin and other locus of enterocyte effacement (LEE)-encoded factors linked to pathogenesis have been extensively reviewed elsewhere (5, 6), we briefly describe some historical information and focus on recent advances in the understanding of LEE-encoded Tir and intimin interactions.

The *eae* gene is part of a larger set of genes that make up the LEE pathogenicity island. The LEE was first described as a 35-kb locus that was conserved in different isolates of EHEC and EPEC but was not present in nonpathogenic strains of *E. coli* (7). Two independent sequencing projects determined the DNA sequence of the LEE and described a locus that was divided into five major operons containing up to 41 genes (8, 9) (Fig. 1). An additional 11 genes were identified in the LEE of EHEC that were not found in EPEC and were associated with a prophage coding region present on the terminal end of the operon (9). Other parallel studies verified the presence of the LEE in a large number of pathogenic *E. coli* strains, all of which were associated with A/E lesion formation, including the *E. coli* RDEC-1 strain found in rabbits, EHEC O26:H–, O15:H–, O103:H2, and *Citrobacter rodentium* (9–12). Further studies identified specific gene products within this island as being required for the colonization of the bovine gut (13, 14), and others verified that the LEE was associated with the development of enteritis in other animal models, such as mice (15), calves (16), sheep (17), and rabbits (18, 19).

A large portion of the LEE represents the coding region for a type 3 secretion system (T3SS), a complex structural system that allows translocation of effector proteins out of the bacterial cell and into the surrounding environment or directly into a host cell. Early studies of EPEC suggested that this pathogen was capable of protein secretion, that these proteins were required for full virulence and the formation of A/E lesions (20, 21), and that this secretion was likely mediated by a T3SS (22). The relationship between the T3SS and the LEE was confirmed by the publication of the aforementioned sequencing projects (8), which helped identify individual components of the T3SS and their role in pathogenesis (23–25).

Since the LEE and its associated effector proteins play such a significant role in the virulence of EHEC, understanding the factors that affect the expression of these genes became a priority. A number of studies have identified a series of transcriptional regulators

Escherichia coli EDL933

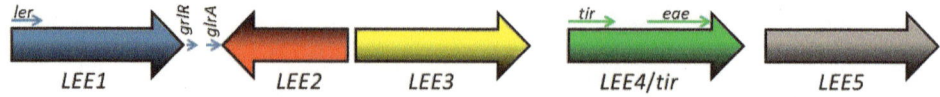

FIGURE 1 Illustration of the EHEC prototype strain EDL933 LEE pathogenicity island. The five LEE operons are depicted, specifically emphasizing gene-encoded proteins associated with adhesion and A/E lesion formation. They include the genes encoding the regulatory proteins Ler, GrlA, and GrlR; the translocated receptor protein Tir; and the adhesin protein intimin. doi:10.1128/microbiolspec.EHEC-0003-2013.f1

that affect the expression of various LEE-associated genes and include several proteins, such as QseA (26), H-NS (27), and LEE-encoded regulators Ler (28), GrlA, and GrlR (29). In addition, there are specific environmental conditions that affect LEE expression, including pH, osmolarity, $Fe(NO_3)_2$, Ca^{2+}, temperature (30), quorum sensing (31, 32), and HCO_3 (30, 33). More generally, LEE expression is also modulated by carbon catabolite repression in vitro (34) and by a combination of factors found in spent media from epithelial cell culture (35) or by posttranscriptional regulators like CsrA (36).

H-NS and Ler as Important Regulators of EHEC Adhesion

Of all the factors involved in the regulation of the LEE, two of the most extensively studied and closely associated with virulence are the negative regulator (silencer) H-NS and the positive regulator (antisilencer) Ler. The Ler protein was first identified as a member of the *LEE1* operon, and it is a transcription factor that positively regulates *LEE2*, *LEE3*, and *LEE4*, and may also contribute to regulation of its own transcription (28, 37–39). As such, Ler and its regulatory activity are required for adherence, A/E lesion formation, and virulence of EHEC (40). Certain *ler* mutants have also shown modified adhesion capabilities and displayed a novel set of fimbriae on the surface of the bacterial cell (39). Ler is involved in multiple regulatory systems, and its transcription is mediated by a large number of factors, including available sugars (41), quorum-sensing associated proteins such as QseA (26), the noncoding RNA DsrA (42), and the universal regulator Hfq (43). Ler is also regulated by GrlA, which completes an autoregulatory loop, as Ler is responsible for initial GrlA activation (44).

Ler imparts its positive regulatory effects by directly interfering with the negative histone-like negative regulator H-NS. This protein regulates a huge number of genes throughout the EHEC chromosome by binding to specifically curved DNA commonly found in promoter regions (45). The Ler protein originates from a similar protein family as H-NS, and the ability of both of these proteins to affect their target promoters requires a capacity to form long oligomers. Recent studies suggest that these two proteins contain similar DNA-binding domains and differ only in the behavior surrounding oligomerization (46). This difference seems to represent the mechanism by which Ler is able to antagonize the binding capacity of H-NS to the target promoter regions (38).

Intimin and Tir

Though H-NS and Ler regulate multiple genes located throughout the chromosome, it is their ability to regulate the proteins intimin and Tir and the long polar fimbriae (discussed below) that highlights their capacity to impact EHEC adhesion. As mentioned, the intimin protein was originally identified in EPEC (47), and researchers quickly identified a homolog in EHEC (2). The EPEC and EHEC intimin proteins share 86% homology at the nucleotide level and 83% homology at the protein level, with most of the divergence between these two proteins at their C-terminal end, which is the domain responsible for target interaction (2, 4). Over the following decade, it became firmly established that intimin was the primary molecule in both EHEC and EPEC associated with intimate bacterial interaction with the epithelia. Cell-culture studies using EHEC confirmed that this protein is essential for intimate adhesion, and disruption of the *eae* gene abolishes this phenotype (4). Furthermore, intimin has been shown to be essential for A/E lesion formation in both gnotobiotic pigs and colostrum-deprived calves (14, 48, 49). Finally, antibodies against intimin inhibit EHEC adherence in a cell-culture model (14, 49, 50). This central role of intimin in the interaction with the intestinal epithelia spurred further research toward understanding this protein's function and role in virulence.

DNA sequencing and structural studies indicated that intimin is a 95-kDa (939 amino acids) protein that can be divided into two functional regions. The N-terminal region (residues 1 to 550) contains a signal peptide region (1 to 39) that is responsible for its interaction with the bacterial Sec pathway and eventual secretion to the bacterial outer membrane, as well as a LysM-type region that allows for integration into peptidoglycan (51). From residues 189 to 550, a β-domain spans the outer membrane of the bacteria and is responsible for promotion of surface exposure for host-cell interaction (52). The C-terminal region contains four smaller domains (D1 to D4), each of which is an independent bacterial immunoglobulin-like domain that forms a surface-exposed rod that binds to Tir and is capped by a C-type lectin domain (53, 54). The interaction between intimin and its binding partner Tir primarily occurs between a "superdomain" formed between domain 3 and domain 4 of intimin and a β-hairpin motif on Tir (55).

The effort to characterize the Tir and intimin interaction originally started as an effort to identify what was thought to be a host-cell receptor protein mediating EHEC adhesion. This receptor was eventually identified in EPEC and called Hp90, whose phosphorylation was shown to mediate the actin rearrangements visible in A/E lesion formation and directly interact with intimin (56). Further studies in EPEC recognized that this was actually the first (and only) example of a bacterial protein being translocated into the host-cell membrane, acting as its own receptor for host-cell interaction (30). This protein used the T3SS to reach the host-cell cytoplasm and was named Tir for "translocated intimin receptor" (55). In EHEC, the previously discovered protein EspE was shown to be a homolog of Tir and was renamed to reflect this identity. Once translocation into the host cell is complete, Tir integrates itself into the host-cell membrane, exposing both terminal ends to the cytoplasm and an extracellular domain for interaction with

intimin (57, 58). Upon integration, however, Tir (EHEC) and Tir (EPEC) behave differently. Tir (EPEC) function is dependent on the phosphorylation of tyrosine 474, which is lacking in EHEC Tir (59–61). Further, EHEC Tir is incapable of complementing EPEC Tir on account of this amino acid difference (60–62). Since Tir (EHEC) lacks tyrosine 474, EHEC-mediated actin rearrangement is instead controlled by a second bacterial translocated virulence factor, EspF$_U$ (formerly TccP) (63, 64). This protein directly interacts with EHEC Tir upon its translocation into the host cell (63), and these two proteins form a larger complex with two host proteins, IRSp53 and IRTKS (64–66). The complex recruits and activates the host proteins N-WASP and Arp2/3, leading to the reorganization of actin within the host cell and the pedestal formation commonly associated with A/E lesion formation (63, 65, 66).

Though this interaction between Tir and intimin is likely the primary means by which EHEC is able to intimately adhere to epithelial cells, other studies attempting to understand the relationship between these two proteins revealed that adhesion is a more complex process than originally realized. Frankel et al. used purified EPEC intimin to determine if and how intimin was interacting with HEp-2 cells in vitro (67). In this scenario, intimin lacks its translocated binding partner Tir. Despite this, fluorescence was observed on the surface of these cells, suggesting that intimin may be capable of interacting with host cells independent of Tir. Follow-up studies using an *eae* mutant were able to identify a single amino acid in the C-terminal 150 residues of intimin that was responsible for this interaction (68). Further, through solid-phase binding assays, intimin was shown to interact with the host cell-surface-located protein $\beta 1$ integrin (69). This interaction, however, is not required for EPEC adhesion, and mutation of the gene encoding $\beta 1$ integrin did not have a negative effect on A/E lesion formation or the attachment of EPEC to the surface of cells in vitro (70). Finally, studies to map the binding domain within intimin identified the previously discussed lectin domain located on the C-terminal end of intimin, and removal of this domain completely abrogated Tir-independent binding (71). Similar studies characterizing Tir-independent binding in EHEC-derived intimin were done by Sinclair et al., who used fluorescent microscopy to colocalize $\beta1$ integrin and intimin (72); a separate study further strengthened this conclusion by abrogating the interaction between EHEC and cultured epithelial monolayers by adding heparin and heparin sulfate (73). These molecules directly interact with integrin and act in a similar fashion as a monoclonal antibody, interfering with the binding of these two proteins.

In addition to the intimin-$\beta1$ integrin interaction, a second Tir-independent interaction was discovered during these initial characterizations. The first studies looking closely at Tir-independent binding in EHEC identified a second binding event occurring between intimin and nucleolin (74). Still associated with the C-terminal end of intimin, the binding of this lectin domain was calculated to have an affinity of around 100 nM to host cells, and through affinity purification and sequencing steps, nucleolin was identified as a binding partner mediating this interaction. More detailed studies on this interaction showed that both EPEC- and EHEC-derived intimin bound to nucleolin with the same affinity but that this interaction was weaker than that between intimin and Tir (75). These results laid the foundation for a hypothesis indicating that the intimin-nucleolin and the intimin-integrin interactions were only secondary binding events during infection and that intimin-Tir interactions are required to keep the pathogen in place while the T3SS and its effector proteins were injected into the host. Finally, some more recent data further illustrated the importance of intimin's interaction with both nucleolin and $\beta1$ integrin through the effects of EHEC-encoded Shiga toxin (Stx). During infection, Stx is internalized via receptor-mediated endocytosis and is capable of modifying the transcription of nucleolin

and $\beta1$ integrin. In the absence of Stx, EHEC adhesion was significantly reduced, suggesting that the pathogen is capable of modifying host transcription patterns to increase its chances of successful adhesion and attachment of epithelial cells (76).

As previously mentioned, intimin contains two functional regions: the N-terminal region required for anchoring the protein into the bacterial membrane, and the C-terminal region responsible for host-cell and Tir interactions. As a result of this extracellular localization, this region was shown to be recognized by the immune system, as antibodies are known to be developed against intimin during the normal course of infection in patients with hemorrhagic colitis and hemolytic-uremic syndrome (77–79). It has been speculated that to avoid recognition by the immune system, both EPEC- and EHEC-derived intimin proteins have shown a high degree of plasticity and subsequent variability in this exposed C-terminal region (67). This variability has allowed allele-specific subtyping of the intimin protein. Overall, 27 total variants of the intimin protein have been identified that are made up of 18 different types and 9 subtypes. They include α, $\alpha2$, $\beta1$, $\beta2$, $\beta3$, $\gamma1$, $\gamma2$, δ, ϵ, $\epsilon2$, $\epsilon3$, $\epsilon4$, ζ, η, $\eta2$, θ, ι, $\iota2$, κ, λ, μ, ν, ξ, o, π, ρ, and σ (80–87). The most common clinically relevant subtypes are those within α-, β-, and γ-type. The α-type is found in multiple EPEC strains, including O127:H6, as is the β-type, with the only exception being EHEC O26:H11. The γ-type includes EHEC O157:H7 and two other EPEC strains, O55:H– and O55:H7 (6).

Because one unique feature between different types of A/E-inducing pathogenic *E. coli* strains relates to tissue tropism, it is logical to propose that these different intimin subtypes are one of the driving forces behind the diverse tropism and variability in EHEC and EPEC interaction with host cells. Indeed, early studies prior to the establishment of the intimin typing system determined that substituting intimin from a γ-subtype (O157:H7) to a β-type (O127:H6) completely changed the target of infection from the large intestine, which is commonly seen in EHEC, to both the small and large intestines, which are more frequently associated with EPEC infections (48). Later studies using in vitro organ cultures supported this hypothesis by showing that restoring different intimin subtypes into a O157:H7 *eae* mutant strain shifted the adhesion from Peyer's patches, common in EHEC, to the small intestine (88). Further, this study validated grouping within the intimin γ-type, as EPEC O55:H7 was also associated with Peyer's patches in the large intestine. Other studies, using subtypes from α-, β-, and γ-types, showed shifting tissue tropism in both in vitro organ-culture models and in vivo infections in mice (89).

Though strong evidence suggests that the emergence of these intimin types and subtypes is a way to predict tissue tropism, these differences may have originally been the result of immune system avoidance. As mentioned, antibodies against intimin are detectable in patients with EHEC and EPEC infection, and these antibodies are generally specific to their serotype (90, 91). The presence of antibodies against intimin, combined with the fact that this protein is extracellular, has made antibody-mediated protection a promising solution to both EPEC and EHEC infections. This protection likely occurs as a result of antibody-mediated interference with intimin interaction with Tir, β1 integrin, or nucleolin. Initial studies have shown that antibodies did significantly limit the amount of adherence observed on the surface of in vitro cultured epithelial cells (50, 92), and vaccination with recombinant intimin was able to induce specific antibody responses (93). Further, a wide range of protection studies using many different types of host organisms and delivery systems have afforded a number of interesting conclusions. First, multiple studies have reinforced the fact that antibody-mediated protection is subtype-specific. For example, γ-intimin-vaccinated mice generated antibodies against EHEC infection and displayed a decreased shedding time (15). Similar results were seen with intimin-α and EPEC (94).

Intimin-based vaccines have also been shown to be effective in rabbits (95) and adult cattle (96), and infant calves were protected by anti-intimin antibodies obtained from colostrum (97). Second, a number of different vector systems have been used to produce intimin, such as developing a plant-derived edible version of intimin (98) or using attenuated *Salmonella* species as the delivery vehicle (99, 100). Ultimately, the ideal vaccine should protect against all major intimin subtypes found in EHEC and EPEC, and it could be used to protect bovine or porcine herds from colonization, thus indirectly diminishing the potential exposure of these categories of *E. coli* strains to humans.

Long-Polar Fimbriae (Lpf1 and Lpf2)

Cumulative evidence supports the fact that although intimin is the primary adhesin in EHEC/STEC, it is by no means the only contributing factor. This became clear when a novel STEC O113:H21 strain was isolated from infected humans and showed to be lacking the intimin gene (101, 102). Not only was this strain infectious, it was able to adhere to in vitro cultured cells, implicating other proteins in contributing to the adhesive capabilities of pathogenic *E. coli*. In addition to this observation, two different EPEC O55 serotypes, O55:H6 and O55:H7, have been shown to express intimin-α and -γ, respectively, which should result in specific tropism to two different regions of the intestine. However, it was observed that these strains are both restricted to interaction with the follicle-associated endothelium (FAE), suggesting that perhaps other factors are contributing to tissue tropism in addition to intimin (103).

Full-genome sequencing and subsequent analysis of two prototype O157:H7 strains identified approximately 1.34 Mbp of DNA sequence present in O157:H7 but absent in the nonpathogenic K-12 strain (104, 105). These regions are distributed throughout the genome as "O islands," and two of these, islands #141 and #154, contained regions with homology to *Salmonella enterica* serovar Typhimurium. The proteins encoded in these regions were predicted to produce the components of the long polar fimbriae (Lpf), originally characterized in *Salmonella* (106) and found to contribute to the pathogen's adherence to the human intestine (Table 1). A mutation in the gene encoding the *Salmonella* major fimbrial subunit (*lpfA*) showed a significant reduction in adhesion when in vitro tissue-cultured cells were used; specifically, this mutation seemed to affect the bacterial interaction with Peyer's patches and M cells (107). As a result of the homology with the *Salmonella lpf* operon, the O157:H7 regions O-141 and O-154 were named *lpf1* and *lpf2*.

The *lpf1* operon consists of five genes with functions predicted to be similar to those described in *Salmonella* serovar Typhimurium (108–110). The first gene, *lpfA*, encodes the major fimbrial subunit protein whereas *lpfB* is a chaperone predicted to participate in proper folding. The third gene, *lpfC*, is the outer membrane usher protein and contains a stop codon within this predicted reading frame, dividing the gene into *lpfC* and *lpfC′*. The *lpfD* gene represents the coding region for a minor fimbrial subunit. Finally, the *lpfE* gene encodes for another predicted subunit of the fimbriae.

The *lpf2* operon contains a duplication of *lpfD* called *lpfD′* and lacks the *lpfE* gene (110). On account of these genetic differences, predicted structural variations have been proposed and may contribute to some small differences in function between Lpf1 and Lpf2. Further, the *lpf* operons are not limited to EHEC, and are seen in a wide range of pathogenic *E. coli* strains and some nonpathogenic commensal strains (109). Similar to intimin, the *lpfA* gene has been divided into allele groups, of which five exist for the *lpfA1* gene and three exist for the *lpfA2* gene (112, 113). Interestingly, there seems to be a correlation between the different Lpf allele group combinations and the subtype of intimin associated with a given *E. coli* strain that may be contributing to tissue tropism within the intestine (112).

In addition to slight differences in their genetic makeup, *lpf1* and *lpf2* are regulated in slightly different ways as well. Studies indicate that transcription of *lpf1* is increased during late log phase in Dulbecco's modified Eagle medium at 37°C, pH of 6.5, and in response to salt concentrations (108). The *lpf2* is positively affected by iron depletion, suggesting that the regulatory protein Fur might participate in this activation (114). At least for the *lpf1* operon, the relationship between the

TABLE 1 Major characteristics of the EHEC fimbrial proteins

EHEC fimbria name	Primary structural gene	Representative strains	Ligand	Cell lines used to show binding
Long polar fimbria 1	*lpfA1*	O157:H7, O55:H7, O127:H6	Fibronectin, collagen IV, laminin	Caco-2, HeLa
Long polar fimbria 2	*lpfA2*	O157:H7	Unknown	Caco-2
Curli	*csgA*	O157:H7, O26:H11	Abiotic surfaces, lettuce leaves	T84
E. coli common pilus	*ecpA*	O157:H7, O26:H11, O32:H37, O104:H4	Unknown	HT-29, HEp-2, HeLa
F9 fimbria	*z2200*	O157:H7, O26:H–, O104:H4, O55:H7, O127:H6	Fibronectin	EBL, HeLa
E. coli laminin-binding fimbria	*ycbQ*	O157:H7, O26:H11, O104:H4	Laminin	HT-29, HEp-2, MDBK
Sorbitol-fermenting fimbria protein	*sfpA*	O157:H–, O157:NM, O165:H25	Unknown	Caco-2, HCT-8
Type 1 fimbria	*fimA*	O26:H11, O118, O157:H7, O157:H–, O55:H–	Abiotic surfaces	REC
Hemorrhagic *E. coli* protein	*hcpA*	O157:H7	Laminin, fibronectin	T84, Caco-2, HeLa, HEp-2, MDBK

environment and its expression is linked to two regulatory proteins: H-NS and Ler. The H-NS and Ler relationship with the *lpf1* operon was originally established by using *E. coli* O157:H7 *hns* and *ler* mutant strains and β-galactosidase assays in which the *lpf1* promoter region was fused to a reporter gene (115). The *hns* mutation showed an increase in β-galactosidase activity, and the *ler* mutant showed a corresponding decrease in activity. The regulatory control of these proteins was further established through quantitative real-time reverse transcription-PCR analysis, primer extension, and electrophoretic mobility shift assays. These experiments also helped identify two σ^{70}-type promoters in the *lpf1* regulatory region and showed that both Ler and H-NS were able to bind those regions (115). Further analysis allowed for the precise definition and localization of three individual binding sites for H-NS (116).

Once conditions are appropriate for the expression of Lpf, these proteins, particularly Lpf1 fimbriae, display characteristics similar to those described for Lpf in *Salmonella* species. Early studies showed that expression of Lpf1 from a plasmid containing the *lpf* operon and transformed into an *E. coli* K-12 strain increased bacterial adhesion to tissue-cultured cells and also increased the number of microcolonies that were formed in vitro (108). Further, the Lpf fimbriae were visible by transmission electron microscopy, and it was determined that they displayed a long, fine structure. In contrast, when the *lpf2* genes were cloned into this same strain of K-12, a reduction in the adherence phenotype was observed (110). Further, a mutation of the *lpfA2* gene in the wild-type O157:H7 did not have a significant effect on the adhesive properties in vitro while interacting with HeLa cells. However, when Caco-2 cells were used, a reduction in adhesion was visible at early time points, suggesting that Lpf2 may contribute to early adhesion whereas Lpf1 may contribute to later steps in the adherence process, perhaps when Ler and H-NS are also regulating the expression of the LEE-encoded

proteins (110). Other studies of EHEC strain O113:H21 showed the presence of a single *lpf* operon located in the same chromosomal position as Lpf2. This operon contained a shortened *lpfD* gene (similar to REPEC O15:H–) that was nonduplicated (making it distinct from Lpf2 from EHEC O157:H7) and also lacked an *lpfE* gene (found in EHEC O157:H7 Lpf1) (109). When this operon was deleted from the genome of O113:H21, the bacteria were significantly less capable of adhering to CHO-K1 epithelial cells, emphasizing the fact that Lpf proteins are critical for adherence in vitro. Finally, to further define the contribution of H-NS and Ler regulation to the function of EHEC Lpf1 and its adhesion characteristics (117), HeLa cells were exposed to EHEC carrying single- or double-knockout mutations in *hns* and *ler*. It was demonstrated that an *hns* mutation caused an increase in adhesion and a *ler* mutation resulted in a decrease. This reduction in adhesion was also observed when anti-Lpf antibodies were added during an in vitro adhesion assay with the wild-type strain. Interestingly, when both *hns* and *ler* genes were deleted, a 6-fold increase in adhesion was observed, suggesting that other positive regulators influencing Lpf1 might exist and have not been described.

Some progress has been made in establishing the host binding partner for Lpf1, as it was shown that wild-type EHEC is able to interact with the extracellular matrix (ECM) proteins, specifically fibronectin, laminin, and collagen IV (118). Further, anti-LpfA1 antibodies were added to these interaction assays, and it was observed that binding to these proteins decreased. Deletion of *lpf1* and *lpf2* reduced binding to the ECM proteins, and deletions of both *hns* and *ler* genes caused a similar decrease in the binding phenotype. It was also noted that purified recombinant LpfA1 was capable of interacting directly with these ECM proteins, and when T84 cells were preincubated with ECM proteins and then incubated with O157:H7 wild-type strains, an increase in bacterial adherence was observed.

The first in vivo models to study the role of Lpf during infection used sheep, conventional pigs, and gnotobiotic piglets infected with the *lpfA1* single or *lpfA1/lpfA2* double mutant strains (119). The study sought to differentiate stages of colonization in lieu of the previously discussed results, suggesting that Lpf1 may play a role early in colonization (110). The early colonization results, in which germ-free pigs were sacrificed 24 h postinfection, indicated that the *lpfA1/lpfA2* double mutant strain colonized significantly less in the spiral colon, but no statistical difference was detected in the cecum. Over an extended period, the *lpf* double mutant also shed significantly less in sheep starting at 2 weeks postinfection, and the *lpfA1/lpfA2* double mutant was not recovered from any intestinal tissue after sacrifice, in contrast to the wild-type strain, which was detected at 100 CFU/g of tissue. A different study used in vitro organ cultures from human intestinal tissue and sought to better understand the effects that Lpf had on tissue tropism (120). This study indicated that the wild-type strain bound only to the FAE of Peyer's patches but was unable to interact with the duodenum, terminal ileum, or the transverse colon. However, when a knockout was introduced into either or both *lpfA1* and *lpfA2* genes, the strains were able to interact with all of these intestinal regions except the transverse colon. This result further emphasized the hypothesis that Lpf may be specifically driving adhesion to the FAE and working with intimin to dictate EHEC tissue tropism. This FAE specificity is especially relevant with the understanding that O157:H7 tends to colonize areas in the intestine where FAE is found, thus providing a role for Lpf during colonization to the human intestine. A study by Fitzhenry et al. identified expression changes in 124 genes between the *lpf* mutants and the wild-type strain (120). The expression changes involved genes encoding some surface-exposed structures (FimG and YagZ) and some transcriptional regulators, suggesting that EHEC contains alternative mechanisms associated with cellular adhesion independent of tissue specificity (120). The presence of these alternative mechanisms has been suggested in another study using the rabbit infant model (18). This model uses rabbit intragastric infections, which are considered to be the most effective way to reproduce EHEC infection in vivo in an animal model. In this study, the *lpfA1/lpfA2* double mutant mixed with the wild-type strain was inoculated into the rabbits. The mutant was outcompeted for colonization in the ileum, cecum, and mid-colon, reinforcing the fact that Lpf was an important contributing adherence factor (121). Unexpectedly, when this double mutant was used in in vitro adhesion assays with colonic epithelial cells (Caco-2), adhesion was observed to increase when compared to wild-type strain. Transmission electron microscopy indicated the presence of a novel structure that appeared to be curli, a thin afimbrial adhesin factor implicated in biofilm formation and induction of inflammation (122). When the major structural subunit of curli (CsgA) was deleted from the parent O157:H7 strain, in vitro adhesion was not significantly decreased, suggesting that curli may only be expressed under conditions in which Lpf expression was decreased or eliminated. This also suggested that there was some sort of common regulatory mechanism between these two surface structures, though changes in the transcription of *csgA* were not observed in the double mutant strain (123).

Curli

Curli has been recognized as an adherence-related factor capable of mediating interactions between EHEC and host cells (Table 1), specifically with the ECM proteins laminin and fibronectin, along with plasminogen and major histocompatibility complex class 1 molecules (124, 125). Curli is formed by thin fibers that aggregate on the surface of the cells and are characterized by their binding to Congo red dye (124, 126). In addition to potentially acting as a compensatory mechanism for adhesion in the EHEC *lpfA* double mutant, curli and Lpf share a number of other characteristics. Curli and Lpf1 genes are both negatively regulated by H-NS and contribute to in vitro adhesion of tissue-cultured cells (122, 127). Curli is transcriptionally influenced by environmental signals, allowing curli transcription to respond to a number of different stimuli, such as pH, temperature, and nutrient limitations (128–130). Curli exists at the genetic level as two distinct and divergently transcribed operons: *csgDEFG* and *csgAB* (127, 131). CsgD is the primary transcriptional activator of *csgA*, and the gene encoding this regulator is under direct control of H-NS, RpoS, Crl, MlrA, Rcs, and the two-component systems OmpR/EnvZ and CpxA/CpxR, IHF, and SdiA (122, 132–136).

In addition to assuming the role of a primary adhesin of EHEC in the absence of Lpf, the primary function of curli is to mediate bacterial interaction with abiotic surfaces and protection of the pathogen from antiseptic chemicals. One study indicated that curli can directly contribute to bacterial attachment to stainless steel and promotes bacterial resistance to chlorine (137). Other studies have identified a large number of other abiotic surfaces to which curli mediates binding, including polystyrene, glass, and rubber (138). Perhaps most important is its role in adhesion of EHEC to vegetable leaves, making curli of particular interest in the investigation of recent outbreaks associated with contaminated spinach and lettuce. Curli-expressing strains of pathogenic *E. coli* have been isolated from multiple outbreaks worldwide, and a recent study indicates that CsgA is up-regulated 20-fold during interactions with lettuce leaves (139). The *csgA* mutant showed a decrease in adhesion to leaf tissue during both short-term and long-term colonization assays. As discussed by Lloyd et al., the difference in positive regulators may explain the difference in expression patterns for Lpf and curli. Whereas Lpf is the default "on" adhesion factor and expressed under conditions mimicking the intestine, curli is strictly controlled by

several regulators, possibly giving EHEC the opportunity to colonize and survive on other surfaces until more favorable environmental conditions are available (121).

Other Fimbrial Adhesin Proteins

ECP (Mat) Fimbriae

In addition to EHEC Lpf1, Lpf2, and curli, a study by Low et al. used genomic data analysis to predict and analyze the expression of 13 more fimbrial operons, many of which have proved to be functionally relevant (Table 1) (140). One of these operons encodes the *E. coli* YagZ homolog (also known as MatB) that was later renamed as *E. coli* common pilus (ECP) (141). The isolate O18:K1:H7 is an *E. coli* strain associated with neonatal meningitis and septicemia (NMEC) and provided the first information regarding Mat/ECP function. This fimbria was first identified after attempts to create a strain of NMEC that was completely devoid of fimbriae, and despite a triple-knockout of known fimbrial proteins, some fimbrial structures were still present on the surface of these cells (142). The fimbriae were purified and identified, and their coding region was verified in a wide range of pathogenic *E. coli* strains, including EHEC, EPEC, enterotoxigenic *E. coli*, enteroaggregative *E. coli*, and nonpathogenic *E. coli* K-12. Despite this, the study was only able to verify Mat/ECP expression in NMEC strains, and expression was only observed at 20°C in rich media and involved both transcriptional and post-transcriptional events. The *mat* (*ecp*) mutant strains showed a loss of Mat/ECP on the surface of their respective strains; these structures could be restored in *trans* via plasmid-expressed Mat/ECP (142). Later studies of EHEC showed that ECP is also critical for adherence to HEp-2 and HeLa cells and that this adherence phenotype can be abolished by mutating *ecpA* (141). The study also assayed for EcpA expression in 176 different strains of *E. coli*, verifying production of the fimbriae in almost 72% of strains tested, which included EHEC O157 and non-O157, as well as non-EHEC strains. In comparison with previous studies, these strains were grown at 26°C in Dulbecco's modified Eagle medium and showed a lack of expression at 37°C and in rich media. This study also showed that ECP expression increases under low oxygen and high CO_2, environmental conditions that are experienced in the intestinal tract. Studies of EPEC also verified this environmental dependence for the expression of *ecpA* (143). Purified EPEC EcpA was able to bind to and interact with the surface of HT-29 epithelial cells, reaffirming that this protein is an adhesin (143).

F9 Fimbriae

Another fimbrial gene cluster analyzed in the study by Low et al. was the F9 fimbriae. This protein had been previously identified as a factor associated with tissue colonization when a series of 59 mutants were obtained by transposon mutagenesis in O157:H7 screened for intestinal colonization in calves (16) and in a second study using O26:H– in cattle (144). A subsequent study of this fimbria performed by Low et al. as a follow-up analysis of one of four candidate adhesins from their original screen showed expression under in vitro conditions (145). Deletion of the *z2200* gene in O157:H7 significantly reduced the pathogen's ability to colonize the calf intestine, but the fimbria was also shown as not essential for bacterial shedding or required for proper tissue tropism (145). F9 fimbria also seems to have an affinity for ECM proteins, as it was shown to interact with fibronectin. Interestingly, when the F9 operon is transformed into an F9-deficient O157 strain, the in vitro adhesion decreased significantly (145). A later study identified CadA as a protein associated with F9 fimbriae expression when a microarray analysis indicated that the F9 fimbriae are up-regulated in a *cadA* mutant (146). The O157 *cadA* mutant resulted in a hyperadherent phenotype in the ileum of an infant rabbit, implying that the F9 fimbriae may also be contributing to tissue-specific adherence during infection.

ELF and Sfp

Another set of genes analyzed by Low et al. was associated with adhesion to host cells, but no extensive work was done to confirm their relationship with adherence or their functional roles (140). One of those proteins, originally identified as "locus 5," was identified as part of the *ycbQRST* fimbrial-like operon and was renamed as ELF for *E. coli* laminin-binding fimbria (147). The ELF fimbria was purified and shown to interact with the ECM protein laminin, and its expression was verified on the surface of O157:H7 during adherence to epithelial cells (147). Like many of the previously discussed fimbriae, this protein is expressed maximally during growth in minimal media and in the presence of host cells, and mutation of the coding region resulted in a loss of adhesion to HEp-2, HT-29, and MDBK (bovine kidney) cells.

Another fimbria described in the EHEC pathogroup is the sorbitol-fermenting fimbria protein (Sfp) that was originally identified in an O157:H– strain from Germany and showed a high degree of similarity to the pyelonephritis-associated pili (*pap*) genes in uropathogenic *E. coli* (148). Sfp was also identified in EHEC O157:NM and O165:H25/NM strains and shown to express poorly under nutrient-rich and aerobic conditions. However, under anaerobic conditions and on media simulating that of the colonic environment, induction was increased and corresponded with an increased level of adhesion to Caco-2 and HCT-8 cells (149, 150).

HCP

Another fimbria recently described is the hemorrhagic *E. coli* pilus (HCP). The fimbria, originally referred to as TFP (type IV pilus) or simply PpdD (predicted protein D), was originally identified in a study that visualized the fimbria on the surface of O157:H7 grown in minimal media (151). This study also showed that HCP was involved in EHEC interactions with human colonic cells (T84 and Caco-2), nonintestinal cells (HeLa and HEp-2), and bovine kidney (MDBK) epithelial cells, as disruption in the *hcp* coding region caused a significant decrease in adhesion to these cell types. HCP was also shown to contribute to adhesion when pig and cow intestinal explants were used, and it was suggested that HCP interaction works in concert with other adhesins and serves as a second binding interaction on the surface of host cells (151). Later studies characterized a series of other functions, some of which are currently unique to HCP, as compared to other EHEC fimbriae, and include involvement in the invasion of host epithelial cells, hemagglutination of erythrocytes, the formation of biofilms, and binding to the ECM proteins laminin and fibronectin (152).

Type 1 Fimbria

The final fimbrial protein discussed here is the type 1 fimbria, which has been extensively studied as a common fimbria in *E. coli*; however, it has also been shown to mediate pathogenic *E. coli* adhesion to epithelial cells in the rumen of cattle (153). Early studies identified the type 1 fimbria in the EHEC strains O26 and O118 (154) after being genetically traced to other EHEC strains, including O157:H7, O157:H–, and O55– (155). Despite the presence of the coding regions in the genome of these strains, type 1 fimbria expression was not observed in these EHEC strains. This phenomenon was attributed to the presence of a "*fim* switch," a 16-bp region in the regulatory region of *fimA*. When absent, the switch is considered to be in the "off" position and will not allow the expression of *fimA*, and when present, is considered to be in the "on" position and expression is allowed to proceed (155, 156). The *fim* switch is turned on in the bovine EHEC strains O26 and O118, and studies indicate that type 1 fimbria is a contributing factor to their virulence in the cattle infection model (156). When the *fim* switch from these strains is cloned into the regulatory region of *fimA* from O157:H7, expression is restored and the adhesion to host cells increases (156). The mutation of the *fimA* gene in non-O157 STEC and O128:H2 was shown to affect adherence to abiotic surfaces, such as polystyrene and glass (157).

Autotransporters

EhaA-D

In addition to the different fimbrial proteins, there is a second group of surface-exposed structures that contribute to EHEC interaction with host cells. These proteins have the capacity to be secreted independently of the conventional cell translocation machinery and are thus referred to as autotransporters. These proteins are listed in Table 2 and generally contain a signal sequence at the N terminus of the polypeptide, a passenger domain in the middle of the protein that determines the activity of the protein, and a C-terminal region that contains a β-barrel transmembrane region that embeds into the outer membrane and provides the passenger domain the ability to be translocated to the extracellular milieu (reviewed in reference 158). These proteins are often associated with the formation and maintenance of biofilms and other virulence traits in pathogens and related commensals (159). One study identified a series of four putative autotransporter-encoding genes, *ehaA–D* present in EHEC O157:H7 (160). EhaA was cloned in *trans* into an avirulent strain of *E. coli* and expressed under control of an arabinose promoter. Upon expression, EhaA was able to induce cellular aggregation as determined by fluorescence microscopy; further, this aggregation was the product of EhaA-EhaA interactions. This protein is also capable

TABLE 2 EHEC autotransporters and their functions

Autotransporter protein	Primary function	Binding target
EhaA	Biofilms	EhaA-EhaA
EhaB	Biofilms	Collagen, laminin
EspP	Protease activity, biofilms	EspP (ropes)
Saa	Host-cell adhesion	HEp-2, Caco-2
Sab	Host-cell adhesion, biofilms	HEp-2
Cah	Autoaggregation, Ca^+ binding, leaf binding	Alfalfa sprouts

of inducing the formation of biofilms, but because the protein is physically shorter than other surface structures, its adhesive capacity can be eliminated by the expression of type 1 fimbria, which is considerably longer than EhaA. Further, 24 of 50 STEC strains were tested for the presence of EhaA and showed expression of this protein under growth in minimal media, perhaps implicating a mechanism where cell-to-cell adhesion is preferred under conditions where type 1 fimbria is not expressed (160). Another study verified a similar set of characteristics for EhaB; expressing this protein in *E. coli* K-12 conferred an ability to form biofilms (161). Expression of EhaB produced an *E. coli* K-12 that was able to interact with the ECM proteins collagen and laminin. When the *ehaB* gene was disrupted in EHEC O157:H7, the mutation did not affect bacterial adhesion to primary epithelial cells derived from the bovine recto-anal junction or to Caco-2 cells. Interestingly, this protein reacted to serum from infected cattle, indicating that it was expressed during infection and is likely to be exposed to the immune system (161).

EspP

Another important autotransporter protein in EHEC is EspP. This protein, a homolog to EspC from EPEC, is a member of the serine protease autotransporters of *Enterobacteriaceae* family of proteins, which, in addition to having adhesive capabilities, also contains protease activity (162). The initial reports indicated that this protein cleaves porcine pepsin A, human coagulation factor V (from the blood coagulation cascade pathway), and apolipoprotein A-1; the purified form of the protein was shown to be cytotoxic to Vero cells (163). Regarding its pathogenic association, this protein is also capable of cleaving the complement proteins C3/C3b and C5, which may help in immunomodulation, and is able to cleave and subsequently inactivate EHEC hemolysin, which might help in controlling and modulating the degree or timing of its own pathological effect (164, 165). Mutations in

espP in O157:H7 abolished adhesion in a calf model and to primary bovine intestinal epithelial cells (164). Addition of exogenous, purified EspP to the in vitro system caused the restoration of the adhesive phenotype of the *espP* mutant strain. A later study investigated the presence of "rope-like fibers" during growth of both EHEC and EPEC strains (167). These fibers were actually made of pure EspP that had oligomerized into fibers measuring up to 2 cm in length. The EspP-based ropes were capable of binding Congo red dye, and it has been proposed that the ropes may partially function in the protection of the bacterial cells from antibiotics and detergents. *E. coli* cells were found to interact directly with this EspP-derived rope, possibly using this protein-based structure as a foundation for biofilm-mediated interactions between bacterial cells (167). This protein was capable of interacting with cultured epithelial cells and with fibronectin, and the interaction with the cultured cells resulted in a cytopathic, but not cytotoxic, effect. Finally, antibodies against EspP were observed in patients with hemolytic-uremic syndrome, indicating that this protein was being expressed during clinical infection and might play a role in the pathogenesis of EHEC (167).

Saa and Sab

Two other autotransporter proteins, Saa and Sab, are both plasmid-encoded adhesins that have been identified primarily in the EHEC LEE-negative O113:H21 strain. Saa was the first adhesin protein to be identified in an LEE-negative strain; the original studies were done in vitro and showed that the expression in *trans* resulted in a nearly 10-fold increase in bacterial adhesion to HEp-2 cells. Conversely, mutation of the *saa* gene resulted in a significant reduction in this binding (168). Multiple variants of the *saa* gene are based on the presence or absence of a series of repeats on the 3′ end. Variations of this 3′ end had been hypothesized as a mechanism for bacterial regulation of the adhesive affinity and shorter variants as less adhesive than longer variants (168, 169). This hypothesis was explored in more depth in a study that verified the presence of the gene in 32 different STEC strains and explored whether these strains adhere to HEp-2 and Caco-2 cells (170). Though there were differences in degrees of adhesion in these strains, the study could not find a correlation between repeat length and adhesive capabilities, nor was any correlation between Saa expression levels and adhesive properties statistically significant (170). Sab has not been characterized as extensively as Saa, but it has been shown to mediate adhesion to HEp-2 cells, and the mutant strains were less adhesive than the wild-type O113 strain (171). Further, the study showed that the protein is present on the surface of these O113 strains, that these proteins could interact with HEp-2 cells in vitro, and that mutant strains were less capable of forming biofilms than their wild-type counterparts (171).

Cah

The final autotransporter protein that has been associated with EHEC infections is the calcium-binding antigen 43 homolog (Cah). When expressed in an *E. coli* K-12 strain, this protein caused bacterial autoaggregation and was also able to bind to Ca^+ ions in solution (172). The *cah* gene is duplicated in EHEC (*cah1* and *cah2*), and their expression is maximized under low nutrient conditions; these proteins were also shown to contribute to the formation of biofilms when grown in minimal media (172). Further studies indicated that this protein was also an important factor in EHEC adherence to alfalfa sprouts and seed coats, a feature of especially high interest when discussing persistence of EHEC in produce (173). These proteins, when provided in *trans* to an *E. coli* K-12 strain, showed increased adhesion to these two alfalfa surfaces. It was hypothesized that this protein may contribute more to a "docking stage" of adhesion, where initial contact is made with the binding surface, than to the "primary stage" of adhesion, where interactions of a higher affinity

promote the long-term adhesion and colonization to a given surface (173).

Flagella

Flagella, proteins associated with bacterial motility, were first characterized as adherence factors in pathogenic *E. coli* in a study that characterized the *fliC* gene in strains O127:H6, O119:H6, O128:H2, and O127:H40 (174). The in vitro studies found an association among flagella, adherence, and the formation of microcolonies on both HeLa cells and HEp-2 cells and also confirmed that these proteins were responsible for the *E. coli* motility phenotype. Flagella were identified as adhesins in EHEC when a *fliC* mutation was found to reduce the degree of virulence in a chick infection model (175). The virulence properties in the intestine are likely a product of a direct interaction with the mucus layer, as EHEC flagella have been shown to interact directly with Muc2, and the EPEC flagella have been shown to interact with Muc1 (176). Interestingly, the EPEC flagellin was capable of interacting with collagen, laminin, and fibronectin, but this interaction was not observed with EHEC flagella (174). The role of flagella as adhesin proteins was further confirmed by a series of standard in vitro adhesion studies showing that a mutation in the *fliC* gene reduced bacterial adhesion to terminal rectum epithelial cells, and complementation of this mutation restored the binding phenotype (177). Purified flagella were capable of interacting with these cells in vitro and of interfering with bacterial binding. Finally, it was shown that the EHEC flagella are actually down-regulated after contact with the epithelium, suggesting that binding may be required only for early stages and provides a means for the T3SS to increase transcription of its components and mediate a more long-term binding event via the A/E lesion (177). Other studies have indicated that flagella are responsible, at least in part, for EHEC binding to leafy greens. Ultrastructural analysis showed the presence of flagella on the surface of EHEC during this bacterium-plant interaction, and mutation of the *fliC* gene negatively affects the binding (152).

Other Adhesin Proteins

Three final adhesins are not included in any of the fimbrial/afimbrial families described above, but their contribution to colonization is significant to EHEC pathogenesis. Two of these, Iha and EibG, are generally regarded as adhesin proteins, though their adhesive characteristics have not been fully demonstrated as others discussed in this review. Iha is a homolog of the *Vibrio cholerae* adhesive protein IrgA and is present in O157:H7 but absent in other strains of EHEC (178). It was originally identified during a random screen, and mutations in this gene showed decreased adhesive capabilities to HeLa and MDBK cells, and subsequently showed increases in binding when added in *trans* to an otherwise non-piliated *E. coli* strain (178). The evidence for its contribution to EHEC adhesion was displayed in two studies. The first of these studies showed that transcription increases under growth conditions simulating short-chain fatty acid concentrations present in the human gut; however, transcription was not regulated by the presence of iron (179). The other study showed that a mutation of the EHEC *iha* gene resulted in a decrease in colonization in ligated pig intestine, but adherence in vitro seemed to be unaffected (180). Despite these data, evidence of Iha directly interacting with cells in vivo or in vitro has not been shown, and the presence of the protein has not been verified over the course of a natural infection, so these effects may be of an indirect nature.

E. coli immunoglobulin-binding protein (EibG) is present in 15% of *eae*-negative strains and exists as three different subtypes (α, β, and γ) (181, 182). Transposon mutagenesis experiments identified an *eibG* mutation in a strain that was no longer able to form its "chain-like adhesion" pattern. This deletion also abrogated in vitro adhesion to HEp-2 cells and eliminated its ability to autoagglutinate red blood cells and to interact with IgG and

IgA (181). Later, in vitro studies using both HCT-8 cells and bovine intestinal epithelial cells indicated that the chain-like adhesion pattern varied in size according to which allele of the *eibG* gene was present in the genome, even within a given subtype (182).

Finally, the role of the outer membrane protein A (OmpA) in adherence was first characterized in EHEC during a transposon mutagenesis screen for a hyperadherent phenotype (183). This screen identified the transcriptional regulator TdcA as a regulator of OmpA. Further investigation led to a specific mutation of *ompA*, which reduced the hyperadherent phenotype in the *tdcA* mutant strain. Antibodies specific to OmpA decreased adhesion by 25% during in vitro adhesion assays with Caco-2 and HeLa cells (184). In addition to being present in EHEC, this protein is commonly present in other commensal and pathogenic strains of *E. coli*, but it is only known to mediate adhesion to brain microvascular cells by NMEC (185–187). Purified OmpA has been crystallized (188) and is known to stimulate dendritic cell migration through polarized epithelial cells (184).

CLOSING REMARKS

As evidenced by the wide scope of this article, a large number of proteins within the EHEC proteome contribute to varying degrees to its interaction with plant leaf surfaces, cattle and human intestine, and abiotic surfaces, such as glass and polystyrene. Figure 2 attempts to graphically depict some of these critical interactions, emphasizing the adhesin proteins that have been demonstrated to be important in intestinal colonization. The variety of surfaces capable of interacting with the bacteria suggests that EHEC possesses a complex regulatory system to control the expression of multiple adhesins, and the temporal regulation adds another layer of complexity in understanding this complex network. Whereas some of these systems may be redundancies built into the pathogen, others likely have a specific function that is not fully characterized. Future

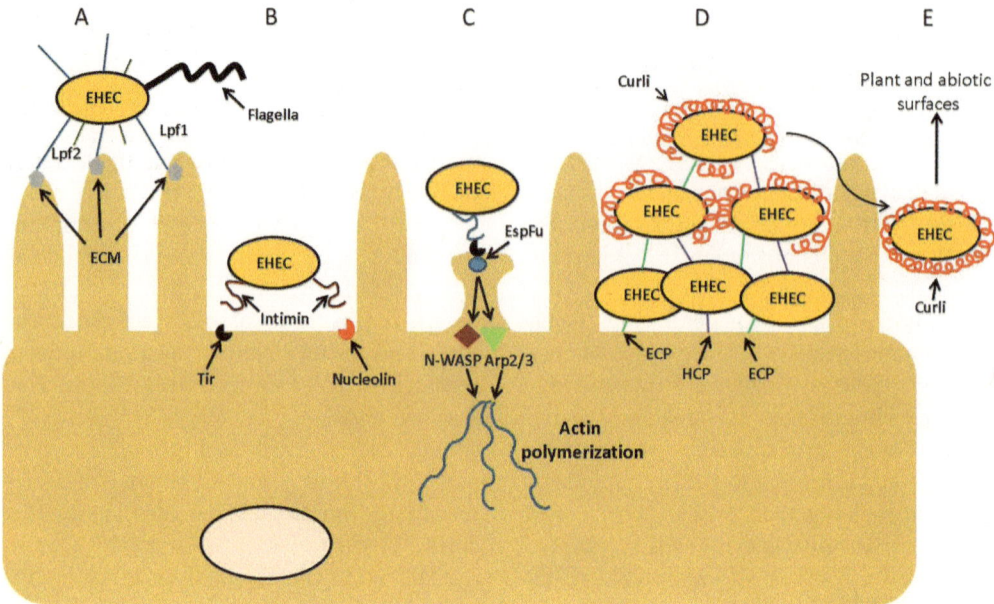

FIGURE 2 Time line and proposed roles of major fimbrial/afimbrial adhesin proteins in EHEC. In (A), Lpf is interacting with the extracellular membrane (ECM) proteins, such as laminin, collagen IV, and fibronectin. After the initial interaction with the intestine is established, EHEC is able to closely attach to the host cell (B) through an interaction of the surface protein intimin with the translocated receptor protein Tir and nucleolin. In (C), the interaction between Tir and intimin is established, initiating a host-cell actin rearrangement via participation of the translocated bacterial protein EspFu and the recruitment of several host proteins. In (D), ECP and HCP are proposed to interact with the surface of host intestinal cells, perhaps strengthening the colonization through the formation of microcolonies and/or biofilms. Further, curli is shown to contribute in the establishment of biofilms, though an interaction with ECP and HCP is not demonstrated. Finally in (E), dominant curli expression has been proposed as detrimental for the survival of EHEC in the intestine, suggesting that this surface structure might become important for the pathogen interaction with abiotic surfaces and during colonization of the surface of vegetable and plant leaves. doi:10.1128/microbiolspec.EHEC-0003-2013.f2

research will attempt to address these questions through structural and functional studies, and it seems probable that new adhesins and their host receptors will be discovered as well. Finally, as we begin to understand EHEC pathogenesis more thoroughly, adhesion models can become more refined and can begin to tease out why, when, and how this hugely diverse set of proteins are mediating EHEC colonization of the intestine or persistence in the environment.

ACKNOWLEDGMENTS

Research work in the AGT laboratory is supported by NIH/NIAID grants AI079154 and AI09956001. The contents are solely the responsibility of the authors and do not necessarily represent the official views of NIAID or NIH.

CITATION

McWilliams BD, Torres AG. 2014. Enterohemorrhagic *Escherichia coli* adhesins. Microbiol Spectrum 2(3):EHEC-0003-2013.

REFERENCES

1. **Francis DH, Collins JE, Duimstra JR.** 1986. Infection of gnotobiotic pigs with an *Escherichia coli* O157:H7 strain associated with an outbreak of hemorrhagic colitis. *Infect Immun* **51:**953–956.
2. **Yu J, Kaper JB.** 1992. Cloning and characterization of the *eae* gene of enterohaemorrhagic

Escherichia coli O157:H7. *Mol Microbiol* **6**:411–417.

3. **Donnenberg MS, Tzipori S, McKee ML, O'Brien AD, Alroy J, Kaper JB.** 1993. The role of the *eae* gene of enterohemorrhagic *Escherichia coli* in intimate attachment in vitro and in a porcine model. *J Clin Invest* **92**:1418–1424.

4. **Donnenberg MS, Tacket CO, James SP, Losonsky G, Nataro JP, Wasserman SS, Kaper JB, Levine MM.** 1993. Role of the *eaeA* gene in experimental enteropathogenic *Escherichia coli* infection. *J Clin Invest* **92**:1412–1417.

5. **Torres AG, Kaper JB.** 2002. Pathogenicity islands of intestinal *E. coli*. Pathogenicity islands (PAIs) and the evolution of pathogenic microbes. *Curr Top Microbiol Immunol* **264**:31–48.

6. **Stevens MP, Wallis TS.** 29 August 2005, posting date Chapter 8.3.2.3, Adhesins of enterohemorrhagic *Escherichia coli*. *In* Böck A, et al. (ed), *EcoSal—Escherichia coli and Salmonella: Cellular and Molecular Biology*. ASM Press, Washington, DC. doi:10.1128/ecosal.8.3.2.3.

7. **McDaniel TK, Jarvis KG, Donnenberg MS, Kaper JB.** 1995. A genetic locus of enterocyte effacement conserved among diverse enterobacterial pathogens. *Proc Natl Acad Sci USA* **92**:1664–1668.

8. **Elliott SJ, Wainwright LA, McDaniel TK, Jarvis KG, Deng YK, Lai LC, McNamara BP, Donnenberg MS, Kaper JB.** 1998. The complete sequence of the locus of enterocyte effacement (LEE) from enteropathogenic *Escherichia coli* E2348/69. *Mol Microbiol* **28**:1–4.

9. **Perna NT, Mayhew GF, Posfai G, Elliott S, Donnenberg MS, Kaper JB, Blattner FR.** 1998. Molecular evolution of a pathogenicity island from enterohemorrhagic *Escherichia coli* O157:H7. *Infect Immun* **66**:3810–3817.

10. **Zhu C, Agin TS, Elliott SJ, Johnson LA, Thate TE, Kaper JB, Boedeker EC.** 2001. Complete nucleotide sequence and analysis of the locus of enterocyte effacement from rabbit diarrheagenic *Escherichia coli* RDEC-1. *Infect Immun* **69**:2107–2115.

11. **Tauschek M, Strugnell RA, Robins-Browne RM.** 2002. Characterization and evidence of mobilization of the LEE pathogenicity island of rabbit-specific strains of enteropathogenic *Escherichia coli*. *Mol Microbiol* **44**:1533–1550.

12. **Deng W, Li Y, Vallance BA, Finlay BB.** 2001. Locus of enterocyte effacement from *Citrobacter rodentium*: sequence analysis and evidence for horizontal transfer among attaching and effacing pathogens. *Infect Immun* **69**:6323–6335.

13. **Bretschneider G, Berberov EM, Moxley RA.** 2007. Isotype-specific antibody responses against *Escherichia coli* O157:H7 locus of enterocyte

effacement proteins in adult beef cattle following experimental infection. *Vet Immunol Immunopathol* **118**:229–238.

14. **Dean-Nystrom EA, Bosworth BT, Moon HW, O'Brien AD.** 1998. *Escherichia coli* O157:H7 requires intimin for enteropathogenicity in calves. *Infect Immun* **66**:4560–4563.

15. **Judge NA, Mason HS, O'Brien AD.** 2004. Plant cell-based intimin vaccine given orally to mice primed with intimin reduces time of *Escherichia coli* O157:H7 shedding in feces. *Infect Immun* **72**:168–175.

16. **Dziva F, van Diemen PM, Stevens MP, Smith AJ, Wallis TS.** 2004. Identification of *Escherichia coli* O157:H7 genes influencing colonization of the bovine gastrointestinal tract using signature-tagged mutagenesis. *Microbiology* **150**:3631–3645.

17. **Cornick NA, Booher SL, Moon HW.** 2002. Intimin facilitates colonization by *Escherichia coli* O157:H7 in adult ruminants. *Infect Immun* **70**:2704–2707.

18. **Ritchie JM, Thorpe CM, Rogers AB, Waldor MK.** 2003. Critical roles for *stx2*, *eae*, and *tir* in enterohemorrhagic *Escherichia coli*-induced diarrhea and intestinal inflammation in infant rabbits. *Infect Immun* **71**:7129–7139.

19. **Abe A, Heczko U, Hegele RG, Brett FB.** 1998. Two enteropathogenic *Escherichia coli* type III secreted proteins, EspA and EspB, are virulence factors. *J Exp Med* **188**:1907–1916.

20. **Ebel F, Deibel C, Kresse AU, Guzman CA, Chakraborty T.** 1996. Temperature- and medium-dependent secretion of proteins by Shiga toxin-producing *Escherichia coli*. *Infect Immun* **64**:4472–4479.

21. **Kenny B, Finlay BB.** 1995. Protein secretion by enteropathogenic *Escherichia coli* is essential for transducing signals to epithelial cells. *Proc Natl Acad Sci USA* **92**:7991–7995.

22. **Jarvis KG, Kaper JB.** 1996. Secretion of extracellular proteins by enterohemorrhagic *Escherichia coli* via a putative type III secretion system. *Infect Immun* **64**:4826–4829.

23. **Hueck CJ.** 1998. Type III protein secretion systems in bacterial pathogens of animals and plants. *Microbiol Mol Biol Rev* **62**:379–433.

24. **Gauthier A, Finlay BB.** 2003. Translocated intimin receptor and its chaperone interact with ATPase of the type III secretion apparatus of enteropathogenic *Escherichia coli*. *J Bacteriol* **185**:6747–6755.

25. **Garmendia J, Frankel G, Crepin VF.** 2005. Enteropathogenic and enterohemorrhagic *Escherichia coli* infections: translocation, translocation, translocation. *Infect Immun* **73**:2573–2585.

26. **Sharp FC, Sperandio V.** 2007. QseA directly activates transcription of LEE1 in enterohemorrhagic *Escherichia coli*. *Infect Immun* **75**:2432–2440.

27. **Atlung T, Ingmer H.** 1997. H-NS: a modulator of environmentally regulated gene expression. *Mol Microbiol* **24:**7–17.

28. **Mellies JL, Elliott SJ, Sperandio V, Donnenberg MS, Kaper JB.** 1999. The Per regulon of enteropathogenic *Escherichia coli*: identification of a regulatory cascade and a novel transcriptional activator, the locus of enterocyte effacement (LEE)-encoded regulator (Ler). *Mol Microbiol* **33:**296–306.

29. **Deng W, Puente JL, Gruenheid S, Li Y, Vallance BA, Vazquez A, Barba J, Ibarra JA, O'Donnell P, Metalnikov P, Ashman K, Lee S, Goode D, Pawson T, Finlay BB.** 2004. Dissecting virulence: systematic and functional analyses of a pathogenicity island. *Proc Natl Acad Sci USA* **101:**3597–3602.

30. **Kenny B, Abe A, Stein M, Finlay BB.** 1997. Enteropathogenic *Escherichia coli* protein secretion is induced in response to conditions similar to those in the gastrointestinal tract. *Infect Immun* **65:**2606–2612.

31. **Bansal T, Jesudhasan P, Pillai S, Wood TK, Jayaraman A.** 2008. Temporal regulation of enterohemorrhagic *Escherichia coli* virulence mediated by autoinducer-2. *Appl Microbiol Biotechnol* **78:**811–819.

32. **Sperandio V, Li CC, Kaper JB.** 2002. Quorum-sensing *Escherichia coli* regulator A: a regulator of the LysR family involved in the regulation of the locus of enterocyte effacement pathogenicity island in enterohemorrhagic *E. coli*. *Infect Immun* **70:**3085–3093.

33. **Abe H, Tatsuno I, Tobe T, Okutani A, Sasakawa C.** 2002. Bicarbonate ion stimulates the expression of locus of enterocyte effacement-encoded genes in enterohemorrhagic *Escherichia coli* O157:H7. *Infect Immun* **70:**3500–3509.

34. **Yin X, Feng Y, Wheatcroft R, Chambers J, Gong J, Gyles CL.** 2011. Adherence of *Escherichia coli* O157:H7 to epithelial cells in vitro and in pig gut loops is affected by bacterial culture conditions. *Can J Vet Res* **75:**81–88.

35. **Bansal T, Kim DN, Slininger T, Wood TK, Jayaraman A.** 2012. Human intestinal epithelial cell-derived molecule(s) increase enterohemorrhagic *Escherichia coli* virulence. *FEMS Immunol Med Microbiol* **66:**399–410.

36. **Bhatt S, Edwards AN, Nguyen HT, Merlin D, Romeo T, Kalman D.** 2009. The RNA binding protein CsrA is a pleiotropic regulator of the locus of enterocyte effacement pathogenicity island of enteropathogenic *Escherichia coli*. *Infect Immun* **77:**3552–3568.

37. **Russell RM, Sharp FC, Rasko DA, Sperandio V.** 2007. QseA and GrlR/GrlA regulation of the locus of enterocyte effacement genes in enterohemorrhagic *Escherichia coli*. *J Bacteriol* **189:**5387–5392.

38. **Bustamante VH, Santana FJ, Calva E, Puente JL.** 2001. Transcriptional regulation of type III secretion genes in enteropathogenic *Escherichia coli*: Ler antagonizes H-NS-dependent repression. *Mol Microbiol* **39:**664–678.

39. **Elliott SJ, Sperandio V, Giron JA, Shin S, Mellies JL, Wainwright L, Hutcheson SW, McDaniel TK, Kaper JB.** 2000. The locus of enterocyte effacement (LEE)-encoded regulator controls expression of both LEE- and non-LEE-encoded virulence factors in enteropathogenic and enterohemorrhagic *Escherichia coli*. *Infect Immun* **68:**6115–6126.

40. **Friedberg D, Umanski T, Fang Y, Rosenshine I.** 1999. Hierarchy in the expression of the locus of enterocyte effacement genes of enteropathogenic *Escherichia coli*. *Mol Microbiol* **34:**941–952.

41. **Njoroge JW, Nguyen Y, Curtis MM, Moreira CG, Sperandio V.** 2012. Virulence meets metabolism: Cra and KdpE gene regulation in enterohemorrhagic *Escherichia coli*. *MBio* **3:**e00280-12.

42. **Laaberki MH, Janabi N, Oswald E, Repoila F.** 2006. Concert of regulators to switch on LEE expression in enterohemorrhagic *Escherichia coli* O157:H7: interplay between Ler, GrlA, HNS and RpoS. *Int J Med Microbiol* **296:**197–210.

43. **Kendall MM, Gruber CC, Rasko DA, Hughes DT, Sperandio V.** 2011. Hfq virulence regulation in enterohemorrhagic *Escherichia coli* O157:H7 strain 86-24. *J Bacteriol* **193:**6843–6851.

44. **Barba J, Bustamante VH, Flores-Valdez MA, Deng W, Finlay BB, Puente JL.** 2005. A positive regulatory loop controls expression of the locus of enterocyte effacement-encoded regulators Ler and GrlA. *J Bacteriol* **187:**7918–7930.

45. **Dorman CJ.** 2004. H-NS: a universal regulator for a dynamic genome. *Nat Rev Microbiol* **2:**391–400.

46. **Garcia J, Cordeiro TN, Prieto MJ, Pons M.** 2012. Oligomerization and DNA binding of Ler, a master regulator of pathogenicity of enterohemorrhagic and enteropathogenic *Escherichia coli*. *Nucleic Acids Res* **40:**10254–10262.

47. **Jerse AE, Yu J, Tall BD, Kaper JB.** 1990. A genetic locus of enteropathogenic *Escherichia coli* necessary for the production of attaching and effacing lesions on tissue culture cells. *Proc Natl Acad Sci USA* **87:**7839–7843.

48. **Tzipori S, Gunzer F, Donnenberg MS, de Montigny L, Kaper JB, Donohue-Rolfe A.** 1995. The role of the *eaeA* gene in diarrhea and neurological complications in a gnotobiotic piglet model of enterohemorrhagic *Escherichia coli* infection. *Infect Immun* **63:**3621–3627.

49. **McKee ML, Melton-Celsa AR, Moxley RA, Francis DH, O'Brien AD.** 1995. Enterohemorrhagic *Escherichia coli* O157:H7 requires intimin

to colonize the gnotobiotic pig intestine and to adhere to HEp-2 cells. *Infect Immun* **63**:3739–3744.

50. **Gansheroff LJ, Wachtel MR, O'Brien AD.** 1999. Decreased adherence of enterohemorrhagic *Escherichia coli* to HEp-2 cells in the presence of antibodies that recognize the C-terminal region of intimin. *Infect Immun* **67**:6409–6417.

51. **Buist G, Steen A, Kok J, Kuipers OP.** 2008. LysM, a widely distributed protein motif for binding to (peptido)glycans. *Mol Microbiol* **68**:838–847.

52. **Touze T, Hayward RD, Eswaran J, Leong JM, Koronakis V.** 2004. Self-association of EPEC intimin mediated by the beta-barrel-containing anchor domain: a role in clustering of the Tir receptor. *Mol Microbiol* **51**:73–87.

53. **Kelly G, Prasannan S, Daniell S, Fleming K, Frankel G, Dougan G, Connerton I, Matthews S.** 1999. Structure of the cell-adhesion fragment of intimin from enteropathogenic *Escherichia coli*. *Nat Struct Biol* **6**:313–318.

54. **Batchelor M, Prasannan S, Daniell S, Reece S, Connerton I, Bloomberg G, Dougan G, Frankel G, Matthews S.** 2000. Structural basis for recognition of the translocated intimin receptor (Tir) by intimin from enteropathogenic *Escherichia coli*. *EMBO J* **19**:2452–2464.

55. **Luo Y, Frey EA, Pfuetzner RA, Creagh AL, Knoechel DG, Haynes CA, Finlay BB, Strynadka NC.** 2000. Crystal structure of enteropathogenic *Escherichia coli* intimin-receptor complex. *Nature* **405**:1073–1077.

56. **Rosenshine I, Ruschkowski S, Stein M, Reinscheid DJ, Mills SD, Finlay BB.** 1996. A pathogenic bacterium triggers epithelial signals to form a functional bacterial receptor that mediates actin pseudopod formation. *EMBO J* **15**:2613–2624.

57. **Campellone KG, Leong JM.** 2003. Tails of two Tirs: actin pedestal formation by enteropathogenic *E. coli* and enterohemorrhagic *E. coli* O157:H7. *Curr Opin Microbiol* **6**:82–90.

58. **Campellone KG, Brady MJ, Alamares JG, Rowe DC, Skehan BM, Tipper DJ, Leong JM.** 2006. Enterohaemorrhagic *Escherichia coli* Tir requires a C-terminal 12-residue peptide to initiate EspF-mediated actin assembly and harbours N-terminal sequences that influence pedestal length. *Cell Microbiol* **8**:1488–1503.

59. **Kenny B.** 1999. Phosphorylation of tyrosine 474 of the enteropathogenic *Escherichia coli* (EPEC) Tir receptor molecule is essential for actin nucleating activity and is preceded by additional host modifications. *Mol Microbiol* **31**:1229–1241.

60. **Ismaili A, Philpott DJ, Dytoc MT, Sherman PM.** 1995. Signal transduction responses following adhesion of verocytotoxin-producing *Escherichia coli*. *Infect Immun* **63**:3316–3326.

61. **DeVinney R, Stein M, Reinscheid D, Abe A, Ruschkowski S, Finlay BB.** 1999. Enterohemorrhagic *Escherichia coli* O157:H7 produces Tir, which is translocated to the host cell membrane but is not tyrosine phosphorylated. *Infect Immun* **67**:2389–2398.

62. **Kenny B, Warawa J.** 2001. Enteropathogenic *Escherichia coli* (EPEC) Tir receptor molecule does not undergo full modification when introduced into host cells by EPEC-independent mechanisms. *Infect Immun* **69**:1444–1453.

63. **de Groot JC, Schluter K, Carius Y, Quedenau C, Vingadassalom D, Faix J, Weiss SM, Reichelt J, Standfuss-Gabisch C, Lesser CF, Leong JM, Heinz DW, Bussow K, Stradal TE.** 2011. Structural basis for complex formation between human IRSp53 and the translocated intimin receptor Tir of enterohemorrhagic *E. coli*. *Structure* **19**:1294–1306.

64. **Garmendia J, Phillips AD, Carlier MF, Chong Y, Schuller S, Marches O, Dahan S, Oswald E, Shaw RK, Knutton S, Frankel G.** 2004. TccP is an enterhemorrhagic *Escherichia coli* O157:H7 type III effector protein that couples Tir to the actin cytoskeleton. *Cell Microbiol* **6**:1167–1183.

65. **Weiss SM, Ladwein M, Schmidt D, Ehinger J, Lommel S, Stading K, Beutling U, Disanza A, Frank R, Jansch L, Scita G, Gunzer F, Rottner K, Stradal TE.** 2009. IRSp53 links the enterohemorrhagic *E. coli* effectors Tir and EspFU for actin pedestal formation. *Cell Host Microbe* **5**:244–258.

66. **Vingadassalom D, Kazlauskas A, Skehan B, Cheng HC, Magoun L, Robbins D, Rosen MK, Saksela K, Leong JM.** 2009. Insulin receptor tyrosine kinase substrate links the *E. coli* O157:H7 actin assembly effectors Tir and EspF(U) during pedestal formation. *Proc Natl Acad Sci USA* **106**:6754–6759.

67. **Frankel G, Candy DC, Everest P, Dougan G.** 1994. Characterization of the C-terminal domains of intimin-like proteins of enteropathogenic and enterohemorrhagic *Escherichia coli*, *Citrobacter freundii*, and *Hafnia alvei*. *Infect Immun* **62**:1835–1842.

68. **Frankel G, Candy DC, Fabiani E, Adu-Bobie J, Gil S, Novakova M, Phillips AD, Dougan G.** 1995. Molecular characterization of a carboxy-terminal eukaryotic-cell-binding domain of intimin from enteropathogenic *Escherichia coli*. *Infect Immun* **63**:4323–4328.

69. **Frankel G, Lider O, Hershkoviz R, Mould AP, Kachalsky SG, Candy DC, Cahalon L, Humphries MJ, Dougan G.** 1996. The cell-binding domain of intimin from enteropathogenic *Escherichia coli* binds to beta1 integrins. *J Biol Chem* **271**:20359–20364.

70. **Liu H, Magoun L, Leong JM.** 1999. beta1-chain integrins are not essential for intimin-mediated host cell attachment and enteropathogenic *Escherichia coli*-induced actin condensation. *Infect Immun* **67:**2045–2049.

71. **Hartland EL, Batchelor M, Delahay RM, Hale C, Matthews S, Dougan G, Knutton S, Connerton I, Frankel G.** 1999. Binding of intimin from enteropathogenic *Escherichia coli* to Tir and to host cells. *Mol Microbiol* **32:**151–158.

72. **Sinclair JF, Dean-Nystrom EA, O'Brien AD.** 2006. The established intimin receptor Tir and the putative eucaryotic intimin receptors nucleolin and beta1 integrin localize at or near the site of enterohemorrhagic *Escherichia coli* O157:H7 adherence to enterocytes in vivo. *Infect Immun* **74:**1255–1265.

73. **Gu L, Wang H, Guo YL, Zen K.** 2008. Heparin blocks the adhesion of *E. coli* O157:H7 to human colonic epithelial cells. *Biochem Biophys Res Commun* **369:**1061–1064.

74. **Sinclair JF, O'Brien AD.** 2002. Cell surface-localized nucleolin is a eukaryotic receptor for the adhesin intimin-gamma of enterohemorrhagic *Escherichia coli* O157:H7. *J Biol Chem* **277:**2876–2885.

75. **Sinclair JF, O'Brien AD.** 2004. Intimin types alpha, beta, and gamma bind to nucleolin with equivalent affinity but lower avidity than to the translocated intimin receptor. *J Biol Chem* **279:**33751–33758.

76. **Liu B, Yin X, Feng Y, Chambers JR, Guo A, Gong J, Zhu J, Gyles CL.** 2010. Verotoxin 2 enhances adherence of enterohemorrhagic *Escherichia coli* O157:H7 to intestinal epithelial cells and expression of {beta}1-integrin by IPEC-J2 cells. *Appl Environ Microbiol* **76:**4461–4468.

77. **Jenkins C, Chart H, Smith HR, Hartland EL, Batchelor M, Delahay RM, Dougan G, Frankel G.** 2000. Antibody response of patients infected with verocytotoxin-producing *Escherichia coli* to protein antigens encoded on the LEE locus. *J Med Microbiol* **49:**97–101.

78. **Karpman D, Bekassy ZD, Sjogren AC, Dubois MS, Karmali MA, Mascarenhas M, Jarvis KG, Gansheroff LJ, O'Brien AD, Arbus GS, Kaper JB.** 2002. Antibodies to intimin and *Escherichia coli* secreted proteins A and B in patients with enterohemorrhagic *Escherichia coli* infections. *Pediatr Nephrol* **17:**201–211.

79. **Li Y, Frey E, Mackenzie AM, Finlay BB.** 2000. Human response to *Escherichia coli* O157:H7 infection: antibodies to secreted virulence factors. *Infect Immun* **68:**5090–5095.

80. **Adu-Bobie J, Frankel G, Bain C, Goncalves AG, Trabulsi LR, Douce G, Knutton S, Dougan G.** 1998. Detection of intimins alpha, beta, gamma, and delta, four intimin derivatives expressed by attaching and effacing microbial pathogens. *J Clin Microbiol* **36:**662–668.

81. **Blanco M, Blanco JE, Blanco J, de Carvalho VM, Onuma DL, Pestana de Castro AF.** 2004. Typing of intimin (*eae*) genes in attaching and effacing *Escherichia coli* strains from monkeys. *J Clin Microbiol* **42:**1382–1383.

82. **China B, Goffaux F, Pirson V, Mainil J.** 1999. Comparison of *eae, tir, espA* and *espB* genes of bovine and human attaching and effacing *Escherichia coli* by multiplex polymerase chain reaction. *FEMS Microbiol Lett* **178:**177–182.

83. **Oswald E, Schmidt H, Morabito S, Karch H, Marches O, Caprioli A.** 2000. Typing of intimin genes in human and animal enterohemorrhagic and enteropathogenic *Escherichia coli*: characterization of a new intimin variant. *Infect Immun* **68:**64–71.

84. **Reid SD, Betting DJ, Whittam TS.** 1999. Molecular detection and identification of intimin alleles in pathogenic *Escherichia coli* by multiplex PCR. *J Clin Microbiol* **37:**2719–2722.

85. **Zhang WL, Kohler B, Oswald E, Beutin L, Karch H, Morabito S, Caprioli A, Suerbaum S, Schmidt H.** 2002. Genetic diversity of intimin genes of attaching and effacing *Escherichia coli* strains. *J Clin Microbiol* **40:**4486–4492.

86. **Ito K, Iida M, Yamazaki M, Moriya K, Moroishi S, Yatsuyanagi J, Kurazono T, Hiruta N, Ratchtrachenchai OA.** 2007. Intimin types determined by heteroduplex mobility assay of intimin gene (*eae*)-positive *Escherichia coli* strains. *J Clin Microbiol* **45:**1038–1041.

87. **Torres AG, Zhou X, Kaper JB.** 2005. Adherence of diarrheagenic *Escherichia coli* strains to epithelial cells. *Infect Immun* **73:**18–29.

88. **Phillips AD, Frankel G.** 2000. Intimin-mediated tissue specificity in enteropathogenic *Escherichia coli* interaction with human intestinal organ cultures. *J Infect Dis* **181:**1496–1500.

89. **Mundy R, Schuller S, Girard F, Fairbrother JM, Phillips AD, Frankel G.** 2007. Functional studies of intimin in vivo and ex vivo: implications for host specificity and tissue tropism. *Microbiology* **153:**959–967.

90. **Ahmed S, Byrd W, Kumar S, Boedeker EC.** 2013. A directed intimin insertion mutant of a rabbit enteropathogenic *Escherichia coli* (REPEC) is attenuated, immunogenic and elicits serogroup specific protection. *Vet Immunol Immunopathol* **152:**146–155.

91. **Guirro M, de Souza RL, Piazza RM, Guth BE.** 2013. Antibodies to intimin and *Escherichia coli*-secreted proteins EspA and EspB in sera of Brazilian children with hemolytic uremic syndrome and healthy controls. *Vet Immunol Immunopathol* **152:**121–125.

92. **McKee ML, O'Brien AD.** 1996. Truncated entero-hemorrhagic *Escherichia coli* (EHEC) O157:H7 intimin (EaeA) fusion proteins promote adherence of EHEC strains to HEp-2 cells. *Infect Immun* 64:2225–2233.

93. **Dean-Nystrom EA, Gansheroff LJ, Mills M, Moon HW, O'Brien AD.** 2002. Vaccination of pregnant dams with intimin(O157) protects suckling piglets from *Escherichia coli* O157:H7 infection. *Infect Immun* 70:2414–2418.

94. **Ghaem-Maghami M, Simmons CP, Daniell S, Pizza M, Lewis D, Frankel G, Dougan G.** 2001. Intimin-specific immune responses prevent bacterial colonization by the attaching-effacing pathogen *Citrobacter rodentium. Infect Immun* 69:5597–5605.

95. **Agin TS, Zhu C, Johnson LA, Thate TE, Yang Z, Boedeker EC.** 2005. Protection against hemorrhagic colitis in an animal model by oral immunization with isogeneic rabbit enteropathogenic *Escherichia coli* attenuated by truncating intimin. *Infect Immun* 73:6608–6619.

96. **Vilte DA, Larzabal M, Cataldi AA, Mercado EC.** 2008. Bovine colostrum contains immunoglobulin G antibodies against intimin, EspA, and EspB and inhibits hemolytic activity mediated by the type three secretion system of attaching and effacing *Escherichia coli. Clin Vaccine Immunol* 15:1208–1213.

97. **Rabinovitz BC, Gerhardt E, Tironi FC, Abdala A, Galarza R, Vilte DA, Ibarra C, Cataldi A, Mercado EC.** 2012. Vaccination of pregnant cows with EspA, EspB, gamma-intimin, and Shiga toxin 2 proteins from *Escherichia coli* O157:H7 induces high levels of specific colostral antibodies that are transferred to newborn calves. *J Dairy Sci* 95:3318–3326.

98. **Amani J, Mousavi SL, Rafati S, Salmanian AH.** 2011. Immunogenicity of a plant-derived edible chimeric EspA, Intimin and Tir of *Escherichia coli* O157:H7 in mice. *Plant Sci* 180:620–627.

99. **Hur J, Lee JH.** 2011. Immune responses to new vaccine candidates constructed by a live attenuated *Salmonella* Typhimurium delivery system expressing *Escherichia coli* F4, F5, F6, F41 and intimin adhesin antigens in a murine model. *J Vet Med Sci* 73:1265–1273.

100. **Oliveira AF, Cardoso SA, Almeida FB, de Oliveira LL, Pitondo-Silva A, Soares SG, Hanna ES.** 2012. Oral immunization with attenuated *Salmonella* vaccine expressing *Escherichia coli* O157:H7 intimin gamma triggers both systemic and mucosal humoral immunity in mice. *Microbiol Immunol* 56:513–522.

101. **Dytoc MT, Ismaili A, Philpott DJ, Soni R, Brunton JL, Sherman PM.** 1994. Distinct binding properties of *eaeA*-negative verocytotoxin-producing *Escherichia coli* of serotype O113:H21. *Infect Immun* 62:3494–3505.

102. **Paton AW, Woodrow MC, Doyle RM, Lanser JA, Paton JC.** 1999. Molecular characterization of a Shiga toxigenic *Escherichia coli* O113:H21 strain lacking *eae* responsible for a cluster of cases of hemolytic-uremic syndrome. *J Clin Microbiol* 37:3357–3361.

103. **Fitzhenry RJ, Reece S, Trabulsi LR, Heuschkel R, Murch S, Thomson M, Frankel G, Phillips AD.** 2002. Tissue tropism of enteropathogenic *Escherichia coli* strains belonging to the O55 serogroup. *Infect Immun* 70:4362–4368.

104. **Hayashi T, Makino K, Ohnishi M, Kurokawa K, Ishii K, Yokoyama K, Han CG, Ohtsubo E, Nakayama K, Murata T, Tanaka M, Tobe T, Iida T, Takami H, Honda T, Sasakawa C, Ogasawara N, Yasunaga T, Kuhara S, Shiba T, Hattori M, Shinagawa H.** 2001. Complete genome sequence of enterohemorrhagic *Escherichia coli* O157:H7 and genomic comparison with a laboratory strain K-12. *DNA Res* 8:11–22.

105. **Perna NT, Plunkett G III, Burland V, Mau B, Glasner JD, Rose DJ, Mayhew GF, Evans PS, Gregor J, Kirkpatrick HA, Posfai G, Hackett J, Klink S, Boutin A, Shao Y, Miller L, Grotbeck EJ, Davis NW, Lim A, Dimalanta ET, Potamousis KD, Apodaca J, Anantharaman TS, Lin J, Yen G, Schwartz DC, Welch RA, Blattner FR.** 2001. Genome sequence of enterohaemorrhagic *Escherichia coli* O157:H7. *Nature* 409:529–533.

106. **Baumler AJ, Heffron F.** 1995. Identification and sequence analysis of *lpfABCDE*, a putative fimbrial operon of *Salmonella typhimurium. J Bacteriol* 177:2087–2097.

107. **Baumler AJ, Tsolis RM, Heffron F.** 1996. The *lpf* fimbrial operon mediates adhesion of *Salmonella typhimurium* to murine Peyer's patches. *Proc Natl Acad Sci USA* 93:279–283.

108. **Torres AG, Giron JA, Perna NT, Burland V, Blattner FR, Avelino-Flores F, Kaper JB.** 2002. Identification and characterization of *lpfABCC'DE*, a fimbrial operon of enterohemorrhagic *Escherichia coli* O157:H7. *Infect Immun* 70:5416–5427.

109. **Doughty S, Sloan J, Bennett-Wood V, Robertson M, Robins-Browne RM, Hartland EL.** 2002. Identification of a novel fimbrial gene cluster related to long polar fimbriae in locus of enterocyte effacement-negative strains of enterohemorrhagic *Escherichia coli. Infect Immun* 70:6761–6769.

110. **Torres AG, Kanack KJ, Tutt CB, Popov V, Kaper JB.** 2004. Characterization of the second long polar (LP) fimbriae of *Escherichia coli* O157:H7 and distribution of LP fimbriae in other pathogenic *E. coli* strains. *FEMS Microbiol Lett* 238:333–344.

111. Osek J, Weiner M, Hartland EL. 2003. Prevalence of the *lpfO113* gene cluster among *Escherichia coli* O157 isolates from different sources. *Vet Microbiol* **96:**259–266.

112. Torres AG, Blanco M, Valenzuela P, Slater TM, Patel SD, Dahbi G, Lopez C, Barriga XF, Blanco JE, Gomes TA, Vidal R, Blanco J. 2009. Genes related to long polar fimbriae of pathogenic *Escherichia coli* strains as reliable markers to identify virulent isolates. *J Clin Microbiol* **47:**2442–2451.

113. Galli L, Torres AG, Rivas M. 2010. Identification of the long polar fimbriae gene variants in the locus of enterocyte effacement-negative Shiga toxin-producing *Escherichia coli* strains isolated from humans and cattle in Argentina. *FEMS Microbiol Lett* **308:**123–129.

114. Torres AG, Milflores-Flores L, Garcia-Gallegos JG, Patel SD, Best A, La Ragione RM, Martinez-Laguna Y, Woodward MJ. 2007. Environmental regulation and colonization attributes of the long polar fimbriae (LPF) of *Escherichia coli* O157:H7. *Int J Med Microbiol* **297:**177–185.

115. Torres AG, Lopez-Sanchez GN, Milflores-Flores L, Patel SD, Rojas-Lopez M, Martinez de la Pena CF, Arenas-Hernandez MM, Martinez-Laguna Y. 2007. Ler and H-NS, regulators controlling expression of the long polar fimbriae of *Escherichia coli* O157:H7. *J Bacteriol* **189:**5916–5928.

116. Rojas-Lopez M, Arenas-Hernandez MM, Medrano-Lopez A, Martinez de la Pena CF, Puente JL, Martinez-Laguna Y, Torres AG. 2011. Regulatory control of the *Escherichia coli* O157:H7 *lpf1* operon by H-NS and Ler. *J Bacteriol* **193:**1622–1632.

117. Torres AG, Slater TM, Patel SD, Popov VL, Arenas-Hernandez MM. 2008. Contribution of the Ler- and H-NS-regulated long polar fimbriae of *Escherichia coli* O157:H7 during binding to tissue-cultured cells. *Infect Immun* **76:**5062–5071.

118. Farfan MJ, Cantero L, Vidal R, Botkin DJ, Torres AG. 2011. Long polar fimbriae of enterohemorrhagic *Escherichia coli* O157:H7 bind to extracellular matrix proteins. *Infect Immun* **79:**3744–3750.

119. Jordan DM, Cornick N, Torres AG, Dean-Nystrom EA, Kaper JB, Moon HW. 2004. Long polar fimbriae contribute to colonization by *Escherichia coli* O157:H7 in vivo. *Infect Immun* **72:**6168–6171.

120. Fitzhenry R, Dahan S, Torres AG, Chong Y, Heuschkel R, Murch SH, Thomson M, Kaper JB, Frankel G, Phillips AD. 2006. Long polar fimbriae and tissue tropism in *Escherichia coli* O157:H7. *Microbes Infect* **8:**1741–1749.

121. Lloyd SJ, Ritchie JM, Rojas-Lopez M, Blumentritt CA, Popov VL, Greenwich JL, Waldor MK, Torres AG. 2012. A double, long polar fimbria mutant of *Escherichia coli* O157:H7 expresses curli and exhibits reduced in vivo colonization. *Infect Immun* **80:**914–920.

122. Barnhart MM, Chapman MR. 2006. Curli biogenesis and function. *Annu Rev Microbiol* **60:**131–147.

123. Fitzhenry RJ, Stevens MP, Jenkins C, Wallis TS, Heuschkel R, Murch S, Thomson M, Frankel G, Phillips AD. 2003. Human intestinal tissue tropism of intimin epsilon O103 *Escherichia coli*. *FEMS Microbiol Lett* **218:**311–316.

124. Olsen A, Jonsson A, Normark S. 1989. Fibronectin binding mediated by a novel class of surface organelles on *Escherichia coli*. *Nature* **338:**652–655.

125. Olsen A, Wick MJ, Morgelin M, Bjorck L. 1998. Curli, fibrous surface proteins of *Escherichia coli*, interact with major histocompatibility complex class I molecules. *Infect Immun* **66:**944–949.

126. Collinson SK, Clouthier SC, Doran JL, Banser PA, Kay WW. 1997. Characterization of the *agfBA* fimbrial operon encoding thin aggregative fimbriae of *Salmonella enteritidis*. *Adv Exp Med Biol* **412:**247–248.

127. Romling U, Bian Z, Hammar M, Sierralta WD, Normark S. 1998. Curli fibers are highly conserved between *Salmonella typhimurium* and *Escherichia coli* with respect to operon structure and regulation. *J Bacteriol* **180:**722–731.

128. Dong T, Schellhorn HE. 2009. Global effect of RpoS on gene expression in pathogenic *Escherichia coli* O157:H7 strain EDL933. *BMC Genomics* **10:**349.

129. Olsen A, Arnqvist A, Hammar M, Sukupolvi S, Normark S. 1993. The RpoS sigma factor relieves H-NS-mediated transcriptional repression of *csgA*, the subunit gene of fibronectin-binding curli in *Escherichia coli*. *Mol Microbiol* **7:**523–536.

130. Lee JH, Cho MH, Lee J. 2011. 3-indolylacetonitrile decreases *Escherichia coli* O157:H7 biofilm formation and *Pseudomonas aeruginosa* virulence. *Environ Microbiol* **13:**62–73.

131. Hammar M, Arnqvist A, Bian Z, Olsen A, Normark S. 1995. Expression of two *csg* operons is required for production of fibronectin- and congo red-binding curli polymers in *Escherichia coli* K-12. *Mol Microbiol* **18:**661–670.

132. Vidal O, Longin R, Prigent-Combaret C, Dorel C, Hooreman M, Lejeune P. 1998. Isolation of an *Escherichia coli* K-12 mutant strain able to form biofilms on inert surfaces: involvement of a new *ompR* allele that increases curli expression. *J Bacteriol* **180:**2442–2449.

133. Dorel C, Vidal O, Prigent-Combaret C, Vallet I, Lejeune P. 1999. Involvement of the Cpx signal transduction pathway of *E. coli* in biofilm formation. *FEMS Microbiol Lett* **178:**169–175.

134. **Brown PK, Dozois CM, Nickerson CA, Zuppardo A, Terlonge J, Curtiss R III.** 2001. MlrA, a novel regulator of curli (AgF) and extracellular matrix synthesis by *Escherichia coli* and *Salmonella enterica* serovar Typhimurium. *Mol Microbiol* **41:**349–363.

135. **Saldana Z, Xicohtencatl-Cortes J, Avelino F, Phillips AD, Kaper JB, Puente JL, Giron JA.** 2009. Synergistic role of curli and cellulose in cell adherence and biofilm formation of attaching and effacing *Escherichia coli* and identification of Fis as a negative regulator of curli. *Environ Microbiol* **11:**992–1006.

136. **Sharma VK, Bearson SM, Bearson BL.** 2010. Evaluation of the effects of *sdiA*, a *luxR* homologue, on adherence and motility of *Escherichia coli* O157:H7. *Microbiology* **156:**1303–1312.

137. **Ryu JH, Beuchat LR.** 2005. Biofilm formation by *Escherichia coli* O157:H7 on stainless steel: effect of exopolysaccharide and curli production on its resistance to chlorine. *Appl Environ Microbiol* **71:**247–254.

138. **Pawar DM, Rossman ML, Chen J.** 2005. Role of curli fimbriae in mediating the cells of enterohaemorrhagic *Escherichia coli* to attach to abiotic surfaces. *J Appl Microbiol* **99:**418–425.

139. **Fink RC, Black EP, Hou Z, Sugawara M, Sadowsky MJ, Diez-Gonzalez F.** 2012. Transcriptional responses of *Escherichia coli* K-12 and O157:H7 associated with lettuce leaves. *Appl Environ Microbiol* **78:**1752–1764.

140. **Low AS, Holden N, Rosser T, Roe AJ, Constantinidou C, Hobman JL, Smith DG, Low JC, Gally DL.** 2006. Analysis of fimbrial gene clusters and their expression in enterohaemorrhagic *Escherichia coli* O157:H7. *Environ Microbiol* **8:**1033–1047.

141. **Rendon MA, Saldana Z, Erdem AL, Monteiro-Neto V, Vazquez A, Kaper JB, Puente JL, Giron JA.** 2007. Commensal and pathogenic *Escherichia coli* use a common pilus adherence factor for epithelial cell colonization. *Proc Natl Acad Sci USA* **104:**10637–10642.

142. **Pouttu R, Westerlund-Wikstrom B, Lang H, Alsti K, Virkola R, Saarela U, Siitonen A, Kalkkinen N, Korhonen TK.** 2001. *matB*, a common fimbrillin gene of *Escherichia coli*, expressed in a genetically conserved, virulent clonal group. *J Bacteriol* **183:**4727–4736.

143. **Saldana Z, Erdem AL, Schuller S, Okeke IN, Lucas M, Sivananthan A, Phillips AD, Kaper JB, Puente JL, Giron JA.** 2009. The *Escherichia coli* common pilus and the bundle-forming pilus act in concert during the formation of localized adherence by enteropathogenic *E. coli*. *J Bacteriol* **191:**3451–3461.

144. **van Diemen PM, Dziva F, Stevens MP, Wallis TS.** 2005. Identification of enterohemorrhagic *Escherichia coli* O26:H– genes required for intestinal colonization in calves. *Infect Immun* **73:**1735–1743.

145. **Low AS, Dziva F, Torres AG, Martinez JL, Rosser T, Naylor S, Spears K, Holden N, Mahajan A, Findlay J, Sales J, Smith DG, Low JC, Stevens MP, Gally DL.** 2006. Cloning, expression, and characterization of fimbrial operon F9 from enterohemorrhagic *Escherichia coli* O157:H7. *Infect Immun* **74:**2233–2244.

146. **Vazquez-Juarez RC, Kuriakose JA, Rasko DA, Ritchie JM, Kendall MM, Slater TM, Sinha M, Luxon BA, Popov VL, Waldor MK, Sperandio V, Torres AG.** 2008. CadA negatively regulates *Escherichia coli* O157:H7 adherence and intestinal colonization. *Infect Immun* **76:**5072–5081.

147. **Samadder P, Xicohtencatl-Cortes J, Saldana Z, Jordan D, Tarr PI, Kaper JB, Giron JA.** 2009. The *Escherichia coli* ycbQRST operon encodes fimbriae with laminin-binding and epithelial cell adherence properties in Shiga-toxigenic *E. coli* O157:H7. *Environ Microbiol* **11:**1815–1826.

148. **Brunder W, Khan AS, Hacker J, Karch H.** 2001. Novel type of fimbriae encoded by the large plasmid of sorbitol-fermenting enterohemorrhagic *Escherichia coli* O157:H(–). *Infect Immun* **69:**4447–4457.

149. **Bielaszewska M, Prager R, Vandivinit L, Musken A, Mellmann A, Holt NJ, Tarr PI, Karch H, Zhang W.** 2009. Detection and characterization of the fimbrial *sfp* cluster in enterohemorrhagic *Escherichia coli* O165:H25/NM isolates from humans and cattle. *Appl Environ Microbiol* **75:**64–71.

150. **Musken A, Bielaszewska M, Greune L, Schweppe CH, Muthing J, Schmidt H, Schmidt MA, Karch H, Zhang W.** 2008. Anaerobic conditions promote expression of Sfp fimbriae and adherence of sorbitol-fermenting enterohemorrhagic *Escherichia coli* O157:NM to human intestinal epithelial cells. *Appl Environ Microbiol* **74:**1087–1093.

151. **Xicohtencatl-Cortes J, Monteiro-Neto V, Ledesma MA, Jordan DM, Francetic O, Kaper JB, Puente JL, Giron JA.** 2007. Intestinal adherence associated with type IV pili of enterohemorrhagic *Escherichia coli* O157:H7. *J Clin Invest* **117:**3519–3529.

152. **Xicohtencatl-Cortes J, Monteiro-Neto V, Saldana Z, Ledesma MA, Puente JL, Giron JA.** 2009. The type 4 pili of enterohemorrhagic *Escherichia coli* O157:H7 are multipurpose structures with pathogenic attributes. *J Bacteriol* **191:**411–421.

153. **Galfi P, Neogrady S, Semjen G, Bardocz S, Pusztai A.** 1998. Attachment of different *Escherichia coli* strains to cultured rumen epithelial cells. *Vet Microbiol* **61:**191–197.

154. **Enami M, Nakasone N, Honma Y, Kakinohana S, Kudaka J, Iwanaga M.** 1999. Expression of

type I pili is abolished in verotoxin-producing *Escherichia coli* O157. *FEMS Microbiol Lett* **179:** 467–472.

155. **Li B, Koch WH, Cebula TA.** 1997. Detection and characterization of the *fimA* gene of *Escherichia coli* O157:H7. *Mol Cell Probes* **11:**397–406.

156. **Roe AJ, Currie C, Smith DG, Gally DL.** 2001. Analysis of type 1 fimbriae expression in verotoxigenic *Escherichia coli*: a comparison between serotypes O157 and O26. *Microbiology* **147:**145–152.

157. **Cookson AL, Cooley WA, Woodward MJ.** 2002. The role of type 1 and curli fimbriae of Shiga toxin-producing *Escherichia coli* in adherence to abiotic surfaces. *Int J Med Microbiol* **292:**195–205.

158. **Leyton DL, Sevastsyanovich YR, Browning DF, Rossiter AE, Wells TJ, Fitzpatrick RE, Overduin M, Cunningham AF, Henderson IR.** 2011. Size and conformation limits to secretion of disulfide-bonded loops in autotransporter proteins. *J Biol Chem* **286:**42283–42291.

159. **Henderson IR, Nataro JP.** 2001. Virulence functions of autotransporter proteins. *Infect Immun* **69:**1231–1243.

160. **Wells TJ, Sherlock O, Rivas L, Mahajan A, Beatson SA, Torpdahl M, Webb RI, Allsopp LP, Gobius KS, Gally DL, Schembri MA.** 2008. EhaA is a novel autotransporter protein of enterohemorrhagic *Escherichia coli* O157:H7 that contributes to adhesion and biofilm formation. *Environ Microbiol* **10:**589–604.

161. **Wells TJ, McNeilly TN, Totsika M, Mahajan A, Gally DL, Schembri MA.** 2009. The *Escherichia coli* O157:H7 EhaB autotransporter protein binds to laminin and collagen I and induces a serum IgA response in O157:H7 challenged cattle. *Environ Microbiol* **11:**1803–1814.

162. **Peterson JH, Tian P, Ieva R, Dautin N, Bernstein HD.** 2010. Secretion of a bacterial virulence factor is driven by the folding of a C-terminal segment. *Proc Natl Acad Sci USA* **107:**17739–17744.

163. **Brunder W, Schmidt H, Karch H.** 1997. EspP, a novel extracellular serine protease of enterohaemorrhagic *Escherichia coli* O157:H7 cleaves human coagulation factor V. *Mol Microbiol* **24:** 767–778.

164. **Orth D, Ehrlenbach S, Brockmeyer J, Khan AB, Huber G, Karch H, Sarg B, Lindner H, Wurzner R.** 2010. EspP, a serine protease of enterohemorrhagic *Escherichia coli*, impairs complement activation by cleaving complement factors C3/C3b and C5. *Infect Immun* **78:**4294–4301.

165. **Brockmeyer J, Aldick T, Soltwisch J, Zhang W, Tarr PI, Weiss A, Dreisewerd K, Muthing J,** **Bielaszewska M, Karch H.** 2011. Enterohaemorrhagic *Escherichia coli* haemolysin is cleaved and inactivated by serine protease EspPalpha. *Environ Microbiol* **13:**1327–1341.

166. **Dziva F, Mahajan A, Cameron P, Currie C, McKendrick IJ, Wallis TS, Smith DG, Stevens MP.** 2007. EspP, a Type V-secreted serine protease of enterohaemorrhagic *Escherichia coli* O157:H7, influences intestinal colonization of calves and adherence to bovine primary intestinal epithelial cells. *FEMS Microbiol Lett* **271:**258–264.

167. **Xicohtencatl-Cortes J, Saldana Z, Deng W, Castaneda E, Freer E, Tarr PI, Finlay BB, Puente JL, Giron JA.** 2010. Bacterial macroscopic rope-like fibers with cytopathic and adhesive properties. *J Biol Chem* **285:**32336–32342.

168. **Paton AW, Srimanote P, Woodrow MC, Paton JC.** 2001. Characterization of Saa, a novel autoagglutinating adhesin produced by locus of enterocyte effacement-negative Shiga-toxigenic *Escherichia coli* strains that are virulent for humans. *Infect Immun* **69:**6999–7009.

169. **Lucchesi PM, Kruger A, Parma AE.** 2006. Distribution of *saa* gene variants in verocytotoxigenic *Escherichia coli* isolated from cattle and food. *Res Microbiol* **157:**263–266.

170. **Toma C, Nakasone N, Miliwebsky E, Higa N, Rivas M, Suzuki T.** 2008. Differential adherence of Shiga toxin-producing *Escherichia coli* harboring *saa* to epithelial cells. *Int J Med Microbiol* **298:**571–578.

171. **Herold S, Paton JC, Paton AW.** 2009. Sab, a novel autotransporter of locus of enterocyte effacement-negative shiga-toxigenic *Escherichia coli* O113:H21, contributes to adherence and biofilm formation. *Infect Immun* **77:**3234–3243.

172. **Torres AG, Perna NT, Burland V, Ruknudin A, Blattner FR, Kaper JB.** 2002. Characterization of Cah, a calcium-binding and heat-extractable autotransporter protein of enterohaemorrhagic *Escherichia coli*. *Mol Microbiol* **45:**951–966.

173. **Torres AG, Jeter C, Langley W, Matthysse AG.** 2005. Differential binding of *Escherichia coli* O157:H7 to alfalfa, human epithelial cells, and plastic is mediated by a variety of surface structures. *Appl Environ Microbiol* **71:**8008–8015.

174. **Giron JA, Torres AG, Freer E, Kaper JB.** 2002. The flagella of enteropathogenic *Escherichia coli* mediate adherence to epithelial cells. *Mol Microbiol* **44:**361–379.

175. **Best A, La Ragione RM, Sayers AR, Woodward MJ.** 2005. Role for flagella but not intimin in the persistent infection of the gastrointestinal tissues of specific-pathogen-free chicks by shiga toxin-negative *Escherichia coli* O157:H7. *Infect Immun* **73:**1836–1846.

176. **Erdem AL, Avelino F, Xicohtencatl-Cortes J, Giron JA.** 2007. Host protein binding and adhesive properties of H6 and H7 flagella of attaching and effacing *Escherichia coli. J Bacteriol* **189:**7426–7435.

177. **Mahajan A, Currie CG, Mackie S, Tree J, McAteer S, McKendrick I, McNeilly TN, Roe A, La Ragione RM, Woodward MJ, Gally DL, Smith DG.** 2009. An investigation of the expression and adhesin function of H7 flagella in the interaction of *Escherichia coli* O157:H7 with bovine intestinal epithelium. *Cell Microbiol* **11:**121–137.

178. **Tarr PI, Bilge SS, Vary JC Jr, Jelacic S, Habeeb RL, Ward TR, Baylor MR, Besser TE.** 2000. Iha: a novel *Escherichia coli* O157:H7 adherence-conferring molecule encoded on a recently acquired chromosomal island of conserved structure. *Infect Immun* **68:**1400–1407.

179. **Herold S, Paton JC, Srimanote P, Paton AW.** 2009. Differential effects of short-chain fatty acids and iron on expression of iha in Shiga-toxigenic *Escherichia coli. Microbiology* **155:**3554–3563.

180. **Yin X, Wheatcroft R, Chambers JR, Liu B, Zhu J, Gyles CL.** 2009. Contributions of O island 48 to adherence of enterohemorrhagic *Escherichia coli* O157:H7 to epithelial cells *in vitro* and in ligated pig ileal loops. *Appl Environ Microbiol* **75:**5779–5786.

181. **Lu Y, Iyoda S, Satou H, Satou H, Itoh K, Saitoh T, Watanabe H.** 2006. A new immunoglobulin-binding protein, EibG, is responsible for the chain-like adhesion phenotype of locus of enterocyte effacement-negative, shiga toxin-producing *Escherichia coli. Infect Immun* **74:**5747–5755.

182. **Merkel V, Ohder B, Bielaszewska M, Zhang W, Fruth A, Menge C, Borrmann E, Middendorf B, Muthing J, Karch H, Mellmann A.** 2010. Distribution and phylogeny of immunoglobulin-binding protein G in Shiga toxin-producing *Escherichia coli* and its association with adherence phenotypes. *Infect Immun* **78:**3625–3636.

183. **Torres AG, Kaper JB.** 2003. Multiple elements controlling adherence of enterohemorrhagic *Escherichia coli* O157:H7 to HeLa cells. *Infect Immun* **71:**4985–4995.

184. **Torres AG, Li Y, Tutt CB, Xin L, Eaves-Pyles T, Soong L.** 2006. Outer membrane protein A of *Escherichia coli* O157:H7 stimulates dendritic cell activation. *Infect Immun* **74:**2676–2685.

185. **Prasadarao NV, Wass CA, Weiser JN, Stins MF, Huang SH, Kim KS.** 1996. Outer membrane protein A of *Escherichia coli* contributes to invasion of brain microvascular endothelial cells. *Infect Immun* **64:**146–153.

186. **Prasadarao NV.** 2002. Identification of *Escherichia coli* outer membrane protein A receptor on human brain microvascular endothelial cells. *Infect Immun* **70:**4556–4563.

187. **Shin S, Lu G, Cai M, Kim KS.** 2005. *Escherichia coli* outer membrane protein A adheres to human brain microvascular endothelial cells. *Biochem Biophys Res Commun* **330:**1199–1204.

188. **Gu J, Ji X, Qi J, Ma Y, Mao X, Zou Q.** 2010. Crystallization and preliminary crystallographic studies of the C-terminal domain of outer membrane protein A from enterohaemorrhagic *Escherichia coli. Acta Crystallogr Sect F Struct Biol Cryst Commun* **66:**929–931.

Animal Models of Enterohemorrhagic *Escherichia coli* Infection

8

JENNIFER M. RITCHIE[1]

INTRODUCTION

Since the first recognized *Escherichia coli* O157:H7 outbreak over 3 decades ago (1), investigators have sought to identify suitable animal hosts that allow study of enterohemorrhagic *E. coli* (EHEC)-mediated disease. The value of any animal infection model ultimately relies on its ability to reproduce the human disease and enable the mechanistic processes that lead to clinical disease, pathogen carriage, and transmission to be examined. As yet, no single animal model mimics the full spectrum of disease caused by EHEC infection. However, since Moxley and Francis's review in 1998 (2), several advances have been made in the field, including the generation of a Shiga toxin (Stx)-producing *Citrobacter rodentium*-murine model, a human intestine xenograft murine model, and a renewed interest in the use of rabbit models. This article reviews what is known about EHEC-mediated disease from human outbreaks and biopsy studies, and within a historical context, describes the features and limitations of EHEC infection models that are based on the three most commonly used species (pigs, rabbits, and mice). Recent new advances are highlighted and discussed in light of mounting evidence for the need to study the biology and virulence strategies of EHEC in the context of its niche within the intestine. The reader is directed elsewhere for excellent reviews on the environmental sources of

[1]School of Biosciences and Medicine, University of Surrey, Guildford, United Kingdom.

Enterohemorrhagic Escherichia coli *and Other Shiga Toxin-Producing* E. coli
Edited by Vanessa Sperandio and Carolyn J. Hovde
© 2015 American Society for Microbiology, Washington, DC
doi:10.1128/microbiolspec.EHEC-0022-2013

EHEC infection (3), EHEC interactions with the intestinal epithelium (4), the molecular basis of pathogenicity (5–8), and the current status of treatment options (9).

OBSERVATIONS FROM HUMAN OUTBREAKS

Most of what is known about EHEC pathogenesis is based on outbreaks caused by *E. coli* O157:H7 (1, 5, 10). From these outbreaks, it is widely accepted that EHEC infection is acquired through the ingestion of feces-contaminated food or water or by hand-to-mouth transmission following contact with infected animals (including humans) or their surroundings. Whereas some individuals who ingest the organism remain asymptomatic, others develop severe abdominal pain and diarrhea typically within 3 to 4 days (11). This can progress to a bloody diarrhea (or hemorrhagic colitis), and in approximately 5 to 15% of infected individuals a potentially fatal hemolytic-uremic syndrome (HUS) develops (12). HUS is defined as a clinical syndrome comprising thrombotic microangiopathy, thrombocytopenia, and hemolytic anemia that leads to acute kidney injury, for which there is no effective treatment. The ubiquitousness of the causative agent, the severity of the clinical sequelae, and the lack of treatment options and effective preventive measures demand a better understanding of how this group of organisms cause disease.

The pathophysiology of disease is largely attributed to the ability of EHEC to colonize the mammalian intestine and produce Shiga-like toxin, a family of potent cytotoxins that are able to cross the epithelial barrier and act at sites distal to the intestine, inhibiting protein synthesis in sensitive target endothelial cells. Consistent with this idea, Stx has been detected on the surface of polymorphonuclear leukocytes (PMNs) circulating in the blood of some infected individuals and in their feces (13, 14). In contrast, bacteremia has not been reported (1, 15), and the organism remains in the intestinal lumen where it is shed in feces. Exactly how Stx crosses the epithelial barrier remains poorly defined, but it is generally believed that distribution of the Stx receptor, globotriaosylceramide (Gb3), determines the site of tissue damage (16, 17). The presence of Gb3 in the endothelial cells that line the blood vessels (vasculature) and the subsequent host cell response to the cellular damage it causes explain the resulting clinical disease (18). Acute renal failure occurs because the highest concentrations of Gb3 are found in this organ (19); however, Gb3 is also detected in the microvasculature of the colon and the cerebellum (20–22). Conflicting views on whether Gb3 is also expressed on the surface of colonic epithelial cells exist in the literature (23–25).

As such, any Stx-producing strain capable of colonizing the mammalian intestine could potentially cause Stx-mediated disease in humans. In reality, *E. coli* O157:H7 strains are the predominant serotype associated with EHEC outbreaks in the United Kingdom and elsewhere (26–28). The reason behind this selection is not clear but may be related to serotype-specific differences in prevalence, persistence, or survival in the environment as well as their inherent virulence. *E. coli* O157:H7 strains usually contain the locus of enterocyte effacement (LEE) pathogenicity island, a virulence attribute deemed to play a critical role in mediating bacterial attachment to the intestinal epithelium (29–31). Among non-O157 strains, bacteria belonging to serogroups O26, O103, O111, O121, O45, and O145 are most commonly reported from infected individuals (28). Some serogroups, notably O113, lack the LEE pathogenicity island, and compared to the LEE-positive O157 strains, relatively little is known about the factors that are important in mediating their attachment to the epithelial surface (32, 33).

Epidemiological studies suggest that non-O157 serogroups cause a similar but somewhat less severe range of clinical manifestations than O157 strains (28, 34, 35). Retrospective analyses of patient clinical profiles from a U.S. hospital revealed similar incidences of bloody diarrhea and HUS in children presenting with either non-O157 (17 patients) or O157 (33 patients) infections (36). In contrast, a larger study using German national surveillance data concluded that other than for the recent O104 outbreak strains, patients presenting with non-O157 infections were half as likely to be hospitalized and one-tenth as likely to die compared to those infected with O157 (37). An unexpectedly high percentage of people developed HUS following the outbreak of Stx-producing enteroaggregative *E. coli* O104:H4 that unfolded in northern Germany during the summer of 2011 (37, 38). In addition, the affected target population consisted of young, healthy females rather than individuals at the extremes of age, who are more likely to be sickened during an *E. coli* O157:H7 outbreak. Thus, clinical outcome may be dependent on the demographic of the affected population as well as the characteristics of the infectious organism. Predicting which strain and host factors led to the progression of a more severe clinical outcome is critical to the development and management of effective control strategies to prevent the loss of human life.

Volunteer Studies

Due to the potentially life-threatening sequelae associated with EHEC infections, volunteer studies using Stx-producing strains are deemed unethical. However, volunteer studies using the closely related diarrheal pathogen enteropathogenic *E. coli* (EPEC) have been performed and demonstrate a clear role in pathogenesis for some of the virulence attributes shared by the two pathotypes of *E. coli*. For example, studies reveal that bacterial adherence to the epithelial surface mediated by Tir-intimin interactions play a major role in the development of diarrhea. Deletion of EspB, a type III secreted protein and required for targeting of Tir to the host cell membrane (39), decreased the incidence and severity

of diarrhea in volunteers who ingested the mutant (40). Similarly, deletion of intimin, a 94-kDa outer membrane protein encoded by the gene *eaeA*, also reduced the severity of infection in volunteers (41). In both studies, however, a few individuals who received the mutant strains still developed diarrhea, albeit producing stools of reduced volume and frequency. Moreover, the numbers of organisms recovered from their stools were similar to those receiving wild-type organisms, at least at around the time of peak organism excretion from the host (41). These results indicate that factors other than Tir and intimin contribute to EHEC colonization in the intestine, and in this regard, the biological significance of Tir-intimin–mediated attachment (and the resultant formation of the characteristic attaching and effacing [A/E] lesion) during disease is not really understood.

Biopsies from Infected Individuals

Intestinal damage

Relatively few studies report on the intestinal changes that occur in individuals infected with EHEC, probably due to the risk of worsening the underlying clinical condition. Colonoscopy examination of infected individuals, which allows the full length of the large intestine to be viewed, found that the most severe disease occurred in the cecum and ascending colon (42, 43). Here, tissue damage consisting of edema, erythema (redness), hemorrhage, and erosion was evident over a wide area and was associated with a marked narrowing of the luminal space. In some individuals, long ulcer-like lesions were also noted on the surface of the tissue. Patchy and less severe areas of inflammatory damage were observed in the transverse colon, descending colon, and sigmoid colon, consistent with descriptions from sigmoidoscopy examinations (44, 45).

Histologic examination of the damaged areas revealed destruction of the surface epithelium and involvement of the lamina propria, while the deeper colonic crypts remained largely intact (42, 43). Focal areas with acute inflammation were evident, with neutrophil infiltration of the lamina propria and, in some cases, the formation of crypt abscesses. Hemorrhage and edema were commonly observed in the lamina propria, as were apoptotic cells in the surface epithelium and colonic crypts. Small fibrin/platelet thrombi formed within the mucosal capillaries. It is interesting to note that adherent bacteria were not observed in the tissue sections, despite the samples being collected during the acute phase of the disease and only a few days after the onset of bloody diarrhea (42, 44). These findings indicate that during EHEC infection the integrity of the epithelial barrier is compromised; whether this occurs due to the direct action of the bacterium or Stx or due to the resulting host response to infection is not clear.

Extraintestinal damage

Readers are referred to two reviews describing the detailed pathological manifestations of Stx-mediated kidney damage in humans (6, 12). Summarized below are the key pathological findings evident in patients who present with diarrhea and HUS. Inward and colleagues reported that the renal pathology consisted of endothelial swelling and glomerular thrombosis with congested rather than ischemic glomeruli (46). In addition, they noted the presence of significant numbers of neutrophils compared to control tissue. Glomerular capillaries appeared to be the main site of damage in the patients. As such, the site of damage mirrors the distribution of Gb3, the Stx receptor, in humans. The highest densities of Gb3 were detected on glomerular endothelial cells, although Gb3 has also been detected in renal proximal tubules (6). Consistent with these findings, Stx was found to bind to both kidney tubular and glomeruli of a pediatric patient with HUS, but only the tubules of a geriatric patient with HUS (16). Whether these findings reflect differences in Gb3 expression, the stage of HUS, or some other factor is not known.

HISTORICAL PERSPECTIVE ON ANIMAL MODELS OF EHEC-MEDIATED DISEASE

Pigs

Nearly 30 years ago, Tzipori and Francis described the first use of gnotobiotic piglets to study the newly emerging group of pathogenic *E. coli* strains that included *E. coli* O157:H7 (47, 48). Gnotobiotic piglets, which are delivered via cesarean section, maintained in germ-free incubators, and artificially reared (aka cesarean-derived colostrum-deprived [CDCD] piglets), had been used previously to study other human enteric pathogens, including cryptosporidium and pathogenic *E. coli* (49, 50). Oral infection of 1-day-old CDCD piglets with high numbers (10^{10} CFU) of *E. coli* O157:H7 caused watery but not bloody diarrhea in most infected animals by 2 to 4 days post inoculation. In these early studies, no other signs of disease were reported. As was found in human infections, histologic abnormalities were detected in the cecum and colon, and primarily consisted of destruction of the mucosal brush border and inflammation (51). These first investigations recognized that *E. coli* O157:H7 caused intestinal lesions that were morphologically similar to those previously described for EPEC (52), and thus EHEC was considered a member of the same pathotype group (53). However, while EHEC and EPEC apparently shared one common mechanism of bacterial attachment, they adhered to different sites in the intestine. EPEC colonized both the small and large intestine, whereas only the large intestine was colonized by EHEC (54).

Neither the pO157 plasmid nor Stx appeared to play a role in lesion formation as mutants deleted for these factors still caused intestinal disease (54, 55). Instead, genes encoded in the now well-described LEE pathogenicity island proved to be critical for intimate bacterial attachment in vitro and intestinal lesion formation in vivo (29, 30). Consistent with findings from volunteer studies, deletion of *eaeA* prevented bacterial attachment and A/E lesion formation, yet allowed similar numbers of wild-type and mutant cells to be recovered in intestinal homogenates obtained from the infected animals (56). These findings suggest that the gnotobiotic piglet intestine, like the human intestine, is permissive for EHEC replication and the organism is capable of surviving in this niche.

In addition to intestinal disease, subsequent studies reported that some piglets exhibit signs of central nervous system (CNS) disease, including ataxia (55). Indeed, support for a role for Stx in extraintestinal disease initially came from a study where gnotobiotic piglets were given either orally administered bacteria or intraperitoneally administered polymyxin B-treated bacterial cell lysates (57). In both instances, piglets that exhibited CNS disease showed evidence of vascular damage in the cerebellum. The vascular damage bore similarities to lesions seen in the brain of a patient who died following *E. coli* O157:H7 infection and in piglets with naturally acquired edema disease (57, 58), infections that are both marked by the presence of Stx. While Stx was found to be important in the development of extraintestinal disease, intimate bacterial attachment to the epithelial surface was not required for this to occur (59).

While early studies focused on the effect of Stx on the CNS, a significant advance in the usefulness of this model was made when investigators reported that gnotobiotic piglets also developed renal lesions following oral EHEC infection (59–61). Gunzer and colleagues were the first to report that the kidneys of infected animals exhibited signs of endothelial cell damage (60). Damage was characterized by diffuse glomerular endothelial swelling and congestion and a narrowing of the capillary vessels. In most instances, morphological signs of thrombotic microangiopathy, which is a hallmark characteristic of HUS in humans, were also noted. Gunzer's findings were confirmed following retrospective examination of kidney samples obtained from multiple EHEC-infected and control

piglets (61). Despite kidney damage, unlike humans, the piglets do not exhibit clinical signs of kidney failure (e.g., elevated creatinine levels) (60). As such, while piglets are one of the few species that show both intestinal and systemic disease following EHEC infection, they do not fully recapitulate all facets of the human disease.

In addition to the inherent biological constraints associated with the use of gnotobiotic animals (e.g., poorly developed immune systems, a lack of normal microflora), most researchers lack the specialized facilities needed to rear animals under sterile conditions. To overcome some of these limitations, conventional piglets were assessed for their susceptibility to EHEC infection. Naturally born piglets, which feed directly from the sow and either remain with the sow for the duration of the study or are subsequently hand-reared and housed in individual microbiological isolators, offer an additional advantage as they enable passive immunization studies to be performed (62). Surprisingly, continuously suckling piglets were found to exhibit more severe neurological disease and more quickly succumbed to EHEC infection than traditional CDCD animals (63). The reason for this finding is not known, but it may be related to the amount of Stx produced under the different conditions in the intestine. Consistent with this idea, Shringi and colleagues found that strains that produced higher amounts of Stx caused earlier and more severe signs of CNS disease in conventionally born piglets (64). Somewhat surprising, no signs of kidney damage were found in these animals, perhaps due to the rapid onset of neurological symptoms. The relationships between animal husbandry practice, the amount of intestinal Stx, and the resulting pathology warrant rigorous investigation in this species to further define this model system.

Rabbits

Rabbits have been used both to study the toxic effects of Stx and the intestinal biology of the organism, and as such, the age, breed, and route of infection affect their response to infection. Suckling New Zealand White (NZW) rabbits are naturally susceptible to oral infection by EHEC and, like piglets, develop diarrhea. Farmer and colleagues first reported that oral inoculation of high numbers (>10^9 CFU) of *E. coli* O157:H7 caused watery diarrhea in young (5 to 10 days old) but not older (20 days old) rabbits. In contrast, the ingestion of control (nonpathogenic) *E. coli* strains failed to cause any signs of morbidity in the animals (65, 66). Large numbers of the bacteria could be recovered from the intestine, suggesting that the organism is able to multiply in this environment (65). Subsequent studies identified the colon as the principal site of EHEC-mediated disease, where histological abnormalities, including edema, hemorrhage, the presence of an inflammatory infiltrate, and mucosal epithelial apoptosis, were observed (66, 67). In these young animals, the development of diarrhea appears to be driven by inflammation-mediated changes in colonic ion transport resulting in decreased Na absorption and increased Cl secretion (68, 69). Unlike humans, suckling rabbits do not develop signs of renal disease; however, the reproduction of diarrhea and colonic inflammation, two key features of EHEC infection in humans, indicates that they may be a useful model for studying the intestinal manifestations of the disease.

The first suckling rabbit studies suggested a role for Stx in the development of diarrhea because natural isolates that produced differing amounts of Stx gave rise to differing severities of disease in the animals (66, 67). We later confirmed a role for Stx in *E. coli* O157:H7 pathogenicity using genetically defined isogenic bacterial mutants (31). Three-day-old NZW rabbits infected with an isogenic mutant that lacked the genes coding for Stx2 developed a transient diarrhea and largely noninflammatory intestinal pathology. Thus, Stx2 appeared to increase the severity and duration of EHEC-induced diarrhea and modulate the host response to infection.

However, Stx2 did not appear to act as a colonization factor because similar numbers of wild-type and mutant organisms were recovered from the colon at various times postinfection. In contrast, deletion of the genes coding for the bacterial outer membrane adhesin, intimin (*eaeA*), and the translocated intimin receptor (*tir*) markedly reduced the ability of the organism to colonize the rabbit intestine (31). Thus, in contrast to piglets, EHEC attachment to the epithelial surface is required for the organism to persist and cause intestinal disease.

One of the strengths of the suckling rabbit model is that it allowed, for the first time, a means to study factors that were important in mediating intestinal colonization and disease. Thus, using suckling rabbits we found that factors other than Tir and intimin influenced the ability of EHEC to colonize the intestine and cause disease. For example, surface-exposed structures such as the long polar fimbriae (Lpf), curli, and the O-antigen capsule all contributed to the ability of EHEC to colonize and survive in the intestine (70–72). Indeed, our studies began to reveal the complexities of regulating colonization factor expression in the intestine. While expression of Lpf contributes to EHEC intestinal colonization, in the absence of these structures, the organism appears to overexpress curli, appendages more usually associated with adherence to abiotic surfaces (73). In addition, we found that bacterial factors (other than Tir), which are translocated into the host cell via the organism's type III secretion system, also modulate intestinal colonization (71). For example, EspFu (also known as TccP) appears to promote the spread of EHEC to new sites in the intestine. Rabbits infected with the *espFu* mutant contained significantly fewer organisms in their ceca and colons at 7 but not 2 days postinfection. Moreover, this mutant covered a significantly smaller area of the intestine than the wild type in the gnotobiotic piglets (74). However, many of the LEE-encoded effector proteins caused only minor changes in the ability of the EHEC to colonize the intestine, most likely because of functional redundancy between the different proteins (71).

Despite being sensitive to the intestinal manifestations of EHEC infection, suckling rabbits do not develop HUS or any other evidence of renal failure. The reasons for this are not clear but may be related to the species-specific distribution of Gb3. Zoja and colleagues reported that Gb3 is present in the tissues of the brain, lung, spinal cord, and intestine of weaned NZW rabbits but is missing from the heart, liver, and kidney (75). While Gb3 homologs have recently been detected in the kidney cortex of NZW and Dutch Belted (DB) rabbits, their role in disease and relative affinity for Stx remain unclear (76). In young rabbits (day 5 of life), Gb3 has been detected in lipid extracts obtained from the colon (77). Conversely, the appearance of Gb3 appears to be maturationally regulated in the small intestine, as jejunal microvillous membranes isolated from rabbits less than 16 days of age did not bind Stx (78, 79). Together, these observations may explain the sensitivity of the rabbit colon to the effects of luminal Stx in suckling animals, and the predominantly neurological manifestations of Stx when given intravenously to adult rabbits (see below).

Following a natural outbreak of bloody diarrhea and sudden death in a group of 10-week-old DB rabbits in which an Stx1-producing *E. coli* O153 isolate was recovered, the possibility was raised that this breed of rabbits may be more sensitive to the systemic effects of Stx than NZW rabbits (80). Experimental infection of 8-week-old weaned DB rabbits with the outbreak strain confirmed that it was able to cause diarrhea and renal vascular and glomerular lesions in the animals (76). As part of this study, the authors also examined the response to infection by an *E. coli* O157:H7 clinical isolate. Rabbits infected with *E. coli* O157 lost less weight (mean ± SE: –4.6% ± 1.6 versus –18.1% ± 4.7) and shed fewer organisms (range 10^3 to 10^5 versus $\geq 10^5$ to 10^9 CFU/g) than O153-infected rabbits, perhaps suggesting that the latter strain is better adapted to colonizing the rabbit intestine. Panda and colleagues observed no significant differences in the response of weaned NZW or DB rabbits to *E. coli* O157 infection (81). However, these authors noted that renal lesions did not appear to be a consistent feature in the infected rabbits. Renal pathology was detected in only 1 of 11 infected rabbits, and as such, the reproducibility of this clinical outcome would need to be improved to enable study of EHEC-induced renal disease or to allow novel therapeutics to be tested.

While suckling and young (10-week-old) rabbits are susceptible to oral administration of EHEC, older rabbits tend to be more recalcitrant to colonization by the organism. Two approaches have been used by investigators to circumvent this issue. The rabbit ligated ileal loop assay originally described by De and colleagues has been widely used to assess the enterotoxicity of pathogenic bacteria, their culture supernatants, or purified toxins (82). Likewise, both EHEC strains and purified Stx1 have been injected into ligated loops to reproduce the disease found in humans (2, 83). However, this "closed" loop model suffers from many limitations when intestinal pathogens such as EHEC are studied, including most notably an interruption to normal gut function (e.g., peristalsis).

The second approach involves the use of a natural rabbit diarrheal pathogen, *E. coli* strain RDEC-1, which, while naturally lacking Stx, contains the LEE pathogenicity island (84). The histological lesions caused by RDEC-1 appeared indistinguishable from those reported for other A/E pathogens, including EHEC (52, 85). To better reproduce the virulence repertoire of EHEC, Sjogren and colleagues generated an Stx-producing variant of RDEC-1 by infecting RDEC-1 with an Stx1-converting bacteriophage (86). By comparing the disease caused by the two strains, the authors observed that Stx-producing RDEC-1 induced a more severe infection than the parent strain did, causing earlier weight loss and higher rates of mortality in the animals. While these studies indicate a role for

Stx in the development of severe disease, the usefulness of this model is somewhat limited by an inability to study whether additional EHEC virulence factors are important to infection.

Intravenous administration of purified Stx causes thrombotic microangiopathy in several organs in rabbits, consistent with the reported distribution of Gb3 (75, 87). The location of the resulting damage appears to depend on the type of Stx used because Stx1 primarily located to the microvasculature of the brain and spinal cord, whereas Stx2 localizes to the microvasculature of the cecum. These differences in tissue tropism also likely explain the greater lethality of Stx1 than Stx2 in rabbits, although this does not reflect the epidemiological data from human outbreaks.

Mice

Mice, the workhorse species of infectious disease research, are naturally resistant to colonization by EHEC and fail to develop signs of intestinal disease following oral infection. In contrast, systemic Stx causes acute kidney damage and death in the animals (88). However, key differences exist between Stx-mediated renal damage in mice and humans (12). Renal damage in mice consists of acute tubular necrosis, which is deemed characteristic of toxin insult, rather than the prothrombotic condition created following Stx-mediated endothelial injury in humans. HUS is viewed as a thrombotic microangiopathy of the glomerular blood vessels that develops with or without tubular cell death (12). This difference in pathology likely reflects species-specific differences in Gb3 distribution. In mice, Gb3 is detected in the microvascular endothelial cells of the brain cortex and pia mater (membranes surrounding the brain and spinal cord) and the capillary vascular endothelial cells surrounding renal tubules (17). As noted earlier, the highest concentrations of Gb3 expression in humans occur in the kidney, specifically on the glomerular endothelium (6).

Despite this fundamental difference in the pathophysiological response to Stx in mice and humans, a rodent model of EHEC-mediated disease remains desirable. As such, many approaches have been taken to render mice more susceptible to oral infection or to present with a more accurate model of HUS. Because of the recent publication of an excellent review on murine models of EHEC and Stx infection by Mohawk and O'Brien (89) to which the readers are directed, this article summarizes only the main features and limitations of murine models before recent advances in the field are discussed.

The natural resistance to EHEC colonization in mice can be reduced by modifying the normal intestinal microflora through pretreatment with antibiotics, use of germ-free animals, or via dietary-induced changes (see Table 1). All lead to increased numbers of EHEC in the intestine and the development of varying levels of renal toxic injury and death in the animals. In addition, the sensitivity of the host (e.g., "lipopolysaccharide [LPS]-responder" mice, MyD88 knockout mice) or the virulence of the bacterium (e.g., host-passaged organisms, treatment with mitomycin C) can be manipulated to further increase clinical severity. Moreover, to more accurately mimic the kidney disease seen during HUS, investigators have also tried coadministering endotoxin (LPS) with Stx. In this situation, LPS acts as an accessory proinflammatory mediator, which primes the host immune response and results in the induction of thrombocytopenia in the animals. However, the etiology of LPS plus Stx challenge differs from that of HUS and does not reflect what occurs during human infection (12).

An alternative approach that circumvents the problems of establishing EHEC colonization in mice is the use of the surrogate bacterium *Citrobacter rodentium*, a natural murine pathogen that induces colonic hyperplasia (90). Both EHEC and *C. rodentium* contain the LEE pathogenicity island that mediates Tir-intimin-based bacterial attachment to the epithelial surface (91). Because of

TABLE 1 Comparative murine models of EHEC-mediated disease

Model	Host	Clinical disease	Key limitations	Reference(s)
Oral infection				
Streptomycin-treated	CD-1 (5- to 8-week old-male mice)	Weight loss, renal tubular damage, mortality	Reduced microflora, artificial colonization, no intestinal disease	88, 122
Streptomycin + mitomycin C treatment	ICR (3 weeks old)	Weight loss, renal tubular damage, mortality	Reduced microflora, no intestinal disease	123
Weaned	BALB/c (17 to 20 days old); host-passaged bacteria	Mortality	No intestinal disease, rapid clearance of the pathogen, no glomerular disease	124
LPS responder	C3H/HeN and C3H/HeJ (8 to 16 weeks old)	Anorexia, ruffled fur, neurological disease	No glomerular disease	125
Intact microflora	BALB/c (6 weeks old)	Weight loss, renal tubular damage, mortality	No glomerular disease	126
Streptomycin + MyD88-deficient mice	MyD88−/− C57BL/6 (6 to 14 weeks old)	Weight loss, kidney toxic insult, mortality	No glomerular disease, host immune system abnormalities	127
Germ-free	Swiss Webster(3 days to 12 weeks old)/IQI (8 weeks old)	Acute renal tubular necrosis, limited glomerular thrombosis, mortality	No natural microflora, poorly developed immune systems	128–130
C. rodentium	C57BL/6 (5 to 8 weeks old)	Watery stools, weight loss, intestinal disease	Not EHEC	94
Injection				
Conventional mice + IP injection of OMV[a]	C57BL/6 (8 to 10 weeks old)	Acute renal tubular necrosis,	Nonbacterial, no thrombosis	131
Conventional mice + multiple IP Stx2	C57Bl/6J (8 to 10 weeks old)	Weight loss, signs of kidney failure, glomerular endothelial damage, neutrophilia	Nonbacterial	132
Conventional + IP or IV Stx1/2	CD-1 (6 to 8 weeks old)	Mild dehydration, renal tubular necrosis	No intestinal disease, no glomerular thrombosis	133
Other				
Mice + human colonic xenograft model	CB-17 SCID mice (6 to 8 weeks old) + 16 maturation	A/E lesion pathology, some mucosal damage	Incomplete immune system, no natural microflora, nonnatural organ environment, no systemic disease	96

[a]Outer membrane vesicles.

this, *C. rodentium* has been used extensively to dissect the function and activity of all LEE-encoded proteins as well as many other non-LEE-encoded proteins that are translocated via the organism's type III secretion system (92, 93). These studies have done much to advance our understanding of how translocated bacterial proteins interfere with host cell processes. However, while *C. rodentium* is useful for dissecting the role and function of type III secreted proteins, *C. rodentium*

does not naturally contain the genes coding for Stx, the principal virulence factor of EHEC. To overcome this limitation, Mallick and colleagues recently developed an Stx-expressing *C. rodentium* strain by lysogenizing the bacterium with an Stx-containing phage (94). The phage taken up by *C. rodentium* was found to carry Stx genes belonging to an Stx2 variant called mucus-activatable Stx2d. Mucus-activatable Stx2d is a particularly toxic subtype of Stx2 that causes high mortality in mice but is usually found in non-LEE-encoding EHEC strains that are less commonly isolated in human oubreaks than EHEC containing Stx2 or Stx2c (95). Compared to wild-type *C. rodentium*, Stx-producing *C. rodentium* causes increased weight loss, severe destruction of the intestinal mucosa, and renal tubular damage. The development of this model allows study of robust intestinal colonization together with Stx-mediated disease, although it still does not overcome the different pathophysiological response to Stx seen in mice and humans. In addition, it does not easily allow study of other EHEC factors that are important to infection.

Another recent advance in the development of murine models has been the construction of human and bovine intestinal xenograft models (96). Immature germ-free human and bovine small intestine or colonic fetal tissues were subcutaneously transferred into SCID mice and the tissue left to develop for up to 16 weeks. The authors demonstrate that after maturation, the xenografted tissue displays appropriate organ-specific tissue architecture, complete with a range of cell types including goblet cells and microvilli. Thereafter, various EHEC strains were injected into the "intestinal lumen" of each organ. Compared to the small intestine xenografts, EHEC strains grew poorly in the lumen of the colonic xenografts. Despite this difference, distinct A/E lesions and mucosal damage were found only on tissue derived from the human colon, and not the human or bovine small intestine segments. Furthermore, A/E lesion formation was shown to be dependent on the expression of a functional type III secretion system, consistent with what has been reported in other model hosts (see above). While xenograft models allow an opportunity to study EHEC interactions with the human intestine without endangering human life, this approach faces several challenges. The "closed" nature of the segment, which does not allow any normal intestinal activity (e.g., peristalsis, food passage) to occur, the lack of a natural microflora, and the incomplete immune system of the host animal may prove to be significant limitations to the acceptance of this model system.

Other Species

Other than pigs, rabbits, and mice, a plethora of animal species have been considered as model hosts of EHEC-induced disease (see Table 2). Like mice, ferrets are naturally resistant to oral infection by EHEC and require pretreatment with antibiotics to establish moderate numbers of the organism in the intestine (97). Despite significant weight loss following infection, none of the animals showed any signs of intestinal disease, and only around 20% showed signs of renal glomerular disease and/or thrombocytopenia. Greyhounds naturally develop cutaneous and renal glomerular vasculopathy, a disease that results in the formation of renal glomerular lesions, which are similar to those seen in human HUS (98, 99). While this model has not been used extensively since its first description, Raife and colleagues have used it to demonstrate that thrombin-dependent mechanisms are important in Stx2-mediated pathogenesis. Treatment of greyhounds with lepirudin, a thrombin inhibitor, reduced HUS-like pathology and enabled the animals to survive beyond the normal course of the experiment (100). Baboons appear to best mimic human HUS symptoms following intravenous injection of Stx (101, 102). In this model, both Stx1 and Stx2 were able to induce HUS in the animals, although the dose, timing, and magnitude of the response differed (103). Rats also develop HUS-like symptoms in response to intraperitoneal challenge of filtered EHEC culture supernatants (104). However, since purified Stx preparations were not used in this model, further investigation is required to dissect the contribution of Stx versus other soluble factors. Finally, the nematode *Caenorhabditis elegans* has beenput forward as a naturally infected and genetic tractable animal host in which to study EHEC infection. As found previously for other enteric pathogens

TABLE 2 Comparative models of EHEC-mediated disease in species other than pigs, rabbits, and mice

Model	Host	Clinical disease	Key limitations	Reference(s)
Ferrets	*Campylobacter*-free female (6 weeks old)	Weight loss, limited renal glomerular pathology	Limited natural microflora, no intestinal disease	97
Greyhound		Bloody diarrhea, microvascular thrombosis, HUS	Nonbacterial	99, 100
Monkeys	*Macaca radiata* (wild-caught adults)	Enteritis with neutrophil infiltration, A/E lesions seen, moderate tubular blood vessel disruption	High dose required to establish infection, no glomerular disease	134
Caenorhabditis elegans fed on EHEC strains	N2	Intestinal colonization, A/E lesion formation, mortality	Not a model of EHEC-mediated disease	106
Rats + IP injection of Stx	Sprague-Dawley (adult male)	Watery diarrhea, kidney failure, thrombocytopenia, hemolytic anemia, leukocytosis	Nonbacterial, crude bacterial supernatants were used	104
Baboons + Stx injection into cephalic vein		Renal failure, thrombocytopenia, glomerular endothelial injury, inflammation, mortality	Nonbacterial, not the natural route of infection	101–103
Chickens + crop cannulation of EHEC	1 day old	Persistent shedding of the organism, A/E lesions observed	No clinical signs	135

(e.g., EPEC) (105), *E. coli* O157:H7 infects and kills *C. elegans* (106). The bacterium is able to multiply and produce A/E lesions in the digestive tract of the nematodes, and killing appears to be at least partly dependent on Stx1 (Stx2 was not important in this model). It remains to be seen whether this model will lead to new information on the pathogenesis and host immune response to infection; however, the availability of powerful genetic analysis methods for *C. elegans* may prove invaluable.

ROLE OF THE HOST IN MODULATING EHEC VIRULENCE

Like other enteric pathogens, EHEC uses many host- and microbiota-derived chemicals to regulate virulence gene expression when in the host (107). To date, only a few of these have been described. For example, it is known that the two-component system encoded by

QseBC acts to recognize and respond to the presence of the host hormones epinephrine and norepinephrine or to the microbiota-derived quorum-sensing molecule AI-3 to activate flagella expression and motility (108). It is thought that signaling through this pathway may alert EHEC to its arrival in the colon, a key site of bacterial attachment. Consistent with this idea, QseC mutants exhibit a reduced ability to colonize the colon of infant rabbits (109). EHEC also senses fucose, a microbiota-derived sugar, whose presence at high concentrations in the intestinal lumen inhibits LEE gene expression until the bacterium reaches the epithelial surface (110). Deletion of FusKR, the fucose-sensing two-component system, reduced EHEC's ability to colonize the rabbit intestine compared to the parental strain. In addition to sugars, EHEC also appears to sense xanthine oxidase, a host defense molecule that is constitutively expressed by intestinal epithelial cells. Physiological concentrations of xanthine oxidase upregulate EHEC virulence factor production only as the bacterium comes into close proximity to the epithelial surface (111, 112). Finally, diet-induced changes to butyrate concentrations in the mouse intestine were found to cause increased mortality in animals following EHEC infection (113). The mechanistic basis for this finding is hypothesized to be due to increased levels of Gb3 expression in the kidneys and intestines of mice fed high-fiber diets. However, mice fed high-fiber diets also contained significantly fewer commensal *Escherichia* spp. As such, this study highlights some of the complexities faced when dealing with intact biological systems, where the host and its associated microbiota are intimately linked.

Over the past decade, there has been a growing realization that the host and its associated microbiota profoundly influence the pathologies caused by noninfectious and infectious disease agents (e.g., see references 114 to 117). Many factors including diet, stress, or the environment alter the physiochemical conditions in the gut and thereby influence the composition and activities of the intestinal microbiota (118). In turn, the intestinal microbiota contributes to numerous biological processes in the host, including metabolic, nutritional, physiological, and immunological function. As such, individual differences in host microbiota characteristics may explain some of the variation in an individual's response to EHEC during an outbreak.

Finally, a recent report suggests that uptake of Stx-encoding genetic material by host cells, rather than the toxic protein, may contribute to Stx-induced pathology in mice (119, 120). These studies raise the question of whether Stx-coding bacteriophages, which are released along with Stx following bacterial lysis in the intestine, play a role in Stx dissemination (121).

CONCLUSIONS

Pigs, rabbits, and mice are the three most common animal species used to study EHEC-mediated disease. Despite recent advances in the field, none reproduce the full clinical spectrum of illness seen during human infection. The choice of which animal model to use must be viewed within the context of the scientific question being asked. However, studies focused on understanding the biology of the organism need to be considered within a model system that reproduces as much as possible the natural state of the intestine.

ACKNOWLEDGMENTS

I declare no conflicts of interest with regard to the manuscript.

CITATION

Ritchie JM. 2014. Animal models of enterohemorrhagic *Escherichia coli* infection. Microbiol Spectrum 2(4):EHEC-0022-2013.

REFERENCES

1. **Riley LW, Remis RS, Helgerson SD, McGee HB, Wells JG, Davis BR, Hebert RJ, Olcott ES, Johnson LM, Hargrett NT, Blake PA, Cohen ML.** 1983. Hemorrhagic colitis associated with a rare *Escherichia coli* serotype. *New Engl J Med* **308:**681–685.
2. **Moxley RA, Francis DH.** 1998. Overview of animal models, p 249–260. *In* Kaper JB, O'Brien AD (ed), *Escherichia coli* O157:H7 and Other Shiga Toxin-Producing *E coli* strains. ASM Press, Washington, DC.
3. **Ferens WA, Hovde CJ.** 2011. *Escherichia coli* O157:H7: animal reservoir and sources of human infection. *Foodborne Pathog Dis* **8:**465–487.
4. **Schuller S.** 2011. Shiga toxin interaction with human intestinal epithelium. *Toxins* **3:**626–639.
5. **Kaper JB, Nataro JP, Mobley HL.** 2004. Pathogenic *Escherichia coli. Nat Rev Microbiol* **2:**123–140.
6. **Obrig TG.** 2010. *Escherichia coli* Shiga toxin mechanisms of action in renal disease. *Toxins* **2:**2769–2794.
7. **Farfan MJ, Torres AG.** 2012. Molecular mechanisms that mediate colonization of Shiga toxin-producing *Escherichia coli* strains. *Infect Immun* **80:**903–913.
8. **Melton-Celsa A, Mohawk K, Teel L, O'Brien A.** 2012. Pathogenesis of Shiga-toxin producing *Escherichia coli. Curr Top Microbiol Immunol* **357:**67–103.

9. **Bitzan M.** 2009. Treatment options for HUS secondary to *Escherichia coli* O157:H7. *Kidney Int Suppl* **112:**S62–S66.

10. **Paton JC, Paton AW.** 1998. Pathogenesis and diagnosis of Shiga toxin-producing *Escherichia coli* infections. *Clin Microbiol Rev* **11:**450–479.

11. **Bell BP, Goldoft M, Griffin PM, Davis MA, Gordon DC, Tarr PI, Bartleson CA, Lewis JH, Barrett TJ, Wells JG, et al.** 1994. A multistate outbreak of *Escherichia coli* O157:H7-associated bloody diarrhea and hemolytic uremic syndrome from hamburgers. The Washington experience. *JAMA* **272:**1349–1353.

12. **Mayer CL, Leibowitz CS, Kurosawa S, Stearns-Kurosawa DJ.** 2012. Shiga toxins and the pathophysiology of hemolytic uremic syndrome in humans and animals. *Toxins* **4:**1261–1287.

13. **Brigotti M, Caprioli A, Tozzi AE, Tazzari PL, Ricci F, Conte R, Carnicelli D, Procaccino MA, Minelli F, Ferretti AV, Paglialonga F, Edefonti A, Rizzoni G.** 2006. Shiga toxins present in the gut and in the polymorphonuclear leukocytes circulating in the blood of children with hemolytic-uremic syndrome. *J Clin Microbiol* **44:**313–317.

14. **te Loo DM, Monnens LA, van Der Velden TJ, Vermeer MA, Preyers F, Demacker PN, van Den Heuvel LP, van Hinsbergh VW.** 2000. Binding and transfer of verocytotoxin by polymorphonuclear leukocytes in hemolytic uremic syndrome. *Blood* **95:**3396–3402.

15. **Johnson WM, Lior H, Bezanson GS.** 1983. Cytotoxic *Escherichia coli* O157:H7 associated with haemorrhagic colitis in Canada. *Lancet* **i:**76.

16. **Chaisri U, Nagata M, Kurazono H, Horie H, Tongtawe P, Hayashi H, Watanabe T, Tapchaisri P, Chongsa-nguan M, Chaicumpa W.** 2001. Localization of Shiga toxins of enterohaemorrhagic *Escherichia coli* in kidneys of paediatric and geriatric patients with fatal haemolytic uraemic syndrome. *Microb Pathog* **31:**59–67.

17. **Okuda T, Tokuda N, Numata S, Ito M, Ohta M, Kawamura K, Wiels J, Urano T, Tajima O, Furukawa K.** 2006. Targeted disruption of Gb3/CD77 synthase gene resulted in the complete deletion of globo-series glycosphingolipids and loss of sensitivity to verotoxins. *J Biol Chem* **281:** 10230–10235.

18. **Ray PE, Liu XH.** 2001. Pathogenesis of Shiga toxin-induced hemolytic uremic syndrome. *Pediatr Nephrol* **16:**823–839.

19. **Boyd B, Lingwood C.** 1989. Verotoxin receptor glycolipid in human renal tissue. *Nephron* **51:**207–210.

20. **Ren J, Utsunomiya I, Taguchi K, Ariga T, Tai T, Ihara Y, Miyatake T.** 1999. Localization of verotoxin receptors in nervous system. *Brain Res* **825:**183–188.

21. **Jacewicz MS, Acheson DW, Binion DG, West GA, Lincicome LL, Fiocchi C, Keusch GT.** 1999. Responses of human intestinal microvascular endothelial cells to Shiga toxins 1 and 2 and pathogenesis of hemorrhagic colitis. *Infect Immun* **67:**1439–1444.

22. **O'Loughlin EV, Robins-Browne RM.** 2001. Effect of Shiga toxin and Shiga-like toxins on eukaryotic cells. *Microbes Infect* **3:**493–507.

23. **Zumbrun SD, Hanson L, Sinclair JF, Freedy J, Melton-Celsa AR, Rodriguez-Canales J, Hanson JC, O'Brien AD.** 2010. Human intestinal tissue and cultured colonic cells contain globotriaosylceramide synthase mRNA and the alternate Shiga toxin receptor globotetraosylceramide. *Infect Immun* **78:**4488–4499.

24. **Schuller S, Frankel G, Phillips AD.** 2004. Interaction of Shiga toxin from *Escherichia coli* with human intestinal epithelial cell lines and explants: Stx2 induces epithelial damage in organ culture. *Cell Microbiol* **6:**289–301.

25. **Holgersson J, Stromberg N, Breimer ME.** 1988. Glycolipids of human large intestine: difference in glycolipid expression related to anatomical localization, epithelial/non-epithelial tissue and the ABO, Le and Se phenotypes of the donors. *Biochimie* **70:**1565–1574.

26. **Griffin PM, Tauxe RV.** 1991. The epidemiology of infections caused by *Escherichia coli* O157:H7, other enterohemorrhagic *E. coli*, and the associated hemolytic uremic syndrome. *Epidemiol Rev* **13:**60–98.

27. **Tarr PI, Gordon CA, Chandler WL.** 2005. Shiga-toxin-producing *Escherichia coli* and haemolytic uraemic syndrome. *Lancet* **365:**1073–1086.

28. **Gould LH, Mody RK, Ong KL, Clogher P, Cronquist AB, Garman KN, Lathrop S, Medus C, Spina NL, Webb TH, White PL, Wymore K, Gierke RE, Mahon BE, Griffin PM.** 2013. Increased recognition of non-O157 Shiga toxin-producing *Escherichia coli* infections in the United States during 2000–2010: epidemiologic features and comparison with *E. coli* O157 infections. *Foodborne Pathog Dis* **10:**453–462.

29. **Donnenberg MS, Tzipori S, McKee ML, O'Brien AD, Alroy J, Kaper JB.** 1993. The role of the *eae* gene of enterohemorrhagic *Escherichia coli* in intimate attachment in vitro and in a porcine model. *J Clin Investig* **92:**1418–1424.

30. **Jerse AE, Yu J, Tall BD, Kaper JB.** 1990. A genetic locus of enteropathogenic *Escherichia coli* necessary for the production of attaching and effacing lesions on tissue culture cells. *Proc Natl Acad Sci USA* **87:**7839–7843.

31. **Ritchie JM, Thorpe CM, Rogers AB, Waldor MK.** 2003. Critical roles for *stx2*, *eae*, and *tir* in enterohemorrhagic *Escherichia coli*-induced di-

arrhea and intestinal inflammation in infant rabbits. *Infect Immun* **71:**7129–7139.

32. **Bonnet R, Souweine B, Gauthier G, Rich C, Livrelli V, Sirot J, Joly B, Forestier C.** 1998. Non-O157:H7 Stx2-producing *Escherichia coli* strains associated with sporadic cases of hemolytic-uremic syndrome in adults. *J Clin Microbiol* **36:** 1777–1780.

33. **Newton HJ, Sloan J, Bulach DM, Seemann T, Allison CC, Tauschek M, Robins-Browne RM, Paton JC, Whittam TS, Paton AW, Hartland EL.** 2009. Shiga toxin-producing *Escherichia coli* strains negative for locus of enterocyte effacement. *Emerg Infect Dis* **15:**372–380.

34. **Hedican EB, Medus C, Besser JM, Juni BA, Koziol B, Taylor C, Smith KE.** 2009. Characteristics of O157 versus non-O157 Shiga toxin-producing *Escherichia coli* infections in Minnesota, 2000–2006. *Clin Infect Dis* **49:**358–364.

35. **Gerber A, Karch H, Allerberger F, Verweyen HM, Zimmerhackl LB.** 2002. Clinical course and the role of shiga toxin-producing *Escherichia coli* infection in the hemolytic-uremic syndrome in pediatric patients, 1997–2000, in Germany and Austria: a prospective study. *J Infect Dis* **186:**493–500.

36. **Hermos CR, Janineh M, Han LL, McAdam AJ.** 2011. Shiga toxin-producing *Escherichia coli* in children: diagnosis and clinical manifestations of O157:H7 and non-O157:H7 infection. *J Clin Microbiol* **49:**955–959.

37. **Preussel K, Hohle M, Stark K, Werber D.** 2013. Shiga toxin-producing *Escherichia coli* O157 is more likely to lead to hospitalization and death than non-O157 serogroups—Except O104. *PloS One* **8:**e78180.

38. **Frank C, Werber D, Cramer JP, Askar M, Faber M, an der Heiden M, Bernard H, Fruth A, Prager R, Spode A, Wadl M, Zoufaly A, Jordan S, Kemper MJ, Follin P, Muller L, King LA, Rosner B, Buchholz U, Stark K, Krause G.** 2011. Epidemic profile of Shiga-toxin-producing *Escherichia coli* O104:H4 outbreak in Germany. *New Engl J Med* **365:**1771–1780.

39. **Foubister V, Rosenshine I, Donnenberg MS, Finlay BB.** 1994. The *eaeB* gene of enteropathogenic *Escherichia coli* is necessary for signal transduction in epithelial cells. *Infect Immun* **62:**3038–3040.

40. **Tacket CO, Sztein MB, Losonsky G, Abe A, Finlay BB, McNamara BP, Fantry GT, James SP, Nataro JP, Levine MM, Donnenberg MS.** 2000. Role of EspB in experimental human enteropathogenic *Escherichia coli* infection. *Infect Immun* **68:**3689–3695.

41. **Donnenberg MS, Tacket CO, James SP, Losonsky G, Nataro JP, Wasserman SS, Kaper JB, Levine MM.** 1993. Role of the *eaeA* gene in experimental enteropathogenic *Escherichia coli* infection. *J Clin Invest* **92:**1412–1417.

42. **Griffin PM, Olmstead LC, Petras RE.** 1990. *Escherichia coli* O157:H7-associated colitis. A clinical and histological study of 11 cases. *Gastroenterology* **99:**142–149.

43. **Shigeno T, Akamatsu T, Fujimori K, Nakatsuji Y, Nagata A.** 2002. The clinical significance of colonoscopy in hemorrhagic colitis due to enterohemorrhagic *Escherichia coli* O157:H7 infection. *Endoscopy* **34:**311–314.

44. **Kelly JK, Pai CH, Jadusingh IH, Macinnis ML, Shaffer EA, Hershfield NB.** 1987. The histopathology of rectosigmoid biopsies from adults with bloody diarrhea due to verotoxin-producing *Escherichia coli*. *Am J Clin Pathol* **88:**78–82.

45. **Ryan CA, Tauxe RV, Hosek GW, Wells JG, Stoesz PA, McFadden HW Jr, Smith PW, Wright GF, Blake PA.** 1986. *Escherichia coli* O157:H7 diarrhea in a nursing home: clinical, epidemiological, and pathological findings. *J Infect Dis* **154:**631–638.

46. **Inward CD, Howie AJ, Fitzpatrick MM, Rafaat F, Milford DV, Taylor CM.** 1997. Renal histopathology in fatal cases of diarrhoea-associated haemolytic uraemic syndrome. British Association for Paediatric Nephrology. *Pediatr Nephrol* **11:**556–559.

47. **Tzipori S, Wachsmuth IK, Chapman C, Birden R, Brittingham J, Jackson C, Hogg J.** 1986. The pathogenesis of hemorrhagic colitis caused by *Escherichia coli* O157:H7 in gnotobiotic piglets. *J Infect Dis* **154:**712–716.

48. **Francis DH, Collins JE, Duimstra JR.** 1986. Infection of gnotobiotic pigs with an *Escherichia coli* O157:H7 strain associated with an outbreak of hemorrhagic colitis. *Infect Immun* **51:**953–956.

49. **Tzipori S, McCartney E, Lawson GH, Rowland AC, Campbell I.** 1981. Experimental infection of piglets with cryptosporidium. *Res Vet Sci* **31:**358–368.

50. **Tzipori S, Chandler D, Smith M, Makin T, Halpin C.** 1982. Experimental colibacillosis in gnotobiotic piglets exposed to 3 enterotoxigenic serotypes. *Aust Vet J* **59:**93–95.

51. **Tzipori S, Robins-Browne RM, Gonis G, Hayes J, Withers M, McCartney E.** 1985. Enteropathogenic *Escherichia coli* enteritis: evaluation of the gnotobiotic piglet as a model of human infection. *Gut* **26:**570–578.

52. **Moon HW, Whipp SC, Argenzio RA, Levine MM, Giannella RA.** 1983. Attaching and effacing activities of rabbit and human enteropathogenic *Escherichia coli* in pig and rabbit intestines. *Infect Immun* **41:**1340–1351.

53. **Levine MM, Kaper JB, Black RE, Clements ML.** 1983. New knowledge on pathogenesis of bacterial enteric infections as applied to vaccine development. *Microbiol Rev* **47:**510–550.

54. **Tzipori S, Gibson R, Montanaro J.** 1989. Nature and distribution of mucosal lesions associated with enteropathogenic and enterohemorrhagic *Escherichia coli* in piglets and the role of plasmid-mediated factors. *Infect Immun* **57:**1142–1150.

55. **Tzipori S, Karch H, Wachsmuth KI, Robins-Browne RM, O'Brien AD, Lior H, Cohen ML, Smithers J, Levine MM.** 1987. Role of a 60-megadalton plasmid and Shiga-like toxins in the pathogenesis of infection caused by enterohemorrhagic *Escherichia coli* O157:H7 in gnotobiotic piglets. *Infect Immun* **55:**3117–3125.

56. **Tzipori S, Gunzer F, Donnenberg MS, de Montigny L, Kaper JB, Donohue-Rolfe A.** 1995. The role of the *eaeA* gene in diarrhea and neurological complications in a gnotobiotic piglet model of enterohemorrhagic *Escherichia coli* infection. *Infect Immun* **63:**3621–3627.

57. **Tzipori S, Chow CW, Powell HR.** 1988. Cerebral infection with *Escherichia coli* O157:H7 in humans and gnotobiotic piglets. *J Clin Pathol* **41:**1099–1103.

58. **Francis DH, Moxley RA, Andraos CY.** 1989. Edema disease-like brain lesions in gnotobiotic piglets infected with *Escherichia coli* serotype O157:H7. *Infect Immun* **57:**1339–1342.

59. **Dean-Nystrom EA, Melton-Celsa AR, Pohlenz JF, Moon HW, O'Brien AD.** 2003. Comparative pathogenicity of *Escherichia coli* O157 and intimin-negative non-O157 Shiga toxin-producing *E coli* strains in neonatal pigs. *Infect Immun* **71:**6526–6533.

60. **Gunzer F, Hennig-Pauka I, Waldmann KH, Sandhoff R, Grone HJ, Kreipe HH, Matussek A, Mengel M.** 2002. Gnotobiotic piglets develop thrombotic microangiopathy after oral infection with enterohemorrhagic *Escherichia coli*. *Am J Clin Pathol* **118:**364–375.

61. **Pohlenz JF, Winter KR, Dean-Nystrom EA.** 2005. Shiga-toxigenic *Escherichia coli*-inoculated neonatal piglets develop kidney lesions that are comparable to those in humans with hemolytic-uremic syndrome. *Infect Immun* **73:**612–616.

62. **Dean-Nystrom EA, Gansheroff LJ, Mills M, Moon HW, O'Brien AD.** 2002. Vaccination of pregnant dams with intimin(O157) protects suckling piglets from *Escherichia coli* O157:H7 infection. *Infect Immun* **70:**2414–2418.

63. **Dean-Nystrom EA, Pohlenz JF, Moon HW, O'Brien AD.** 2000. *Escherichia coli* O157:H7 causes more-severe systemic disease in suckling piglets than in colostrum-deprived neonatal piglets. *Infect Immun* **68:**2356–2358.

64. **Shringi S, Garcia A, Lahmers KK, Potter KA, Muthupalani S, Swennes AG, Hovde CJ, Call DR, Fox JG, Besser TE.** 2012. Differential virulence of clinical and bovine-biased enterohemorrhagic *Escherichia coli* O157:H7 genotypes in piglet and Dutch belted rabbit models. *Infect Immun* **80:**369–380.

65. **Farmer JJ 3rd, Potter ME, Riley LW, Barrett TJ, Blake PA, Bopp CA, Cohen ML, Kaufmann A, Morris GK, Remis RS, Thomason BM, Wells JG.** 1983. Animal models to study *Escherichia coli* O157:H7 isolated from patients with haemorrhagic colitis. *Lancet* **i:**702–703.

66. **Potter ME, Kaufmann AF, Thomason BM, Blake PA, Farmer JJ 3rd.** 1985. Diarrhea due to *Escherichia coli* O157:H7 in the infant rabbit. *J Infect Dis* **152:**1341–1343.

67. **Pai CH, Kelly JK, Meyers GL.** 1986. Experimental infection of infant rabbits with verotoxin-producing *Escherichia coli*. *Infect Immun* **51:**16–23.

68. **Elliott E, Li Z, Bell C, Stiel D, Buret A, Wallace J, Brzuszczak I, O'Loughlin E.** 1994. Modulation of host response to *Escherichia coli* O157:H7 infection by anti-CD18 antibody in rabbits. *Gastroenterology* **106:**1554–1561.

69. **Li Z, Bell C, Buret A, Robins-Browne R, Stiel D, O'Loughlin E.** 1993. The effect of enterohemorrhagic *Escherichia coli* O157:H7 on intestinal structure and solute transport in rabbits. *Gastroenterology* **104:**467–474.

70. **Lloyd SJ, Ritchie JM, Rojas-Lopez M, Blumentritt CA, Popov VL, Greenwich JL, Waldor MK, Torres AG.** 2012. A double, long polar fimbria mutant of *Escherichia coli* O157:H7 expresses Curli and exhibits reduced in vivo colonization. *Infect Immun* **80:**914–920.

71. **Ritchie JM, Waldor MK.** 2005. The locus of enterocyte effacement-encoded effector proteins all promote enterohemorrhagic *Escherichia coli* pathogenicity in infant rabbits. *Infect Immun* **73:**1466–1474.

72. **Shifrin Y, Peleg A, Ilan O, Nadler C, Kobi S, Baruch K, Yerushalmi G, Berdichevsky T, Altuvia S, Elgrably-Weiss M, Abe C, Knutton S, Sasakawa C, Ritchie JM, Waldor MK, Rosenshine I.** 2008. Transient shielding of intimin and the type III secretion system of enterohemorrhagic and enteropathogenic *Escherichia coli* by a group 4 capsule. *J Bacteriol* **190:**5063–5074.

73. **Lloyd SJ, Ritchie JM, Torres AG.** 2012. Fimbriation and curliation in *Escherichia coli* O157:H7: a paradigm of intestinal and environmental colonization. *Gut Microbes* **3:**272–276.

74. **Ritchie JM, Brady MJ, Riley KN, Ho TD, Campellone KG, Herman IM, Donohue-Rolfe**

A, Tzipori S, Waldor MK, Leong JM. 2008. EspFU, a type III-translocated effector of actin assembly, fosters epithelial association and late-stage intestinal colonization by *E. coli* O157:H7. *Cell Microbiol* **10**:836–847.

75. Zoja C, Corna D, Farina C, Sacchi G, Lingwood C, Doyle MP, Padhye VV, Abbate M, Remuzzi G. 1992. Verotoxin glycolipid receptors determine the localization of microangiopathic process in rabbits given verotoxin-1. *J Lab Clin Med* **120**:229–238.

76. Garcia A, Bosques CJ, Wishnok JS, Feng Y, Karalius BJ, Butterton JR, Schauer DB, Rogers AB, Fox JG. 2006. Renal injury is a consistent finding in Dutch Belted rabbits experimentally infected with enterohemorrhagic *Escherichia coli*. *J Infect Dis* **193**:1125–1134.

77. Stone SM, Thorpe CM, Ahluwalia A, Rogers AB, Obata F, Vozenilek A, Kolling GL, Kane AV, Magun BE, Jandhyala DM. 2012. Shiga toxin 2-induced intestinal pathology in infant rabbits is A-subunit dependent and responsive to the tyrosine kinase and potential ZAK inhibitor imatinib. *Front Cell Infect Microbiol* **2**:135.

78. Mobassaleh M, Donohue-Rolfe A, Jacewicz M, Grand RJ, Keusch GT. 1988. Pathogenesis of shigella diarrhea: evidence for a developmentally regulated glycolipid receptor for shigella toxin involved in the fluid secretory response of rabbit small intestine. *J Infect Dis* **157**:1023–1031.

79. Mobassaleh M, Gross SK, McCluer RH, Donohue-Rolfe A, Keusch GT. 1989. Quantitation of the rabbit intestinal glycolipid receptor for Shiga toxin. Further evidence for the developmental regulation of globotriaosylceramide in microvillus membranes. *Gastroenterology* **97**:384–391.

80. Garcia A, Marini RP, Feng Y, Vitsky A, Knox KA, Taylor NS, Schauer DB, Fox JG. 2002. A naturally occurring rabbit model of enterohemorrhagic *Escherichia coli*-induced disease. *J Infect Dis* **186**:1682–1686.

81. Panda A, Tatarov I, Melton-Celsa AR, Kolappaswamy K, Kriel EH, Petkov D, Coksaygan T, Livio S, McLeod CG, Nataro JP, O'Brien AD, DeTolla LJ. 2010. *Escherichia coli* O157:H7 infection in Dutch belted and New Zealand white rabbits. *Comp Med* **60**:31–37.

82. De SN, Chatterje DN. 1953. An experimental study of the mechanism of action of *Vibrio cholerae* on the intestinal mucous membrane. *J Pathol Bacteriol* **66**:559–562.

83. Keenan KP, Sharpnack DD, Collins H, Formal SB, O'Brien AD. 1986. Morphologic evaluation of the effects of Shiga toxin and *E coli* Shiga-like toxin on the rabbit intestine. *Am J Pathol* **125**:69–80.

84. Agin TS, Cantey JR, Boedeker EC, Wolf MK. 1996. Characterization of the *eaeA* gene from rabbit enteropathogenic *Escherichia coli* strain RDEC-1 and comparison to other *eaeA* genes from bacteria that cause attaching-effacing lesions. *FEMS Microbiol Lett* **144**:249–258.

85. Cantey JR, Lushbaugh WB, Inman LR. 1981. Attachment of bacteria to intestinal epithelial cells in diarrhea caused by *Escherichia coli* strain RDEC-1 in the rabbit: stages and role of capsule. *J Infect Dis* **143**:219–230.

86. Sjogren R, Neill R, Rachmilewitz D, Fritz D, Newland J, Sharpnack D, Colleton C, Fondacaro J, Gemski P, Boedeker E. 1994. Role of Shiga-like toxin I in bacterial enteritis: comparison between isogenic *Escherichia coli* strains induced in rabbits. *Gastroenterology* **106**:306–317.

87. Richardson SE, Rotman TA, Jay V, Smith CR, Becker LE, Petric M, Olivieri NF, Karmali MA. 1992. Experimental verocytotoxemia in rabbits. *Infect Immun* **60**:4154–4167.

88. Wadolkowski EA, Sung LM, Burris JA, Samuel JE, O'Brien AD. 1990. Acute renal tubular necrosis and death of mice orally infected with *Escherichia coli* strains that produce Shiga-like toxin type II. *Infect Immun* **58**:3959–3965.

89. Mohawk KL, O'Brien AD. 2011. Mouse models of *Escherichia coli* O157:H7 infection and Shiga toxin injection. *J Biomed Biotechnol* **2011**:258185.

90. Luperchio SA, Schauer DB. 2001. Molecular pathogenesis of *Citrobacter rodentium* and transmissible murine colonic hyperplasia. *Microbes Infect* **3**:333–340.

91. Deng W, Li Y, Vallance BA, Finlay BB. 2001. Locus of enterocyte effacement from *Citrobacter rodentium*: sequence analysis and evidence for horizontal transfer among attaching and effacing pathogens. *Infect Immun* **69**:6323–6335.

92. Deng W, Puente JL, Gruenheid S, Li Y, Vallance BA, Vazquez A, Barba J, Ibarra JA, O'Donnell P, Metalnikov P, Ashman K, Lee S, Goode D, Pawson T, Finlay BB. 2004. Dissecting virulence: systematic and functional analyses of a pathogenicity island. *Proc Natl Acad Sci USA* **101**:3597–3602.

93. Deng W, de Hoog CL, Yu HB, Li Y, Croxen MA, Thomas NA, Puente JL, Foster LJ, Finlay BB. 2010. A comprehensive proteomic analysis of the type III secretome of *Citrobacter rodentium*. *J Biol Chem* **285**:6790–6800.

94. Mallick EM, McBee ME, Vanguri VK, Melton-Celsa AR, Schlieper K, Karalius BJ, O'Brien AD, Butterton JR, Leong JM, Schauer DB. 2012. A novel murine infection model for Shiga toxin-producing *Escherichia coli*. *J Clin Invest* **122**:4012–4024.

95. Bielaszewska M, Friedrich AW, Aldick T, Schurk-Bulgrin R, Karch H. 2006. Shiga toxin activatable by intestinal mucus in *Escherichia coli* isolated from humans: predictor for a severe clinical outcome. *Clin Infect Dis* **43**:1160–1167.

96. Golan L, Gonen E, Yagel S, Rosenshine I, Shpigel NY. 2011. Enterohemorrhagic *Escherichia coli* induce attaching and effacing lesions and hemorrhagic colitis in human and bovine intestinal xenograft models. *Dis Model Mech* **4**:86–94.

97. Woods JB, Schmitt CK, Darnell SC, Meysick KC, O'Brien AD. 2002. Ferrets as a model system for renal disease secondary to intestinal infection with *Escherichia coli* O157:H7 and other Shiga toxin-producing *E. coli*. *J Infect Dis* **185**:550–554.

98. Cowan LA, Hertzke DM, Fenwick BW, Andreasen CB. 1997. Clinical and clinicopathologic abnormalities in greyhounds with cutaneous and renal glomerular vasculopathy: 18 cases (1992–1994). *J Am Vet Med Assoc* **210**:789–793.

99. Fenwick B, Cowan L. 1998. Canine model of hemolytic-uremic syndrome, p 268–277. *In* Kaper JB, O'Brien AD (ed), *Escherichia coli* O157:H7 and Other Shiga Toxin-Producing *E coli* strains. ASM Press, Washington, DC.

100. Raife T, Friedman KD, Fenwick B. 2004. Lepirudin prevents lethal effects of Shiga toxin in a canine model. *Thromb Haemost* **2**:387–393.

101. Siegler RL, Obrig TG, Pysher TJ, Tesh VL, Denkers ND, Taylor FB. 2003. Response to Shiga toxin 1 and 2 in a baboon model of hemolytic uremic syndrome. *Pediatr Nephrol* **18**:92–96.

102. Taylor FB Jr, Tesh VL, DeBault L, Li A, Chang AC, Kosanke SD, Pysher TJ, Siegler RL. 1999. Characterization of the baboon responses to Shiga-like toxin: descriptive study of a new primate model of toxic responses to Stx-1. *Am J Pathol* **154**:1285–1299.

103. Stearns-Kurosawa DJ, Collins V, Freeman S, Tesh VL, Kurosawa S. 2010. Distinct physiologic and inflammatory responses elicited in baboons after challenge with Shiga toxin type 1 or 2 from enterohemorrhagic *Escherichia coli*. *Infect Immun* **78**:2497–2504.

104. Zotta E, Lago N, Ochoa F, Repetto HA, Ibarra C. 2008. Development of an experimental hemolytic uremic syndrome in rats. *Pediatr Nephrol* **23**:559–567.

105. Anyanful A, Dolan-Livengood JM, Lewis T, Sheth S, Dezalia MN, Sherman MA, Kalman LV, Benian GM, Kalman D. 2005. Paralysis and killing of *Caenorhabditis elegans* by enteropathogenic *Escherichia coli* requires the bacterial tryptophanase gene. *Mol Microbiol* **57**:988–1007.

106. Chou TC, Chiu HC, Kuo CJ, Wu CM, Syu WJ, Chiu WT, Chen CS. 2013. Enterohaemorrhagic *Escherichia coli* O157:H7 Shiga-like toxin 1 is required for full pathogenicity and activation of the p38 mitogen-activated protein kinase pathway in *Caenorhabditis elegans*. *Cell Microbiol* **15**:82–97.

107. Hernandez-Doria JD, Sperandio V. 2013. Nutrient and chemical sensing by intestinal pathogens. *Microbes Infect* **15**:759–764.

108. Clarke MB, Hughes DT, Zhu C, Boedeker EC, Sperandio V. 2006. The QseC sensor kinase: a bacterial adrenergic receptor. *Proc Natl Acad Sci USA* **103**:10420–10425.

109. Rasko DA, Moreira CG, Li de R, Reading NC, Ritchie JM, Waldor MK, Williams N, Taussig R, Wei S, Roth M, Hughes DT, Huntley JF, Fina MW, Falck JR, Sperandio V. 2008. Targeting QseC signaling and virulence for antibiotic development. *Science* **321**:1078–1080.

110. Pacheco AR, Curtis MM, Ritchie JM, Munera D, Waldor MK, Moreira CG, Sperandio V. 2012. Fucose sensing regulates bacterial intestinal colonization. *Nature* **492**:113–117.

111. Crane J. 2013. Role of host xanthine oxidase in infection due to enteropathogenic and Shiga-toxigenic *Escherichia coli*. *Gut Microbes* **4**:388–391.

112. Crane JK, Naeher TM, Broome JE, Boedeker EC. 2013. Role of host xanthine oxidase in infection due to enteropathogenic and Shiga-toxigenic *Escherichia coli*. *Infect Immun* **81**:1129–1139.

113. Zumbrun SD, Melton-Celsa AR, Smith MA, Gilbreath JJ, Merrell DS, O'Brien AD. 2013. Dietary choice affects Shiga toxin-producing *Escherichia coli* (STEC) O157:H7 colonization and disease. *Proc Natl Acad Sci USA* **110**:E2126–E2133.

114. Willing BP, Vacharaksa A, Croxen M, Thanachayanont T, Finlay BB. 2011. Altering host resistance to infections through microbial transplantation. *PLoS One* **6**:e26988.

115. Ferreira RB, Gill N, Willing BP, Antunes LC, Russell SL, Croxen MA, Finlay BB. 2011. The intestinal microbiota plays a role in *Salmonella*-induced colitis independent of pathogen colonization. *PLoS One* **6**:e20338.

116. Sekirov I, Finlay BB. 2009. The role of the intestinal microbiota in enteric infection. *J Physiol* **587**:4159–4167.

117. Wen L, Ley RE, Volchkov PY, Stranges PB, Avanesyan L, Stonebraker AC, Hu C, Wong FS, Szot GL, Bluestone JA, Gordon JI, Chervonsky AV. 2008. Innate immunity and intestinal microbiota in the development of Type 1 diabetes. *Nature* **455**:1109–1113.

118. Sekirov I, Russell SL, Antunes LC, Finlay BB. 2010. Gut microbiota in health and disease. *Physiol Rev* **90**:859–904.

119. **Bentancor LV, Bilen MF, Mejias MP, Fernandez-Brando RJ, Panek CA, Ramos MV, Fernandez GC, Isturiz M, Ghiringhelli PD, Palermo MS.** 2013. Functional capacity of Shiga-toxin promoter sequences in eukaryotic cells. *PloS One* **8:**e57128.

120. **Bentancor LV, Mejias MP, Pinto A, Bilen MF, Meiss R, Rodriguez-Galan MC, Baez N, Pedrotti LP, Goldstein J, Ghiringhelli PD, Palermo MS.** 2013. Promoter sequence of Shiga toxin 2 (Stx2) is recognized in vivo, leading to production of biologically active Stx2. *MBio* **4:**e00501-13.

121. **Lengeling A, Mahajan A, Gally DL.** 2013. Bacteriophages as pathogens and immune modulators? *MBio* **4:**e00868-13.

122. **Wadolkowski EA, Burris JA, O'Brien AD.** 1990. Mouse model for colonization and disease caused by enterohemorrhagic *Escherichia coli* O157:H7. *Infect Immun* **58:**2438–2445.

123. **Fujii J, Kita T, Yoshida S, Takeda T, Kobayashi H, Tanaka N, Ohsato K, Mizuguchi Y.** 1994. Direct evidence of neuron impairment by oral infection with verotoxin-producing *Escherichia coli* O157:H- in mitomycin-treated mice. *Infect Immun* **62:**3447–3453.

124. **Fernandez-Brando RJ, Miliwebsky E, Mejias MP, Baschkier A, Panek CA, Abrey-Recalde MJ, Cabrera G, Ramos MV, Rivas M, Palermo MS.** 2012. Shiga toxin-producing *Escherichia coli* O157 : H7 shows an increased pathogenicity in mice after the passage through the gastrointestinal tract of the same host. *J Med Microbiol* **61:**852–859.

125. **Karpman D, Connell H, Svensson M, Scheutz F, Alm P, Svanborg C.** 1997. The role of lipopolysaccharide and Shiga-like toxin in a mouse model of *Escherichia coli* O157:H7 infection. *J Infect Dis* **175:**611–620.

126. **Mohawk KL, Melton-Celsa AR, Zangari T, Carroll EE, O'Brien AD.** 2010. Pathogenesis of *Escherichia coli* O157:H7 strain 86-24 following oral infection of BALB/c mice with an intact commensal flora. *Microb Pathog* **48:**131–142.

127. **Calderon Toledo C, Rogers TJ, Svensson M, Tati R, Fischer H, Svanborg C, Karpman D.** 2008. Shiga toxin-mediated disease in MyD88-deficient mice infected with *Escherichia coli* O157:H7. *Am J Pathol* **173:**1428–1439.

128. **Eaton KA, Friedman DI, Francis GJ, Tyler JS, Young VB, Haeger J, Abu-Ali G, Whittam TS.** 2008. Pathogenesis of renal disease due to enterohemorrhagic *Escherichia coli* in germ-free mice. *Infect Immun* **76:**3054–3063.

129. **Tyler JS, Beeri K, Reynolds JL, Alteri CJ, Skinner KG, Friedman JH, Eaton KA, Friedman DI.** 2013. Prophage induction is enhanced and required for renal disease and lethality in an EHEC mouse model. *PLoS Pathog* **9:**e1003236.

130. **Taguchi H, Takahashi M, Yamaguchi H, Osaki T, Komatsu A, Fujioka Y, Kamiya S.** 2002. Experimental infection of germ-free mice with hyper-toxigenic enterohaemorrhagic *Escherichia coli* O157:H7, strain 6. *J Med Microbiol* **51:**336–343.

131. **Kim SH, Lee YH, Lee SH, Lee SR, Huh JW, Kim SU, Chang KT.** 2011. Mouse model for hemolytic uremic syndrome induced by outer membrane vesicles of *Escherichia coli* O157:H7. *FEMS Immunol Med Microbiol* **63:**427–434.

132. **Sauter KA, Melton-Celsa AR, Larkin K, Troxell ML, O'Brien AD, Magun BE.** 2008. Mouse model of hemolytic-uremic syndrome caused by endotoxin-free Shiga toxin 2 (Stx2) and protection from lethal outcome by anti-Stx2 antibody. *Infect Immun* **76:**4469–4478.

133. **Tesh VL, Burris JA, Owens JW, Gordon VM, Wadolkowski EA, O'Brien AD, Samuel JE.** 1993. Comparison of the relative toxicities of Shiga-like toxins type I and type II for mice. *Infect Immun* **61:**3392–3402.

134. **Kang G, Pulimood AB, Koshi R, Hull A, Acheson D, Rajan P, Keusch GT, Mathan VI, Mathan MM.** 2001. A monkey model for enterohemorrhagic *Escherichia coli* infection. *J Infect Dis* **184:**206–210.

135. **Beery JT, Doyle MP, Schoeni JL.** 1985. Colonization of chicken cecae by *Escherichia coli* associated with hemorrhagic colitis. *Appl Environ Microbiol* **49:**310–315.

Enterohemorrhagic *Escherichia coli* Virulence Gene Regulation

9

JAY L. MELLIES[1] and EMILY LORENZEN[2]

As a species, *Escherichia coli* is highly successful, adapting to inhabit the lower intestine of warm-blooded animals. Commensal *E. coli*, part of the normal biota, resides harmlessly in the gut, producing vitamin K. However, *E. coli* also causes three types of disease in humans: urinary tract infections, sepsis in newborns, and diarrheal disease. Enterohemorrhagic *E. coli* (EHEC) plays a prominent role in the third type of illness. It has been estimated that, for the pan genome of *E. coli*, the nonpathogenic and pathogenic strains only contain a core set of genes comprising approximately 20% of any one genome (1, 2). Much of the horizontally acquired genetic information is clustered within genomic islands in pathogens. As for EHEC, this has allowed the organism to not only attach and colonize the large intestine of humans and other animals, to outcompete commensal *E. coli* and other bacteria at the site of infection, but also to cause serious disease.

Horizontally acquired genetic information in EHEC results in evolution into a specific pathotype—genotype dictates phenotype. How this genetic information is controlled is of equal importance for the success and virulence of the organism. Indeed, Abu-Ali et al. (3) investigated differences in virulence gene regulation in two distinct EHEC isolate lineages, clade 8 and clade 2. A clade is a group of EHEC isolates with one ancestor and all its descendants. Eight clades of *E. coli* O157 isolates were defined by single nucleotide polymorphism (SNP)

[1]Department of Biology, Reed College, Portland, OR 97202; [2]Rockefeller University, New York, NY 10065.

Enterohemorrhagic Escherichia coli *and Other Shiga Toxin-Producing* E. coli
Edited by Vanessa Sperandio and Carolyn J. Hovde
© 2015 American Society for Microbiology, Washington, DC
doi:10.1128/microbiolspec.EHEC-0004-2013

analyses, where clade 8 was a group of hypervirulent bacteria compared to the other seven clades (4). By examining multiple strains per lineage, the investigators found increased expression of horizontally acquired virulence genes in clade 8 versus clade 2. Genes expressed to higher levels in clade 8, which is associated with a greater number of cases of *E. coli* hemorrhagic disease compared to those in clade 2, included several virulence factors: *rpoS*, 29 of the 41 locus of enterocyte effacement (LEE)-encoded genes, *gadX* involved in acid tolerance, and the pO157 plasmid-encoded *stcE* adhesin and *hlyA* hemolysin. These data provide evidence that genetic regulation correlates with EHEC virulence.

However, our understanding of the regulatory network controlling EHEC pathogenesis remains incomplete. Some of the questions pursued by researchers include the following: What are the molecular signals perceived by EHEC strains in the human host and within cattle that allow the bacteria to properly express virulence traits? What are the signals and bacterial responses required to pass through the acid environment of the stomach, to ultimately reside in the large intestine, and to cause disease in humans but to attach to and colonize harmlessly the recto-anal junction (RAJ) of cattle (5). How are expression of the type III secretion system (T3SS), attachment to host cell surfaces, the effector molecules destined for translocation into host cells, secretion itself, and the Shiga toxin leading to serious disease controlled? This article summarizes our current knowledge of EHEC virulence gene regulation, indicating that spatiotemporal control of pathogenesis in humans and carriage in cattle is coming to light.

ACID TOLERANCE

As for any intestinal pathogen that causes disease, EHEC must survive the acidic environment of the stomach. Three acid resistance systems have been described: a glucose-repressed, or oxidative, system and glutamate- and arginine-dependent systems (for review see reference 6). Evidence suggests that the glucose-repressed system is used for EHEC survival in acidic food items (7). For the glutamate-dependent system, glutamate decarboxylases GadA and GadB convert glutamate to γ-amino butyric acid, or GABA (8–10). The arginine decarboxylase AdiA converts arginine to agmatine (11). The Gad system is regulated by a number of environmental conditions and the global regulatory proteins H-NS and CRP and alternate sigma factors RpoS and RpoN (8–10, 12, 13) (Fig. 1). For both amino acid-dependent systems, pH homeostasis is maintained by displacing the α-carboxyl group of the amino acids with a proton that is transported from the environment to the cytoplasm. Cytoplasmic glutamate and arginine are restored by their respective

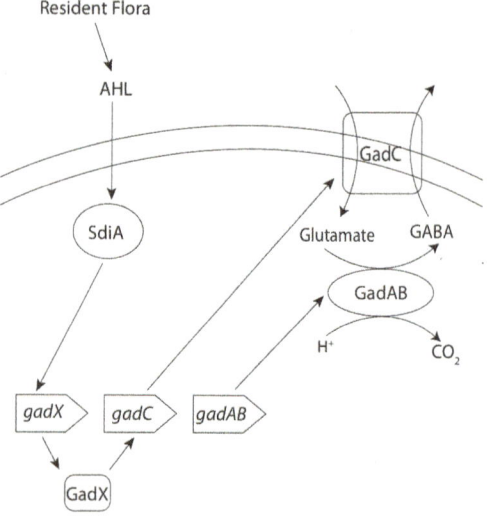

FIGURE 1 Control of acid tolerance by EHEC. Resident flora produce AHL signaling molecules that stimulate expression of the glutamate-dependent acid resistance system (14). The GAD system removes excess protons by exchanging the alpha-carboxyl groups of glutamate with a proton. The resulting GABA molecules are transported out of the cell in exchange for additional glutamate (143).
doi:10.1128/microbiolspec.EHEC-0004-2013.f1

antiporters, GadC and AdiC. The glutamate-dependent Gad system provides a high level of acid tolerance and is necessary for passage through the stomachs and for colonization of the RAJ in cattle (7).

In cattle, EHEC passes through the rumen to eventually colonize the RAJ, and the LuxR homolog SdiA is involved in this process. Although EHEC does not possess the LuxI *N*-acyl homoserine lactone (AHL) synthase, SdiA perceives the oxo-C6-homoserine lactone produced by other bacteria in the rumen (Fig. 1). In turn, SdiA increases *gadX* expression (14), which is a regulator of the *gad* genes, encoding glutamate-dependent acid resistance. Using a cattle model, investigators demonstrated that wild-type EHEC outcompeted the *sdiA* deletion strain, which was defective in colonization of the RAJ over the 6-day assay. The finding that an estimated 70 to 80% of cattle herds in the United States carry EHEC is a major step forward in our understanding of the infection process in the EHEC reservoir.

MOTILITY

Flagellar motility is important in the initial stages of infection for many bacterial pathogens. For EHEC, motility and taxis are stimulated through the two-component system, QseC and QseB, which perceives the AI-3 signal and the interkingdom communication molecules epinephrine and norepinephrine (15–18) (Fig. 2). QseBC signaling activates expression of the *flhDC* master regulators and FliA, an alternate sigma factor, ultimately turning on a number of operons necessary for flagellar biosynthesis and motility. Furthermore, motility and biofilm formation are enhanced in the presence of epinephrine and norepinephrine (19). It is important to note that the AI-3 quorum-sensing molecule is distinct from AI-2, which was initially thought to control motility and type III secretion (20–22). Unlike AI-2, AI-3 is not dependent on LuxS for synthesis. Deletion of the gene encoding the AI-3 and epinephrine- and norepinephrine-

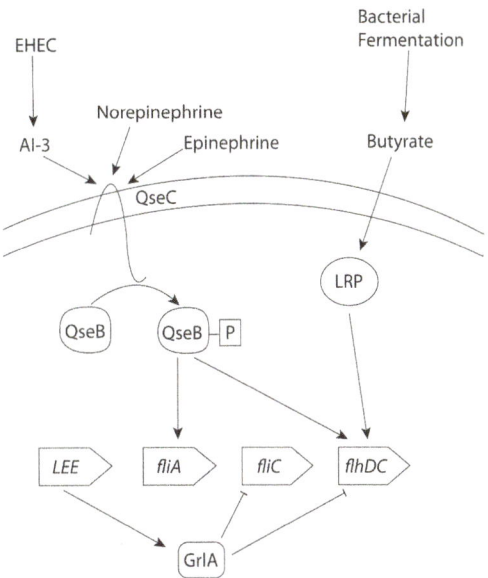

FIGURE 2 Regulation of flagellar biosynthesis and motility. Extracellular signals norepinephrine, epinephrine, and AI-3 stimulate expression of *fliA* (motility) and *flhDC* (biosynthesis) through the two-component system QseBC (16). Following induction of the *LEE* operons, *grlA* inhibits expression of *fliC* and *flhDC* (28). Blunted arrow indicates negative control. doi:10.1128/microbiolspec.EHEC-0004-2013.f2

recognizing QseC sensor kinase attenuated EHEC virulence in a rabbit model of infection. Collectively, these data clearly demonstrate that flagellar motility, and the associated signaling, is necessary for virulence (15, 16).

Additional molecules are known to stimulate flagellar motility, and data indicate temporal control correlates with the infection process. Three classes of promoters are responsible for transcribing the flagellar operons (23). The *flhDC* operon is designated class I and is required for class II promoter activity, which includes the *fliA* operon. In turn, class II promoters are required for class III expression. Butyrate, a short-chain fatty acid found in the large intestine, increases expression of *flhDC* (24) (Fig. 2). Regulation of *flhDC* by butyrate is leucine-responsive regulatory protein (LRP)-dependent, whereas induction of *fliA* by this short-chain fatty acid is LRP-independent. This regulation does not occur in the presence of propionate or acetate. However, propionate and acetate do increase *fliC* (class III) expression but not through *flhDC*. Short-chain fatty acids are found in concentrations ranging from 20 to 140 mM in the large bowel, and recognition of this signal, along with the AI-3 and interkingdom signaling molecules, likely contributes to EHEC niche recognition and adaptation. Some evidence also suggests that the flagellum acts as an initial adhesin to epithelial cells from the bovine terminal rectum (25).

Once motility is no longer needed, or prior to its necessity for colonization, the flagella are downregulated, an important step toward avoiding immune recognition. One observed mechanism is that the LuxR homolog SdiA downregulates FliC expression (26). In another mechanism, mucin, produced by epithelial tissues, diminishes expression of flagellar genes. Mucin on agar plates repressed motility, and transcriptome analysis and quantitative PCR confirmed these data (27). In addition, the GrlA regulator, expressed from the LEE, decreased flagella biosynthesis by downregulating *fliC* and *flhDC* (28) (Fig. 2). Consistent with flagella not being needed after surface adherence, Tobe et al. observed that after 5 h of attachment to epithelial cells, flagella were downregulated in a GrlA-dependent manner. Furthermore, in a neonatal meningitis-causing *E. coli* strain, the *E. coli* common pilus (ECP), found in pathogens and nonpathogens alike, is necessary for attachment and biofilm formation. The regulator of ECP, EcpR, also downregulates flagellar motility after adherence occurs (29). By downregulating the flagellar master regulators *flhDC* by GrlA in EHEC, spatiotemporal regulation controls pathogenesis in coordination with the transition from a planktonic to an adhesive lifestyle.

CONTINUING THE INTESTINAL JOURNEY

Carbohydrate recognition and metabolism are important for EHEC niche adaptation (30) and ensure that EHEC virulence proteins are expressed only at the appropriate site of infection. A glycolytic environment, at 0.4% glucose, inhibits *ler* and LEE expression, whereas 0.1% glucose mimicking gluconeogenesis conditions enhances their expression (31). The LEE pathogenicity island (PAI) encodes T3SS and is organized into five major polycistronic operons and a number of bicistronic and monocistronic genes (32, 33). In gluconeogenesis, instead of oxidizing glucose, glucose is produced from two to three carbon molecules, such as acetate, succinate, or pyruvate. Consistently, glucose and glycerol inhibit *ler* expression, while succinate stimulates *ler*, mimicking glycolysis and gluconeogenesis, respectively (Fig. 3). This regulation occurs through the proteins KdpE, a response regulator that also senses osmotic stress, and Cra, both of which bind to *ler* regulatory DNA.

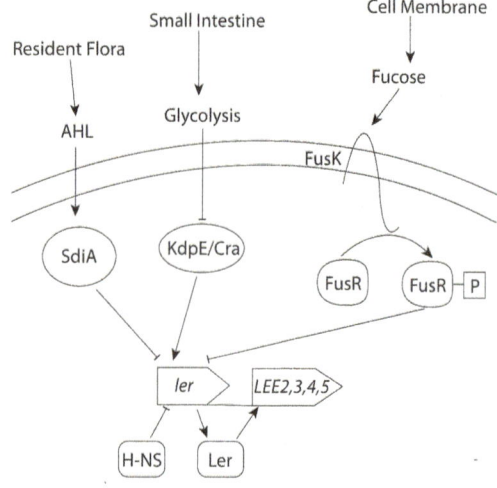

FIGURE 3 Inhibition of EHEC effector molecules in the cattle rumen and human small intestine. Resident flora in the rumen produce the signaling molecule AHL that causes SdiA to fold and inhibit transcription of *ler* (14). Glycolytic conditions in the small intestine inhibit the *ler* transcriptional activators KdpE and Cra (31). Fucose, a component of cell-surface glycans, signals through the two-component system FusKR to inhibit Ler expression (36). The transcriptional silencer H-NS maintains Ler downregulation (48). Blunted arrows indicate negative control. doi:10.1128/microbiolspec.EHEC-0004-2013.f3

Deletion of *kdpE* and *cra* results in ablation of attaching and effacing (A/E) lesion formation by EHEC in vitro(31).

In the gut, EHEC also perceives and responds to the sugar fucose, which enhances EHEC colonization (34). In the mammalian intestine, fucose is abundant, reaching concentrations of ~100 ×M (35), because it is cleaved from host mucin glycoproteins by fucosidases produced by the commensal bacterium *Bacterioides thetaiotamicron*. EHEC perceives the fucose signal via the two-component regulatory pair FusK, the sensor kinase, and FusR, the response regulator (36) (Fig. 3). Transcriptional activity of the *LEE1* operon-encoded *ler* is increased in Δ*fusK* and Δ*fusR* strains. Thus these data indicate that the fucose-sensing, two-component system represses Ler and ultimately T3SS expression. Regulation is direct, as purified FusR binds to *ler* regulatory sequences, and binding is enhanced when the protein is in the phosphorylated state. This carbohydrate signaling is independent of using fucose as a carbon source. The importance of fucose signaling in virulence is highlighted by the findings that in an infant rabbit model of infection, *fusK* and *fusR* deletion mutants are outcompeted by the wild-type strain (36). In sum, *ler* and the T3SS are repressed by high glucose concentrations, and the presence of mucus-derived fucose, while stimulated under gluconeogenesis conditions, is mimicked by the presence of molecules such as succinate, pyruvate, acetate, and propionate.

EHEC virulence gene regulation has been correlated further with intestinal physiology. Butyrate, found in the large intestine, increases expression of the T3SS (37) (Fig. 4). More recently, ethanolamine, a bacterial and animal cell membrane component that is released into the lumen of the gut upon normal epithelial cell turnover, has been implicated as a molecule important in EHEC virulence gene regulation. Ethanolamine is perceived by EHEC, leading to increased expression of the quorum-sensing regulators QseA, QseC, and QseE and Ler (38). In addition, the *LEE* operons 1 through 5 are thus stimulated in the presence of ethanolamine. Though there are 17 ethanolamine utilization genes, *eut*, necessary for the use of this compound as a nitrogen source, the observed signaling in EHEC is independent of ethanolamine metabolism, but partially dependent on the EutR transcriptional regulator that binds ethanolamine (38) (Fig. 4). Furthermore, there is evidence that ethanolamine confers a growth advantage to EHEC during the stationary phase of growth, when this compound is used as a nitrogen source in media mimicking bovine intestinal contents (39). Ethanolamine is a part of normal physiology of the human and bovine large intestine membranes, and for EHEC, it provides a growth advantage and signals for expression of the T3SS as part of ecological niche recognition and adaptation. This signaling most likely coincides with close proximity or attachment to the host epithelium.

T3SS EXPRESSION AND ATTACHMENT

LEE

Elaboration of the T3SS of EHEC allows for attachment to the intestinal epithelium in cattle, specifically to the epithelial cells of the RAJ (5, 40). Disruption or deletion of genes encoding components of this apparatus reduces colonization in the bovine host (41, 42). Furthermore, the translocated intimin receptor (Tir) and intimin proteins, facilitators of the tight attachment of the A/E intestinal lesions, play a critical role in colonization of a neonatal calf model of infection (43). In addition to binding to the bacterial-derived Tir molecule, intimin also binds to integrins and nucleolin on the host cell surface. The molecular syringe, the apparatus necessary for altering signaling events in the formation of actin-rich pedestals, ultimately injects ~50 distinct effector molecules into host cells (44). Assembly of the apparatus, expression of both LEE and non-LEE encoded effectors,

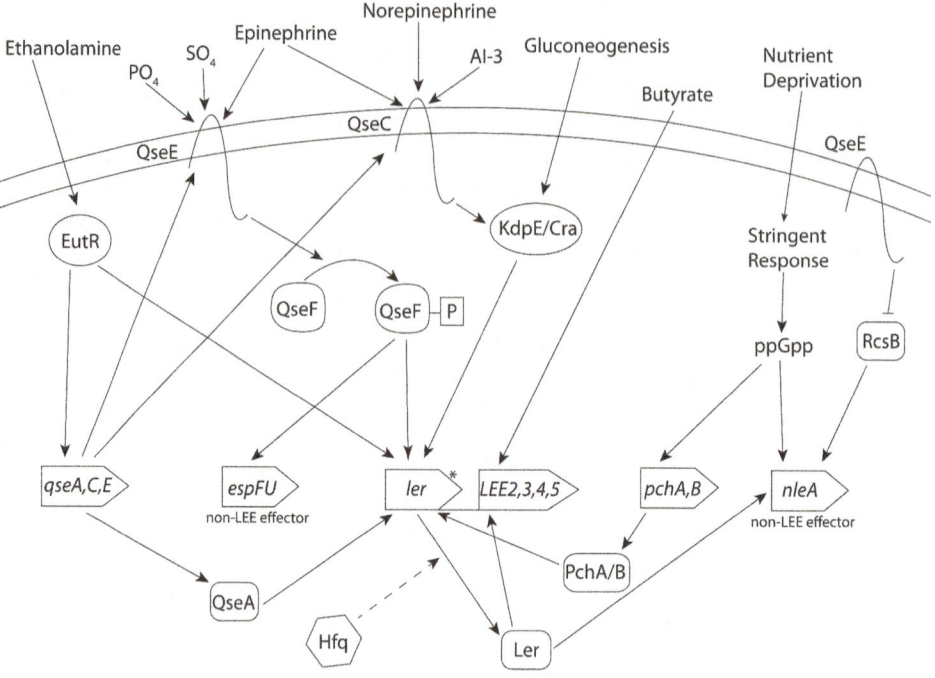

FIGURE 4 Stimulation of LEE and non-LEE effector molecules required for infection by EHEC. In the large intestine of humans and RAJ, the host-produced hormones norepinephrine and epinephrine, EHEC-derived signaling molecule AI-3, and sulfate and phosphate trigger the two-component sensors QseC,F and/or QseEF to upregulate *ler* and *espFu* transcription (17, 85). In addition, ethanolamine, a cell membrane component, stimulates *ler* production through EutR and QseC (38). Gluconeogenic conditions of the large intestine activate the *ler* activators KdpE and Cra, and butyrate directly activates LEE transcriptional activity (37, 125). The RNA chaperone Hfq affects LEE expression through interactions with Ler mRNA but has negative or positive effects depending on the strain of EHEC, as indicated by a dashed arrow (71–73). Finally, nutrient deprivation associated with the infection site activates the stringent response leading to production of ppGpp, which promotes expression of the LEE transcriptional activators *pchA* and *pchB* and the non-LEE effector NleA (68, 117). The asterisk indicates that *ler* (*LEE1*) expression is upregulated by several other factors, including temperature, pH, iron, ammonium, calcium, bicarbonate, and the regulatory proteins IHF, Fis, BipA, PerC, and GadX (previously reviewed in reference 45). Blunted arrow indicates negative control.
doi:10.1128/microbiolspec.EHEC-0004-2013.f4

and their translocation must be regulated in a spatiotemporally correct manner to establish infection and to avoid immune detection in both humans and cattle.

The LEE PAI, encoding the T3SS, is silenced by H-NS (for review see reference 45). Navarre et al. demonstrated that the function of H-NS, conserved across the *Enterobacteriaceae*, is to silence horizontally transferred genetic elements, indicated by a lower GC or higher AT content than the resident genome. Along with *LEE1*, multiple *LEE* operons are

directly silenced by H-NS (46–48), and thus, much of the regulatory network described for the LEE directly involves relieving H-NS-mediated repression. The *LEE1*-encoded Ler protein, an H-NS homolog, is a key regulator of the LEE and acts as an antisilencer (32). As most H-NS-controlled genes are regulated in response to environmental adaptation (49), numerous environmental inputs and regulatory proteins control *ler* gene expression.

To demonstrate the many inputs controlling *ler* gene expression, researchers working with both EHEC and enteropathogenic *E. coli* (EPEC) found that Ler is stimulated in response to environmental signals such as temperature, pH, iron, ammonium, calcium, bicarbonate (50–54), and quorum-sensing signaling (55, 56), as well as the proteins IHF (57), Fis (58), BipA (59), PerC homologs, or Pch (60), GrlRA (61, 62), and GadX (63) (Fig. 4). The GrlRA regulators constitute a complicated feedback loop whereby GrlA directly activates *ler* expression, and GrlR acts posttranslationally to reduce the cellular quantity of the GrlA protein (28, 61, 64, 65). Once expressed, Ler, in turn, acts as an H-NS antisilencer to increase the expression of *LEE2, LEE3, LEE4,* and *LEE5* operons (Fig. 4).

LEE gene expression is stimulated as EHEC transitions into stationary phase, controlled by quorum-sensing signaling, and also in response to nutrient deprivation, much through direct regulation of *ler* (21) (Fig. 4). Nutrient deprivation, the starvation for a number of nutritional requirements such as amino acids, carbon, nitrogen, or phosphorus, induces the stringent response (for review see reference 66). The stringent response produces the signaling molecule ppGpp, a process dependent on the synthase RelA and SpoT, a hydrolase and synthase (67). Expression of the EHEC LEE and adherence to Caco-2 cells in culture are increased in response to nutrient deprivation in a RelA-dependent manner, whereby increased pools of ppGpp are created (68). The protein DksA, which interacts with RNA polymerase in the transcription complex, is also involved in this regulation. Regulation of EspB and Tir, and thus the *LEE4* and *LEE5* operons, is controlled, in part, by ppGpp interaction with the *ler, pchA,* and *pchB* genes. Thus the stringent response rapidly activates expression of the T3SS components when nutrients are limiting and upon transitioning into stationary phase, conditions EHEC is predicted to encounter when entering the lower intestine.

A number of posttranscriptional control mechanisms add to the complexity and highlight the importance of coordinating expression of LEE genes within the host. RNaseE-dependent RNA processing occurs at the *LEE4* operon, resulting in differential control of expression of the SepL protein. This RNA processing is thought to be involved in the contact regulation that controls the switch from translocator to effector type III secretion (62, 69) and the regulation of EspADB proteins necessary for forming the filament and pore that ultimately embed in the host cell membrane (70). In EHEC strain 86-24, serotype O157:H7, the RNA chaperone Hfq acts positively on the expression of *ler* and *LEE* operons 2 through 5 in all phases of growth (71) (Fig. 4). This study shows that Hfq increases the expression of the quorum-sensing, two-component system QseBC. QseC also acts to increase Ler expression and thus illustrates coordinated regulation. Other researchers, though working with strain EDL933 as opposed to 86-24, observed that the RNA chaperone Hfq negatively affects the expression of the LEE (72, 73). Shakhnovich et al. observed that Hfq negatively regulates expression of the LEE in a Ler-dependent manner and also controls expression of a number of non-LEE effector molecules. The *ler,* or *LEE1* transcript, was a direct target of Hfq regulation. Congruently, Hansen and Kaper observed that Hfq affects Ler expression by negatively affecting the GrlRA regulators through posttranscriptional control, the stability of the *grlRA* transcript, while negatively affecting LEE expression in stationary phase in a manner independent of *grlRA*. Consistently with this set of observations, researchers also demonstrated that Hfq negatively affects LEE expression in the related AE pathogen EPEC (73). Thus, LEE genes are apparently regulated by Hfq in an opposite manner in the O157:H7 strains 86-24 and EDL933. Collectively, these regulatory observations illustrate the intricate and variable nature of virulence gene control in related but distinct isolates associated with outbreaks of EHEC disease and the evolution of regulatory networks associated with niche adaptation.

Additionally, a number of O-island regulators contribute to the expression of the LEE. O-islands are horizontally acquired clusters of genes that reside in the EHEC genome. The regulators EtrA and EivF are encoded in a cryptic, second T3SS gene cluster in EHEC, and negatively regulate the LEE PAI and secretion (74). Encoded within O-island 51, the regulator RgdR is a positive regulator of the LEE and secretion, through the Ler regulatory cascade (75). Thus expression of the T3SS is controlled not only by LEE-encoded regulators and regulatory elements that are part of the core *E. coli* genome but also by genes acquired by horizontal gene transfer.

PerC is an important regulator of the EPEC LEE, though a *perC*-like gene does not exist on the pO157 virulence plasmid of EHEC (33, 46, 76). However, in the O157:H7 Sakai stain, Iyoda and Watanabe (60) identified five chromosomally encoded PerC homologs (*pch*). Of those identified, *pchA*, *pchB*, and *pchC*, but not *pchD* or *pchE*, were confirmed to regulate the LEE when expressed in *trans*. Using double mutations, the authors demonstrated that, in combination, *pchA* and *pchB*, or *pchA* and *pchC* resulted in significantly decreased expression of the LEE, through the global regulator Ler, and reduction of adherence to HEp-2 epithelial cells in culture (Fig. 4). Interestingly, the *pchABC* genes of EHEC and *perC* of EPEC are interchangeable in their ability to activate expression of the *LEE1* operons of both organisms (77). Thus the EHEC *pch* genes, with the exception of *pchD* and *pchE*, all encoded within phage-like elements, are involved in pathogenesis.

Intense investigation has been directed toward cell-to-cell communication or quorum-sensing signaling of the LEE and its cognate effector molecules. The best understood of the autoinducers, AHLs, are not produced by *E. coli*. Thus EHEC strains do not have the LuxI AHL synthase, though they do possess a LuxR-type transcriptional regulator SdiA (78, 79), and structural studies indicate that SdiA properly folds only in the presence of AHLs (80). Therefore, regulation through SdiA occurs by perceiving AHLs produced by other bacteria, leading to alterations in EHEC gene expression.

AHL molecules are produced by the resident flora in the rumen of the bovine intestine, indicated by an *Agrobacterium* reporter strain, but are not found in the terminal rectum, the site of EHEC colonization (14). Through SdiA sensing of the oxo-C6-AHL signal LEE genes, including *ler* located in *LEE1* and *LEE4* operons, are downregulated (Fig. 3). It has been proposed that quorum sensing facilitates EHEC escaping the rumen en route to the site of colonization, the RAJ in cattle, by downregulation and avoiding inappropriate expression of the LEE, while inducing the *gad* acid tolerances system, as described above (Fig. 1).

Elucidation of the interkingdom signaling systems of EHEC controlling virulence has revealed how these organisms communicate with members of their own species, resident bacteria, and the host organism. The QseA quorum-sensing regulator acts directly on the expression of the *LEE1* operon, specifically at the P1 promoter (81, 82) (Fig. 4). Indeed, multiple two-component regulators perceive chemical signals and the EHEC-derived, aromatic AI-3 molecule to control expression of the T3SS. The sensor kinase QseE responds to sulfate, phosphate, and host-derived epinephrine to phosphorylate its cognate response regulator QseF that ultimately stimulates *ler* expression (83) (Fig. 4). The sensor kinase QseC perceives the catecholamines epinephrine and norepinephrine and the EHEC-produced AI-3 molecule to increase *ler* transcription through phosphorylation of KdpE (17), leading to the elaboration of the type III system. To add emphasis to the importance of quorum-sensing signaling controlling virulence, in an infant rabbit model of infection, deletion or inhibition of *qseC* in rabbit EPEC severely attenuated virulence (15, 84), and a *qseF* deletion mutant of EHEC does not form A/E lesions (85). Thus the ability to respond to host catecholamines and AI-3 is required for EHEC to stimulate T3SS expression and to establish infection.

Non-LEE

A number of non-T3SS adhesins, either demonstrated to be necessary for or hypothesized to be involved in attachment to host cells, have been described (Table 1). Upon sequencing the *E. coli* O157:H7 genome of strain EDL933, it was noted that this organism has at least 10 fimbrial gene clusters and 13 regions that encode nonfimbrial adhesins, some of which were not found in nonpathogenic *E. coli* (86, 87). Extracellular structures include the *E. coli* YcbQ laminin-binding fimbriae (ELF) (88), two long polar fimbriae (LPF) (89, 90), the F9 fimbriae (91), a type IV pilus called "hemorrhagic *coli* pilus" (HCP) (92), curli (93, 94), OmpA (95), the EHEC factor for adherence (Efa1) (96), the IgrA homolog adhesin (Iha) (97), the ECP, the autotransporter protein EhaG (98), and the pO157 virulence plasmid-encoded StcE (99, 100). Surely these molecules contribute to adherence to surfaces, both biotic and abiotic, but a comprehensive understanding of the role of these adhesins in human disease, carriage in cattle, and survival and propagation in food does not exist.

However, the long polar fimbria 1 (Lpf1) of *E. coli* O157:H7 is a well-characterized example of one such adhesion. Lpf1 facilitates binding of EHEC to not only epithelial cells but also extracellular matrix proteins, including fibronectin, laminin, and collagen IV (101). Lpf1 is tightly regulated, with maximal expression in late exponential phase of growth in iron-deprived and slightly acidic environments (102) (Table 1). Two *lpf* loci exist in *E. coli* O157:H7, and when one or both are deleted, colonization in animal infection models is altered (103, 104) (Table 1). The Lpf1 fimbria is coordinately regulated with the LEE because the operon is silenced by H-NS, while Ler functions as its antisilencer (105).

ECP, common to both pathogens and non-pathogens, has also been described. Deletion of the *ecp* genes resulted in decreased adherence of both O157:H7 strain EDL933 and commensal *E. coli* to human epithelial cells in culture (106). EcpR, a LuxR-like regulator containing a helix-turn-helix DNA-binding motif, controls expression of the ECP (107). The EcpR protein was shown to bind to a TTCCT sequence upstream of the *ecp* operon, and deletion of *ecpR* resulted in decreased adherence by EHEC. H-NS silences *ecp*, and antisilencing occurs by EcpR, assisted by the protein IHF. Thus the ECP is most likely involved in colonization by pathogens and nonpathogens alike. Similarly, a trimeric autotransporter protein, EhaG, of EHEC has been described (98). As for EcpR, these conserved proteins are found in both pathogens and nonpathogens, and in EHEC EhaG is involved in autoaggregation, biofilm formation, and adherence to extracellular matrix proteins and colorectal epithelial cells. Not surprisingly, expression is, in part, controlled by H-NS (Table 1). Finally, the StcE zinc metalloprotease encoded on the pO157 virulence plasmid enhances pedestal formation and is predicted to facilitate adherence to

TABLE 1 Regulation of non-LEE adhesins and factors involved in EHEC adherence

Adhesin	Type	Regulation	Reference(s)
YcbQ	Laminin-binding fimbriae		88
Lpf1	Long-polar fimbriae	H-NS, Ler, late exponential phase, low pH	89, 90, 101, 103–105, 146
F9	Fimbriae		91
Curli	Fimbriae	Fis, Hha, RcsB, zinc, heat shock	93, 94, 147–149
OmpA	Outer membrane protein A	Hfq, nitrogen	95, 150–152
Efa1	Toxin		96, 153
Iha	IgrA homolog adhesin	Temperature, iron	97, 154, 155
ECP	*E. coli* common pilus	EcpR, H-NS, IHF	106, 107
EhaG	Trimeric autotransporter protein	H-NS	98
StcE	Zinc metalloprotease	H-NS, Ler	99, 100, 108

epithelial cells in the host by cleavage of glycoproteins (99, 108). The *stcE* gene is coordinately regulated with the LEE through Ler and H-NS. Clearly there is much work to be done to understand the role of non-T3SS adhesins in niche recognition, pathogenesis in general, and how their expression is coordinated with other virulence factors.

INJECTION OF EFFECTORS

Effector molecules slated for injection by the T3SS into host cells are encoded within the LEE, O-islands, phage, and other integrative elements located in the genome. Those located within the LEE, including Tir, EspF, Map, EspG, EspH, and EspZ, are of course coordinately regulated with the expression of the secretion apparatus itself. Evidence demonstrates that a number of non-LEE-encoded effectors, including EspJ, NleB, NleE, NleF, and NleH, are involved in EHEC colonization and survival, though experiments demonstrating these phenotypes were conducted in a number of different animal infection models (109–111). Thus further experimentation is necessary to determine the role of these non-LEE effectors in EHEC disease and carriage, particularly in colonization of the RAJ in cattle.

Assuming that the non-LEE effectors are important for disease in humans and carriage in cattle, they should be coordinately expressed with LEE-encoded effectors. Investigators have begun to unravel their regulation. In particular, several studies have focused on the NleA (also named EspI) effector because of evidence that it plays an important role in pathogenesis. After translocation into the host cell, NleA colocalizes with the Golgi apparatus, and has been associated with severe disease in the *Citrobacter rodentium* A/E mouse model (112). NleA is conserved in many A/E pathogens (113, 114), and evidence indicates that it affects protein trafficking and secretion through the rough endoplasmic reticulum by binding to the COPII coat protein (112, 113).

NleA is secreted into the EHEC extracellular milieu (115) and is coordinately regulated with the LEE. As for the LEE, the regulatory proteins H-NS, Ler, and GrlA control expression of the phage-encoded *nleA* gene (115–117). Ler acts directly, binding to the regulatory region upstream of the *nleA* promoter. As with *LEE1*, *nleA* is regulated by quorum-sensing signaling, whereby QseE controls expression through inhibition of the positive-acting RcsB response regulator (118). Expression of NleA is induced by osmolarity, response to starvation signals, and RecA-dependent DNA-damage signaling, the latter also controlling LEE expression (117, 119) (Fig. 4).

EspFu (also named TccP) is another non-LEE effector important for the EHEC A/E histopathology. Unlike EPEC where all genes required for A/E lesion formation are found within the LEE, EspFu is necessary for this phenotype in EHEC and is encoded within phage U (120–122). EspFu binds to the GTP-binding domain of N-WASP, mimicking the eukaryotic SH2/SH3 adapter protein and facilitates actin polymerization, and is subject to regulation by environmental signals (123). The genes *espFu* and *espJ* are influenced by changes in temperature, pH, osmolarity, and oxygen pressure (124). The regulators KdpE and Cra act to increase expression of *espFu* (125). The QseEF two-component, quorum-sensing system controls expression of EspFu through AI-3, epinephrine, and norepinephrine signaling (85) (Fig. 4). Thus the non-LEE effectors NleA and EspFu, important for virulence in EHEC, are coordinately controlled with the LEE by overlapping signaling: *nleA* expression is Ler-dependent and, in part, induced by the SOS response, while expression of *espFu* is controlled by the QseA and QseEF quorum-sensing pathways that directly stimulate the LEE (Fig. 4).

Data are emerging to show that there are at least two modes of regulation of the non-LEE effector molecules. The study by Garcia-Angulo et al. (116), addressing whether a common mechanism for the non-LEE effector molecules exists, is an important investigation because to date a common regulatory network linking their expression had not been established. Though performing much of their experimentation on the related A/E pathogen EPEC, investigators identified a 13-bp inverted repeat, with a 5-bp spacer upstream of many of the genes encoding the non-LEE effector molecules. They found in EPEC, by deletion of these sequences and mutagenesis, that the repeats were essential for transcription of the *nleH1* and *nleB2* genes. The authors termed these sequences NRIR, for *nle* regulatory inverted repeat. Transferring this information to EHEC virulence gene regulation, the authors found a number of *nle* genes in the EDL933 and Sakai EHEC strains that contain NRIR sequences in positions predicted to affect transcription by in silico analysis. These loci included, in both strains EDL933 and Sakai, *nleH1-nleF1*, *nleB2-nleC*, *espX*, a number of *nleG* genes, and *espFu*. The authors concluded that the NRIR sequences were involved in coordinated expression of the non-LEE effector molecules in a second mode of regulation, independent of the LEE regulators Ler and GrlA. Future work will undoubtedly involve identifying protein components, as well as additional genetic elements necessary for this regulatory network, and determining how these effector molecules are coordinately regulated with the type III system required for their translocation into host cells.

TOXIN EXPRESSION AND PHAGE INDUCTION

As with many of the non-LEE effector molecules and most major bacterial exotoxins, the EHEC *stx1* and *stx2* genes are encoded within prophages. The Shiga toxins Stx1 and Stx2 are responsible for bloody diarrhea associated with EHEC infection and the serious complication hemolytic-uremic syndrome. Toxin production occurs upon phage induction. The genes *stx1* and *stx2* are encoded in separate prophages, and both are expressed during the phage lytic cycle (126). Initial studies indicated

that *stx1* is also regulated by low levels of iron as a mechanism to acquire this micronutrient (127–129) (Fig. 5). As for induction of other lambdoid phage, Stx production is induced by DNA damage and regulated by RecA. Antibiotics that target DNA synthesis, such as quinolones and mitomycin C, stimulate Stx production (130). Similarly, hydrogen peroxide induces *stx* expression, most likely due to activation of the SOS response (131). In a seemingly contradictory study, the reactive oxygen species NO inhibited Stx phage induction and Stx production in the presence of the DNA-damaging agent mitomycin C (132). Here the authors claim that iNOS expressed in enterocytes, simulated by type I cytokines producing NO, leads to a bacterial NsrR-related stress response, sensitizing and protecting EHEC from DNA damage. Nonetheless, because the only known mechanism

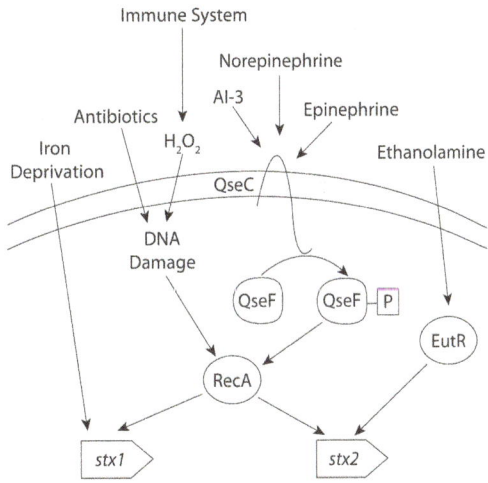

FIGURE 5 Stimulation of Shiga toxin expression. Low iron levels lead to upregulation of *stx1* (127, 128). Hydrogen peroxide and antibiotics targeting DNA synthesis, such as mytomycin C and quinolines, lead to DNA damage and activate RecA to induce Stx1 production (130, 144, 145). Sulfate, phosphate, and epinephrine molecules signal through the two-component system QseCF to RecA, thus leading activation of Stx2 production. Ethanolamine also upregulates *stx2* through the transcription factor EutR (38). doi:10.1128/microbiolspec.EHEC-0004-2013.f5

of Stx release is by lysing of the bacterium, combined with the now obvious reasons that DNA-damaging antibiotics lead to increased risk of hemolytic-uremic syndrome, there is a need for a more complete understanding of how *stx* genes are regulated.

More recent data correlate Stx production with signal molecules associated with specific regions of the human and bovine intestine. The two-component sensor kinase QseC regulates Stx2 expression through the SOS response, RecA cleavage of the lambda repressor cI (17) (Fig. 5). Consistently, *recA* expression was decreased in a *qseC* mutant. QseC responds to the interkingdom signaling molecules epinephrine and norepinephrine and the related AI-3 quorum-sensing molecule produced by *E. coli*. Several studies indicated that Stx production is controlled by AI-3. Another two-component pair, QseE, the sensor kinase, and QseF, the response regulator, affects Stx expression (85, 133). QseF can be phosphorylated by either QseE, responding to epinephrine, sulfate, and phosphate, or QseC to stimulate *stx2*. Ethanolamine, released from epithelium, also stimulates Stx2 production in EHEC (38). The genetic pathway necessary for the signaling involves the ethanolamine utilization regulator *eutR*, as deletion of the gene encoding this protein resulted in decreased expression of *stx2* by microarray (Fig. 5). Thus we begin to see a clearer picture of Stx regulation in EHEC, correlating expression to bacteria-derived autoinducers, host-produced signal molecules, and stress-related signals, and ultimately the genetic pathways necessary for the bacterial response.

Because of the necessity of iron for colonization and pathogenesis within the animal host, the EhxA enterohemolysin is also an important EHEC virulence factor. EhxA is a member of the RTX family, and its function is to lyse red blood cells. The *ehxCABD* operon, found on the pO157 virulence plasmid (134), is positively regulated by DsrA and RpoS at host body temperature (135). DsrA is a small RNA that acts by allowing translation to proceed

(136). The expression of *ehx* is also positively regulated by GrlA and Ler (64, 137), and thus the enterohemolysin is coordinately regulated with the T3SS.

A REGULATED INFECTION CONTINUUM

The work of many research groups leads to a model of EHEC pathogenesis correlating with gut physiology, bacterial metabolism along the route of infection, and host-microbe and microbe-microbe signaling. For both the human and bovine hosts, EHEC must pass through the acidic environment of the stomach and the rumen, respectively. Three acid tolerance systems assist the bacterium in this process. GadX is induced by AHL molecules produced by resident flora of the bovine rumen (14) and perceived by the LuxR homolog SdiA (Fig. 1). The LEE is downregulated by SdiA in this environment because it is not the normal site of attachment, allowing the bacterium to travel toward the RAJ in cattle (Fig. 3). Consistently, in a separate study, data indicated that flagellar motility is upregulated by acid stress while expression of Stx remains unchanged and the T3SS is downregulated by acute acid stress (138).

In the large intestines of humans, conditions mimicking gluconeogenic versus glycolytic metabolism are predicted to stimulate expression of the EHEC T3SS (31, 139) (Fig. 4). This observation is also understandable in colonization of the RAJ of cattle, where nutrients, particularly carbohydrates, are less abundant compared to the rumen. One might posit that EHEC avoids the small intestine in humans by downregulating expression of LEE genes in the presence of high-glucose, glycolytic conditions (Fig. 3). It is curious why EPEC, which infects the small intestine, also downregulates LEE genes under glycolytic conditions, whereas, identical to EHEC, LEE genes are stimulated under low-glucose conditions. This observation might be explained by the specificity of the infection process, that perception of multiple signals is necessary for proper niche recognition.

Carbohydrate metabolism and signaling are clearly linked to virulence, and several studies have indicated that fucose is important for EHEC colonization (34, 140). Resident *Bacterioides* spp. cleave fucose from the glycans found in the intestine, and suppression of Ler, T3SS function, and A/E lesions ensues from the fucose signal perceived by FusKR (36) (Fig. 3). EHEC can use a number of mucus-derived carbohydrates, in particular, mannose, N-acetylglucosamine, and N-acetylglucosamine, and galactose catabolism confers a competitive advantage against commensal bacteria, most likely assisting safe passage through the small intestine (39, 141). In both humans and cattle, EHEC expresses the T3SS in response to membrane-derived ethanolamine, clearly encountered at the sites of infection in both hosts. The catecholamine and AI-3 signals also contribute to elaboration of the secretion apparatus (Fig. 4). Why would epinephrine and norepinephrine exist here when the adrenergic receptors are not located on the apical side of the intestinal lumen, but these molecules were detected in the lumen of the intestine of mice containing specific pathogen-free microbiota (142). It is also plausible that the epinephrine/norepinephrine signal becomes more available upon damage to the host epithelium, releasing these hormones into the lumen of the gut, enhancing expression of the T3SS and firmly establishing infection (Fig. 4). Nonetheless, researchers have established that three important signals, ethanolamine, epinephrine/norepinephrine, combined with the bacteria-derived AI-3, correlated with the sites of infection.

A number of signals, including the catecholamines, also control the expression of *stx* genes, mediated through two-component, quorum-sensing proteins QseC and QseE (Fig. 5). Membrane-derived ethanolamine and the well-studied RecA-dependent SOS response stimulate Stx production. EHEC is expected to encounter all of these conditions during niche recognition at the sites of at-tachment in humans and cattle. These observations are consistent with the presumption that substantive expression of Stx will occur when colonization has commenced.

CONCLUSIONS

Researchers have made significant progress toward understanding the genetic regulation controlling EHEC pathogenesis. How EHEC gains safe passage through the acidic environment of the stomach, how flagellar motility is turned on and turned off, and what signals and metabolites are necessary for outcompeting commensal flora prior to arriving at the sites of infection for humans and cattle are now recognized on a basic level. Researchers have discovered myriad regulatory networks and environmental signals that control type III secretion, and we are beginning to understand how effector molecules slated for translocation into host cells are coordinately regulated with the T3SS itself and how the dangerous Shiga toxin is expressed. Though the LEE-encoded effectors and non-LEE NleA and EspFU are controlled by Ler and quorum-sensing signaling, the majority of those proteins targeted for secretion are regulated by different mechanisms. The NRIR sequences found upstream of many of the phage- and O-island-encoded effectors most likely coordinate their expression, but this has yet to be established experimentally for EHEC. Similarly, with the exception of Lpf1 and StcE, we do not know if, or how, expression of most non-LEE adhesins is coordinated with that of the LEE (Table 1). Finally, EHEC tissue tropism is not well understood, but the regulatory networks described herein most likely play an important role in niche recognition in human and animal hosts.

ACKNOWLEDGMENTS

We declare no conflicts of interest with regard to the manuscript.

CITATION

Mellies JL, Lorenzen E. 2014. Enterohemorrhagic *Escherichia coli* virulence gene regulation. Microbiol Spectrum 2(4):EHEC-0004-2013.

REFERENCES

1. **Lukjancenko O, Wassenaar TM, Ussery DW.** 2010. Comparison of 61 sequenced *Escherichia coli* genomes. *Microb Ecol* **60:**708–720.

2. **Zhaxybayeva O, Doolittle WF.** 2011. Lateral gene transfer. *Curr Biol* **21:**R242–R246.

3. **Abu-Ali GS, Ouellette LM, Henderson ST, Lacher DW, Riordan JT, Whittam TS, Manning SD.** 2010. Increased adherence and expression of virulence genes in a lineage of *Escherichia coli* O157:H7 commonly associated with human infections. *PLoS ONE* **5:**e10167.

4. **Manning SD, Motiwala AS, Springman AC, Qi W, Lacher DW, Ouellette LM, Mladonicky JM, Somsel P, Rudrik JT, Dietrich SE, Zhang W, Swaminathan B, Alland D, Whittam TS.** 2008. Variation in virulence among clades of *Escherichia coli* O157:H7 associated with disease outbreaks. *Proc Natl Acad Sci USA* **105:**4868–4873.

5. **Naylor SW, Low JC, Besser TE, Mahajan A, Gunn GJ, Pearce MC, McKendrick IJ, Smith DG, Gally DL.** 2003. Lymphoid follicle-dense mucosa at the terminal rectum is the principal site of colonization of enterohemorrhagic *Escherichia coli* O157:H7 in the bovine host. *Infect Immun* **71:**1505–1512.

6. **Foster JW.** 2004. *Escherichia coli* acid resistance: tales of an amateur acidophile. *Nat Rev Microbiol* **2:**898–907.

7. **Price SB, Wright JC, DeGraves FJ, Castanie-Cornet MP, Foster JW.** 2004. Acid resistance systems required for survival of *Escherichia coli* O157:H7 in the bovine gastrointestinal tract and in apple cider are different. *Appl Environ Microbiol* **70:**4792–4799.

8. **Castanie-Cornet MP, Penfound TA, Smith D, Elliott JF, Foster JW.** 1999. Control of acid resistance in *Escherichia coli*. *J Bacteriol* **181:**3525–3535.

9. **De Biase D, Tramonti A, Bossa F, Visca P.** 1999. The response to stationary-phase stress conditions in *Escherichia coli*: role and regulation of the glutamic acid decarboxylase system. *Mol Microbiol* **32:**1198–1211.

10. **Lin J, Smith MP, Chapin KC, Baik HS, Bennett GN, Foster JW.** 1996. Mechanisms of acid resistance in enterohemorrhagic *Escherichia coli*. *Appl Environ Microbiol* **62:**3094–3100.

11. **Gong S, Richard H, Foster JW.** 2003. YjdE (AdiC) is the arginine:agmatine antiporter essential for arginine-dependent acid resistance in *Escherichia coli*. *J Bacteriol* **185:**4402–4409.

12. **Mitra A, Fay PA, Morgan JK, Vendura KW, Versaggi SL, Riordan JT.** 2012. Sigma factor N, liaison to an ntrC and rpoS dependent regulatory pathway controlling acid resistance and the LEE in enterohemorrhagic *Escherichia coli*. *PLoS ONE* **7:**e46288.

13. **Tramonti A, Visca P, De Canio M, Falconi M, De Biase D.** 2002. Functional characterization and regulation of gadX, a gene encoding an AraC/XylS-like transcriptional activator of the *Escherichia coli* glutamic acid decarboxylase system. *J Bacteriol* **184:**2603–2613.

14. **Hughes DT, Terekhova DA, Liou L, Hovde CJ, Sahl JW, Patankar AV, Gonzalez JE, Edrington TS, Rasko DA, Sperandio V.** 2010. Chemical sensing in mammalian host-bacterial commensal associations. *Proc Natl Acad Sci USA* **107:**9831–9836.

15. **Clarke MB, Hughes DT, Zhu C, Boedeker EC, Sperandio V.** 2006. The QseC sensor kinase: a bacterial adrenergic receptor. *Proc Natl Acad Sci USA* **103:**10420–10425.

16. **Clarke MB, Sperandio V.** 2005. Transcriptional regulation of flhDC by QseBC and sigma (FliA) in enterohaemorrhagic *Escherichia coli*. *Mol Microbiol* **57:**1734–1749.

17. **Hughes DT, Clarke MB, Yamamoto K, Rasko DA, Sperandio V..** 2009. The QseC adrenergic signaling cascade in enterohemorrhagic *E. coli* (EHEC). *PLoS Pathog* **5:**e1000553.

18. **Sperandio V, Torres AG, Kaper JB.** 2002. Quorum sensing *Escherichia coli* regulators B and C (QseBC): a novel two-component regulatory system involved in the regulation of flagella and motility by quorum sensing in *E. coli*. *Mol Microbiol* **43:**809–821.

19. **Bansal T, Englert D, Lee J, Hegde M, Wood TK, Jayaraman A.** 2007. Differential effects of epinephrine, norepinephrine, and indole on *Escherichia coli* O157:H7 chemotaxis, colonization, and gene expression. *Infect Immun* **75:**4597–4607.

20. **Kendall MM, Rasko DA, Sperandio V.** 2007. Global effects of the cell-to-cell signaling molecules autoinducer-2, autoinducer-3, and epinephrine in a luxS mutant of enterohemorrhagic *Escherichia coli*. *Infect Immun* **75:**4875–4884.

21. **Sperandio V, Torres AG, Jarvis B, Nataro JP, Kaper JB.** 2003. Bacteria-host communication: the language of hormones. *Proc Natl Acad Sci USA* **100:**8951–8956.

22. **Walters M, Sircili MP, Sperandio V.** 2006. AI-3 synthesis is not dependent on luxS in *Escherichia coli*. *J Bacteriol* **188:**5668–5681.

23. **Kutsukake K, Ohya Y, Iino T.** 1990. Transcriptional analysis of the flagellar regulon of *Salmonella typhimurium*. *J Bacteriol* **172:**741–747.

24. **Tobe T, Nakanishi N, Sugimoto N.** 2011. Activation of motility by sensing short-chain fatty acids via two steps in a flagellar gene regulatory cascade in enterohemorrhagic *Escherichia coli*. *Infect Immun* **79:**1016–1024.

25. **Mahajan A, Currie CG, Mackie S, Tree J, McAteer S, McKendrick I, McNeilly TN, Roe A, La Ragione RM, Woodward MJ, Gally DL, Smith DG.** 2009. An investigation of the expression and adhesin function of H7 flagella in the interaction of *Escherichia coli* O157:H7 with bovine intestinal epithelium. *Cell Microbiol* **11:**121–137.

26. **Sharma VK, Bearson SM, Bearson BL.** 2010. Evaluation of the effects of sdiA, a luxR homologue, on adherence and motility of *Escherichia coli* O157:H7. *Microbiology* **156:**1303–1312.

27. **Kim JC, Yoon JW, Kim CH, Park MS, Cho SH.** 2012. Repression of flagella motility in enterohemorrhagic *Escherichia coli* O157:H7 by mucin components. *Biochem Biophys Res Commun* **423:**789–792.

28. **Iyoda S, Koizumi N, Satou H, Lu Y, Saitoh T, Ohnishi M, Watanabe H.** 2006. The GrlR-GrlA regulatory system coordinately controls the expression of flagellar and LEE-encoded type III protein secretion systems in enterohemorrhagic *Escherichia coli*. *J Bacteriol* **188:**5682–5692.

29. **Lehti TA, Bauchart P, Dobrindt U, Korhonen TK, Westerlund-Wikstrom B.** 2012. The fimbriae activator MatA switches off motility in *Escherichia coli* by repression of the flagellar master operon flhDC. *Microbiology* **158:**1444–1455.

30. **Miranda RL, Conway T, Leatham MP, Chang DE, Norris WE, Allen JH, Stevenson SJ, Laux DC, Cohen PS.** 2004. Glycolytic and gluconeogenic growth of *Escherichia coli* O157:H7 (EDL933) and *E. coli* K-12 (MG1655) in the mouse intestine. *Infect Immun* **72:**1666–1676.

31. **Njoroge JW, Nguyen Y, Curtis MM, Moreira CG, Sperandio V.** 2012. Virulence meets metabolism: Cra and KdpE gene regulation in enterohemorrhagic *Escherichia coli*. *MBio* **3:**e00280-12.

32. **Elliott SJ, Sperandio V, Giron JA, Shin S, Mellies JL, Wainwright L, Hutcheson SW, McDaniel TK, Kaper JB.** 2000. The locus of enterocyte effacement (LEE)-encoded regulator controls expression of both LEE- and non-LEE-encoded virulence factors in enteropathogenic and enterohemorrhagic *Escherichia coli*. *Infect Immun* **68:**6115–6126.

33. **Mellies JL, Elliott SJ, Sperandio V, Donnenberg MS, Kaper JB.** 1999. The Per regulon of enteropathogenic *Escherichia coli*: identification of a regulatory cascade and a novel transcriptional activator, the locus of enterocyte effacement (LEE)-encoded regulator (Ler). *Mol Microbiol* **33:**296–306.

34. **Snider TA, Fabich AJ, Conway T, Clinkenbeard KD.** 2009. *E. coli* O157:H7 catabolism of intestinal mucin-derived carbohydrates and colonization. *Vet Microbiol* **136:**150–154.

35. **Jaswal VM, Babbar HS, Mahmood A.** 1988. Changes in sialic acid and fucose contents of enterocytes across the crypt-villus axis in developing rat intestine. *Biochem Med Metab Biol* **39:**105–110.

36. **Pacheco AR, Curtis MM, Ritchie JM, Munera D, Waldor MK, Moreira CG, Sperandio V.** 2012. Fucose sensing regulates bacterial intestinal colonization. *Nature* **492:**113–117.

37. **Nakanishi N, Tashiro K, Kuhara S, Hayashi T, Sugimoto N, Tobe T.** 2009. Regulation of virulence by butyrate sensing in enterohaemorrhagic *Escherichia coli*. *Microbiology* **155:**521–530.

38. **Kendall MM, Gruber CC, Parker CT, Sperandio V.** 2012. Ethanolamine controls expression of genes encoding components involved in interkingdom signaling and virulence in enterohemorrhagic *Escherichia coli* O157:H7. *MBio* **3:**e00050-12.

39. **Bertin Y, Girardeau JP, Chaucheyras-Durand F, Lyan B, Pujos-Guillot E, Harel J, Martin C.** 2011. Enterohaemorrhagic *Escherichia coli* gains a competitive advantage by using ethanolamine as a nitrogen source in the bovine intestinal content. *Environ Microbiol* **13:**365–377.

40. **Low JC, McKendrick IJ, McKechnie C, Fenlon D, Naylor SW, Currie C, Smith DG, Allison L, Gally DL.** 2005. Rectal carriage of enterohemorrhagic *Escherichia coli* O157 in slaughtered cattle. *Appl Environ Microbiol* **71:**93–97.

41. **Dziva F, van Diemen PM, Stevens MP, Smith AJ, Wallis TS.** 2004. Identification of *Escherichia coli* O157:H7 genes influencing colonization of the bovine gastrointestinal tract using signature-tagged mutagenesis. *Microbiology* **150:**3631–3645.

42. **Naylor SW, Roe AJ, Nart P, Spears K, Smith DG, Low JC, Gally DL.** 2005. *Escherichia coli* O157:H7 forms attaching and effacing lesions at the terminal rectum of cattle and colonization requires the LEE4 operon. *Microbiology* **151:**2773–2781.

43. **Dean-Nystrom EA, Bosworth BT, Moon HW, O'Brien AD.** 1998. *Escherichia coli* O157:H7 requires intimin for enteropathogenicity in calves. *Infect Immun* **66:**4560–4563.

44. **Tobe T, Beatson SA, Taniguchi H, Abe H, Bailey CM, Fivian A, Younis R, Matthews S, Marches O, Frankel G, Hayashi T, Pallen MJ.** 2006. An extensive repertoire of type III secretion effectors

in *Escherichia coli* O157 and the role of lambdoid phages in their dissemination. *Proc Natl Acad Sci USA* **103**:14941–14946.

45. **Mellies JL, ABarron AM, Carmona AM.** 2007. Enteropathogenic and enterohemorrhagic *Escherichia coli* virulence gene regulation. *Infect Immun* **75**:4199–4210.

46. **Bustamante VH, Santana FJ, Calva E, Puente JL.** 2001. Transcriptional regulation of type III secretion genes in enteropathogenic *Escherichia coli*: Ler antagonizes H-NS-dependent repression. *Mol Microbiol* **39**:664–678.

47. **Haack KR, Robinson CL, Miller KJ, Fowlkes JW, Mellies JL.** 2003. Interaction of Ler at the *LEE5* (*tir*) operon of enteropathogenic *Escherichia coli*. *Infect Immun* **71**:384–392.

48. **Umanski T, Rosenshine I, Friedberg D.** 2002. Thermoregulated expression of virulence genes in enteropathogenic *Escherichia coli*. *Microbiology* **148**:2735–2744.

49. **Hommais F, Krin E, Laurent-Winter C, Soutourina O, Malpertuy A, Le Caer JP, Danchin A, Bertin P.** 2001. Large-scale monitoring of pleiotropic regulation of gene expression by the prokaryotic nucleoid-associated protein, H-NS. *Mol Microbiol* **40**:20–36.

50. **Abe H, Tatsuno I, Tobe T, Okutani A, Sasakawa C.** 2002. Bicarbonate ion stimulates the expression of locus of enterocyte effacement-encoded genes in enterohemorrhagic *Escherichia coli* O157:H7. *Infect Immun* **70**:3500–3509.

51. **Beltrametti F, Kresse AU, Guzman CA.** 1999. Transcriptional regulation of the *esp* genes of enterohemorrhagic *Escherichia coli*. *J Bacteriol* **181**:3409–3418.

52. **Ide T, Michgehl S, Knappstein S, Heusipp G, Schmidt MA.** 2003. Differential modulation by Ca2+ of type III secretion of diffusely adhering enteropathogenic *Escherichia coli*. *Infect Immun* **71**:1725–1732.

53. **Kenny B, Abe A, Stein M, Finlay BB.** 1997. Enteropathogenic *Escherichia coli* protein secretion is induced in response to conditions similar to those in the gastrointestinal tract. *Infect Immun* **65**:2606–2612.

54. **Kenny B, Finlay BB.** 1995. Protein secretion by enteropathogenic *Escherichia coli* is essential for transducing signals to epithelial cells. *Proc Natl Acad Sci USA* **92**:7991–7995.

55. **Sperandio V, Mellies JL, Nguyen W, Shin S, Kaper JB.** 1999. Quorum sensing controls expression of the type III secretion gene transcription and protein secretion in enterohemorrhagic and enteropathogenic *Escherichia coli*. *Proc Natl AcadSci USA* **96**:15196–15201.

56. **Sperandio V, Torres AG, Giron JA, Kaper JB.** 2001. Quorum sensing is a global regulatory mechanism in enterohemorrhagic *Escherichia coli* O157:H7. *J Bacteriol* **183**:5187–5197.

57. **Friedberg D, Umanski T, Fang Y, Rosenshine I.** 1999. Hierarchy in the expression of the locus of enterocyte effacement genes of enteropathogenic *Escherichia coli*. *Mol Microbiol* **34**:941–952.

58. **Goldberg MD, Johnson M, Hinton JC, Williams PH.** 2001. Role of the nucleoid-associated protein Fis in the regulation of virulence properties of enteropathogenic *Escherichia coli*. *Mol Microbiol* **41**:549–559.

59. **Grant AJ, Farris M, Alefounder P, Williams PH, Woodward MJ, O'Connor CD.** 2003. Coordination of pathogenicity island expression by the BipA GTPase in enteropathogenic *Escherichia coli* (EPEC). *Mol Microbiol* **48**:507–521.

60. **Iyoda S, Watanabe H.** 2004. Positive effects of multiple *pch* genes on expression of the locus of enterocyte effacement genes and adherence of enterohaemorrhagic *Escherichia coli* O157:H7 to HEp-2 cells. *Microbiology* **150**:2357–2571.

61. **Barba J, Bustamante VH, Flores-Valdez MA, Deng W, Finlay BB, Puente JL.** 2005. A positive regulatory loop controls expression of the locus of enterocyte effacement-encoded regulators Ler and GrlA. *J Bacteriol* **187**:7918–7930.

62. **Deng W, Puente JL, Gruenheid S, Li Y, Vallance BA, Vazquez A, Barba J, Ibarra JA, O'Donnell P, Metalnikov P, Ashman K, Lee S, Goode D, Pawson T, Finlay BB.** 2004. Dissecting virulence: systematic and functional analyses of a pathogenicity island. *Proc Natl Acad Sci USA* **101**:3597–3602.

63. **Shin S, Castanie-Cornet MP, Foster JW, Crawford JA, Brinkley C, Kaper JB.** 2001. An activator of glutamate decarboxylase genes regulates the expression of enteropathogenic *Escherichia coli* virulence genes through control of the plasmid-encoded regulator, Per. *Mol Microbiol* **41**:1133–1150.

64. **Iyoda S, Honda N, Saitoh T, Shimuta K, Terajima J, Watanabe H, Ohnishi M.** 2011. Coordinate control of the locus of enterocyte effacement and enterohemolysin genes by multiple common virulence regulators in enterohemorrhagic *Escherichia coli*. *Infect Immun* **79**:4628–4637.

65. **Russell RM, Sharp FC, Rasko DA, Sperandio V.** 2007. QseA and GrlR/GrlA regulation of the locus of enterocyte effacement genes in enterohemorrhagic *Escherichia coli*. *J Bacteriol* **189**:5387–5392.

66. **Boutte CC, Crosson S.** 2013. Bacterial lifestyle shapes stringent response activation. *Trends Microbiol* **21**:174–180.

67. **Hernandez VJ, Bremer H.** 1991. *Escherichia coli* ppGpp synthetase II activity requires spoT. *J Biol Chem* **266**:5991–5999.

68. **Nakanishi N, Abe H, Ogura Y, Hayashi T, Tashiro K, Kuhara S, Sugimoto N, Tobe T.** 2006. ppGpp with DksA controls gene expression in the locus of enterocyte effacement (LEE) pathogenicity island of enterohaemorrhagic *Escherichia coli* through activation of two virulence regulatory genes. *Mol Microbiol* **61:**194–205.

69. **Bhatt S, Edwards AN, Nguyen HT, Merlin D, Romeo T, Kalman D.** 2009. The RNA binding protein CsrA is a pleiotropic regulator of the locus of enterocyte effacement pathogenicity island of enteropathogenic *Escherichia coli*. *Infect Immun* **77:**3552–3568.

70. **Lodato PB, Kaper JB.** 2009. Post-transcriptional processing of the LEE4 operon in enterohaemorrhagic *Escherichia coli*. *Mol Microbiol* **71:**273–290.

71. **Kendall MM, Gruber CC, Rasko DA, Hughes DT, Sperandio V.** 2011. Hfq virulence regulation in enterohemorrhagic *Escherichia coli* O157:H7 strain 86-24. *J Bacteriol* **193:**6843–6851.

72. **Hansen AM, Kaper JB.** 2009. Hfq affects the expression of the LEE pathogenicity island in enterohaemorrhagic *Escherichia coli*. *Mol Microbiol* **73:**446–465.

73. **Shakhnovich EA, Davis BM, Waldor MK.** 2009. Hfq negatively regulates type III secretion in EHEC and several other pathogens. *Mol Microbiol* **74:**347–363.

74. **Zhang L, Chaudhuri RR, Constantinidou C, Hobman JL, Patel MD, Jones AC, Sarti D, Roe AJ, Vlisidou I, Shaw RK, Falciani F, Stevens MP, Gally DL, Knutton S, Frankel G, Penn CW, Pallen MJ.** 2004. Regulators encoded in the *Escherichia coli* type III secretion system 2 gene cluster influence expression of genes within the locus for enterocyte effacement in enterohemorrhagic E. coli O157:H7. *Infect Immun* **72:**7282–7293.

75. **Flockhart AF, Tree JJ, Xu X, Karpiyevich M, McAteer SP, Rosenblum R, Shaw DJ, Low CJ, Best A, Gannon V, Laing C, Murphy KC, Leong JM, Schneiders T, La Ragione R, Gally DL.** 2012. Identification of a novel prophage regulator in *Escherichia coli* controlling the expression of type III secretion. *Mol Microbiol* **83:**208–223.

76. **Gomez-Duarte OG, Kaper JB.** 1995. A plasmid-encoded regulatory region activates chromosomal eaeA expression in enteropathogenic *Escherichia coli*. *Infect Immun* **63:**1767–1776.

77. **Porter ME, Mitchell P, Free A, Smith DG, Gally DL.** 2005. The *LEE1* promoters from both enteropathogenic and enterohemorrhagic *Escherichia coli* can be activated by PerC-like proteins from either organism. *J Bacteriol* **187:**458–472.

78. **Ahmer BM.** 2004. Cell-to-cell signalling in *Escherichia coli* and *Salmonella enterica*. *Mol Microbiol* **52:**933–945.

79. **Michael B, Smith JN, Swift S, Heffron F, Ahmer BM.** 2001. SdiA of *Salmonella enterica* is a LuxR homolog that detects mixed microbial communities. *J Bacteriol* **183:**5733–5742.

80. **Yao Y, Martinez-Yamout MA, Dickerson TJ, Brogan AP, Wright PE, Dyson HJ.** 2006. Structure of the *Escherichia coli* quorum sensing protein SdiA: activation of the folding switch by acyl homoserine lactones. *J Mol Biol* **355:**262–273.

81. **Sharp FC, Sperandio V.** 2007. QseA directly activates transcription of LEE1 in enterohemorrhagic *Escherichia coli*. *Infect Immun* **75:**2432–2440.

82. **Sperandio V, Li CC, Kaper JB.** 2002. Quorum-sensing *Escherichia coli* regulator A: a regulator of the LysR family involved in the regulation of the locus of enterocyte effacement pathogenicity island in enterohemorrhagic E. coli. *Infect Immun* **70:**3085–3093.

83. **Reading NC, Rasko D, Torres AG, Sperandio V.** 2010. A transcriptome study of the QseEF two-component system and the QseG membrane protein in enterohaemorrhagic *Escherichia coli* O157:H7. *Microbiology* **156:**1167–1175.

84. **Rasko DA, Moreira CG, Li de R, Reading NC, Ritchie JM, Waldor MK, Williams N, Taussig R, Wei S, Roth M, Hughes DT, Huntley JF, Fina MW, Falck JR, Sperandio V.** 2008. Targeting QseC signaling and virulence for antibiotic development. *Science* **321:**1078–1080.

85. **Reading NC, Torres AG, Kendall MM, Hughes DT, Yamamoto K, Sperandio V.** 2007. A novel two-component signaling system that activates transcription of an enterohemorrhagic *Escherichia coli* effector involved in remodeling of host actin. *J Bacteriol* **189:**2468–2476.

86. **Hayashi T, Makino K, Ohnishi M, Kurokawa K, Ishii K, Yokoyama K, Han CG, Ohtsubo E, Nakayama K, Murata T, Tanaka M, Tobe T, Iida T, Takami H, Honda T, Sasakawa C, Ogasawara N, Yasunaga T, Kuhara S, Shiba T, Hattori M, Shinagawa H.** 2001. Complete genome sequence of enterohemorrhagic *Escherichia coli* O157:H7 and genomic comparison with a laboratory strain K-12. *DNA Res* **8:**11–22.

87. **Perna NT, Plunkett G 3rd, Burland V, Mau B, Glasner JD, Rose DJ, Mayhew GF, Evans PS, Gregor J, Kirkpatrick HA, Posfai G, Hackett J, Klink S, Boutin A, Shao Y, Miller L, Grotbeck EJ, Davis NW, Lim A, Dimalanta ET, Potamousis KD, Apodaca J, Anantharaman TS, Lin J, Yen G, Schwartz DC, Welch RA, Blattner FR.** 2001. Genome sequence of enterohaemorrhagic *Escherichia coli* O157:H7. *Nature* **409:**529–533.

88. **Samadder P, Xicohtencatl-Cortes J, Saldana Z, Jordan D, Tarr PI, Kaper JB, Giron JA.** 2009. The *Escherichia coli* ycbQRST operon encodes fimbriae with laminin-binding and epithelial cell adherence properties in Shiga-toxigenic *E. coli* O157:H7. *Environ Microbiol* **11**:1815–1826.

89. **Doughty S, Sloan J, Bennett-Wood V, Robertson M, Robins-Browne RM, Hartland EL..** 2002. Identification of a novel fimbrial gene cluster related to long polar fimbriae in locus of enterocyte effacement-negative strains of enterohemorrhagic *Escherichia coli*. *Infect Immun* **70**: 6761–6769.

90. **Tatsuno I, Mundy R, Frankel G, Chong Y, Phillips AD, Torres AG, Kaper JB.** 2006. The *lpf* gene cluster for long polar fimbriae is not involved in adherence of enteropathogenic *Escherichia coli* or virulence of *Citrobacter rodentium*. *Infect Immun* **74**:265–272.

91. **Low AS, Holden N, Rosser T, Roe AJ, Constantinidou C, Hobman JL, Smith DG, Low JC, Gally DL.** 2006. Analysis of fimbrial gene clusters and their expression in enterohaemorrhagic *Escherichia coli* O157:H7. *Environ Microbiol* **8**:1033–1047.

92. **Xicohtencatl-Cortes J, Monteiro-Neto V, Saldana Z, Ledesma MA, Puente JL, Giron JA.** 2009. The type 4 pili of enterohemorrhagic *Escherichia coli* O157:H7 are multipurpose structures with pathogenic attributes. *J Bacteriol* **191**:411–421.

93. **Kim SH, Kim YH.** 2004. *Escherichia coli* O157:H7 adherence to HEp-2 cells is implicated with curli expression and outer membrane integrity. *J Vet Sci* **5**:119–124.

94. **Saldana Z, Xicohtencatl-Cortes J, Avelino F, Phillips AD, Kaper JB, Puente JL, Giron JA.** 2009. Synergistic role of curli and cellulose in cell adherence and biofilm formation of attaching and effacing *Escherichia coli* and identification of Fis as a negative regulator of curli. *Environ Microbiol* **11**:992–1006.

95. **Torres AG, Li Y, Tutt CB, Xin L, Eaves-Pyles T, Soong L.** 2006. Outer membrane protein A of *Escherichia coli* O157:H7 stimulates dendritic cell activation. *Infect Immun* **74**:2676–2685.

96. **Nicholls L, Grant TH, Robins-Browne RM.** 2000. Identification of a novel genetic locus that is required for in vitro adhesion of a clinical isolate of enterohaemorrhagic *Escherichia coli* to epithelial cells. *Mol Microbiol* **35**:275–288.

97. **Tarr PI, Bilge SS, Vary JCJ, Jelacic S, Habeeb RL, Ward TR, Baylor MR, Besser TE.** 2000. Iha: a novel *Escherichia coli* O157:H7 adherence-conferring molecule encoded on a recently acquired chromosomal island of conserved structure. *Infect Immun* **68**:1400–1407.

98. **Totsika M, Wells TJ, Beloin C, Valle J, Allsopp LP, King NP, Ghigo JM, Schembri MA.** 2012. Molecular characterization of the EhaG and UpaG trimeric autotransporter proteins from pathogenic *Escherichia coli*. *Appl Environ Microbiol* **78**:2179–2189.

99. **Grys TE, Siegel MB, Lathem WW, Welch RA.** 2005. The StcE protease contributes to intimate adherence of enterohemorrhagic *Escherichia coli* O157:H7 to host cells. *Infect Immun* **73**:1295–1303.

100. **Lathem WW, Grys TE, Witowski SE, Torres AG, Kaper JB, Tarr PI, Welch RA.** 2002. StcE, a metalloprotease secreted by *Escherichia coli* O157:H7, specifically cleaves C1 esterase inhibitor. *Mol Microbiol* **45**:277–288.

101. **Farfan MJ, Cantero L, Vidal R, Botkin DJ, Torres AG.** 2011. Long polar fimbriae of enterohemorrhagic *Escherichia coli* O157:H7 bind to extracellular matrix proteins. *Infect Immun* **79**: 3744–3750.

102. **Torres AG, Milflores-Flores L, Garcia-Gallegos JG, Patel SD, Best A, La Ragione RM, Martinez-Laguna Y, Woodward MJ.** 2007. Environmental regulation and colonization attributes of the long polar fimbriae (LPF) of *Escherichia coli* O157:H7. *Int J Med Microbiol* **297**:177–185.

103. **Jordan DM, Cornick N, Torres AG, Dean-Nystrom EA, Kaper JB, Moon HW.** 2004. Long polar fimbriae contribute to colonization by *Escherichia coli* O157:H7 in vivo. *Infect Immun* **72**:6168–6171.

104. **Lloyd SJ, Ritchie JM, Rojas-Lopez M, Blumentritt CA, Popov VL, Greenwich JL, MWaldor MK, Torres AG.** 2012. A double, long polar fimbria mutant of *Escherichia coli* O157:H7 expresses Curli and exhibits reduced in vivo colonization. *Infect Immun* **80**:914–920.

105. **Torres AG, Lopez-Sanchez GN, Milflores-Flores L, Patel SD, Rojas-Lopez M, Martinez de la Pena CF, Arenas-Hernandez MM, Martinez-Laguna Y.** 2007. Ler and H-NS, regulators controlling expression of the long polar fimbriae of *Escherichia coli* O157:H7. *J Bacteriol* **189**:5916–5928.

106. **Rendon MA, Saldana Z, Erdem AL, Monteiro-Neto V, Vazquez A, Kaper JB, Puente JL, Giron JA.** 2007. Commensal and pathogenic *Escherichia coli* use a common pilus adherence factor for epithelial cell colonization. *Proc Natl Acad Sci USA* **104**:10637–10642.

107. **Martinez-Santos VI, Medrano-Lopez A, Saldana Z, Giron JA, Puente JL.** 2012. Transcriptional regulation of the *ecp* operon by EcpR, IHF, and H-NS in attaching and effacing *Escherichia coli*. *J Bacteriol* **194**:5020–5033.

108. **Lathem WW, Bergsbaken T, Welch RA.** 2004. Potentiation of C1 esterase inhibitor by StcE, a metalloprotease secreted by *Escherichia coli* O157:H7. *J Exp Med* **199:**1077–1087.

109. **Echtenkamp F, Deng W, Wickham ME, Vazquez A, Puente JL, Thanabalasuriar A, Gruenheid S, Finlay BB, Hardwidge PR.** 2008. Characterization of the NleF effector protein from attaching and effacing bacterial pathogens. *FEMS Microbiol Lett* **281:**98–107.

110. **Hemrajani C, Marches O, Wiles S, Girard F, Dennis A, Dziva F, Best A, Phillips AD, Berger CN, Mousnier A, Crepin VF, Kruidenier L, Woodward MJ, Stevens MP, La Ragione RM, MacDonald TT, Frankel G.** 2008. Role of NleH, a type III secreted effector from attaching and effacing pathogens, in colonization of the bovine, ovine, and murine gut. *Infect Immun* **76:**4804–4813.

111. **Kelly M, Hart E, Mundy R, Marches O, Wiles S, Badea L, Luck S, Tauschek M, Frankel G, Robins-Browne RM, Hartland EL.** 2006. Essential role of the type III secretion system effector NleB in colonization of mice by *Citrobacter rodentium*. *Infect Immun* **74:**2328–2337.

112. **Gruenheid S, Sekirov I, Thomas NA, Deng W, O'Donnell P, Goode D, Li Y, Frey EA, Brown NF, Metalnikov P, Pawson T, Ashman K, Finlay BB.** 2004. Identification and characterization of NleA, a non-LEE-encoded type III translocated virulence factor of enterohaemorrhagic *Escherichia coli* O157:H7. *Mol Microbiol* **51:**1233–1249.

113. **Creuzburg K, Schmidt H.** 2007. Molecular characterization and distribution of genes encoding members of the type III effector nleA family among pathogenic *Escherichia coli* strains. *J Clin Microbiol* **45:**2498–2507.

114. **Mundy R, Jenkins C, Yu J, Smith H, Frankel G.** 2004. Distribution of espI among clinical enterohaemorrhagic and enteropathogenic *Escherichia coli* isolates. *J Med Microbiol* **53:**1145–1149.

115. **Roe AJ, Tysall L, Dransfield T, Wang D, Fraser-Pitt D, Mahajan A, Constandinou C, Inglis N, Downing A, Talbot R, Smith DG, Gally DL.** 2007. Analysis of the expression, regulation and export of NleA-E in *Escherichia coli* O157:H7. *Microbiology* **153:**1350–1360.

116. **Garcia-Angulo VA, Martinez-Santos VI, Villasenor T, Santana FJ, Huerta-Saquero A, Martinez LC, Jimenez R, Lara-Ochoa C, Tellez-Sosa J, Bustamante VH, Puente JL.** 2012. A distinct regulatory sequence is essential for the expression of a subset of *nle* genes in attaching and effacing *Escherichia coli*. *J Bacteriol* **194:**5589–5603.

117. **Schwidder M, Hensel M, Schmidt H.** 2011. Regulation of nleA in Shiga toxin-producing *Escherichia coli* O84:H4 strain 4795/97. *J Bacteriol* **193:**832–841.

118. **Njoroge J, Sperandio V.** 2012. Enterohemorrhagic *Escherichia coli* virulence regulation by two bacterial adrenergic kinases, QseC and QseE. *Infect Immun* **80:**688–703.

119. **Mellies JL, Haack KR, Galligan DC.** 2007. SOS regulation of the type III secretion system of enteropathogenic *Escherichia coli*. *J Bacteriol* **189:**2863–2872.

120. **Campellone KG, Robbins D, Leong JM.** 2004. EspFU is a translocated EHEC effector that interacts with Tir and N-WASP and promotes Nck-independent actin assembly. *Dev Cell* **7:**217–228.

121. **Garmendia J, Phillips AD, Carlier MF, Chong Y, Schuller S, Marches O, Dahan S, Oswald E, Shaw RK, Knutton S, Frankel G.** 2004. TccP is an enterohaemorrhagic *Escherichia coli* O157:H7 type III effector protein that couples Tir to the actin-cytoskeleton. *Cell Microbiol* **6:**1167–1183.

122. **Ritchie JM, Brady MJ, Riley KN, Ho TD, Campellone KG, Herman IM, Donohue-Rolfe A, Tzipori S, Waldor MK, Leong JM.** 2008. EspFU, a type III-translocated effector of actin assembly, fosters epithelial association and late-stage intestinal colonization by *E. coli* O157:H7. *Cell Microbiol* **10:**836–847.

123. **Sallee NA, Rivera GM, Dueber JE, Vasilescu D, Mullins RD, Mayer BJ, Lim WA.** 2008. The pathogen protein EspF(U) hijacks actin polymerization using mimicry and multivalency. *Nature* **454:**1005–1008.

124. **Garmendia J, Frankel G, Crepin VF.** 2005. Enteropathogenic and enterohemorrhagic *Escherichia coli* infections: translocation, translocation, translocation. *Infect Immun* **73:**2573–2585.

125. **Njoroge JW, Gruber C, Sperandio V.** 2013. The interacting Cra and KdpE regulators are involved in the expression of multiple virulence factors in enterohemorrhagic *Escherichia coli*. *J Bacteriol* **195:**2499–2508.

126. **Tyler JS, Mills MJ, Friedman DI.** 2004. The operator and early promoter region of the Shiga toxin type 2-encoding bacteriophage 933W and control of toxin expression. *J Bacteriol* **186:**7670–7679.

127. **O'Brien AD, LaVeck GD, Thompson MR, Formal SB.** 1982. Production of *Shigella dysenteriae* type 1-like cytotoxin by *Escherichia coli*. *J Infect Dis* **146:**763–769.

128. **Wagner PL, Livny J, Neely MN, Acheson DW, Friedman DI, Waldor MK.** 2002. Bacteriophage control of Shiga toxin 1 production and release by *Escherichia coli*. *Mol Microbiol* **44:**957–970.

129. **Waldor MK, Friedman DI.** 2005. Phage regulatory circuits and virulence gene expression. *Curr Opin Microbiol* **8**:459–465.

130. **Zhang X, McDaniel AD, Wolf LE, Keusch GT, Waldor MK, Acheson DW.** 2000. Quinolone antibiotics induce Shiga toxin-encoding bacteriophages, toxin production, and death in mice. *J Infect Dis* **181**:664–670.

131. **Los JM, Los M, Wegrzyn A, Wegrzyn G.** 2010. Hydrogen peroxide-mediated induction of the Shiga toxin-converting lambdoid prophage ST2-8624 in *Escherichia coli* O157:H7. *FEMS Immunol Med Microbiol* **58**:322–329.

132. **Vareille M, de Sablet T, Hindre T, Martin C, Gobert AP.** 2007. Nitric oxide inhibits Shigatoxin synthesis by enterohemorrhagic *Escherichia coli*. *Proc Natl Acad Sci USA* **104**:10199–10204.

133. **Reading NC, Rasko DA, Torres AG, Sperandio V.** 2009. The two-component system QseEF and the membrane protein QseG link adrenergic and stress sensing to bacterial pathogenesis. *Proc Natl Acad Sci USA* **106**:5889–5894.

134. **Ogura Y, Ooka T, Iguchi A, Toh H, Asadulghani M, Oshima K, Kodama T, Abe H, Nakayama K, Kurokawa K, Tobe T, Hattori M, Hayashi T.** 2009. Comparative genomics reveal the mechanism of the parallel evolution of O157 and non-O157 enterohemorrhagic *Escherichia coli*. *Proc Natl AcadSci USA* **106**:17939–17944.

135. **Li H, Granat A, Stewart V, Gillespie JR.** 2008. RpoS, H-NS, and DsrA influence EHEC hemolysin operon (ehxCABD) transcription in *Escherichia coli* O157:H7 strain EDL933. *FEMS Microbiol Lett* **285**:257–262.

136. **Gottesman S.** 2004. The small RNA regulators of *Escherichia coli*: roles and mechanisms. *Annu Rev Microbiol* **58**:303–328.

137. **Saitoh T, Iyoda S, Yamamoto S, Lu Y, Shimuta K, Ohnishi M, Terajima J, Watanabe H.** 2008. Transcription of the *ehx* enterohemolysin gene is positively regulated by GrlA, a global regulator encoded within the locus of enterocyte effacement in enterohemorrhagic *Escherichia coli*. *J Bacteriol* **190**:4822–4830.

138. **House B, Kus JV, Prayitno N, Mair R, Que L, Chingcuanco F, Gannon V, Cvitkovitch DG, Barnett Foster D.** 2009. Acid-stress-induced changes in enterohaemorrhagic *Escherichia coli* O157:H7 virulence. *Microbiology* **155**:2907–2918.

139. **Calderon VE, Chang Q, McDermott M, Lytle MB, McKee G, Rodriguez K, Rasko DA, Sperandio V, Torres AG.** 2010. Outbreak caused by cad-negative Shiga toxin-producing *Escherichia coli* O111, Oklahoma. *Foodborne Pathog Dis* **7**:107–109.

140. **Fabich AJ, Jones SA, Chowdhury FZ, Cernosek A, Anderson A, Smalley D, McHargue JW, Hightower GA, Smith JT, Autieri SM, Leatham MP, Lins JJ, Allen RL, Laux DC, Cohen PS, Conway T.** 2008. Comparison of carbon nutrition for pathogenic and commensal *Escherichia coli* strains in the mouse intestine. *Infect Immun* **76**:1143–1152.

141. **Kamada N, Kim Y-G, Sham HP, Vallance BA, Puente JL, Martens EC, Núñez G.** 2012. Regulated virulence controls the ability of a pathogen to compete with the gut microbiota. *Science* **336**:1325–1329.

142. **Asano Y, Hiramoto T, Nishino R, Aiba Y, Kimura T, Yoshihara K, Koga Y, Sudo N.** 2012. Critical role of gut microbiota in the production of biologically active, free catecholamines in the gut lumen of mice. *Am J Physiol Gastrointest Liver Physiol* **303**:G1288–G1295.

143. **Hersh BM, Farooq FT, Barstad DN, Blankenhorn DL, Slonczewski JL.** 1996. A glutamate-dependent acid resistance gene in *Escherichia coli*. *J Bacteriol* **178**:3978–3981.

144. **McGannon CM, Fuller CA, Weiss AA.** 2010. Different classes of antibiotics differentially influence Shiga toxin production. *Antimicrob Agents Chemother* **54**:3790–3798.

145. **Wagner PL, Neely MN, Zhang X, Acheson DW, Waldor MK, Friedman DI.** 2001. Role for a phage promoter in Shiga toxin 2 expression from a pathogenic *Escherichia coli* strain. *J Bacteriol* **183**:2081–2085.

146. **Torres AG, Slater TM, Patel SD, Popov VL, Arenas-Hernandez MM.** 2008. Contribution of the Ler- and H-NS-regulated long polar fimbriae of *Escherichia coli* O157:H7 during binding to tissue-cultured cells. *Infect Immun* **76**:5062–5071.

147. **Carter MQ, Parker CT, Louie JW, Huynh S, Fagerquist CK, Mandrell RE.** 2012. RcsB contributes to the distinct stress fitness among *Escherichia coli* O157:H7 curli variants of the 1993 hamburger-associated outbreak strains. *Appl Environ Microbiol* **78**:7706–7719.

148. **Lim J, Lee KM, Kim SH, Kim Y, Kim SH, Park W, Park S.** 2011. YkgM and ZinT proteins are required for maintaining intracellular zinc concentration and producing curli in enterohemorrhagic *Escherichia coli* (EHEC) O157:H7 under zinc deficient conditions. *Int J Food Microbiol* **149**:159–170.

149. **Sharma VK, Bearson BL.** 2013. Hha controls *Escherichia coli* O157:H7 biofilm formation by differential regulation of global transcriptional regulators FlhDC and CsgD. *Appl Environ Microbiol* **79**:2384–2396.

150. **Baev MV, Baev D, Radek AJ, Campbell JW.** 2006. Growth of *Escherichia coli* MG1655 on LB medium: monitoring utilization of amino acids, peptides, and nucleotides with transcriptional

microarrays. *Appl Microbiol Biotechnol* **71:**317–322.

151. **Douchin V, Bohn C, Bouloc P.** 2006. Down-regulation of porins by a small RNA bypasses the essentiality of the regulated intramembrane proteolysis protease RseP in *Escherichia coli. J Biol Chem* **281:**12253–12259.

152. **Udekwu KI, Darfeuille F, Vogel J, Reimegard J, Holmqvist E, Wagner EG.** 2005. Hfq-dependent regulation of OmpA synthesis is mediated by an antisense RNA. *Genes Dev* **19:**2355–2366.

153. **Stevens MP, Roe AJ, Vlisidou I, van Diemen PM, La Ragione RM, Best A, Woodward MJ, Gally DL, Wallis TS.** 2004. Mutation of toxB and a truncated version of the efa-1 gene in *Escherichia coli* O157:H7 influences the expression and secretion of locus of enterocyte effacement-encoded proteins but not intestinal colonization in calves or sheep. *Infect Immun* **72:**5402–5411.

154. **Rashid RA, Tabata TA, Oatley MJ, Besser TE, Tarr PI, Moseley SL.** 2006. Expression of putative virulence factors of *Escherichia coli* O157:H7 differs in bovine and human infections. *Infect Immun* **74:**4142–4148.

155. **Rashid RA, Tarr PI, Moseley SL.** 2006. Expression of the *Escherichia coli* IrgA homolog adhesin is regulated by the ferric uptake regulation protein. *Microb Pathog* **41:**207–217.

INCIDENCE, EPIDEMIOLOGY, AND ECOLOGY

Shiga Toxin (Verotoxin)-Producing *Escherichia coli* in Japan

10

JUN TERAJIMA,[1] SUNAO IYODA,[1] MAKOTO OHNISHI,[1] and
HARUO WATANABE[1]

A series of outbreaks of infection with Shiga toxin (or verotoxin [VT])-producing *Escherichia coli* or enterohemorrhagic *E. coli* (EHEC) O157:H7 occurred in Japan in 1996, the largest outbreak occurring in primary schools in Sakai City, Osaka Prefecture, where more than 7,500 cases were reported (1). Although the reason for the sudden increase in the number of reports of EHEC isolates in 1996 is not known, the number of reports has grown to more than 3,000 cases per year since 1996 from an average of 105 cases reported each year during the previous 5-year period (1991–1995) (2). Despite control measures instituted since 1996, including designating EHEC infection as a notifiable disease, and the disease being monitored effectively through nationwide surveillance, the number of reports remains high, around 3,800 cases per year (Fig. 1). Serogroup O157 predominates over other EHEC serogroups, but isolation frequency of non-O157 EHEC has gone up slightly over the past few years. Non-O157 EHEC has caused outbreaks where consumption of a raw beef dish was the source of the infection and some fatal cases were occurred. Laboratory surveillance consisting of prefectural and municipal public health institutes (PHIs) and the National Institute of Infectious Diseases has contributed to finding not only multiprefectural outbreaks but also recognizing sporadic cases that could have been missed as an outbreak without the aid of molecular

[1]Department of Bacteriology I, National Institute of Infectious Diseases, Tokyo, Japan.
Enterohemorrhagic Escherichia coli *and Other Shiga Toxin-Producing* E. coli
Edited by Vanessa Sperandio and Carolyn J. Hovde
© 2015 American Society for Microbiology, Washington, DC
doi:10.1128/microbiolspec.EHEC-0011-2013

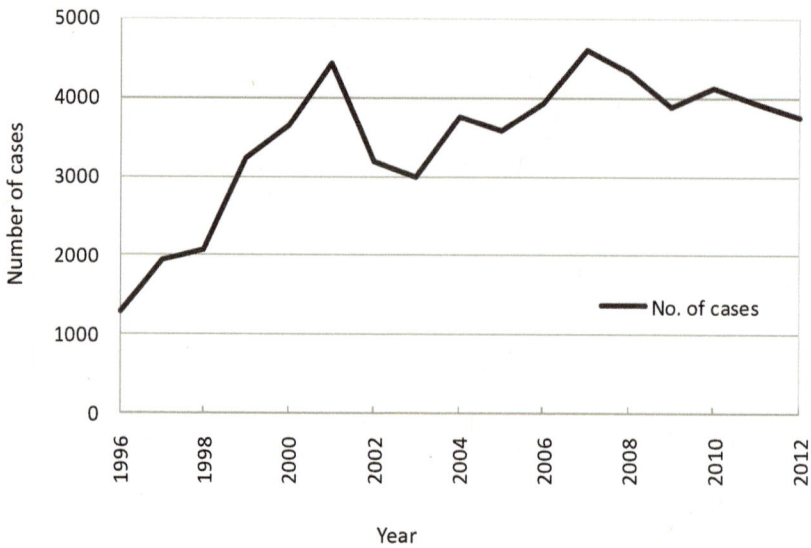

FIGURE 1 Reported cases of EHEC infection, 1996 to 2012. doi:10.1128/microbiolspec.EHEC-0011-20013.f1

subtyping of EHEC isolates. This short overview presents recent information on the surveillance of EHEC infections in Japan.

SURVEILLANCE AND REPORTING

In 1966, EHEC infection was designated as a disease requiring mandatory reporting regardless of source. EHEC infection is one of category III notifiable infectious diseases along with other bacterial infections caused by *Vibrio cholerae* O1 or O139, *Shigella* species, *Salmonella enterica* serovar Typhi, and *Salmonella* serovar Paratyphi A in the National Epidemiological Surveillance of Infectious Diseases (NESID) under the Law Concerning the Prevention of Infectious Diseases and Medical Care for Patients of Infections (the Infectious Diseases Control Law) enacted in April 1999. Under a surveillance system for food poisoning based on the Food Sanitation Law, an EHEC infection is reported as food poisoning by physicians or judged as such by the director of the health center and reported as such by the local government to the Ministry of Health, Labour and Welfare (MHLW). Active

surveillance of the population surrounding the outbreak, including checking food-intake and stool specimens of infected cases and persons engaged in food preparation, revealed that approximately 35% of EHEC infections were asymptomatic and that the proportion of asymptomatic infection was high among the middle-aged group whereas symptomatic cases were more frequent in young and old age groups. The public health importance of symptomatic shedding (3) and asymptomatic shedding (4) in transmission among preschool children is well established, but asymptomatic shedding in adults is not yet well understood though high rates of culture-positive, asymptomatic adults with EHEC O157:H7 infection were reported in 1998 (5), and the tendency remains similar with EHEC infection currently (6). Apart from NESID, results of characterization (serotypes, VT types, etc.) of EHEC isolates at prefectural and municipal PHIs are reported to the Infectious Disease Surveillance Center (IDSC) at the National Institute of Infectious Diseases. The summary showed that EHEC O157 serogroup is the predominant one, followed by O26, O111, O103, O121, O145, and others. However, as seen in the United States

(7), the percentage of non-EHEC O157 serogroups among all EHEC isolates from human infection has been growing slightly; the rate of isolation frequency of EHEC O157 has declined from approximately 70% of all EHEC isolates in 2000 to less than 60% in 2012, though the situation of EHEC infection in Japan is not exactly the same as in continental Europe where non-O157 EHEC serotypes are more common than infection with O157:H7 EHEC (8, 9). The number of EHEC isolates reported to IDSC is about half the number of EHEC infection cases reported to NESID, because only a small proportion of isolates in hospitals or commercial laboratories are sent to PHIs.

LABORATORY SURVEILLANCE

Stool samples from suspected cases of diarrhea should be routinely screened for EHEC by plating on sorbitol-MacConkey agar containing cefixime and tellurite in addition to conventional *E. coli* isolation agar. Rhamnose/sorbose-MacConkey agar containing cefixime and tellurite and chromogenic media could also be used for O26/O111 isolates as these serogroups are major ones following O157. The use of an enrichment culture step or immunomagnetic separation available for *E. coli* O157, O26, O111, O103, and O145 facilitates isolation of these pathogens. It is essential that *E. coli* isolates be examined for production of VT or the presence of VT genes regardless of serogroup of the isolates since these are prerequisite results for diagnosis of EHEC infection. It could be important to examine production of VT or the presence of VT genes in all isolates on selective agar plates with and without cefixime and tellurite to detect various serogroups of EHEC (10). Molecular subtyping techniques, including pulsed-field gel electrophoresis (PFGE), have sufficient sensitivity and discriminatory power for use in epidemiological investigations. The establishment of the molecular subtyping network, PulseNet, in the United States (11) encouraged us to construct an equivalent molecular

subtyping network based on PFGE analysis, PulseNet Japan (12). It participates in the Asia and Pacific regional network, PulseNet Asia Pacific, which constitutes a part of PulseNet International (13), and contributed in investigations of domestic outbreaks (14) and an international EHEC O157 outbreak that occurred between Japan and the United States (15). Although PFGE is still in wide use in outbreak investigations for its high discrimination power, there are additional DNA-based methods for strain analysis that are simpler, faster to perform, and equivalent to PFGE in strain discrimination. Among these, multilocus sequence typing was shown to have insufficient discriminatory power for EHEC O157:H7 (16, 17). On the other hand, multiple-locus variable-number tandem-repeat analysis (MLVA) has proven to be useful for genetic fingerprinting of EHEC O157:H7 (13, 18–21), O26, and O111 (22) and pathogenic bacteria (23). For strain analysis, in addition to MLVA, an insertion element (IS)-printing system that is a PCR-based strain-typing method for EHEC O157 (24) has been applied to outbreak investigations. The IS-printing system showed that it had equivalent capacity to subtype the isolates in the outbreak. It is based on the variability in genomic location of IS*629* among EHEC O157 strains and can produce the results in a much shorter time than other subtyping methods such as PFGE. The method has another advantage: if standard protocol is established and quality assurance is achieved, then the results are suitable for creating a database since PCR results can be expressed as presence or absence of the amplicon for the sample, which is easily digitized, and the information can readily be shared with other laboratories involved in the outbreak investigation. Characterization of EHEC isolates using DNA-based methods could be critical for linking simultaneously occurring sporadic cases that would otherwise have been unrecognized as a diffuse outbreak (14). It is important to note that a prompt epidemiological investigation is necessary to confirm that a cluster of genetically indistinguishable isolates may represent an outbreak with a common source. Also, it is noteworthy that epidemiological links are not established for all the clusters; many remain unresolved.

HEMOLYTIC-UREMIC SYNDROME (HUS)

The case definition of EHEC infection was partly amended in April 2006 as follows: if VT is detected in feces, or O-antigen agglutinating antibody or anti-VT antibody is detected in serum of a patient with HUS, the patient should be reported as having EHEC infection. From 2006 to 2012, the average annual number of HUS cases (including serodiagnosed cases) was 100 and the incidence rate of HUS (HUS cases/symptomatic cases) was 3.7% (Fig. 2) (25–29). Increased rates of HUS in children less than 10 years old and the elderly (30, 31) are shown in Fig. 2, and a rise in incidence rate of HUS for the age group older than 65 in 2012 reflects an outbreak of EHEC O157 where five HUS cases occurred in a facility for the elderly (32). A rise in incidence

rate for the age group 15 to 64 years in 2011 was due to a large outbreak of EHEC O111 in which the age of 21 HUS cases ranged from 15 to 64 years among 34 HUS cases in the outbreak. From 2006 to 2012, 67% of the 605 HUS cases were culture-confirmed by laboratory testing, and the rest of the cases were diagnosed by detecting antilipopolysaccharide antibody of *E. coli* in the serum of the patients or VT in the stool samples of the patients. EHEC O157 was the predominant serogroup, occupying 85% of all isolates in culture-confirmed HUS cases, followed by O111 (5%), O121 (2.1%), O26 (1.7%), O165 (1%), O145 (1%), and the rest of the O serogroups, including O55, O174, O183, and unknown serogroup samples. Although non-O157 serogroup strains were isolated in the culture-confirmed HUS cases, 94% of all EHEC isolates in the culture-confirmed HUS cases were either VT2 or both VT1 and VT2 producers, which is consistent with epidemiological evidence that VT2-containing EHEC O157:H7 strains are more frequently associated with HUS than the strains containing VT1 (33, 34). Some non-O157 EHEC strains that were also classified

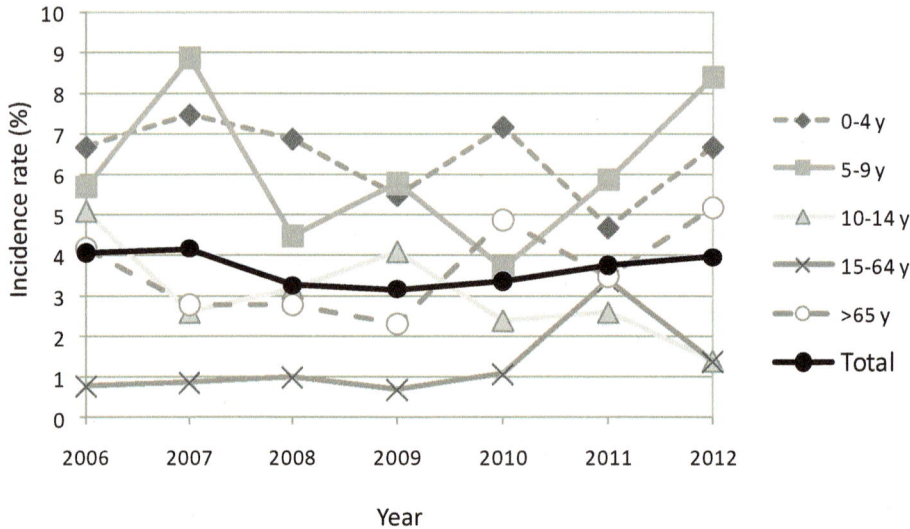

FIGURE 2 Incidence rate of HUS in EHEC infection by age groups in Japan, 2006 to 2012. Incidence rate (%) was calculated as (number of patients with HUS) ÷ (number of symptomatic cases) × 100%. doi:10.1128/microbiolspec.EHEC-0011-20013.f2

as enteroaggregative *E. coli* have been isolated in HUS cases due to O104:H4 infection in Germany (35, 36) as well as an outbreak due to O111:H2 in France (37), but there has been only one such case reported in Japan, where a 3-year-old boy in Kagoshima Prefecture developed HUS, developed encephalopathy, and died due to the infection of VT2-producing and enteroaggregative *E. coli* O86:HNM (38).

OUTBREAKS

In the 197 outbreaks with more than 10 culture-positive patients, which were reported to IDSC from 2000 to 2012 (6, 32, 39–50), the main mode of transmission of the infection was person to person (41%), food borne (29%), and water borne (3%,), and in about one-third of the outbreaks, the mode of transmission of the infection remains unknown. The most frequent serogroup of EHEC associated with the outbreaks was EHEC O157 (47%), followed by O26 (35%), O111 (8%), O103 (4%), O121 (3%), O145 (2%), and others. The predominance of EHEC O157 serogroup and other representative serogroups that had caused the outbreaks is fairly consistent with the laboratory surveillance results obtained by PHIs for EHEC infection that includes outbreaks and sporadic cases.

Among 197 outbreaks, 101 outbreaks have occurred in nursery schools, which may account for the high proportion of person-to-person transmissions of the infection in the outbreaks. Interestingly, the most prevalent serogroup of EHEC in the outbreaks in nursery schools was O26 (52%), followed by O157 (27%), O111 (9%), O103 (4%), O121 (4%), O145 (3%), and OUT, which may account for relatively mild clinical manifestations, including asymptomatic cases and, consequently, frequent person-to-person transmission. More O26-associated cases may have occurred; however, such cases would not be detected as sporadic due to mild symptoms among adults and children. Larger outbreaks resulted from consumption of contaminated foods. Between 2000 and 2012, there were 12 EHEC outbreaks that had more than 100 culture-positive cases (Table 1). All 12 outbreaks appear to have resulted from consumption of contaminated foods, and in some of these outbreaks, microbiological testing confirmed the implicated foods. These included beef products (39); lightly salted cucumber (40); Koumi-ae consisting of boiled spinach and steamed chicken meat seasoned with welsh onion, ginger, and soy sauce (40); boxed meals (43); lettuce (44); school lunch (45); Yukhoe (raw beef) (6), and Japanese rice cakes (6). The infectious dose of EHEC O157 is very low, probably fewer than 100 organisms (51, 52), and this is a critical factor in the transmission of the EHEC when people consume raw or lightly cooked foods such as sushi and vegetables. Because sushi and raw or lightly cooked meat are favorite food items in Japan, outbreaks were associated with consumption of salmon roe sushi in 1998 (53) and "rare" roast beef contaminated with EHEC O157 in 2001 (39). In 2011, a large EHEC O111 outbreak due to consumption of Yukhoe, a Korean dish of raw beef and egg yolk, was identified at Yakiniku chain restaurants. EHEC O111:H8 was isolated from 85 of 181 patients (median age 20 years); in 34 of those patients HUS developed; encephalopathy developed in 21 patients; and 5 patients died (6). EHEC O111: H8 was also isolated from the conserved part of the original meat preparations distributed to the chain restaurants. In response to persistent food poisonings caused by raw beef, MHLW revised the standards of beef product quality for raw-eating and put them into operation in October 2011. Further, after the EHEC O157 was detected in the inner part of cattle livers, MHLW banned marketing of cattle liver for raw-eating. Probably as a consequence of these preventive measures, the incidence of EHEC O157 cases related to consumption of raw meat decreased by almost half in 1 year from 2011 to 2012 (32).

Some of the outbreaks were associated with consumption of vegetables. Among four outbreaks associated with consumption of

TABLE 1 Characterization of 12 EHEC outbreaks with more than 100 culture-positive cases between 2000 and 2012

Year	Prefecture/city	Setting (reference)	Serotype	VT type	Symptomatic cases	Culture positives	Likely mode of transmission
2001	Chiba P.	Patient's home (39)	O157:H7	VT1 & 2	195	257	Beef products[a]
2002	Fukuoka C.	Nursery school (40)	O157:H-	VT2	74	112	Lightly salted cucumber[a]
2002	Utsunomiya C.	Hospital and home for the elderly (40)	O157:H7	VT1 & 2	123	111	Koumi-ae[a]
2003	Yokohama C.	Kindergarten (41)	O26:H11	VT1	141	449	Food borne
2004	Ishikawa P.	High school (42)	O111:H-	VT1 & 2	110	103	Food borne
2007	Tokyo M.	School refectory (43)	O157:H7	VT2	467	204	Food borne
2007	Miyagi P., Sendai C., and Akita C.	Restaurant (43)	O157:H7	VT1 & 2	314	173	Boxed meals[a]
2009	Saga P.	Nursery school (44)	O26:H11	VT1	ND	133	Lettuce[a]
2010	Mie P.	High school (45)	O157:H7	VT2	138	164	School lunch[a]
2011	Toyama P.	Chain restaurants (6)	O111:H8	VT2, VT–	181	102	Yukhoe (raw beef)[a]
			O157:H7	VT1, VT2, VT1 & 2		38	
2011	Yamagata P.	Festival (6)	O157:H7	VT1 & 2	287	189	Japanese rice cakes[a]
2012	Osaka C.	Nursery school (32)	O26:H-	VT1	68	115	Food borne

[a]Confirmed microbiologically; ND, no data.

vegetables in 2011 (6), EHEC O26:H11 was isolated from cabbage in an outbreak and EHEC O157:H7 was isolated from pickled eggplant and green perilla, green perilla served with grated radish, and cucumber in three outbreaks, respectively. In 2012, there was a large EHEC O157:H7 outbreak due to consumption of a brand of lightly salted vegetables from a company in Sapporo, Hokkaido (32). EHEC O157:H7 was isolated from the implicated product. Because of the products' wide distribution, 169 patients were reported from five facilities for the elderly, hotels, restaurants, and families in Hokkaido and included four cases in different prefectures from which EHEC O157:H7 was isolated. Among 169 patients, EHEC O157:H7 was isolated from 73 patients, 8 of whom, mostly elderly, died.

ROUTES OF TRANSMISSION OF EHEC INFECTIONS

Consumption of contaminated foods is the most common source of EHEC infection, and because cattle are considered to be major reservoirs of EHEC, consumption of raw or undercooked foods of bovine origin has been the most common means of transmitting EHEC infection. Laboratory investigations have played a critical role in the investigation of EHEC infections since some cases were geographically dispersed, and without the aid of DNA-based fingerprinting of the isolates, it was difficult to recognize whole cases as a diffuse outbreak. For example, in 2009, there was a diffuse outbreak from a restaurant chain that spread all over Japan. Implicated restaurants were providing cubically assembled meat, a type of processed meat made by mixing and filling different parts of chopped meat, that is frozen and cut into cubes to make the appearance of cube-cut steak. EHEC O157 contaminating the process probably survived in the product, and its intake caused illness. Initially, some cases were reported as sporadic cases by different prefectures, but the results of PFGE and MLVA of EHEC O157 isolates of the cases as well as the product showed identical genotype and strongly suggested that these isolates were part of the outbreak (44). Other examples are the EHEC O157 infections due to consumption of raw beef liver in Aichi Prefecture in 2010 (45). There were four successive food poisoning incidents due to EHEC O157 in Nagoya City, Aichi Prefecture, and all were related to consumption of raw beef liver. During the same period, food poisonings also caused by EHEC O157 occurred in multiple restaurants serving grilled meat in Aichi Prefecture. Laboratory investigation of the cases showed identical genotypes of the isolates, and investigation of the marketing routes revealed that the apparently diffuse incidents were actually the same-sourced food poisoning cases affecting a wide area.

Prevalence of EHEC strains in beef cattle in Japan between November 2007 and March 2008 was reported to be 8.9% and 0.4% for EHEC O157 and O26, respectively, among 2,436 beef cattle reared on 406 farms (54). In another study, prevalence of EHEC strains in 932 healthy dairy cows from 123 farms was 12%, and 31 different O-serogroups, including O26 but not O157, were identified (55). Using *stx*-PCRs for screening, the same study also found that the prevalence of the *stx* gene-positive samples among the dairy cows was 30.4%. Although EHEC O157 was found in retail meat at a low frequency (0.1%) in a national food surveillance (56), relatively high prevalence of EHEC in cattle on farms could account for sporadic cases of EHEC and outbreaks when people have contact with these animals at farms (57–59).

Raw vegetables, such as lettuce (44, 46), cabbage (6), and cucumber (6), have been implicated in EHEC outbreaks. Pickled vegetables (6) and lightly salted vegetables (32, 39, 40), which were fresh as salad, have also been implicated in the outbreaks. Trace-back studies of an EHEC O104 outbreak associated with consumption of fenugreek sprouts in Germany (60) and France (61) showed that two outbreaks shared fenugreek seeds imported from Egypt as the most likely common link (62), and studies of a spinach outbreak due to EHEC O157 in the United States also showed that environmental samples including river water, cattle feces, and wild pig feces from the field adjacent to the spinach-growing field were contaminated with identical pulsotype isolates of EHEC O157 (63); however, few trace-back studies have been conducted in Japanese outbreaks associated with consumption of vegetables.

Person-to-person transmission of EHEC is the most likely mode of transmission in the outbreaks reported from nursery schools, and prolonged fecal shedding of EHEC that was similar to that reported in the literature (64–66) was seen in the outbreaks due to EHEC O26 (67), O157 (68), and O103 (69). Because both symptomatic and asymptomatic patients are usually among members of the staff of the daycare and family members of the children at daycare, extended stool culture examination of all children attending daycare (4) and family members would facilitate in taking control measures to prevent further dissemination of EHEC infection.

CONCLUSIONS

The large outbreak in Sakai City and the subsequent outbreaks that occurred domestically in 1996 illustrate some of the serious public health problems associated with EHEC O157 infections. The government initiated a series of preventive control measures against the disease, including enacting the new infectious disease control law in 1999. Since then, however, a gradual increase in infection caused by EHEC has been reported, partly because there have been improvements in methodologies for isolating these organisms and more laboratories were practicing enhanced screening for these organisms. Although EHEC O157 has been isolated predominantly in the EHEC infection, increased rates of non-O157 EHEC cases are being reported with a variety of non-O157 EHEC serotypes. Especially among the outbreaks between 2000 and 2012, one-half of the outbreaks were due to EHEC O157 and the other half were due to O26, O111, O121, O103, O145, and other serogroups; some serious cases were caused by O111. Precautions need to be taken not only for non-O157 EHEC in addition to EHEC O157 but also for emerging types of Shiga toxin-producing enteroaggregative *E. coli*, such as O104:H4 in Germany; a fatal case due to O86:HNM with similar phenotype to the O104 *E. coli* was reported in Japan in 1999 (38). Nursery schools have become frequent outbreak settings where EHEC infection was prolonged by person-to-person transmission involving family members of the child. To reduce the risk of outbreaks, it is important to improve the sanitary controls in the nursery schools according to the guideline presented by MHLW to prevent EHEC dissemination.

Laboratory investigations employing DNA-based methods are critically important components in finding and controlling diffuse outbreaks as early as possible while simultaneously conducting epidemiological investigations of the outbreak. It is also important for relevant organizations and bodies to communicate effectively and work together to control and prevent EHEC infection.

ACKNOWLEDGMENTS

We thank all investigators of prefectural and municipal public health institutes for providing us samples and information, and staffs in IDSC for data collection and analysis.

We declare no conflicts of interest with regard to the manuscript.

CITATION

Terajima J, Iyoda S, Ohnishi M, Watanabe H. 2014. Shiga toxin (verotoxin)-producing *Escherichia coli* in Japan. Microbiol Spectrum 2(5): EHEC-0011-2013.

REFERENCES

1. **National Institute of Infectious Diseases and Tuberculosis and Infectious Diseases Control Division, Ministry of Health, Labour and Welfare.** 1998. Enterohemorrhagic *E. coli* (verocytotoxin-producing *E. coli*) infection, 1996–April 1998. *Infec Agen Surv Rep* **19:**122–123.
2. **National Institute of Infectious Diseases and Infectious Diseases Control Division, Ministry of Health and Welfare of Japan.** 1997. Verocytotoxin-producing *Escherichia coli* (Enterohemorrhagic *E. coli*) infections, Japan, 1996–June 1997. *Infec Agen Surv Rep* **18:**153–154.
3. **Belongia EA, Osterholm MT, Soler JT, Ammend DA, Braun JE, MacDonald KL.** 1993. Transmission of *Escherichia coli* O157:H7 infection in Minnesota child day-care facilities. *JAMA* **269:**883–888.
4. **Gilbert M, Monk C, Wang HL, Diplock K, Landry L.** 2008. Screening policies for daycare attendees: lessons learned from an outbreak of *E. coli* O157:H7 in a daycare in Waterloo, Ontario. *Can J Public Health* **99:**281–285.
5. **Terajima J, Izumiya H, Wada A, Tamura K, Watanabe H.** 1999. Shiga toxin-producing *Escherichia coli* O157:H7 in Japan. *Emerg Infect Dis* **5:**301–302.
6. **National Institute of Infectious Diseases and Tuberculosis and Infectious Diseases Control Division, Ministry of Health, Labour and Welfare.** 2012. Enterohemorrhagic *Escherichia coli* infection in Japan as of April 2012. *Infec Agen Surv Rep* **33:**115–116.
7. **Gould LH, Mody RK, Ong KL, Clogher P, Cronquist AB, Garman KN, Lathrop S, Medus C, Spina NL, Webb TH, White PL, Wymore K, Gierke RE, Mahon BE, Griffin PM for the Emerging Infections Program Foodnet Working Group PM.** 2013. Increased recognition of non-O157 Shiga toxin-producing *Escherichia coli* infections in the United States During 2000–2010: Epidemiologic features and comparison with *E. coli* O157 infections. *Foodborne Pathog Dis* **10:**453–460.
8. **Caprioli A, Tozzi AE.** 1998. Epidemiology of Shiga toxin-producing *Escherichia coli* infections in continental Europe, p 38–48. *In* Kaper JB, O'Brien AD (ed), Escherichia coli *O157:H7 and Other Shiga Toxin-Producing* E. coli *Strains.* ASM Press, Washington, DC.
9. **Blanco JE, Blanco M, Alonso MP, Mora A, Dahbi G, Coira MA, Blanco J.** 2004. Serotypes, virulence genes, and intimin types of Shiga toxin (verotoxin)-producing *Escherichia coli* isolates from human patients: prevalence in Lugo, Spain, from 1992 through 1999. *J Clin Microbiol* **42:**311–319.
10. **Seto K, Taguchi M, Kobayashi K, Kozaki S.** 2007. Biochemical and molecular characterization of minor serogroups of Shiga toxin-producing *Escherichia coli* isolated from humans in Osaka Prefecture. *J Vet Med Sci* **69:**1215–1222.
11. **Swaminathan B, Barrett TJ, Hunter SB, Tauxe RV.** 2001. PulseNet: the molecular subtyping network for foodborne bacterial disease surveillance, United States. *Emerg Infect Dis* **7:**382–389.

12. **Watanabe H, Terajima J, Izumiya H, Iyoda S, Tamura K.** 2002. [PulseNet Japan: surveillance system for the early detection of diffuse outbreak based on the molecular epidemiological method]. *Kansenshogaku Zasshi* **76:**842–848 (In Japanese).

13. **Swaminathan B, Gerner-Smidt P, Ng LK, Lukinmaa S, Kam KM, Rolando S, Gutierrez EP, Binsztein N.** 2006. Building PulseNet International: an interconnected system of laboratory networks to facilitate timely public health recognition and response to foodborne disease outbreaks and emerging foodborne diseases. *Foodborne Pathog Dis* **3:**36–50.

14. **Terajima J, Izumiya H, Iyoda S, Mitobe J, Miura M, Watanabe H.** 2006. Effectiveness of pulsed-field gel electrophoresis for the early detection of diffuse outbreaks due to Shiga toxin-producing *Escherichia coli* in Japan. *Foodborne Pathog Dis* **3:**68–73.

15. **Centers for Disease Control and Prevention.** 2005. *Escherichia coli* O157:H7 infections associated with ground beef from a U.S. military installation—Okinawa, Japan, February 2004. *Morbid Mortal Weekly Rep* **54:**40–42.

16. **Foley SL, Simjee S, Meng J, White DG, McDermott PF, Zhao S.** 2004. Evaluation of molecular typing methods for *Escherichia coli* O157:H7 isolates from cattle, food, and humans. *J Food Prot* **67:**651–657.

17. **Noller AC, McEllistrem MC, Stine OC, Morris JG Jr, Boxrud DJ, Dixon B, Harrison LH.** 2003. Multilocus sequence typing reveals a lack of diversity among *Escherichia coli* O157:H7 isolates that are distinct by pulsed-field gel electrophoresis. *J Clin Microbiol* **41:**675–679.

18. **Hyytia-Trees E, Smole SC, Fields PA, Swaminathan B, Ribot EM.** 2006. Second generation subtyping: a proposed PulseNet protocol for multiple-locus variable-number tandem repeat analysis of Shiga toxin-producing *Escherichia coli* O157 (STEC O157). *Foodborne Pathog Dis* **3:**118–131.

19. **Lindstedt BA, Vardund T, Kapperud G.** 2004. Multiple-locus variable-number tandem-repeats analysis of *Escherichia coli* O157 using PCR multiplexing and multi-colored capillary electrophoresis. *J Microbiol Methods* **58:**213–222.

20. **Keys C, Kemper S, Keim P.** 2005. Highly diverse variable number tandem repeat loci in the *E. coli* O157:H7 and O55:H7 genomes for high-resolution molecular typing. *J Appl Microbiol* **98:**928–940.

21. **Noller AC, McEllistrem MC, Pacheco AG, Boxrud DJ, Harrison LH.** 2003. Multilocus variable-number tandem repeat analysis distinguishes outbreak and sporadic *Escherichia coli* O157:H7 isolates. *J Clin Microbiol* **41:**5389–5397.

22. **Izumiya H, Pei Y, Terajima J, Ohnishi M, Hayashi T, Iyoda S, Watanabe H.** 2010. New system for multilocus variable-number tandem-repeat analysis of the enterohemorrhagic *Escherichia coli* strains belonging to three major serogroups: O157, O26, and O111. *Microbiol Immunol* **54:**569–577.

23. **Lindstedt BA.** 2005. Multiple-locus variable number tandem repeats analysis for genetic fingerprinting of pathogenic bacteria. *Electrophoresis* **26:**2567–2582.

24. **Ooka T, Terajima J, Kusumoto M, Iguchi A, Kurokawa K, Ogura Y, Asadulghani M, Nakayama K, Murase K, Ohnishi M, Iyoda S, Watanabe H, Hayashi T.** 2009. Development of a multiplex PCR-based rapid typing method for enterohemorrhagic *Escherichia coli* O157 strains. *J Clin Microbiol* **47:**2888–2894.

25. **National Institute of Infectious Diseases and Tuberculosis and Infectious Diseases Control Division, Ministry of Health, Labour and Welfare.** 2009. Enterohemorrhagic *Escherichia coli* infection 2006, 2007. *Infec Dis Weekly Rep* **11:**16–22 (In Japanese).

26. **National Institute of Infectious Diseases and Tuberculosis and Infectious Diseases Control Division, Ministry of Health, Labour and Welfare.** 2009. Enterohemorrhagic *Escherichia coli* infection 2008. *Infec Dis Weekl. Rep* **11:**15–23 (In Japanese).

27. **National Institute of Infectious Diseases and Tuberculosis and Infectious Diseases Control Division, Ministry of Health, Labour and Welfare.** 2010. Enterohemorrhagic *Escherichia coli* infection 2009. *Infec Dis Weekly Rep* **12:**11–19 (In Japanese).

28. **National Institute of Infectious Diseases and Tuberculosis and Infectious Diseases Control Division, Ministry of Health, Labour and Welfare.** 2011. Enterohemorrhagic *Escherichia coli* infection 2010. *Infec Dis Weekly Rep* **13:**11–20 (In Japanese).

29. **National Institute of Infectious Diseases and Tuberculosis and Infectious Diseases Control Division, Ministry of Health, Labour and Welfare.** 2012. Enterohemorrhagic *Escherichia coli* infection 2011. *Infec Dis Weekly Rep* **14:**9–19 (In Japanese).

30. **Tarr PI, Gordon CA, Chandler WL.** 2005. Shiga-toxin-producing *Escherichia coli* and haemolytic uraemic syndrome. *Lancet* **365:**1073–1086.

31. **Boyce TG, Swerdlow DL, Griffin PM.** 1995. *Escherichia coli* O157:H7 and the hemolytic-uremic syndrome. *N Engl J Med* **333:**364–368.

32. **National Institute of Infectious Diseases and Tuberculosis and Infectious Diseases Control Division, Ministry of Health, Labour and Welfare.** 2013. Enterohemorrhagic *Escherichia*

coli infection in Japan as of April 2013. *Infec Agen Surv Rep* **34**:123–124.

33. **Ostroff SM, Tarr PI, Neill MA, Lewis JH, Hargrett-Bean N, Kobayashi JM.** 1989. Toxin genotypes and plasmid profiles as determinants of systemic sequelae in *Escherichia coli* O157:H7 infections. *J Infect Dis* **160**:994–998.

34. **Scotland SM, Willshaw GA, Smith HR, Rowe B.** 1987. Properties of strains of *Escherichia coli* belonging to serogroup O157 with special reference to production of Vero cytotoxins VT1 and VT2. *Epidemiol Infect* **99**:613–624.

35. **Scheutz F, Nielsen EM, Frimodt-Moller J, Boisen N, Morabito S, Tozzoli R, Nataro JP, Caprioli A.** 2011. Characteristics of the enteroaggregative Shiga toxin/verotoxin-producing *Escherichia coli* O104:H4 strain causing the outbreak of haemolytic uraemic syndrome in Germany, May to June 2011. *Euro Surveill* **16**:5–10.

36. **Morabito S, Karch H, Mariani-Kurkdjian P, Schmidt H, Minelli F, Bingen E, Caprioli A.** 1998. Enteroaggregative, Shiga toxin-producing *Escherichia coli* O111:H2 associated with an outbreak of hemolytic-uremic syndrome. *J Clin Microbiol* **36**:840–842.

37. **Frank C, Werber D, Cramer JP, Askar M, Faber M, an der Heiden M, Bernard H, Fruth A, Prager R, Spode A, Wadl M, Zoufaly A, Jordan S, Kemper MJ, Follin P, Muller L, King LA, Rosner B, Buchholz U, Stark K, Krause G, Team HUSI.** 2011. Epidemic profile of Shiga-toxin-producing *Escherichia coli* O104:H4 outbreak in Germany. *N Engl J Med* **365**:1771–1780.

38. **Iyoda S, Tamura K, Itoh K, Izumiya H, Ueno N, Nagata K, Togo M, Terajima J, Watanabe H.** 2000. Inducible *stx2* phages are lysogenized in the enteroaggregative and other phenotypic *Escherichia coli* O86:HNM isolated from patients. *FEMS Microbiol Lett* **191**:7–10.

39. **National Institute of Infectious Diseases and Tuberculosis and Infectious Diseases Control Division, Ministry of Health, Labour and Welfare.** 2002. Enterohemorrhagic *Escherichia coli* infection as of April 2002. *Infec Agen Surv Rep* **23**:137–138.

40. **National Institute of Infectious Diseases and Tuberculosis and Infectious Diseases Control Division, Ministry of Health, Labour and Welfare.** 2003. Enterohemorrhagic *Escherichia coli* infection as of May 2003. *Infec Agen Surv Rep* **24**:129–130.

41. **National Institute of Infectious Diseases and Tuberculosis and Infectious Diseases Control Division, Ministry of Health, Labour and Welfare.** 2004. Enterohemorrhagic *Escherichia coli* infection as of May 2004, Japan. *Infec Agen Surv Rep* **25**:138–139.

42. **National Institute of Infectious Diseases and Tuberculosis and Infectious Diseases Control Division, Ministry of Health, Labour and Welfare.** 2005. Enterohemorrhagic *Escherichia coli* infection as of May 2005. *Infec Agen Surv Rep* **26**:137–138.

43. **National Institute of Infectious Diseases and Tuberculosis and Infectious Diseases Control Division, Ministry of Health, Labour and Welfare.** 2008. Enterohemorrhagic *Escherichia coli* infection in Japan as of April 2008. *Infec Agen Surv Rep* **29**:117–118.

44. **National Institute of Infectious Diseases and Tuberculosis and Infectious Diseases Control Division, Ministry of Health, Labour and Welfare.** 2010. Enterohemorrhagic *Escherichia coli* infection in Japan as of May 2010. *Infec Agen Surv Rep* **31**:152–153.

45. **National Institute of Infectious Diseases and Tuberculosis and Infectious Diseases Control Division, Ministry of Health, Labour and Welfare.** 2011. Enterohemorrhagic *Escherichia coli* infection in Japan as of April 2011. *Infec Agen Surv Rep* **32**:125–126.

46. **National Institute of Infectious Diseases and Tuberculosis and Infectious Diseases Control Division, Ministry of Health, Labour and Welfare.** 2001. Enterohemorrhagic *Escherichia coli* infection in Japan as of April 2001. *Infec Agen Surv Rep* **22**:135–136.

47. **National Institute of Infectious Diseases and Tuberculosis and Infectious Diseases Control Division, Ministry of Health, Labour and Welfare.** 2006. Enterohemorrhagic *Escherichia coli* infection in Japan as of May 2006. *Infec Agen Surv Rep* **27**:141–142.

48. **National Institute of Infectious Diseases and Tuberculosis and Infectious Diseases Control Division, Ministry of Health, Labour and Welfare.** 2007. Enterohemorrhagic *Escherichia coli* infection in Japan as of April 2007. *Infec Agen Surv Rep* **28**:131–132.

49. **National Institute of Infectious Diseases and Tuberculosis and Infectious Diseases Control Division, Ministry of Health, Labour and Welfare.** 2009. Enterohemorrhagic *Escherichia coli* infection in Japan as of April 2009. *Infec Agen Surv Rep* **30**:119–120.

50. **Kato K, Shimoura R, Nashimura K, Yoshifuzi K, Shiroshita K, Sakurai N, Kodama H, Kuramoto S.** 2005. Outbreak of enterohemorrhagic *Escherichia coli* O111 among high school participants in excursion to Korea. *Jpn J Infect Dis* **58**:332–333.

51. **Tilden J Jr, Young W, McNamara AM, Custer C, Boesel B, Lambert-Fair MA, Majkowski J, Vugia D, Werner SB, Hollingsworth J, Morris JG Jr.** 1996. A new route of transmission for

Escherichia coli: infection from dry fermented salami. *Am J Public Health* **86**:1142–1145.

52. Paton AW, Ratcliff RM, Doyle RM, Seymour-Murray J, Davos D, Lanser JA, Paton JC. 1996. Molecular microbiological investigation of an outbreak of hemolytic-uremic syndrome caused by dry fermented sausage contaminated with Shiga-like toxin-producing *Escherichia coli*. *J Clin Microbiol* **34**:1622–1627.

53. Terajima J, Izumiya H, Iyoda S, Tamura K, Watanabe H. 1999. Detection of a multi-prefectual *E. coli* O157:H7 outbreak caused by contaminated Ikura-Sushi ingestion. *Jpn J Infect Dis* **52**:52–53.

54. Sasaki Y, Tsujiyama Y, Kusukawa M, Murakami M, Katayama S, Yamada Y. 2011. Prevalence and characterization of Shiga toxin-producing *Escherichia coli* O157 and O26 in beef farms. *Vet Microbiol* **150**:140–145.

55. Kobayashi H, Kanazaki M, Ogawa T, Iyoda S, Hara-Kudo Y. 2009. Changing prevalence of O-serogroups and antimicrobial susceptibility among STEC strains isolated from healthy dairy cows over a decade in Japan between 1998 and 2007. *J Vet Med Sci* **71**:363–366.

56. Hara-Kudo Y, Konuma H, Kamata Y, Miyahara M, Takatori K, Onoue Y, Sugita-Konishi Y, Ohnishi T. 2012. Prevalence of the main food-borne pathogens in retail food under the national food surveillance system in Japan. *Food Addit Contam Part A Chem Anal Control Expo Risk Assess* **30**:1450–1458.

57. Waguri A, Sakuraba M, Sawada Y, Abe K, Onishi M, Tanaka J, Kudo Y, Saito K. 2007. Enterohemorrhagic *Escherichia coli* O157 infection presumably caused by contact with infected cows, Aomori Prefecture, Japan. *Jpn J Infect Dis* **60**:321–322.

58. Muto T, Matsumoto Y, Yamada M, Ishiguro Y, Kitazume H, Sasaki K, Toba M. 2008. Outbreaks of enterohemorrhagic *Escherichia coli* O157 in-fections among children with animal contact at a dairy farm in Yokohama City, Japan. *Jpn J Infect Dis* **61**:161–162.

59. Akiba Y, Kimura T, Takagi M, Akimoto T, Mitsui Y, Ogasawara Y, Omichi M. 2005. Out-break of enterohemorrhagic *Escherichia coli* O121 among school children exposed to cattle in a ranch for public education on dairy farming. *Jpn J Infect Dis* **58**:190–192.

60. Buchholz U, Bernard H, Werber D, Bohmer MM, Remschmidt C, Wilking H, Delere Y, an der Heiden M, Adlhoch C, Dreesman J, Ehlers J, Ethelberg S, Faber M, Frank C, Fricke G, Greiner M, Hohle M, Ivarsson S, Jark U, Kirchner M, Koch J, Krause G, Luber P, Rosner B, Stark K, Kuhne M. 2011. German outbreak of *Escherichia coli* O104:H4 associated with sprouts. *N Engl J Med* **365**:1763–1770.

61. Gault G, Weill FX, Mariani-Kurkdjian P, Jourdan-da Silva N, King L, Aldabe B, Charron M, Ong N, Castor C, Mace M, Bingen E, Noel H, Vaillant V, Bone A, Vendrely B, Delmas Y, Combe C, Bercion R, d'Andigne E, Desjardin M, de Valk H, Rolland P. 2011. Outbreak of haemolytic uraemic syndrome and bloody diar-rhoea due to *Escherichia coli* O104:H4, south-west France, June 2011. *Euro Surveill* **16**:1–3.

62. European Food Safety Authority. 2011. *Tracing seeds, in particular fenugreek (Trigonella foenum-graecum) seeds, in relation to the Shiga toxin-producing E. coli (STEC) O104:H4 2011 outbreaks in Germany and France*. Technical Report of EFSA. European Food Safety Authority, Parma, Italy.

63. California Food Emergency Response Team, California Department of Health Services, US Food and Drug Administration. 2007. *Investi-gation of an Escherichia coli O157:H7 outbreak associated with Dole pre-packaged spinach*. En-vironmental Investigation Report. California De-partment of Public Health, Sacramento, CA.

64. Swerdlow DL, Griffin PM. 1997. Duration of faecal shedding of *Escherichia coli* O157:H7 among children in day-care centres. *Lancet* **349**:745–746.

65. O'Donnell JM, Thornton L, McNamara EB, Prendergast T, Igoe D, Cosgrove C. 2002. Out-break of Vero cytotoxin-producing *Escherichia coli* O157 in a child day care facility. *Commun Dis Public Health* **5**:54–58.

66. Galanis E, Longmore K, Hasselback P, Swann D, Ellis A, Panaro L. 2003. Investigation of an *E. coli* O157:H7 outbreak in Brooks, Alberta, June–July 2002: the role of occult cases in the spread of infection within a daycare setting. *Can Commun Dis Rep* **29**:21–28.

67. Kanazawa Y, Ishikawa T, Shimizu K, Inaba S. 2007. Enterohemorrhagic *Escherichia coli* outbreaks in nursery and primary schools. *Jpn J Infect Dis* **60**:326–327.

68. Sugiyama A, Iwade Y, Akachi S, Nakano Y, Matsuno Y, Yano T, Yamauchi A, Nakayama O, Sakai H, Yamamoto K, Nagasaka Y, Nakano T, Ihara T, Kamiya H. 2005. An outbreak of Shigatoxin-producing *Eshcherichia coli* O157:H7 in a nursery school in Mie Prefecture. *Jpn J Infect Dis* **58**:398–400.

69. Muraoka R, Okazaki M, Fujimoto Y, Jo N, Yoshida R, Kiyoyama T, Oura Y, Hirakawa K, Jyukurogi M, Kawano K, Okada M, Shioyama Y, Iryoda K, Wakamatu H, Kawabata N. 2007. An enterohemorrhagic *Escherichia coli* O103 outbreak at a nursery school in Miyazaki Prefec-ture, Japan. *Jpn J Infect Dis* **60**:410–411.

Animal Reservoirs of Shiga Toxin-Producing *Escherichia coli*

11

ANIL K. PERSAD[1] and JEFFREY T. LeJEUNE[1]

Escherichia coli strains that carry Shiga toxin genes are commonly isolated from the gastrointestinal tract of a wide variety of animal species (Table 1). Intestinal carriage of most Shiga toxin-producing *E. coli* (STEC) strains by domestic and wild animals has little clinical relevance to either the animal hosts or humans. Most animals lack receptors for Shiga toxin, and in humans, the presence of additional virulence factors, in addition to the *stx* gene, is associated with disease outcomes (1–3). However, animals may harbor STEC strains that are pathogenic to humans. This article focuses on the role of animals as reservoirs for infection or as spillover hosts. Within the animal, these bacteria may be resident or transient in the gastrointestinal tract. Determining whether STEC is resident in flora or transient is not possible during cross-sectional observational epidemiological studies when only one sample is collected from an animal and there is no serial testing. Even under experimental conditions it is difficult to determine if repeated isolation from the feces over time is a result of replication of the organism in the animal or repeated exposure.

Animals capable of maintaining STEC carriage in the absence of continuous exposure or those that frequently are reexposed to STEC from environmental sources can serve as potential sources of interspecies and intraspecies infection. Cattle are regarded as the natural reservoir of STEC (1), but other ruminant

[1]Food Animal Health Research Program, Ohio Agriculture Research and Development Center, The Ohio State University, Wooster, OH 44691.

Enterohemorrhagic Escherichia coli *and Other Shiga Toxin-Producing* E. coli
Edited by Vanessa Sperandio and Carolyn J. Hovde
© 2015 American Society for Microbiology, Washington, DC
doi:10.1128/microbiolspec.EHEC-0027-2014

TABLE 1 Animal hosts of Shiga toxin-producing *E. coli*

Common name	Scientific name	Reference(s)
Cattle	*Bos taurus*	1, 7, 8, 10, 19, 21–23, 27, 29–33
Goats	*Capra aegagrus hircus*	34, 39, 40, 43, 44, 48, 49, 53
Sheep	*Ovis aries*	1, 35, 39, 43–47
Water buffalo	*Bubalus bubalis*	53, 54, 61
White-tailed deer	*Odocoileus virginianus*	62–64, 67–71
Bison	*Bison bison*	74–77
Elk	*Cervus canadensis*	72, 73, 80
Llamas	*Lama glama*	191
Alpaca	*Lama pacos*	83, 192
Yak	*Bos grunniens*	83
Eland	*Taurotragus oryx*	83
Antelope	*Antilope cervicapra*	83
Mountain goat	*Oreamnos americanus*	84
Guanaco	*Lama guanicoe*	79
Horses	*Equus ferus caballus*	85–88, 91
Donkey	*Equus africanus asinus*	84, 89, 90
Domestic swine	*Sus domesticus*	1, 92, 94–96, 101, 102
Feral swine	*Sus scrofa*	103–105
Chicken	*Gallus gallus domesticus*	92, 94, 125, 126
Turkeys	*Meleagris gallopavo*	92, 126
Pigeon	*Columba livia*	111, 116
Starling	*Sturnus vulgaris*	110, 112–114
Geese	*Branta canadensis*	107, 119
Turtle dove	*Streptopelia turtur*	112
Barn swallow	*Hirundo rustica*	112
Dogs	*Canis lupus familiaris*	39, 163, 165
Cats	*Felis catus*	166, 170, 171
Coyote	*Canis latrans*	84
Fox	*Vulpes vulpes*	84
Rabbit	*Oryctolagus cuniculus*	143, 144
Raccoon	*Procyon lotor*	152
Fish and shellfish		129–132
Norway rats	*Rattus norvegicus*	108, 137, 138
Ground hog	*Marmota monax*	84
Patagonian cavy	*Dolichotis patagonus*	83
Frogs		193
Ferrets[a]	*Mustela putorius furo*	172
Mice[a]	*Mus* spp.	114, 142, 180

[a]Experimental infections only.

species such as sheep, goats, and deer may also act as reservoirs. Animals may also be categorized as spillover hosts. Similar to reservoirs, these animals are susceptible to colonization and may transmit disease; however, once they are no longer exposed to a source of STEC, they do not maintain this colonization. This inability to maintain STEC colonization in the absence of exposure is the critical factor that differentiates these animals from reservoirs. Epidemiological evidence indicates that birds, swine, dogs, and horses may be spillover hosts. Dead-end hosts, as the name suggests, are incapable of transmitting STEC naturally to other animals. In the absence of evidence that aquatic species such as finfish and shellfish transmit the organism to other animals, they may act as dead-end hosts for STEC, only transmitting STEC when they are consumed (4–6).

The factors governing the prevalence and number of bacteria present in the digestive tract of animals are poorly understood, even for the best-studied species, bovines. The prevalence and magnitude of STEC infection in animals is dependent on a complex interaction of external and internal conditions: the frequency and dose of exposure, the host's susceptibility to infection, and the duration of shedding. Moreover, these factors may vary considerably among species and even among the same species as a function of age, immunity, housing, diet, climate, and sanitation.

ANIMAL SPECIES OF IMPORTANCE IN THE EPIDEMIOLOGY OF STEC

Cattle

Cattle are recognized as a primary reservoir for STEC strains, especially the serogroup O157 (1). Like humans, cattle are exposed to STEC through contaminated food and water or by exposure to the feces of other animals shedding the organism. The infectious dose in cattle is estimated to be as low as 300 CFU

(7). STEC colonization in cattle is usually asymptomatic due to the absence of vascular receptors for Shiga toxins (8). The absence of these globotriaosylceramide-3 (Gb3) vascular receptors, especially in the intestinal vasculature, means Shiga toxins cannot be endocytosed and transported to other organs that may be sensitive to Shiga toxins (9). The terminal part of the large intestine, the recto-anal junction (RAJ), is the main site of STEC colonization in cattle (10). The increased production of factors associated with environmental survival among cattle isolates of STEC O157, compared to human-origin isolates, combined with the low infectious dose, may provide a selection bias for these organisms to recolonize cattle and maintain the organism in the bovine population (11). Improper feed storage facilities or poorly designed feeding troughs can result in feed being contaminated with the feces of wild or domestic animals.

Livestock drinking water contamination can occur at its source or at the farm. Surface water and groundwater sources may be contaminated from effluent runoff from farms and urban areas. Leaching from pastures may also result in groundwater contamination (12, 13). At the farm, improperly designed water troughs can be contaminated by animal feces. LeJeune et al. (14) showed that 1.3% of 473 water troughs sampled in three U.S. states were positive for STEC O157. STEC has also been demonstrated to persist for more than 4 months in contaminated water troughs (15).

Other management practices may also affect the incidence of STEC in animal populations. Flushing alleyways with water increased the incidence of STEC in animals compared to other manure removal strategies (16). Animals housed on sawdust were also found to have a higher incidence of STEC than animals housed on sand-based bedding (17). Movement of animals to and from farms also increases the risk of STEC transmission: Animals carried to animal exhibitions have a greater likelihood of contracting STEC than

animals not carried to shows (18). These animals, on returning to the farm, can then shed STEC, thus exposing other animals to infection or colonization.

In the United States, STEC O157 is found on almost all cattle farms, with the organism being shed intermittently by most animals (19). STEC is shed mainly through the feces of colonized animals; however, Shiga toxin genes have been detected from *E. coli* strains isolated from the milk of mastitic cows (20). Although a rare occurrence, milk from these animals can be a potential source of STEC infection to nursing calves, animals fed waste milk, and the human population. Most milk-borne STEC cases are, however, due to postmilking contamination and the subsequent consumption of these products without pasteurization.

The prevalence of STEC in cattle populations is highly variable, with peaking and dropping at unpredictable times. At any specific time, the global prevalence of STEC O157 in cattle has been reported to range from 0 to 71% (21), and the herd infection rate has been reported to be up to 100% in some studies (22). In the United States, the herd prevalence of STEC may range between 10 and 20% (23). The global prevalence of STEC O157 has been reported to range from 0.2 to 48.8% in dairy cattle and 0.2 to 27.8% in beef animals whereas the global prevalence of non-O157 STEC may range from 0.4 to 74% in dairy cattle and 2.1 to 70.1% for beef animals, respectively, as reported in two independent studies (24, 25).

Colonized cattle can shed STEC O157 at levels as high as 1.1×10^5 CFU/g feces (26) and for as long as 10 weeks (27). The average duration of STEC O157 carriage is 30 days; however, in rare cases animals may be colonized for up to 1 year (28). Animals excreting greater than 10^4 CFU/g feces are termed "super-shedding animals" (29, 30). Longitudinal studies, however, indicate super-shedding represents a phase or stage of colonization of all cattle and can be typically observed in a small fraction of animals in a population at any given time (31). Nevertheless, it is not debatable that animals excreting these high levels of STEC are responsible for the majority of environmental contamination (30, 32). Calves tend to shed STEC at the lowest levels before weaning; however, the highest shedding is exhibited in the period immediately post weaning (33). The shedding of STEC also tends to be higher in the warmer months, with peak prevalence being in summer and early fall with a drastic decrease in prevalence during the winter months (19).

Small Ruminants

Small ruminants, particularly sheep and goats, are important reservoirs of STEC O157 (34). Considerable research has focused on the role of sheep in the epidemiology of STEC infections; however, there is limited published research on the role of goats (34). Although cattle have been identified as the major reservoir for STEC in the United States, small ruminants play a greater role in the epidemiology of STEC infections in other countries. For example, sheep have been identified as the host of significance in Australia (1) and have also been recognized as an important reservoir of STEC O26 in Norway (35). In addition to STEC serogroups O157 and O26, sheep have been cited as reservoirs for more than 100 other serotypes of STEC, including O115, O128, and O130 (34, 35)

Transmission of STEC to small ruminants occurs through the same pathway as in cattle. The site of STEC colonization, however, may be different. Unlike cattle, tropism for RAJ has not been described for all small ruminants (34). Following exposure to STEC O157, in some studies few attachment and effacement lesions were visible on the intestinal mucosa and the entire distal intestine, including the cecum, colon, and rectum, was colonized, not only the RAJ (34, 36). However, in mature sheep given a single oral dose of a human clinical isolate of STEC O157:H7, analysis of digesta and intestinal mucosa showed colonization occurs exclusively at the RAJ mucosa

(37; CJ Hovde, University of Idaho, personal communication). STEC O157 shedding patterns between rectally and orally inoculated sheep were found to be similar, thus indicating that STEC O157 may be able to effectively colonize the terminal rectum (38). However, inefficient attachment of large numbers of STEC to the RAJ may account for reduced shedding periods compared to those in cattle.

Similar to cattle, small ruminants tend to be asymptomatic shedders of STEC. This trait was demonstrated when the screening of "healthy animals" in Berlin reported that 66.6% of the sheep and 56.1% of the goats tested were found to be STEC carriers (39). Similar results were obtained in Spain where 47% of healthy goats tested positive for shedding STEC (40). The asymptomatic feature of STEC carriage is possibly also due to the lack of Shiga toxin vascular receptors in small ruminants.

Many direct-contact human infections are attributed to contact with sheep and goats at petting zoos and open farms (41, 42). One study investigating the prevalence of zoonotic agents on small city farms in southern Germany found that 100% of the sheep and 89.3% of the goats tested positive for STEC (43). Small ruminants, especially goats, generally exhibit inquisitive behavior and thus may have greater contact with humans, increasing the potential for transmission to humans (34). Human infections have also been linked to the consumption of unpasteurized milk and cheese made from contaminated goat or sheep milk (40, 44).

Sheep are the primary reservoir for STEC in Australia, with the serotype of importance being O26; however, the risk of human infection was deemed insignificant due to low prevalence rates (1, 45). Although the within-herd prevalence was low, previous research reported that 90% of Australian sheep farms had animals testing positive for STEC (46). In Norway, however, the risk of human infection from sheep was much more significant since almost 50% of the sheep O26 isolates had multiple-locus variable number tandem repeat analysis profiles similar to that found in human clinical cases (35). The importance of sheep in STEC epidemiology was also demonstrated by Oporto et al. (47), who reported that greater than 50% of sheep herds in Spain had animals shedding non-O157 STEC compared to 20.7% in dairy cattle and 46% in beef cattle.

In the United States, Jacob et al. reported that 11.1% of goat fecal samples collected at slaughter had STEC O157 and 14.5% had at least one non-O157 STEC serotype (48, 49). The STEC O157 flock prevalence in Spain was reported to be 8.7% and individual prevalence, 7.8% (47). A similarly low STEC O157 prevalence of 5.8% was also reported in Scotland (50). Low STEC O157 prevalences were also reported in the United Kingdom and Holland, with the prevalence being 0.1% and 4.0%, respectively (51, 52). Lesser developed countries have also reported the presence of STEC in their small ruminant population. In Vietnam, 100% of the goat farms surveyed had animals shedding STEC, and the within-herd shedding was dramatically higher than that reported elsewhere, with up to 65% of animals shedding STEC (53). In Bangladesh, almost 10% of the small ruminants being slaughtered tested positive for STEC O157 (54). As evidenced by past outbreaks, STEC in animals from lesser developed countries can potentially be a serious threat to food safety, since in those countries there may not be strict hygienic slaughter practices; thus contaminated meat could easily enter the food chain (55, 56).

The shedding of STEC in small ruminants has been demonstrated to be age and season dependent. Younger animals tend to have a lower prevalence of STEC than older animals do (40, 57–59). A longitudinal study spanning 6 months in the United States demonstrated a peak in STEC prevalence during summer (60). This trend was also observed in Italy, where animals screened during the warmer months of the year had a higher prevalence of STEC O157 (58).

Other Ruminants

Water Buffalo (*Bubalus bubalis*)

In addition to cattle, sheep, and goats, other ruminant species have also been identified as shedders of STEC. Water buffalo (*Bubalus bubalis*) has been identified as an important reservoir of STEC O157 in many countries (61). The water buffalo is reared in many countries because of its ability to serve a dual purpose, as both a milk and meat producer. Buffaloes are also able to thrive much better on poor quality forages than the *Bos taurus* species, thus making them suitable for subsistence farming. There are large commercial meat and milk water buffalo herds in Asia and South America, while in Europe water buffalo is primarily reared for milk production. In Bangladesh, STEC colonies were isolated from 38% of the buffaloes sampled before slaughter. Almost half of these isolates were identified as being O157 (54). Galiero et al. reported an almost a similar prevalence in Italy, with 14.5% of the animals shedding O157 (61). In Vietnam, 27% of the buffaloes screened were found to be positive for STEC. Serotyping of the isolates, however, revealed that none of the isolates were O157 (53).

Deer

There are an estimated 30 million white-tailed deer in the United States (194). The role of white-tailed deer (*Odocoileus virginianus*) as a potential reservoir for STEC was first reported in 1999 when almost 2.4% of deer sampled tested positive (62). The presence of STEC O157:H7 in deer feces was later confirmed by Renter et al. (68), who found the STEC prevalence in Nebraska white-tailed deer to be 0.25%. Similarly, low STEC prevalences of 0.2% were reported in hunter-harvested captive deer in Louisiana (63) and 3.3% in farm-raised deer in Ohio (64). Other species of deer, including red deer (*Cervus elaphus*), fallow deer (*Dama dama*), and roe deer (*Capreolus capreolus*), have also been identified as capable of shedding STEC serotypes (65, 66). Almost 50% of Pennsylvanian white-tailed deer fecal samples screened tested positive for *stx* genes; however, only 8% possessed the *eae* gene, which is necessary for colonization of the human intestine (67).

Feral deer are known to share pastures with cattle and can also be found in close proximity to many dairy farms. The close association between deer and livestock implies that deer can serve to maintain and disseminate STEC between and within cattle herds (68, 69).

Human STEC O157 infections were first associated with venison in 1997 when six persons became ill as a result of consuming jerky made from venison (70). Since then, there have been numerous other cases associated with venison, with one of the most recent published reports being an outbreak of non-O157 STEC among high school students that was associated with consumption of venison they had killed and processed. Two STEC serotypes, O103:H2 and O145:NM, were isolated from the samples analyzed; however, the O145:NM serotype was found to be Shiga toxin negative (71).

Elk (*Cervus canadensis*)

Similar to deer, elk have also been associated with numerous food-borne disease outbreaks (72). Gilbreath et al. (80) reported that over 22% of wild Idaho elk screened were positive for STEC. A slightly lower prevalence (7%) of *stx* genes was detected in fecal pellets collected from elk in Colorado (73). In this study none of the animals were found to shed serogroup O157, but serogroups O103 and O146 were detected. Interestingly in both studies, the incidence of STEC in the elk feces was found to be higher than in mule deer, which shared the same grazing grounds.

Bison (*Bison bison*)

Another potentially important animal reservoir of STEC is the American bison (*Bison bison*). This potential is supported by the fact that both cattle and bison share similar RAJ

morphological characteristics (74). In the United States consumption of bison meat has increased, thus increasing the risk of transmission from bison to humans (75). This risk was exemplified in 2010 when a multistate outbreak of STEC O157:H7 was associated with bison meat consumption. The prevalence of STEC O157:H7 in bison has been reported to be as high as 42% (76). STEC O157 has also been isolated from the carcass of slaughtered bison at a prevalence of 1.13% (77). Non-O157 STEC serotypes including O45, O103, O111, O113, O121, and O145 have also been isolated from bison carcasses; however, none of these isolates possessed *stx* genes (78).

STEC O157 and non-O157 STEC strains have been isolated from other captive and wild nondomesticated ruminant species. They include llamas, moose, alpacas, antelopes, and yaks (79–84). These animals can transmit STEC to humans directly by contact at petting zoos or indirectly through fecal deposition in water sources, vegetable fields, or recreational areas or on meat. Further research is required to determine the role these animals may have in the epidemiology of human STEC infections.

Equine

Published data on the epidemiology of STEC carriage in horses are limited. There are also no published case reports describing the clinical features of STEC infection in horses. The available published data on the prevalence of STEC in horses (85–88) and donkeys (84, 89, 90) indicate that they are not major reservoirs of STEC and may instead be spillover hosts. Only one of 400 horse fecal samples screened in Germany was positive for STEC. The serotype isolated was O113: H21 (88). A similarly low prevalence of STEC has also been detected in the equine population in the United States. Only one of 242 horse fecal samples from Ohio tested positive for STEC O157:H7. Interestingly, this case shared housing accommodations with a goat that also shed STEC O157. The isolates from both animals had indistinguishable multiple-locus variable number tandem repeat analysis patterns (87). A similarly low prevalence has been reported by Hancock et al., who reported that 1% of horses sampled (n = 90) tested positive for O157:H7 (85). Screening of fecal samples from horses located in the Sacramento Valley revealed a slightly higher prevalence than that recorded in Ohio. Four of 156 samples (2.6%) tested positive for the Shiga toxin 2 gene (86). Notably, as was seen in Ohio, all the positive horses were also housed on farms containing ruminants. Despite the low STEC prevalence in horses, there are reported human clinical cases associated with infection from horse contact (91), and one must be aware of this potential source of infection.

Swine

Swine can be colonized with various serotypes of STEC, including O157; however, the risk of causing human disease is low (12, 92, 93). The prevalence of STEC O157:H7 in domestic swine has been reported to range from 0 up to 10%, with the prevalence in the United States usually being less than 2% (12, 92, 94, 95). As in humans, STEC O157:H7 can be highly pathogenic in pigs (92). Unlike ruminants, pigs possess *stx*-sensitive vascular receptors, and edema disease can develop post intestinal colonization with STEC strains producing Stx2e (8, 96). Stx2e is the most frequent subtype of Stx2 found in porcine feces (97). The receptor affinity of Stx2e is different from Stx1a and Stx2a, since the primary receptor targets are not Gb3 receptors but rather globotetraosylceramide (Gb4) receptors (98, 99). Recently weaned pigs are most susceptible to edema disease, and the clinical signs include subcutaneous and submucosal edema, ataxia, incoordination, stupor, and recumbency (98, 100). While morbidity in the herd may be low, the case fatality rate for edema disease is high, and surviving pigs may have neurologic deficits.

Though a relatively low prevalence of STEC O157:H7 has been reported, swine have been shown to harbor and shed STEC for up to 2 months post infection (101). Non-O157 STEC serotypes have also been isolated from pigs; however, many of these isolates lack the virulence factors required to cause human disease (1). Despite a low prevalence of pathogenic STEC serotypes, the potential for human infection from swine exists. This risk is exemplified by a recent Canadian outbreak of STEC O157:H7 associated with consumption of pork, with infected persons having the identical STEC O157:H7 isolate to that found in the pork meal served (102)

Feral Swine

Feral swine is another wildlife species that has been associated with STEC disease in the human population. There are approximately 5 million feral swine in the United States, and they can be found in more than 35 states (195). These animals are highly adaptable to varying environmental conditions and can serve as a vector for disease between livestock farms and as a source of contamination of vegetable production fields.

In the United States, feral swine was first identified as a reservoir for STEC O157:H7 in 2007 in California (103). In that study, STEC O157:H7 was isolated from 14.9% of the swine specimens tested, and these isolates were found to be indistinguishable from STEC O157:H7 isolates obtained from an outbreak in the human population associated with the consumption of spinach. Interestingly, all cattle, feral swine, and environmental samples from the region where the spinach was cultivated had the same STEC isolate O157 (103). STEC was also detected in feral swine from Sweden, Switzerland, and Spain. Approximately 9% of the tonsil samples screened (n = 153) in Switzerland were positive for STEC O157, but none of the corresponding fecal samples were positive (104). A similar prevalence of 8% was reported for Spanish feral swine fecal samples (105). The isolates were serotyped, and 3.3% of the animals were identified as shedding STEC O157:H7 and 5.2% of the animals as shedding non-O157 STEC.

The identification of STEC from feral swine samples indicates that they can play a role in the epidemiology of STEC infections. As such, their ability to potentially contaminate vegetable production fields and serve as vectors for STEC transmission between livestock must be recognized, and measures employed to mitigate this risk

Birds

Birds are capable of harboring many bacterial organisms in their gastrointestinal tract and are capable of acting as spillover hosts for STEC. Wild birds were first identified as a potential source of STEC infection in 1997 (106). Since then, STEC has been isolated from starlings, pigeons, sparrows, and other avian species (106, 107). Many species of wild birds can be found in close proximity to livestock operations and waste disposal landfill sites. These birds are attracted to farms since they can easily obtain a food source from animal feed. Nielsen et al. identified that 2% of the wild bird fecal samples collected in close proximity to farms contained *stx* genes (108). Similar results were also obtained in England where 1.5% of wild bird samples had the *stx1* gene, 7.9% the *stx2* gene, and 4.9% the *eae* gene (109). Similarly, low prevalence rates of STEC O157:H7 in the starling (*Sturnus vulgaris*) population in Ohio and other wild bird species in Scotland and Japan have also been reported (110–112). Though the STEC prevalence levels are reportedly low, the potential of these birds to transmit STEC to other birds and contaminate the environment is of serious risk. Studies have shown that once colonized, a starling may shed STEC O157 at levels greater than 100 CFU/g of feces for up to 13 days post colonization (113).

The migratory pattern of birds and the fact they can traverse long distances in a single day mean they can serve as a mode of transmission of STEC between and within farms. This was demonstrated by Williams et al. (114), who reported that starlings and cattle on different farms had molecularly indistinguishable subtypes of STEC O157:H7, thus confirming that starlings were able to transmit STEC to different farms. Bolstering the role of starlings in STEC epidemiology is the fact that the number of starlings per milking dairy cow was also found to be significantly associated with the presence of STEC O157:H7 in bovine fecal pats (115).

Migratory birds can also interact with peridomestic birds such as pigeons and thus propagate the transmission of STEC. Pigeons and finches have been identified as two species that can potentially serve as a source of human infection since these birds inhabit buildings, parks, and other recreational areas and are in close association with the human population (111, 116). Fecal depositions by these birds in areas frequented by humans increase the exposure potential to STEC.

Water fowl, including geese and ducks, are identified as a source of surface water and pasture contamination (107, 117, 118) and implicated as the source of numerous food-borne pathogens (119–122). One goose is reportedly capable of producing up to 5 pounds of feces per day, and this can result in mass contamination, since these birds are usually found in flocks (123). These birds are also able to travel large distances per day and can disperse pathogens over a wide area. Geese are known to forage within vegetable fields and also inhabit ponds and other surface water sources used for irrigation (124). The birds can thus contaminate produce when they defecate within vegetable fields and irrigation water sources.

STEC carriage has also been reported in domestic poultry. The prevalence of STEC O157:H7 in domestic chicken is relatively low, ranging from 0 to 1.5%, depending on the geographic location sampled (92, 94, 125, 126). Interestingly, the prevalence in turkeys was higher than that in chickens, with up to 7.5% of fecal samples testing positive (92, 126). Experimentally colonized chickens have been reported to harbor and shed STEC O157:H7 in their feces for periods in excess of 11 months (127). Pet birds such as canaries (*Serinus canaria domestica*) have also been reported to be capable of harboring and shedding STEC (128).

That both wild and domestic birds are able to harbor, transmit, and shed STEC is of serious concern since they potentially are a major risk to human health and disease, and as such, precautions should be taken to limit human or animal exposure to the excrement from these birds.

Fish and Shellfish

Fish and shellfish can be exposed to STEC when their aquatic environment becomes contaminated with mammalian fecal matter. These species do not act as reservoirs of infection or spillover hosts but rather dead-end hosts. Fish residing in close proximity or downstream of animal livestock facilities have also been found to be contaminated with STEC (129). Shellfish, due to their filter feeding ability, pose a significant risk to human infection since they can concentrate and retain pathogens (4, 130). Numerous studies have reported the recovery of both O157 and non-O157 STEC from the carcass of fish and shellfish offered for sale (131–135). The detection of STEC in these carcasses highlights the potential for human STEC infection through the consumption of under-cooked or raw fish and shellfish.

Rodents

Rodents have also been identified as being capable of harboring STEC within their gastrointestinal flora (108, 136–138). STEC O157 and non-O157 STEC have been recovered from *Rattus* spp. living in urban areas and on farms (137, 138). Cizek et al. demonstrated that Norway rats (*Rattus novegicus*) were

capable of shedding STEC O157:H7 for up to 11 days post exposure to high doses of STEC (10^9 CFU) and 5 days post exposure to lower doses of STEC (10^5 CFU) (139). This shedding ability indicates that while rats may not be long-term reservoirs of STEC, they are certainly capable of transmitting STEC between and within farms. This transmission potential was highlighted by Nielsen et al. (108), who found that STEC recovered from Norwegian rat fecal pellets was identical to that shed by cattle on the same farm. Contaminated rat feces may also be capable of harboring STEC O157:H7 for up to 9 months post inoculation, thus increasing the risk of transmission to other animals (139). Although rodents, especially rats, are not regarded as having a major role in the epidemiology of STEC (69, 85, 140), their potential to harbor and transmit STEC exists. Unlike for rats, there is little published information describing the role of mice in the epidemiology of STEC. Mice have, however, been used as animal models to study STEC infection in humans (141, 142).

Rabbits

Rabbits have been identified as potential vehicles for the transmission of both O157 and non-O157 STEC (143–145). Rabbits have also been used as a possible animal model to study STEC infection in humans, since they demonstrate enteric and renal lesions when challenged with STEC (146). Globally, consumption of rabbit meat is increasing; over 1 million tons of rabbit meat are consumed annually (147), thus increasing the potential for food-borne infection. Rabbits are also popular at petting zoos, and more than 6 million rabbits are kept as pets in United States (148). Similar to dogs and cats, their close association with humans may lead to the exchange of microbiota between species and thus STEC transmission. Wild rabbits are able to traverse long distances and may inhabit both urban and agricultural areas and potentially serve as vectors for the transmission of STEC from farm environments to the human population (144).

Raccoons

Raccoons are of particular interest since they can reside in a wide range of habitats, including agricultural, forested, and urban areas. Raccoons have been identified as a reservoir for numerous pathogens, including *Salmonella, Leptospira*, and *Campylobacter* species (149–151); however, there is only one report of STEC being isolated from raccoon feces. This animal had been residing within the hay barn of a dairy farm (152) and thus may be a spillover host. Despite an extensive literature search, no other reports of STEC raccoon colonization could be found (153, 154).

Insects

Insects can be important vectors in the transmission and dissemination of STEC in the environment. STEC O157:H7 has been recovered from houseflies (*Musca domestica*), dump flies (*Hydrotaea aenescens*), and dung beetles (*Catharsius molossus*) residing on animal farms and at animal fairs (155–158). Houseflies, in addition to being a mechanical vector, may also be involved in bioenhanced transmission (159). Kobayabashi et al. (159) suggested this additional role because STEC O15H:H7 could be detected within the alimentary tract of inoculated houseflies for at least 3 days post inoculation. The ability of houseflies to transmit STEC O157:H7 to animals was demonstrated by Ahmad et al. (160), who exposed naïve calves to houseflies inoculated with STEC; within 24 hours, STEC could be recovered from the feces of all eight calves in the experiment (160). Houseflies have also been demonstrated to transfer STEC onto the surface of vegetable produce (161). Houseflies are able to travel greater than 4 miles (162), and given their ability to transmit STEC, one has to be cognizant of the role they may play in the epidemiology of STEC infection.

Pets

Pets, especially dogs and cats, are capable of shedding a diverse range of STEC serotypes in their feces (163–166). Interestingly, although both O157 and non-O157 STEC have been recovered from dogs, there are no published reports indicating that O157 STEC has ever been recovered from cat feces. Dogs and cats have historically had close interaction with humans, with exchange of microbiota resulting in the possible transmission of STEC between species. These animals can be asymptomatic shedders of STEC, as demonstrated by Beutin et al. (39), who reported that up to 12% of healthy dogs shed STEC in their feces. In addition to household dogs, farm dogs can be a vector for the transmission of STEC. These dogs move freely among animals and humans, thus potentiating the spread of enterohemorrhagic *E. coli* (167). Dogs have also been reported to shed non-O157 STEC serotypes in their feces (163). Human infections due to canine exposure were also reported; one outbreak in Sweden resulted in 50 cases in humans after they attended a dog show (168). STEC has also been recovered from the feces of wild canids (169).

A highly virulent strain of STEC O146:NM has been isolated from the feces of an asymptomatic cat in Argentina (170). In Germany, there is also evidence of a cat and its owner shedding the same STEC O146:NM serotype (171). In this case, the source of the infection could not be determined, nor which animal was the index case.

Animal Models

Animal models have been used to study the in vivo pathogenesis of STEC. Numerous animal models have been developed, including mice, rats, chickens, rabbits, cows, greyhounds, baboons, and macaques (141, 173–178). Although these models do not fully replicate all aspects of the STEC infection in humans, they provide valuable insight into intestinal colonization, STEC pathogenesis, immune response, and efficacy of possible treatment regimens (179, 180).

Compared to other animal models, the mouse models are preferred for in vivo STEC studies because of their small size, low cost, ease of care and maintenance, availability of numbers, and varying genetic backgrounds (<141). There are at least four mouse models, including streptomycin-treated, streptomycin and mitocyin C/ciprofloxacin-treated, intragastric-fed but not streptomycin-treated, and the malnourished mouse models (181). The two most popular models are, however, the streptomycin-treated mouse and axenic mice. The models have reduced or no gastrointestinal flora and are also susceptible to STEC or enterohemorrhagic *E. coli* colonization (141). The response of mouse models to STEC exposure is, however, dependent on the method of infection, the strain of STEC, and the type of mouse model used (181).

Humans

Although animals are generally regarded as the main reservoir of STEC, humans can also be STEC reservoirs and may play a much larger role in the epidemiology of STEC infections than previously thought (182, 183). Asymptomatic infections in the human population can result in dissemination STEC and further propagation of outbreaks (182). Given that food contamination is the source of almost 40% of STEC outbreaks and almost 30% are of unknown origin (184), it is possible that contamination by asymptomatic humans may the source of many of these outbreaks.

Human STEC infections can present a wide spectrum of clinical signs, ranging from symptomatic infections to severe clinical syndromes such as hemorrhagic colitis and possibly hemolytic-uremic syndrome and thrombotic thrombocytopenic purpura in approximately 7% of the cases. Elderly persons, young children, and immune-compromised persons are at greatest risk (185).

Food and environmental contamination with STEC can occur as result of shedding by asymptomatic workers. Approximately 12% of dairy families in a Canadian study tested positive for O157 antibodies, and STEC isolates were recovered from 6% of the fecal samples, yet none of these positive cases could recall any clinical signs associated with STEC (186). Similarly, 1.1% of farm workers in Italy were found to be shedding STEC O157 asymptomatically in their feces. Contamination of meat carcasses at the abattoir during slaughter is also a possibility. One study reported that 1.3% of abattoir workers sampled were actively shedding STEC in their feces (187). Asymptomatic children can also shed STEC serotypes for up to 30 days post detection, whereas adults in the recent German O104:H4 outbreak shed the organism for up to 13 weeks (182, 188).

Person-to-person or secondary transmission is important in propagation of outbreaks and can account for 15 to 20% of cases within outbreaks (184, 189). At particular risk of disease due to secondary transmission are children (1 to 6 years of age) due to close contact, their immature immune systems, reduced personal hygiene, and prolonged shedding time (184). An analysis of STEC outbreaks occurring between 1982 and 2006 as a result of person-to-person transmission showed that 45% of these outbreaks occurred due to transmission at home, 11% at nurseries, and 10% at recreational water sources (190).

CONCLUSION

Most warm-blooded animals are capable of acting as reservoirs (symptomatic and asymptomatic), spillover hosts, or dead-end hosts of STEC. Animals are exposed to STEC by direct or indirect contact with the feces of a shedding animal. Cattle are recognized as the main reservoir of STEC, however, and other livestock species, including goats, sheep, bison, horses, pigs, and water buffalo, have been demonstrated to be capable of harboring these organisms. Wild birds and animals pose a unique risk in their ability to travel long distances, increasing the dissemination of STEC in the environment and thus potentiating its spread. Domestic pets are also capable of harboring STEC, and thus serve as a source of contamination within the household. Given the demonstrated ability of STEC to colonize the gastrointestinal tract of a wide variety of animals, it is expected that numerous other unreported species may also be potential sources of contamination or transmission of STEC. Recognition of these potential novel animal sources of transmission and propagation of STEC is essential when conducting epidemiological investigations and developing proper risk mitigation strategies. Despite the widespread carriage of STEC by a variety of animal species, it is important to consider that the presence of the *stx* gene alone is not an indication of pathogenicity in the human host. Assessment of the complement of virulence factors present in STEC recovered from animal hosts is therefore important to develop risk models. The understanding of why certain pathoserotypes or strains of STEC have a predilection for different animal species may provide valuable insight in the design of interventions to control the organism in the live animals that have impacts on human health.

ACKNOWLEDGMENTS

We declare no conflicts of interest with regard to the manuscript.

CITATION

Persad AK, LeJeune JT. 2014. Animal reservoirs of Shiga toxin-producing *Escherichia coli*. Microbiol Spectrum 2(4):EHEC-0027-2014.

REFERENCES

1. **Gyles CL.** 2007. Shiga toxin-producing *Escherichia coli*: an overview. *J Anim Sci* **85**:E45–E62.
2. **Etcheverria AI, Padola NL.** 2013. Shiga toxin-producing *Escherichia coli*: factors involved in virulence and cattle colonization. *Virulence* **4**:366–372.
3. **Friedrich AW, Bielaszewska M, Zhang WL, Pulz M, Kuczius T, Ammon A, Karch H.** 2002. *Escherichia coli* harboring Shiga toxin 2 gene variants: frequency and association with clinical symptoms. *J Infect Dis* **185**:74–84.
4. **Bennani M, Badri S, Baibai T, Oubrim N, Hassar M, Cohen N, Amarouch H.** 2011. First detection of Shiga toxin-producing *Escherichia coli* in shellfish and coastal environments of Morocco. *Appl Biochem Biotechnol* **165**:290–299.
5. **Manna SK, Das R, Manna C.** 2008. Microbiological quality of finfish and shellfish with special reference to shiga toxin-producing *Escherichia coli* O157. *J Food Sci* **73**:M283–M286.
6. **Gourmelon M, Montet MP, Lozach S, Le Mennec C, Pommepuy M, Beutin L, Vernozy-Rozand C.** 2006. First isolation of Shiga toxin 1d producing *Escherichia coli* variant strains in shellfish from coastal areas in France. *J Appl Microbiol* **100**:85–97.
7. **Besser TE, Richards BL, Rice DH, Hancock DD.** 2001. *Escherichia coli* O157:H7 infection of calves: infectious dose and direct contact transmission. *Epidemiol Infect* **127**:555–560.
8. **Pruimboom-Brees IM, Morgan TW, Ackermann MR, Nystrom ED, Samuel JE, Cornick NA, Moon HW.** 2000. Cattle lack vascular receptors for *Escherichia coli* O157:H7 Shiga toxins. *Proc Natl Acad Sci USA* **97**:10325–10329.
9. **Nguyen Y, Sperandio V.** 2012. Enterohemorrhagic *E. coli* (EHEC) pathogenesis. *Front Cell Infect Microbiol* **2**:90.
10. **Naylor SW, Low JC, Besser TE, Mahajan A, Gunn GJ, Pearce MC, McKendrick IJ, Smith DG, Gally DL.** 2003. Lymphoid follicle-dense mucosa at the terminal rectum is the principal site of colonization of enterohemorrhagic *Escherichia coli* O157:H7 in the bovine host. *Infect Immun* **71**:1505–1512.
11. **Vanaja SK, Springman AC, Besser TE, Whittam TS, Manning SD.** 2010. Differential expression of virulence and stress fitness genes between *Escherichia coli* O157:H7 strains with clinical or bovine-biased genotypes. *Appl Environ Microbiol* **76**:60–68.
12. **Fairbrother JM, Nadeau E.** 2006. *Escherichia coli*: on-farm contamination of animals. *Rev Sci Tech* **25**:555–569.
13. **Gagliardi JV, Karns JS.** 2000. Leaching of *Escherichia coli* O157:H7 in diverse soils under various agricultural management practices. *Appl Environ Microbiol* **66**:877–883.
14. **LeJeune JT, Besser TE, Merrill NL, Rice DH, Hancock DD.** 2001. Livestock drinking water microbiology and the factors influencing the

quality of drinking water offered to cattle. *J Dairy Sci* **84:**1856–1862.

15. **LeJeune JT, Besser TE, Hancock DD.** 2001. Cattle water troughs as reservoirs of *Escherichia coli* O157. *Appl Environ Microbiol* **67:**3053–3057.

16. **Garber L, Wells S, Schroeder-Tucker L, Ferris K.** 1999. Factors associated with fecal shedding of verotoxin-producing *Escherichia coli* O157 on dairy farms. *J Food Prot* **62:**307–312.

17. **Lejeune JT, Kauffman MD.** 2005. Effect of sand and sawdust bedding materials on the fecal prevalence of *Escherichia coli* O157:H7 in dairy cows. *Appl Environ Microbiol* **71:**326–330.

18. **Cernicchiaro N, Pearl DL, Ghimire S, Gyles CL, Johnson RP, LeJeune JT, Ziebell K, McEwen SA.** 2009. Risk factors associated with *Escherichia coli* O157:H7 in Ontario beef cow-calf operations. *Prev Vet Med* **92:**106–115.

19. **Hancock D, Besser T, Lejeune J, Davis M, Rice D.** 2001. The control of VTEC in the animal reservoir. *Int J Food Microbiol* **66:**71–78.

20. **Lira WM, Macedo C, Marin JM.** 2004. The incidence of Shiga toxin-producing *Escherichia coli* in cattle with mastitis in Brazil. *J Appl Microbiol* **97:**861–866.

21. **Cerqueira AM, Guth BE, Joaquim RM, Andrade JR.** 1999. High occurrence of Shiga toxin-producing *Escherichia coli* (STEC) in healthy cattle in Rio de Janeiro State, Brazil. *Vet Microbiol* **70:**111–121.

22. **Farrokh C, Jordan K, Auvray F, Glass K, Oppegaard H, Raynaud S, Thevenot D, Condron R, De Reu K, Govaris A, Heggum K, Heyndrickx M, Hummerjohann J, Lindsay D, Miszczycha S, Moussiegt S, Verstraete K, Cerf O.** 2013. Review of Shiga-toxin-producing *Escherichia coli* (STEC) and their significance in dairy production. *Int J Food Microbiol* **162:**190–212.

23. **Hancock DD, Besser TE, Kinsel ML, Tarr PI, Rice DH, Paros MG.** 1994. The prevalence of *Escherichia coli* O157.H7 in dairy and beef cattle in Washington state. *Epidemiol Infect* **113:**199–207.

24. **Hussein HS, Bollinger LM.** 2005. Prevalence of Shiga toxin-producing *Escherichia coli* in beef cattle. *J Food Prot* **68:**2224–2241.

25. **Hussein HS, Sakuma T.** 2005. Prevalence of shiga toxin-producing *Escherichia coli* in dairy cattle and their products. *J Dairy Sci* **88:**450–465.

26. **Fegan N, Vanderlinde P, Higgs G, Desmarchelier P.** 2004. The prevalence and concentration of *Escherichia coli* O157 in faeces of cattle from different production systems at slaughter. *J Appl Microbiol* **97:**362–370.

27. **Widiasih DA, Ido N, Omoe K, Sugii S, Shinagawa K.** 2004. Duration and magnitude of faecal shedding of Shiga toxin-producing *Escherichia coli* from naturally infected cattle. *Epidemiol Infect* **132:**67–75.

28. **Lim JY, Li J, Sheng H, Besser TE, Potter K, Hovde CJ.** 2007. *Escherichia coli* O157:H7 colonization at the rectoanal junction of long-duration culture-positive cattle. *Appl Environ Microbiol* **73:**1380–1382.

29. **Chase-Topping M, Gally D, Low C, Matthews L, Woolhouse M.** 2008. Super-shedding and the link between human infection and livestock carriage of *Escherichia coli* O157. *Nat Rev Microbiol* **6:**904–912.

30. **Omisakin F, MacRae M, Ogden ID, Strachan NJ.** 2003. Concentration and prevalence of *Escherichia coli* O157 in cattle feces at slaughter. *Appl Environ Microbiol* **69:**2444–2447.

31. **Williams ML, Pearl DL, Bishop KE, Lejeune JT.** 2013. Use of multiple-locus variable-number tandem repeat analysis to evaluate *Escherichia coli* O157 subtype distribution and transmission dynamics following natural exposure on a closed beef feedlot facility. *Foodborne Pathog Dis* **10:**827–834.

32. **Matthews L, Low JC, Gally DL, Pearce MC, Mellor DJ, Heesterbeek JA, Chase-Topping M, Naylor SW, Shaw DJ, Reid SW, Gunn GJ, Woolhouse ME.** 2006. Heterogeneous shedding of *Escherichia coli* O157 in cattle and its implications for control. *Proc Natl Acad Sci USA* **103:**547–552.

33. **Nielsen EM, Tegtmeier C, Andersen HJ, Gronbaek C, Andersen JS.** 2002. Influence of age, sex and herd characteristics on the occurrence of verocytotoxin-producing *Escherichia coli* O157 in Danish dairy farms. *Vet Microbiol* **88:**245–257.

34. **La Ragione RM, Best A, Woodward MJ, Wales AD.** 2009. *Escherichia coli* O157:H7 colonization in small domestic ruminants. *FEMS Microbiol Rev* **33:**394–410.

35. **Brandal LT, Sekse C, Lindstedt BA, Sunde M, Lobersli I, Urdahl AM, Kapperud G.** 2012. Norwegian sheep are an important reservoir for human-pathogenic *Escherichia coli* O26:H11. *Appl Environ Microbiol* **78:**4083–4091.

36. **Grauke LJ, Kudva IT, Yoon JW, Hunt CW, Williams CJ, Hovde CJ.** 2002. Gastrointestinal tract location of *Escherichia coli* O157:H7 in ruminants. *Appl Environ Microbiol* **68:**2269–2277.

37. **Woodward MJ, Best A, Sprigings KA, Pearson GR, Skuse AM, Wales A, Hayes CM, Roe JM, Low JC, La Ragione RM.** 2003. Non-toxigenic *Escherichia coli* O157:H7 strain NCTC12900 causes attaching-effacing lesions and eae-dependent persistence in weaned sheep. *Int J Med Microbiol* **293:**299–308.

38. Best A, Clifford D, Crudgington B, Cooley WA, Nunez A, Carter B, Weyer U, Woodward MJ, La Ragione RM. 2009. Intermittent *Escherichia coli* O157:H7 colonisation at the terminal rectum mucosa of conventionally-reared lambs. *Vet Res* **40:**9.

39. Beutin L, Geier D, Steinruck H, Zimmermann S, Scheutz F. 1993. Prevalence and some properties of verotoxin (Shiga-like toxin)-producing *Escherichia coli* in seven different species of healthy domestic animals. *J Clin Microbiol* **31:** 2483–2488.

40. Cortes C, De la Fuente R, Blanco J, Blanco M, Blanco JE, Dhabi G, Mora A, Justel P, Contreras A, Sanchez A, Corrales JC, Orden JA. 2005. Serotypes, virulence genes and intimin types of verotoxin-producing *Escherichia coli* and enteropathogenic *E. coli* isolated from healthy dairy goats in Spain. *Vet Microbiol* **110:**67–76.

41. Heuvelink AE, van Heerwaarden C, Zwartkruis-Nahuis JT, van Oosterom R, Edink K, van Duynhoven YT, de Boer E. 2002. *Escherichia coli* O157 infection associated with a petting zoo. *Epidemiol Infect* **129:**295–302.

42. LeJeune JT, Davis MA. 2004. Outbreaks of zoonotic enteric disease associated with animal exhibits. *J Am Vet Med Assoc* **224:**1440–1445.

43. Schilling AK, Hotzel H, Methner U, Sprague LD, Schmoock G, El-Adawy H, Ehricht R, Wohr AC, Erhard M, Geue L. 2012. Zoonotic agents in small ruminants kept on city farms in southern Germany. *Appl Environ Microbiol* **78:**3785–3793.

44. Espie E, Vaillant V, Mariani-Kurkdjian P, Grimont F, Martin-Schaller R, De Valk H, Vernozy-Rozand C. 2006. *Escherichia coli* O157 outbreak associated with fresh unpasteurized goats' cheese. *Epidemiol Infect* **134:**143–146.

45. Duffy LL, Small A, Fegan N. 2010. Concentration and prevalence of *Escherichia coli* O157 and *Salmonella* serotypes in sheep during slaughter at two Australian abattoirs. *Aust Vet J* **88:**399–404.

46. Djordjevic SP, Hornitzky MA, Bailey G, Gill P, Vanselow B, Walker K, Bettelheim KA. 2001. Virulence properties and serotypes of Shiga toxin-producing *Escherichia coli* from healthy Australian slaughter-age sheep. *J Clin Microbiol* **39:**2017–2021.

47. Oporto B, Esteban JI, Aduriz G, Juste RA, Hurtado A. 2008. *Escherichia coli* O157:H7 and non-O157 Shiga toxin-producing *E. coli* in healthy cattle, sheep and swine herds in Northern Spain. *Zoonoses Public Health* **55:**73–81.

48. Jacob ME, Foster DM, Rogers AT, Balcomb CC, Shi X, Nagaraja TG. 2013. Evidence of non-O157 Shiga toxin-Producing *Escherichia coli* in the feces of meat goats at a U.S. slaughter plant. *J Food Prot* **76:**1626–1629.

49. Jacob ME, Foster DM, Rogers AT, Balcomb CC, Sanderson MW. 2013. Prevalence and relatedness of *Escherichia coli* O157:H7 strains in the feces and on the hides and carcasses of U.S. meat goats at slaughter. *Appl Environ Microbiol* **79:**4154–4158.

50. Solecki O, MacRae M, Strachan N, Lindstedt BA, Ogden I. 2009. *E. coli* O157 from sheep in northeast Scotland: prevalence, concentration shed, and molecular characterization by multilocus variable tandem repeat analysis. *Foodborne Pathog Dis* **6:**849–854.

51. Milnes AS, Stewart I, Clifton-Hadley FA, Davies RH, Newell DG, Sayers AR, Cheasty T, Cassar C, Ridley A, Cook AJ, Evans SJ, Teale CJ, Smith RP, McNally A, Toszeghy M, Futter R, Kay A, Paiba GA. 2008. Intestinal carriage of verocytotoxigenic *Escherichia coli* O157, *Salmonella*, thermophilic *Campylobacter* and *Yersinia enterocolitica*, in cattle, sheep and pigs at slaughter in Great Britain during 2003. *Epidemiol Infect* **136:**739–751.

52. Heuvelink AE, van den Biggelaar FL, de Boer E, Herbes RG, Melchers WJ, Huis in 't Veld JH, Monnens LA. 1998. Isolation and characterization of verocytotoxin-producing *Escherichia coli* O157 strains from Dutch cattle and sheep. *J Clin Microbiol* **36:**878–882.

53. Vu-Khac H, Cornick NA. 2008. Prevalence and genetic profiles of Shiga toxin-producing *Escherichia coli* strains isolated from buffaloes, cattle, and goats in central Vietnam. *Vet Microbiol* **126:**356–363.

54. Islam MA, Mondol AS, de Boer E, Beumer RR, Zwietering MH, Talukder KA, Heuvelink AE. 2008. Prevalence and genetic characterization of shiga toxin-producing Escherichia coli isolates from slaughtered animals in Bangladesh. *Appl Environ Microbiol* **74:**5414–5421.

55. Effler E, Isaacson M, Arntzen L, Heenan R, Canter P, Barrett T, Lee L, Mambo C, Levine W, Zaidi A, Griffin PM. 2001. Factors contributing to the emergence of *Escherichia coli* O157 in Africa. *Emerg Infect Dis* **7:**812–819.

56. Cunin P, Tedjouka E, Germani Y, Ncharre C, Bercion R, Morvan J, Martin P. 1999. An epidemic of bloody diarrhea: *Escherichia coli* O157 emerging in Cameroon? *Emerg Infect Dis* **5:**285.

57. Battisti A, Lovari S, Franco A, Di Egidio A, Tozzoli R, Caprioli A, Morabito S. 2006. Prevalence of *Escherichia coli* O157 in lambs at slaughter in Rome, central Italy. *Epidemiol Infect* **134:**415–419.

58. Franco A, Lovari S, Cordaro G, Di Matteo P, Sorbara L, Iurescia M, Donati V, Buccella C, Battisti A. 2009. Prevalence and concentration of verotoxigenic *Escherichia coli* O157:H7 in adult

sheep at slaughter from Italy. *Zoonoses Public Health* **56:**215–220.

59. **Orden JA, Cortes C, Horcajo P, De la Fuente R, Blanco JE, Mora A, Lopez C, Blanco J, Contreras A, Sanchez A, Corrales JC, Dominguez-Bernal G.** 2008. A longitudinal study of verotoxin-producing *Escherichia coli* in two dairy goat herds. *Vet Microbiol* **132:**428–434.

60. **Kudva IT, Hatfield PG, Hovde CJ.** 1996. *Escherichia coli* O157:H7 in microbial flora of sheep. *J Clin Microbiol* **34:**431–433.

61. **Galiero G, Conedera G, Alfano D, Caprioli A.** 2005. Isolation of verocytotoxin-producing *Escherichia coli* O157 from water buffaloes (*Bubalus bubalis*) in southern Italy. *Vet Rec* **156:**382–383.

62. **Sargeant JM, Hafer DJ, Gillespie JR, Oberst RD, Flood SJ.** 1999. Prevalence of *Escherichia coli* O157:H7 in white-tailed deer sharing rangeland with cattle. *J Am Vet Med Assoc* **215:**792–794.

63. **Dunn JR, Keen JE, Moreland D, Alex T.** 2004. Prevalence of *Escherichia coli* O157:H7 in white-tailed deer from Louisiana. *J Wildl Dis* **40:**361–365.

64. **French E, Rodriguez-Palacios A, LeJeune JT.** 2010. Enteric bacterial pathogens with zoonotic potential isolated from farm-raised deer. *Foodborne Pathog Dis* **7:**1031–1037.

65. **Bardiau M, Gregoire F, Muylaert A, Nahayo A, Duprez JN, Mainil J, Linden A.** 2010. Enteropathogenic (EPEC), enterohaemorragic (EHEC) and verotoxigenic (VTEC) *Escherichia coli* in wild cervids. *J Appl Microbiol* **109:**2214–2222.

66. **Sanchez S, Garcia-Sanchez A, Martinez R, Blanco J, Blanco JE, Blanco M, Dahbi G, Mora A, Hermoso de Mendoza J, Alonso JM, Rey J.** 2009. Detection and characterisation of Shiga toxin-producing *Escherichia coli* other than *Escherichia coli* O157:H7 in wild ruminants. *Vet J* **180:**384–388.

67. **Kistler WM, Mulugeta S, Mauro SA.** 2011. Detection of *stx* and *stx* genes in Pennsylvanian white-tailed deer. *Toxins* (Basel) **3:**640–646.

68. **Renter DG, Sargeant JM, Hygnstorm SE, Hoffman JD, Gillespie JR.** 2001. *Escherichia coli* O157:H7 in free-ranging deer in Nebraska. *J Wildl Dis* **37:**755–760.

69. **Rice DH, Hancock DD, Besser TE.** 2003. Faecal culture of wild animals for *Escherichia coli* O157: H7. *Vet Rec* **152:**82–83.

70. **Keene WE, Sazie E, Kok J, Rice DH, Hancock DD, Balan VK, Zhao T, Doyle MP.** 1997. An outbreak of *Escherichia coli* O157:H7 infections traced to jerky made from deer meat. *JAMA* **277:**1229–1231.

71. **Rounds JM, Rigdon CE, Muhl LJ, Forstner M, Danzeisen GT, Koziol BS, Taylor C, Shaw BT, Short GL, Smith KE.** 2012. Non-O157 Shiga toxin-producing *Escherichia coli* associated with venison. *Emerg Infect Dis* **18:**279–282.

72. **Laidler MR, Tourdjman M, Buser GL, Hostetler T, Repp KK, Leman R, Samadpour M, Keene WE.** 2013. *Escherichia coli* O157:H7 infections associated with consumption of locally grown strawberries contaminated by deer. *Clin Infect Dis* **57:**1129–1134.

73. **Franklin AB, Vercauteren KC, Maguire H, Cichon MK, Fischer JW, Lavelle MJ, Powell A, Root JJ, Scallan E.** 2013. Wild ungulates as disseminators of Shiga toxin-producing *Escherichia coli* in urban areas. *PloS One* **8:**e81512.

74. **Kudva IT, Stasko JA.** 2013. Bison and bovine rectoanal junctions exhibit similar cellular architecture and *Escherichia coli* O157 adherence patterns. *BMC Vet Res* **9:**266.

75. **Johnson K.** 2011. Plains giants have foothold on tables. *The New York Times*, Jan. 23:A16. http://www.nytimes.com/2011/01/23/us/23buffalo.html?_r=0.

76. **Reinstein S, Fox JT, Shi X, Alam MJ, Nagaraja TG.** 2007. Prevalence of *Escherichia coli* O157:H7 in the American bison (*Bison bison*). *J Food Prot* **70:**2555–2560.

77. **Li Q, Sherwood JS, Logue CM.** 2004. The prevalence of *Listeria, Salmonella, Escherichia coli* and *E. coli* O157:H7 on bison carcasses during processing. *Food Microbiol* **21:**791–799.

78. **Magwedere K, Dang HA, Mills EW, Cutter CN, Roberts EL, Debroy C.** 2013. Incidence of Shiga toxin-producing *Escherichia coli* strains in beef, pork, chicken, deer, boar, bison, and rabbit retail meat. *J Vet Diagn Invest* **25:**254–258.

79. **Mercado EC, Rodriguez SM, Elizondo AM, Marcoppido G, Parreño V.** 2004. Isolation of Shiga toxin-producing *Escherichia coli* from a South American camelid (*Lama guanicoe*) with diarrhea. *J Clin Microbiol* **42:**4809–4811.

80. **Gilbreath JJ, Shields MS, Smith RL, Farrell LD, Sheridan PP, Spiegel KM.** 2009. Shiga toxins, and the genes encoding them, in fecal samples from native Idaho ungulates. *Appl Environ Microbiol* **75:**862–865.

81. **Cid D, Martín-Espada C, Maturrano L, García A, Luna L, Rosadio R.** 2012. Diarrheagenic *Escherichia coli* strains isolated from neonatal Peruvian alpacas (*Vicugnapacos*) with diarrhea, p 223–228. *In* Pérez-Cabal MA, Gutiérrez JP, Cervantes I, Alcalde MJ (ed), *Fibre Production in South American Camelids and Other Fibre Animals.* Wageningen Academic Publishers, Wageningen, The Netherlands.

82. **Mohammed Hamzah A, Mohammed Hussein A, Mahmoud Khalef J.** 2013. Isolation of *Escherichia coli* O157:H7 strain from fecal samples of zoo animal. *Scientific World Journal* **2013:**843968.

83. Leotta GA, Deza N, Origlia J, Toma C, Chinen I, Miliwebsky E, Iyoda S, Sosa-Estani S, Rivas M. 2006. Detection and characterization of Shiga toxin-producing *Escherichia coli* in captive non-domestic mammals. *Vet Microbiol* **118:**151–157.

84. Chandran A, Mazumder A. 2013. Prevalence of diarrhea-associated virulence genes and genetic diversity in *Escherichia coli* isolates from fecal material of various animal hosts. *Appl Environ Microbiol* **79:**7371–7380.

85. Hancock DD, Besser TE, Rice DH, Ebel ED, Herriott DE, Carpenter LV. 1998. Multiple sources of *Escherichia coli* O157 in feedlots and dairy farms in the northwestern USA. *Prev Vet Med* **35:**11–19.

86. Larson DG. 2009. *Prevalence of Shiga toxin-producing Escherichia coli in the Sacramento Valley equine population.* M.S. thesis, California State University, Sacramento.

87. Lengacher B, Kline TR, Harpster L, Williams ML, Lejeune JT. 2010. Low prevalence of *Escherichia coli* O157:H7 in horses in Ohio, USA. *J Food Prot* **73:**2089–2092.

88. Pichner R, Sander A, Steinruck H, Gareis M. 2005. [Occurrence of *Salmonella* spp. and shigatoxin-producing *Escherichia coli* (STEC) in horse faeces and horse meat products]. *Berl Munch Tierarztl Wochenschr* **118:**321–325 (In German).

89. Pritchard GC, Smith R, Ellis-Iversen J, Cheasty T, Willshaw GA. 2009. Verocytotoxigenic *Escherichia coli* O157 in animals on public amenity premises in England and Wales, 1997 to 2007. *Vet Rec* **164:**545–549.

90. Momtaz H, Farzan R, Rahimi E, Safarpoor Dehkordi F, Souod N. 2012. Molecular characterization of Shiga toxin-producing *Escherichia coli* isolated from ruminant and donkey raw milk samples and traditional dairy products in Iran. *Scientific World Journal* **2012:**231–341.

91. Chalmers RM, Salmon RL, Willshaw GA, Cheasty T, Looker N, Davies I, Wray C. 1997. Vero-cytotoxin-producing *Escherichia coli* O157 in a farmer handling horses. *Lancet* **349:**1816.

92. Ferens WA, Hovde CJ. 2011. *Escherichia coli* O157:H7: animal reservoir and sources of human infection. *Foodborne Pathog Dis* **8:**465–487.

93. Kaufmann M, Zweifel C, Blanco M, Blanco JE, Blanco J, Beutin L, Stephan R. 2006. *Escherichia coli* O157 and non-O157 Shiga toxin-producing *Escherichia coli* in fecal samples of finished pigs at slaughter in Switzerland. *J Food Prot* **69:**260–266.

94. Chapman PA, Siddons CA, Gerdan Malo AT, Harkin MA. 1997. A 1-year study of *Escherichia coli* O157 in cattle, sheep, pigs and poultry. *Epidemiol Infect* **119:**245–250.

95. Richards HA, Perez-Conesa D, Doane CA, Gillespie BE, Mount JR, Oliver SP, Pangloli P, Draughon FA. 2006. Genetic characterization of a diverse *Escherichia coli* O157:H7 population from a variety of farm environments. *Foodborne Pathog Dis* **3:**259–265.

96. Waddell TE, Coomber BL, Gyles CL. 1998. Localization of potential binding sites for the edema disease verotoxin (VT2e) in pigs. *Can J Vet Res* **62:**81–86.

97. Fratamico PM, Bagi LK, Bush EJ, Solow BT. 2004. Prevalence and characterization of shiga toxin-producing *Escherichia coli* in swine feces recovered in the National Animal Health Monitoring System's Swine 2000 study. *Appl Environ Microbiol* **70:**7173–7178.

98. Imberechts H, De Greve H, Lintermans P. 1992. The pathogenesis of edema disease in pigs. A review. *Vet Microbiol* **31:**221–233.

99. Müthing J, Meisen I, Zhang W, Bielaszewska M, Mormann M, Bauerfeind R, Schmidt MA, Friedrich AW, Karch H. 2012. Promiscuous Shiga toxin 2e and its intimate relationship to Forssman. *Glycobiology* **22:**849–862.

100. Meisen I, Rosenbrück R, Galla HJ, Hüwel S, Kouzel IU, Mormann M, Karch H, Müthing J. 2013. Expression of Shiga toxin 2e glycosphingolipid receptors of primary porcine brain endothelial cells and toxin-mediated breakdown of the blood-brain barrier. *Glycobiology* **23:**745–759.

101. Booher SL, Cornick NA, Moon HW. 2002. Persistence of *Escherichia coli* O157:H7 in experimentally infected swine. *Vet Microbiol* **89:**69–81.

102. Trotz-Williams LA, Mercer NJ, Walters JM, Maki AM, Johnson RP. 2012. Pork implicated in a Shiga toxin-producing *Escherichia coli* O157:H7 outbreak in Ontario, Canada. *Can J Public Health* **103:**e622–e326.

103. Jay MT, Cooley M, Carychao D, Wiscomb GW, Sweitzer RA, Crawford-Miksza L, Farrar JA, Lau DK, O'Connell J, Millington A, Asmundson RV, Atwill ER, Mandrell RE. 2007. *Escherichia coli* O157:H7 in feral swine near spinach fields and cattle, central California coast. *Emerg Infect Dis* **13:**1908–1911.

104. Wacheck S, Fredriksson-Ahomaa M, Konig M, Stolle A, Stephan R. 2010. Wild boars as an important reservoir for foodborne pathogens. *Foodborne Pathog Dis* **7:**307–312.

105. Sanchez S, Martinez R, Garcia A, Vidal D, Blanco J, Blanco M, Blanco JE, Mora A, Herrera-Leon S, Echeita A, Alonso JM, Rey J. 2010. Detection and characterisation of O157:H7 and non-O157 Shiga toxin-producing *Escherichia coli* in wild boars. *Vet Microbiol* **143:**420–423.

106. **Wallace JS, Cheasty T, Jones K.** 1997. Isolation of vero cytotoxin-producing *Escherichia coli* O157 from wild birds. *J Appl Microbiol* **82:**399–404.

107. **Pedersen K, Clark L.** 2007. A review of Shiga toxin *Escherichia coli* and *Salmonella enterica* in cattle and free-ranging birds: potential association and epidemiological links. *Human–Wildlife Conflicts* **1:**68–77.

108. **Nielsen EM, Skov MN, Madsen JJ, Lodal J, Jespersen JB, Baggesen DL.** 2004. Verocytotoxin-producing *Escherichia coli* in wild birds and rodents in close proximity to farms. *Appl Environ Microbiol* **70:**6944–6947.

109. **Hughes LA, Bennett M, Coffey P, Elliott J, Jones TR, Jones RC, Lahuerta-Marin A, McNiffe K, Norman D, Williams NJ, Chantrey J.** 2009. Risk factors for the occurrence of *Escherichia coli* virulence genes *eae*, *stx1* and *stx2* in wild bird populations. *Epidemiol Infect* **137:**1574–1582.

110. **LeJeune J, Homan J, Linz G, Pearl DL.** 2008. Role of the European starling in the transmission of *E. coli* O157 on dairy farms, p 31–34. *In* Timm RM, Madon MB (ed), *Proceedings of the Twenty-Third Vertebrate Pest Conference.* University of California, Davis, CA.

111. **Foster G, Evans J, Knight HI, Smith AW, Gunn GJ, Allison LJ, Synge BA, Pennycott TW.** 2006. Analysis of feces samples collected from a wild-bird garden feeding station in Scotland for the presence of verocytotoxin-producing *Escherichia coli* O157. *Appl Environ Microbiol* **72:**2265–2267.

112. **Kobayashi H, Kanazaki M, Hata E, Kubo M.** 2009. Prevalence and characteristics of *eae*- and *stx*-positive strains of *Escherichia coli* from wild birds in the immediate environment of Tokyo Bay. *Appl Environ Microbiol* **75:**292–295.

113. **Kauffman MD, LeJeune J.** 2011. European starlings (*Sturnus vulgaris*) challenged with *Escherichia coli* O157 can carry and transmit the human pathogen to cattle. *Lett Appl Microbiol* **53:**596–601.

114. **Williams ML, Pearl DL, Lejeune JT.** 2011. Multiple-locus variable-nucleotide tandem repeat subtype analysis implicates European starlings as biological vectors for *Escherichia coli* O157:H7 in Ohio, USA. *J Appl Microbiol* **111:**982–988.

115. **Cernicchiaro N, Pearl DL, McEwen SA, Harpster L, Homan HJ, Linz GM, Lejeune JT.** 2012. Association of wild bird density and farm management factors with the prevalence of *E. coli* O157 in dairy herds in Ohio (2007–2009). *Zoonoses Public Health* **59:**320–329.

116. **Morabito S, Dell'Omo G, Agrimi U, Schmidt H, Karch H, Cheasty T, Caprioli A.** 2001. Detection and characterization of Shiga toxin-producing *Escherichia coli* in feral pigeons. *Vet Microbiol* **82:**275–283.

117. **Moriarty EM, Weaver L, Sinton LW, Gilpin B.** 2012. Survival of *Escherichia coli*, enterococci and *Campylobacter jejuni* in Canada goose faeces on pasture. *Zoonoses Public Health* **59:**490–497.

118. **Kirschner AK, Zechmeister TC, Kavka GG, Beiwl C, Herzig A, Mach RL, Farnleitner AH.** 2004. Integral strategy for evaluation of fecal indicator performance in bird-influenced saline inland waters. *Appl Environ Microbiol* **70:**7396–7403.

119. **Kullas H, Coles M, Rhyan J, Clark L.** 2002. Prevalence of *Escherichia coli* serogroups and human virulence factors in faeces of urban Canada geese (*Branta canadensis*). *Int J Environ Health Res* **12:**153–162.

120. **Murinda SE, Nguyen LT, Nam HM, Almeida RA, Headrick SJ, Oliver SP.** 2004. Detection of sorbitol-negative and sorbitol-positive Shiga toxin-producing *Escherichia coli*, *Listeria monocytogenes*, *Campylobacter jejuni*, and *Salmonella* spp. in dairy farm environmental samples. *Foodborne Pathog Dis* **1:**97–104.

121. **Rutledge ME, Siletzky RM, Gu W, Degernes LA, Moorman CE, DePerno CS, Kathariou S.** 2013. Characterization of *Campylobacter* from resident Canada geese in an urban environment. *J Wildl Dis* **49:**1–9.

122. **Siembieda JL, Miller WA, Byrne BA, Ziccardi MH, Anderson N, Chouicha N, Sandrock CE, Johnson CK.** 2011. Zoonotic pathogens isolated from wild animals and environmental samples at two California wildlife hospitals. *J Am Vet Med Assoc* **238:**773–783.

123. **Bedard J, Gauthier G.** 1986. Assessment of faecal output in geese. *J Appl Ecol* **23:**77–90.

124. **Ankney CD.** 1996. An embarrassment of riches: too many geese. *J Wildl Manaset* **60:**217–223.

125. **Doyle MP, Schoeni JL.** 1987. Isolation of *Escherichia coli* O157:H7 from retail fresh meats and poultry. *Appl Environ Microbiol* **53:**2394–2396.

126. **Doane CA, Pangloli P, Richards HA, Mount JR, Golden DA, Draughon FA.** 2007. Occurrence of *Escherichia coli* O157:H7 in diverse farm environments. *J Food Prot* **70:**6–10.

127. **Schoeni JL, Doyle MP.** 1994. Variable colonization of chickens perorally inoculated with *Escherichia coli* O157:H7 and subsequent contamination of eggs. *Appl Environ Microbiol* **60:**2958–2962.

128. **Gholami-Ahangaran M, Zia-Jahromi N.** 2012. Identification of Shiga toxin and intimin genes in *Escherichia coli* detected from canary (*Serinus canaria domestica*). *Toxicol Ind Health* Oct 9.

129. **Tuyet DT, Yassibanda S, Nguyen Thi PL, Koyenede MR, Gouali M, Bekondi C, Mazzi J,**

Germani Y. 2006. Enteropathogenic *Escherichia coli* o157 in Bangui and N'Goila, Central African Republic: a brief report. *Am J Trop Med Hyg* **75:**513–515.

130. Guyon R, Dorey F, Collobert JF, Foret J, Goubert C, Mariau V, Malas JP. 2000. Detection of Shiga toxin-producing *Escherichia coli* O157 in shellfish (*Crassostrea gigas*). *Sciences des Aliments* **20:**457–466.

131. Surendraraj A, Thampuran N, Joseph TC. 2010. Molecular screening, isolation, and characterization of enterohemorrhagic *Escherichia coli* O157: H7 from retail shrimp. *J Food Prot* **73:**97–103.

132. Sanath Kumar H, Otta SK, Karunasagar I, Karunasagar I. 2001. Detection of Shigatoxigenic *Escherichia coli* (STEC) in fresh seafood and meat marketed in Mangalore, India by PCR. *Lett Appl Microbiol* **33:**334–338.

133. Thampuran N, Surendraraj A, Surendran PK. 2005. Prevalence and characterization of typical and atypical *Escherichia coli* from fish sold at retail in Cochin, India. *J Food Prot* **68:**2208–2211.

134. Gupta B, Ghatak S, Gill JPS. 2013. Incidence and virulence properties of *E. coli* isolated from fresh fish and ready-to-eat fish products. *Vet World* **6:**5–9.

135. Gourmelon M, Montet MP, Lozach S, Le Mennec C, Pommepuy M, Beutin L, Vernozy-Rozand C. 2006. First isolation of Shiga toxin 1d producing *Escherichia coli* variant strains in shellfish from coastal areas in France. *J Appl Microbiol* **100:**85–97.

136. Kilonzo C, Li X, Vivas EJ, Jay-Russell MT, Fernandez KL, Atwill ER. 2013. Fecal shedding of zoonotic food-borne pathogens by wild rodents in a major agricultural region of the central California coast. *Appl Environ Microbiol* **79:**6337–6344.

137. Cizek A, Alexa P, Literak I, Hamrik J, Novak P, Smola J. 1999. Shiga toxin-producing *Escherichia coli* O157 in feedlot cattle and Norwegian rats from a large-scale farm. *Lett Appl Microbiol* **28:**435–439.

138. Blanco Crivelli X, Rumi MV, Carfagnini JC, Degregorio O, Bentancor AB. 2012. Synanthropic rodents as possible reservoirs of shigatoxigenic *Escherichia coli* strains. *Front Cell Infect Microbiol* **2:**134.

139. Cizek A, Literak I, Scheer P. 2000. Survival of *Escherichia coli* O157 in faeces of experimentally infected rats and domestic pigeons. *Lett Appl Microbiol* **31:**349–352.

140. Nkogwe C, Raletobana J, Stewart-Johnson A, Suepaul S, Adesiyun A. 2011. Frequency of detection of *Escherichia coli*, *Salmonella* spp., and *Campylobacter* spp. in the faeces of wild rats (*Rattus* spp.) in Trinidad and Tobago. *Vet Med Int* **2011:**686–923.

141. Mohawk KL, O'Brien AD. 2011. Mouse models of *Escherichia coli* O157:H7 infection and Shiga toxin injection. *J Biomed Biotechnol* **2011:**258185.

142. Wadolkowski EA, Burris JA, O'Brien AD. 1990. Mouse model for colonization and disease caused by enterohemorrhagic *Escherichia coli* O157: H7. *Infect Immun* **58:**2438–2445.

143. Pritchard GC, Williamson S, Carson T, Bailey JR, Warner L, Willshaw G, Cheasty T. 2001. Wild rabbits—a novel vector for verocytotoxigenic *Escherichia coli* O157. *Vet Rec* **149:**567.

144. Scaife HR, Cowan D, Finney J, Kinghorn-Perry SF, Crook B. 2006. Wild rabbits (*Oryctolagus cuniculus*) as potential carriers of verocytotoxin-producing *Escherichia coli*. *Vet Rec* **159:**175–178.

145. Garcia A, Fox JG. 2003. The rabbit as a new reservoir host of enterohemorrhagic *Escherichia coli*. *Emerg Infect Dis* **9:**1592–1597.

146. Panda A, Tatarov I, Melton-Celsa AR, Kolappaswamy K, Kriel EH, Petkov D, Coksaygan T, Livio S, McLeod CG, Nataro JP, O'Brien AD, DeTolla LJ. 2010. *Escherichia coli* O157:H7 infection in Dutch belted and New Zealand white rabbits. *Comp Med* **60:**31–37.

147. Dalle Zotte A, Szendrő Z. 2011. The role of rabbit meat as functional food. *Meat Sci* **88:**319–331.

148. Cook AJ, McCobb E. 2012. Quantifying the shelter rabbit population: an analysis of Massachusetts and Rhode Island animal shelters. *J Appl Anim Welf Sci* **15:**297–312.

149. Compton JA, Baney JA, Donaldson SC, Houser BA, San Julian GJ, Yahner RH, Chmielecki W, Reynolds S, Jayarao BM. 2008. *Salmonella* infections in the common raccoon (*Procyon lotor*) in western Pennsylvania. *J Clin Microbiol* **46:**3084–3086.

150. Lee K, Iwata T, Nakadai A, Kato T, Hayama S, Taniguchi T, Hayashidani H. 2011. Prevalence of *Salmonella*, *Yersinia* and *Campylobacter* spp. in feral raccoons (*Procyon lotor*) and masked palm civets (*Paguma larvata*) in Japan. *Zoonoses Public Health* **58:**424–431.

151. Beltrán-Beck B, García F, Gortázar C. 2012. Raccoons in Europe: disease hazards due to the establishment of an invasive species. *Eur J Wildl Res* **58:**5–15.

152. Shere JA, Bartlett KJ, Kaspar CW. 1998. Longitudinal study of *Escherichia coli* O157:H7 dissemination on four dairy farms in Wisconsin. *Appl Environ Microbiol* **64:**1390–1399.

153. Renter DG, Sargeant JM, Oberst RD, Samadpour M. 2003. Diversity, frequency, and persistence of *Escherichia coli* O157 strains from range cattle environments. *Appl Environ Microbiol* **69:**542–547.

154. Jay-Russell M, Atwill E, Cooley M, Carychao D, Vivas E, Chandler S, Orthmeyer D, Lii X,

Mandrell R. 2010, p 23–27. Occurrence of *Escherichia coli* O157:H7 in wildlife in a major produce production region in California. Poster. 110th Gen Meet Am Soc for Microbiol. American Society for Microbiology, Washington, DC.

155. **Alam MJ, Zurek L.** 2004. Association of *Escherichia coli* O157:H7 with houseflies on a cattle farm. *Appl Environ Microbiol* **70:**7578–7580.

156. **Keen JE, Wittum TE, Dunn JR, Bono JL, Durso LM.** 2006. Shiga-toxigenic *Escherichia coli* O157 in agricultural fair livestock, United States. *Emerg Infect Dis* **12:**780–786.

157. **Xu J, Liu Q, Jing H, Pang B, Yang J, Zhao G, Li H.** 2003. Isolation of *Escherichia coli* O157: H7 from dung beetles *Catharsius molossus*. *Microbiol Immunol* **47:**45–49.

158. **Szalanski A, Owens C, McKay T, Steelman C.** 2004. Detection of *Campylobacter* and *Escherichia coli* O157: H7 from filth flies by polymerase chain reaction. *Med Vet Entomol* **18:**241–246.

159. **Kobayashi M, Sasaki T, Saito N, Tamura K, Suzuki K, Watanabe H, Agui N.** 1999. Houseflies: not simple mechanical vectors of enterohemorrhagic *Escherichia coli* O157:H7. *Am J Trop Med Hyg* **61:**625–629.

160. **Ahmad A, Nagaraja TG, Zurek L.** 2007. Transmission of *Escherichia coli* O157:H7 to cattle by house flies. *Prev Vet Med* **80:**74–81.

161. **Wasala L, Talley JL, Desilva U, Fletcher J, Wayadande A.** 2013. Transfer of *Escherichia coli* O157:H7 to spinach by house flies, *Musca domestica* (Diptera: Muscidae). *Phytopathology* **103:**373–380.

162. **Pickens L, Morgan N, Hartsock J, Smith J.** 1967. Dispersal patterns and populations of the house fly affected by sanitation and weather in rural Maryland. *J Econ Entomol* **60:**1250–1255.

163. **Roopnarine RR, Ammons D, Rampersad J, Adesiyun AA.** 2007. Occurrence and characterization of verocytotoxigenic *Escherichia coli* (VTEC) strains from dairy farms in Trinidad. *Zoonoses Public Health* **54:**78–85.

164. **LeJeune JT, Hancock DD.** 2001. Public health concerns associated with feeding raw meat diets to dogs. *J Am Vet Med Assoc* **219:**1222–1225.

165. **Hogg RA, Holmes JP, Ghebrehewet S, Elders K, Hart J, Whiteside C, Willshaw GA, Cheasty T, Kay A, Lynch K, Pritchard GC.** 2009. Probable zoonotic transmission of verocytotoxigenic *Escherichia coli* O 157 by dogs. *Vet Rec* **164:**304–305.

166. **Beutin L.** 1999. *Escherichia coli* as a pathogen in dogs and cats. *Vet Res* **30:**285–298.

167. **Bentancor A, Rumi MV, Carbonari C, Gerhardt E, Larzabal M, Vilte DA, Pistone-Creydt V, Chinen I, Ibarra C, Cataldi A, Mercado EC.** 2012. Profile of Shiga toxin-producing *Escherichia*

coli strains isolated from dogs and cats and genetic relationships with isolates from cattle, meat and humans. *Vet Microbiol* **156:**336–342.

168. **Anonymous.** 2011. Dog show sparks new Swedish EHEC outbreak, *The Local*, June 16 http://www.thelocal.se/20110616/34384.

169. **Mora A, Lopez C, Dhabi G, Lopez-Beceiro AM, Fidalgo LE, Diaz EA, Martinez-Carrasco C, Mamani R, Herrera A, Blanco JE, Blanco M, Blanco J.** 2012. Seropathotypes, phylogroups, Stx subtypes, and intimin types of wildlife-carried, Shiga toxin-producing *Escherichia coli* strains with the same characteristics as human-pathogenic isolates. *Appl Environ Microbiol* **78:** 2578–2585.

170. **Rumi MV, Irino K, Deza N, Huguet MJ, Bentancor AB.** 2012. First isolation in Argentina of a highly virulent Shiga toxin-producing *Escherichia coli* O145:NM from a domestic cat. *J Infect Dev Ctries* **6:**358–363.

171. **Busch U, Hormansdorfer S, Schranner S, Huber I, Bogner KH, Sing A.** 2007. Enterohemorrhagic *Escherichia coli* excretion by child and her cat. *Emerg Infect Dis* **13:**348–349.

172. **Woods JB, Schmitt CK, Darnell SC, Meysick KC, O'Brien AD.** 2002. Ferrets as a model system for renal disease secondary to intestinal infection with *Escherichia coli* O157:H7 and other Shiga toxin-producing *E. coli*. *J Infect Dis* **185:**550–554.

173. **Zotta E, Lago N, Ochoa F, Repetto HA, Ibarra C.** 2008. Development of an experimental hemolytic uremic syndrome in rats. *Pediatr Nephrol* **23:**559–567.

174. **Garcia A, Bosques CJ, Wishnok JS, Feng Y, Karalius BJ, Butterton JR, Schauer DB, Rogers AB, Fox JG.** 2006. Renal injury is a consistent finding in Dutch belted rabbits experimentally infected with enterohemorrhagic *Escherichia coli*. *J Infect Dis* **193:**1125–1134.

175. **Sueyoshi M, Nakazawa M.** 1994. Experimental infection of young chicks with attaching and effacing *Escherichia coli*. *Infect Immun* **62:**4066–4071.

176. **Tzipori S, Wachsmuth IK, Chapman C, Birden R, Brittingham J, Jackson C, Hogg J.** 1986. The pathogenesis of hemorrhagic colitis caused by *Escherichia coli* O157:H7 in gnotobiotic piglets. *J Infect Dis* **154:**712–716.

177. **Taylor FB Jr, Tesh VL, DeBault L, Li A, Chang AC, Kosanke SD, Pysher TJ, Siegler RL.** 1999. Characterization of the baboon responses to Shiga-like toxin: descriptive study of a new primate model of toxic responses to Stx-1. *Am J Pathol* **154:**1285–1299.

178. **Kang G, Pulimood AB, Koshi R, Hull A, Acheson D, Rajan P, Keusch GT, Mathan VI, Mathan MM.** 2001. A monkey model for

enterohemorrhagic *Escherichia coli* infection. *J Infect Dis* **184**:206–210.

179. **Mohawk KL, Melton-Celsa AR, Zangari T, Carroll EE, O'Brien AD.** 2010. Pathogenesis of *Escherichia coli* O157:H7 strain 86-24 following oral infection of BALB/c mice with an intact commensal flora. *Microb Pathog* **48**:131–142.

180. **Mayer CL, Leibowitz CS, Kurosawa S, Stearns-Kurosawa DJ.** 2012. Shiga toxins and the pathophysiology of hemolytic uremic syndrome in humans and animals. *Toxins* (Basel) **4**:1261–1287.

181. **Melton-Celsa AR, O'Brien AD.** 2003. Animal models for STEC-mediated disease. *Methods Mol Med* **73**:291–305.

182. **Beutin L, Martin A.** 2012. Outbreak of Shiga toxin-producing *Escherichia coli* (STEC) O104:H4 infection in Germany causes a paradigm shift with regard to human pathogenicity of STEC strains. *J Food Prot* **75**:408–418.

183. **Karch H, Tarr PI, Bielaszewska M.** 2005. Enterohaemorrhagic *Escherichia coli* in human medicine. *Int J Med Microbiol* **295**:405–418.

184. **Snedeker KG, Shaw DJ, Locking ME, Prescott RJ.** 2009. Primary and secondary cases in *Escherichia coli* O157 outbreaks: a statistical analysis. *BMC Infect Dis* **9**:144.

185. **Mead PS, Griffin PM.** 1998. *Escherichia coli* O157:H7. *Lancet* **352**:1207–1212.

186. **Wilson JB, Clarke RC, Renwick SA, Rahn K, Johnson RP, Karmali MA, Lior H, Alves D, Gyles CL, Sandhu KS, McEwen SA, Spika JS.** 1996. Vero cytotoxigenic *Escherichia coli* infection in dairy farm families. *J Infect Dis* **174**:1021–1027.

187. **Wilson JB, Johnson RP, Clarke RC, Rahn K, Renwick SA, Alves D, Karmali MA, Michel P,** Orrbine E, Spika JS. 1997. Canadian perspectives on verocytotoxin-producing *Escherichia coli* infection. *J Food Prot* **60**:1451–1453.

188. **Miliwebsky E, Deza N, Chinen I, Martinez Espinosa E, Gomez D, Pedroni E, Caprile L, Bashckier A, Manfredi E, Leotta G.** 2007. Prolonged fecal shedding of Shiga toxin-producing *Escherichia coli* among children attending day-care centers in Argentina. *Rev Argent Microbiol* **39**:90–92.

189. **Rangel JM, Sparling PH, Crowe C, Griffin PM, Swerdlow DL.** 2005. Epidemiology of *Escherichia coli* O157:H7 outbreaks, United States, 198–2002. *Emerg Infect Dis* **11**:603–609.

190. **Pennington H.** 2010. *Escherichia coli* O157. *Lancet* **376**:1428–1435.

191. **Mohammed Hamzah A, Mohammed Hussein A, Mahmoud Khalef J.** 2013. Isolation of *Escherichia coli* O157: H7 strain from fecal samples of zoo animal. *Scientific World Journal* **2013**.

192. **Featherstone CA, Foster AP, Chappell SA, Carson T, Pritchard GC.** 2011. Verocytotoxigenic *Escherichia coli* O157 in camelids. *Vet Rec* **168**:194–195.

193. **Dipineto L, Gargiulo A, Russo TP, De Luca Bossa LM, Borrelli L, d'Ovidio D, Sensale M, Menna LF, Fioretti A.** 2010. Survey of *Escherichia coli* O157 in captive frogs. *J Wildl Dis* **46**:944–946.

194. **Johnson M.** 2005. Deer eating away at forests nationwide. *The Associated Press*, January 18, 2005.

195. **United States Department of Agriculture.** 2011. *Feral swine: damage and disease threat.* http://www.aphis.usda.gov/publications/wildlife_damage/content/printable_version/feral_swine.pdf.

Shiga Toxin-Producing *Escherichia coli* in Fresh Produce: A Food Safety Dilemma

12

PETER FENG[1]

INTRODUCTION

The worldwide trends for healthier lifestyles to reduce obesity and other complications arising from unhealthy diets have greatly increased the consumption of fresh fruits and vegetables. This increased demand, coupled with ever busier consumer lifestyles, also stimulated the growth of a "convenience" food industry and popularized the concept of bagged salad vegetables and fruits. It has been estimated that several millions of bags of fresh produce are sold daily in the United States. Bagged produce, also referred to as "fresh cut" or "precut," is often regarded as ready-to-eat (RTE) and consumed without further intervention steps. However, because produce is predominantly cultivated in soil in open fields, it is susceptible to contamination and can contain high levels of complex microbial populations, occasionally including bacterial pathogens. As a result, increases in fresh produce demand and consumption coupled with changes in production practices have also contributed to increases in incidents of foodborne illness. In the United States, about 0.7% of the infections in the 1970s were attributed to fresh produce, but this increased to 6% in the 1990s (1). Since "fresh cut" products are often mass produced, broadly distributed, and marketed worldwide, a single pathogen contamination event can have broadly impacting consequences, and several large, produce-related outbreaks have

[1]Division of Microbiology, U.S. Food and Drug Administration, College Park, MD 20740-3835.

Enterohemorrhagic Escherichia coli *and Other Shiga Toxin-Producing* E. coli
Edited by Vanessa Sperandio and Carolyn J. Hovde
© 2015 American Society for Microbiology, Washington, DC
doi:10.1128/microbiolspec.EHEC-0010-2013

occurred in many countries (2, 3). In 2006, a large multistate outbreak in the United States that infected more than 200 persons was traced to bagged spinach contaminated with *Escherichia coli* O157:H7 (4). Several months later, another O157:H7 outbreak in a fast-food restaurant chain had initially implicated green onions but appeared to have been due to bagged lettuce. At about the same time, bagged lettuce was implicated in another O157:H7 outbreak that affected three states (5). Likewise, increased consumption of sprouts caused several outbreaks of *Salmonella* sp., *E. coli* O157:H7, and other Shigatoxin-producing *E. coli* (STEC) strains. STEC serotype O26:H11 strains caused an outbreak with alfalfa sprouts and, more recently, with clover sprouts, and the large outbreak of O104:H4 in 2011 in the European Union also implicated the consumption of sprouts (6). These large produce-related outbreak incidents worldwide have greatly raised concerns about the safety of fresh produce and about the microbiological and sanitary quality of fresh produce.

MICROBIOLOGICAL QUALITY OF FRESH PRODUCE

Total Microbial Populations

The basic fresh produce production practices have remained essentially unchanged as most produce is still cultivated in soil in a field and irrigated with available sources of water. This constant exposure of produce plants to the environment makes them susceptible to contamination, which can come from many sources, including soil, water, compost, animal wastes, and other environmental sources (http://www.fda.gov/Food/RecallsOutbreaks Emergencies/Outbreaks/ucm235477.htm).

As a result, it is not unusual to find high microbial counts on fresh produce in the field. Several studies examined the microbial content of preharvest produce plants in the field and observed that total bacterial counts

ranged from 10^4 to 10^7 CFU/g (7–9). Bagged, "fresh cut" produce undergoes processing and is washed multiple times before and after shredding and before bagging and, therefore, would be expected to have lower microbial contents. However, it was surprising that these finished products can also contain high levels of bacteria. Microbiological surveys from several countries showed that total bacterial counts of bagged produce were similar to levels observed in plants in the field and ranged widely from 10^3 to 10^7 CFU/g (10). Furthermore, data from various countries showed that it was not unusual to have mean total microbial counts of 10^7 CFU/g or higher in both bagged salads and sprouts (11–14). Recent FDA survey studies of bagged produce in the United States showed similar results and wide ranges, and in some cases, total counts as high as 10^8 CFU/g were observed in some bags of spinach (15, 16). Analysis of these microflora populations showed that the most frequently identified bacteria were *Bacillus subtilis*, *Pseudomonas fluorescens*, *Pantoea agglomerans*, and *Sphingomonas paucimobilis* (16).

Coliform and Generic *E. coli* Populations

Coliform and *E. coli* have been used as indirect indicators of fecal contamination for more than 100 years, but there seems to be little correlation between the presence or the levels of indicator bacteria in food with the presence of pathogens. Still, these indicators continue to be of use in monitoring general sanitary conditions. For instance, some produce industries test their products to establish a baseline indicator level. In subsequent testing, the detection of any large deviations from baseline may be indicative of abnormal lots or processing conditions that warrant investigation.

Coliform and *E. coli* are ubiquitous in the cultivation environments and, therefore, are often found on produce plants. Surveys of produce in the field showed that coliform and generic *E. coli* levels often ranged from 10 to 10^4 CFU/g (7–9), and analogous to the findings with total bacterial counts, high levels of these indicator bacteria can persist in bagged finished products as well.

There are no microbial limits for coliforms and *E. coli* for bagged produce in the United States, but guidelines and limits exist in other countries. The Brazilian standard for minimally processed RTE vegetables has a fecal coliform limit of 100 CFU/g (13). The Public Health Laboratory Service of the United Kingdom has established microbiological quality guidelines for RTE foods, which include bagged produce, and has set an *E. coli* limit of <20/g as satisfactory, 20 to <100/g as acceptable, and ≥100/g as unsatisfactory (17). Indicator bacteria are often enumerated by using the most probable number (MPN) method, and surveys from various countries showed that, like total counts, coliform levels in bagged salad varied greatly, ranging from undetected (<3.0 MPN/g) to 10^3 or 10^4 MPN/g (10, 15, 16). Similarly, generic *E. coli* may or may not be present, but, if found, the prevalence rate and the levels also varied greatly (10, 12, 15, 16). For example, surveys of lettuce samples at restaurants and university cafeterias in Spain showed that the prevalence rate of *E. coli* varied from 6.6% to ~26% (10, 18). A more startling finding is the study from Brazil that showed that 73% of the minimally processed vegetable salads examined had exceeded the fecal coliform limit of 100 CFU/g (13). Likewise, *E. coli* can be prevalent in sprouts and at high levels, as a study from Spain showed that 40% of the sprout samples tested positive and, in several, the counts were >10^3 MPN/g (11). Two large surveys in the United Kingdom tested approximately 3,000 samples of organic and conventional salad vegetables and showed that only 0.5% of the samples exceeded the *E. coli* limit of >100 CFU/g and, therefore, were unsatisfactory (19, 20). Two surveys of bagged lettuce and spinach in the United States showed that *E. coli* was not detected in one study (16), and in the other, 16% of the samples had *E. coli*, but all were at levels of <10 MPN/g (15). So, with a few exceptions

where high levels were reported, *E. coli* levels in most bagged produce samples seem to be low and within acceptable limits (17). These findings show that coliforms are too prevalent in produce and their levels are too variable to be able to generate a useful baseline. However, *E. coli* does not appear to be part of normal flora in fresh produce, and because its presence and levels are usually low, it may be feasible to establish an *E. coli* baseline, where spikes in *E. coli* levels detected may be indicative of unsanitary or abnormal processing conditions.

These surveys showed that total bacteria and coliforms are prevalent in produce, and interestingly, some studies showed little differences in counts between organic versus conventional produce (9) or between imported and domestic produce (8). But one observation that is consistent is that the levels of total and coliform counts in produce vary greatly and can be unpredictable. One study showed that even seemingly identical products, of the same brand and "use by" dates and tested on the same day, can have as much as 2 to 3 log differences in counts (15). Such discrepancies have also been noted in RTE produce in other countries (10, 11, 13). Whether these large variations are due to the produce types, the complexity of produce microflora, seasonal or regional variations, or perhaps to inconsistencies or variations in processing parameters remains uncertain.

PRESENCE OF PATHOGENIC *E. coli* IN FRESH PRODUCE

Considering the huge quantities of fresh produce that are harvested and consumed daily, the frequency of pathogen contamination in produce is very low. In response to increasing concerns about food-borne outbreaks associated with the consumption of fresh produce, the Agricultural Marketing Service of the U.S. Department of Agriculture initiated the Microbiological Data Program (MDP) in 2001 to conduct surveillance of fresh produce samples collected from wholesale distribution centers across the country. On the average, about 10,000 to 15,000 produce samples of many varieties were tested yearly for the presence of *Salmonella* sp., enterotoxigenic *E. coli* (ETEC), *E. coli* serotype O157:H7, and other STEC strains. MDP analyses showed that many of these pathogens are found in various types of fresh produce, and the program provided one of the largest, publicly available databases on the presence of pathogens in fresh produce. Unfortunately, the program was affected by budget cuts and terminated in 2012. The yearly MDP reports are available at http://www.ams.usda.gov/AMSv1.0/mdp.

Enterotoxigenic *E. coli*

ETEC is commonly known as the causative agent of traveler's diarrhea; however, it is also an important diarrheal pathogen in infants. The two trait virulence factors of ETEC are the plasmid-encoded, heat-labile (LT) and heat-stable (ST) enterotoxins. The two serologically distinct LT types are designated LT-I and LT-II, but the latter is produced mostly by animal isolates and has not been associated with illness. There are also two distinct ST types; STa is produced by both human and porcine ETEC strains, but STb production is limited mainly to porcine strains (21).

ETEC infections are most often caused by the consumption of contaminated food and water, and so ETEC strains can be found in various foods, including produce and produce used in street-vendor food in Mexico (22). MDP analyses of produce showed that ETEC was isolated almost every year from samples including lettuce, parsley, cilantro, alfalfa sprouts, and spinach. Typically, only a handful of strains are isolated a year, but in 2009, 11 ETEC isolates were found and 4 came from cilantro and spinach samples, respectively. Considering that about 2,200 samples of each were tested yearly, the estimated ETEC prevalence rate in these products is ~0.18% (23). Most of these ETEC isolates had STb, which is associated with porcine strains, but a few

also had STa or LT. Although not surprising to find ETEC in fresh produce, ETEC has a fairly high infectious dose, which, based on volunteer feeding studies, has been estimated to be 10^8 to 10^{10} cells. As a result, quantifying the levels of ETEC present in produce would be more useful than the presence or absence data in assessing the health risks of ETEC in fresh produce.

In the MDP produce testing scheme, a multiplex PCR assay was used to simultaneously screen produce samples for the presence of ETEC and STEC (24). Interestingly, this assay detected some *E. coli* strains that carried both ETEC and STEC virulence factors. Although many of these strains were only partially identified and characterized, two strains from the 2004 MDP analysis were studied in detail. These two *E. coli* strains, found to carry both Stx1 and ST toxin genes, had an untypeable O antigen and H52 flagellar antigen and were isolated from fresh cilantro and cantaloupe, respectively. Analysis by pulsed-field gel electrophoresis showed that the XbaI profiles of both strains were similar and did not resemble those of selected Stx1-producing STEC or ST-producing ETEC strains examined, but they shared >90% XbaI profile similarity to two clinical Ont:H52 strains that had identical phenotypes and traits (24). Furthermore, multilocus sequence typing of all four strains showed that they had sequence type (ST) 274, which did not fall into any of the defined STEC clonal groups and, therefore, appeared to be a unique clone. The Stx1 and ST genes reside on phage and plasmid, respectively, and can be transferred, so it is uncertain if these Ont:H52 strains in produce were ETEC that was infected by the Stx_1 phage or, conversely, a STEC that acquired the ST-encoding plasmid.

Shiga Toxin-Producing *E. coli*

STEC is characterized by the production of Stx, and there are two main types, designated Stx1 and Stx2. Within each are many subtypes, and currently, there are three known Stx1

(Stx1a, Stx1c, and Stx1d) and seven known Stx2 (Stx2a, Stx2b, Stx2c, Stx2d, Stx2e, Stx2f, and Stx2g) subtypes (25). There are estimated to be 300 to 400 known STEC serotypes that can produce any of the Stx or combination of Stx subtypes, but some subtypes seem to be found mostly in environmental or animal strains and have not affected humans (26); thus, not all STEC strains appear to cause illness. STEC has been found in various forms of wildlife (27) and in environmental sources like water and soil and is also common in farm or agricultural areas (http://www.fda.gov/downloads/Food/FoodSafety/Foodborne Illness/UCM235923.pdf).

STEC strains can also be found in meats and other foods (27–29), and fresh produce is no exception. During 2004 to 2009, European Union member states conducted a microbiological survey of produce and found that of the ~6,000 samples tested, 11 (0.18%) had STEC (http://www.efsa.europa.eu/en/supporting/doc/166e.pdf). Similarly, 10 years of MDP analyses showed that STEC strains were isolated from various types of produce in the United States, especially, spinach, lettuce, and cilantro (http://www.ams.usda.gov/AMSv1.0/mdp). From MDP statistics on the number of STEC isolations per year in relation to the ~2,200 samples of each product tested yearly, it is estimated that STEC was present in 0.5 to 0.6% of the spinach, 0.3 to 0.5% of the cilantro, and 0.04 to 0.18% of the lettuce samples. These estimates, however, may not reflect the overall trend. For instance, the prevalence rate obtained for STEC in lettuce was based on MDP data from the past few years, which had low isolation rates, but there had been higher numbers of STEC isolations from lettuce in other years. These observations suggest that STEC prevalence may vary from year to year and depends on many factors, including, perhaps, regional and seasonal variations.

Spinach was included in the MDP testing scheme in 2008 in response to the 2006 spinach outbreak with *E. coli* O157:H7. Since then, this product has accounted for most of the STEC isolations, and the prevalence rate has remained fairly steady. For example, from 2009 to 2011, 11 to 14 STEC strains were isolated from spinach yearly, giving a prevalence rate of ~0.5 to 0.6%. But in 2012, there were 32 STEC isolations, 21 of which came from spinach (0.95%). This fairly consistent prevalence of STEC strains in spinach suggests that, perhaps, there may be some correlation between STEC and spinach plants or with spinach cultivation practices.

All STEC strains isolated by MDP were serotyped by the *E. coli* Reference Laboratory at Penn State University. However, more than 50% of the isolates could only be partially serotyped, when either the O or the H antigens or both could not be identified. Among those that were serotyped, there were strains from diverse serogroups, most of which were unremarkable, with no history of having caused human illness (Table 1). But there were strains from recognized pathogenic serotypes, including O157:H7, O26:H11, O121:H19, and O113:H21, as well as other serotypes like O165:H25 and O91:H21 (Table 2) (30) that, historically, have been implicated with severe human illness (31, 32).

STEC—CHALLENGES IN PRODUCE TESTING AND IN RISK ANALYSIS

Testing for the presence of any pathogen in fresh produce is challenging. The heterogeneous distribution of pathogen contamination and the many types and varieties of fresh produce present problems to sampling and sample processing procedures. Furthermore, fresh produce can contain very high levels of microflora, which can mask or overwhelm the sporadic presence of often very low levels of pathogens. To complicate matters further, produce can contain inhibitors that interfere with assays; hence, fresh produce samples often have to be incubated in broth media containing antibiotics or other inhibitory substances to "enrich" for the target bacteria and to suppress the growth of background flora. Typical enrichment steps take 1 to 2 days, so

TABLE 1 Selected STEC strains and serotypes isolated from various produce commodities[a]

Product	No.	Serotypes[b]
Cantaloupe	3	Ont[c]:H11; Ont:H52; O88:H38
Cilantro	18	Ont:H16, H31, H49; O1:H+; O8:H16, H28; O113:H5; O139:H1; O153:H21; O168:H8; others
Coriander	3	Ont:H7; O2:H25; O119:H4
Hot peppers	3	O8:H9; O24:H11; O180:H14
Lettuce	28	Ont:Hnt, O6:H49; O8:H28; Ont:H2,H8; O136:H16; O143:H34; O163:H19; O168:H-, H8; O181:H49; others
Parsley	1	Ont:H38
Spinach	70	O8:H–, H28; O11:H15; O21:Hnt; O76:H+; O88:Hnt; O98:H36; O107:Hnt; O113:H36; O130:H11; O159:H19; O181:H49; many Ont with various H types
Sprouts (alfalfa)	3	Ont:H28, Hnt; O36:H14
Tomatoes	3	Ont:Hnt
Total	132	

[a]Table modified from reference 30.
[b]Excludes pathogenic serotypes listed in Table 2.
[c]nt, not typeable.

these prerequisite produce sample preparation steps before testing are media-, time-, and labor-intensive.

Once the produce samples are media-enriched, they are often screened by antibody-based or PCR or real-time PCR (qPCR) assays that are specific for particular targets. Since STEC strains are characterized solely by the production of Stx, anti-Stx assays or the use of *stx*-specific PCR or qPCR assays is effective in screening produce enrichments. Although these assays are adequate to determine the presence of STEC in produce, such data alone are inadequate for making risk-assessment decisions.

To determine whether a bacterium poses health risk is difficult as it depends on many variables such as virulence factors, infectious dosage, the pathogenicity of the organism, etc., but it also depends on the health conditions

and the susceptibilities of the human host. For instance, even generic *E. coli* strains are deemed as opportunistic pathogens as they can cause infections in immunocompromised hosts and, therefore, cannot be regarded as being "risk free." Risk assessment for STEC is much more difficult due to the complexity of STEC pathogenesis, plus the group comprises a large diversity of strains, from several hundred serotypes, that can produce any of the

TABLE 2 Serotype and pathotype of selected produce STEC strains[a]

Commodity	Serotype[b]	Pathotype(s)[c]
Cherry tomato	O8:H19[d]	stx1, stx2, ehxA, subA/B
Cilantro	O20:H19[d]	stx1, stx2, saa, ehxA
	O26:H11	stx1, eae, ehxA
	O165:H25	stx1, stx2, ehxA, eae
	Untyped	stx1, eae, ehxA
Lettuce	O2:H27[e]	stx2, ehxA
	O121:H19	stx2, ehxA, eae
	O157:H7	stx2, eae
	O157:H7	stx2, eae, ehxA
	O163:H19[d]	stx2, ehxA, subA/B
	O165:H25	stx1, stx2, ehxA, eae
	O174:H21[d]	stx2
Spinach	**O157:H7**[f]	stx2, eae, ehxA
	O8:H19[d]	stx2, ehxA
	O8:H19[d]	stx2, ehxA
	O26:H11	stx1, eae, ehxA
	O82:H8[e]	stx2, ehxA, saa
	O82:H8[e]	stx2, ehxA, saa
	O91:H21	stx2, ehxA, saa
	O98:H36[g]	stx1, eae, ehxA
	O113:H21	stx2, ehxA, saa, subAB
	O113:H21	stx2, ehxA, saa, subAB
	O113:H21	stx2, ehxA, saa, subAB
	O116:H21[h]	stx2, ehxA, saa, subAB
	O157:H7	stx2, eae, ehxA
	O157:H7	stx2, eae, ehxA
	O174:H2[d]	stx2, ehxA, saa
	O174:H21[d]	stx1, stx2, ehxA, saa, subAB

[a]Table modified from reference 30.
[b]Known pathogenic serotypes are shown in bold type.
[c]stx1, Shiga toxin 1; stx2, Shiga toxin 2; eae, intimin; ehxA, enterohemolysin; saa, STEC agglutinating adhesin; subAB, subtilase cytotoxin.
[d]Serotype reported to have history of causing HUS (31).
[e]Serotype reported to have history of causing other illnesses (31).
[f]From the spinach outbreak of 2006. Included as reference.
[g]Serotype has no history of causing illness (31), but has eae.
[h]Serotype reported to have history of causing HUS (32).

many Stx subtypes or combinations of subtypes. Though some of these subtypes such as Stx1e do not seem to affect humans, and infections by STEC strains with Stx1c tend to range from asymptomatic to mild diarrhea (33), nevertheless, they may only be regarded as low or minimal risk, but not without risk. On the other hand, infections by other STEC strains can cause severe diseases, such as hemorrhagic colitis (HC) that can lead to life-threatening complications like hemolytic-uremic syndrome (HUS). In these instances, the production of Stx alone is deemed to be insufficient, and an adherence factor that enables the pathogen to attach to epithelial cells is also needed for severe disease to occur. Other general information on pathogenic *E. coli* can be found in a consumer-oriented report titled "*E. coli*: Good, Bad & Deadly" published by the American Academy of Microbiology and is available at http://academy.asm.org/images/stories/documents/EColi.pdf. The most notable adherence factor among pathogenic STEC is the intimin protein, encoded by the *eae* gene that resides on the locus for enterocyte effacement pathogenicity islands. The presence of *eae* and *stx2* has been found to be a good predictor that the STEC strain may cause HC or HUS (34). Hence, the term enterohemorrhagic *E. coli* has been used by some to designate a subset of STEC that comprises pathogenic strains, and most of these have *eae*. Among EHEC strains, serotype O157:H7 is the prototypic strain, but others in the serogroup O26, O111, O103, O145, to name a few, have also caused severe human illnesses. Also, other EHEC strains, such as strains in the serotype O113:H21, O91:H21, O104:H4, etc., do not have *eae* but have caused HUS (35, 36), so these pathogens appear to have other means of attachment. For example, the O104:H4 strain that caused the large HUS outbreak in the European Union in 2011 did not have *eae*, but it was an enteroaggregative *E. coli* strain, and its ability to attach and aggregate on epithelial cells, coupled with Stx2 production, is postulated to have caused the severity of infections.

The mechanisms of attachment used by other *eae*-negative EHEC strains are less certain. Analysis of the *eae*-negative O113:H21 strain that caused an outbreak of HUS in Australia identified the plasmid-borne *saa* gene that encodes for STEC agglutinating adhesin (Saa) (36). Saa was determined to be an adherence protein as a plasmid-cured, *saa*-negative O113:H21 mutant showed reduced adherence compared to wild type, and purified Saa protein enhanced adherence to HEp-2 cells (37). However, studies that examined the distribution of the *saa* genes found that over half of the STEC strains isolated from healthy cattle were positive for *saa* (38), and although it was also found in some clinical STEC strains, there was no significant correlation between the presence of *saa* and HUS (39).

Among the other factors that are often associated with pathogenic STEC strains are the plasmid-encoded enterohemolysin and subtilase cytotoxin. The *ehxA* gene that encodes for enterohemolysin has been found to be very common among STEC strains isolated from all sources, including clinical, environmental, and foods. The prevalence rate can range from 30% in STEC isolated from wildlife to as many as 70% of the STEC strains isolated from ground meats (40). However, the prevalence of *ehxA* is not limited to STEC, as analysis of ~300 environmental isolates of generic *E. coli* from surface waters showed that almost all carried and expressed *ehxA* and none were STEC (41). Moreover, the role of enterohemolysin in STEC pathogenesis is uncertain. A study of ~300 clinical STEC isolates showed that 77% had the *ehxA* gene, but its presence could not be correlated with the occurrence of HUS or bloody diarrhea (34). In addition, the sorbitol-fermenting variants of O157:H7 (SFO157) do not express *ehxA* but are highly pathogenic and have caused many outbreaks of HUS in Europe (42). Despite the uncertainty of its role in pathogenesis, the expression of enterohemolysin has become a useful marker in STEC isolation procedures (see below).

The subtilase cytotoxin encoded by the *subAB* gene and produced predominantly by *eae*-negative STEC strains has been determined to be a potent toxin that is even more cyto- toxic to Vero cells than Stx (43). The *subAB* gene has been found in ~50% of the STEC strains isolated from ground meats (40). An- other study showed that 72% and 86% of the *eae*-negative STEC strains from patients with diarrhea and healthy sheep, respectively, had *subAB* (44), so it is very prevalent among STEC strains. However, other *eae*-negative STEC strains, like O91:H21 and O22:H8, have been implicated in severe illnesses, but ground beef isolates of these serotypes did not have *subAB* (40). So the role of this toxin in the pathogenesis of *eae*-negative STEC strains also remains elusive.

A recent study characterized ~130 STEC strains isolated from a variety of fresh produce samples over a period of about 10 years. The isolates were tested by PCR for the presence of *eae*, *ehxA*, *saa*, and *subAB* genes (30). The study showed that *eae* was present in about 8% of the isolates, and most of these were recognized pathogenic serotypes (Table 2) (30). The other genes were even more prev- alent among STEC strains in produce, as ~60% of the STEC strains had *ehxA*, and 35% and 32% of the strains had the *saa* and *subAB* genes, respectively (30). Thus, considering the complexities of STEC pathogenesis, the variety of STEC serotypes that exist, and the wide distribution of various virulence and putative virulence genes among STEC strains in produce, it is clear that much more in- formation beyond the production of Stx or the presence of *stx* is needed for STEC risk assessment.

Fresh produce has a short shelf life, so it is crucial to obtain relevant risk analysis data as quickly as possible. One such testing strategy is to use multiplex PCR assays to simulta- neously screen for a combination of virulence and putative virulence factor genes plus other relevant markers directly in produce enrich- ments. Others use sequential strategies, where samples positive for virulence factor genes are

followed up by other trait-specific assays. The key STEC virulence genes tested by almost everyone worldwide are *stx* and *eae*, and if the samples are positive, they are often tested with assays specific for certain serotypes. While these strategies may seem straightfor- ward, the selection of target serotypes is not. Studies showed that there are regional variations in the STEC serotypes that cause infections (45); hence, the follow-up serotype- specific assays may vary geographically. For example, serotype O157:H7 continues to be regarded as the most important EHEC path- ogen, and so it is almost always included in all screening strategies. But other non-O157 STEC strains are increasingly being impli- cated in food-borne infections worldwide. In the United States, strains from serotypes O26, O111, O121, O103, O145, and O45 have been isolated most frequently from clinical infec- tions, and so these six, commonly referred to as the "big 6," have emerged as being impor- tant and, recently, declared as adulterants in meats. These strains are equally important in fresh produce as strains of O111, O121, and O145 have caused outbreaks in lettuce and O26 strains in alfalfa and clover sprouts. However, focusing on the "big 6" may be in- sufficient in produce testing, as other known pathogenic STEC serotypes have been isolated from fresh produce (30). Hence, for risk as- sessment of STEC in produce, it is essential that samples found to be positive for key vir- ulence factors by initial screening be further tested to obtain additional relevant data to determine health risk.

While multiplex assays can generate many data points in one assay, there are drawbacks to using this approach to screen complex, mixed bacterial population samples, as all the gene targets detected may not be within the same bacteria. For example, a survey for *stx*, *eae*, and the "big 6" serotype-specific genes in another flora-complex food like ground meats showed that it was not uncommon to find bacterial strains that carried each of these gene targets independently (40). But when they were all present in the same sample, it gave positive multiplex results for all three targets and, thereby, the misleading impression that a pathogen with those attributes was present in the sample (40). Produce can contain equally complex, mixed microbial populations; hence, it is imperative that samples initially screened and found to be positive are plated onto selective or differential media to isolate the STEC strain and to verify that all the relevant genes targeted are within the same organism.

Without a doubt, isolation and confirmation of STEC strains from produce enrichments are time- and labor-intensive. Fortunately, these steps are actually easier and straightforward for O157:H7 strains due to the unique phenotypes expressed by this pathogen. Unlike typical *E. coli*, O157:H7 strains do not ferment sorbitol or express β-glucuronidase activity, so both traits are used extensively in plating agar media to differentiate O157:H7 colonies from others. Presence of the O157 and the H7 antigens has also eased identification, as simple latex agglutination tests can be used to identify O157:H7 strains quickly (21). An example of a method that uses these traits to identify O157:H7 is outlined in the FDA Bacteriological Analytical Manual (http://www.fda.gov/Food/FoodScienceResearch/LaboratoryMethods/ucm2006949.htm).

In contrast, isolation and confirmation of the non-O157 STEC strains are much more complex, laborious, and time-consuming and, undoubtedly, the most problematic step in STEC testing. As mentioned, there are ~300 STEC serotypes that comprise diverse strains and serotypes, and not all appear to affect humans. Those strains that are known pathogens may share common virulence traits but do not uniformly exhibit unique phenotypes that can be used in differentiation and isolation. As a result, there are no methods that can select and isolate all the diverse strains existing within the pathogenic STEC group. Many have used various chromogenic substrates and combinations of phenotypes to develop plating media for differentiation of STEC strains. However, these media can be

costly, are not very inclusive, and are useful only for selected serotypes and strains. Also, the high levels and the complex microflora present in produce samples can often mask or mimic pathogenic colonies on these media, thereby interfering with differentiation of STEC colonies. Similarly, to facilitate the isolation of selected serotypes from enrichment broths, specific antibodies have been used in immunomagnetic separation (IMS) methods to capture specific organisms. But the capture efficiency of IMS can vary greatly, depending on the antibodies used and on the type of produce samples being tested. In addition, the antibodies used in IMS often target serogroups rather than serotypes. For example, an anti-O157 IMS method is not specific for O157:H7, as it will capture O157 strains regardless of H type. There are many O157 strains that have H antigens other than H7, and they are not STEC nor are they pathogenic, but can be found in foods. As a result of these limitations and logistical problems, isolating a STEC strain from produce enrichments often entails the laborious process of picking and pooling numbers of colonies from the plate, screening it for Stx or *stx* by antibody or PCR assays, and repeating the process until a pure STEC strain is isolated for serotyping and virulence testing. This process is not only labor-intensive and costly, it is time-consuming and does not provide timely data for making risk-assessment decisions.

The availability of good differential plating media would greatly simplify the process of STEC isolation from produce enrichments. As mentioned, even though its role in pathogenesis is uncertain, the majority of STEC strains produce enterohemolysin. Strains of O157:H7 almost always carry *ehxA* and express enterohemolysin, as do many of the other pathogenic STEC strains. Enterohemolysin activity can be visualized as a faint zone of hemolysis around the colonies on washed blood agar (WBA) medium, which is made up of tryptose blood agar base supplemented with 10 mM of $CaCl_2$ and 5% of defibrinated sheep blood (46). The WBA medium was later modified to become the WBMA medium by the addition of 0.5 μg/ml of mitomycin C, which has been shown to induce and enhance the visualization of enterohemolytic activity and also increased the detection rate of STEC strains (47). A similar medium called SHIBAM, which uses heart infusion agar base, is described in the FDA Bacteriological Analytical Manual. Although not all STEC strains express *ehxA*, the use of WBMA or SHIBAM has eased the recognition of STEC strains by the zones of hemolysis around the colonies and, thereby, facilitated the selection and isolation of STEC strains from complex, mixed-flora food enrichments like produce samples.

Once a STEC strain in produce has been isolated, characterized, and serotyped, subjectivity remains in making risk-assessment decisions. To illustrate these uncertainties, one study characterized ~130 STEC strains isolated from produce samples over a 10-year period (30). In some cases, risk-assessment decisions were straightforward. For example, some produce samples had O157:H7 strains, and these, without a doubt, are of serious safety concern. A few atypical O157:H7 strains were also isolated from produce, including strains that did not produce enterohemolysin or Stx. Both factors are encoded by mobile genetic elements that can be lost or transferred. The role of enterohemolysin in pathogenesis is uncertain and the SFO157 strains do not express enterohemolysin but are highly virulent, so an O157:H7 strain that is enterohemolysin negative should not be regarded lightly. The Stx encoding phage can be lost during culture and isolation, so the possibility that a toxigenic strain exists in the product cannot be ruled out. Hence, in both instances, the presence of these atypical O157:H7 strains should also be regarded as being a health risk. Similarly, if a STEC strain is found to have *eae* and of a known pathogenic serotype, it is also a serious health risk; a handful of STEC strains in produce fit these criteria (Table 2). However, even if the serotype is unknown, a STEC strain that had *eae* should

be regarded as a public health concern. Of the ~130 strains in produce tested in that study, 11 strains (~8%) had *eae*, and except for a strain of O98:H36 serotype and another that was untyped, all the other *eae*-bearing strains belonged to well-recognized pathogenic serotypes and, therefore, were of safety concern (Table 2) (30).

There are EHEC strains that do not have *eae* but have caused severe illness, and some of these, like strains of O113:H21 and O91:H21 serotype, have been found by MDP in fresh spinach samples. Characterization and comparison of the O113:H21 strains from spinach to strains that caused infections in Australia showed that although the produce strains had a different sequence type (ST223), they had identical traits and phenotypes as the pathogenic strains (48). A more detailed study used a microarray to analyze 41 pathogenic *E. coli* markers in 65 O113:H21 strains isolated from food, spinach and environmental sources worldwide and, showed that they could not be distinguished from the O113:H21 strains that caused HUS and therefore, are most likely pathogenic as well (49). Produce samples also contained several strains that belonged to serotypes like O2:H27, O82:H8, O116:H21, O174:H21, etc., that have been reported to have been isolated from human infections (31, 32). None of these strains carried *eae*, but many had *ehxA*, *saa*, and *subAB* (Table 2). In spite of the uncertain role of these genes in STEC pathogenesis, the historical association of these serotypes with illness should be considered when making risk-assessment decisions. Excluding the ~30 STEC strains in produce that fit the criteria mentioned and discussed above, risk assessment on the remaining ~100 STEC isolates from produce is more complex and difficult. These strains did not have *eae*, but ~60% had *ehxA*, 35% had *saa*, and 32% had *subAB*. Moreover, almost half of these only had partial O and H type data or could not be serologically typed. Knowing the serotype of the STEC isolates is important, but it is difficult to determine due to the large numbers and the complex combinations of O and H antigen types that can exist among *E. coli* strains. The limitations of using antibodies for serotyping, however, may be remedied in the future with genotypic assays, as a microarray was able to identify the serotype of many untypeable STEC strains isolated from fresh produce (50). In the meantime, considering the complexities and uncertainties of these putative virulence factors in pathogenesis and the incomplete or lack of serotype identity, it is almost impossible to determine if these STEC strains in produce are of safety concern.

Thus, risk assessment on STEC isolates from produce can be made systematically and with some rationality in some cases, but most assessments tend to be inconclusive due to partial identity or the uncertain role of putative virulence factors detected. Even more critical is the inability to get timely data to make risk-assessment decisions on such short-shelf-life products as fresh produce. Occasionally, the ease of isolation and identification of O157:H7 makes it feasible to obtain data in time to decide to withdraw the product from the market. For most of the other STEC strains, by the time the strain is isolated, characterized, and serotyped, the product has either been consumed or passed its expiration date and has been taken off the market and discarded.

CONCLUSION

Fresh produce is constantly exposed to the environment and, therefore, can contain complex microflora populations at levels that sometimes exceed 10^8 CFU/g. It is also common for produce to contain indicator bacteria, and the levels of coliforms can vary from undetected to as high as 10^4 CFU/g. The levels of generic *E. coli* in produce are much lower and, if *E. coli* is present, the level is usually <10 CFU/g, but in occasional samples, it may exceed 100 CFU/g. Pathogenic *E. coli* can be found in produce samples but seems to be more prevalent in spinach where prevalence rates of ~0.2% and 0.5% for ETEC and STEC, respectively, have been observed in some years. Screening for STEC in produce is straightforward, but there are no effective isolation methods; hence the confirmation process to isolate, characterize, and serotype the strains is time- and labor-intensive. Produce contains many different serotypes of STEC and many strains that could only be partially serotyped or are untyped. Only a handful of the STEC strains isolated from fresh produce carried the *eae* gene that encodes for the adherence protein, intimin, and can be regarded as a health risk. They included O157:H7 strains, a few "big 6" serotype strains like O121:H19 and O26:H11, but also included other pathogenic serotypes such as O165:H25. The majority of the STEC isolates from produce did not have *eae*, but some of them were of serotype O113:H21 and O91:H21 strains, which historically have been implicated with severe illness and, therefore, may also be a safety concern. Risk assessment for the other STEC strains in produce is more difficult as many of these strains carried putative virulence factor genes that had uncertain roles in pathogenesis, only had partial serological identity, or belonged to serotypes that had no history of causing human illness. Thus, the logistical problems associated with testing for a complex group of bacteria like STEC in microflora-complex, short-shelf-life products like fresh produce have greatly limited the ability to obtain timely and relevant data for making risk-assessment decisions.

ACKNOWLEDGMENT

I declare no conflict of interest with regard to the manuscript.

CITATION

Feng P. 2014. Shiga toxin-producing *Escherichia coli* in fresh produce—a food safety dilemma. Microbiol Spectrum 2(4):EHEC-0010-2013.

REFERENCES

1. **Sivapalasingam S, Friedman CR, Cohen L, Tauxe RV.** 2004. Fresh produce: a growing cause of outbreaks of foodborne illness in the United States, 1973 through 1997. *J Food Prot* **67:**2342–2353.

2. **Lynch MF, Tauxe RV, Hedberg CW.** 2009. The growing burden of foodborne outbreaks due to contaminated fresh produce: risks and opportunities. *Epidemiol Infect* **137:**307–315.

3. **Berger CN, Sodha SV, Shaw RK, Griffin PM, Pink D, Hand P, Frankel G.** 2010. Fresh fruit and vegetables as vehicles for the transmission of human pathogens. *Environ Microbiol* **12:**2385–2397.

4. **Centers for Disease Control and Prevention.** 2006. Ongoing multistate outbreak of *Escherichia coli* serotype O157:H7 infections associated with consumption of fresh spinach—United States, September 2006. *Morb Mortal Wkly Rep* **55:**1045–1046.

5. **Doyle MP, Erickson MC.** 2008. Summer meeting 2007—the problem with fresh produce: an overview. *J Appl Microbiol* **105:**317–330.

6. **Beutin L, Martin A.** 2012. Outbreak of Shiga toxin-producing *Escherichia coli* (STEC) O104:H4 infection in Germany causes a paradigm shift with regard to human pathogenicity of STEC strains. *J Food Prot* **75:**408–418.

7. **Johnston LM, Jaykus LA, Moll D, Martinez MC, Anciso J, Mora B, Moe CL.** 2005. A field study of the microbiological quality of fresh produce. *J Food Prot* **68:**1840–1847.

8. **Johnston LM, Jaykus LA, Moll D, Anciso J, Mora B, Moe CL.** 2006. A field study of the microbiological quality of fresh produce of domestic and Mexican origin. *Int J Food Microbiol* **112:**83–95.

9. **Mukherjee A, Speh D, Jones AT, Buesing KM, Diez-Gonzalez F.** 2006. Longitudinal microbiological survey of fresh produce grown by farmers in the upper Midwest. *J Food Prot* **69:**1928–1936.

10. **Soriano JM, Rico H, Moltó JC, Mañes J.** 2006. Assessment of the microbiological quality and wash treatments of lettuce served in University restaurants. *Int J Food Microbiol* **58:**123–128.

11. **Abadias M, Usall J, Anguera M, Solsona C, Viñas I.** 2008. Microbiological quality of fresh, minimally-processed fruit and vegetables, and sprouts from retail establishments. *Int J Food Microbiol* **123:**121–129.

12. **Caponigro V, Ventura M, Chiancone I, Amato L, Parente E, Piro F.** 2010. Variation of microbial load and visual quality of ready-to-eat salads by vegetable type, season, processor and retailer. *Food Microbiol* **27:**1071–1077.

13. **Fröder H, Martins CG, De Souza KL, Landgraf M, Franco BD, Destro MT.** 2007. Minimally processed vegetable salads: microbial quality evaluation. *J Food Prot* **70:**1277–1280.

14. **Hagenmeaier RD, Baker RA.** 1998. A survey of the microbial population and ethanol content of bagged salad. *J Food Prot* **61:**357–359.

15. **Valentin-Bon I, Jacobson A, Monday S, Feng P.** 2008. Microbiological quality of bagged cut spinach and lettuce mixes. *Appl Environ Microbiol* **74:**1240–1242.

16. **Kase JA, Borenstein S, Feng PCH.** 2012. Microbial quality of bagged baby spinach and romaine lettuce—effects of top vs. bottom sampling. *J Food Prot* **75:**132–136.

17. **Sagoo SK, Little CL, Mitchell RT.** 2003. Microbiological quality of open ready-to-eat salad vegetables: effectiveness of food hygiene training of management. *J Food Prot* **66:**1581–1586.

18. **Sospedra I, Rubert J, Soriano JM, Mañes J.** 2013. Survey of microbial quality of plant-based foods served in restaurants. *Food Control* **30:**418–422.

19. **Sagoo SK, Little CL, Mitchell RT.** 2001. The microbiological examination of ready-to-eat organic vegetables from retail establishments in the United Kingdom. *Lett Appl Microbiol* **33:**434–439.

20. **Sagoo SK, Little CL, Ward L, Gillespie IA, Mitchell RT.** 2003. Microbiological study of ready-to-eat salad vegetables from retail establishments uncovers a national outbreak of salmonellosis. *J Food Prot* **66:**403–409.

21. **Feng P, Strockbine N, Fratamico P.** 2012. Methods of detection and characterization of pathogenic *Escherichia coli*. *In* UNESCO-EOLSS Joint Committee (ed), Food Quality and Standards, in *Encyclopedia of Life Support Systems (EOLSS)*. Developed under the Auspices of the UNESCO. Eolss Publishers, Oxford, UK. (http://www.eolss.net)

22. **Lopez-Saucedo C, Cerna JF, Estrada-Garcia T.** 2010. Non-O157 Shiga toxin-producing *Escherichia coli* is the most prevalent diarrheagenic *E. coli* pathotype in street-vended taco dressings in Mexico City. *Clin Infect Dis* **50:**450–451.

23. **Feng PCH, Reddy SP.** 2014. Prevalence and diversity of enterotoxigenic *Escherichia coli* strains in fresh produce. *J Food Prot* **77:**820–823.

24. **Monday SR, Keys C, Hansen P, Shen Y, Whittam TS, Feng P.** 2006. Produce isolates of *Escherichia coli* Ont:H52 serotype that carry both Shiga toxin 1 and StableToxin genes. *Appl Environ Microbiol* **72:**3062–3065.

25. **Scheutz F, Teel LD, Beutin L, Piérard D, Buvens G, Karch H, Mellmann A, Caprioli A, Tozzoli R, Morabito S, Strockbine NA, Melton-Celsa AR, Sanchez M, Persson S, O'Brien AD.**

2012. Multicenter evaluation of a sequence-based protocol for subtyping Shiga toxins and standardizing Stx nomenclature. *J Clin Microbiol* **50:**2951–2963.

26. **Martin A, Beutin L.** 2011. Characteristics of Shiga toxin-producing *Escherichia coli* from meat and milk products of different origins and association with food producing animals as main contamination sources. *Int J Food Microbiol* **146:**99–104.

27. **Monaghan A, Byrne B, Fanning S, Sweeney T, McDowell D, Bolton DJ.** 2011. Serotypes and virulence profiles of non-O157 Shiga toxin-producing *Escherichia coli* isolates from bovine farms. *Appl Environ Microbiol* **77:**8662–8668.

28. **Mora A, López C, Dhabi G, López-Beceiro AM, Fidalgo LE, Díaz EA, Martínez-Carrasco C, Mamani R, Herrera A, Blanco JE, Blanco M, Blanco J.** 2012. Seropathotypes, phylogroups, Stx subtypes, and intimin types of wildlife-carried, Shiga toxin-producing *Escherichia coli* strains with same characteristics as human-pathogenic isolates. *Appl Environ Microbiol* **78:**2578–2585.

29. **García-Aljaro C, Muniesa M, Blanco JE, Blanco M, Blanco J, Jofre J, Blanch AR.** 2005. Characterization of Shiga toxin-producing *Escherichia coli* isolated from aquatic environments. *FEMS Microbiol Lett* **246:**55–65.

30. **Feng PCH, Reddy S.** 2013. Prevalences of Shiga toxin subtypes and selected other virulence factors among Shiga-toxigenic *Escherichia coli* strains isolated from fresh produce. *Appl Environ Microbiol* **79:**6917–6923.

31. **Hussein HS.** 2007. Prevalence and pathogenicity of Shiga toxin-producing *Escherichia coli* in beef cattle and their products. *J Anim Sci* **85**(13 Suppl): E63–E72.

32. **Newton HJ, Sloan J, Bulach DM, Seemann T, Allison CC, Tauschek M, Robins-Browne RM, Paton JC, Whittam TS, Paton AW, Hartland EL.** 2009. Shiga toxin producing *Escherichia coli* strains negative for locus of enterocyte effacement. *Emerg Infect Dis* **15:**372–380.

33. **Friedrich AW, Borell J, Bielaszewska M, Fruth A, Tschape H, Karch H.** 2003. Shiga toxin 1c-producing *Escherichia coli* strains: phenotypic and genetic characterization and association with human disease. *J Clin Microbiol* **41:**2248–2453.

34. **Ethelberg S, Olsen KE, Scheutz F, Jensen C, Schiellerup P, Enberg J, Olesen B, Gerner-Smidt P, Mølbak K.** 2004. Virulence factors for hemolytic uremic syndrome, Denmark. *Emerg Infect Dis* **10:**842–847.

35. **Karmali MA, Petric M, Lim C, Fleming PC, Arbus GS, Lior H.** 1985. The association between idiopathic hemolytic uremic syndrome and infection by verocytotoxin-producing *Escherichia coli. J Infect Dis* **151:**775–782.

36. **Paton AW, Woodrow MC, Doyle RM, Lanser JA, Paton JC.** 1999. Molecular characterization of a Shiga toxigenic *Escherichia coli* O113:H21 strain lacking *eae* responsible for a cluster of cases of hemolytic-uremic syndrome. *J Clin Microbiol* **37:**3357–3361.

37. **Paton AW, Srimanote P, Woodrow MC, Paton JC.** 2001. Characterization of Saa, a novel auto-agglutinating adhesin produced by locus of enterocyte effacement-negative Shiga-toxigenic *Escherichia coli* strains that are virulent for humans. *Infect Immun* **69:**6999–7009.

38. **Galli L, Miliwebsky E, Irino K, Leotta G, Rivas M.** 2010. Virulence profile comparison between LEE-negative Shiga toxin-producing *Escherichia coli* (STEC) strains isolated from cattle and humans. *Vet Microbiol* **143:**307–313.

39. **Jenkins C, Perry NT, Cheasty T, Shaw DJ, Frankel G, Dougan G, Gunn GJ, Smith HR, Paton AW, Paton JC.** 2003. Distribution of the *saa* gene in strains of Shiga toxin-producing *Escherichia coli* of human and bovine origins. *J Clin Microbiol* **41:**1775–1778.

40. **Bosilevac JM, Koohmaraie M.** 2011. Prevalence and characterization of non-O157 Shiga toxin producing *Escherichia coli* isolated from commercial ground beef in the United States. *Appl Environ Microbiol* **77:**2103–2112.

41. **Boczek LA, Johnson CH, Rice EW, Kinkle BK.** 2006. The widespread occurrence of the enterohemolysin gene *ehlyA* among environmental strains of *Escherichia coli. FEMS Microbiol Lett* **254:**281–284.

42. **Karch H, Bielaszewska M.** 2001. Sorbitol-fermenting Shiga toxin-producing *Escherichia coli* O157:H⁻ strains: epidemiology, phenotypic and molecular characteristic, and microbiological diagnosis. *J Clin Microbiol* **39:**2043–2049.

43. **Paton AW, Beddoe T, Thorpe CM, Whisstock JC, Wilce MC, Rossjohn J, Talbot UM, Paton JC.** 2006. AB5 subtilase cytotoxin inactivates the endoplasmic reticulum chaperone BiP. *Nature* **443:** 548–552.

44. **Michelacci V, Tozzoli R, Caprioli A, Martínez R, Scheutz F, Grande L, Sánchez S, Morabito S.** 2013. A new pathogenicity island carrying an allelic variant of the Subtilase cytotoxin is common among Shiga toxin producing *Escherichia coli* of human and ovine origin. *Clin Microbiol Infect* **19:** E149–E156.

45. **Johnson KE, Thorpe CM, Sears CL.** 2006. The emerging clinical importance of non-O157 Shiga toxin-producing *Escherichia coli. Clin Infect Dis.* **43:**1587–1595.

46. **Beutin L, Montenegro MA, Orskov I, Prada J, Zimmerman S, Stephan R.** 1989. Close association of verotoxin (Shiga-like toxin) production

with enterohemolysin production in strains of *Escherichia coli*. *J Clin Microbiol* **27:**2559–2564.

47. **Sugiyama K, Inoue K, Sakazaki R.** 2001. Mitomycin-supplemented washed blood agar for the isolation of Shiga toxin-producing *Escherichia coli* other than O157:H7. *Lett Appl Microbiol* **33:**193–195.

48. **Feng PCH, Councell T, Key C, Monday SR.** 2011. Virulence characterization of Shigatoxigenic *Escherichia coli* serotypes isolated from wholesale produce. *Appl Environ Microbiol* **77:**343–345.

49. **Feng PCH, Delannoy S, Lacher DW, dos Santos LF, Beutin L, Fach P, Rivas M, Hartland EL, Paton AW, Guth BEC.** 2014. Genetic diversity and virulence potential of Shiga toxin-producing *Escherichia coli* O113:H21 strains isolated from clinical, environmental, and food sources. *Appl Environ Microbiol* **80:**4757–4763.

50. **Lacher DW, Gangiredla J, Jackson SA, Elkins CA, Feng PCH.** 2014. A novel microarray design for molecular serotyping of Shiga toxin-producing *Escherichia coli* isolated from fresh produce. *Appl Environ Microbiol* **80:**4677–4682.

Public Health Microbiology of Shiga Toxin-Producing *Escherichia coli*

13

ALFREDO CAPRIOLI,[1] GAIA SCAVIA[1] and STEFANO MORABITO[1]

Shiga toxin-producing *Escherichia coli* (STEC) represents a major issue for public health because of the capability to cause large outbreaks and the severity of the associated illnesses (1). STEC strains are the only pathogenic group of *E. coli* that has a definite zoonotic origin, with ruminants and, in particular, cattle being recognized as the major reservoir for human infections (2). Most human infections are food borne, but the routes of transmission include direct contact with animals and a wide variety of environment-related exposures (3). Therefore, STEC public health microbiology spans the fields of medical, veterinary, food, water, and environmental microbiology, requiring a "One Health" perspective (4) and laboratory scientists with the ability to work effectively across disciplines. Public health microbiology laboratories play a central role in the surveillance of STEC infections, as well as in the preparedness for responding to outbreaks and in providing scientific evidence for the implementation of prevention and control measures. This article reviews in depth (i) how the integration of surveillance of STEC infections and monitoring of these pathogens in animal reservoirs and potential food vehicles may contribute to their control; (ii) the role of reference laboratories; and (iii) the public health perspectives, including those related to regulatory issues in both the European Union and the United States.

[1]European Union Reference Laboratory for *E. coli*, Dipartimento di Sanità Pubblica Veterinaria e Sicurezza Alimentare, Istituto Superiore di Sanità, Rome, Italy.

Enterohemorrhagic Escherichia coli *and Other Shiga Toxin-Producing* E. coli
Edited by Vanessa Sperandio and Carolyn J. Hovde
© 2015 American Society for Microbiology, Washington, DC
doi:10.1128/microbiolspec.EHEC-0014-2013

SURVEILLANCE AND MONITORING OF STEC INFECTIONS

Surveillance of STEC Infections in Humans

In the medical field, dedicated surveillance systems of human STEC infections have been developed in most of the industrialized areas of the world because of the public health importance of these infections. Such systems are of utmost importance for the prompt recognition and management of outbreaks and the implementation of specific strategies to control the spread of the infections in the community.

Surveillance systems should record not only the epidemic outbreaks but also sporadic cases of infection. In particular, cases of severe illness, such as bloody diarrhea and the hemolytic-uremic syndrome (HUS), may reveal a wider circulation of STEC strains in a community and should be considered as possible syndromic sentinel events of possible underlying clusters of infections (5).

The data sets that should be part of surveillance systems of STEC infections include laboratory and clinical data, as well as information on risk factors and the possible association with other cases of infection. Surveillance systems are usually based on "laboratory-confirmed" case definitions that are dependent on the methods applied for laboratory diagnosis. Such methods may vary considerably. In many settings, they are limited to the detection of STEC O157, leading to an underestimation of the incidence of STEC non-O157 infections (6). The information on the clinical outcome of the reported cases of STEC infection is very important, in particular when HUS develops, since it is crucial to estimate the burden of disease due to these pathogens and to define which STEC serotypes are consistently associated with severe disease.

In Europe, the surveillance of STEC infections is embedded in the Food- and Waterborne Diseases and Zoonoses (FWD) surveillance system (http://ecdc.europa.eu/en/activities/diseaseprogrammes/fwd/Pages/index.aspx) coordinated by the European Centre for Disease Prevention and Control (ECDC). FWD is a passive surveillance system, collecting data on STEC infections from the European Union and the European Economic Area (EEA) countries, based on case definitions that include laboratory-confirmed cases, probable cases, and possible cases. For laboratory-confirmed cases, collected data include the serotype and the main virulence genes (*stx1*, *stx2*, *eae*) of the infecting strain, together with the clinical manifestation, in particular, the development of HUS. The data on STEC infections are published yearly in the *European Union Summary Reports on Trends and Sources of Zoonoses, Zoonotic Agents and Food-borne Outbreaks* and provide information on the trend of STEC infections in the European Union, according to the serogroup of the infecting strains and the development of HUS. Figure 1 shows the number of STEC infections reported to the ECDC-FWD surveillance system in the period 2007 to 2011 (7–11). An increasing trend of reported cases was observed, likely due to a general improvement of the surveillance systems in the participating countries. This hypothesis is supported by the sharp increase in the cases caused by STEC O157 and STEC non-O157 other than O104 notified in 2011, when the large outbreak of STEC O104:H4 infections occurred in Germany and other European countries (12). Such an increase in the reporting was likely due to the enhanced attention toward STEC infections raised by the outbreak. At the same time, a decrease in the number of cases for which the infecting strain was not serotyped was observed, probably due to the enhanced rate of submission of STEC strains to reference laboratories for confirmation and typing.

The serogroups other than O157 and O104 most frequently associated with STEC infections in the European Union in the period 2007 to 2011 are reported in Fig. 2. STEC O26 was the serogroup most frequently reported throughout the period, followed by O103, O91, O145, and O111.

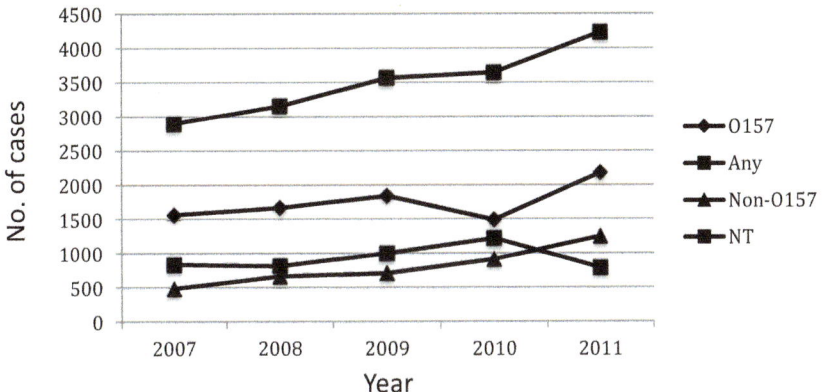

FIGURE 1 Number of STEC infections reported in the period 2007–2011 to the FWD surveillance system co-ordinated by the ECDC. The 2011 data do not include the cases due to STEC O104, which occurred in the framework of the large outbreak that occurred in Germany and other European countries. NT, cases with no information available on the serogroup of the STEC infecting strain.
doi:10.1128/microbiolspec.EHEC-0014-2013.f1

Unfortunately, not all cases reported in the ECDC-FWD databases had the complete set of data, and the information on the development of HUS was unknown for 30 to 40% of them (13). Despite this lack of information, a number of HUS cases ranging from 103 in 2007 to 277 in 2011 (not including those associated with the STEC O104 outbreak) were reported. Most cases occurred in children, with about 60% occurring in the age group 0 to 4 years. The STEC serogroups consistently associated with HUS were O157, O26, O103, O145, and O111, whereas the syndrome was rarely observed among patients with STEC O91 infection.

In the United States, the Centers for Disease Control and Prevention (CDC) established a surveillance network for cases of STEC infections and other food-borne diseases (FoodNet), with the aim of determining

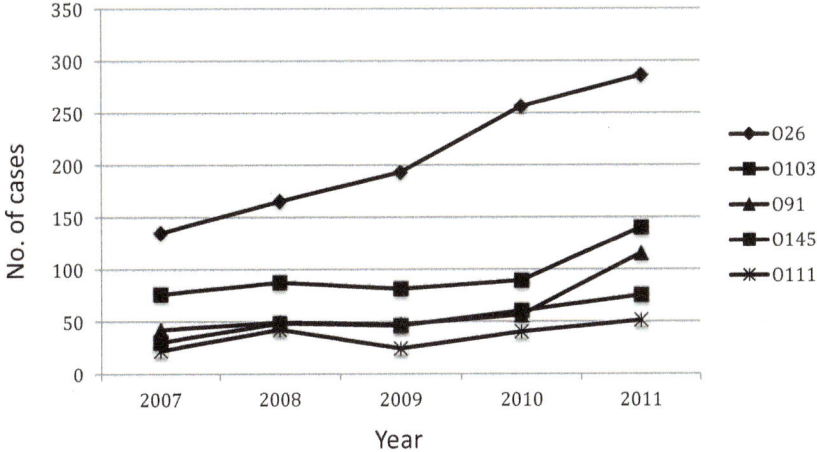

FIGURE 2 Number of STEC infections in the European Union associated with the serogroups other than O157 and O104 most frequently reported in the period 2007–2011 to the FWD surveillance system coordinated by the ECDC. doi:10.1128/microbiolspec.EHEC-0014-2013.f2

the incidence of laboratory-confirmed infections for bacterial pathogens transmitted commonly through food and attributing illnesses to specific sources and settings (www.cdc.gov/foodnet/). FoodNet was established in 1995 in collaboration with the U.S. Department of Agriculture's Food Safety and Inspection Service and the Food and Drug Administration. It involves 10 state health departments, with a surveillance area including 15% of the U.S. population (47 million persons). Differently from the FWD, FoodNet is an active surveillance system, with public health officials routinely communicating with the clinical laboratories to identify new cases. Cases of HUS are specifically recorded through a network of pediatric nephrologists and infection-control practitioners on the basis of clinical diagnosis (14). Infections with STEC O157 and non-O157 are included in the FoodNet surveillance system, and interestingly, similar incidence values (0.97 and 1.10 per 100,000, respectively) were recorded for the two groups in 2011 (15). However, among the patients with HUS tested with appropriate laboratory methods, the prevalence of STEC O157 infections was much higher than that of STEC non-O157. In general, among the STEC infections with the serogroup identified, the most common were O157 (47%), O26 (14%), and O103 (11%).

The FoodNet network is pulled alongside with PulseNet, a network performing standardized molecular subtyping of STEC and other food-borne pathogens by pulsed-field gel electrophoresis to detect case clusters and to facilitate the early identification of common-source outbreaks (www.cdc.gov/pulsenet/). PulseNet is also coordinated by the CDC and originated as a North American network of laboratories, then extended to other similar networks around the world (16).

Surveillance activities should be maintained over time to allow evaluating the trends of STEC infections and understanding the dynamics of the circulation of serotypes and even specific clones. In Italy, a surveillance system for HUS has been in place since 1988

(17) and has pinpointed significant changes in the prevalence of HUS-associated serogroups over time. As shown in Fig. 3, STEC O157 was associated with more than 50% of cases in the first decade of the surveillance. After this first period, infections with STEC O26 and O145 increased, outnumbering those associated with STEC O157 in the decade 1998 to 2007. Finally, the distribution of cases due to STEC O157 and STEC O26 reached a more balanced relative figure in the current period (unpublished data from the Italian Registry for HUS, 1988–2007). Since the methods for laboratory diagnosis of STEC infection used in the surveillance (17) did not change over time, the observed variations in the prevalence of the HUS-associated serogroups likely reflect a varying exposure of the population to the sources and reservoirs of these organisms. This hypothesis might have important implications from a public health perspective and seems to be supported by the observation that patients with STEC O157 infections were older (median, 32 months) than those with non-O157 infections (median, 22 months) and showed a more clear summer peak (unpublished data from the Italian Registry for HUS, 1988–2007).

Serogroup O26 has been strongly associated with HUS also in other European countries (7–11) and the United States (18). In Europe, most of the STEC O26 pathogens causing

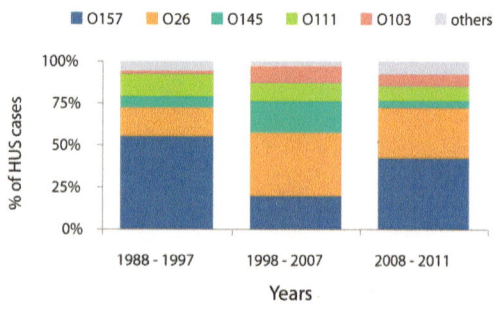

FIGURE 3 STEC serogroups associated with the hemolytic-uremic syndrome in Italy, 1988–2011. Data from the Italian Registry for HUS.
doi:10.1128/microbiolspec.EHEC-0014-2013.f3

HUS belong to a highly virulent clone, which seems to have emerged in the mid-1990s (19), with the first reports from Germany (20) and Italy (21). This O26 clone belongs to a particular multilocus sequence type, ST29, and harbors the *stx2a* subtype of Shiga toxin gene (19). It has been speculated (19) that its evolutionary success might be due to the ability to lose the *stx2a*-harboring bacteriophage (21) without entering the lytic cycle. As a matter of fact, such an ability might allow the avoidance of bacterial lysis following stimulation to release the *stx2a*-harboring bacteriophage in the human gut and might also account for a prolonged survival in the environment. Such *stx*-negative STEC O26 variants also represent targets for lysogenization by other *stx*-harboring phages, explaining their greater diffusion with respect to other lytic clones.

Monitoring of STEC Along the Food Chain

Monitoring the presence of STEC in the animal reservoirs and the potential food vehicles can provide useful information to identify which animals and foodstuffs are the main sources of human infections. However, data on the prevalence of STEC in food and animals from different investigations could be difficult to compare due to differences in the sampling strategies and analytical methods employed. As International Organization for Standardization Technical Specification (ISO/TS) 13136:2012, the international standard for the detection of STEC in food, was published only in late 2012, a variety of detection methods for STEC have been used in monitoring programs and official control plans, based on two main approaches: (i) the detection of any STEC strain present in the sample, assessed by the production of Stx and/or the presence of *stx* genes; and (ii) a serogroup-specific detection strategy, aimed in most cases at the detection of *E. coli* O157 and, more recently, a few other serogroups that have been consistently associated with HUS and are capable to cause outbreaks.

These serogroups include O26, O103, O111, and O145 and have been classified in the seropathotype B group of the scheme proposed by Karmali et al. (23) (see also "A Proactive Approach to Food Control: Which STEC Should Be Considered Pathogenic?" below). In the United States, such a serogroup-specific approach has been extended to serogroups O121 and O45 that have been considered epidemiologically relevant in that country (18).

In the European Union, STEC is included among the zoonotic agents for which monitoring activities are mandatory, according to Directive 2003/99/EC, which obligates the European Union member states to collect relevant and possibly comparable data on zoonoses and zoonotic agents. Data are collected by the European Food Safety Authority (EFSA) and published in the *European Union Summary Report on Trends and Sources of Zoonoses, Zoonotic Agents and Food-borne Outbreaks* (7–11) together with data on human STEC infections. EFSA issued recommendations for the monitoring of STEC in animals and food (24) that gave priority to STEC O157 as the STEC serogroup most frequently associated with severe human infections, in particular HUS, in Europe. Monitoring should then be extended to other STEC serogroups (e.g., O26, O103, O91, O145, and O111) indicated by the periodic analysis of the data on human infections in Europe. EFSA also issued technical specifications for STEC monitoring activities (25). For animals, the sampling of the hide of young cattle and the fleeces of sheep at slaughter was indicated. For food, sampling of commodities that are perceived to be sources of STEC infections was suggested, including beef products that could be eaten with minimal cooking, ready-to-eat fermented meats, fresh produce, raw and low-heat-treated milk, and derived dairy products.

The monitoring data reported in the last published *European Union Summary Report on Zoonoses, Zoonotic Agents and Food-borne Outbreaks* (11) confirmed that STEC is mainly found in ruminants and products thereof, meat and raw milk. The proportion of STEC-positive samples can vary widely among countries, but these differences could be due to the differences in the sampling strategies and analytical methods. The reported prevalence of STEC contamination in vegetables and fruits was very low, but this probably reflects the sampling plans adopted so far; these matrices included numbers of tested samples much lower than those of foods of animal origin.

Despite the problems in the standardization of sampling and detection methods, the *European Union Summary Reports on Trends and Sources of Zoonoses, Zoonotic Agents and Food-borne Outbreaks* represent an important example of integrated medical, veterinary, and food data and can provide a valuable contribution to the source attribution of the burden of STEC infections due the serotypes and clones in humans and to the evaluation of cost-effective control measures.

The Role of Reference Laboratories

The microbiology of STEC infections is particularly complex because of the difficulties in distinguishing STEC from the other ubiquitous and generally harmless types of *E. coli*. Moreover, the physiologic and genomic ductility of these microorganisms may favor the emergence of new pathogenic clones, hindering the development of reliable schemes for their characterization. Therefore, establishing specific reference laboratories is of utmost importance for the prevention and control of STEC infections.

Reference laboratories, whether established within a normative framework or not, are usually appointed to provide reference diagnostics, reference materials, external quality assessment (EQA), scientific advice, collaboration on research and monitoring, and participation in alert and response activities. Reliable STEC reference laboratories are pivotal for supporting surveillance activities and for enabling preparedness for the threats caused by emerging pathogenic clones and epidemic outbreaks.

Public Health Reference Laboratories

Networks of STEC national reference laboratories (NRLs) have been established in the industrialized countries and operate in different contexts according to their geographic distribution.

The services provided by STEC NRLs should include confirmation, serotyping, and molecular typing of suspected *E. coli* strains isolated by front-line clinical microbiology laboratories. NRLs should also participate in surveillance activities, research, and dissemination of information and provide technical training and reference strains to front-line laboratories.

In the European Union, the public health field, differently from the food and veterinary sector, does not refer to specific norms regulating the asset of the networks of reference laboratories. Nevertheless, a web of STEC NRLs has been established over time, composed of scientific institutions historically collaborating on this topic (26) and, since 2007, connected within the framework of the ECDC-FWD program described earlier. In such a framework, the European *E. coli* NRLs provide the data on STEC infections to the ECDC-FWD database. In turn, the ECDC supports the STEC NRL network through the standardization of identification and typing methods, and the provision of reference materials (control strains) and EQA, organized to evaluate the performance of laboratories, to identify areas for improvement in laboratory methods and to ensure that identification and typing of STEC are carried out uniformly and that the results provided to the FWD database are comparable. The EQA usually includes the identification of STEC by detection and typing of the main virulence genes (*stx1, stx2, eae*) and O:H serotyping (www.ecdc.europa.eu/en/publications/_layouts/forms/Publication_DispForm.aspx?List=4f55ad51-4aed-4d32-b960-af70113dbb90&ID=1041).

In the United States, the FoodNet network supports the 650 clinical laboratories that provide data on STEC infections with laboratory protocols and recommendations and conducts periodic surveys to understand the current diagnostic practices and monitor their changes over time (27).

Veterinary/Food Reference Laboratories

The role of reference laboratories is particularly important in the veterinary/food sector. Differently from clinical microbiology, where the isolation of any STEC strain from a case of infection displaying compatible symptoms is sufficient to establish an etiologic diagnosis, in food microbiology the need to assign a rank of risk to the STEC isolates for the humans' health does not benefit from a clear-cut definition of which STEC types have to be considered as a hazard (see "A Proactive Approach to Food Control: Which STEC Should Be Considered Pathogenic?" below). Moreover, most diagnostic tests for the detection of these organisms in food vehicles and animal reservoirs have been specifically developed for *E. coli* O157, taking advantage of its particular metabolic and antigenic features (24). Conversely, the development of assays to distinguish non-O157 STEC from nonpathogenic *E. coli* remains challenging.

In the food safety field, the duties of reference laboratories include the development, evaluation, and validation of the test methods for a reliable detection of STEC in foods, the provision of reference materials, the organization of EQA programs for the laboratories involved in food control, and the scientific and technical assistance to the public health authorities, particularly in risk assessment exercises.

In the European Union, a network of *E. coli* NRLs operates within the framework of Regulation (EC) 882/2004, which lays out the official controls on food and also establishes the nomination of European Union Reference Laboratories (EU-RLs) to face specific food and feed hazards or specific animal diseases. According to the Regulation (EC) 882/2004 prescripts, each European Union member state must designate its own NRL for each

established EU-RL to create laboratory networks on the specific hazards. The NRLs collaborate with the EU-RL and coordinate the official laboratories responsible for food analyses in their country, also through the organization of national EQA. All laboratories included in the network at either the national or the European Union level must be accredited according to the norm EN ISO IEC 17025:2005. The final aim of this cascade system is that the official controls conducted on any produced or imported foodstuff are carried out using the same state-of-art methods and with comparable levels of proficiency throughout the European Union territory.

The EU-RL for *E. coli* was established in 2006, and at present it coordinates a network of 34 NRLs designated by the European Union and EEA member states (www.iss.it/vtec).

The first aim of the EU-RL was the harmonization of the identification and typing methods for STEC strains used by the veterinary/food NRLs with those used within the network of public health NRLs participating in the ECDC-FWD surveillance program to allow the comparison of data referring to STEC strains isolated from human infections and from food and animal sources. This was achieved by the organization of five EQA schemes on STEC strain identification and typing, two of which are conducted jointly with the ECDC-FWD network. The methods used in these EQA studies and the results obtained are available at the EU-RL website (www.iss.it/vtec). The results referring to the detection of the STEC main virulence genes and to the identification of the main STEC serogroups are summarized in Fig. 4 and Fig. 5, respectively.

As far as the methods for the detection of STEC in food are concerned, the EU-RL developed and published on its website several operating procedures destined to the NRL network (www.iss.it/vtec). Moreover, it coordinated the collaborative development of an international standard for the detection of these microorganisms in food upon mandate of the European Committee for Standardization.

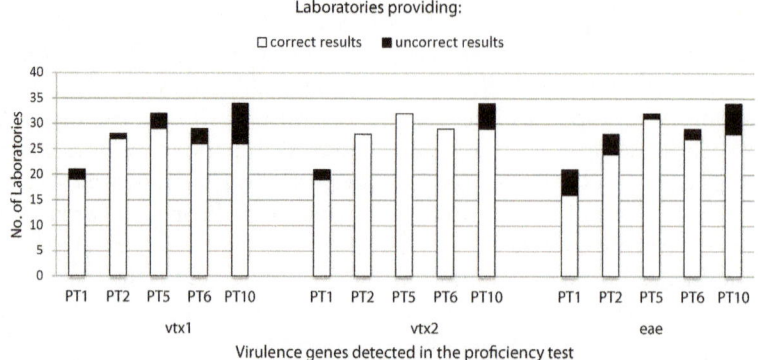

FIGURE 4 External quality assessment organized by the EU-RL for *E. coli* on the identification of STEC strains by detection of their main virulence genes by PCR. For each gene, white bars represent the number of laboratories that obtained correct results for all the strains included in the test and black bars the number of laboratories that provided incorrect results or did not perform the assay. doi:10.1128/microbiolspec.EHEC-0014-2013.f4

The developed standard is a horizontal method based on the real-time PCR screening of food enrichment cultures for the presence of the virulence genes (*stx1*, *stx2*, and *eae*) and genes specific for five STEC serogroups widely involved in severe human infections in Europe: O157, O26, O103, O111, and O145. Samples positive for *stx* genes are submitted to a further

step aimed at isolating the STEC strain responsible for the positive PCR reactions (see also "Monitoring of STEC Along the Food Chain" above).

The standard has been published by the International Organization for Standardization in November 2012 as a technical specification (ISO/TS 13136:2012). Starting in 2009,

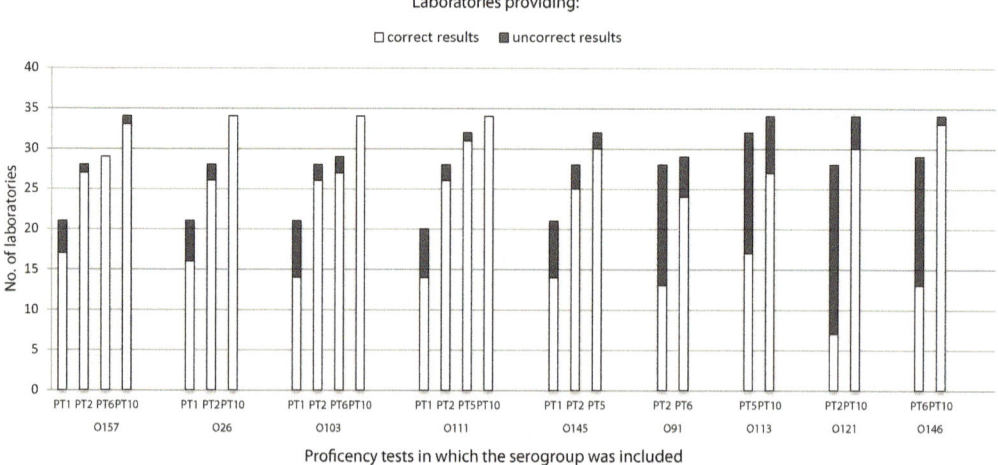

FIGURE 5 External quality assessment organized by the EU-RL for *E. coli* on the identification of the STEC serogroups most involved in human disease in Europe. For each serogroup, white bars represent the number of laboratories that obtained correct results for all the strains included in the test and black bars the number of laboratories that provided incorrect results or did not perform the assay. doi:10.1128/microbiolspec.EHEC-0014-2013.f5

the analytical approach described by ISO/TS 13136:2012 was evaluated in five proficiency-testing schemes organized by the EU-RL and conducted on different artificially contaminated matrices. The reports of such EQA studies are available at the EU-RL website (www.iss.it/vtec), and a synopsis of the aggregated results is shown in Fig. 6, which shows an increasing trend in the number of participating laboratories and in their analytical performances.

The existence of networks of reference laboratories reveals its pivotal value during food safety crises involving different countries, when testing food with standardized, rapid, and reliable methods is essential to provide the competent authorities the data needed to plan appropriate control measures and to inform consumers correctly. In this respect, the European Union network of *E. coli* reference laboratories was challenged in 2011 by the occurrence of the major outbreak of STEC O104:H4 infections (12). As usually occurs during food safety crises due to food-borne outbreaks, testing food for the presence of the outbreak strains was urgently required. Because of the activities carried out in the previous years, the European laboratory network already had a suitable screening method to exclude the presence of any type of STEC in food, tested by several rounds of EQA. An additional standard operating procedure specific for detecting the STEC O104:H4 outbreak strain was developed by the EU-RL and released through the EU-RL website 3 days after the occurrence of the outbreak was communicated (28). DNA samples to be used as positive control in the molecular assays for the detection of STEC O104:H4 were also prepared and distributed to the NRLs in the following days. During the entire period of the crisis, the EU-RL provided continuous scientific and technical support to the European Union structures involved in the management of the crisis. The experience of the STEC O104:H4 outbreak confirmed that the activities carried out within the networks of reference laboratories can provide an important contribution in terms of preparedness to face food safety crises.

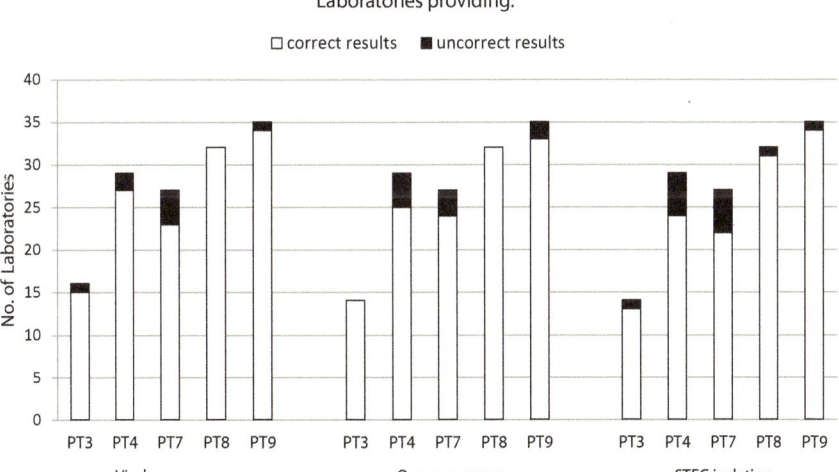

FIGURE 6 External quality assessment organized by the EU-RL for *E. coli* on the detection in food of the STEC serogroups most involved in human disease in Europe, using the real-time PCR-based ISO/TS 13136 method. For each step of the procedure, white bars represent the number of laboratories that obtained correct results for all the samples included in the test and black bars the number of laboratories that provided incorrect results or did not perform the assay. doi:10.1128/microbiolspec.EHEC-0014-2013.f6

PERSPECTIVES IN PUBLIC HEALTH

A Proactive Approach to Food Control: Which STEC Should Be Considered Pathogenic?

Adequate risk assessments are crucial for securing the microbiological safety of food, and the characterization of the microbial hazards represents a milestone in such processes. A precise definition of the pathogens provides the basis for their detection and allows the assessment of their prevalence and setting the objectives for the reduction of the associated risk for human health. The characterization of STEC as a food-borne microbial hazard is complex and, at present, a definition of the characteristics associated with STEC pathogenicity remains a matter of discussion. Whether all STEC strains are pathogenic has been disputed among scientists and risk assessors for a long time; the dispute has been partially linked to the broad spectrum of symptoms associated with STEC infections, with a clinical manifestation ranging from the severe forms of HUS and hemorrhagic colitis (HC) to mild diarrhea and asymptomatic carriage. This variability in the clinical outcome, together with the STEC heterogeneity at the genomic level and the plausible effect of the general health status of the patient on the development of the disease, makes it difficult to define unambiguously the features and the genetic background of STEC that might, respectively, cause severe disease, milder symptoms, or no disease at all.

Since the 1980s, it has been proposed that the STEC strains involved in HC and HUS, also termed enterohemorrhagic *E. coli*, were restricted to particular serogroups and were characterized by the ability to induce a typical lesion in the intestine, termed "attaching and effacing" (A/E) and governed by the locus of enterocyte effacement (LEE) pathogenicity island (29). The first attempt to build a coherent classification system for pathogenic STEC was made in the early 2000s, and it is represented by the scheme developed by Karmali and coworkers (23). Such a scheme was based on the evaluation of the virulence and serological features of the strains combined with their association with severe disease and epidemic outbreaks. The integrated analysis of such data introduced the concept of "seropathotype" and led to the construction of a paradigm allocating STEC strains isolated from either human disease or the animal reservoir (Table 1).

The analysis of human cases of STEC infection notified to the ECDC-FWD surveillance system between 2007 and 2011 showed that STEC strains belonging to serogroups O157, O26, O111, O103, and O145 were responsible for roughly 90% of the cases of HUS occurring yearly in the European Union, confirming the applicability of Karmali's scheme (7–11, 13). A similar analysis performed in the United States led to the same observation, with the exception that the most represented

TABLE 1 Classification of STEC serotypes into seropathotypes[a]

Seropathotype	Relative incidence	Frequency of involvement in outbreaks[b]	Association with severe disease	Serotypes[c]
A	High	Common	Yes	O157:H7, O157:NM
B	Moderate	Uncommon	Yes	O26:H11, O103:H2, O111:NM, O121:H19, O145:NM
C	Low	Rare	Yes	O91:H21, O104:H21, O113:H21, other
D	Low	Rare	No	Multiple
E	Nonhuman only	NA	NA	Multiple

[a]Adapted from reference 23.
[b]NA, not applicable.
[c]NM, nonmotile.

serogroups also included O121 and O45 (15; www.cdc.gov/foodnet). However, while the seropathotype model is effective in accommodating the strains from human infections, it presents limitations when a STEC strain isolated from a potential vehicle of infection has to be assigned to a seropathotype or has to be generally evaluated for its wherewithal of representing a hazard. As a matter of fact, STEC strains possessing the LEE pathogenicity island are commonly found in animals, the environment, or food samples, but they can belong to serotypes different from those indicated in the seropathotype scheme. Similarly, STEC strains belonging to the serotypes included in groups A and B, those allocating the most hazardous STEC strains, can possess a partial asset of virulence genes, and, theoretically, they could be considered to be less or not pathogenic at all. Finally, given the high genomic plasticity of *E. coli*, new STEC variants may emerge, with characteristics completely different from those included in the seropathotype concept (28, 30).

In spite of the limitations in categorizing the level of danger associated with STEC from nonhuman sources, the seropathotype concept raised a large consensus in the scientific community and was endorsed by EFSA, which recommended focusing food testing for STEC on the seropathotype A and B groups (24, 25). However, the massive outbreak caused by an enteroaggregative STEC O104:H4 strain in 2011 in Germany (12) has permanently questioned the efficacy of the seropathotype scheme to categorize STEC from food for the adoption of intervention measures. As a matter of fact, the STEC O104 outbreak strain was undoubtedly the most pathogenic STEC ever described and yet it did not fit the seropathotype A and B groups, being negative for the LEE and belonging to a serotype not included in the scheme (28, 31).

The crisis resulting from the STEC O104 outbreak forced the risk managers to reconsider the pathogenicity assessment for STEC, and EFSA conducted a new risk assessment exercise at the request of some European

Union member states (13). A thorough evaluation of the STEC characteristics recorded in the ECDC and EFSA databases in the past 5 years led to the acknowledgment that there is not a combination of genoserotypes that identifies all pathogenic STEC strains, even when the field is narrowed to the STEC causing HC or HUS. It was recognized that some combination of virulence genes, such as particular *stx2* gene subtypes (32), together with the LEE pathogenicity island, might be associated with HUS (33, 34). However, while this observation seems to be consistent for the *stx2* subtypes, the role of the LEE in the colonization of the human intestinal mucosa could be taken over by the action of other adhesion factors, as demonstrated by the STEC O104:H4 outbreak in 2011 (28, 31). Moreover, it was highlighted that patient-associated factors, such as age, immune status, and the administration of antibiotic therapy in the early course of infection (13), have an important role in the final outcome of the STEC infection. Based on these considerations, it was recognized that neither the seropathotype concept nor the analysis of surveillance data allows a certain definition of the pathogenicity of a STEC strain or its level of danger to human health, serving as a proactive tool to protect humans' health. Such uncertainty also applies to molecular risk assessments based on the evaluation of the genetic asset only, since there are no specific genetic markers that, alone or in combination, can assign unambiguously the status of food-borne microbiological hazard to a STEC strain (13).

Regulatory Perspectives in the European Union and the United States

The uncertainties of STEC pathogenicity discussed above hindered the development of control measures to limit the presence of STEC in food and the issuing of specific microbiological criteria for food safety. Nevertheless, some steps toward food-safety rules have been made, often as a reaction to dramatic events such as outbreaks causing deaths and with large mass media impact. The first official measure regarding STEC was taken in the United States at the end of the 20th century. In October 1994, the U.S. Food Safety and Inspection Service declared *E. coli* O157:H7 an adulterant in ground beef in response to the multistate outbreak caused in 1993 by undercooked hamburgers contaminated with STEC O157 (35). The pronouncement was extended in 1999 to all nonintact raw beef products (36). In the 1990s, it was a popular opinion in the United States that *E. coli* O157 was the only STEC serotype causing human disease, and such a viewpoint lasted for more than 10 years. However, the growing number of infections and outbreaks caused by STEC non-O157 led to the assessment that *E. coli* O157 was not the only STEC representing a hazard (18) and the Food Safety and Inspection Service decided on the implementation of sampling and testing of beef manufacturing trimmings for STEC non-O157 starting June 4, 2012 (37). STEC strains belonging to serogroups O26, O111, O103, O145, O121, and O45 were declared as adulterant in these food commodities and included in the sampling plans in addition to *E. coli* O157. Reactions included the withdrawal from the market of the positive batches.

In the European Union, the introduction of food-safety rules related to STEC was incited by the high impact of the STEC O104:H4 outbreak (12), linked to the consumption of contaminated sprouts (38). The high number of casualties and HUS cases, together with the attention of public opinion and the backlash affecting the trade of vegetable products, forced the European Commission to take measures against the possibility that other STEC crises could occur in the European Union. EFSA was asked to perform a risk assessment exercise on the presence of STEC and other pathogenic microorganisms in sprouts and seeds intended for sprouting (38). At the same time, a technical working group, involving experts from the European Union member states and the EU-RL for *E. coli*, discussed the issues related to the definition

of microbiological criteria for STEC in sprouts and the methodology to be adopted for the conformity assessment of this food commodity. The entire process took about 1 year and resulted in the issuance of Regulation (EU) 209/2013, containing the microbiological criteria for STEC in sprouts and amending Regulation (EC) 2073/2005, which lists all microbiological criteria for the assessment of the safety of food and the verification of the process hygiene criteria in the European Union. Regulation (EU) 209/2013 introduced for the first time in the European Union legislation a specific criterion for STEC regarding the presence in sprouts of the five STEC serogroups included in the seropathotypes A and B of Karmali's scheme (23) plus STEC O104:H4. One of the regulation's recitals explained that the reasons for the restriction to certain STEC groups resided in the observation that STEC O157, O26, O103, O111, and O145 are recognized as causing most of the HUS cases in the European Union, whereas STEC O104:H4 caused the large 2011 outbreak. Such a criterion might appear in opposition to the conclusions of the newly released EFSA opinion on STEC pathogenicity assessment (13) (see also "A Proactive Approach to Food Control: Which STEC Should Be Considered Pathogenic?" above). However, in agreement with such an opinion, the same regulation recital stated: "It cannot be excluded that other STEC serogroups may be pathogenic to humans as well. In fact, such STEC may cause less severe forms of disease such as diarrhoea and or bloody diarrhoea or may also cause HUS and therefore represent a hazard for the consumer's health." This last sentence widens the concept of pathogenicity to all STEC and refers to the food business operator the choice of releasing on the market sprouts positive for the presence of STEC that do not fit the microbiological criterion. This approach, apparently contradicting the role of the competent authorities in ensuring food safety, is in agreement with one of the main principles laid down in the general food law [Regulation (EC) 178/2004] that assigns to the food business operators the responsibility to place safe food on the market.

In conclusion, securing food safety is a complex matter that becomes even more complicated when dealing with STEC. As a matter of fact, STEC represents one of the most elusive pathogens in terms of phenotypic characteristics and genomic arrangement, and a continuous challenge for the laboratories in charge of assessing the food safety by applying analytical controls. Furthermore, the food market is a dynamic entity, always chasing the most available and convenient sources of food commodities capable of satisfying the need for cheap food and consumers' demand for "exotic" flavors. This often results in providers of food and raw materials from developing countries (39) introducing, from time to time, pathogens such as the STEC O104:H4 that caused the German outbreak in 2011, bringing into question the food safety and public health systems.

Coping with such complex challenges requires the interplay of the different health care sectors. In fact, it is crucial that all roles of the risk assessment, evaluation, and management referring to both the public health and the food and veterinary fields collaborate, enforcing the concept of "One Health."

ACKNOWLEDGMENT

We declare no conflicts of interest with regard to the manuscript.

CITATION

Caprioli A, Scavia G, Morabito S. 2014. Public health microbiology of Shiga toxin-producing *Escherichia coli*. Microbiol Spectrum 2(4): EHEC-0014-2013.

REFERENCES

1. **Tarr PI, Gordon CA, Chandler WL.** 2005. Shiga-toxin-producing *Escherichia coli* and haemolytic uraemic syndrome. *Lancet* **365:**1073–1086.

2. **Ferens WA, Hovde CJ.** 2011. *Escherichia coli* O157:H7: animal reservoir and sources of human infection. *Foodborne Pathog Dis* **8:**465–487.

3. **Caprioli A, Morabito S, Brugere H, Oswald E.** 2005. Enterohaemorrhagic *Escherichia coli*: emerging issues on virulence and modes of transmission. *Vet Res* **36:**289–311.

4. **Garcia A, Fox JG, Besser TE.** 2010. Zoonotic enterohemorrhagic *Escherichia coli*: a One Health perspective. *ILAR J* **51:**221–232.

5. **Ong KL, Apostal M, Comstock N, Hurd S, Webb TH, Mickelson S, Scheftel J, Smith G, Shiferaw B, Boothe E, Gould LH.** 2012. Strategies for surveillance of pediatric hemolytic uremic syndrome: Foodborne Diseases Active Surveillance Network (FoodNet), 2000–2007. *Clin Infect Dis* **54**(Suppl 5):S424–S431.

6. **Lockary VM, Hudson RF, Ball CL.** 2007. Shiga toxin-producing *Escherichia coli*, Idaho. *Emerg Infect Dis* **13:**1262–1264.

7. **European Food Safety Authority, European Centre for Disease Prevention and Control.** 2009. The Community Summary Report on trends and sources of zoonoses and zoonotic agents in the European Union in 2007. *EFSA Journal* **7:**223.

8. **European Food Safety Authority, European Centre for Disease Prevention and Control.** 2010. The Community Summary Report on trends and sources of zoonoses, zoonotic agents and food-borne outbreaks in the European Union in 2008. *EFSA Journal* **8:**1496.

9. **European Food Safety Authority, European Centre for Disease Prevention and Control.** 2011. The European Union Summary Report on trends and sources of zoonoses, zoonotic agents and food-borne outbreaks in 2009. *EFSA Journal* **9:**2090.

10. **European Food Safety Authority, European Centre for Disease Prevention and Control.** 2012. The European Union Summary Report on trends and sources of zoonoses, zoonotic agents and food-borne outbreaks in 2010. *EFSA Journal* **10:**2597.

11. **European Food Safety Authority, European Centre for Disease Prevention and Control.** 2013. The European Union Summary Report on trends and sources of zoonoses, zoonotic agents and food-borne outbreaks in 2011. *EFSA Journal* **11:**3129.

12. **Buchholz U, Bernard H, Werber D, Bohmer MM, Remschmidt C, Wilking H, Delere Y, an der Heiden M, Adlhoch C, Dreesman J, Ehlers J, Ethelberg S, Faber M, Frank C, Fricke G, Greiner M, Hohle M, Ivarsson S, Jark U, Kirchner M, Koch J, Krause G, Luber P, Rosner B, Stark K, Kuhne M.** 2011. German outbreak of

Escherichia coli O104:H4 associated with sprouts. *N Engl J Med* **365**:1763–1770.

13. **European Food Safety Authority, Panel on Biological Hazards (BIOHAZ).** 2013. Scientific Opinion on VTEC-seropathotype and scientific criteria regarding pathogenicity assessment. *EFSA Journal* **11**(4).

14. **Gould LH, Demma L, Jones TF, Hurd S, Vugia DJ, Smith K, Shiferaw B, Segler S, Palmer A, Zansky S, Griffin PM.** 2009. Hemolytic uremic syndrome and death in persons with *Escherichia coli* O157:H7 infection, foodborne diseases active surveillance network sites, 2000–2006. *Clin Infect Dis* **49**:1480–1485.

15. **Centers for Disease Control and Prevention.** *Foodborne Diseases Active Surveillance Network (FoodNet): FoodNet Surveillance Report for 2011 (Final Report), 2012.* U.S. Department of Health and Human Services, Centers for Disease Control and Prevention, Atlanta, GA. Available from http://www.cdc.gov/foodnet/PDFs/2011_annual_report_508c.pdf.

16. **Swaminathan B, Gerner-Smidt P, Ng LK, Lukinmaa S, Kam KM, Rolando S, Gutierrez EP, Binsztein N.** 2006. Building PulseNet International: an interconnected system of laboratory networks to facilitate timely public health recognition and response to foodborne disease outbreaks and emerging foodborne diseases. *Foodborne Pathog Dis* **3**:36–50.

17. **Tozzi AE, Caprioli A, Minelli F, Gianviti A, De Petris L, Edefonti A, Montini G, Ferretti A, De Palo T, Gaido M, Rizzoni G.** 2003. Shiga toxin-producing *Escherichia coli* infections associated with hemolytic uremic syndrome, Italy, 1988–2000. *Emerg Infect Dis* **9**:106–108.

18. **US Department of Agriculture, Food Safety and Inspection Service.** *Risk Profile for Pathogenic Non-O157 Shiga Toxin-Producing* Escherichia coli *(non-O157 STEC).* 2012. Available from http://www.fsis.usda.gov/wps/wcm/connect/92de038d-c30e-4037-85a6-065c3a709435/Non_O157_STEC_Risk_Profile_May2012.pdf?MOD=AJPERES.

19. **Bielaszewska M, Mellmann A, Bletz S, Zhang W, Kock R, Kossow A, Prager R, Fruth A, Orth-Holler D, Marejkova M, Morabito S, Caprioli A, Pierard D, Smith G, Jenkins C, Curova K, Karch H.** 2013. Enterohemorrhagic *Escherichia coli* O26:H11/H–: a new virulent clone emerges in Europe. *Clin Infect Dis* **56**:1373–1381.

20. **Zhang WL, Bielaszewska M, Liesegang A, Tschape H, Schmidt H, Bitzan M, Karch H.** 2000. Molecular characteristics and epidemiological significance of Shiga toxin-producing *Escherichia coli* O26 strains. *J Clin Microbiol* **38**:2134–2140.

21. **Ricotti GC, Buonomini MI, Merlitti A, Karch H, Luzzi I, Caprioli A.** 1994. A fatal case of hemorrhagic colitis, thrombocytopenia, and renal failure associated with verocytotoxin-producing, non-O157 *Escherichia coli. Clin Infect Dis* **19**:815–816.

22. **Bielaszewska M, Prager R, Kock R, Mellmann A, Zhang W, Tschape H, Tarr PI, Karch H.** 2007. Shiga toxin gene loss and transfer in vitro and in vivo during enterohemorrhagic *Escherichia coli* O26 infection in humans. *Appl Environ Microbiol* **73**:3144–3150.

23. **Karmali MA, Mascarenhas M, Shen S, Ziebell K, Johnson S, Reid-Smith R, Isaac-Renton J, Clark C, Rahn K, Kaper JB.** 2003. Association of genomic O island 122 of *Escherichia coli* EDL 933 with verocytotoxin-producing *Escherichia coli* seropathotypes that are linked to epidemic and/or serious disease. *J Clin Microbiol* **41**:4930–4940.

24. **European Food Safety Authority, Panel on Biological Hazards (BIOHAZ).** 2007. Scientific Opinion of the Panel on Biological Hazards on a request from EFSA on monitoring of verotoxigenic *Escherichia coli* (VTEC) and identification of human pathogenic VTEC types. *EFSA Journal* **579**:1–61.

25. **European Food Safety Authority, Panel on Biological Hazards (BIOHAZ).** 2009. Technical specifications for the monitoring and reporting of verotoxigenic *Escherichia coli* (VTEC) on animals and food (VTEC surveys on animals and food) on request of EFSA. *EFSA Journal* **7**:1366.

26. **Fisher IS.** 1999. The Enter-net international surveillance network—how it works. *Euro Surveill* **4**:52–55.

27. **Hoefer D, Hurd S, Medus C, Cronquist A, Hanna S, Hatch J, Hayes T, Larson K, Nicholson C, Wymore K, Tobin-D'Angelo M, Strockbine N, Snippes P, Atkinson R, Griffin PM, Gould LH.** 2011. Laboratory practices for the identification of Shiga toxin-producing *Escherichia coli* in the United States, FoodNet sites, 2007. *Foodborne Pathog Dis* **8**:555–560.

28. **Scheutz F, Nielsen EM, Frimodt-Moller J, Boisen N, Morabito S, Tozzoli R, Nataro JP, Caprioli A.** 2011. Characteristics of the enteroaggregative Shiga toxin/verotoxin-producing *Escherichia coli* O104:H4 strain causing the outbreak of haemolytic uraemic syndrome in Germany, May to June 2011. *Euro Surveill* **16**(24).

29. **Levine MM.** 1987. *Escherichia coli* that cause diarrhea: enterotoxigenic, enteropathogenic, enteroinvasive, enterohemorrhagic, and enteroadherent. *J Infect Dis* **155**:377–389.

30. **Colic E, Dieperink H, Titlestad K, Tepel M.** 2011. Management of an acute outbreak of diarrhoea-associated haemolytic uraemic syndrome

with early plasma exchange in adults from southern Denmark: an observational study. *Lancet* **378:**1089–1093.

31. **Bielaszewska M, Mellmann A, Zhang W, Kock R, Fruth A, Bauwens A, Peters G, Karch H.** 2011. Characterisation of the *Escherichia coli* strain associated with an outbreak of haemolytic uraemic syndrome in Germany, 2011: a microbiological study. *Lancet Infect Dis* **11:**671–676.

32. **Scheutz F, Teel LD, Beutin L, Pierard D, Buvens G, Karch H, Mellmann A, Caprioli A, Tozzoli R, Morabito S, Strockbine NA, Melton-Celsa AR, Sanchez M, Persson S, O'Brien AD.** 2012. Multicenter evaluation of a sequence-based protocol for subtyping Shiga toxins and standardizing Stx nomenclature. *J Clin Microbiol* **50:**2951–2963.

33. **Friedrich AW, Bielaszewska M, Zhang WL, Pulz M, Kuczius T, Ammon A, Karch H.** 2002. *Escherichia coli* harboring Shiga toxin 2 gene variants: frequency and association with clinical symptoms. *J Infect Dis* **185:**74–84.

34. **Persson S, Olsen KE, Ethelberg S, Scheutz F.** 2007. Subtyping method for *Escherichia coli* Shiga toxin (verocytotoxin) 2 variants and correlations to clinical manifestations. *J Clin Microbiol* **45:** 2020–2024.

35. **Barrett TJ, Lior H, Green JH, Khakhria R, Wells JG, Bell BP, Greene KD, Lewis J, Griffin PM.** 1994. Laboratory investigation of a multistate food-borne outbreak of *Escherichia coli* O157:H7 by using pulsed-field gel electrophoresis and phage typing. *J Clin Microbiol* **32:**3013–3017.

36. **US Department of Agriculture, Food Safety and Inspection Service.** 1999. *FSIS Policy on Nonintact Raw Beef Products Contaminated with E. coli O157:H7.* Available from http://www.fsis.usda.gov/Oa/background/O157policy.htm.

37. **US Department of Agriculture, Food Safety and Inspection Service.** 2011. Shiga toxin-producing *Escherichia coli* in certain raw beef products. *Fed Regist* **76:**58157–58165.

38. **European Food Safety Authority, Panel on Biological Hazards (BIOHAZ).** 2011. Scientific Opinion on the risk posed by Shiga toxin-producing *Escherichia coli* (STEC) and other pathogenic bacteria in seeds and sprouted seeds. *EFSA Journal* **9:**2424.

39. **European Food Safety Authority.** 2011. *Tracing seeds, in particular fenugreek (Trigonella foenum-graecum) seeds, in relation to the Shiga toxin-producing E. coli (STEC) O104:H4 2011 outbreaks in Germany and France.* Available from http://www.efsa.europa.eu/it/supporting/doc/176e.pdf.

DIAGNOSIS, DETECTION, AND STRAIN CHARACTERIZATION

Detection of Shiga Toxin-Producing *Escherichia coli* from Nonhuman Sources and Strain Typing

14

LOTHAR BEUTIN[1] and PATRICK FACH[2]

INTRODUCTION

Shiga toxin (verotoxin)-producing *Escherichia coli* (STEC) strains were first described in 1977 by their ability to cause cytotoxic effects on Vero cells (1). In the early days, production of Shiga toxin (Stx) by *E. coli* was thought to be associated with certain human enteropathogenic *E. coli* (EPEC) strains (1–3). STEC was recognized as a zoonotic agent when the first known outbreaks occurred in 1982. STEC O157:H7, a rare serotype of *E. coli*, was isolated from patients developing hemorrhagic colitis (HC) after they ingested undercooked beef in restaurant chains (4). STEC O157 was isolated from the incriminated beef, indicating a possible transmission from a bovine reservoir. In the following years cattle were identified as a worldwide natural reservoir for STEC O157 and non-O157 strains and as an important source of food contamination (5–8). Repeated sampling of cattle revealed that the agent was occasionally present in the majority of cattle farms in Europe and America (9). However, recent findings on the epidemiology of enteroaggregative Shiga toxin-producing *E. coli* (EAEC-STEC) O104:H4 indicate that not all STEC strains have a zoonotic origin (10, 11).

[1]National Reference Laboratory for *Escherichia coli*, Department of Biological Safety, Federal Institute for Risk Assessment (BfR), D-12277 Berlin, Germany; [2]Food Safety Laboratory, ANSES (French Agency for Food, Environmental and Occupational Health and Safety), F-94706 Maisons-Alfort, France.

Enterohemorrhagic Escherichia coli *and Other Shiga Toxin-Producing* E. coli
Edited by Vanessa Sperandio and Carolyn J. Hovde
© 2015 American Society for Microbiology, Washington, DC
doi:10.1128/microbiolspec.EHEC-0001-2013

STEC in Food, Animals, and the Environment

With the improvements in STEC detection methods, the search for natural STEC reservoirs and for *E. coli* serotypes associated with Stx production intensified. STEC was identified as a frequent colonizer of domestic and wild ruminants such as cattle, sheep, goats, and deer (9, 12–16). Some nonruminant domestic and wild animals such as pigs, wild boar, and hares were also identified as important sources and natural reservoirs of serologically diverse STEC types (12, 15, 16).

Today, it is known that STEC is present as a resident of the gut flora or transiently carried by many species of mammals, as well as birds, insects, mollusks, and fish (17). Contamination of food of animal origin occurs frequently at the critical stages of food production, such as milking, slaughtering, and evisceration. Accordingly, numerous types of STEC are found in meat and milk products from domestic and wildanimals (8, 18). Fecal shedding of STEC from infected animals results in wide contamination of farmland, water, and fresh produce (19). Wastewater processing plants do not eliminate STEC completely from treated water (20). As a consequence, STEC may be present in irrigation water, resulting in surface contamination of herbal foodstuffs such as vegetables and fruits. Effluents from farmland containing STEC contaminate coastal waters and shellfish (21, 22). Birds and insects play an important role in disseminating STEC over long distances, thus allowing the spread of the agent over larger geographical areas (17). Direct and indirect transmission of STEC from animal to animal and from animal to humans was observed for O157 and other STEC types (17, 23–25). STEC of various serotypes was found to persist on contaminated farmland, pen floors, and the skin of animals, and evidence for horizontal dissemination of STEC clones and recontamination of animals was found (26). Once in the environment, STEC may survive in soil and water for weeks and months, depending on biotic factors and the physicochemical conditions in general (27). The dissemination of STEC is also promoted by the occurrence of free *stx* phage particles in the environment (28). Transfer of *stx* genes to *E. coli* was found to occur by free Stx phages in food and water, giving numerous possibilities for the generation and spread of STEC strains (29).

The ecology of STEC in animals and its relative stability to physicochemical stress conditions have the effect that this agent is found in practically all kinds of foodstuffs, animals, and environments. This also includes geographical areas that are not used for agricultural activities.

Diversity of STEC Types

After the first outbreaks with Stx-producing O157:H7 became known, the search for STEC concentrated mainly on this particular serotype. However, investigations of animal reservoirs and human patients infected with STEC revealed an enormous diversity of *E. coli* serotypes associated with Stx production. A compilation published in 2005 of STEC strains isolated from human patients listed more than 400 serotypes (30). It is likely that the number of STEC serotypes from patients has increased. A review article from 2007 listed 171 *E. coli* O groups as associated with Stx production. The STEC strains were from diseased humans and healthy cattle and sheep (31). In summary, with a few exceptions, *E. coli* strains of virtually all known O serogroups were already found associated with Stx production. In addition, many STEC strains with untypeable (unknown) O antigens were isolated from patients, animals, and food, indicating that the serotype diversity of STEC is beyond the current serotype scheme, including more than 182 O antigens and 53 H antigens (18, 30, 31, 33).

On the other hand, epidemiological data on STEC serotypes isolated from humans indicate that a small number of STEC serotypes predominate in patients with severe illness (34–36). Most of these STEC strains belong to the group of enterohemorrhagic *E. coli* (EHEC) strains and mainly to serogroups O26, O45, O103, O111, O121, O145, and O157. EHEC is most frequently implicated in severe disease such as HC and hemolytic-uremic syndrome (HUS) (33, 36). In contrast to other STEC strains, EHEC strains were found to be similar to each other for the presence of genes involved in the intestinal attaching and effacing (LEE) mechanism and for non-LEE (*nle*) effector genes. Accordingly, a molecular risk assessment (MRA) concept was developed to assess the relative pathogenicity of STEC strains according to the presence of virulence attributes and their frequency in human infections (37–40).

Other STEC strains lacking LEE and non-LEE effectors were also found to cause HC and HUS in humans. Among these, STEC strains of serogroups O91, O104, O113, O128, and O146 are most frequently mentioned (33–36, 41). In contrast to EHEC, these STEC strains are less well defined for the molecular basis of their human pathogenicity (42). However, the outbreak with EAEC-STEC O104:H4 strains in Europe showed that effective colonization of the human intestine together with production of Stx2a is sufficient to cause HC and HUS (11, 43). Therefore, it is likely that other serotypes of STEC strains might emerge as severe human pathogens in the future. As a consequence, the search for STEC should a priori not be restricted to a limited range of *E. coli* serotypes. Accordingly, this report concentrates on both STEC detection procedures independent of the serotype and methods specifically developed for detection of EHEC O157:H7 and major non-O157 EHEC serogroups (O26, O45, O103, O111, O121, and O145) that are associated with severe illness in humans worldwide. Methods specifically developed for the detection of EHEC O157:H7 only are not addressed here.

STEC DETECTION STRATEGIES

Because of the variety of STEC serotypes and phenotypes, the identification of Shiga toxins or the underlying genes is the only way to detect virtually all STEC types from any kind of sample (44, 45). Over the last 3 decades, Stx detection assays were developed. These assays can be grouped into tissue culture cytotoxicity assays, immunological assays, and DNA-based methodologies.

Cytotoxicity assays and immunological assays are phenotypic assays indicating the synthesis and the relative amount of Stx produced by the bacterium. DNA-based methodologies deal with the molecular detection of *stx* genes. Shiga toxins comprise a growing family of genes with enormous type diversity (46). The Stx family splits into two major branches, Stx1 and Stx2, which are immunologically not cross-reactive and show about 55% difference in their amino acid sequences (47). The Stx1 family divides into four major subtypes (Stx1, Stx1a, Stx1c, and Stx1d), which further split into a number of genetic variants. The Stx2 family branches into seven major subtypes (Stx2a, Stx2b, Stx2c, Stx2d, Stx2e, Stx2f, and Stx2g), which subdivide into a total of 93 genetic variants (46). Because of the enormous type diversity of Stx1 and Stx2 family toxins, problems can arise in detecting all toxin subtypes with both serological and genetic assays.

Cytoxicity Assays for Stx

Cytotoxicity tests are regarded as the "gold standard" for Stx testing as only these tests indicate the cytotoxic effect on mammalian cells (48). They are commonly performed with Vero cells but require neutralization tests with Stx1- and Stx2-specific antisera to confirm the specificity of the result (49). HeLa cells might also be used but were shown to be less sensitive to certain Stx variants (50). Cytotoxicity tests are suitable for demonstrating the biological activity of Stx or the expression of the *stx* genes detected by serological or DNA-based assays, respectively (51, 52). Comparative studies have shown that cytotoxicity tests are more sensitive for detection of Stx than immunological assays (48). The detection limits of any test system can be critical if the samples contain low amounts of Stx or if the assay fails to detect certain subtypes of Stx1 or Stx2 (51, 53, 54). Cytotoxicity tests can be used for direct Stx testing from fecal samples where immunological methods can fail (53). However, fecal, food, and environmental samples may contain mixtures of bacteria that can produce enterotoxins, cytolysins (hemolysins), or other cytotoxins different from Stx but equally active on Vero or HeLa cells (50). Without Stx-neutralization assays, these kinds of samples may produce confounding results.

On the other hand, test samples containing glycolipids (Gb3 + Gb4) that bind to the Stx receptor (B-subunit) can neutralize the cytotoxic effect in cell culture assays (55, 56). For routine screening of test samples, cytotoxicity tests are less frequently used because they are labor-intensive, need specific equipment and personnel skills, and require 72 h before final conclusive reading (48).

Immunological Assays for Stx

A large number of immunological assays for the detection of Stx have been developed, and many are available commercially. These serological assays are enzyme-linked immunosorbent assays (ELISAs), reverse passive latex agglutination assays, immunochromatographic lateral flow test systems, and the colony immunoblot. Capture of Stx from the sample may occur through the P1 glycoprotein, Gb3 receptor, or monoclonal antibodies directed against Stx1 and Stx2 (50).

Some commercially available tests were evaluated for their sensitivity and specificity (51, 53, 54, 57–59). A number of serological assays gave weak reactions or did not detect some strains expressing Stx variants such as Stx2c, Stx2e, Stx2f, and Stx2g (51, 57, 59, 60). Significant differences in the performance of some commercial Stx ELISAs were revealed by an interlaboratory study performed on beef samples inoculated with STEC strains producing different kinds of Stx (61).

Stx ELISA and an immunochromatographic lateral flow system were used to compare representative strains covering all known toxin types, and all Stx variants, except Stx2f, were detectable. Except for Stx2f, the false-negative results were assumed to be due to low toxin production, which was below the detection limit of the serological assay (59). Stx ELISA and the immunochromatographic flow system detected <10 CFU of EHEC (O26, O103, O111, O118, O121, O145, and O157) per 25 g in enrichment cultures grown from artificially contaminated salad samples (59).

Detection limits of different serological tests were reported as between 0.5 ng/ml to 126 ng/ml, depending on the test system and the Stx types analyzed (58, 62, 63). Stx2f is genetically most distant from all other Shiga toxins (46). It is therefore possible that the Stx2-specific antibodies used in most immunological assays do not react with the Stx2f variant (59, 60). The combination of a serological test with a PCR assay (Immuno-PCR) was shown to decrease the detection limit for Stx2 from 1 ng/ml to 10 pg/ml (64). PCR-ELISA was reported to increase 100-fold the sensitivity of a conventional Stx PCR assay (65). A PCR-ELISA technique was successfully used to screen and isolate STEC from naturally contaminated dairy products and retail minced-beef samples (66, 67).

Immunological assays, such as the verocytotoxigenic *E. coli* (VTEC) reverse passive latex agglutination assay, were found sufficiently sensitive to identify single Stx-producing colonies from agar plates after treatment with polymyxin B (68). When larger numbers of colonies need to be screened for the presence of STEC, the colony immunoblot offers the possibility of detection and isolation in a one-step procedure (44, 69–72) (Fig. 1). Commercially available hybridoma cell lines for Stx1 (ATCC CRL-1794) and Stx2 (ATCC CRL-1907) proved to be successful for identification of different STEC strains from food and fecal samples (69, 70, 73). Food inspection laboratories in Germany widely use the colony immunoblot since, according to German legislation,

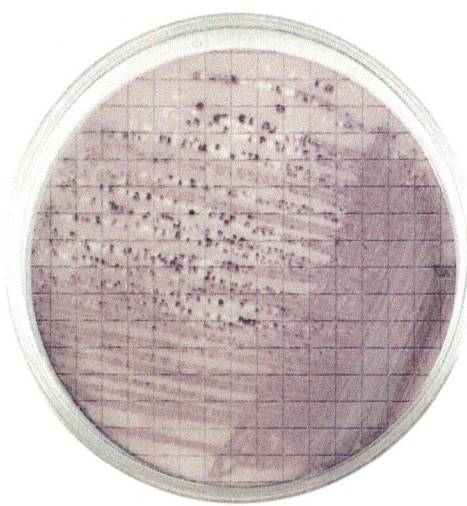

FIGURE 1 Identification and isolation of STEC from mixed samples of bacteria with the Stx colony immunoblot. Mauve-stained Stx2-positive STEC colonies are detected from a sample containing STEC mixed with Stx-negative bacteria.
doi:10.1128/microbiolspec.EHEC-0001-2013.f1

all STEC types present in food must be regarded as potentially hazardous (61).

DNA-Based Assays for Stx and Other Virulence Markers of STEC

DNA-based assays to detect STEC are rapid to perform and have the advantage of being independent of the availability of specific Stx antisera and cell culture facilities. Universal Stx1 and Stx2 gene probes (74, 75) were developed and found useful for simultaneous detection and isolation of STEC in colony hybridization assays with different kinds of samples (50, 76–80). For general detection of STEC from all types of samples, common stx PCR assays covering the different stx variant genes were developed (67, 81–84). Considerable differences in specificities were found when some PCR and PCR-restriction fragment length polymorphism (RFLP) methods were compared (85). PCR-RFLP protocols use specific differences in the restriction endonuclease patterns of stx PCR products for further toxin subtyping. PCR-RFLP was also helpful in detecting multiple copies of stx_2 variant genes, which are frequently present in certain STEC strains (86, 87). As universal PCR primers fail to detect some Stx subtypes such as stx_{2f}, a minimum of one primer pair for the Stx1 family and two primer pairs for the Stx2 family were found necessary to cover all known Shiga toxin variants (46, 88).

Conventional multiplex PCR assays were developed to detect STEC and were successfully tested on bacterial isolates (89). Conventional PCR assays require separation of amplified DNA molecules by gel electrophoresis followed by interpretation of the results from agarose gels. The interpretation of PCR results can cause problems in assays performed with DNA obtained from mixed cultures of bacteria. Results were ambiguous in testing enrichment cultures from meat samples with a common primer system for stx_1 and stx_2 (90) (Fig. 2). The same stx primers were used in a study of STEC in fecal samples of human healthy volunteers, resulting in 13.9% stx-PCR-positive samples; however, STEC could be detected only in 1 of the 23 stx-PCR-positive samples (91). New rapid PCR assays that do not need gel electrophoresis, such as loop-mediated isothermal amplification, were used to detect 1 to 20 CFU of EHEC/25 g by detecting serogroup-associated genes in spiked meat and produce samples and could be an alternative to conventional PCR assays (92).

Real-time PCR assays overcome the gel electrophoresis step and have the advantage that the PCR is combined with gene probe hybridization, thus ensuring a quantitative, more specific, and very rapid detection assay. In the last decade, real-time PCR systems for universal STEC screening became more important for examining clinical, food, and environmental samples (93–97). Two sets of primers were successfully used for real-time PCR using SYBR Green to detect different variants of Stx1 and Stx2 genes (98). In an evaluation study on a large number of STEC and non-STEC strains, two degenerated

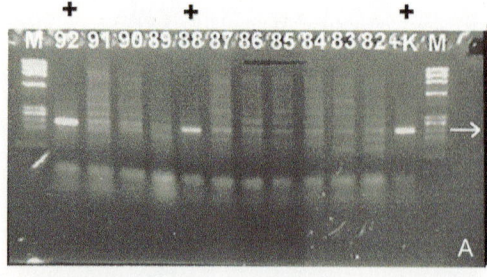

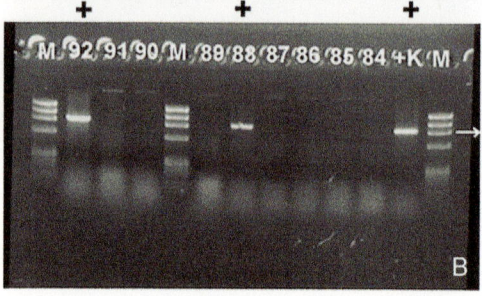

FIGURE 2 Suitability of conventional PCRs for detection of *stx* genes in enriched samples from minced meat. The same panel of enrichment cultures (#84–92) was tested with two PCR systems using the MK1/MK2 primer system (A) and the Lin-up/Lin-down primer system (B) (90). Primers MK1/MK2 generate 230-bp-long PCR products (arrow), and a ladder of nonspecific bands of different sizes is visible in almost all meat samples. (B) Lin-up/Lin-down generates 900-bp size PCR products (arrow), and only two STEC-positive meat samples (#88+92) give corresponding PCR products. M, molecular weight standard, K, positive STEC control; arrow, position of the specific PCR product.
doi:10.1128/microbiolspec.EHEC-0001-2013.f2

primer pairs and two labeled gene probes were found sufficient to detect all stx_1 and stx_2 variant genes except for stx_{2f} (99). The same primers and probes were found to detect STEC from enrichment cultures of ground beef samples inoculated with 2 to 20 CFU of bacteria/25 g (100). Alternative real-time PCR detection systems using common stx_1 and stx_2 primer sets and probes have been successfully tested on ground beef samples artificially contaminated with EHEC (1 to 10 CFU/25 g) belonging to the most prominent serogroups (101). Two sets of PCR primers and TaqMan minor groove binder probes were successfully employed to detect EHEC O26, O111, O118,

O121, O145, and O157 (1 to 10 CFU/25 g) in spiked salad samples and in sprouted seeds naturally contaminated with STEC O104:H4 (102). A large number of tests based on real-time PCR amplifications have been developed, and many are available commercially. However, a comparison of different sets of commercialized real-time PCR assays for the detection of STEC revealed significant differences in their specificity and sensitivity values (103).

Most promising with real-time PCR systems is the simultaneous detection of the characteristic virulence markers and phenotypic traits of EHEC strains, permitting their rapid identification from mixed cultures in a multiplex assay. Two reference methods based on real-time PCR detection of the most important EHEC types of strains from food have been developed. One has been standardized at European (Commission for European Normalization, CEN) and international (International Organization for Standardization, ISO) levels (104). The other is an official method used in the United States for meat products (MLG 5B.01) (105). Both methods rely on the detection of the most important EHEC types. STEC strains are considered highly pathogenic when they combine the presence of stx_1 or stx_2 and *eae* genes and belong to one of the serotypes O157:H7, O26:H11, O145:H28, O111:H8, and O103:H2 ("big five") or their nonmotile variants (104, 106). The method used in the United States includes the same panel of EHEC serotypes and, in addition, EHEC of serogroups O45 and O121 ("big seven group") (105). Interestingly, EHEC O45:H2 strains are isolated from patients in North America but are not yet notified as disease agents in Europe. In contrast, EHEC O121:H19 strains were repeatedly isolated from patients with HUS on the European and North American continents (34–36, 107, 108).

The European and U.S. EHEC screening methods first investigate the presence of *stx* (97) and *eae* (109) genes. In the case of a positive result, the presence of genes encoding the lipopolysaccharide (LPS) of the above-mentioned EHEC serogroups is tested (88, 104, 105). A list of the real-time PCR detectors used in the European CEN/ISO method was published by Stephan et al. (88). Both procedures were published and became effective in 2012.

This sequential procedure can provide four different results. When *stx* and *eae* genes are not detected, one concludes that STEC strains are absent. When an *stx* gene is detected but no *eae* gene, one presumes the presence of an STEC strain with no or low risk for public health. In both cases, no further action is taken. If *stx* and *eae* genes are detected, the sample contains STEC strains potentially pathogenic to humans. In that case, the presence of genes coding for the above-mentioned EHEC serogroups is investigated. If one of these serogroups is detected, the test portion may contain STEC highly pathogenic to humans, and a confirmation is required by isolating the supposed EHEC strain from the sample. Bacteria from samples positive for *stx*, *eae,* and one or more of the investigated LPS-encoding genes are isolated using serotype-specific immunomagnetic assays to confirm the presence of these markers in the same strain (104, 105). The single colonies thus isolated are investigated with the previously described genetic markers to confirm the presence of all pathogenicity markers in the same strain.

The isolation of the EHEC strain is important, as samples from feces, food, and water may contain mixtures of *eae*-negative STEC strains, *stx*-negative strains belonging to the EHEC serogroups, and *eae*-positive EPEC strains not producing Stx (110–114). Without doubt, the adaptation of these or alternative (115) real-time PCR protocols will be very helpful for detecting most EHEC strains pathogenic to humans from food samples.

However, it is still not completely clear which other factors contribute to the human pathogenicity of STEC strains not belonging to the "big five" or "big seven" EHEC groups that are targeted by sequential assays described above. As a consequence of such restrictions, there is always a risk that other serotypes of STEC that might be highly pathogenic for humans as well are not taken into account. For example, a considerable number of clinical isolates of STEC strains different from the "big five" and "big seven" groups were isolated from patients with HUS in Germany and are listed in the HUSEC collection (108). A number of *eae* -and *stx*-positive STEC strain types serologically different from the "big seven" group were recently found to be similar to the "big seven" EHEC strains according to their *nle* gene profiles (40). These "emerging EHEC" strains belong to serotypes that were previously shown to be associated with HC and HUS in humans (O5:H–, O15:H2, O55:H7, O103:H11, O103:H25, O118:H16, O123, O165:H25, O172:H25, and O177:H–) (34, 116–119).

The outbreak in Europe with EAEC-STEC O104:H4 in summer 2011 caused a paradigm shift with regard to human pathogenicity of STEC strains (10, 11). The presence of the *eae*-negative EAEC-STEC O104:H4 strain in a food matrix would not have been considered as dangerous according to the criteria based on the European Union and U.S. guidelines described above (11). It is therefore important to adopt such procedures for the current situation by including new significant gene targets such as *aggR* for identification of EAEC-STEC strains (11, 102). Recently, PCR tests specific for EAEC-STEC O104:H4 (120) were stipulated in the new regulation on sprouts and seeds published in Europe in March 2013.

Severe disease such as bloody diarrhea cannot be attributed simply to the *stx* genotype or certain serotypes of STEC (36, 121–123) and is not only restricted to infections with typical EHEC strains. Besides O104:H4, other types of LEE (*eae*)-negative STEC strains can play a role as severe pathogens of humans (42). Only half of human STEC infections in the European Union were caused by "big five" strains (107). This indicates that searching for STEC potentially pathogenic to humans in animals, food, and the environment should not be limited to the set of *stx*- and LEE (*eae*)-positive STEC strains.

Cultural Enrichment of STEC from Nonhuman Samples

STEC is generally present in relatively low numbers in environmental and food matrices and in feces of ruminant animals that are the natural hosts for these strains. For STEC screening, it is therefore important to include cultural enrichment steps that allow growth of STEC to detectable numbers by either method. Enrichment in liquid media by sample dilution (generally 1:10) reduces possible inhibitors of STEC growth that might be present in the sample. A second enrichment step of plating aliquots from liquid enrichment cultures grown on solid agar plates and investigating DNA prepared from the lawn of bacteria can enhance the sensitivity of STEC detection (99, 102).

For detection of Stx by serological assays including the colony immunoblot, media containing substances that enhance production of Stx such as mitomycin C are employed (50, 53, 59, 72). Mitomycin C can boost drastically Stx production and toxin release by bacteria (124, 125). Certain STEC strains do not produce detectable quantities of Stx without induction (52, 125). Apart from mitomycin C, which is a highly toxic substance, different classes of low-toxicity antibiotics were found to induce the production of Stx (126–128). These antibiotics may present alternatives as toxin enhancers for detection of Stx in phenotypic assays (125).

The addition of Stx enhancers is not needed for STEC detection by DNA-based methods. Nevertheless, microbial enrichment of test samples is always recommended and is a critical step in any protocol for detection of STEC from all kinds of samples (88, 129). Ideally, enrichment favors growth of STEC and disfavors growth of other organisms present in the sample. However, apart from protocols developed for STEC O157:H7 (130, 131), there is no standard method for specific enrichment of non-O157 STEC strains from food and other samples (129, 132). As a consequence, numerous enrichment media and growth conditions for STEC from food, fecal, and environmental samples were developed (129, 133, 134) (Table 1). However, most of these have not been evaluated for their relative efficacy (135).

Testing growth of STEC strains in different types of commonly used enrichment media revealed strain-specific differences, depending on incubation temperature and media supplements (136). Enrichment media containing novobiocin that are used in standard protocols for enrichment of STEC O157 strains (130, 131) were found to inhibit growth of some non-O157 STEC (136) and freeze-injured STEC O157 strains (61, 88, 137). A preenrichment step in a nonselective medium is therefore generally recommended for growing stressed STEC from food and other samples (88, 129, 131).

Enrichment of STEC from suspected food samples is also important to avoid false-positive results for STEC. Real-time PCR showed high CT values (>35), indicating the presence of a very low number of STEC, in 7.5% of meat and 16.0% of cheese samples after cultural enrichment (138). Very high CT values after enrichment may indicate that nonenrichable *stx* sequences are present that might correspond to *stx* phages or nonculturable STEC strains (138, 139) (Fig. 3).

TABLE 1 Different types of enrichment media for STEC[a]

Tryptose soy broth (TSB)
Buffered peptone water (BPW)
Gram-negative broth (GN)
E. coli broth (EC)
Mineral modified glutamate broth (MMGB)
Brilliant green bile broth (BGBB)
Lauryl sulfate tryptose broth (LSTB)
MacConkey broth (MCB)
Enterohemorrhagic *E. coli* broth (EHEC)
Brain heart infusion broth (BHI)

[a]The following substances are used (singly or in combination) as supplements to increase selectivity: novobiocin, vancomycin, cefixime, acriflavine, cefsulodin, cefixime, K-tellurite, and bile salts. Adapted from reference 134.

Apart from the composition of the enrichment medium, the incubation time for enrichment (generally 6 to 24 h) and the temperature (between 35 and 41.5°C) are critical. STEC O157 was found to grow best between 30 and 42°C (140). The incubation temperature of 41.5 ± 1°C, as recommended by ISO 16654 (130), might hamper the recovery of injured bacteria (88). The composition of the growth medium together with the incubation temperature (37°C versus 42°C) has a significant influence and may cause multiple effects on the growth behavior of STEC belonging to different serotypes (136).

The chosen temperature can be very important for cultural enrichment of STEC from samples where high numbers of a bacterial flora from the environment are present.

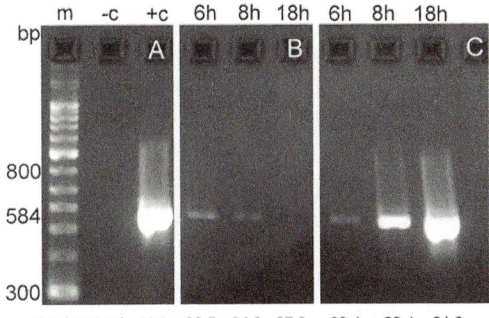

FIGURE 3 Importance of cultural enrichment for the detection of viable STEC in meat samples by conventional and real-time PCR. Agarose gel showing Stx2-specific PCR products in enrichment cultures. The PCR was performed with common stx2 PCR primers LP43/LP44 as described in reference 86. The corresponding CT values from real-time PCR are indicated below. (A) Lanes: m, molecular weight standard; –c, Stx-negative *E. coli* control strain; +c, Stx2-positive *E. coli* control strain. (B) Sample containing nonamplifiable *stx* genes. PCR and real-time PCR are performed after 6, 8, and 18 h of enrichment. *Stx* gene-specific signals are decreasing following prolonged enrichment with both conventional and real-time PCR. (C) Sample containing amplifiable *stx* genes. PCR and real-time PCR are performed after 6, 8, and 18 h of enrichment. *Stx* gene-specific signals are increasing following prolonged enrichment with both conventional and real-time PCR.
doi:10.1128/microbiolspec.EHEC-0001-2013.f3

Vegetables such as ready-to-eat salads may contain about 10^7 CFU/g of total aerobic bacteria, and bacterial growth continues at retail storage temperatures of 6 to 10°C (102, 141). More than 90% of the total microbial counts in ready-to-eat salads are represented by *Pseudomonas* and *Enterobacteriaceae* (142, 143). As these bacteria are generally inhibited by a growth temperature of 44°C, a pre-enrichment step at 37°C for 6 h in liquid medium (buffered peptone water or brilliant green bile broth) (Table 1) followed by a second enrichment step by growth on selective agar plates incubated at 44°C for 18 to 22 h was found suitable to recover an initial inoculum of 1 to 10 CFU of EHEC/25 g from ready-to-eat salad samples (102). The choice of the best enrichment media, enrichment time, and temperature may vary according to the sample matrix (food, environment, or feces),which has to be investigated for STEC; no general recommendation can be given at the present state.

ISOLATION OF STEC FROM Stx-POSITIVE SAMPLES

The isolation and further typing of STEC from incriminated samples are required to confirm positive results from Stx screening and to uncover chains of infection in STEC outbreaks and sporadic cases (50). Detecting Stx production or the underlying genes in any kind of sample without attempting to isolate the STEC strain is incomplete and should be considered as presumptive (50, 132).

In most samples, STEC is present in low numbers, and isolation may involve testing large numbers of colonies (132). Thus isolation of STEC is the most laborious part of STEC screening as the majority of STEC strains show phenotypic properties very similar to those of commensal *E. coli* and other *Enterobacteriaceae,* which can be present in any kind of sample.

The most straightforward way to isolate STEC from a sample is direct screening for

Stx-positive colonies through the Stx colony immunoblot or colony hybridization (50). Kits for DNA hybridization and a Stx colony immunoblot kit are commercially available, and these methods for STEC detection are officially recommended for food inspection laboratories in Germany (61). However, as these procedures are labor-intensive and time-consuming, they reach their limits when larger numbers of samples have to be investigated in time.

Because many countries concentrate only on a limited number of EHEC serogroups (O157, O103, O111, O26, and O145), commercially available immunomagnetic separation (IMS) assays are used for enrichment of suspected colonies before cultural detection (100, 104, 105, 137, 144). Studies have shown that the IMS enrichment increases the isolation rate of the targeted bacteria (45, 50, 131). IMS beads directed to other EHEC serogroups can be prepared with commercially available reagents if the appropriate antisera for coating are available (101). However, the use of IMS is limited to the targeted *E. coli* serogroups, and serologically O-rough derivatives of these STEC strains will not be captured specifically by O-antigen-directed IMS beads.

Independent of the enrichment procedure used, STEC colonies must be grown on solid agar plates to obtain pure cultures for storage and further characterization. Solid media for STEC isolation can be divided into nonselective media (such as brain heart infusion agar) (145, 146), indicator media for certain phenotypic properties of STEC strains (such as enterohemolysin agar) (57), and selective media which frequently contain tellurite or antibiotics (such as CT-SMAC) (50). Very frequently, indicator media are modified by the addition of selective agents or antibiotics to increase their selectivity (101, 147–151).

Nonselective media are suggested for isolation of presumptive EHEC according to the CEN/ISO method. After IMS enrichment, a minimum of 50 colonies on nonselective medium (nutrient agar) followed by *stx* PCR investigation of pooled and single colonies is proposed for isolation of STEC from positively tested samples (104). The USDA/FSIS method uses a chromogenic, selective indicator medium (modified Rainbow Agar) for the detection of major EHEC-type strains after IMS and a serological slide agglutination test for identification, followed by PCR detection of relevant genes for confirmation (105).

A number of indicator media and selective media have been developed for detection and selection of STEC strains (102, 129, 146, 150). Most of these media were specifically developed for the detection of non-sorbitol-fermenting STEC O157:H7 strains. Addition of antibiotics such as novobiocin, vancomycin, cefixime, and cefsulodin to inhibit Gram-positive organisms, plus *Proteus* and *Pseudomonas* species, was shown to increase the selectivity of growth media for STEC strains (129, 145, 147). However, the efficacy was found to depend largely on the type and the concentration of the antibiotics used (129, 136, 146, 152). Lower recovery of plated STEC strains was found with some commercial selective media when compared to nonselective brain heart infusion agar (146). Which additives are used for the selectivity of the medium also depends on the background microflora, which may vary between different sample types.

Chromogenic media such as modified Rainbow Agar, Chromagar-STEC, and non-commercial formulas are widely used to identify STEC strains by observing colony growth, morphology, and color (100, 102, 105, 150, 151, 153). These media commonly contain tellurite, which is used for counterselection to suppress growth of accompanying non-STEC bacteria. Tellurite is highly selective for growing non-sorbitol-fermenting *E. coli* O157 and certain other EHEC strains belonging to the "big seven" EHEC serogroups. Resistance to tellurite is associated with a cluster of genes encoded by the ter operon (102, 154–156). However, about half of the EHEC O103:H2 strains, all sorbitol-fermenting EHEC O157:[H7] strains, and single representatives belonging to other EHEC serogroups (O121:H19, O145:H28) were found to lack ter genes and showed no growth on media containing tellurite (102, 155). The maximum concentrations of tellurite permitting growth were found to be significantly different when STEC strains were compared, and spontaneous recombinants of EHEC O157 strains developing sensitivity to tellurite were found (146, 157). Most STEC strains not belonging to the "big seven" group were found to be negative for *ter* genes and were inhibited by tellurite in growth media (102, 156).

On the other hand, many *Enterobacteriaceae* from the environment that are found in vegetables and surface water grow on media containing tellurite (Beutin L, unpublished data). By screening a large number of *E. coli* strains, 12% of apathogenic *E. coli* (#150) and 12.5% of EPEC (#287) were found to carry *ter* genes and showed growth on media containing tellurite (102). In summary, tellurite supplementation of growth media proved to be suitable for selection of the most important EHEC strains, but absence of growth on these media does not indicate that EHEC is absent from the sample.

Selective agar plates were found to grow generally lower portions of inoculated STEC when compared with unselective brain heart infusion agar (146). The choice of a nonselective indicator medium, such as enterohemolysin (washed sheep blood) agar, may thus serve as a supplement or alternative to selective media used in isolation of STEC. Enterohemolysin (also called EHEC-hemolysin) is encoded by genes (*ehxA/ehlyA*) located on large plasmids present in EHEC and STEC strains (158, 159). In contrast to *E. coli* alpha-hemolysin, enterohemolysin is only visible on blood agar plates containing washed sheep blood erythrocytes (57, 160). Hemolysis zones caused by enterohemolysin are generally narrow and turbid and appear after longer incubation, generally overnight, of bacteria plated on enterohemolysin agar (57). They are easily discerned from hemolysis zones caused by alpha-hemolysin, which is frequently found in uropathogenic *E. coli* and enterotoxigenic *E. coli* from animals, but not in EHEC and very rarely in STEC strains (160–162). Alpha-hemolysin is produced in the exponential growth phase, and lysis zones appear after 3 to 4 h of incubation of inoculated plates at 37°C. The lysis zones are generally larger compared to those of enterohemolysin. The detection of low quantities of enterohemolysin-positive (Ehly$^+$) STEC from mixed cultures of bacteria can be improved by plating dilutions of bacterial enrichment cultures. Single Ehly$^+$ colonies indicating STEC can thus be easily detected in a lawn of morphologically different bacteria (57) (Fig. 4).

Enterohemolysin agar is commercially available, and its selectivity can be improved by the addition of antibiotics (57, 147, 149) (Fig. 5). About 90% of EHEC and 40% of STEC strains carry the *ehxA/ehlyA* genes, which are rarely found in non-STEC strains (40, 163). However, some of the important EHEC types, such as sorbitol-fermenting O157 and O104:H4, do not express or do not carry genes for *ehxA/ehlyA*, respectively (43, 164). On the other hand, *ehxA/ehlyA* genes are frequent in derivatives of EHEC that have lost the *stx* genes and in some EPEC O26:H11 strains as well (165–167). In STEC from human patients, the presence of *ehxA* genes and an enterohemolytic phenotype was found more associated with strains possessing the *eae* gene (96.2%) than with *eae*-negative STEC strains (65.2%) (34, 168). Enterohemolysin genes were found in 37.9 to 55% in STEC from food specimens (18, 86, 87, 169–171). Moreover, significant differences for the presence of *ehlyA* genes were found between STEC strains according to their animal food origin (52, 172). *ehly/ehxA* genes were also found in more than 92% of STEC strains isolated from the environment of dairy farms (26) and were significantly associated with STEC from food derived from cattle, wild boar, and red deer (18). As some *E. coli* strains produce enterohemolysin without being STEC, all isolates of presumptive STEC taken from enterohemolysin agar have to be confirmed for Stx production or the presence of *stx* genes.

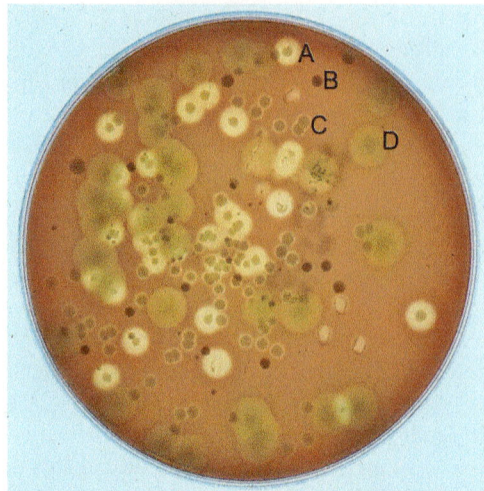

FIGURE 4 Stool sample of a human patient with HUS infected with EHEC O157:H7 grown on enterohemolysin agar (57). Morphological differences between the hemolysis zones on enterohemolysin agar facilitate the detection of enterohemolytic, *ehlyA*-positive STEC strains. Plating of 10-fold dilutions from stool enrichment cultures grown in tryptic soy broth for detection of enterohemolytic, presumptive STEC colonies. Four morphologically different types of bacteria designated from A to D were detected: A, alpha-hemolysin producing, Stx-negative *E. coli*; B, hemolysin-negative *Enterobacteriaceae*; C, enterohemolytic (*ehlyA*)-positive STEC O157:H7; D, *Pseudomonas aeruginosa*.
doi:10.1128/microbiolspec.EHEC-0001-2013.f4

STEC STRAIN TYPING METHODS

Bacterial subtyping includes microbiological and molecular methods that are useful for differentiation below the species level. Microbiological methods such as serotyping, biotyping, colicin typing, phage lysotyping, and antibiotic susceptibility typing are widely used for *E. coli* and other bacterial species. A growing number of molecular methods are described for typing of STEC strains (50, 173), but only some have gained broader acceptance. An overview of genotyping methods for STEC was presented recently, and the most promising molecular methods for STEC genotyping were compared for their discriminatory power; ease of standardization, performance, and interpretation; availability; and costs (173).

Many typing methods are not generally applicable to all strains of the STEC group but were found useful for the identification and characterization of certain STEC strains, subtypes, and subclones within a given serotype or serogroup. Biotyping was found important for rapid screening of clinically and epidemiologically important subtypes among strains belonging to the same O:H type such as O157:[H7] and O26:[H11] (166, 174–176). Bacteriophage lysotyping was used for subtyping clinically and epidemiologically important STEC O157:H7 strains, but its use is restricted to a few laboratories and is not successfully employed on non-O157 STEC strains (50, 177, 178). Resistance to certain antimicrobials such as sulfamethoxazole, tetracycline, and streptomycin is frequent in many STEC and EHEC strains (179). However, an extended antimicrobial resistance typing can be very helpful for the identification and characterization of subclones within STEC strains and types (43, 50, 118, 180). Most STEC strains carrying plasmids and plasmid profiles can be heterogeneous within a given serotype (181, 182). Analysis of plasmid profiles can nevertheless be helpful for describing clinically important subclones among certain STEC types (183).

Multi-locus sequence typing (MLST) has replaced multi-locus enzyme electrophoresis and is now widely used for analysis of phylogenetic relationships among *E. coli* strains (173, 184). However, MLST can be also employed for characterization of epidemiologically and pathogenetically important subtypes among serotypes and clones of *E. coli* strains. Conventional and molecular serotyping was found to provide useful information for outbreak investigations, for defining seropathotypes, and for all kinds of epidemiological studies on reservoirs and spread of certain STEC types. The complete O:H serotype of STEC strains can be used as an indicator for the clonal type of certain STEC strains, as shown by multi-locus enzyme electrophoresis and MLST analysis

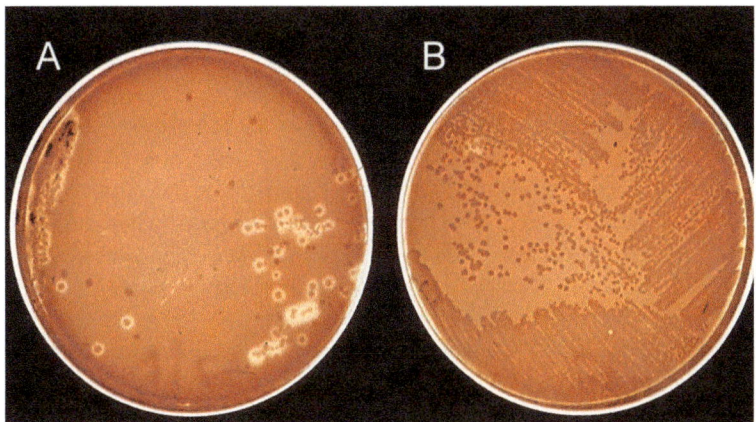

FIGURE 5 Enhancement of selectivity for isolation of STEC from bovine fecal samples on enterohemolysin agar supplemented with vancomycin. (A) Growth of enterohemolytic STEC colonies from a sample of bovine feces on enterohemolysin agar supplemented with 8 mg/liter of vancomycin. The gram-positive flora is suppressed. (B) Growth of enterohemolytic STEC colonies from the same sample of bovine feces on nonselective enterohemolysin agar. The abundant gram-positive flora overgrows STEC present in the sample. doi:10.1128/microbiolspec.EHEC-0001-2013.f5

(118, 185, 186). In a similar way, other molecular typing methods such as pulsed-field gel electrophoresis (PFGE) gave more meaningful results when strains belonging to the same serotypes were compared instead of randomly sampled STEC isolates (187, 188).

Phenotypical and Molecular O and H Serotyping

A serotyping scheme of *E. coli* O (LPS) and H (flagella) antigens was established in the 1940s by Fritz Kauffmann and further extended by the International Escherichia and Klebsiella Centre in Copenhagen, Denmark, in collaboration with national *E. coli* reference laboratories (189). At present, a panel of *E. coli* O antigens (O1 to O181) and H antigens (H1 to H56) is described (190), and four additional O antigens (O182 to O186) are currently being investigated in some reference laboratories (18, 191, 192). Complete serotyping of *E. coli* O and H antigens is performed in a few laboratories that participate in quality assurance tests (193). STEC belongs to more than 400 serotypes (see "Diversity of STEC Types" above), and many diagnostic laboratories

concentrate on the identification of the most important EHEC serogroups. A large panel of *E. coli* antisera for O and H serotyping is commercially available and can be used to screen diagnostically interesting types. Complete serotyping by agglutination of O and H antigens is laborious and time-consuming, as slide agglutination tests with live bacteria have to be confirmed by agglutination in tubes with heat-treated (O antigen detection) and formaldehyde-inactivated motile cultures of bacteria (H antigen detection) (189). A number of serological tests for detecting *E. coli* O157 and H7 antigens and conjugated antibodies directed to important EHEC serogroups are commercially available to design and perform rapid serological assays.

However, a considerable number of STEC and EHEC isolates from clinical, food, and other sources are not typeable for the O and H antigens, as O-rough (spontaneously agglutinating) and nonmotile (NM) strains are frequently isolated. About 18% (n=107) of 593 STEC strains from different types of food were untypeable (O-rough or unknown O type) for their O antigens using O antisera from O1 to O186 (18). In a study of 677 STEC strains from human patients, 221 (32.6%) were found to be NM and thus not serotypeable for their H antigens (34). The NM strains frequently include representatives of the "big seven" EHEC O groups (30, 34, 164).

Most STEC strains showing an NM phenotype were nevertheless found positive for the *fliC* gene, which encodes most of the flagellar antigens in *E. coli* (194). In those cases, H serotyping can be substituted or accompanied by molecular typing of the *fliC* gene in *E. coli*. A number of PCR and RFLP methods were described for *fliC* genotyping (195–197). In some cases, *fliC* genotyping was found to be more specific for detection of genetic subtypes among serologically cross-reacting H antigens and for possible detection of new flagellar types in *E. coli* (198–200). The *fliC* genes representing 43 H antigens in *E. coli* have been sequenced, and the sequence information can be used for confirmation of H serotyping and *fliC* typing results (194).

Numerous molecular serotyping methods have been developed as alternatives to traditional serotyping and identify most often the genes associated with the somatic (O) and flagellar (H) antigens or detect particular serotype-associated sequences in housekeeping or virulence genes. These methods have the advantage of also being applicable on O-rough strains that cannot be serotyped phenotypically. Molecular tools developed for the determination of the O antigen are mainly based on the detection of the *rfbE* (also named *per*), *wzx*, and *wzy* genes (95, 97, 201–206). Ballmer and coworkers (207) developed a serotyping microarray to identify the 24 most epidemiologically relevant serogroups based on oligonucleotides designed on the *wzx* and *wzy* genes. In addition, this array tube-based assay targeted 47 different H antigens based on the identification of variations on the *fliC* gene that code for the flagellar monomer (207, 208). Lin and coworkers (209, 210) fabricated a TaqMan real-time multiplex PCR assay and a microbead-based suspension array that identify the 10 most clinically relevant STEC serogroups (O157, O26, O45, O91, O103, O111,

O113, O121, O128, and O145), targeting the *wzx* and *wzy* genes of the *rfb* locus. An alternative molecular O serotyping method based on sequence variations of the *gnd* gene was presented by Gilmour and coworkers (211). The *gnd* gene is located close to the O antigen cluster. By sequencing this gene, they demonstrated the presence of unique alleles for each of the examined STEC serogroups. Furthermore, this approach differentiates between STEC and non-STEC strains of serogroups O157, O26, O55, O6, and O117 (https://www.corefacility.ca/ecoli_typer/). Recently Norman et al. (212) identified single nucleotide polymorphisms in the O-antigen operon of serogroups O26, O45, O103, O111, O121, and O145 that could be used to differentiate between strains of these serogroups that contain STEC-associated virulence factors and those that do not. Their study supports the idea that differences in the genetic sequence of the O-antigen operon correspond well with differences in the virulence gene profiles and provide evidence of separate clustering for STEC and non-STEC strains. These findings may open the development of new detection tests and thus represent a significant improvement in the identification of EHEC strains. Another recent approach described by Delannoy et al. (120, 213) is based on the genetic diversity of the clustered, regularly interspaced, short palindromic repeats (CRISPR) regions of EHEC to design simplex real-time PCR assays for each of the "big seven" EHEC serogroups and for STEC O104:H4. The identification of EHEC serotype-specific CRISPR sequences was found to be more specific than the mere identification of O antigen gene sequences as it is used in current PCR protocols for the detection of EHEC strains (104, 105).

PFGE and Other Genotyping Methods

PFGE of total bacterial DNA digested with infrequently cutting restriction enzymes produces DNA patterns that allow comparison of strains at a genomic level. PFGE is considered the gold standard of subtyping methods because of its discriminatory power and its use-fulness in epidemiological investigations (173). PFGE is widely used for typing of EHEC O157:H7, and the electronic submission of DNA fingerprints stored in databases allows world-wide survey of these pathogens (214, 215).

Less frequently, PFGE is also used for typing of non-O157 EHEC strains. Most data are available from non-O157 EHEC "big five" strains, but other STEC types also were in-vestigated (15, 87, 187, 216–219). A number of studies were performed on *E. coli* O26 strains from patients, food, and animals (166, 176, 182, 220, 221). PFGE typing was compared with other genotyping methods such as variable number of tandem-repeats analysis (MLVA) (166). IS621 multiplex PCR-based fingerprint-ing was found to be less discriminatory than PFGE in a study on O26 isolates from different sources, but epidemiologically related strains were identified with both methods (221). MLST, PFGE, and MLVA were found suit-able to discriminate between major clusters of O26:[H11] EPEC and EHEC strains (166, 222). PFGE was used for typing STEC and other *E. coli* strains belonging to serogroup O103 and compared with other typing methods such as P gene profiling and ribotyping. Clonal types were reported to be best defined by a combination of PFGE and P gene profiles (177). Stx phage PCR-RFLP was found to be less discriminative than PFGE for typing a series of STEC strains isolated in Japan, in-cluding O157, O26, O111, and others (223). PFGE and MLST with seven housekeeping genes were compared in an analysis on 54 *E. coli* O103 strains from different sources. PFGE was very useful for identification of genetically closely related subgroups among strains belonging to the same MLST type, such as Stx2-producing EHEC O103:H2 strains from patients with HUS (224). The suitability of PFGE for identification and subtyping clin-ically important STEC strains was also dem-onstrated with serogroup O145 strains (225).

Moreover, PFGE was successfully employed to verify transmission of STEC strains between animals, from animals to humans, and in the food chain (217, 218, 226–229). PFGE analysis of STEC strains belonging to the same sero-types was also helpful to show that wild and domestic animals form a common reservoir for certain STEC and EHEC strains that con-taminate the respective food products and are finally found in human patients (87, 166, 216).

The ability of Stx phages to integrate at dis-tinct sites in chromosomes of *E. coli* (230, 231) was employed as a molecular typing method for STEC O157:H7 strains with regard to their virulence for humans. Sixteen genotypes of STEC O157:H7 strains were defined on the basis of Stx type and occupation of phage in-tegration sites in the chromosome of STEC O157:H7 (232). The majority (95%) of STEC O157:H7 clinical isolates from humans clus-tered into genotypes 1, 2, and 3 in contrast to only 51.3% of the bovine STEC O157:H7 strains (232). Human-associated STEC O157:H7 genotype strains were characterized by in-creased expression of EHEC virulence genes (pO157 and LEE), which could explain their prevalence in disease (233). Human STEC O157:H7 strains are characterized by the pre-sence of Stx2a phages whereas Stx2c phages are present in typical bovine O157:H7 strains, which could also explain the virulence dif-ferences observed (234).

Subtyping of STEC- and EHEC-Related Virulence Genes

Virulence markers such as bacteriophage-encoded *stx* genes, the LEE pathogenicity is-land-located *eae* (intimin) genes, and some non-LEE effector genes (*nle*) such as *nleA* were shown to be genetically very heteroge-neous, splitting in various subtypes (46, 235–239). Genotyping of *eae*, *stx*, and some *nle* genes was thus employed for typing STEC strains, to describe STEC clones, for defining pathotypes, and for investigating animal host reservoirs. Subtyping of these genes is per-formed with real-time and conventional specific PCRs, restriction endonuclease di-gestion of relevant PCR products, and nucle-otide sequencing.

Subtyping of *stx* genes below the Stx1 and Stx2 level becomes increasingly important for describing potentially human pathogenic STEC strains. Certain subtypes of Stx1 and Stx2 are related to the virulence of STEC for humans and are frequently associated with EHEC strains (46, 240–242). Some genotypes of Stx1 and Stx2 are typical for certain serotypes of STEC strains. Significant relationships among the food-producing animal host species, STEC serotypes, and *stx* genotypes could be established (18). In detail, Stx2e, as the key factor in porcine edema disease principle, is frequently found in STEC from pigs, wild boar, and their meat products but rarely in other animal species. Most likely, Stx2e is not associated with human pathogenicity (18, 52, 243, 244). Stx2f is associated with certain serotypes of STEC strains from birds and with strains of *Escherichia albertii* (245, 246). STEC producing Stx2f was found positive for LEE genes (encoding the attaching and effacing phenotype) and was isolated as an agent of nonbloody diarrhea in humans (241, 247). Stx1c and Stx2b are frequent among STEC strains from sheep, goats, red deer, and their food products (18, 216, 248–250). Stx1c and Stx2b are more associated with nonbloody diarrhea in humans but not with HC and HUS (34, 241, 242). Stx1a, Stx2a, Stx2c, and Stx2d were found to be significantly associated with STEC from bovines and food of bovine origin (18, 144, 251, 252). Stx2a is highly associated with STEC strains causing HUS in humans, such as O157:H7 and O104:H4 (43, 234, 253). Stx2a, Stx2c, and Stx2d genes show little differences in their genotypes (46), and PCR-RFLP typing was found useful for detecting multiple types of these genes, which are frequently present in single STEC strains (46, 86). A close association of certain serotypes, *stx* subtypes, and the species of the food-producing animals was found in a study of 597 STEC strains from different food categories (18). Accordingly, multiple types of *stx* variants are associated with STEC isolated from food (18, 86, 87, 254). A similar diversity of Stx types was also found in STEC

isolated from the farm environment (255, 256).

Some, but not all, subtypes of Stx1 and Stx2 are associated with strains carrying the LEE-encoded *eae* genes involved in intestinal adhesion of EPEC and EHEC strains. About 35 variants of the *eae* gene have been described in *E. coli* (257) and in other *Enterobacteriaceae* (246, 258). The "big seven" EHEC strains express different genetic variants of the *eae* gene, which can be analyzed by PCR, PCR-RFLP, multiplex real-time PCR, and nucleotide sequencing (109, 144, 236, 259, 260). Typing of *eae* genes was found to be useful for characterizing EHEC from food and animals and for discriminating between EPEC and EHEC strains belonging to serogroups O103, O111, O145, and O157 (113, 144, 172, 216, 220, 224, 225). Typing of *eae* genes in combination with analysis of *nle* genes can also be useful for discriminating between EPEC and EHEC derivative strains that have lost their *stx* genes and for characterizing new types of emerging EHEC strains (34, 163, 169, 261, 262).

RISK ASSESSMENT OF STEC FOR HUMANS

With the growing number of different STEC serotypes isolated from different sources, attempts to classify these strains according to their virulence for humans have increased. The term EHEC was coined in 1987 for those STEC strains that cause HC and HUS, and a DNA probe derived from the O157 virulence plasmid (pO157) was developed for typing purposes (263). Certain genetic variants of Stx, such as Stx1a and Stx2a, were found more associated with HC and HUS in patients than others (46). However, the mere presence of stx_1 or stx_2 genes in an STEC strain does not itself indicate that it causes severe disease in humans.

The relation between Stx type and human pathogenicity of an STEC strain becomes statistically clear if the presence of other virulence genes encoded by virulence plasmids (*ehlyA*) and the LEE-encoded *eae* genes are considered (11, 37, 264, 265). Efficient adhesion of an STEC strain in the human intestine seems to be a prerequisite for its capacity to cause severe disease such as HC and HUS. The LEE-encoded attaching and effacing system is a typical colonization mechanism of the "big seven" EHEC strains. On the other hand, the outbreak with EAEC O104:H4 has demonstrated that the LEE system can be substituted by other effective intestinal adherence mechanisms such as those present in EAEC strains (11, 43). A number of other types of fimbrial and afimbrial adhesions, which can play a role in intestinal adhesion of STEC strains, have been described (266). Further analysis of STEC strains from human patients will show whether the presence of one or more of these adhesins is specifically associated with severe illness.

A growing number of genes located on virulence plasmids of STEC and on genomic islands (LEE, *nle* genes) were associated with strains causing disease in humans (42). Elucidation of genomic sequences of major EHEC strains allowed further research on the distribution of various effector genes located on prophages and genomic islands in strains belonging to the "big seven" EHEC types (267–269). It thus became possible to establish an MRA approach (39) to define STEC according to its pathogenicity for humans. The first MRA approach was established by Karmali and coworkers in 2003 (37). STEC strains were classified as seropathotypes A to E according to their incidence, frequency in outbreaks, and association with HC and HUS. Seropathotypes A and B comprise the EHEC "big five" plus O121 strains whereas seropathotype C includes LEE-negative STEC strains that were already associated with HC and HUS (O91:H21, O113:H21, O104:H21, and others). Strains of seropathotype D encompass all STEC strains that were already involved in cases of diarrhea but not HC and HUS. Finally, strains of seropathotype E have not been associated with human infections. The MRA approach was adapted in further studies revealing links between severe illness,

outbreaks, and the presence of genomic islands, such as OI-122 and OI-71 (39, 270). Bugarel and coworkers (40, 163, 271) studied the distribution of coding sequences found on genomic islands OI-71 and OI-122 in a collection of more than 500 EHEC, EPEC, STEC, and nonpathogenic *E. coli* strains. Both genomic islands code for Nle type III effectors that are potentially involved in STEC virulence. The results revealed an association between the distribution of *nle* genes harbored on OI-71 (*nleF*, *nleH1-2*, and *nleA*) and OI-122 (*ent/sen/espL2*, *nleB*, and *nleE*) with the "big seven" EHEC serotypes and with emerging EHEC serologically different from the "big seven" that had been already associated with severe illness in humans.

EHEC strains show either one of two predominant profiles, which differ according to their composition of *nle* genes. EHEC O157: [H7], O111:[H8], O26:[H11], and O121:H19 generally have the complete set of the six *nle* genes located on islands OI-71 and OI-122, as found with emerging EHEC serotype of O5: NM, O103:H25, and O55:H7 strains. On the other hand, EHEC O103:H2 and O145:[H28] strains showed an altered profile composed of OI-122 *nle* genes *ent/sen/espL2*, *nleB*, *nleE*, and the OI-71-located *nleH1-2* gene. *nleF* and *nleA* genes (OI-71) were not detected in O103:H2 and O145:[H28] strains using the PCR assays developed in this study. However, the possibility that these strains carry genetic variants of these *nle* genes, as previously reported for *nleA*, cannot be excluded (235).

The "big seven" EHEC and some of the emerging EHEC types are characterized by four variants of the *eae* genes present in these strains (163, 169, 271). The *eae*-gamma variant is characteristic for EHEC O157:H7, O55:H7, and O145:H28. The *eae*-epsilon variant is a hallmark of EHEC O103:H2, O45:H2, and O121:H19 strains (163, 236). The *eae*-theta variant is associated with EHEC strain O111:[H8] and with the emerging EHEC O103: H25 strain that caused an HUS outbreak in Norway (272). Finally, the *eae*-beta variant is

TABLE 2 Serotypes of STEC isolated from human patients[a]

O1:H–	O8:H21	O25:K2:H2	O52:H23	O83:H1	O103:H21	O114:H?	O127	O148:H28	O172:H–
O1:H1	O8:H25	O25:H14	O52:H25	O84:H–	O103:H25	O115:H10	O128:H–	O150:H–	O172:H?
O1:H2	O9ab:H–	O26:H–	O54:H21	O84:H2	O103:HNT	O115:H18	O128:H2	O150:H8	O173:H2
O1:H7	O9:H7	O26:H2	O55:H–	O84:H20	O104:H–	O116:H–	O128:H7	O150:H10	O174:H–
O1:H20	O9:H21	O26:H8	O55:H6	O85:H–	O104:H2	O116:H4	O128:H8	O152:H4	O174:H2
O1:HNT	O11:H–	O26:H11	O55:H7	O85:H10	O104:H7	O116:H10	O128:H10	O153:H2	O174:H8
O2:H–	O11:H2	O26:H12	O55:H9	O85:H23	O104:H16	O116:H19	O128:H12	O153:H11	O174:H21
O2:H1	O11:H8	O26:H32	O55:H10	O86:H–	O104:H21	O117:H–	O128:H25	O153:H12	O175:H16
O2:K1:H2	O11:H49	O26:H46	O55:H19	O86:H10	O105acH18	O117:H4	O128:H31	O153:H21	O176:H–
O2:H5	O12:H–	O27:H–	O55:H?	O86:H40	O105:H19	O117:H7	O128:H45	O153:H25	O177:H–
O2:H6	O14:H–	O27:H30	O60:H–	O87:H6	O105:H20	O117:K1:H7	O129:H–	O153:H30	O177:H11
O2:H7	O15:H–	O28ab:H–	O64:H25	O88:H–	O106	O117:H8	O130:H11	O153:H33	O178:H7
O2:H11	O15:H2	O28:H25	O65:H16	O88:H25	O107:H27	O117:H19	O131:H4	O154:H–	O179:H8
O2:H27	O15:H8	O28:H35	O68:H–	O89:H–	O109:H2	O117:H28	O132:H–	O154:H4	O181:H15
O2:H29	O15:H27	O30:H2	O69:H–	O90:H–	O109:H16	O118:H–	O133:H–	O154:H19/20	O181:H49
O2:H44	O16:H–	O30:H21	O69:H11	O91:H–	O110:H–	O118:H2	O133:H53	O156:H–	
O3:H10	O16:H6	O30:H41	O70:H11	O91:H4	O110:H19	O118:H12	O134:H25	O156:H4	
O4:H–	O16:H21	O37:H41	O71:H–	O91:H10	O110:H28	O118:H16	O137:H6	O156:H7	
O4:H5	O17:H18	O38:H21	O73:H34	O91:H14	O111:H–	O118:H30	O137:H41	O156:H25	
O4:H10	O17:H41	O38:H26	O74	O91:H15	O111:H2	O119:H–	O138:H2	O156:H27	
O4:H40	O18:H–	O39:H4	O75:H–	O91:H21	O111:H7	O119:H5	O141:H–	O156:HNT	

O5:H–	**O18:H7**	O39:H8	O75:H1	O91:H40	O111:H8	O119:H6	O141:H2	**O157:H7**
O5:H16	O18:H12	O39:H28	O75:H5	O91:HNT	O111:H11	O119:H25	O141:H8	**O157:NM**
O6:H–	O18:H15	O40:H2	O75:H8	O92:H3	O111:H21	O120:H19	O142	O160:H?
O6:H1	O18:H?	O40:H8	O76:H7	O92:H11	O111:H30	O121:H–	O143:H–	O161:H–
O6:H2	O20:H–	O41:H2	**O76:H19**	O95:H–	O111:H34	O121:H8	O144:H–	O162:H4
O6:H4	O20:H7	O41:H26	O77:H–	**O96:H10**	O111:H40	O121:H11	O145:H–	**O163:H–**
O6:H12	O20:H19	O44	O77:H4	O98:H–	O111:H49	O121:H19	O145:H4	**O163:H19**
O6:H28	O21:H5	O45:H–	O77:H7	O98:H8	O111:H?	O123:H19	O145:H8	O163:H25
O6:H29	O21:H8	O45:H2	O77:H18	O100:H25	O112abcH2	O123:H49	O145:H16	O165:H–
O6:H31	**O21:H?**	O45:H7	O77:H41	O100:H32	O112:H19	O124:H–	O145:H25	O165:H10
O6:H34	**O22:H–**	O46:H2	O78:H–	O101:H–	O112:H21	O125:H–	O145:H26	O165:H19
O6:H49	O22:H1	O46:H31	O79:H7	O101:H9	O113:H2	O125:H8	**O145:H28**	O165:H21
O7:H4	O22:H5	**O46:H38**	**O79:H14**	O102:H6	**O113:H4**	O125:H?	O145:H46	O165:H25
O7:H8	**O22:H8**	O48:H21	O79:H23	O103:H–	O113:H5	O126:H–	O145:HNT	O166:H12
O8:H–	**O22:H16**	O49:H–	**O80:H–**	**O103:H2**	O113:H7	O126:H2	**O146:H–**	O166:H15
O8:H2	O22:H40	O49:H10	O81:H?	O103:H4	**O113:H21**	O126:H8	L	**O160:H28**
O8:H9	O23:H7	O50:H–	O82:H–	O103:H6	O113:H32	O126:H11	O146:H11	O168:H–
O8:H11	O23:H16	O50:H7	O82:H5	O103:H7	O113:H53	**O126:H20**	O146:H14	O169:H–
O8:H14	O23:H21	O51:H49	**O82:H8**	O103:H11	O114:H4	O126:H21	**O146:H21**	O171:H–
O8:H19	O25:H–	O52:H19	O83:H–	O103:H18	O114:H48	O126:H27	**O146:H28**	**O171:H2**

aSerotypes in bold were identified as food-borne STEC in Germany (593 strains isolated between 2005 and 2009) (18) and correspond to 19.5% of the 375 STEC serotypes from humans listed in this table. Adapted from reference 30.

present in EHEC and EPEC O26:H11 (167) and in emerging EHEC strains of serotype O5:NM and O118:H16.

In a given STEC strain, the presence of *ent/sen/espL2, nleB, nleE,* and *nleH1-2* effectors, when associated with one of the four genetic variants of the *eae* gene—*eae*-beta, *eae*-gamma, *eae*-epsilon, or *eae*-theta—constitutes the characteristic signature of typical and emerging EHEC strains that have been identified over the past few years in Europe. Detection of these genetic markers constitutes an innovative DNA-based approach in the MRA of STEC strains.

As these *nle* genes and *eae* variants can be shared by Stx-negative EPEC strains as well (271), identification of genetic markers allowing a more targeted screening of EHEC strains in complex food samples is still challenging, and genetic targets that may support such an approach still have to be clarified. Monitoring EHEC in foods requires, in particular, selection of genetic markers able to discriminate clearly EHEC from EPEC strains. A recent study pointed out that genes carried by genomic O island 57 (OI-57) may be associated with increased virulence of STEC strains in humans (273). Open reading frames (ORFs) inside OI-57 such as Z2097, Z2098, Z2121, and Z2149 seem to be associated preferentially with the EHEC strains (273). Recently, Delannoy and coworkers (32) identified two ORFs on OI-57 called Z2098 and Z2099 as suitable markers to discriminate between EHEC and EPEC strains. Z2098 and Z2099 show a higher specificity for the "big seven" EHEC serotype strains than does the *eae* gene and the other *nle* genes that were investigated so far. Accordingly, Z2098 and Z2099 were rarely found in EPEC (10% and 12%, respectively), STEC (2% and 15%), and apathogenic *E. coli* (1% each), in contrast to EHEC (87% and 91% positive) and *stx*-negative EHEC derivative strains.

Apart from the "big seven" EHEC and some emerging EHEC serotypes, there is no MRA approach to classify the broad number of LEE-negative and *nle*-negative STEC strains that constitute the major part of STEC isolated from animals, food, and the environment. Nevertheless, 16 of the 41 strains of the HUSEC collection are LEE-negative, pointing out that at least some of these STEC strains are able to cause severe disease in humans (108). Some serogroups of these LEE-negative STEC strains (O91, O63, O113, O128, and O146) accounted for 21.4 to 25.7% of serotypeable, non-O157 STEC strains isolated from human patients in the European Union (107). Most of these strains are frequently found in food of different origins and can thus be transmitted as food-borne infections. Food-borne, LEE-negative STEC strains were found to group in the same MLST clusters as LEE-negative strains that were isolated from patients with HUS (108, 123). Serotyping of 593 food-borne STEC strains isolated in Germany between 2005 and 2009 revealed 73 serotypes that were already found associated with human infections (18, 50) (Table 2). A future challenge will be to establish MRA concepts that enable a better definition of human pathogenic and nonpathogenic STEC of the diverse group of LEE-negative STEC strains.

ACKNOWLEDGMENT

We declare no conflicts of interest with regard to the manuscript.

CITATION

Beutin L, Fach P. 2014. Detection of Shiga toxin-producing *Escherichia coli* from nonhuman sources and strain typing. Microbiol Spectrum 2(3):EHEC-0001-2013.

REFERENCES

1. **Konowalchuk J, Speirs JI, Stavric S.** 1977. Vero response to a cytotoxin of *Escherichia coli. Infect Immun* **18:**775–779.
2. **Wade WG, Thom BT, Evans N.** 1979. Cytotoxic enteropathogenic *Escherichia coli. Lancet* **ii:**1235–1236.
3. **Wilson MW, Bettelheim KA.** 1980. Cytotoxic *Escherichia coli* serotypes. *Lancet* **i:**201.

4. **Riley LW, Remis RS, Helgerson SD, McGee HB, Wells JG, Davis BR, Hebert RJ, Olcott ES, Johnson LM, Hargrett NT, Blake PA, Cohen ML.** 1983. Hemorrhagic colitis associated with a rare *Escherichia coli* serotype. *N Engl J Med* **308:** 681–685.

5. **Orskov F, Orskov I, Villar JA.** 1987. Cattle as reservoir of verotoxin-producing *Escherichia coli* O157:H7. *Lancet* **ii:**276.

6. **Martin ML, Shipman LD, Wells JG, Potter ME, Hedberg K, Wachsmuth IK, Tauxe RV, Davis JP, Arnoldi J, Tilleli J.** 1986. Isolation of *Escherichia coli* O157:H7 from dairy cattle associated with two cases of haemolytic uraemic syndrome. *Lancet* **ii:**1043.

7. **Borczyk AA, Karmali MA, Lior H, Duncan LM.** 1987. Bovine reservoir for verotoxin-producing *Escherichia coli* O157:H7. *Lancet* **i:**98.

8. **Mathusa EC, Chen Y, Enache E, Hontz L.** 2010. Non-O157 Shiga toxin-producing *Escherichia coli* in foods. *J Food Prot* **73:**1721–1736.

9. **Blanco J, Blanco M, Blanco J, Mora A, Alonso MP, Gonzalez EA, Bernardez MI.** 2001. Epidemiology of verocytotoxigenic *Escherichia coli* (VTEC) in ruminants, p 113–148. *In* Duffy G, Garvey P, McDowell DA (ed), *Verocytotoxigenic E. coli.* Food & Nutrition Press, Inc, Trumbull, CT.

10. **Karch H, Denamur E, Dobrindt U, Finlay BB, Hengge R, Johannes L, Ron EZ, Tonjum T, Sansonetti PJ, Vicente M.** 2012. The enemy within us: lessons from the 2011 European *Escherichia coli* O104:H4 outbreak. *EMBO Mol Med* **4:**841–848.

11. **Beutin L, Martin A.** 2012. Outbreak of Shiga toxin-producing *Escherichia coli* (STEC) O104:H4 infection in Germany causes a paradigm shift with regard to human pathogenicity of STEC strains. *J Food Prot* **75:**408–418.

12. **Sanchez S, Garcia-Sanchez A, Martinez R, Blanco J, Blanco JE, Blanco M, Dahbi G, Mora A, Hermoso DM, Alonso JM, Rey J.** 2009. Detection and characterisation of Shiga toxin-producing *Escherichia coli* other than *Escherichia coli* O157:H7 in wild ruminants. *Vet J* **180:**384–388.

13. **Beutin L, Geier D, Steinruck H, Zimmermann S, Scheutz F.** 1993. Prevalence and some properties of verotoxin (Shiga-like toxin)-producing *Escherichia coli* in seven different species of healthy domestic animals. *J Clin Microbiol* **31:**2483–2488.

14. **Caprioli A, Morabito S, Brugère H, Oswald E.** 2005. Enterohaemorrhagic *Escherichia coli*: emerging issues on virulence and modes of transmission. *Vet Res* **36:**289–311.

15. **Mora A, Herrera A, Lopez C, Dahbi G, Mamani R, Pita JM, Alonso MP, Llovo J, Bernardez MI, Blanco JE, Blanco M, Blanco J.** 2011.

Characteristics of the Shiga-toxin-producing enteroaggregative *Escherichia coli* O104:H4 German outbreak strain and of STEC strains isolated in Spain. *Int Microbiol* **14:**121–141.

16. **Asakura H, Makino S, Shirahata T, Tsukamoto T, Kurazono H, Ikeda T, Takeshi K.** 1998. Detection and genetical characterization of Shiga toxin-producing *Escherichia coli* from wild deer. *Microbiol Immunol* **42:**815–822.

17. **Ferens WA, Hovde CJ.** 2011. *Escherichia coli* O157:H7: animal reservoir and sources of human infection. *Foodborne Pathog Dis* **8:**465–487.

18. **Martin A, Beutin L.** 2011. Characteristics of Shiga toxin-producing *Escherichia coli* from meat and milk products of different origins and association with food producing animals as main contamination sources. *Int J Food Microbiol* **146:**99–104.

19. **Franz E, van Bruggen AH.** 2008. Ecology of *E. coli* O157:H7 and *Salmonella enterica* in the primary vegetable production chain. *Crit Rev Microbiol* **34:**143–161.

20. **Loukiadis E, Kerouredan M, Beutin L, Oswald E, Brugere H.** 2006. Characterization of Shiga toxin gene (*stx*)-positive and intimin gene (*eae*)-positive *Escherichia coli* isolates from wastewater of slaughterhouses in France. *Appl Environ Microbiol* **72:**3245–3251.

21. **Gourmelon M, Montet MP, Lozach S, Le Mennec C, Pommepuy M, Beutin L, Vernozy-Rozand C.** 2006. First isolation of Shiga toxin 1d producing *Escherichia coli* variant strains in shellfish from coastal areas in France. *J Appl Microbiol* **100:**85–97.

22. **Bennani M, Badri S, Baibai T, Oubrim N, Hassar M, Cohen N, Amarouch H.** 2011. First detection of Shiga toxin-producing *Escherichia coli* in shellfish and coastal environments of Morocco. *Appl Biochem Biotechnol* **165:**290–299.

23. **Kauffman MD, LeJeune J.** 2011. European starlings (*Sturnus vulgaris*) challenged with *Escherichia coli* O157 can carry and transmit the human pathogen to cattle. *Lett Appl Microbiol* **53:**596–601.

24. **Ihekweazu C, Carroll K, Adak B, Smith G, Pritchard GC, Gillespie IA, Verlander NQ, Harvey-Vince L, Reacher M, Edeghere O, Sultan B, Cooper R, Morgan G, Kinross PT, Boxall NS, Iversen A, Bickler G.** 2011. Large outbreak of verocytotoxin-producing *Escherichia coli* O157 infection in visitors to a petting farm in South East England, 2009. *Epidemiol Infect*1–14.

25. **Henderson H.** 2008. Direct and indirect zoonotic transmission of Shiga toxin-producing *Escherichia coli. J Am Vet Med Assoc* **232:**848–859.

26. **Fremaux B, Raynaud S, Beutin L, Rozand CV.** 2006. Dissemination and persistence of Shiga toxin-producing *Escherichia coli* (STEC) strains on French dairy farms. *Vet Microbiol* **117:**180–191.

27. **Fremaux B, Prigent-Combaret C, Vernozy-Rozand C.** 2008. Long-term survival of Shiga toxin-producing *Escherichia coli* in cattle effluents and environment: an updated review. *Vet Microbiol* **132:**1–18.

28. **Muniesa M, Serra-Moreno R, Jofre J.** 2004. Free Shiga toxin bacteriophages isolated from sewage showed diversity although the *stx* genes appeared conserved. *Environ Microbiol* **6:**716–725.

29. **Imamovic L, Jofre J, Schmidt H, Serra-Moreno R, Muniesa M.** 2009. Phage-mediated Shiga toxin 2 gene transfer in food and water. *Appl Environ Microbiol* **75:**1764–1768.

30. **Scheutz F, Strockbine NA.** 2005. Genus I. *Escherichia*, p 607–624. *In* Garrity GM, Brenner DJ, Krieg NR, Staley JT (ed), *Bergey's Manual of Systematic Bacteriology*, 2nd ed. Springer, New York, NY.

31. **Bettelheim KA.** 2007. The non-O157 Shiga-toxigenic (verocytotoxigenic) *Escherichia coli*; under-rated pathogens. *Crit Rev Microbiol* **33:** 67–87.

32. **Delannoy S, Beutin L, Fach P.** 2013. Towards a molecular definition of enterohemorrhagic *Escherichia coli* (EHEC): detection of genes located on O island 57 as markers to distinguish EHEC from closely related enteropathogenic *E. coli* strains. *J Clin Microbiol* **51:**1083–1088.

33. **European Centre for Disease Prevention and Control and European Food Safety Authority.** 2011. *Shiga toxin/verotoxin-producing Escherichia coli* in humans, food and animals in the EU/EEA, with special reference to the German outbreak strain STEC O104. European Centre for Disease Prevention and Control, Stockholm, Sweden, and European Food Safety Authority, Parma, Italy.

34. **Beutin L, Krause G, Zimmermann S, Kaulfuss S, Gleier K.** 2004. Characterization of Shiga toxin-producing *Escherichia coli* strains isolated from human patients in Germany over a 3-year period. *J Clin Microbiol* **42:**1099–1108.

35. **Kappeli U, Hachler H, Giezendanner N, Cheasty T, Stephan R.** 2010. Shiga toxin-producing *Escherichia coli* O157 associated with human infections in Switzerland, 2000–2009. *Epidemiol Infect* **28:** 1–8.

36. **Brooks JT, Sowers EG, Wells JG, Greene KD, Griffin PM, Hoekstra RM, Strockbine NA.** 2005. Non-O157 Shiga toxin-producing *Escherichia coli* infections in the United States, 1983–2002. *J Infect Dis* **192:**1422–1429.

37. **Karmali MA, Mascarenhas M, Shen S, Ziebell K, Johnson S, Reid-Smith R, Isaac-Renton J, Clark C, Rahn K, Kaper JB.** 2003. Association of genomic O island 122 of *Escherichia coli* EDL 933 with verocytotoxin-producing *Escherichia coli*

seropathotypes that are linked to epidemic and/or serious disease. *J Clin Microbiol* **41:**4930–4940.

38. **Konczy P, Ziebell K, Mascarenhas M, Choi A, Michaud C, Kropinski AM, Whittam TS, Wickham M, Finlay B, Karmali MA.** 2008. Genomic O island 122, locus for enterocyte effacement, and the evolution of virulent verocytotoxin-producing *Escherichia coli*. *J Bacteriol* **190:**5832–5840.

39. **Coombes BK, Wickham ME, Mascarenhas M, Gruenheid S, Finlay BB, Karmali MA.** 2008. Molecular analysis as an aid to assess the public health risk of non-O157 Shiga toxin-producing *Escherichia coli* strains. *Appl Environ Microbiol* **74:**2153–2160.

40. **Bugarel M, Beutin L, Martin A, Gill A, Fach P.** 2010. Micro-array for the identification of Shiga toxin-producing *Escherichia coli* (STEC) seropathotypes associated with hemorrhagic colitis and hemolytic uremic syndrome in humans. *Int J Food Microbiol* **142:**318–329.

41. **Centers for Disease Control and Prevention.** 1995. Outbreak of acute gastroenteritis attributable to *Escherichia coli* serotype O104:H21—Helena, Montana, 1994. *Morb Mortal Wkly Rep* **44:**501–503.

42. **Kaper JB, Nataro JP, Mobley HL.** 2004. Pathogenic *Escherichia coli*. *Nat Rev Microbiol* **2:**123–140.

43. **Bielaszewska M, Mellmann A, Zhang W, Kock R, Fruth A, Bauwens A, Peters G, Karch H.** 2011. Characterisation of the *Escherichia coli* strain associated with an outbreak of haemolytic uraemic syndrome in Germany, 2011: a microbiological study. *Lancet Infect Dis* **11:**671–676.

44. **Bettelheim KA, Beutin L.** 2003. Rapid laboratory identification and characterization of verocytotoxigenic (Shiga toxin producing) *Escherichia coli* (VTEC/STEC). *J Appl Microbiol* **95:**205–217.

45. **Paton JC, Paton AW.** 2003. Methods for detection of STEC in humans. An overview. *Methods Mol Med* **73:**9–26.

46. **Scheutz F, Teel LD, Beutin L, Pierard D, Buvens G, Karch H, Mellmann A, Caprioli A, Tozzoli R, Morabito S, Strockbine NA, Melton-Celsa AR, Sanchez M, Persson S, O'Brien AD.** 2012. Multicenter evaluation of a sequence-based protocol for subtyping shiga toxins and standardizing stx nomenclature. *J Clin Microbiol* **50:**2951–2963.

47. **Muthing J, Schweppe CH, Karch H, Friedrich AW.** 2009. Shiga toxins, glycosphingolipid diversity, and endothelial cell injury. *Thromb Haemost* **101:**252–264.

48. **Paton JC, Paton AW.** 1998. Pathogenesis and diagnosis of Shiga toxin-producing *Escherichia coli* infections. *Clin Microbiol Rev* **11:**450–479.

49. **Smith HR, Scotland SM.** 1988. Vero cytotoxin-producing strains of *Escherichia coli. J Med Microbiol* **26:**77–85.

50. **Strockbine NA, Wells JG, Bopp CA, Barrett TJ.** 1998. Overview of detection and subtyping methods, p 331–356. *In* Kaper JB, O'Brien AD (ed), *Escherichia coli O157:H7 and Other Shiga Toxin-Producing E. coli Strains.* American Society for Microbiology, Washington, DC.

51. **Beutin L, Steinruck H, Krause G, Steege K, Haby S, Hultsch G, Appel B.** 2007. Comparative evaluation of the Ridascreen®Verotoxin enzyme immunoassay for detection of Shiga-toxin producing strains of *Escherichia coli* (STEC) from food and other sources. *J Appl Microbiol* **102:**630–639.

52. **Beutin L, Kruger U, Krause G, Miko A, Martin A, Strauch E.** 2008. Evaluation of major types of Shiga toxin 2e producing *Escherichia coli* present in food, pigs and in the environment as potential pathogens for humans. *Appl Environ Microbiol* **74:**4806–4816.

53. **Staples M, Jennison AV, Graham RM, Smith HV.** 2012. Evaluation of the Meridian Premier EHEC assay as an indicator of Shiga toxin presence in direct faecal specimens. *Diagn Microbiol Infect Dis* **73:**322–325.

54. **Segura-Alvarez M, Richter H, Conraths FJ, Geue L.** 2003. Evaluation of enzyme-linked immunosorbent assays and a PCR test for detection of shiga toxins for Shiga toxin-producing *Escherichia coli* in cattle herds. *J Clin Microbiol* **41:** 5760–5763.

55. **Watarai S, Tana, Inoue K, Kushi Y, Isogai E, Yokota K, Naka K, Oguma K, Kodama H.** 2001. Inhibition of Vero cell cytotoxic activity in *Escherichia coli* O157:H7 lysates by globotriaosylceramide, Gb3, from bovine milk. *Biosci Biotechnol Biochem* **65:**414–419.

56. **Gallegos KM, Conrady DG, Karve SS, Gunasekera TS, Herr AB, Weiss AA.** 2012. Shiga toxin binding to glycolipids and glycans. *PLoS One* **7:**e30368.

57. **Beutin L, Zimmermann S, Gleier K.** 1996. Rapid detection and isolation of Shiga-like toxin (verocytotoxin)-producing *Escherichia coli* by direct testing of individual enterohemolytic colonies from washed sheep blood agar plates in the VTEC-RPLA assay. *J Clin Microbiol* **34:**2812–2814.

58. **Karmali MA, Petric M, Bielaszewska M.** 1999. Evaluation of a microplate latex agglutination method (Verotox-F assay) for detecting and characterizing verotoxins (Shiga toxins) in *Escherichia coli. J Clin Microbiol* **37:**396–399.

59. **Burgos Y, Beutin L.** 2012. Evaluation of an immuno-chromatographic detection system for Shiga toxins and the *E. coli* O157 antigen, p 29–40.

In Abuelzein E (ed), *Trends in Immunolabelled and Related Techniques,* 1st ed. InTech, Rijeka, Croatia.

60. **Feng PC, Jinneman K, Scheutz F, Monday SR.** 2011. Specificity of PCR and serological assays in the detection of *Escherichia coli* Shiga toxin subtypes. *Appl Environ Microbiol* **77:**6699–6702.

61. **Beutin L, Martin A, Krause G, Steege K, Haby S, Pries K, Albrecht N, Miko A, Jahn S.** 2010. Ergebnisse, Schlussfolgerungen und Empfehlungen aus zwei Ringversuchen zum Nachweis und zur Isolierung von Shiga (Vero) Toxin bildenden *Escherichia coli* (STEC) aus Hackfleischproben. *J Verbrauch Lebensm* **5:**21–34.

62. **Parma YR, Chacana PA, Lucchesi PM, Roge A, Granobles Velandia CV, Kruger A, Parma AE, Fernandez-Miyakawa ME.** 2012. Detection of Shiga toxin-producing *Escherichia coli* by sandwich enzyme-linked immunosorbent assay using chicken egg yolk IgY antibodies. *Front Cell Infect Microbiol* **2:**84.

63. **Kehl SC.** 2002. Role of the laboratory in the diagnosis of enterohemorrhagic *Escherichia coli* infections. *J Clin Microbiol* **40:**2711–2715.

64. **Zhang W, Bielaszewska M, Pulz M, Becker K, Friedrich AW, Karch H, Kuczius T.** 2008. A new immuno-PCR assay for the detection of low concentrations of Shiga toxin 2 and its variants. *J Clin Microbiol* **46:**1292–1297.

65. **Ge B, Zhao S, Hall R, Meng J.** 2002. A PCR-ELISA for detecting Shiga toxin-producing *Escherichia coli. Microbes Infect* **4:**285–290.

66. **Auvray F, Lecureuil C, Tache J, Leclerc V, Deperrois V, Lombard B.** 2007. Detection, isolation and characterization of Shiga toxin-producing *Escherichia coli* in retail-minced beef using PCR-based techniques, immunoassays and colony hybridization. *Lett Appl Microbiol* **45:**646–651.

67. **Fach P, Perelle S, Dilasser F, Grout J.** 2001. Comparison between a PCR-ELISA test and the vero cell assay for detecting Shiga toxin-producing *Escherichia coli* in dairy products and characterization of virulence traits of the isolated strains. *J Appl Microbiol* **90:**809–818.

68. **Beutin L, Zimmermann S, Gleier K.** 2002. Evaluation of the VTEC-screen "Seiken" test for detection of different types of Shiga toxin (verotoxin)-producing *Escherichia coli* (STEC) in human stool samples. *Diagn Microbiol Infect Dis* **42:**1–8.

69. **Klie H, Timm M, Richter H, Gallien P, Perlberg KW, Steinruck H.** 1997. Detection and occurrence of verotoxin-forming and/or shigatoxin producing *Escherichia coli* (VTEC and/or STEC) in milk. *Berl Munch Tierarztl Wochenschr* **110:** 337–341 (In German).

70. **Richter H, Klie H, Timm M, Gallien P, Steinruck H, Perlberg KW, Protz D.** 1997. Verotoxin-producing *E. coli* (VTEC) in feces from cattle slaughtered in Germany. *Berl Munch Tierarztl Wochenschr* **110:**121–127 (In German).

71. **Desrosiers A, Fairbrother JM, Johnson RP, Desautels C, Letellier A, Quessy S.** 2001. Phenotypic and genotypic characterization of *Escherichia coli* verotoxin-producing isolates from humans and pigs. *J Food Prot* **64:**1904–1911.

72. **Hull AE, Acheson DW, Echeverria P, Donohue-Rolfe A, Keusch GT.** 1993. Mitomycin immunoblot colony assay for detection of Shiga-like toxin-producing *Escherichia coli* in fecal samples: comparison with DNA probes. *J Clin Microbiol* **31:**1167–1172.

73. **Timm M, Klie H, Richter H, Perlberg KW.** 1996. A method for specific isolation of verotoxin-producing *Escherichia coli* colonies. *Berl Munch Tierarztl Wochenschr* **109:**270–272 (In German).

74. **Thomas A, Smith HR, Willshaw GA, Rowe B.** 1991. Non-radioactively labelled polynucleotide and oligonucleotide DNA probes, for selectively detecting *Escherichia coli* strains producing Vero cytotoxins VT1, VT2 and VT2 variant. *Mol Cell Probes* **5:**129–135.

75. **Karch H, Meyer T.** 1989. Evaluation of oligonucleotide probes for identification of Shiga-like-toxin-producing *Escherichia coli*. *J Clin Microbiol* **27:**1180–1186.

76. **Tokhi AM, Peiris JS, Scotland SM, Willshaw GA, Smith HR, Cheasty T.** 1993. A longitudinal study of Vero cytotoxin producing *Escherichia coli* in cattle calves in Sri Lanka. *Epidemiol Infect* **110:**197–208.

77. **Todd EC, Szabo RA, MacKenzie JM, Martin A, Rahn K, Gyles C, Gao A, Alves D, Yee AJ.** 1999. Application of a DNA hybridization-hydrophobic-grid membrane filter method for detection and isolation of verotoxigenic *Escherichia coli*. *Appl Environ Microbiol* **65:**4775–4780.

78. **Montenegro MA, Bulte M, Trumpf T, Aleksic S, Reuter G, Bulling E, Helmuth R.** 1990. Detection and characterization of fecal verotoxin-producing *Escherichia coli* from healthy cattle. *J Clin Microbiol* **28:**1417–1421.

79. **Dorn CR, Scotland SM, Smith HR, Willshaw GA, Rowe B.** 1989. Properties of Vero cytotoxin-producing *Escherichia coli* of human and animal origin belonging to serotypes other than O157:H7. *Epidemiol Infect* **103:**83–95.

80. **Kobayashi H, Shimada J, Nakazawa M, Morozumi T, Pohjanvirta T, Pelkonen S, Yamamoto K.** 2001. Prevalence and characteristics of shiga toxin-producing *Escherichia coli* from healthy cattle in Japan. *Appl Environ Microbiol* **67:**484–489.

81. **Lin Z, Kurazono H, Yamasaki S, Takeda Y.** 1993. Detection of various variant verotoxin genes in *Escherichia coli* by polymerase chain reaction. *Microbiol Immunol* **37:**543–548.

82. **Karch H, Meyer T.** 1989. Single primer pair for amplifying segments of distinct Shiga-like-toxin genes by polymerase chain reaction. *J Clin Microbiol* **27:**2751–2757.

83. **Paton AW, Paton JC.** 1998. Detection and characterization of Shiga toxigenic *Escherichia coli* by using multiplex PCR assays for *stx1*, *stx2*, *eaeA*, enterohemorrhagic *E. coli hlyA*, *rfb*O111, and *rfb*O157. *J Clin Microbiol* **36:**598–602.

84. **Gannon VP, King RK, Kim JY, Thomas EJ.** 1992. Rapid and sensitive method for detection of Shiga-like toxin-producing *Escherichia coli* in ground beef using the polymerase chain reaction. *Appl Environ Microbiol* **58:**3809–3815.

85. **Ziebell KA, Read SC, Johnson RP, Gyles CL.** 2002. Evaluation of PCR and PCR-RFLP protocols for identifying Shiga toxins. *Res Microbiol* **153:**289–300.

86. **Beutin L, Miko A, Krause G, Pries K, Haby S, Steege K, Albrecht N.** 2007. Identification of human-pathogenic strains of Shiga toxin-producing *Escherichia coli* from food by a combination of serotyping and molecular typing of Shiga toxin genes. *Appl Environ Microbiol* **73:**4769–4775.

87. **Miko A, Pries K, Haby S, Steege K, Albrecht N, Krause G, Beutin L.** 2009. Assessment of Shiga toxin-producing *Escherichia coli* isolates from wildlife meat as potential pathogens for humans. *Appl Environ Microbiol* **75:**6462–6470.

88. **Stephan R, Zweifel C, Fach P, Morabito S, Beutin L.** 2011. Shiga toxin-producing *Escherichia coli* in food, p 229–239. *In* Hoorfar J (ed), *Rapid Detection, Characterization, and Enumeration of Foodborne Pathogens.* ASM Press, Washington, DC.

89. **Wang G, Clark CG, Rodgers FG.** 2002. Detection in *Escherichia coli* of the genes encoding the major virulence factors, the genes defining the O157:H7 serotype, and components of the type 2 Shiga toxin family by multiplex PCR. *J Clin Microbiol* **40:**3613–3619.

90. **Peitz R, Weber H, Gleier K, Zimmermann S, Beutin L.** 2000. Detection of enterohemorrhagic *Escherichia coli* (EHEC) in raw meat and raw meat sausages. *Fleischwirtschaft* **3:**71–74.

91. **Urdahl AM, Solheim HT, Vold L, Hasseltvedt V, Wasteson Y.** 2013. Shiga toxin-encoding genes (*stx* genes) in human faecal samples. *APMIS* **121:**202–210.

92. **Wang F, Jiang L, Yang Q, Prinyawiwatkul W, Ge B.** 2012. Rapid and specific detection of *Escherichia coli* serogroups O26, O45, O103, O111, O121, O145, and O157 in ground beef, beef trim,

and produce by loop-mediated isothermal amplification. *Appl Environ Microbiol* **78:**2727–2736.

93. **Reischl U, Youssef MT, Kilwinski J, Lehn N, Zhang WL, Karch H, Strockbine NA.** 2002. Real-time fluorescence PCR assays for detection and characterization of Shiga toxin, intimin, and enterohemolysin genes from Shiga toxin-producing *Escherichia coli*. *J Clin Microbiol* **40:**2555–2565.

94. **Sekse C, Solberg A, Petersen A, Rudi K, Wasteson YB.** 2005. Detection and quantification of Shiga toxin-encoding genes in sheep faeces by real-time PCR. *Mol Cell Probes* **19:**363–370.

95. **Stefan A, Scaramagli S, Bergami R, Mazzini C, Barbanera M, Perelle S, Fach P.** 2007. Real-time PCR and enzyme-linked fluorescent assay methods for detecting Shiga-toxin-producing *Escherichia coli* in mincemeat samples. *Can J Microbiol* **53:**337–342.

96. **Perelle S, Dilasser F, Grout J, Fach P.** 2007. Screening food raw materials for the presence of the world's most frequent clinical cases of Shiga toxin-encoding *Escherichia coli* O26, O103, O111, O145 and O157. *Int J Food Microbiol* **113:**284–288.

97. **Perelle S, Dilasser F, Grout J, Fach P.** 2004. Detection by 5′-nuclease PCR of Shiga-toxin producing *Escherichia coli* O26, O55, O91, O103, O111, O113, O145 and O157:H7, associated with the world's most frequent clinical cases. *Mol Cell Probes* **18:**185–192.

98. **Chassagne L, Pradel N, Robin F, Livrelli V, Bonnet R, Delmas J.** 2009. Detection of *stx*1, *stx*2, and *eae* genes of enterohemorrhagic *Escherichia coli* using SYBR Green in a real-time polymerase chain reaction. *Diagn Microbiol Infect Dis* **64:**98–101.

99. **Beutin L, Jahn S, Fach F.** 2009. Evaluation of the "GeneDisc" real-time PCR system for detection of enterohaemorrhagic *Escherichia coli* (EHEC) O26, O103, O111, O145 and O157 strains according to their virulence markers and their O- and H-antigen-associated genes. *J Appl Microbiol* **106:**1122–1132.

100. **Fratamico PM, Bagi LK.** 2012. Detection of Shiga toxin-producing *Escherichia coli* in ground beef using the GeneDisc real-time PCR system. *Front Cell Infect Microbiol* **2:**152.

101. **Fratamico PM, Bagi LK, Cray WC Jr, Narang N, Yan X, Medina M, Liu Y.** 2011. Detection by multiplex real-time polymerase chain reaction assays and isolation of Shiga toxin-producing *Escherichia coli* serogroups O26, O45, O103, O111, O121, and O145 in ground beef. *Foodborne Pathog Dis* **8:**601–607.

102. **Tzschoppe M, Martin A, Beutin L.** 2012. A rapid procedure for the detection and isolation of enterohaemorrhagic *Escherichia coli* (EHEC) serogroup O26, O103, O111, O118, O121, O145 and O157 strains and the aggregative EHEC O104:H4 strain from ready-to-eat vegetables. *Int J Food Microbiol* **152:**19–30.

103. **Chui L, Couturier MR, Chiu T, Wang G, Olson AB, McDonald RR, Antonishyn NA, Horsman G, Gilmour MW.** 2010. Comparison of Shiga toxin-producing *Escherichia coli* detection methods using clinical stool samples. *J Mol. Diagn* **12:**469–475.

104. **International Organization for Standardization (ISO).** 2012. *ISO/TS 13136:2012, Microbiology of food and animal feed—Real-time polymerase chain reaction (PCR)-based method for the detection of food-borne pathogens—Horizontal method for the detection of Shiga toxin-producing Escherichia coli (STEC) and the determination of O157, O111, O26, O103 and O145 serogroups. (ISO/TS 13136:2012).* 11-7-2012. International Organization for Standardization, ISO Central Secretariat, Geneva, Switzerland.

105. **USDA Working Group.** 2011. Detection and isolation of non-O157 Shiga-toxin producing *Escherichia coli* strains (STEC) from meat products, p 1–16. *In* Laboratory QA/QC Division (ed), *Laboratory Guidebook.* U.S. Department of Agriculture, Laboratory QA/QC Division, Athens, GA.

106. **European Food Safety Authority.** 2012. *Manual for Reporting on Zoonoses, Zoonotic Agents and Antimicrobial Resistance in the Framework of Directive 2003/99/EC and of Some Other Pathogenic Microbiological Agents for Information Derived from the Year 2011.* European Food Safety Authority, Parma, Italy.

107. **European Food Safety Authority.** 2012. The European Union summary report on trends and sources of zoonoses, zoonotic agents and foodborne outbreaks in 2010. *EFSA J* **10:**2597.

108. **Mellmann A, Bielaszewska M, Kock R, Friedrich AW, Fruth A, Middendorf B, Harmsen D, Schmidt MA, Karch H.** 2008. Analysis of collection of hemolytic uremic syndrome-associated enterohemorrhagic *Escherichia coli*. *Emerg Infect Dis* **14:**1287–1290.

109. **Nielsen EM, Andersen MT.** 2003. Detection and characterization of verocytotoxin-producing *Escherichia coli* by automated 5′ nuclease PCR assay. *J Clin Microbiol* **41:**2884–2893.

110. **Shelton DR, Karns JS, Higgins JA, Van Kessel JA, Perdue ML, Belt KT, Russell-Anelli J, Debroy C.** 2006. Impact of microbial diversity on rapid detection of enterohemorrhagic *Escherichia coli* in surface waters. *FEMS Microbiol Lett* **261:**95–101.

111. **Krause G, Zimmermann S, Beutin L.** 2005. Investigation of domestic animals and pets as a reservoir for intimin (*eae*) gene positive *Escherichia coli* types. *Vet Microbiol* **106:**87–95.

112. Kozub-Witkowski E, Krause G, Frankel G, Kramer D, Appel B, Beutin L. 2008. Serotypes and virutypes of enteropathogenic and enterohaemorrhagic *Escherichia coli* strains from stool samples of children with diarrhoea in Germany. *J Appl Microbiol* **104:**403–410.

113. Feng PC, Keys C, Lacher D, Monday SR, Shelton D, Rozand C, Rivas M, Whittam T. 2010. Prevalence, characterization and clonal analysis of *Escherichia coli* O157: non-H7 serotypes that carry *eae* alleles. *FEMS Microbiol Lett* **308:**62–67.

114. Alonso MZ, Padola NL, Parma AE, Lucchesi PM. 2011. Enteropathogenic *Escherichia coli* contamination at different stages of the chicken slaughtering process. *Poult Sci* **90:**2638–2641.

115. Anklam KS, Kanankege KS, Gonzales TK, Kaspar CW, Dopfer D. 2012. Rapid and reliable detection of Shiga toxin-producing *Escherichia coli* by real-time multiplex PCR. *J Food Prot* **75:**643–650.

116. Sekse C, O'Sullivan K, Granum PE, Rorvik LM, Wasteson Y, Jorgensen HJ. 2009. An outbreak of *Escherichia coli* O103. *Int J Food Microbiol* **133:**259–264.

117. Iguchi A, Iyoda S, Ohnishi M. 2012. Molecular characterization reveals three distinct clonal groups among clinical shiga toxin-producing *Escherichia coli* strains of serogroup O103. *J Clin Microbiol* **50:**2894–2900.

118. Maidhof H, Guerra B, Abbas S, Elsheikha HM, Whittam TS, Beutin L. 2002. A multiresistant clone of Shiga toxin-producing *Escherichia coli* O118:[H16] is spread in cattle and humans over different European countries. *Appl Environ Microbiol* **68:**5834–5842.

119. Bielaszewska M, Prager R, Vandivinit L, Musken A, Mellmann A, Holt NJ, Tarr PI, Karch H, Zhang W. 2009. Detection and characterization of the fimbrial sfp cluster in enterohemorrhagic *Escherichia coli* O165:H25/NM isolates from humans and cattle. *Appl Environ Microbiol* **75:**64–71.

120. Delannoy S, Beutin L, Burgos YK, Fach P. 2012. Specific detection of enteroaggregative hemorrhagic *Escherichia coli* O104:H4 strains using the CRISPR locus as target for a diagnostic real-time PCR. *J Clin Microbiol* **50:**3485–3492.

121. Jelacic JK, Damrow T, Chen GS, Jelacic S, Bielaszewska M, Ciol M, Carvalho HM, Melton-Celsa AR, O'Brien AD, Tarr PI. 2003. Shiga toxin-producing *Escherichia coli* in Montana: bacterial genotypes and clinical profiles. *J Infect Dis* **188:**719–729.

122. Paton AW, Srimanote P, Woodrow MC, Paton JC. 2001. Characterization of Saa, a novel autoagglutinating adhesin produced by locus of enterocyte effacement-negative Shiga-toxigenic *Escherichia coli* strains that are virulent for humans. *Infect Immun* **69:**6999–7009.

123. Hauser E, Mellmann A, Semmler T, Stoeber H, Wieler LH, Karch H, Kuebler N, Fruth A, Harmsen D, Weniger T, Tietze E, Schmidt H. 2013. Phylogenetic and molecular analysis of food-borne Shiga toxin-producing *Escherichia coli*. *Appl Environ Microbiol* **79:**2731–2740.

124. Shimizu T, Ohta Y, Noda M. 2009. Shiga toxin 2 is specifically released from bacterial cells by two different mechanisms. *Infect Immun* **77:**2813–2823.

125. Rocha LB, Piazza RM. 2007. Production of Shiga toxin by Shiga toxin-expressing *Escherichia coli* (STEC) in broth media: from divergence to definition. *Lett Appl Microbiol* **45:**411–417.

126. McGannon CM, Fuller CA, Weiss AA. 2010. Different classes of antibiotics differentially influence shiga toxin production. *Antimicrob Agents Chemother* **54:**3790–3798.

127. Zhang X, McDaniel AD, Wolf LE, Keusch GT, Waldor MK, Acheson DW. 2000. Quinolone antibiotics induce Shiga toxin-encoding bacteriophages, toxin production, and death in mice. *J Infect Dis* **181:**664–670.

128. Bielaszewska M, Idelevich EA, Zhang W, Bauwens A, Schaumburg F, Mellmann A, Peters G, Karch H. 2012. Effects of antibiotics on Shiga toxin 2 production and bacteriophage induction by epidemic *Escherichia coli* O104:H4 strain. *Antimicrob Agents Chemother* **56:**3277–3282.

129. Hussein HS, Bollinger LM. 2008. Influence of selective media on successful detection of Shiga toxin-producing *Escherichia coli* in food, fecal, and environmental samples. *Foodborne Pathog Dis* **5:**227–244.

130. International Organization for Standardization. 2009. *ISO 16654: Microbiology of food and animal feeding stuffs—horizontal method for the detection of Escherichia coli O157*. International Organization for Standardization, Geneva, Switzerland.

131. de Boer E, Heuvelink AE. 2000. Methods for the detection and isolation of Shiga toxin-producing *Escherichia coli*. *Symp Ser Soc Appl Microbiol* **29:**133S–143S.

132. World Health Organization. 1998. *Zoonotic non-O157 Shiga toxin-producing Escherichia coli (STEC)*. Report of a WHO Scientific Working Group Meeting, Berlin, Germany, 23–26 June 1998. WHO/CSR/APH/98.8, 1–30. 1998. World Health Organization, Geneva, Switzerland.

133. Vimont A, Vernozy-Rozand C, Delignette-Muller ML. 2006. Isolation of *E. coli* O157:H7 and non-O157 STEC in different matrices: review of the most commonly used enrichment protocols. *Lett Appl Microbiol* **42:**102–108.

134. **Bolton DJ, O'Sullivan J, Duffy G, Baylis CL, Tozzoli R, Wasteson Y, Lofdahl S (ed).** 2007. *Methods for Detection and Molecular Characterisation of Pathogenic Escherichia coli.* Ashtown Food Research Centre, Ashtown, Ireland.

135. **Vimont A, Vernozy-Rozand C, Montet MP, Lazizzera C, Bavai C, Delignette-Muller ML.** 2006. Modeling and predicting the simultaneous growth of *Escherichia coli* O157:H7 and ground beef background microflora for various enrichment protocols. *Appl Environ Microbiol* **72:**261–268.

136. **Baylis CL.** 2008. Growth of pure cultures of verocytotoxin-producing *Escherichia coli* in a range of enrichment media. *J Appl Microbiol* **105:**1259–1265.

137. **Kanki M, Seto K, Harada T, Yonog S, Kumeda Y.** 2011. Comparison of four enrichment broths for the detection of non-O157 Shiga-toxin-producing *Escherichia coli* O91, O103, O111, O119, O121, O145 and O165 from pure culture and food samples. *Lett Appl Microbiol* **53:**167–173.

138. **Auvray F, Lecureuil C, Dilasser F, Tache J, Derzelle S.** 2009. Development of a real-time PCR assay with an internal amplification control for the screening of Shiga toxin-producing *Escherichia coli* in foods. *Lett Appl Microbiol* **48:**554–559.

139. **Jahn S, Weber H, Beutin L.** 2008. Comparison of enzyme immunoassay and quantitiver real time PCR as proof of Shigatoxin producing *Escherichia coli* (STEC) in mincemeat. *J Verbrauch Lebensm* **3:**385–395 (In German).

140. **Doyle MP, Schoeni JL.** 1984. Survival and growth characteristics of *Escherichia coli* associated with hemorrhagic colitis. *Appl Environ Microbiol* **48:**855–856.

141. **Lack WK, Becker B, Holzapfel WH.** 1996. Hygienischer Status frischer vorverpackter Mischsalate im Jahr 1995. *Arch Lebensmittelhyg* **47:**129–152.

142. **Klepzig I, Teufel P, Schott W, Hildebrandt G.** 1999. Auswirkungen einer Unterbrechung der Kühlkette auf die mikrobiologische Beschaffenheit von vorzerkleinerten Mischsalaten. *Arch Lebensmittelhyg* **50:**95–104.

143. **Sagoo SK, Little CL, Mitchell RT.** 2003. Microbiological quality of open ready-to-eat salad vegetables: effectiveness of food hygiene training of management. *J Food Prot* **66:**1581–1586.

144. **Madic J, Vingadassalon N, Peytavin de Garam C, Marault M, Scheutz F, Brugère H, Jamet E, Auvray F.** 2011. Detection of Shiga toxin-producing *Escherichia coli* (STEC) O26:H11, O103:H2, O111:H8, O145:H28 and O157:H7 in rawmilk cheeses by using multiplex real-time PCR. *Appl Environ Microbiol* **77:**2035–2041.

145. **Hussein HS, Bollinger LM, Hall MR.** 2008. Growth and enrichment medium for detection and isolation of Shiga toxin-producing *Escherichia coli* in cattle feces. *J Food Prot* **71:**927–933.

146. **Gill A, Martinez-Perez A, McIlwham S, Blais B.** 2012. Development of a method for the detection of verotoxin-producing *Escherichia coli* in food. *J Food Prot* **75:**827–837.

147. **Lehmacher A, Meier H, Aleksic S, Bockemuhl J.** 1998. Detection of hemolysin variants of Shiga toxin-producing *Escherichia coli* by PCR and culture on vancomycin-cefixime-cefsulodin blood agar. *Appl Environ Microbiol* **64:**2449–2453.

148. **Lin A, Nguyen L, Clotilde LM, Kase JA, Son I, Lauzon CR.** 2012. Isolation of Shiga toxin-producing *Escherichia coli* from fresh produce using STEC heart infusion washed blood agar with mitomycin-C. *J Food Prot* **75:**2028–2030.

149. **Sugiyama K, Inoue K, Sakazaki R.** 2001. Mitomycin-supplemented washed blood agar for the isolation of Shiga toxin-producing *Escherichia coli* other than O157:H7. *Lett Appl Microbiol* **33:**193–195.

150. **Posse B, De Zutter L, Heyndrickx M, Herman L.** 2008. Novel differential and confirmation plating media for Shiga toxin-producing *Escherichia coli* serotypes O26, O103, O111, O145 and sorbitol-positive and -negative O157. *FEMS Microbiol Lett* **282:**124–131.

151. **Posse B, De Zutter L, Heyndrickx M, Herman L.** 2008. Quantitative isolation efficiency of O26, O103, O111, O145 and O157 STEC serotypes from artificially contaminated food and cattle faeces samples using a new isolation protocol. *J Appl Microbiol* **105:**227–235.

152. **Vimont A, Delignette-Muller ML, Vernozy-Rozand C.** 2007. Supplementation of enrichment broths by novobiocin for detecting Shiga toxin-producing *Escherichia coli* from food: a controversial use. *Lett Appl Microbiol* **44:**326–331.

153. **Gouali M, Ruckly C, Carle I, Lejay-Collin M, Weill FX.** 2013. Evaluation of CHROMagar STEC and STEC O104 chromogenic agar media for detection of Shiga toxin-producing *Escherichia coli* in stool specimens. *J Clin Microbiol* **51:**894–900.

154. **Taylor DE, Rooker M, Keelan M, Ng LK, Martin I, Perna NT, Burland NT, Blattner FR.** 2002. Genomic variability of O islands encoding tellurite resistance in enterohemorrhagic *Escherichia coli* O157:H7 isolates. *J Bacteriol* **184:**4690–4698.

155. **Bielaszewska M, Tarr PI, Karch H, Zhang W, Mathys W.** 2005. Phenotypic and molecular analysis of tellurite resistance among enterohemorrhagic *Escherichia coli* O157:H7 and sorbitol-fermenting O157:NM clinical isolates. *J Clin Microbiol* **43:**452–454.

156. **Orth D, Grif K, Dierich MP, Wurzner R.** 2007. Variability in tellurite resistance and the ter gene cluster among Shiga toxin-producing *Escherichia coli* isolated from humans, animals and food. *Res Microbiol* **158:**105–111.

157. **Bielaszewska M, Middendorf B, Tarr PI, Zhang W, Prager R, Aldick T, Dobrindt U, Karch H, Mellmann A.** 2011. Chromosomal instability in enterohaemorrhagic *Escherichia coli* O157:H7: impact on adherence, tellurite resistance and colony phenotype. *Mol Microbiol* **79:**1024–1044.

158. **Schmidt H, Karch H, Beutin L.** 1994. The large-sized plasmids of enterohemorrhagic *Escherichia coli* O157 strains encode hemolysins which are presumably members of the *E. coli* alpha-hemolysin family. *FEMS Microbiol Lett* **117:**189–196.

159. **Brunder W, Schmidt H, Frosch M, Karch H.** 1999. The large plasmids of Shiga-toxin-producing *Escherichia coli* (STEC) are highly variable genetic elements. *Microbiology* **145**(Pt 5):1005–1014.

160. **Beutin L, Montenegro MA, Orskov I, Prada J, Zimmermann S, Stephan R.** 1989. Close association of verotoxin (Shiga-like toxin) production with enterohemolysin production in strains of *Escherichia coli. J Clin Microbiol* **27:**2559–2564.

161. **Burgos Y, Beutin L.** 2010. Common origin of plasmid encoded alpha-hemolysin genes in *Escherichia coli. BMC Microbiol* **10:**193.

162. **Prada J, Baljer G, De Rycke J, Steinruck H, Zimmermann S, Stephan R, Beutin L.** 1991. Characteristics of alpha-hemolytic strains of *Escherichia coli* isolated from dogs with gastroenteritis. *Vet Microbiol* **29:**59–73.

163. **Bugarel M, Beutin L, Fach P.** 2010. Low-density macroarray targeting non-locus of enterocyte effacement effectors (*nle* genes) and major virulence factors of Shiga toxin-producing *Escherichia coli* (STEC): a new approach for molecular risk assessment of STEC isolates. *Appl Environ Microbiol* **76:**203–211.

164. **Karch H, Bielaszewska M.** 2001. Sorbitol-fermenting Shiga toxin-producing *Escherichia coli* O157:H(−) strains: epidemiology, phenotypic and molecular characteristics, and microbiological diagnosis. *J Clin Microbiol* **39:**2043–2049.

165. **Bielaszewska M, Kock R, Friedrich AW, von Eiff C, Zimmerhackl LB, Karch H, Mellmann A.** 2007. Shiga toxin-mediated hemolytic uremic syndrome: time to change the diagnostic paradigm? *PLoS One* **2:**e1024.

166. **Miko A, Lindstedt BA, Brandal LT, Lobersli I, Beutin L.** 2010. Evaluation of multiple-locus variable number of tandem-repeats analysis (MLVA) as a method for identification of clonal groups among enteropathogenic, enterohaemorrhagic and avirulent *Escherichia coli* O26 strains. *FEMS Microbiol Lett* **303:**137–146.

167. **Bugarel M, Beutin L, Scheutz F, Loukiadis E, Fach P.** 2011. Identification of genetic markers for differentiation of Shiga toxin-producing, enteropathogenic and avirulent strains of *Escherichia coli* O26. *Appl Environ Microbiol* **77:**2275–2281.

168. **Gyles C, Johnson R, Gao A, Ziebell K, Pierard D, Aleksic S, Boerlin P.** 1998. Association of enterohemorrhagic *Escherichia coli* hemolysin with serotypes of shiga-like-toxin-producing *Escherichia coli* of human and bovine origins. *Appl Environ Microbiol* **64:**4134–4141.

169. **Aidar-Ugrinovich L, Blanco J, Blanco M, Blanco JE, Leomil L, Dahbi G, Mora A, Onuma DL, Silveira WD, Pestana de Castro AF.** 2007. Serotypes, virulence genes, and intimin types of Shiga toxin-producing *Escherichia coli* (STEC) and enteropathogenic *E. coli* (EPEC) isolated from calves in Sao Paulo, Brazil. *Int J Food Microbiol* **115:**297–306.

170. **Mora A, Blanco M, Blanco JE, Dahbi G, Lopez C, Justel P, Alonso MP, Echeita A, Bernardez MI, Gonzalez EA, Blanco J.** 2007. Serotypes, virulence genes and intimin types of Shiga toxin (verocytotoxin)-producing *Escherichia coli* isolates from minced beef in Lugo (Spain) from 1995 through 2003. *BMC Microbiol* **7:**13.

171. **Pradel N, Livrelli V, Champs C, Palcoux JB, Reynaud A, Scheutz F, Sirot J, Joly B, Forestier C.** 2000. Prevalence and characterization of Shiga toxin-producing *Escherichia coli* isolated from cattle, food, and children during a one-year prospective study in France. *J Clin Microbiol* **38:**1023–1031.

172. **Kaufmann M, Zweifel C, Blanco M, Blanco JE, Blanco J, Beutin L, Stephan RB.** 2006. *Escherichia coli* O157 and non-O157 Shiga toxin-producing *Escherichia coli* in fecal samples of finished pigs at slaughter in Switzerland. *J Food Prot* **69:**260–266.

173. **Karama M, Gyles CL.** 2010. Methods for genotyping verotoxin-producing *Escherichia coli. Zoonoses Public Health* **57:**447–462.

174. **Karch H, Bohm H, Schmidt H, Gunzer F, Aleksic S, Heesemann J.** 1993. Clonal structure and pathogenicity of Shiga-like toxin-producing, sorbitol-fermenting *Escherichia coli* O157:H−. *J Clin Microbiol* **31:**1200–1205.

175. **Feng P, Lampel KA, Karch H, Whittam TS.** 1998. Genotypic and phenotypic changes in the emergence of *Escherichia coli* O157:H7. *J Infect Dis* **177:**1750–1753.

176. **Murinda SE, Batson SD, Nguyen LT, Gillespie BE, Oliver SP.** 2004. Phenotypic and genetic markers for serotype-specific detection of Shiga toxin-producing *Escherichia coli* O26 strains from North America. *Foodborne Pathog Dis* **1:**125–135.

177. **Prager R, Liesegang A, Voigt W, Rabsch W, Fruth A, Tschape H.** 2002. Clonal diversity of Shiga toxin-producing *Escherichia coli* O103:H2/H(-) in Germany. *Infect Genet Evol* **1:**265–275.

178. **Willshaw GA, Smith HR, Cheasty T, O'Brien SJ.** 2001. Use of strain typing to provide evidence for specific interventions in the transmission of VTEC O157 infections. *Int J Food Microbiol* **66:**39–46.

179. **Schroeder CM, Meng J, Zhao S, Debroy C, Torcolini J, Zhao C, McDermott PF, Wagner DD, Walker RD, White DG.** 2002. Antimicrobial resistance of *Escherichia coli* O26, O103, O111, O128, and O145 from animals and humans. *Emerg Infect Dis* **8:**1409–1414.

180. **Ziebell K, Johnson RP, Kropinski AM, Reid-Smith R, Ahmed R, Gannon VP, Gilmour M, Boerlin P.** 2011. Gene cluster conferring strepto-mycin, sulfonamide, and tetracycline resistance in *Escherichia coli* O157:H7 phage types 23, 45, and 67. *Appl Environ Microbiol* **77:**1900–1903.

181. **Souza MR, Klassen G, Toni FD, Rigo LU, Henkes C, Pigatto CP, Dalagassa CB, Fadel-Picheth CM.** 2010. Biochemical properties, enterohaemolysin production and plasmid car-riage of Shiga toxin-producing *Escherichia coli* strains. *Mem Inst Oswaldo Cruz* **105:**318–321.

182. **Zhang WL, Bielaszewska M, Liesegang A, Tschape H, Schmidt H, Bitzan M, Karch H.** 2000. Molecular characteristics and epidemio-logical significance of Shiga toxin-producing *Escherichia coli* O26 strains. *J Clin Microbiol* **38:**2134–2140.

183. **Pradel N, Boukhors K, Bertin Y, Forestier C, Martin C, Livrelli V.** 2001. Heterogeneity of Shiga toxin-producing *Escherichia coli* strains isolated from hemolytic-uremic syndrome patients, cattle, and food samples in central France. *Appl Environ Microbiol* **67:**2460–2468.

184. **Chaudhuri RR, Henderson IR.** 2012. The evo-lution of the *Escherichia coli* phylogeny. *Infect Genet Evol* **12:**214–226.

185. **Whittam TS, Wolfe ML, Wachsmuth IK, Orskov F, Orskov I, Wilson RA.** 1993. Clonal relationships among *Escherichia coli* strains that cause hemorrhagic colitis and infantile diarrhea. *Infect Immun* **61:**1619–1629.

186. **Abu-Ali GS, Lacher DW, Wick LM, Qi W, Whittam TS.** 2009. Genomic diversity of patho-genic *Escherichia coli* of the EHEC 2 clonal complex. *BMC Genomics* **10:**296.

187. **Nielsen EM, Scheutz F, Torpdahl M.** 2006. Continuous surveillance of Shiga toxin-producing *Escherichia coli* infections by pulsed-field gel electrophoresis shows that most infections are sporadic. *Foodborne Pathog Dis* **3:**81–87.

188. **Keskimaki M, Eklund M, Pesonen H, Heiskanen T, Siitonen A.** 2001. EPEC, EAEC and STEC in stool specimens: prevalence and molec-ular epidemiology of isolates. *Diagn Microbiol Infect Dis* **40:**151–156.

189. **Orskov F, Orskov I.** 1984. Serotyping of *Esche-richia coli*, p 43–112. *In* Bergan T (ed), *Methods in Microbiology*. Academic Press, London, United Kingdom.

190. **Scheutz F, Cheasty T, Woodward D, Smith HR.** 2004. Designation of O174 and O175 to temporary O groups OX3 and OX7, and six new *E. coli* O groups that include verocytotoxin-producing *E. coli* (VTEC): O176, O177, O178, O179, O180 and O181. *APMIS* **112:**569–584.

191. **Blanco M, Padola NL, Krüger A, Sanz ME, Blanco JE, Gonzalez EA, Dahbi G, Mora A, Bernardez MI, Etcheverria AI, Arroyo GH, Lucchesi PMA, Parma AE, Blanco J.** 2004. Virulence genes and intimin types of Shiga-toxin-producing *Escherichia coli* isolated from cattle and beef products in Argentina. *Int Microbiol* **7:**269–276.

192. **Much P, Pichler J, Kasper SS, Allerberger F.** 2009. Foodborne outbreaks, Austria 2007. *Wien Klin Wochenschr* **121:**77–85.

193. **Scheutz F.** 2012. *Technical report: external quality assurance scheme for typing of verocytotoxin-pro-ducing E. coli (VTEC)*. 1–28. 1-4-2012. Stockholm, European Centre for Disease Prevention and Control.

194. **Wang L, Rothemund D, Curd H, Reeves PR.** 2003. Species-wide variation in the *Escherichia coli* flagellin (H-antigen) gene. *J Bacteriol* **185:**2936–2943.

195. **Machado J, Grimont F, Grimont PA.** 2000. Identification of *Escherichia coli* flagellar types by restriction of the amplified *fliC* gene. *Res Microbiol* **151:**535–546.

196. **Prager R, Strutz U, Fruth A, Tschape H.** 2003. Subtyping of pathogenic *Escherichia coli* strains using flagellar (H)-antigens: serotyping versus fliC polymorphisms. *Int J Med Microbiol* **292:** 477–486.

197. **Fields PI, Blom K, Hughes HJ, Helsel LO, Feng P, Swaminathan B.** 1997. Molecular char-acterization of the gene encoding H antigen in *Escherichia coli* and development of a PCR-re-striction fragment length polymorphism test for identification of *E. coli* O157:H7 and O157:NM. *J Clin Microbiol* **35:**1066–1070.

198. **Beutin L, Strauch E.** 2007. Identification of se-quence diversity in the *Escherichia coli fliC* genes encoding flagellar types H8 and H40 and its use in typing of Shiga toxin-producing *E. coli* O8, O22, O111, O174, and O179 strains. *J Clin Microbiol* **45:**333–339.

199. **Ayala CO, Ramos Moreno AC, Martinez MB, Burgos YK, Pestana de Castro AF, Bando SY.** 2012. Determination of flagellar types by PCR-RFLP analysis of enteropathogenic *Escherichia coli* (EPEC) and Shiga toxin-producing *E. coli* (STEC) strains isolated from animals in Sao Paulo, Brazil. *Res Vet Sci* **92**:18–23.

200. **Wang L, Rothemund D, Curd H, Reeves PR.** 2000. Sequence diversity of the *Escherichia coli* H7 *fliC* Genes: implication for a DNA-based typing scheme for *E. coli* O157:H7. *J Clin Microbiol* **38**:1786–1790.

201. **Desmarchelier PM, Bilge SS, Fegan N, Mills L, Vary JC Jr, Tarr PI.** 1998. A PCR specific for *Escherichia coli* O157 based on the *rfb* locus encoding O157 lipopolysaccharide. *J Clin Microbiol* **36**:1801–1804.

202. **Fratamico PM, Debroy C, Miyamoto T, Liu Y.** 2009. PCR detection of enterohemorrhagic *Escherichia coli* O145 in food by targeting genes in the *E. coli* O145 O-antigen gene cluster and the shiga toxin 1 and shiga toxin 2 genes. *Foodborne Pathog Dis* **6**:605–611.

203. **O'Hanlon KA, Catarame TM, Duffy G, Blair IS, and McDowell DA.** 2004. RAPID detection and quantification of *E. coli* O157/O26/O111 in minced beef by real-time PCR. *J Appl Microbiol* **96**:1013–1023.

204. **Paton AW, Paton JC.** 1999. Molecular characterization of the locus encoding biosynthesis of the lipopolysaccharide O antigen of *Escherichia coli* serotype O113. *Infect Immun* **67**:5930–5937.

205. **Perelle S, Dilasser F, Grout J, Fach P.** 2005. Detection of *Escherichia coli* serogroup O103 by real-time polymerase chain reaction. *J Appl Microbiol* **98**:1162–1168.

206. **Valadez AM, Debroy C, Dudley E, Cutter CN.** 2011. Multiplex PCR detection of Shiga toxin-producing *Escherichia coli* strains belonging to serogroups O157, O103, O91, O113, O145, O111, and O26 experimentally inoculated in beef carcass swabs, beef trim, and ground beef. *J Food Prot* **74**:228–239.

207. **Ballmer K, Korczak BM, Kuhnert P, Slickers P, Ehricht R, Hachler H.** 2007. Fast DNA serotyping of *Escherichia coli* by use of an oligonucleotide microarray. *J Clin Microbiol* **45**:370–379.

208. **Anjum MF, Mafura M, Slickers P, Ballmer K, Kuhnert P, Woodward MJ, Ehricht R.** 2007. Pathotyping *Escherichia coli* by using miniaturized DNA microarrays. *Appl Environ Microbiol* **73**:5692–5697.

209. **Lin A, Nguyen L, Lee T, Clotilde LM, Kase JA, Son I, Carter JM, Lauzon CR.** 2011. Rapid O serogroup identification of the ten most clinically relevant STECs by Luminex microbead-based suspension array. *J Microbiol Methods* **87**:105–110.

210. **Lin A, Sultan O, Lau HK, Wong E, Hartman G, Lauzon CR.** 2011. O serogroup specific real time PCR assays for the detection and identification of nine clinically relevant non-O157 STECs. *Food Microbiol* **28**:478–483.

211. **Gilmour MW, Olson AB, Andrysiak AK, Ng LK, Chui L.** 2007. Sequence-based typing of genetic targets encoded outside of the O-antigen gene cluster is indicative of Shiga toxin-producing *Escherichia coli* serogroup lineages. *J Med Microbiol* **56**:620–628.

212. **Norman KN, Strockbine NA, Bono JL.** 2012. Association of nucleotide polymorphisms within the O-antigen gene cluster of *Escherichia coli* O26, O45, O103, O111, O121, and O145 with serogroups and genetic subtypes. *Appl Environ Microbiol* **78**:6689–6703.

213. **Delannoy S, Beutin L, Fach P.** 2012. Use of clustered regularly interspaced short palindromic repeat sequence polymorphisms for specific detection of enterohemorrhagic *Escherichia coli* strains of serotypes O26:H11, O45:H2, O103:H2, O111:H8, O121:H19, O145:H28, and O157:H7 by real-time PCR. *J Clin Microbiol* **50**:4035–4040.

214. **Gerner-Smidt P, Hise K, Kincaid J, Hunter S, Rolando S, Hyytia-Trees E, Ribot EM, Swaminathan B.** 2006. PulseNet USA: a five-year update. *Foodborne Pathog Dis* **3**:9–19.

215. **Gerner-Smidt P, Scheutz F.** 2006. Standardized pulsed-field gel electrophoresis of Shiga toxin-producing *Escherichia coli*: the PulseNet Europe feasibility study. *Foodborne Pathog Dis* **3**:74–80.

216. **Mora A, Lopez C, Dhabi G, Lopez-Beceiro AM, Fidalgo LE, Diaz EA, Martinez-Carrasco C, Mamani R, Herrera A, Blanco JE, Blanco M, Blanco J.** 2012. Seropathotypes, phylogroups, Stx subtypes, and intimin types of wildlife-carried, Shiga toxin-producing *Escherichia coli* strains with the same characteristics as human-pathogenic isolates. *Appl Environ Microbiol* **78**:2578–2585.

217. **Beutin L, Geier D, Zimmermann S, Aleksic S, Gillespie HA, Whittam TS.** 1997. Epidemiological relatedness and clonal types of natural populations of *Escherichia coli* strains producing Shiga toxins in separate populations of cattle and sheep. *Appl Environ Microbiol* **63**:2175–2180.

218. **Thomas KM, McCann MS, Collery MM, Logan A, Whyte P, McDowell DA, Duffy G.** 2012. Tracking verocytotoxigenic *Escherichia coli* O157, O26, O111, O103 and O145 in Irish cattle. *Int J Food Microbiol* **153**:288–296.

219. **Welinder-Olsson C, Badenfors M, Cheasty T, Kjellin E, Kaijser B.** 2002. Genetic profiling of enterohemorrhagic *Escherichia coli* strains in

relation to clonality and clinical signs of infection. *J Clin Microbiol* **40:**959–964.

220. **Vaz TMI, Irino K, Nishimura LS, Cergole-Novella MC, Guth BEC.** 2006. Genetic heterogeneity of Shiga toxin-producing *Escherichia coli* strains isolated in Sao Paulo, Brazil, from 1976 through 2003, as revealed by pulsed-field gel electrophoresis. *J Clin Microbiol* **44:**798–804.

221. **Mainil JG, Bardiau M, Ooka T, Ogura Y, Murase K, Etoh Y, Ichihara S, Horikawa K, Buvens G, Pierard D, Itoh T, Hayashi T.** 2011. Typing of O26 enterohaemorrhagic and enteropathogenic *Escherichia coli* isolated from humans and cattle with IS621 multiplex PCR-based fingerprinting. *J Appl Microbiol* **111:**773–786.

222. **Leomil L, Pestana de Castro AF, Krause G, Schmidt H, Beutin L.** 2005. Characterization of two major groups of diarrheagenic *Escherichia coli* O26 strains which are globally spread in human patients and domestic animals of different species. *FEMS Microbiol Lett* **249:**335–342.

223. **Sugimoto N, Shima K, Hinenoya A, Asakura M, Matsuhisa A, Watanabe H, Yamasaki S.** 2011. Evaluation of a PCR-restriction fragment length polymorphism (PCR-RFLP) assay for molecular epidemiological study of Shiga toxin-producing *Escherichia coli*. *J Vet Med Sci* **73:**859–867.

224. **Beutin L, Kaulfuss S, Herold S, Oswald E, Schmidt H.** 2005. Genetic analysis of enteropathogenic and enterohemorrhagic *Escherichia coli* serogroup O103 strains by molecular typing of virulence and housekeeping genes and pulsed-field gel electrophoresis. *J Clin Microbiol* **43:**1552–1563.

225. **Sonntag AK, Prager R, Bielaszewska M, Zhang W, Fruth A, Tschape H, Karch H.** 2004. Phenotypic and genotypic analyses of enterohemorrhagic *Escherichia coli* O145 strains from patients in Germany. *J Clin Microbiol* **42:**954–962.

226. **Louie M, Read S, Louie L, Ziebell K, Rahn K, Borczyk A, Lior H.** 1999. Molecular typing methods to investigate transmission of *Escherichia coli* O157:H7 from cattle to humans. *Epidemiol Infect* **123:**17–24.

227. **Beutin L, Bulte M, Weber A, Zimmermann S, Gleier K.** 2000. Investigation of human infections with verocytotoxin-producing strains of *Escherichia coli* (VTEC) belonging to serogroup O118 with evidence for zoonotic transmission. *Epidemiol Infect* **125:**47–54.

228. **Thomas KM, McCann MS, Collery MM, Moschonas G, Whyte P, McDowell DA, Duffy G.** 2013. Transfer of verocytotoxigenic *Escherichia coli* O157, O26, O111, O103 and O145 from fleece to carcass during sheep slaughter in an Irish export abattoir. *Food Microbiol* **34:**38–45.

229. **Centers for Disease Control and Prevention.** 2012. Outbreak of Shiga toxin-producing *Escherichia coli* O111 infections associated with a correctional facility diary—Colorado 2010. *Morb Mortal Wkly Rep* **61:**149–152.

230. **Serra-Moreno R, Jofre J, Muniesa M.** 2007. Insertion site occupancy by stx2 bacteriophages depends on the locus availability of the host strain chromosome. *J Bacteriol* **189:**6645–6654.

231. **Shaikh N, Tarr PI.** 2003. *Escherichia coli* O157:H7 Shiga toxin-encoding bacteriophages: integrations, excisions, truncations, and evolutionary implications. *J Bacteriol* **185:**3596–3605.

232. **Besser TE, Shaikh N, Holt NJ, Tarr PI, Konkel ME, Malik-Kale P, Walsh CW, Whittam TS, Bono JL.** 2007. Greater diversity of Shiga toxin-encoding bacteriophage insertion sites among *Escherichia coli* O157:H7 isolates from cattle than in those from humans. *Appl Environ Microbiol* **73:**671–679.

233. **Vanaja SK, Springman AC, Besser TE, Whittam TS, Manning SD.** 2010. Differential expression of virulence and stress fitness genes between *Escherichia coli* O157:H7 strains with clinical or bovine-biased genotypes. *Appl Environ Microbiol* **76:**60–68.

234. **Shringi S, Schmidt C, Katherine K, Brayton KA, Hancock DD, Besser TE.** 2012. Carriage of stx2a differentiates clinical and bovine-biased strains of *Escherichia coli* O157. *PLoS One* **7:**e51572.

235. **Creuzburg K, Schmidt H.** 2007. Molecular characterization and distribution of genes encoding members of the type III effector *nleA* family among pathogenic *Escherichia coli* strains. *J Clin Microbiol* **45:**2498–2507.

236. **Oswald E, Schmidt H, Morabito S, Karch H, Marches O, Caprioli A.** 2000. Typing of intimin genes in human and animal enterohemorrhagic and enteropathogenic *Escherichia coli*: characterization of a new intimin variant. *Infect Immun* **68:**64–71.

237. **Jores J, Rumer L, Wieler LH.** 2004. Impact of the locus of enterocyte effacement pathogenicity island on the evolution of pathogenic *Escherichia coli*. *Int J Med Microbiol* **294:**103–113.

238. **Mora A, Blanco M, Yamamoto D, Dahbi G, Blanco JE, Lopez C, Alonso MP, Vieira MA, Hernandes RT, Abe CM, Piazza RM, Lacher DW, Elias WP, Gomes TA, Blanco J.** 2009. HeLa-cell adherence patterns and actin aggregation of enteropathogenic *Escherichia coli* (EPEC) and Shiga-toxin-producing *E. coli* (STEC) strains carrying different *eae* and tir alleles. *Int Microbiol* **12:**243–251.

239. **Creuzburg K, Heeren S, Lis CM, Kranz M, Hensel M, Schmidt H.** 2011. Genetic background and mobility of variants of the gene nleA in attaching and effacing *Escherichia coli*. *Appl Environ Microbiol* **77:**8705–8713.

240. **Persson S, Olsen KE, Ethelberg S, Scheutz F.** 2007. Subtyping method for *Escherichia coli* shiga toxin (verocytotoxin) 2 variants and correlations to clinical manifestations. *J Clin Microbiol* **45**:2020–2024.

241. **Friedrich AW, Bielaszewska M, Zhang WL, Pulz M, Kuczius T, Ammon A, Karch H.** 2002. *Escherichia coli* harboring Shiga toxin 2 gene variants: frequency and association with clinical symptoms. *J Infect Dis* **185**:74–84.

242. **Friedrich AW, Borell J, Bielaszewska M, Fruth A, Tschape H, Karch H.** 2003. Shiga toxin 1c-producing *Escherichia coli* strains: phenotypic and genetic characterization and association with human disease. *J Clin Microbiol* **41**:2448–2453.

243. **Fratamico PM, Bagi LK, Bush EJ, Solow BT.** 2004. Prevalence and characterization of shiga toxin-producing *Escherichia coli* in swine feces recovered in the National Animal Health Monitoring System's Swine 2000 study. *Appl Environ Microbiol* **70**:7173–7178.

244. **Muthing J, Meisen I, Zhang W, Bielaszewska M, Mormann M, Bauerfeind R, Schmidt MA, Friedrich AW, Karch H.** 2012. Promiscuous Shiga toxin 2e and its intimate relationship to Forssman. *Glycobiology* **22**:849–862.

245. **Morabito S, Dell'Omo G, Agrimi U, Schmidt H, Karch H, Cheasty T, Caprioli A.** 2001. Detection and characterization of Shiga toxin-producing *Escherichia coli* in feral pigeons. *Vet Microbiol* **82**:275–283.

246. **Ooka T, Seto K, Kawano K, Kobayashi H, Etoh Y, Ichihara S, Kaneko A, Isobe J, Yamaguchi K, Horikawa K, Gomes TA, Linden A, Bardiau M, Mainil JG, Beutin L, Ogura Y, Hayashi T.** 2012. Clinical significance of *Escherichia albertii*. *Emerg Infect Dis* **18**:488–492.

247. **Prager R, Fruth A, Siewert U, Strutz U, Tschape H.** 2009. *Escherichia coli* encoding Shiga toxin 2f as an emerging human pathogen. *Int J Med Microbiol* **299**:343–353.

248. **Hofer E, Cernela N, Stephan R.** 2012. Shiga toxin subtypes associated with Shiga toxin-producing *Escherichia coli* strains isolated from red deer, roe deer, chamois, and ibex. *Foodborne Pathog Dis* **9**:792795.

249. **Brett KN, Ramachandran V, Hornitzky MA, Bettelheim KA, Walker MJ, Djordjevic SP.** 2003. stx1c Is the most common Shiga toxin 1 subtype among Shiga toxin-producing *Escherichia coli* isolates from sheep but not among isolates from cattle. *J Clin Microbiol* **41**:926–936.

250. **Eggert M, Stuber E, Heurich M, Fredriksson-Ahomaa M, Burgos Y, Beutin L, Martlbauer E.** 2012. Detection and characterization of Shiga toxin-producing *Escherichia coli* in faeces and lymphatic tissue of free-ranging deer. *Epidemiol Infect* **28**:1–9.

251. **Kumar A, Taneja N, Kumar Y, Sharma M.** 2012. Detection of Shiga toxin variants among Shiga toxin-forming *Escherichia coli* isolates from animal stool, meat and human stool samples in India. *J Appl Microbiol* **113**:1208–1216.

252. **Masana MO, D'Astek BA, Palladino PM, Galli L, del Castillo LL, Carbonari C, Leotta GA, Vilacoba E, Irino K, Rivas M.** 2011. Genotypic characterization of non-O157 Shiga toxin-producing *Escherichia coli* in beef abattoirs of Argentina. *J Food Prot* **74**:2008–2017.

253. **D'Astek BA, del Castillo LL, Miliwebsky E, Carbonari C, Palladino PM, Deza N, Chinen I, Manfredi E, Leotta GA, Masana MO, Rivas M.** 2012. Subtyping of *Escherichia coli* O157:H7 strains isolated from human infections and healthy cattle in Argentina. *Foodborne Pathog Dis* **9**:457–464.

254. **Slanec T, Fruth A, Creuzburg K, Schmidt H.** 2009. Molecular analysis of virulence profiles and Shiga toxin genes in food-borne Shiga toxin-producing *Escherichia coli*. *Appl Environ Microbiol* **75**:6187–6197.

255. **Vernozy-Rozand C, Montet MP, Bertin Y, Trably F, Girardeau JP, Martin C, Livrelli V, Beutin L.** 2004. Serotyping, stx2 subtyping, and characterization of the locus of enterocyte effacement island of Shiga toxin-producing *Escherichia coli* and *E. coli* O157:H7 strains isolated from the environment in France. *Appl Environ Microbiol* **70**:2556–2559.

256. **Polifroni R, Etcheverria AI, Sanz ME, Cepeda RE, Kruger A, Lucchesi PM, Fernandez D, Parma AE, Padola NL.** 2012. Molecular characterization of Shiga toxin-producing *Escherichia coli* isolated from the environment of a dairy farm. *Curr Microbiol* **65**:337–343.

257. **Wasilenko JL, Fratamico PM, Narang N, Tillman GE, Ladely S, Simmons M, Cray WC Jr.** 2012. Influence of primer sequences and DNA extraction method on detection of non-O157 Shiga toxin-producing *Escherichia coli* in ground beef by real-time PCR targeting the *eae*, *stx*, and serogroup-specific genes. *J Food Prot* **75**:1939–1950.

258. **Deng W, Puente JL, Gruenheid S, Li Y, Vallance BA, Vazquez A, Barba J, Ibarra JA, O'Donnell P, Metalnikov P, Ashman K, Lee S, Goode D, Pawson T, Finlay BB.** 2004. Dissecting virulence: systematic and functional analyses of a pathogenicity island. *Proc Natl Acad Sci USA* **101**:3597–3602.

259. **Tarr CL, Whittam TS.** 2002. Molecular evolution of the intimin gene in O111 clones of pathogenic *Escherichia coli*. *J Bacteriol* **184**:479–487.

260. **Zhang WL, Kohler B, Oswald E, Beutin L, Karch H, Morabito S, Caprioli A, Suerbaum S, Schmidt H.** 2002. Genetic diversity of intimin genes of attaching and effacing *Escherichia coli* strains. *J Clin Microbiol* **40:**4486–4492.

261. **Jores J, Zehmke K, Eichberg J, Rumer L, Wieler LH.** 2003. Description of a novel intimin variant (type zeta) in the bovine O84:NM verotoxin-producing *Escherichia coli* strain 537/89 and the diagnostic value of intimin typing. *Exp Biol Med (Maywood)* **228:**370–376.

262. **Creuzburg K, Middendorf B, Mellmann A, Martaler T, Holz C, Fruth A, Karch H, Schmidt H.** 2011. Evolutionary analysis and distribution of type III effector genes in pathogenic *Escherichia coli* from human, animal and food sources. *Environ Microbiol* **13:**439–452.

263. **Levine MM.** 1987. *Escherichia coli* that cause diarrhea: enterotoxigenic, enteropathogenic, enteroinvasive, enterohemorrhagic, and enteroadherent. *J Infect Dis* **155:**377–389.

264. **Boerlin P, McEwen SA, Boerlin-Petzold F, Wilson JB, Johnson RP, Gyles CL.** 1999. Associations between virulence factors of Shiga toxin-producing *Escherichia coli* and disease in humans. *J Clin Microbiol* **37:**497–503.

265. **Ethelberg S, Olsen KE, Scheutz F, Jensen C, Schiellerup P, Enberg J, Petersen AM, Olesen B, Gerner-Smidt P, Molbak K.** 2004. Virulence factors for hemolytic uremic syndrome, Denmark. *Emerg Infect Dis* **10:**842–847.

266. **Farfan MJ, Torres AG.** 2012. Molecular mechanisms that mediate colonization of Shiga toxin-producing *Escherichia coli* strains. *Infect Immun* **80:**903–913.

267. **Perna NT, Plunkett G III, Burland V, Mau B, Glasner JD, Rose DJ, Mayhew GF, Evans PS, Gregor J, Kirkpatrick HA, Posfai G, Hackett J, Klink S, Boutin A, Shao Y, Miller L, Grotbeck EJ, Davis NW, Lim A, Dimalanta ET, Potamousis KD, Apodaca J, Anantharaman TS, Lin J, Yen G, Schwartz DC, Welch RA, Blattner FR.** 2001. Genome sequence of enterohaemorrhagic *Escherichia coli* O157:H7. *Nature* **409:**529–533.

268. **Ogura Y, Ooka T, Iguchi A, Toh H, Asadulghani M, Oshima K, Kodama T, Abe H, Nakayama K, Kurokawa K, Tobe T, Hattori M, Hayashi T.** 2009. Comparative genomics reveal the mechanism of the parallel evolution of O157 and non-O157 enterohemorrhagic *Escherichia coli*. *Proc Natl Acad. Sci USA* **106:**17939–17944.

269. **Ogura Y, Ooka T, Asadulghani, Terajima J, Nougayrede JP, Kurokawa K, Tashiro K, Tobe T, Nakayama K, Kuhara S, Oswald E, Watanabe H, Hayashi T.** 2007. Extensive genomic diversity and selective conservation of virulence-determinants in enterohemorrhagic *Escherichia coli* strains of O157 and non-O157 serotypes. *Genome Biol* **8:**R138.

270. **Wickham ME, Lupp C, Mascarenhas M, Vazquez A, Coombes BK, Brown NF, Coburn BA, Deng W, Puente JL, Karmali MA, Finlay BB.** 2006. Bacterial genetic determinants of non-O157 STEC outbreaks and hemolytic-uremic syndrome after infection. *J Infect Dis* **194:**819–827.

271. **Bugarel M, Martin A, Fach P, Beutin L.** 2011. Virulence gene profiling of enterohemorrhagic (EHEC) and enteropathogenic (EPEC) *Escherichia coli* strains: a basis for molecular risk assessment of typical and atypical EPEC strains. *BMC Microbiol* **11:**142.

272. **Schimmer B, Nygard K, Eriksen HM, Lassen J, Lindstedt BA, Brandal LT, Kapperud G, Aavitsland P.** 2008. Outbreak of haemolytic uraemic syndrome in Norway caused by stx2-positive *Escherichia coli* O103:H25 traced to cured mutton sausages. *BMC Infect Dis* **8:**41.

273. **Imamovic L, Tozzoli R, Michelacci V, Minelli F, Marziano ML, Caprioli A, Morabito S.** 2010. OI-57, a genomic island of *Escherichia coli* O157, is present in other seropathotypes of Shiga toxin-producing *E. coli* associated with severe human disease. *Infect Immun* **78:**4697–4704.

CLINICAL, PATHOLOGICAL, AND PATHOPHYSIOLOGICAL ASPECTS

Shiga Toxin/Verocytotoxin-Producing *Escherichia coli* Infections: Practical Clinical Perspectives

15

T. KEEFE DAVIS,[1] NICOLE C. A. J. VAN DE KAR,[2] and PHILLIP I. TARR[3]

INTRODUCTION

Shiga toxin (Stx)-producing *Escherichia coli* (STEC) cause illness with a spectrum of severity ranging from mild (even asymptomatic) carriage to life-threatening disease (1–3). STEC infections are relatively uncommon; in the United States, extrapolation of data from FoodNet (4) to a nationwide population that exceeds 300,000,000 indicates there are fewer than 4,000 diagnosed cases of *E. coli* O157:H7 infection per annum. *E. coli* O157:H7 remains the near-exclusive cause of hemolytic-uremic syndrome (HUS) throughout most of the world, and the single serotype on which most data have been generated. Therefore, we emphasize this particular pathogen in this article. The European Food Safety Authority and the European Centre for Disease Prevention and Control report similar epidemiology: 4,000 confirmed infections caused by Stx-producing *E. coli* strains (mostly belonging to the O157 serogroup) in 27 European Union member states. The number of reported infections attributed to *E. coli* strains that produce Shiga toxins has increased since 2008 (5).

[1]Division of Nephrology, Department of Pediatrics, Washington University School of Medicine, St. Louis, MO 63110; [2]Division of Nephrology, Department of Pediatrics, Radboud University Medical Centre, Nijmegen, The Netherlands; [3]Division of Gastroenterology, Hepatology, and Nutrition, Department of Pediatrics, and Department of Molecular Microbiology, Washington University School of Medicine, St. Louis, MO 63110.

Enterohemorrhagic Escherichia coli *and Other Shiga Toxin-Producing* E. coli
Edited by Vanessa Sperandio and Carolyn J. Hovde
© 2015 American Society for Microbiology, Washington, DC
doi:10.1128/microbiolspec.EHEC-0025-2014

Despite their low overall incidence, human infections are medically and epidemiologically actionable. The rarity with which Shiga toxin-producing *E. coli* infections occur, barriers to timely microbial diagnosis, consequences of missed diagnoses, and the many difficulties in attempts to generate high-quality evidence on which to justify treatments pose challenges for clinicians and public health systems. In this review, we focus on clinical aspects (i) early in illness; (ii) in the intermediate stage of illness as HUS evolves in the approximately 15 to 20% of infected children in whom this complication occurs; and (iii) during the HUS phase and its aftermath. When data exist, we cite the appropriate literature, but in other circumstances, we rely on our cumulative experience, as noted.

DEFINITIONS

This field of study has been vexed by multiple nomenclature issues. In this review and in our own papers and practice, we describe the toxins produced by enteric pathogens that cause HUS as Shiga toxins (Stxs). This term is synonymous with verocytotoxins (VT), named for the toxic effect of these proteins on Vero cells, as originally described by Konawalchuk et al. (6) and identified as the key phenotype of these pathogens by Karmali et al. (7). *E. coli* strains that produce Stx are termed Stx-producing *E. coli* (STEC), which is synonymous with VT-producing *E. coli* (VTEC). However, in this text we use the term STEC/VTEC when describing bacteria that produce Stx/VT. This combined term reflects our current misgivings about the term Stx to refer to the cardinal virulence trait of these pathogens. These misgivings are rooted in a practical matter: many physicians, on learning that their patient is infected with a Shiga toxin-producing organism, assume erroneously that the laboratory is describing an infection with *Shigella* sp. (4). Such misconceptions are, in our experience, preludes to the inappropriate administration of antibiotics.

Enterohemorrhagic *E. coli* (EHEC) is another term for STEC/VTEC strains that cause human disease. This term is also problematic, because it implies that the diarrhea stools contain visible blood, which is sometimes not the case in *E. coli* O157:H7 infections, and frequently not the case in infections caused by STEC/VTEC strains belonging to other serotypes. Furthermore, a small subset of patients with HUS will have no diarrhea, but their stool will nevertheless contain *E. coli* O157:H7 (8, 9).

We also have preferences for clinical descriptors related to STEC/VTEC infection. In lieu of the time-honored term hemorrhagic colitis (coined in the first outbreak report) (10), we prefer the more encompassing concept of "bloody diarrhea." For HUS, we urge a urinalysis-independent, stringent case definition, consisting of the simultaneous presence of nonimmune hemolytic anemia (hematocrit/packed cell volume <30% with smear evidence of hemolysis and a negative Coombs test), thrombocytopenia (platelet count <150,000 mm³), and azotemia (creatinine > upper limit of normal for age) (11). There is much hazard and little benefit to be gained from using less stringent clinical definitions of this complication of STEC/VTEC infections. For example, reliance on an abnormal urinalysis to define HUS, especially if the serum creatinine is normal, risks consideration of incorrect diagnoses such as urinary tract infections, especially as the possibility of contamination with fecal material is high in the setting of diarrhea. Moreover, these widely available blood tests enable physicians to relate their patient's course to those described in many other studies during the past 3 decades from multiple countries (12–31).

HISTORY

HUS was first described in the mid-1950s by Gasser et al. (32). In that series of 10 fatal illnesses, cases 3 and 4 had "Brechdurchfall," which is vomiting plus diarrhea (case 3 had these signs throughout the entire illness, while the diarrhea of case 4 occurred only preterminally). None of the other clinical courses suggested enteric illnesses, and notably none of the reports used the terms bloody diarrhea or dysentery. However, we have found earlier papers describing cases in which diarrhea or dysentery preceded renal failure and death within a time frame that closely resembles that of *E. coli*-related HUS (33–37).

E. coli O157:H7 and the closely related pathogen *E. coli* O157:H⁻ were estimated to have split from the same progenitor about 7,000 years ago (38). This common ancestor acquired a gene encoding Stx2/VT2 before that. However, it was not until the 1970s when *E. coli* that had been isolated from food was reported to produce Stx/VT (6). This phenotype preceded by several years the first description of STEC/VTEC strains as causes of disease in 1983. In that year, near-simultaneous publications introduced these pathogens to the medical community: Riley et al. described a hamburger-associated outbreak of *E. coli* O157:H7 infections (10), and Karmali et al. linked fecal STEC/VTEC to HUS (7). Also in 1983, additional investigators demonstrated the production of Stx/VT by *E. coli* O157:H7 (39–42).

DISTINCTION BETWEEN STEC/VTEC BELONGING TO SEROTYPE O157:H7 AND THOSE BELONGING TO ALL OTHER SEROTYPES

There are important practical reasons to differentiate STEC/VTEC strains that express the O (somatic) 157 and flagellar (H) 7 antigens from all other serotypes, which we collectively term non-O157:H7 STEC/VTEC. Most compellingly, *E. coli* O157:H7 is the STEC/VTEC that remains to this day the near-exclusive cause of postdiarrheal HUS (8, 12, 16, 21, 43–49). This serotype is also the one most strongly associated with outbreaks (though most infections are sporadic). If stool specimens are handled adroitly (i.e., immediately transported to the laboratory and inoculated on sorbitol-MacConkey agar on receipt), the microbiologist can often inform the clinician of a presumptive positive or negative for this serotype within 18 to 24 h. That simple piece of information provides valuable clarity to the management of patients with acute diarrhea. However, because a proportion of STEC/VTEC infections are caused by non-O157:H7 STEC/VTEC, there is considerable merit to also determining their presence, but it is much more imperative to exert the greatest effort to confirm or refute the presence of *E. coli* O157:H7 in a stool culture. One important exception to this statement exists: sorbitol-fermenting *E. coli* O157:H⁻ is as virulent, or possibly more virulent, than *E. coli* O157:H7 (50), and this clone remains endemic in Germany. These organisms, which are closely related to *E. coli* O157:H7 (38), are not detected by sorbitol-MacConkey agar screening.

E. coli O157:H7 is best detected in stool by using sorbitol-MacConkey agar with or without cefixime-tellurite (51), because unlike most commensal *E. coli* and non-O157:H7 STEC/VTEC strains, *E. coli* O157:H7 does not ferment sorbitol after overnight incubation. Hence, the presence of a colorless colony on sorbitol-MacConkey agar that agglutinates with an appropriate serologic reagent enables the microbiologist to make a confident and timely presumptive diagnosis. For inexplicable reasons, *E. coli* O157:H7 is more easily detected by sorbitol-MacConkey agar plating than by toxin testing of broth cultures of stool (1, 16, 52–56). Because of the greater sensitivity of agar plating, the critical importance of making a diagnosis of *E. coli* O157:H7 infection as rapidly as possible, and the recognition that a small subset of non-O157:H7 STEC/VTEC infections can be severe, we agree with the guidance of the Centers for Disease Control and Prevention that advises the simultaneous testing for *E. coli* O157:H7 (on agar plates) and non-O157:H7 STEC/VTEC (using, in most cases, a toxin enzyme immunoassay [EIA]) (57). We strongly disagree with detection algorithms that assume that the EIA can be used as a screen with sorbitol-MacConkey agar plating only for positives. Such protocols underdetect *E. coli* O157:H7 and delay answering an important question: is the patient infected with *E. coli* O157:H7 or not?

Though non-O157:H7 STEC/VTEC can cause HUS, the likelihood that any non-O157:H7 STEC/VTEC infection will result in serious kidney injury is extremely low. As noted above, *E. coli* O157:H7 is the overwhelming cause of postdiarrheal HUS (8, 12, 16, 21, 43–49). If the stool of a patient with HUS does not contain *E. coli* O157:H7, the most likely explanation is that the specimen was first cultured for this organism at a point in illness when the pathogen had been eliminated (44).

Table 1 summarizes the *E. coli* O157:H7/non-O157:H7 STEC/VTEC acuity pyramid derived from several defined populations (the HUS studies are globally distributed, but the non-HUS cohorts are from the United States). As illness severity increases, inferred from the setting of acquisition of culture (large geographic region to emergency facility to cohorts with HUS), the ratio of non-O157:H7 STEC/VTEC to *E. coli* O157:H7 diminishes. Findings from several of these studies are particularly worth mentioning. Of 229 Connecticut patients with infections caused by non-O157:H7 STEC/VTEC who were studied over a decade-long interval (48), HUS developed in only one, while HUS developed in 45 of the 434 patients in this series who were infected with *E. coli* O157. In a United States-

wide study of 83 patients with HUS (12), 70 patients had stool cultures in which bacteria grew. These specimens were obtained a median of 8 days after illness onset. Of the 30 STEC/VTEC strains identified in these 30 specimens, 25 were *E. coli* O157. Of the five patients whose stools yielded non-O157:H7 STEC/VTEC, four had serologic testing in convalescence and three of them had antibody evidence of recent exposure to the O157 lipopolysaccharide antigen. Hence, an EIA (or, increasingly, PCR) (58, 59) to detect an STEC/VTEC strain, when added to sorbitol MacConkey agar screening, is likely to lead to the diagnosis of cases that are at much lower risk of HUS developing. Nonetheless, making a diagnosis in at least a subset of these cases is worthwhile, if for no other reason than to provide etiologic clarity to an illness that is usually of greater severity than most gastroenteritides. However, data do not exist to calculate the overall value or cost of such policies, but, again, sorbitol-MacConkey agar screening is most crucial to include when performing stool cultures for bacterial pathogens.

CLINICAL MANAGEMENT OF STEC/VTEC INFECTIONS: PRE-HUS PHASE

Our approach to STEC/VTEC infections is based on three overarching principles:

1. Early detection is critical: time is not on your side when treating STEC/VTEC infections.
2. Early and vigorous volume expansion is associated with avoidance of the most severe renal injury.
3. Highly strategic test selection avoids generating misleading and potentially harmful results.

Early identification (and hospitalization) of infected patients is critical because it lowers the risk of secondary cases (60), avoids diagnostic misadventures (for example, we have seen patients started on steroids because of

TABLE 1 Acuity pyramid[a]

Setting and study years	*E. coli* O157:H7	Non-O157:H7 STEC/VTEC
Wide geographic areas (1998–2009) (48, 141)	Montana: 38% Connecticut: 42%	Montana: 62% Connecticut: 58%
Pediatric emergency facilities (1991–2005) (1, 16, 142)	71%	29%
HUS (1984–2010) (8, 12, 16, 21, 43–49)	95–99%	1–5%

[a]Frequency of recovery of *E. coli* O157:H7 and non-O157:H7 STEC/VTEC, according to the setting of acquisition of specimen.

presumed fulminant ulcerative colitis), and facilitates the commencement of intravenous volume expansion. Several lines of evidence suggest that renal perfusion is threatened prior to and during HUS and that diminished kidney blood flow increases the likelihood of severe renal injury. First, there is evidence of prothrombotic abnormalities in the infected host before HUS (and even if HUS does not ensue). Factor 1.2 (the prothrombin activation peptide), D-dimers, plasminogen activator inhibitor, and platelet-activating factor are each elevated during *E. coli* O157:H7 infections, and von Willebrand's factor is also sheared (indicating flow-related rheological stress, probably caused by nascent thrombi) (13, 17, 61). These prothrombotic abnormalities, which are demonstrated at a point in illness when the blood counts are normal, probably produce some degree of renal ischemia, even before there is smear evidence of microangiopathy. Second, if HUS develops, dehydration at presentation (manifest as elevated hemoglobin) is associated with less favorable short-term (62–64) and long-term (65) outcomes. Third, intravenous volume expansion early in illness, starting as soon as possible after presentation, is associated with less severe (i.e., nonanuric) HUS, if HUS ensues (20, 21).

The details of our fluid management protocols are provided in reference 66. We have slightly changed our recommendations (articulated in reference 66) to now suggest complete blood counts every 12 h until there is assurance that the hemoglobin is falling with volume expansion, because we have not found other indices of circulating blood volume (BUN:creatinine ratio, skin turgor, or vital signs) to be reliable in this setting (authors' personal experience). We aim for a decrement in the hemoglobin of 0.5 g/dl per each 12-h period over the first 1 or 2 days. It can be difficult to accurately assess host volume status, and there is a risk of overload with vigorous volume expansion, so we stress the importance of assiduous monitoring of infected children in centers that are adept at pediatric care (20, 66).

Antibiotics were first suggested as potentially increasing the risk of HUS developing in the initial report linking STEC/VTEC infection to this disorder (7). No credible evidence has emerged since then that supports the concept that antibiotics administered to children or adults early in illness reduce the likelihood of HUS subsequently developing. In fact, extensive data from multiple studies, including more than 1,000 patients infected in epidemics and sporadically, demonstrate that antibiotics are at best neutral and quite likely increase the risk of HUS developing (Table 2). Indeed, the largest risk is demonstrated in the studies with the most robust data: large cohorts, interviewed prospectively, with extensive analysis of timing of administration of antibiotics, and representing infections with multiple different strains.

We also urge against the use of narcotics and antidiarrheal agents in patients with infections that could be caused by STEC/VTEC because of their association with higher rates of HUS or neurological sequelae (67, 68). We also do not endorse nonsteroidal anti-inflammatory agents, because, in our experience, they have no value and because of their nephrotoxic potential (69), which might be exacerbated in the dehydrated state (70).

Early in HUS Prognostic Factors

HUS occurs in 15 to 20% of children who are culture positive for *E. coli* O157:H7. Several indicators apparent early in the course of HUS are associated with a severe course of HUS. A combination of hypocalcemia (≤ 2 mmol/liter) plus oliguria (urine output <0.4 ml kg^{-1} h^{-1} for 24 h) within 48 h of hospitalization had the highest predictive value for negative outcomes (death, need for dialysis, hypertension requiring therapy, or central nervous system sequelae at discharge) (71). Multiple seizures, coma, retinal hemorrhages, hyperkalemia (>7.5 mmol/liter), acidosis (bicarbonate <8 mmol/liter), or a diastolic blood pressure >90 mm Hg are also suggested to be early indicators of poor outcome in HUS (72).

TABLE 2 Summary of antibiotic experience in multiple case control studies of children and adults[a]

Study	Year performed, setting	Ages of patients	Predominant antibiotics given	Details and comments	HUS rate in group receiving antibiotics	HUS rate in group not receiving antibiotics
Carter (143)	1985 Outbreak analysis, Canada	16–67 yr	Amoxicillin, tetracycline	Timing not specified. Outbreak characterized by two phases: primary, contaminate food; secondary, person-to-person transmission. Antibiotic therapy within the 2 days before food exposure (primary phase) did not have increased risk of HUS developing. However, those on antibiotics during the outbreak (secondary phase) had a 10.3 relative risk of HUS developing.	Does not specify[b]	Does not specify[b]
Pavia et al. (144)	1988 Outbreak, case-control study, Utah	6–39 yr	Predominantly trimethoprim-sulfamethoxazole	Timing not specified. Comment: All antimicrobial agents were begun with 72 h after onset of diarrhea.	5/8 (63%)	0/7 (0%)
Proulx (145)	1989–1990 Randomized controlled trial, Canada, antibiotics administered late in illness	5 ± 4 yr (average)	Trimethoprim-sulfamethoxazole (1)	Yes	2/22 (8%)	4/25 (16%)
Bell et al. (67)	1993 Outbreak, retrospective cohort, Washington State	<16 yr	Trimethoprim-sulfamethoxazole (62%), ampicillin or amoxicillin (26%), cephalosporins (12%), metronidazole (8%)	Yes	8/50 (16%)	28/218 (13%)
Wong et al. (146) (superseded by reference 31 described below)	1997–1999 Multistrain, prospective cohort study, four states	<10 yr	Trimethoprim-sulfamethoxazole (2/5), β-lactams (3/5)	Yes	5/9 (56%)	5/62 (8%)

Study	Study design	Age	Antibiotic agents	Timing of antibiotic administration	HUS in treated	HUS in untreated
Dundas et al. (147)	1996 Outbreak, retrospective cohort study, Scotland	18 mo to 94 yr; Mean = 63	Ciprofloxacin	Timing not specified. Comment: HUS developed in 8 (57%) of the 14 patients who received any antibiotic in the 4 wk prior to the outbreak. HUS developed in 7 (47%) of 15 cases treated with ciprofloxacin ≤4 days after symptom onset compared to 25% of the 104 cases that did not receive antibiotic treatment (the difference was not statistically significant).	8/14 (57%) treated with antibiotics in the 4 wk before illness onset; 7/15 (47%) treated with antibiotics within 4 days after illness onset	26/104 (25%)
Wong et al. (31) (extended cohort analysis of reference 146 described above)	1997–2006 Multistrain, prospective cohort study, five states	<10 yr	Trimethoprim-sulfamethoxazole (9/25), β-lactams (9/25), metronidazole (3/25), azithromycin (4/25)	Yes	9/25 (36%)	27/234 (12%)
Smith et al. (148).	1996–2002 Multistrain, age matched, case-case comparison	<20 yr	β-lactams (22% case, 4% control), sulfonamides (14% case, 24% control), metronidazole (6% case, 2% control)	Timing partly specified. Subjects received antibiotics in two specific periods: within the first 3 days after diarrhea onset and in the first 7 days after diarrhea onset.	27/63c (43%)	38/125 (30%)[c]
Cimolai et al. (68)	1984–1989 Multistrain, sporadic cases, retrospective cohort study, British Columbia, Canada	Age range not reported. HUS cohort: mean = 49 mo Gastroenteritis cohort: mean = 83 mo	Agents not specified, but were characterized as "appropriate" if antimicrobial was recognized to be effective in the treatment of shigellosis and if isolate was susceptible in vitro testing.	Timing not specified. Duration of antibiotic use termed "short" if ≤24 h or prolonged if >24 h.	14.3%[d]	4.4%[d]

[a]Modified from reference 93 with permission.
[b]Risk ratio in lieu of HUS rate was provided, and was 8.5 (95% confidence interval 2.7–27.5) in favor of antibiotics associated with HUS development.
[c]Results report the exposure of antibiotics within the first 7 days.
[d]Results limited to "appropriate" antibiotics administered for short terms.

An elevated polymorphonuclear leukocyte count in diarrhea-associated HUS is also a risk factor for poor outcome (73, 74). Unfortunately, these risk factors and biomarkers were measured at different points of illness, often after poor outcomes are becoming self-evident. Nonetheless, the greatest determinant of short- (75) and long-term (62, 76–87) outcome of *E. coli*-related acute kidney injury remains oligoanuria.

HUS with and without Oligoanuria

There are two categories of HUS: oligoanuric and nonoligoanuric. We emphasize the importance of averting oligoanuria to the extent possible, because of the repeated associations between chronic renal sequelae and presence and duration of oligoanuria during HUS (62, 76–87). In reality, oligoanuric HUS is equivalent to anuric HUS (though there can be a day of oliguria before renal shutdown is complete). Anuria probably reflects acute tubular necrosis. The mechanism of acute tubular necrosis in STEC/VTEC HUS is not completely established but could represent either the effect of Shiga toxin on renal tubules (88–91) or ischemia secondary to thrombotic occlusion of the renal vasculature (92). In view of the abundant evidence of prothrombotic activation before azotemia ensues, and in consideration of examples of diminished renal blood flow preceding anuric renal response in many other clinical situations, we have tended to favor the occlusive/ischemic mechanism as the cause of anuria in HUS. As noted above, oligoanuric HUS has categorically worse short- and long-term implications for patients.

HUS that requires dialysis occurs in up to 71% of patients, according to a summary of HUS series over the past 4 years (Table 3). The median length of stay after the case definition of HUS is attained is 12 days for patients with oligoanuria versus 6 days for patients with nonoligoanuric renal failure (20). Dialysis should be instituted soon after anuria onset to prevent cardiopulmonary overload, avoid electrolyte disturbances, and treat hypertension. Early initiation of dialysis if anuria

develops allows the provision of nutrition without exacerbating the above complications.

In our institutions, peritoneal dialysis is the most commonly used modality although intermittent hemodialysis and continuous renal replacement therapies are equally effective. From our perspective, there several reasons to use peritoneal dialysis in HUS, including avoiding unnecessary care in the intensive care unit (decreasing cost) and allowing direct access to peritoneal fluid, which is helpful if the possibility of bowel perforation is raised. If necessary, home renal replacement can be used in the event of delayed recovery.

Renal replacement therapies, i.e., peritoneal and hemodialysis, are the chief supportive modalities in oligoanuric HUS. A review of these interventions is beyond the scope of this chapter. However, some complications of dialysis seem to be relatively frequent during HUS. First, as we recently reviewed (93), infectious complications of peritoneal dialysis are common. Catheter malfunction, probably related to bowel wall and mesenteric edema, also complicates peritoneal dialysis during HUS. Catheter failure typically presents when a catheter infuses dialysate but does not drain. Catheter malfunction often obligates surgical replacement or conversion to hemodialysis. For hemodialysis, we recommend a dual-lumen catheter of age-appropriate size, preferably in the internal jugular vein. The authors generally use regional citrate anticoagulation of the extracorporeal circuit, but systemic heparin anticoagulation can also be used. Invasive procedures, such as peritoneal dialysis catheter and central vascular line placements, can be performed safely without excessive bleeding during the thrombocytopenia of HUS; platelet transfusions are rarely necessary (94, 95).

Fluid and Electrolyte Abnormalities during HUS

There are numerous electrolyte disturbances associated with acute HUS. Hyponatremia, hyperkalemia or hypokalemia, hypocalcemia,

TABLE 3 Severity of HUS in series identified in PubMed published between 2009 and 2013, using search terms hemolytic-uremic syndrome AND children, on September 4, 2013[a]

Year of cases (reference)	Site	Age group	Dialysis rate	Fatalities	Comments
1997–2006 (31)	Washington, Oregon, Idaho, Wyoming, and Missouri	N = 36, <10 yr	31%	0	Many of these patients were well hydrated (i.e., a subset were among those in a single center series [20] at the onset of HUS, which could account for the low dialysis rate).
2007–2008 (21)	California, Washington, Missouri, Ohio, Wisconsin, Arkansas, Indiana, Glasgow, New Mexico, Tennessee	N = 50 <18 yr	68%	0	
1994–2010 (149)	Alberta, Canada	N = 124 <18 yr	43%	2%	This case series employed a case definition of HUS that did not obligate azotemia. Hence, the low dialysis rate might reflect patients who would not have been considered to have had HUS in other series, and demonstrates another reason to avoid urinalysis-dependent definitions of HUS.
1998–2008 (150)	Buenos Aires, Argentina	N = 365 Ages not reported	43%	(Not reported)	94% of patients underwent peritoneal dialysis; 24% peritonitis rate
1995–2011 (62)	Buenos Aires, Argentina	N = 137 Ages not reported	52%	(Not reported)	Better hydration during the prodromal phase was associated with lower frequencies of oligoanuria and need for dialysis.
2011 (30)	Hamburg, Germany	N = 90 <18 yr	71%	1.1%	Outbreak of HUS caused by *E. coli* O104:H4. Outcomes and severity resembled those of *E. coli* O157:H7 HUS.

[a]Only articles with diarrhea associated HUS and dialysis rates are included.

and hypoalbuminemia are common. Notably, however, these abnormalities usually by themselves do not obligate dialysis if urine is still being produced (96). Fatalities in the absence of anuria are exceptionally rare, and recent retrospective data suggesting the value of early renal replacement therapy if children are >10% overloaded probably do not apply to the still urinating child with HUS (97). Hyperuricemia is common, as is elevated lactate, and could be related to diminished renal flow, impaired clearance, and increased production (98).

Hypertension

Hypertension during and after acute HUS is common, with up to 70% of patients affected (99). The mechanism of HUS-induced hypertension is multifactorial and likely related to volume overload and to endothelial and vascular injury. Renovascular hypertension has been suggested to play a role, but plasma renin activity during HUS is below age-appropriate norms (100) (but renal vein renin concentrations have not been determined in this situation). We have had excellent success using calcium channel blockers in acute HUS.

If hypertension is present at discharge, we use angiotensin-converting enzyme inhibition even if the creatinine has not yet normalized, provided the serum creatinine concentration is falling and the patient is not on dialysis.

Hematologic Complications during HUS

Almost all patients with HUS require erythrocyte transfusions because of hemolysis. The basis for the hemolysis is presumably physical shearing as red cells course through small vessels in which fibrin thrombi are abundant. Transfusion requirements appear independent of the severity of the renal injury (authors' personal observations). We use cardiopulmonary compromise or tachycardia as an indication for transfusion, rather than an arbitrary hemoglobin concentration, though in reality it is common to factor in the rate of hemolysis, time of day, vascular access, and point in illness. The transfusion requirement can continue several days beyond resolution of thrombocytopenia and return of creatinine to normal (authors' personal observations).

Several additional elements should be considered when pondering the need for erythrocyte transfusion. First, erythrocyte life span is short in HUS, ranging between 8 and 24 days (101), and the fibrin debris presumably recedes on a day-by-day basis. Hence, transfused red cells might last longer if transfusion can be delayed until needed. Second, we try to use an entire unit at each transfusion. Third, we have frequently noted hypertension immediately following transfusion, so antihypertensive medications should be readily available. After transfusion needs abate, we usually do not provide iron to correct the residual anemia because the total body iron is not low; reticulocyte counts might be helpful in this situation.

We also rarely transfuse platelets because the underlying process leading to thrombocytopenia is most likely entrapment of platelets in thrombi, and thrombocytes have short circulating half-lives in HUS (101). Also, HUS is a thrombotic process, which is not well served by platelet transfusions. It is also concerning that most HUS-related strokes are thrombotic and not hemorrhagic (102, 103). Fibrinogen turnover is not increased in HUS as it is in classic consumptive coagulopathies (101). We therefore recommend against requesting disseminated intravascular coagulation laboratory tests as they are not likely to provide relevant information.

Neurologic Complications during HUS

HUS can be associated with a variety of bona fide neurologic lesions, and signs and symptoms of central nervous system dysfunction have been reported in 20 to 50% of cases. HUS has an apparent predilection for causing basal ganglia lesions (104), but every structure of the brain can be affected (105, 106). The most common serious neurologic complications of HUS are coma, convulsions, and strokes. These complications are usually poor prognostic signs but are rarely by themselves lethal. Possible mechanisms for these complications include endothelial injury with microthrombotic formation and hypoxia. Neurological dysfunction may be further exacerbated by hyponatremia, hypertension, and uremia. Although involvement of the central nervous system might portend a poor prognosis, it must be appreciated that neurologic recovery can be delayed for weeks or months, even after exceptionally severe HUS (107–110). Indeed, in comprehensive studies of survivors of HUS who were infected during the 2011 outbreak of *E. coli* O104:H4 infections, children and adults usually made complete neurologic recoveries even after exceptionally severe neurologic abnormalities during the acute phase (30, 111, 112).

Patients infected with STEC/VTEC often are irritable, lethargic, and jittery, and we do not treat these signs. It is possible that the around-the-clock defecations during the pre-HUS phase contribute to their occurrence. The use of sedatives to prevent patient movement is not recommended, because the mental status of patients with HUS is often altered and such sedation might confound clinical assessment. Acetaminophen and, if necessary, fentanyl are our preferred analgesics. Morphine should be avoided because of the neurotoxic effects of its metabolites, which are cleared by the kidneys (113, 114).

Additional nonnephrologic complications of HUS are summarized in Table 4.

Chronic Renal-Related Sequelae of HUS

The precise risk of long-term consequences of HUS is difficult to gauge. A meta-analysis of 49 papers including 3,476 patients from 18 countries estimated incidences of death at 9% and end-stage renal disease at 3%, but most of these two outcomes occurred during acute

HUS (86). In the same paper, a meta-analysis of 2,372 patients with a minimum of 1 year follow-up estimated 8%, 6%, and 1.8% of patients who have recovered from HUS will have a glomerular filtration rate (per 1.73 m^2) of 60–80, 30–59, and 5–29 ml min^{-1}. In this same group, 10% had hypertension and 15% had proteinuria. Meta-regression analysis indicated the severity of illness and the presence of central nervous system symptoms were associated with worse outcomes. Full renal recovery was not achieved if dialysis exceeded 4 weeks, but we have seen patients who have made nearly complete recoveries after such prolonged anuria. There are profound selection biases in computing chronic sequelae rates if the cohorts are limited

TABLE 4 Selected nonnephrologic, nonhematologic complications of HUS

Complication	Comments	Reference(s)
Pancreatitis	Do not pursue mild ("chemical") pancreatitis by extensive investigation or withholding oral intake. Hyperlipasemia and hyperamylasemia could be related to intestinal and not biliary injury, emesis, and diminished renal clearance.	151
Diabetes mellitus	Insulin dependence can be transient during acute HUS, or persist following renal improvement, or rarely present in the convalescent phase.	152–157
Intestinal perforation and necrosis	These complications are often difficult to identify. Acidosis that fails to resolve with dialysis suggests a severe intestinal complication warranting a laparotomy.	158
Biliary lithiasis	This is usually apparent in the several weeks after HUS resolves and is manifest as right upper quadrant pain and rarely biliary obstruction. This is probably caused by massive hemolysis and subsequent pigment load in the biliary system during HUS.	152, 153
Irritable bowel syndrome	This can occur after STEC/VTEC infections, as with other bacterial enteric infections. Its prognosis is good, often resolving within a year.	159
Elevated transaminases	These might reflect liver injury or, possibly, arise from hemolysis. There is rarely actionable liver injury during HUS.	160
Bowel obstruction	Post-HUS strictures can occur in the small or large bowel. One of the authors is aware of a postinfectious stricture occurring in an adult in whom HUS did not develop, but this complication usually is manifest in the several weeks or months after HUS resolves. These are best detected by contrast studies (small bowel follow-through or barium enema studies).	161, 162
Cardiac ischemia and/or myocarditis	This complication is unusual but can complicate HUS. Troponin determination and cardiac ultrasound might be helpful. We have noted late in illness, i.e., as HUS resolves, complications in elderly patients with HUS.	9, 163
Retinal hemorrhages	Ocular abnormalities are rarely sought in young children, so frequency might be higher than has been appreciated. Long-term complications included decreased visual acuity and optic nerve atrophy.	164, 165
Acute respiratory distress syndrome and pulmonary hemorrhage	Pulmonary hemorrhage is a poor prognostic factor.	9, 64
Sudden death	This complication is rare. Case reports occurring during the acute phase of illness.	163, 166, 167

to those patients who are still returning to follow-up years after HUS. It is also important to note that it is not clear how clinically relevant many of these sequelae (such as microalbuminuria) are, and in our experience, chronic renal failure later in life after normalization of the serum creatinine is exceedingly unusual. However, it is important to note that a recurring set of data suggests that presence and duration of oligoanuria are major predictors of chronic renal sequelae (62, 76–87), reinforcing the need to avoid this complication during acute HUS, if at all possible.

Use of Plasmapheresis and Eculizumab

Eculizumab administration and therapeutic plasma exchange during the large German outbreak in 2011 ignited debate about using these modalities in *E. coli*-related HUS. Eculizumab prevents formation of the membrane attack complex by inhibiting C5 function, and its use has been prompted by a letter to the *New England Journal of Medicine* (115). However, HUS in each of the patients described in that letter was already resolving (decreasing lactate dehydrogenase and/or rising platelet counts) when eculizumab was started. Several in vitro and animal experiments suggest activation of complement after exposure to Stx/VT (116–118), but these experiments employed STEC/VTEC concentrations several orders of magnitude higher than the levels that have ever been documented in humans (119). In contrast, a primate model of lethal STEC/VTEC challenge demonstrated no evidence of complement activation (120). Finding evidence of alternate complement pathway activation in children with HUS is not evidence of a pathogenic role for this branch of the innate immune system (121), as complement is often activated in multisystem organ injury, such as trauma (122). Finally, there are potentially deleterious effects of inhibiting complement during HUS, most notably sepsis (123).

There is similarly no justification for therapeutic plasma exchange in *E. coli*-related HUS. There is no credible evidence of deficiency in functional ADAMTS13, the enzyme that catalyzes shearing of von Willebrand's factor into less thrombogenic forms, inhibitors of this enzyme (i.e., an antibody), or circulating ultra-large von Willebrand's factor multimers, which suggests a lesion that might be treated with plasma therapies (124–126). Analyses of the *E. coli* O104:H4 outbreak provided no evidence of the value of eculizumab and plasma exchange (127–129). Risks of plasma exchange include complications of catheter insertion, hypocalcemia, and exposure to blood products (130). It is important to remember that patients with HUS often deteriorate after they are initially diagnosed, that the median number of days of anuria (among those whose urine output ceases) is 8 (20), and that gradual spontaneous resolution of the microangiopathy and recovery of renal and neurologic function are the rule and not the exceptions. Creative treatments offer no benefit to standard, assiduous, intensive care monitoring but do carry risks of adverse events. Such interventions should not be conducted outside the context of controlled trials, and only if sufficient data exist to suggest a state of clinical equipoise as to their potential value. Therapeutic plasma exchange and anticomplement therapies do not meet this standard in *E. coli*-related HUS.

Clinical Pitfalls in HUS

Postdiarrheal HUS overwhelmingly occurs on a tightly choreographed trajectory (131): diarrhea (usually painful) evolves into grossly bloody diarrhea (80 to 85% of the time). All three criteria for HUS are usually met between the 5th and 13th day of illness, with the first day of diarrhea assigned to be the first day of illness. However, opportunities for Type 1 and 2 diagnostic errors arise because many different microangiopathic disorders have at least some laboratory and historic elements in common with *E. coli*-related HUS, and microbial diagnosis of *E. coli*-related HUS is often elusive.

Type 1 diagnostic errors (i.e., falsely assuming a patient has *E. coli*-related HUS when the patient's microangiopathy is caused by another process) generally occur in patients with aberrant prodromes to renal failure (minimal or no diarrhea, or chronic diarrhea), or laboratory values that are inconsistent with a diagnosis of *E. coli*-related HUS. This problem is compounded by look-alike disorders that do not have highly stereotypical presentations so are more difficult to recognize. Type 2 diagnostic errors (i.e., incorrectly assuming a patient does not have *E. coli*-related HUS) generally relate to lost microbiologic diagnostic opportunities. To avoid Type 1 errors, we search for other etiologies of microangiopathy if there is a persistent documented fever in a health care setting; exceptionally long (>10 days) or short (<5 days) prodromal illnesses; prominent respiratory symptoms or findings; hypotension/shock; family history (especially of distant past episodes) (132); the patient is under 6 months of age, uses specific medications (e.g., oral contraceptives, cyclosporine), is pregnant; or there are discordances between renal injury (severe) and hematologic abnormalities (minimal anemia or thrombocytopenia). The gastrointestinal symptoms that accompany thrombotic thrombocytopenic purpura (133) or atypical HUS caused by complement regulatory proteins (134, 135) rarely resemble those of the enteric prodrome of STEC/VTEC-related HUS, and for this reason, we do not routinely seek other disorders if the presentation is typical. Adjunctive tests, which should be requested and interpreted with circumspection, include chest X rays, blood and urine cultures (and testing for Shiga toxin production if an *E. coli* bacterium is isolated), assays for ADAMTS13 activity, and complement regulatory protein gene sequencing (135–137). Tests that are not helpful diagnostically and often misleading include serum C3, C4, and total hemolytic complement. In our experience with HUS, Type 2 diagnostic errors are more common than Type 1 errors, and are often based on the misconception that failing to find evidence of an STEC/VTEC infection proves that such an etiologic agent is absent. It is important to note that microbiology testing early in illness has the highest yield for STEC/VTEC and that many children with HUS are culture negative at the time of presentation with HUS (44). We strive to increase the microbiologic yield by performing a rectal swab culture on admission of all children with HUS, attempting to take possession of specimens the earliest in illness (usually agar plates), and seeking STEC/VTEC in these specimens in our own centers' microbiology laboratories. Serology, i.e., seeking evidence of circulating immunoglobulins to the O157 lipopolysaccharide (or to the lipopolysaccharide of several other serogroups), can also be used to assign etiology to cases of HUS in patients in whom a pathogen has not been recovered from the stool (138, 139). Antibodies to VT/Stx are less frequently sought, but newer enzyme immunoassays might offer greater ease of performance (140). However, serologic testing is not widely available. Also, absence of diarrhea does not exclude the possibility of an STEC/VTEC infection (8, 9), and finding such a pathogen can avert a much more extensive evaluation and therapeutic misadventures. Therefore, in any atypical presentation of a microangiopathic disorder we nevertheless exclude, to the best our ability, an etiologic agent by either culture or serologic investigation.

SUMMARY

E. coli O157:H7 remains the most exceptional pathogen among the STEC/VTEC group, in view of its enduring association with HUS worldwide, its leading frequency in case series of STEC/VTEC infections (compared to any other serotype), and its ability to cause epidemics as well as sporadic cases. Agar plating of all incoming stools is the best way to detect this pathogen. Early in illness, aggressive volume expansion is associated with reduced renal injury. Specific therapies directed at this pathogen or its products are either harmful (antibiotics) or unlikely to work (the toxemia is short lived). Therapeutic plasma exchange and complement inhibition are not justified by credible data.

ACKNOWLEDGMENTS

Cepheid has paid P. I. Tarr an honorarium for a lecture at its headquarters on this topic. Some work in his laboratory is supported by a grant to another investigator at Washington University from Alexion Corporation (manufacturers of eculizumab). N.C. A. J. van de Kar is a member of the International Advisory Board of aHUS, Alexion Corporation.

We are grateful to families, patients, and collaborators for teaching us much about gut infections and HUS over the past 3 decades. We thank Ariana Jasarevic for expert manuscript assistance, Alexander Weymann for translating original literature from German to English, and Vikas Dharnidharka for providing helpful comments during the creation of this manuscript.

CITATION

Davis TK, Van De Kar NCAJ, Tarr PI. 2014. Shiga toxin/verocytotoxin-producing *Escherichia coli* infections: practical clinical perspectives. Microbiol Spectrum 2(4):EHEC-0025-2014.

REFERENCES

1. **Denno DM, Shaikh N, Stapp JR, Qin X, Hutter CM, Hoffman V, Mooney JC, Wood KM, Stevens HJ, Jones R, Tarr PI, Klein EJ.** 2012. Diarrhea etiology in a pediatric emergency department: a case control study. *Clin Infect Dis* **55:**897–904.

2. **Watanabe H, Terajima J, Izumiya H, Wada A, Tamura K.** 1999. Molecular analysis of enterohemorrhagic *Escherichia coli* isolates in Japan and its application to epidemiological investigation. *Pediatr Int* **41:**202–208.

3. **Stephan R, Untermann F.** 1999. Virulence factors and phenotypical traits of verotoxin-producing *Escherichia coli* strains isolated from asymptomatic human carriers. *J Clin Microbiol* **37:**1570–1572.

4. **Gould LH, Mody RK, Ong KL, Clogher P, Cronquist AB, Garman KN, Lathrop S, Medus C, Spina NL, Webb TH, White PL, Wymore K, Gierke RE, Mahon BE, Griffin PM, Emerging Infections Program Foodnet Working Group.** 2013. Increased recognition of non-O157 Shiga toxin-producing *Escherichia coli* infections in the United States during 2000–2010: epidemiologic features and comparison with *E. coli* O157 infections. *Foodborne Pathog Dis* **10:**453–460.

5. **European Food Safety Authority.** 2012. The European Union Summary Report on Trends and Sources of Zoonoses, Zoonotic Agents and Foodborne Outbreaks in 2010. *EFSA Journal* **10:**161–188.

6. **Konowalchuk J, Speirs JI, Stavric S.** 1977. Vero response to a cytotoxin of *Escherichia coli*. *Infect Immun* **18:**775–779.

7. **Karmali MA, Steele BT, Petric M, Lim C.** 1983. Sporadic cases of haemolytic-uraemic syndrome associated with faecal cytotoxin and cytotoxin-producing *Escherichia coli* in stools. *Lancet* **1:**619–620.

8. **Miceli S, Jure MA, de Saab OA, de Castillo MC, Rojas S, de Holgado AP, de Nader OM.** 1999. A clinical and bacteriological study of children suffering from haemolytic uraemic syndrome in Tucuman, Argentina. *Jpn J Infect Dis* **52:**33–37.

9. **Brandt JR, Fouser LS, Watkins SL, Zelikovic I, Tarr PI, Nazar-Stewart V, Avner ED.** 1994. *Escherichia coli* O 157:H7-associated hemolytic-uremic syndrome after ingestion of contaminated hamburgers. *J Pediatr* **125:**519–526.

10. **Riley LW, Remis RS, Helgerson SD, McGee HB, Wells JG, Davis BR, Hebert RJ, Olcott ES, Johnson LM, Hargrett NT, Blake PA, Cohen ML.** 1983. Hemorrhagic colitis associated with a rare *Escherichia coli* serotype. *N Engl J Med* **308:**681–685.

11. **Meites S, Buffone GJ.** 1989. *Pediatric Clinical Chemistry: Reference (Normal) Values.* AACC Press, Washington, DC.

12. **Banatvala N, Griffin PM, Greene KD, Barrett TJ, Bibb WF, Green JH, Wells JG, Hemolytic Uremic Syndrome Study Collaborators.** 2001. The United States National Prospective Hemolytic Uremic Syndrome Study: microbiologic, serologic, clinical, and epidemiologic findings. *J Infect Dis* **183:**1063–1070.

13. **Tsai HM, Chandler WL, Sarode R, Hoffman R, Jelacic S, Habeeb RL, Watkins SL, Wong CS, Williams GD, Tarr PI.** 2001. von Willebrand factor and von Willebrand factor-cleaving metalloprotease activity in *Escherichia coli* O157:H7-associated hemolytic uremic syndrome. *Pediatr Res* **49:**653–659.

14. **Jelacic S, Wobbe CL, Boster DR, Ciol MA, Watkins SL, Tarr PI, Stapleton AE.** 2002. ABO and P1 blood group antigen expression and stx genotype and outcome of childhood *Escherichia coli* O157:H7 infections. *J Infect Dis* **185:** 214–219.

15. **Cornick NA, Jelacic S, Ciol MA, Tarr PI.** 2002. *Escherichia coli* O157:H7 infections: discordance between filterable fecal shiga toxin and disease outcome. *J Infect Dis* **186:**57–63.

16. **Klein EJ, Stapp JR, Clausen CR, Boster DR, Wells JG, Qin X, Swerdlow DL, Tarr PI.** 2002. Shiga toxin-producing *Escherichia coli* in children with diarrhea: a prospective point-of-care study. *J Pediatr* **141:**172–177.

17. **Smith JM, Jones F, Ciol MA, Jelacic S, Boster DR, Watkins SL, Williams GD, Tarr PI, Henderson WR Jr.** 2002. Platelet-activating factor and *Escherichia coli* O157:H7 infections. *Pediatr Nephrol* **17:**1047–1052.

18. **Thayu M, Chandler WL, Jelacic S, Gordon CA, Rosenthal GL, Tarr PI.** 2003. Cardiac ischemia during hemolytic uremic syndrome. *Pediatr Nephrol* **18:**286–289.

19. **Jelacic JK, Damrow T, Chen GS, Jelacic S, Bielaszewska M, Ciol M, Carvalho HM, Melton-Celsa AR, O'Brien AD, Tarr PI.** 2003. Shiga toxin-producing *Escherichia coli* in Montana: bacterial genotypes and clinical profiles. *J Infect Dis* **188:**719–729.

20. **Ake JA, Jelacic S, Ciol MA, Watkins SL, Murray KF, Christie DL, Klein EJ, Tarr PI.** 2005. Relative nephroprotection during *Escherichia coli* O157:H7 infections: association with intravenous volume expansion. *Pediatrics* **115:** e673–e680.

21. **Hickey CA, Beattie TJ, Cowieson J, Miyashita Y, Strife CF, Frem JC, Peterson JM, Butani L, Jones DP, Havens PL, Patel HP, Wong CS, Andreoli SP, Rothbaum RJ, Beck AM, Tarr PI.** 2011. Early volume expansion during diarrhea and relative nephroprotection during subsequent hemolytic uremic syndrome. *Arch Pediatr Adolesc Med* **165:**884–889.

22. **Zimmerhackl LB, Rosales A, Hofer J, Riedl M, Jungraithmayr T, Mellmann A, Bielaszewska M, Karch H.** 2010. Enterohemorrhagic *Escherichia coli* O26:H11-associated Hemolytic uremic syndrome: bacteriology and clinical presentation. *Semin Thromb Hemost* **36:**586–593.

23. **Gerber A, Karch H, Allerberger F, Verweyen HM, Zimmerhackl LB.** 2002. Clinical course and the role of shiga toxin-producing *Escherichia coli* infection in the hemolytic-uremic syndrome in pediatric patients, 1997–2000, in Germany and Austria: a prospective study. *J Infect Dis* **186:**493–500.

24. **Bielaszewska M, Middendorf B, Kock R, Friedrich AW, Fruth A, Karch H, Schmidt MA, Mellmann A.** 2008. Shiga toxin-negative attaching and effacing *Escherichia coli*: distinct clinical associations with bacterial phylogeny and virulence traits and inferred in-host pathogen evolution. *Clin Infect Dis* **47:**208–217.

25. **Bielaszewska M, Kock R, Friedrich AW, von Eiff C, Zimmerhackl LB, Karch H, Mellmann A.** 2007. Shiga toxin-mediated hemolytic uremic syndrome: time to change the diagnostic paradigm? *PLoS One* **2:**e1024.

26. **Friedrich AW, Zhang W, Bielaszewska M, Mellmann A, Kock R, Fruth A, Tschape H, Karch H.** 2007. Prevalence, virulence profiles, and clinical significance of Shiga toxin-negative variants of enterohemorrhagic *Escherichia coli* O157 infection in humans. *Clin Infect Dis* **45:**39–45.

27. **Bielaszewska M, Friedrich AW, Aldick T, Schurk-Bulgrin R, Karch H.** 2006. Shiga toxin activatable by intestinal mucus in *Escherichia coli* isolated from humans: predictor for a severe clinical outcome. *Clin Infect Dis* **43:**1160–1167.

28. **Sonntag AK, Prager R, Bielaszewska M, Zhang W, Fruth A, Tschape H, Karch H.** 2004. Phenotypic and genotypic analyses of enterohemorrhagic *Escherichia coli* O145 strains from patients in Germany. *J Clin Microbiol* **42:**954–962.

29. **Ikeda K, Ida O, Kimoto K, Takatorige T, Nakanishi N, Tatara K.** 1999. Effect of early fosfomycin treatment on prevention of hemolytic uremic syndrome accompanying *Escherichia coli* O157:H7 infection. *Clin Nephrol* **52:**357–362.

30. **Loos S, Ahlenstiel T, Kranz B, Staude H, Pape L, Hartel C, Vester U, Buchtala L, Benz K, Hoppe B, Beringer O, Krause M, Muller D, Pohl M, Lemke J, Hillebrand G, Kreuzer M, Konig J, Wigger M, Konrad M, Haffner D, Oh J, Kemper MJ.** 2012. An outbreak of Shiga toxin-producing *Escherichia coli* O104:H4 hemolytic uremic syndrome in Germany: presentation and short-term outcome in children. *Clin Infect Dis* **55:**753–759.

31. **Wong CS, Mooney JC, Brandt JR, Staples AO, Jelacic S, Boster DR, Watkins SL, Tarr PI.** 2012. Risk factors for the hemolytic uremic syndrome in children infected with *Escherichia coli* O157:H7: a multivariable analysis. *Clin Infect Dis* **55:**33–41.

32. **Gasser C, Gautier E, Steck A, Siebenmann RE, Oechslin R.** 1955. [Hemolytic-uremic syndrome: bilateral necrosis of the renal cortex in acute acquired hemolytic anemia]. *Schweiz Med Wochenschr* **85:**905–909.

33. **Fahr T.** 1925. Kreislaufstörungen in der Niere, p. 121–155. *In* Fahr T, Gruber G, Koch M, Lubarsch O, Stoerk O (ed.), *Harnorgane Männliche Geschlechtsorgane*, vol. **6**. Springer, Vienna, Austria.

34. **Bamforth J.** 1923. A case of symmetrical cortical necrosis of the kidneys occurring in an adult man. *J Pathol Bacteriol* **26**:40–45.

35. **Campbell AC, Henderson JL.** 1949. Symmetrical cortical necrosis of the kidneys in infancy and childhood. *Arch Dis Child* **24**:269–285, illust.

36. **Dunn JS, Montgomery GL.** 1941. Acute necrotising glomerulonephritis. *J Pathol Bacteriol* **52**:1–16.

37. **Hensley WJ.** 1952. Haemolytic anaemia in acute glomerulonephritis. *Australas Ann Med* **1**:180–185.

38. **Leopold SR, Magrini V, Holt NJ, Shaikh N, Mardis ER, Cagno J, Ogura Y, Iguchi A, Hayashi T, Mellmann A, Karch H, Besser TE, Sawyer SA, Whittam TS, Tarr PI.** 2009. A precise reconstruction of the emergence and constrained radiations of *Escherichia coli* O157 portrayed by backbone concatenomic analysis. *Proc Natl Acad Sci USA* **106**:8713–8718.

39. **O'Brien AO, Lively TA, Chen ME, Rothman SW, Formal SB.** 1983. *Escherichia coli* O157:H7 strains associated with haemorrhagic colitis in the United States produce a Shigella dysenteriae 1 (SHIGA) like cytotoxin. *Lancet* **1**:702.

40. **O'Brien AD, LaVeck GD, Thompson MR, Formal SB.** 1982. Production of Shigella dysenteriae type 1-like cytotoxin by *Escherichia coli*. *J Infect Dis* **146**:763–769.

41. **Johnson WM, Lior H, Bezanson GS.** 1983. Cytotoxic *Escherichia coli* O157:H7 associated with haemorrhagic colitis in Canada. *Lancet* **1**:76.

42. **O'Brien AD, LaVeck GD.** 1983. Purification and characterization of a Shigella dysenteriae 1-like toxin produced by *Escherichia coli*. *Infect Immun* **40**:675–683.

43. **Neill MA, Tarr PI, Clausen CR, Christie DL, Hickman RO.** 1987. *Escherichia coli* O157:H7 as the predominant pathogen associated with the hemolytic uremic syndrome: a prospective study in the Pacific Northwest. *Pediatrics* **80**:37–40.

44. **Tarr PI, Neill MA, Clausen CR, Watkins SL, Christie DL, Hickman RO.** 1990. *Escherichia coli* O157:H7 and the hemolytic uremic syndrome: importance of early cultures in establishing the etiology. *J Infect Dis* **162**:553–556.

45. **Rowe PC, Orrbine E, Lior H, Wells GA, Yetisir E, Clulow M, McLaine PN.** 1998. Risk of hemolytic uremic syndrome after sporadic *Escherichia coli* O157:H7 infection: results of a Canadian collaborative study. Investigators of the Canadian Pediatric Kidney Disease Research Center. *J Pediatr* **132**:777–782.

46. **Rivas M, Miliwebsky E, Chinen I, Roldan CD, Balbi L, Garcia B, Fiorilli G, Sosa-Estani S, Kincaid J, Rangel J, Griffin PM, Case-Control Study Group.** 2006. Characterization and epidemiologic subtyping of Shiga toxin-producing *Escherichia coli* strains isolated from hemolytic uremic syndrome and diarrhea cases in Argentina. *Foodborne Pathog Dis* **3**:88–96.

47. **Pollock KG, Young D, Beattie TJ, Todd WT.** 2008. Clinical surveillance of thrombotic microangiopathies in Scotland, 2003-2005. *Epidemiol Infect* **136**:115–121.

48. **Hadler JL, Clogher P, Hurd S, Phan Q, Mandour M, Bemis K, Marcus R.** 2011. Ten-year trends and risk factors for non-O157 Shiga toxin-producing *Escherichia coli* found through Shiga toxin testing, Connecticut, 2000-2009. *Clin Infect Dis* **53**:269–276.

49. **Mody RK, Luna-Gierke RE, Jones TF, Comstock N, Hurd S, Scheftel J, Lathrop S, Smith G, Palmer A, Strockbine N, Talkington D, Mahon BE, Hoekstra RM, Griffin PM.** 2012. Infections in pediatric postdiarrheal hemolytic uremic syndrome: factors associated with identifying shiga toxin-producing *Escherichia coli*. *Arch Pediatr Adolesc Med* **166**:902–909.

50. **Werber D, Bielaszewska M, Frank C, Stark K, Karch H.** 2011. Watch out for the even eviler cousin-sorbitol-fermenting *E coli* O157. *Lancet* **377**:298–299.

51. **Chapman PA, Siddons CA.** 1996. A comparison of immunomagnetic separation and direct culture for the isolation of verocytotoxin-producing *Escherichia coli* O157 from cases of bloody diarrhoea, non-bloody diarrhoea and asymptomatic contacts. *J Med Microbiol* **44**:267–271.

52. **Fey PD, Wickert RS, Rupp ME, Safranek TJ, Hinrichs SH.** 2000. Prevalence of non-O157:H7 shiga toxin-producing *Escherichia coli* in diarrheal stool samples from Nebraska. *Emerg Infect Dis* **6**:530–533.

53. **Carroll KC, Adamson K, Korgenski K, Croft A, Hankemeier R, Daly J, Park CH.** 2003. Comparison of a commercial reversed passive latex agglutination assay to an enzyme immunoassay for the detection of Shiga toxin-producing *Escherichia coli*. *Eur J Clin Microbiol Infect Dis* **22**:689–692.

54. **Manning SD, Madera RT, Schneider W, Dietrich SE, Khalife W, Brown W, Whittam TS, Somsel P, Rudrik JT.** 2007. Surveillance for Shiga toxin-producing *Escherichia coli*, Michigan, 2001-2005. *Emerg Infect Dis* **13**:318–321.

55. **Teel LD, Daly JA, Jerris RC, Maul D, Svanas G, O'Brien AD, Park CH.** 2007. Rapid detection of Shiga toxin-producing *Escherichia coli* by optical immunoassay. *J Clin Microbiol* **45**:3377–3380.

56. **Park CH, Kim HJ, Hixon DL, Bubert A.** 2003. Evaluation of the duopath verotoxin test for detection of shiga toxins in cultures of human stools. *J Clin Microbiol* **41**:2650–2653.

57. **Gould LH, Bopp C, Strockbine N, Atkinson R, Baselski V, Body B, Carey R, Crandall C, Hurd S, Kaplan R, Neill M, Shea S, Somsel P, Tobin-D'Angelo M, Griffin PM, Gerner-Smidt P, Centers for Disease Control and Prevention (CDC).** 2009. Recommendations for diagnosis of shiga toxin–producing *Escherichia coli* infections by clinical laboratories. *MMWR Recomm Rep* **58:**1–14.

58. **Grys TE, Sloan LM, Rosenblatt JE, Patel R.** 2009. Rapid and sensitive detection of Shiga toxin-producing *Escherichia coli* from non-enriched stool specimens by real-time PCR in comparison to enzyme immunoassay and culture. *J Clin Microbiol* **47:**2008–2012.

59. **Vallieres E, Saint-Jean M, Rallu F.** 2013. Comparison of three different methods for detection of Shiga toxin-producing *Escherichia coli* in a tertiary pediatric care center. *J Clin Microbiol* **51:**481–486.

60. **Werber D, Mason BW, Evans MR, Salmon RL.** 2008. Preventing household transmission of Shiga toxin-producing *Escherichia coli* O157 infection: promptly separating siblings might be the key. *Clin Infect Dis* **46:**1189–1196.

61. **Chandler WL, Jelacic S, Boster DR, Ciol MA, Williams GD, Watkins SL, Igarashi T, Tarr PI.** 2002. Prothrombotic coagulation abnormalities preceding the hemolytic-uremic syndrome. *N Engl J Med* **346:**23–32.

62. **Balestracci A, Martin SM, Toledo I, Alvarado C, Wainsztein RE.** 2012. Dehydration at admission increased the need for dialysis in hemolytic uremic syndrome children. *Pediatr Nephrol* **27:**1407–1410.

63. **Coad NA, Marshall T, Rowe B, Taylor CM.** 1991. Changes in the postenteropathic form of the hemolytic uremic syndrome in children. *Clin Nephrol* **35:**10–16.

64. **Oakes RS, Siegler RL, McReynolds MA, Pysher T, Pavia AT.** 2006. Predictors of fatality in postdiarrheal hemolytic uremic syndrome. *Pediatrics* **117:**1656–1662.

65. **Ojeda JM, Kohout I, Cuestas E.** 2013. Dehydration upon admission is a risk factor for incomplete recovery of renal function in children with haemolytic uremic syndrome. *Nefrologia* **33:**372–376.

66. **Holtz LR, Neill MA, Tarr PI.** 2009. Acute bloody diarrhea: a medical emergency for patients of all ages. *Gastroenterology* **136:**1887–1898.

67. **Bell BP, Griffin PM, Lozano P, Christie DL, Kobayashi JM, Tarr PI.** 1997. Predictors of hemolytic uremic syndrome in children during a large outbreak of *Escherichia coli* O157:H7 infections. *Pediatrics* **100:**E12.

68. **Cimolai N, Basalyga S, Mah DG, Morrison BJ, Carter JE.** 1994. A continuing assessment of risk factors for the development of *Escherichia coli* O157:H7-associated hemolytic uremic syndrome. *Clin Nephrol* **42:**85–89.

69. **Misurac JM, Knoderer CA, Leiser JD, Nailescu C, Wilson AC, Andreoli SP.** 2013. Nonsteroidal anti-inflammatory drugs are an important cause of acute kidney injury in children. *J Pediatr* **162:**1153–1159, 1159 e1.

70. **John CM, Shukla R, Jones CA.** 2007. Using NSAID in volume depleted children can precipitate acute renal failure. *Arch Dis Child* **92:**524–526.

71. **Havens PL, O'Rourke PP, Hahn J, Higgins J, Walker AM.** 1988. Laboratory and clinical variables to predict outcome in hemolytic-uremic syndrome. *Am J Dis Child* **142:**961–964.

72. **Vitacco M, Sanchez Avalos J, Gianantonio CA.** 1973. Heparin therapy in the hemolytic-uremic syndrome. *J Pediatr* **83:**271–275.

73. **Walters MD, Matthei IU, Kay R, Dillon MJ, Barratt TM.** 1989. The polymorphonuclear leucocyte count in childhood haemolytic uraemic syndrome. *Pediatr Nephrol* **3:**130–134.

74. **Milford DV, Staten J, MacGreggor I, Dawes J, Taylor CM, Hill FG.** 1991. Prognostic markers in diarrhoea-associated haemolytic-uraemic syndrome: initial neutrophil count, human neutrophil elastase and von Willebrand factor antigen. *Nephrol Dial Transplant* **6:**232–237.

75. **Balestracci A, Martin SM, Toledo I, Alvarado C, Wainsztein RE.** 2014. Laboratory predictors of acute dialysis in hemolytic uremic syndrome. *Pediatr Int* **56:**234–239.

76. **Dolislager D, Tune B.** 1978. The hemolytic-uremic syndrome: spectrum of severity and significance of prodrome. *Am J Dis Child* **132:**55–58.

77. **Gianantonio CA, Vitacco M, Mendilaharzu F, Gallo GE, Sojo ET.** 1973. The hemolytic-uremic syndrome. *Nephron* **11:**174–192.

78. **Gianantonio CA, Vitacco M, Mendilaharzu F, Gallo G.** 1968. The hemolytic-uremic syndrome. Renal status of 76 patients at long-term follow-up. *J Pediatr* **72:**757–765.

79. **Tonshoff B, Sammet A, Sanden I, Mehls O, Waldherr R, Scharer K.** 1994. Outcome and prognostic determinants in the hemolytic uremic syndrome of children. *Nephron* **68:**63–70.

80. **Siegler RL, Pavia AT, Christofferson RD, Milligan MK.** 1994. A 20-year population-based study of postdiarrheal hemolytic uremic syndrome in Utah. *Pediatrics* **94:**35–40.

81. **Mizusawa Y, Pitcher LA, Burke JR, Falk MC, Mizushima W.** 1996. Survey of haemolytic-uraemic syndrome in Queensland 1979–1995. *Med J Aust* **165:**188–191.

82. **Spizzirri FD, Rahman RC, Bibiloni N, Ruscasso JD, Amoreo OR.** 1997. Childhood hemolytic

uremic syndrome in Argentina: long-term follow-up and prognostic features. *Pediatr Nephrol* **11:** 156–160.

83. **Mencia Bartolome S, Martinez de Azagra A, de Vicente Aymat A, Monleon Luque M, Casado Flores J.** 1999. [Uremic hemolytic syndrome. Analysis of 43 cases]. *An Esp Pediatr* **50:**467–470 (In Spanish).

84. **Huseman D, Gellermann J, Vollmer I, Ohde I, Devaux S, Ehrich JH, Filler G.** 1999. Long-term prognosis of hemolytic uremic syndrome and effective renal plasma flow. *Pediatr Nephrol* **13:** 672–677.

85. **Loirat C.** 2001. [Post-diarrhea hemolytic-uremic syndrome: clinical aspects]. *Arch Pediatr* **8** Suppl 4:776s–784s (In French).

86. **Garg AX, Suri RS, Barrowman N, Rehman F, Matsell D, Rosas-Arellano MP, Salvadori M, Haynes RB, Clark WF.** 2003. Long-term renal prognosis of diarrhea-associated hemolytic uremic syndrome: a systematic review, meta-analysis, and meta-regression. *JAMA* **290:**1360–1370.

87. **Oakes RS, Kirkham JK, Nelson RD, Siegler RL.** 2008. Duration of oliguria and anuria as predictors of chronic renal-related sequelae in post-diarrheal hemolytic uremic syndrome. *Pediatr Nephrol* **23:**1303–1308.

88. **Nestoridi E, Kushak RI, Duguerre D, Grabowski EF, Ingelfinger JR.** 2005. Up-regulation of tissue factor activity on human proximal tubular epithelial cells in response to Shiga toxin. *Kidney Int* **67:**2254–2266.

89. **Hughes AK, Stricklett PK, Kohan DE.** 1998. Cytotoxic effect of Shiga toxin-1 on human proximal tubule cells. *Kidney Int* **54:**426–437.

90. **Taguchi T, Uchida H, Kiyokawa N, Mori T, Sato N, Horie H, Takeda T, Fujimoto J.** 1998. Verotoxins induce apoptosis in human renal tubular epithelium derived cells. *Kidney Int* **53:** 1681–1688.

91. **Kaneko K, Kiyokawa N, Ohtomo Y, Nagaoka R, Yamashiro Y, Taguchi T, Mori T, Fujimoto J, Takeda T.** 2001. Apoptosis of renal tubular cells in Shiga-toxin-mediated hemolytic uremic syndrome. *Nephron* **87:**182–185.

92. **Abuelo JG.** 2007. Normotensive ischemic acute renal failure. *N Engl J Med* **357:**797–805.

93. **Davis TK, McKee R, Schnadower D, Tarr PI.** 2013. Treatment of Shiga toxin–Producing *Escherichia coli* infections. *Infect Dis Clin North Am* **27:**577–597.

94. **Weil BR, Andreoli SP, Billmire DF.** 2010. Bleeding risk for surgical dialysis procedures in children with hemolytic uremic syndrome. *Pediatr Nephrol* **25:**1693–1698.

95. **Balestracci A, Martin SM, Toledo I, Alvarado C, Wainsztein RE.** 2013. Impact of platelet transfusions in children with post-diarrheal hemolytic uremic syndrome. *Pediatr Nephrol* **28:** 919–925.

96. **Schulman SL, Kaplan BS.** 1996. Management of patients with hemolytic uremic syndrome demonstrating severe azotemia but not anuria. *Pediatr Nephrol* **10:**671–674.

97. **Sutherland SM, Zappitelli M, Alexander SR, Chua AN, Brophy PD, Bunchman TE, Hackbarth R, Somers MJ, Baum M, Symons JM, Flores FX, Benfield M, Askenazi D, Chand D, Fortenberry JD, Mahan JD, McBryde K, Blowey D, Goldstein SL.** 2010. Fluid overload and mortality in children receiving continuous renal replacement therapy: the prospective pediatric continuous renal replacement therapy registry. *Am J Kidney Dis* **55:**316–325.

98. **Kaplan BS, Thomson PD.** 1976. Hyperuricemia in the hemolytic-uremic syndrome. *Am J Dis Child* **130:**854–856.

99. **Robson WL, Leung AK, Brant R.** 1993. Relationship of the recovery in the glomerular filtration rate to the duration of anuria in diarrhea-associated hemolytic uremic syndrome. *Am J Nephrol* **13:**194–197.

100. **Proesmans W, VanCauter A, Thijs L, Lijnen P.** 1994. Plasma renin activity in haemolytic uraemic syndrome. *Pediatr Nephrol* **8:**444–446.

101. **Katz J, Krawitz S, Sacks PV, Levin SE, Thomson P, Levin J, Metz J.** 1973. Platelet, erythrocyte, and fibrinogen kinetics in the hemolytic-uremic syndrome of infancy. *J Pediatr* **83:**739–748.

102. **Nakahata T, Tanaka H, Tateyama T, Ueda T, Suzuki K, Osari S, Kasai M, Waga S.** 2001. Thrombotic stroke in a child with diarrhea-associated hemolytic-uremic syndrome with a good recovery. *Tohoku J Exp Med* **193:**73–77.

103. **Trevathan E, Dooling EC.** 1987. Large thrombotic strokes in hemolytic-uremic syndrome. *J Pediatr* **111:**863–866.

104. **Steinborn M, Leiz S, Rudisser K, Griebel M, Harder T, Hahn H.** 2004. CT and MRI in haemolytic uraemic syndrome with central nervous system involvement: distribution of lesions and prognostic value of imaging findings. *Pediatr Radiol* **34:**805–810.

105. **Nathanson S, Kwon T, Elmaleh M, Charbit M, Launay EA, Harambat J, Brun M, Ranchin B, Bandin F, Cloarec S, Bourdat-Michel G, Pietrement C, Champion G, Ulinski T, Deschenes G.** 2010. Acute neurological involvement in diarrhea-associated hemolytic uremic syndrome. *Clin J Am Soc Nephrol* **5:**1218–1228.

106. **Weissenborn K, Donnerstag F, Kielstein JT, Heeren M, Worthmann H, Hecker H, Schmitt R, Schiffer M, Pasedag T, Schuppner R, Tryc**

AB, Raab P, Hartmann H, Ding XQ, Hafer C, Menne J, Schmidt BM, Bultmann E, Haller H, Dengler R, Lanfermann H, Giesemann AM. 2012. Neurologic manifestations of *E coli* infection-induced hemolytic-uremic syndrome in adults. *Neurology* **79**:1466–1473.

107. Signorini E, Lucchi S, Mastrangelo M, Rapuzzi S, Edefonti A, Fossali E. 2000. Central nervous system involvement in a child with hemolytic uremic syndrome. *Pediatr Nephrol* **14**:990–992.

108. Steele BT, Murphy N, Chuang SH, McGreal D, Arbus GS. 1983. Recovery from prolonged coma in hemolytic uremic syndrome. *J Pediatr* **102**:402–404.

109. Steinberg A, Ish-Horowitcz M, el-Peleg O, Mor J, Branski D. 1986. Stroke in a patient with hemolytic-uremic syndrome with a good outcome. *Brain Dev* **8**:70–72.

110. Kahn SI, Tolkan SR, Kothari O, Garella S. 1982. Spontaneous recovery of the hemolytic uremic syndrome with prolonged renal and neurological manifestations. *Nephron* **32**:188–191.

111. Magnus T, Rother J, Simova O, Meier-Cillien M, Repenthin J, Moller F, Gbadamosi J, Panzer U, Wengenroth M, Hagel C, Kluge S, Stahl RK, Wegscheider K, Urban P, Eckert B, Glatzel M, Fiehler J, Gerloff C. 2012. The neurological syndrome in adults during the 2011 northern German *E. coli* serotype O104:H4 outbreak. *Brain* **135**:1850–1859.

112. Braune SA, Wichmann D, von Heinz MC, Nierhaus A, Becker H, Meyer TN, Meyer GP, Muller-Schulz M, Fricke J, de Weerth A, Hoepker WW, Fiehler J, Magnus T, Gerloff C, Panzer U, Stahl RA, Wegscheider K, Kluge S. 2013. Clinical features of critically ill patients with Shiga toxin-induced hemolytic uremic syndrome. *Crit Care Med* **41**:1702–1710.

113. Murphy EJ. 2005. Acute pain management pharmacology for the patient with concurrent renal or hepatic disease. *Anaesth Intensive Care* **33**:311–322.

114. Niscola P, Scaramucci L, Vischini G, Giovannini M, Ferrannini M, Massa P, Tatangelo P, Galletti M, Palumbo R. 2010. The use of major analgesics in patients with renal dysfunction. *Curr Drug Targets* **11**:752–758.

115. Lapeyraque AL, Malina M, Fremeaux-Bacchi V, Boppel T, Kirschfink M, Oualha M, Proulx F, Clermont MJ, Le Deist F, Niaudet P, Schaefer F. 2011. Eculizumab in severe Shiga-toxin-associated HUS. *N Engl J Med* **364**:2561–2563.

116. Orth D, Khan AB, Naim A, Grif K, Brockmeyer J, Karch H, Joannidis M, Clark SJ, Day AJ, Fidanzi S, Stoiber H, Dierich MP, Zimmerhackl LB, Wurzner R. 2009. Shiga toxin activates complement and binds factor H: evidence for an active role of complement in hemolytic uremic syndrome. *J Immunol* **182**:6394–6400.

117. Morigi M, Galbusera M, Gastoldi S, Locatelli M, Buelli S, Pezzotta A, Pagani C, Noris M, Gobbi M, Stravalaci M, Rottoli D, Tedesco F, Remuzzi G, Zoja C. 2011. Alternative pathway activation of complement by Shiga toxin promotes exuberant C3a formation that triggers microvascular thrombosis. *J Immunol* **187**:172–180.

118. Stahl AL, Sartz L, Karpman D. 2011. Complement activation on platelet-leukocyte complexes and microparticles in enterohemorrhagic *Escherichia coli*-induced hemolytic uremic syndrome. *Blood* **117**:5503–5513.

119. Lopez EL, Contrini MM, Glatstein E, Ayala SG, Santoro R, Ezcurra G, Teplitz E, Matsumoto Y, Sato H, Sakai K, Katsuura Y, Hoshide S, Morita T, Harning R, Brookman S. 2012. An epidemiologic surveillance of Shiga-like toxin-producing *Escherichia coli* infection in Argentinean children: risk factors and serum Shiga-like toxin 2 values. *Pediatr Infect Dis J* **31**:20–24.

120. Lee BC, Mayer CL, Leibowitz CS, Stearns-Kurosawa DJ, Kurosawa S. 2013. Quiescent complement in nonhuman primates during *E coli* Shiga toxin-induced hemolytic uremic syndrome and thrombotic microangiopathy. *Blood* **122**:803–806.

121. Thurman JM, Marians R, Emlen W, Wood S, Smith C, Akana H, Holers VM, Lesser M, Kline M, Hoffman C, Christen E, Trachtman H. 2009. Alternative pathway of complement in children with diarrhea-associated hemolytic uremic syndrome. *Clin J Am Soc Nephrol* **4**:1920–1924.

122. Ganter MT, Brohi K, Cohen MJ, Shaffer LA, Walsh MC, Stahl GL, Pittet JF. 2007. Role of the alternative pathway in the early complement activation following major trauma. *Shock* **28**:29–34.

123. Römer S, Schaumburg R, Karch H, Schwindt W, Waltenberger J, Lebiedz P. 2012. Septic shock and ARDS after Eculizumab application in Shiga-toxin associated HUS during the outbreak of Shiga toxin-producing enterohaemorrhagic *E. coli* O104:H4. *Medizinische Klinik, Intensivmedizin und Notfallmedizin* **107**:329–330.

124. Tarr PI. 2009. Shiga toxin-associated hemolytic uremic syndrome and thrombotic thrombocytopenic purpura: distinct mechanisms of pathogenesis. *Kidney Int Suppl* :S29–S32.

125. Colic E, Dieperink H, Titlestad K, Tepel M. 2011. Management of an acute outbreak of diarrhoea-associated haemolytic uraemic syndrome with early plasma exchange in adults from southern Denmark: an observational study. *Lancet* **378**:1089–1093.

126. Veyradier A, Obert B, Haddad E, Cloarec S, Nivet H, Foulard M, Lesure F, Delattre P,

Lakhdari M, Meyer D, Girma JP, Loirat C. 2003. Severe deficiency of the specific von Willebrand factor-cleaving protease (ADAMTS 13) activity in a subgroup of children with atypical hemolytic uremic syndrome. *J Pediatr* **142**:310–317.

127. Kielstein JT, Beutel G, Fleig S, Steinhoff J, Meyer TN, Hafer C, Kuhlmann U, Bramstedt J, Panzer U, Vischedyk M, Busch V, Ries W, Mitzner S, Mees S, Stracke S, Nurnberger J, Gerke P, Wiesner M, Sucke B, Abu-Tair M, Kribben A, Klause N, Schindler R, Merkel F, Schnatter S, Dorresteijn EM, Samuelsson O, Brunkhorst R, Collaborators of the DGfN STEC-HUS registry. 2012. Best supportive care and therapeutic plasma exchange with or without eculizumab in Shiga-toxin-producing *E. coli* O104:H4 induced haemolytic-uraemic syndrome: an analysis of the German STEC-HUS registry. *Nephrol Dial Transplant* **27**:3807–3815.

128. Samuelsson O, Follin P, Rundgren M, Rylander C, Selga D, Stahl A. 2012. [The HUS epidemic in the summer of 2011 was severe. German and Swedish experiences of the EHEC outbreak]. *Lakartidningen* **109**:1230–1234 (In Swedish).

129. Nitschke M, Sayk F, Hartel C, Roseland RT, Hauswaldt S, Steinhoff J, Fellermann K, Derad I, Wellhoner P, Buning J, Tiemer B, Katalinic A, Rupp J, Lehnert H, Solbach W, Knobloch JK. 2012. Association between azithromycin therapy and duration of bacterial shedding among patients with Shiga toxin-producing enteroaggregative *Escherichia coli* O104:H4. *JAMA* **307**:1046–1052.

130. Som S, Deford CC, Kaiser ML, Terrell DR, Kremer Hovinga JA, Lammle B, George JN, Vesely SK. 2012. Decreasing frequency of plasma exchange complications in patients treated for thrombotic thrombocytopenic purpura-hemolytic uremic syndrome, 1996 to 2011. *Transfusion* **52**:2525–2532; quiz 2524.

131. Havens PL, Hoffman GM, Shith KJ. 1989. The homogeneous nature of the hemolytic uremic syndrome. *Clin Pediatr (Phila)* **28**:482–483.

132. Kaplan BS, Chesney RW, Drummond KN. 1975. Hemolytic uremic syndrome in families. *N Engl J Med* **292**:1090–1093.

133. Karpac CA, Li X, Terrell DR, Kremer Hovinga JA, Lammle B, Vesely SK, George JN. 2008. Sporadic bloody diarrhoea-associated thrombotic thrombocytopenic purpura-haemolytic uraemic syndrome: an adult and paediatric comparison. *Br J Haematol* **141**:696–707.

134. Geerdink LM, Westra D, van Wijk JA, Dorresteijn EM, Lilien MR, Davin JC, Komhoff M, Van Hoeck K, van der Vlugt A, van den Heuvel LP, van de Kar NC. 2012. Atypical hemolytic uremic syndrome in children: complement mutations and clinical characteristics. *Pediatr Nephrol* **27**:1283–1291.

135. Noris M, Caprioli J, Bresin E, Mossali C, Pianetti G, Gamba S, Daina E, Fenili C, Castelletti F, Sorosina A, Piras R, Donadelli R, Maranta R, van der Meer I, Conway EM, Zipfel PF, Goodship TH, Remuzzi G. 2010. Relative role of genetic complement abnormalities in sporadic and familial aHUS and their impact on clinical phenotype. *Clin J Am Soc Nephrol* **5**:1844–1859.

136. Westra D, Wetzels JF, Volokhina EB, van den Heuvel LP, van de Kar NC. 2012. A new era in the diagnosis and treatment of atypical haemolytic uraemic syndrome. *Neth J Med* **70**:121–129.

137. Ariceta G, Besbas N, Johnson S, Karpman D, Landau D, Licht C, Loirat C, Pecoraro C, Taylor CM, Van de Kar N, Vandewalle J, Zimmerhackl LB, European Paediatric Study Group for HUS. 2009. Guideline for the investigation and initial therapy of diarrhea-negative hemolytic uremic syndrome. *Pediatr Nephrol* **24**:687–696.

138. Espie E, Grimont F, Mariani-Kurkdjian P, Bouvet P, Haeghebaert S, Filliol I, Loirat C, Decludt B, Minh NN, Vaillant V, de Valk H. 2008. Surveillance of hemolytic uremic syndrome in children less than 15 years of age, a system to monitor O157 and non-O157 Shiga toxin-producing *Escherichia coli* infections in France, 1996-2006. *Pediatr Infect Dis J* **27**:595–601.

139. Chart H, Cheasty T. 2008. Human infections with verocytotoxin-producing *Escherichia coli* O157—10 years of *E. coli* O157 serodiagnosis. *J Med Microbiol* **57**:1389–1393.

140. Fernandez-Brando RJ, Bentancor LV, Mejias MP, Ramos MV, Exeni A, Exeni C, Laso Mdel C, Exeni R, Isturiz MA, Palermo MS. 2011. Antibody response to Shiga toxins in Argentinean children with enteropathic hemolytic uremic syndrome at acute and long-term follow-up periods. *PLoS One* **6**:e19136.

141. Close RM, Ejidokun OO, Verlander NQ, Fraser G, Meltzer M, Rehman Y, Muir P, Ninis N, Stuart JM. 2011. Early diagnosis model for meningitis supports public health decision making. *J Infect* **63**:32–38.

142. Bokete TN, O'Callahan CM, Clausen CR, Tang NM, Tran N, Moseley SL, Fritsche TR, Tarr PI. 1993. Shiga-like toxin-producing *Escherichia coli* in Seattle children: a prospective study. *Gastroenterology* **105**:1724–1731.

143. Carter AO, Borczyk AA, Carlson JA, Harvey B, Hockin JC, Karmali MA, Krishnan C, Korn DA, Lior H. 1987. A severe outbreak of *Escherichia coli* O157:H7-associated hemorrhagic colitis in a nursing home. *N Engl J Med* **317**:1496–1500.

144. **Pavia AT, Nichols CR, Green DP, Tauxe RV, Mottice S, Greene KD, Wells JG, Siegler RL, Brewer ED, Hannon D.** 1990. Hemolytic-uremic syndrome during an outbreak of *Escherichia coli* O157:H7 infections in institutions for mentally retarded persons: clinical and epidemiologic observations. *J Pediatr* **116:**544–551.

145. **Proulx F, Turgeon JP, Delage G, Lafleur L, Chicoine L.** 1992. Randomized, controlled trial of antibiotic therapy for *Escherichia coli* O157:H7 enteritis. *J Pediatr* **121:**299–303.

146. **Wong CS, Jelacic S, Habeeb RL, Watkins SL, Tarr PI.** 2000. The risk of the hemolytic-uremic syndrome after antibiotic treatment of *Escherichia coli* O157:H7 infections. *N Engl J Med* **342:**1930–1936.

147. **Dundas S, Todd WT, Stewart AI, Murdoch PS, Chaudhuri AK, Hutchinson SJ.** 2001. The central Scotland *Escherichia coli* O157:H7 outbreak: risk factors for the hemolytic uremic syndrome and death among hospitalized patients. *Clin Infect Dis* **33:**923–931.

148. **Smith KE, Wilker PR, Reiter PL, Hedican EB, Bender JB, Hedberg CW.** 2012. Antibiotic treatment of *Escherichia coli* O157 infection and the risk of hemolytic uremic syndrome, Minnesota. *Pediatr Infect Dis J* **31:**37–41.

149. **Grisaru S, Morgunov MA, Samuel SM, Midgley JP, Wade AW, Tee JB, Hamiwka LA.** 2011. Acute renal replacement therapy in children with diarrhea-associated hemolytic uremic syndrome: a single center 16 years of experience. *Int J Nephrol* **2011:**930539.

150. **Adragna M, Balestracci A, Garcia Chervo L, Steinbrun S, Delgado N, Briones L.** 2012. Acute dialysis-associated peritonitis in children with D+ hemolytic uremic syndrome. *Pediatr Nephrol* **27:**637–642.

151. **Grodinsky S, Telmesani A, Robson WL, Fick G, Scott RB.** 1990. Gastrointestinal manifestations of hemolytic uremic syndrome: recognition of pancreatitis. *J Pediatr Gastroenterol Nutr* **11:**518–524.

152. **Brandt JR, Joseph MW, Fouser LS, Tarr PI, Zelikovic I, McDonald RA, Avner ED, McAfee NG, Watkins SL.** 1998. Cholelithiasis following *Escherichia coli* O157:H7-associated hemolytic uremic syndrome. *Pediatr Nephrol* **12:**222–225.

153. **Schweighofer S Jr, Primack WA, Slovis TL, Fleischmann LE, Slovis TL, Hight DW.** 1980. Cholelithiasis associated with the hemolytic-uremic syndrome. *Am J Dis Child* **134:**622.

154. **Suri RS, Mahon JL, Clark WF, Moist LM, Salvadori M, Garg AX.** 2009. Relationship between *Escherichia coli* O157:H7 and diabetes mellitus. *Kidney Int Suppl* **:**S44–S46.

155. **US Census Bureau.** 2010. *2010 Census Population and Housing Tables (CPH-Ts)*. http://quickfacts. census.gov/qfd/states/00000.html. Accessed on June 18, 2014.

156. **Suri RS, Clark WF, Barrowman N, Mahon JL, Thiessen-Philbrook HR, Rosas-Arellano MP, Zarnke K, Garland JS, Garg AX.** 2005. Diabetes during diarrhea-associated hemolytic uremic syndrome: a systematic review and meta-analysis. *Diabetes Care* **28:**2556–2562.

157. **Andreoli SP, Bergstein JM.** 1982. Development of insulin-dependent diabetes mellitus during the hemolytic-uremic syndrome. *J Pediatr* **100:**541–545.

158. **Tapper D, Tarr P, Avner E, Brandt J, Waldhausen J.** 1995. Lessons learned in the management of hemolytic uremic syndrome in children. *J Pediatr Surg* **30:**158–163.

159. **Dundas S, Todd WT.** 2000. Clinical presentation, complications and treatment of infection with verocytotoxin-producing *Escherichia coli*. Challenges for the clinician. *Symp Ser Soc Appl Microbiol* **:**24S–30S.

160. **Whitington PF, Friedman AL, Chesney RW.** 1979. Gastrointestinal disease in the hemolytic-uremic syndrome. *Gastroenterology* **76:**728–733.

161. **Yates RS, Osterholm RK.** 1980. Hemolytic-uremic syndrome colitis. *J Clin Gastroenterol* **2:**359–363.

162. **Sawaf H, Sharp MJ, Youn KJ, Jewell PA, Rabbani A.** 1978. Ischemic colitis and stricture after hemolytic-uremic syndrome. *Pediatrics* **61:**315–317.

163. **Abu-Arafeh I, Gray E, Youngson G, Auchterlonie I, Russell G.** 1995. Myocarditis and haemolytic uraemic syndrome. *Arch Dis Child* **72:**46–47.

164. **Sturm V, Menke MN, Landau K, Laube GF, Neuhaus TJ.** 2010. Ocular involvement in paediatric haemolytic uraemic syndrome. *Acta Ophthalmol* **88:**804–807.

165. **Siegler RL, Brewer ED, Swartz M.** 1988. Ocular involvement in hemolytic-uremic syndrome. *J Pediatr* **112:**594–597.

166. **Manton N, Smith NM, Byard RW.** 2000. Unexpected childhood death due to hemolytic uremic syndrome. *Am J Forensic Med Pathol* **21:**90–92.

167. **Robson WL, Leung AK, Montgomery MD.** 1991. Causes of death in hemolytic uremic syndrome. *Child Nephrol Urol* **11:**228–233.

The Inflammatory Response during Enterohemorrhagic *Escherichia coli* Infection

16

JACLYN S. PEARSON[1] and ELIZABETH L. HARTLAND[1,2]

INTRODUCTION

The mammalian host is equipped with two major types of immune response, innate and adaptive, that are essential for effective control and elimination of infectious agents. The innate immune system is the first line of host defense against invading microbial pathogens and is promptly activated by the recognition of pathogen-associated molecular patterns, such as lipopolysaccharide (LPS), flagellin, peptidoglycan, and CpG DNA (1). Pathogen-associated molecular patterns are recognized by specialized germline-encoded pattern recognition receptors (PRRs) expressed by immune cells. To date, a number of PRR families have been described, including the Toll-like receptors (TLRs), retinoic acid-inducible gene-I-like receptors, and nucleotide-binding oligomerization domain (NOD)-like receptors (NLRs) (2, 3). The first and most significant consequence of PRR-mediated pathogen recognition in the host is the rapid production of proinflammatory cytokines that stimulates the innate immune response (2, 3). In addition, the innate immune response directs the development of the more specific and long-term adaptive response to a particular pathogen, mediated by B and T cells.

[1]Department of Microbiology and Immunology, University of Melbourne at the Peter Doherty Institute for Infection and Immunity, Victoria 3000, Australia; [2]Murdoch Children's Research Institute, Royal Children's Hospital, Parkville, Victoria 3052, Australia.

Enterohemorrhagic Escherichia coli *and Other Shiga Toxin-Producing* E. coli
Edited by Vanessa Sperandio and Carolyn J. Hovde
© 2015 American Society for Microbiology, Washington, DC
doi:10.1128/microbiolspec.EHEC-0012-2013

The primary site of infection with the attaching and effacing (A/E) pathogens, enterohemorrhagic and enteropathogenic *Escherichia coli* (EHEC/EPEC), is the epithelium lining the mucosal surface of the gastrointestinal tract (4). Not only do these gastrointestinal epithelial cells play a pivotal role in ion transport, fluid uptake, and secretion that are critical to the homeostatic state of the digestive system, they also coordinate the expression and upregulation of specific antimicrobial products in response to infection, including cytokines with proinflammatory (tumor necrosis factor [TNF] and interleukin-1 [IL-1]) and chemoattractant (IL-8, macrophage inflammatory protein 1-alpha [MIP1-α], monocyte chemoattractant protein-1 [MCP-1]) functions (5). In vitro, the initial and most potent activation of inflammation by EHEC/EPEC occurs by TLR5 recognition of flagellin and, to a lesser extent, TLR4 recognition of LPS (6–8). Infection studies conducted in gnotobiotic piglets show EHEC and EPEC induce extensive inflammatory cell infiltration in the lamina propria as well as transmigration of inflammatory cells across the intestinal epithelium into the intestinal lumen (9–11). Depletion of neutrophils in mice infected with the mouse A/E pathogen *Citrobacter rodentium* results in elevated bacterial loads in the liver and spleen, suggesting an important role for neutrophils in controlling infection with A/E pathogens (12). Similarly, patients with EHEC infection consistently have significantly increased numbers of leukocytes in their feces, more so than with *Salmonella* or *Shigella* infections (13), whereas antibodies against the leukocyte adhesion molecule CD18 reduce clinical symptoms and pathology of EHEC infections in rabbits (14).

Neutrophils are a type of polymorphonuclear (PMN) leukocyte that play a key regulatory role in acute inflammation (15–18). They represent the first leukocytes recruited to the site of infection with a pathogen and can mediate killing by phagocytosis, degranulation (release of antibacterial proteins), or, in the case of extracellular pathogens, the release of neutrophil extracellular traps. Transmigration of neutrophils from the vasculature to the lamina propria and into epithelial tissues occurs in response to chemoattractive factors released by either resident sentinel leukocytes or the epithelium itself upon recognition of a pathogen (17–19). IL-8 is a potent chemoattractant for neutrophils, and a number of early studies showed that EHEC (20), EPEC (21), and other bacterial enteric pathogens stimulate epithelium-derived IL-8 expression during infection (22–24). Furthermore, increased IL-8 levels, along with neopterin and IL-10, are markers for increased risk of hemolytic-uremic syndrome (HUS) in EHEC-infected children. Despite this, it is not entirely clear to what extent epithelium-derived IL-8 contributes to mucosal defense and the inflammatory response generated against A/E pathogens.

NF-κB AND MAPK SIGNALING

The most crucial factor for initiation of IL-8 gene expression in the host is activation of the key transcriptional regulator of innate immune signaling, NF-κB (nuclear factor-κB) (25). NF-κB proteins are transcription factors that control gene expression during inflammation and are activated rapidly in response to various stimuli, including pathogens, stress signals, and proinflammatory cytokines such as TNF and IL-1β (26). There are five members of the NF-κB/Rel family of proteins: p65 (RelA), p50 (NF-κB1), p52 (NF-κB2), c-Rel, and RelB, which all share an N-terminal Rel homology domain that mediates DNA binding, dimerization, and nuclear translocation (27, 28). The p65, c-Rel, and RelB subunits harbor an additional C-terminal transactivation domain, which strongly activates transcription from NF-κB-binding sites in target genes. The p50 and p52 subunits lack the transactivation domain but still bind to NF-κB consensus sites and act as transcriptional repressors (29).

NF-κB activation is generally categorized as canonical, noncanonical, or alternative, depending on the stimuli. Canonical NF-κB signaling is representative of the general scheme of NF-κB signaling predominantly considered in this article, and is triggered by TLR ligands, proinflammatory cytokines, pathogens, and engagement of the T-cell receptor by antigen (30). Upon ligand recognition, the cognate receptor such as TNFR1, IL-1R, or PRRs triggers signaling events that result in activation of the IκB kinase (IKK) complex, comprising IKKβ (IKK2), IKKα (IKK1), and IKKγ (NEMO). Activated IKK in turn phosphorylates the inhibitory protein of NF-κB (IκB), followed by subsequent ubiquitylation and degradation of IκB by the host cell proteasome (31). In a resting cell IκB is bound to p50/p65 dimers, which are released upon proteosomal degradation of IκB and transported into the nucleus through the nuclear pore complex where they can promote expression of multiple cytokine genes, including *IL8* (26).

In addition to NF-κB, the IL-8 promoter region contains binding sites for several other transcription factors, including NF-IL-6, AP-1, and AP-3 (32). Several studies have demonstrated AP-1-dependent IL-8 production during EHEC and EPEC infection in cultured intestinal epithelial cells (20, 33, 34). AP-1 activation is regulated by mitogen-activated protein kinases (MAPK), which are highly conserved host proteins that play a central role in a number of cell responses, including regulation of cytokine expression, stress responses, and cytoskeletal reorganization (35, 36). The MAPKs comprise three subfamilies of serine/threonine kinases, including the extracellular signal-related protein kinases ERK and two stress-activated protein kinases, p38 and JNK (c-Jun amino-terminal kinase). These kinases regulate cellular processes through phosphorylation of target protein substrates, including other protein kinases, phospholipases, transcription factors, and cytoskeletal proteins. Phosphorylation by MAPKs acts as an on/off switch for the activity of target substrates, a process that can be reversed by cellular protein phosphatases (37).

PROINFLAMMATORY RESPONSES DURING EHEC INFECTION

EHEC produces a number of factors that potentially upregulate inflammatory cytokine production by intestinal epithelial cells. For example, early studies showed that purified Shiga toxin (Stx) could stimulate low-level production of IL-8 by cultured colon cancer cell lines (38–40). However, subsequent studies using primary human colonic epithelial cells showed that TLR5 recognition of H7 flagellin was the major factor inducing IL-8 production during EHEC infection, and not Stx signaling through the Gb3 receptor (7, 41). Indeed, in vitro the initial and most potent activation of inflammation by EHEC/EPEC occurs by TLR5 recognition of flagellin and, to a lesser extent, TLR4 recognition of LPS (6–8). Culture supernatants from the prototype EPEC strain E2348/69, but not from an EPEC flagellin (fliC) mutant, induced significant IL-8 production from cultured colonic epithelial cells (42). In contrast, EHEC O157: H7 lacking Stx potently activates p38 and ERK1/2 MAPKs and NF-κB, and induced significant production of IL-8 by human intestinal epithelial cells (41). Flagellin/TLR5 signaling thus seems important to induce inflammatory signaling in vitro. In vivo studies using EHEC O157:H7-infected rabbits showed a marked increase in neutrophil infiltration in the colonic epithelium resulting in diarrhea and disruption of solute transport, none of which was dependent on Stx expression (43, 44). Although TLR5 is not generally located on the apical surface of enterocytes (45), A/E pathogens may gain access to basolateral receptors by compromising the epithelial barrier and cell polarity (46, 47). TLR5 is also expressed by lamina propria dendritic cells, which may sample luminal flagellin (48). In addition, some EHEC strains appear to have a predilection for the follicle-associated epithelium of the gastrointestinal tract that expresses TLR5 (49, 50). Hence it is likely that flagellin is sensed during EHEC infection.

Despite the fact that Stx alone induces relatively low levels of proinflammatory cytokine production, some studies have suggested that the loss of barrier function induced by neutrophil transmigration allows increased translocation of Stx from the lumen into the circulation (51), promoting a systemic cytokine response. Indeed, patients with HUS have increased levels of circulating proinflammatory cytokines, including IL-8, TNF, IL-6, IL-1β, and IL-10. Although no studies to date have been able to detect circulating Stx in patients with toxin-producing EHEC infections, others have demonstrated Stx bound to circulating neutrophils in patients with HUS (52). A recent study in nonhuman primates used purified Stx1 and Stx2 to evaluate their relative contribution to inflammatory cytokine production. Both toxins induced a significant increase in expression of IL-8 (CXCL8), MCP-1 (CCL2), and MIP1-α (CCL3) in kidney tissue (53), whereas TNF and IL-12p35 levels were comparably low. Urine analysis indicated a substantial increase in the production of IL-6 and vascular endothelial growth factor in response to Stx1 compared to Stx2 48 h after challenge, which was consistent with an increase in leukocyte infiltration in the kidneys. Although both toxins were chemotactic in the kidneys, Stx1 induced a more rapid inflammatory response and time to euthanasia compared with Stx2 (53).

The type IV pilus, hemorrhagic coli pilus (HCP), of EHEC O157:H7 is another factor capable of inducing an inflammatory response in vitro. Production of IL-8 and TNF is significantly increased at the basolateral surface of polarized T84 and HT-29 and in nonpolarized HeLa cells during infection with HCP-expressing EHEC strains (54). Deletion of the HCP major pilin subunit *hcpA* significantly reduced IL-8 production in cultured intestinal epithelial cells during EHEC O157:H7 strain EDL933 infection, but not to the same extent as a *fliC/hcpA* double mutant,

confirming the importance of EHEC flagellin in eliciting an inflammatory response. NF-κB and AP-1 activation were increased in the presence of purified HCP and after 3 h of infection with EHEC O157:H7 strain EDL933 in cultured intestinal epithelial cells (54).

Overall, the initial induction of proinflammatory cytokines in the intestinal mucosa during EHEC infection appears to result from activation of MAPK and NF-κB signaling by a variety of factors. This in turn drives the increased production of cytokines by intestinal epithelial cells and subsequent PMN transmigration. Intestinal tissue damage is sustained as a result of the large number of infiltrating PMNs releasing proteolytic enzymes, reactive oxygen intermediates, and phospholipid derivatives, contributing to diarrheal disease and the release of luminal contents (including Stx) into the circulation.

C. rodentium AND IMMUNE RESPONSES IN MICE

Studying the in vivo inflammatory response intensively in the human host is not a viable option for EHEC and EPEC infection. Therefore, animal models of infection have been invaluable in understanding the type of immune response elicited, including which cytokines and chemokines are released, and how they might contribute to clearance or pathogenesis. Rabbits, calves, and lambs are the primary animals used to test the characteristics of infection and pathology with EHEC strains, whereas the closely related mouse pathogen *C. rodentium* is also used as a model of EHEC and EPEC colonization. Besides the advantage of infecting a small laboratory animal, *C. rodentium* shares most key virulence factors of EHEC and EPEC, including a type III secretion system (T3SS) and effector proteins encoded by the locus of enterocyte effacement (LEE) pathogenicity island and the non-LEE effectors encoded on various genomic islands and integrative elements. The key virulence factors that *C. rodentium* lacks compared to EHEC are genes encoding Stx and flagellin. Despite these limitations, *C. rodentium* infection of mice has allowed two critically important research questions in the field to be partially addressed, namely, (i) does innate immunity provide protection against A/E pathogens or does it just induce tissue damage and (ii) what are the relative contributions of innate and adaptive immunity in controlling infection with A/E pathogens.

Infection with *C. rodentium* causes colitis characterized by inflammatory cell infiltration, hyperplasia of crypt cells, loss of goblet cells, and significant intestinal barrier disruption (55, 56). Both innate and adaptive responses contribute to the control of *C. rodentium* infection, dissemination, and elimination. For instance, mice lacking B cells, T cells, or both have greater pathogen loads in their colonic and peripheral tissues, and despite a severe disease phenotype in these mice, a significant percentage of mice survive, suggesting more than an adaptive response is involved (57, 58). For example, in the absence of neutrophils, mice infected with *C. rodentium* also have a high bacterial load, which highlights the importance of the inflammatory response in the early stages of infection (12). Activation of NF-κB has been demonstrated in vivo during *C. rodentium* infection (59); however, unlike EHEC and EPEC, *C. rodentium* is nonflagellated and does not activate TLR5 (60). Therefore, it is likely that activation is either T3SS-dependent and/or a result of TLR4 recognition of LPS.

Mice with an intact adaptive system but deficient in mast cells display increased colonic inflammation and increased production of proinflammatory cytokines (61). These mice also experience systemic infection, and most die within 4 to 7 days of *C. rodentium* infection (61), which suggests that mast cells play a role in control and clearance of *C. rodentium* rather than regulating inflammation.

A strong Th1/Th17 response, characterized by an increase in the expression of IL-1β, TNF, IL-12, gamma interferon (IFNγ), IL-17, and IL-22, is elicited in the colon of *C. rodentium*-infected mice and produces pathology similar to that seen in mouse models of inflammatory bowel disease (IBD) (62–64). Th17 cells are a subset of T-helper cells that have an innate function and secrete large amounts of the proinflammatory cytokines IL-17 and IL-22. *C. rodentium* infection induces a Th17 response within 2 weeks of infection, and this is needed for clearance of the pathogen (64). The induction of this response requires NOD1/NOD2 signaling and the pro-inflammatory cytokine IL-6 (63). IL-6-deficient mice develop severe mucosal ulcerations and increased crypt cell apoptosis, suggesting a protective role for the cytokine during gut infection (65). Although it is not clear if all these immune and inflammatory factors are required for mucosal defense during human infection with EHEC, some studies have highlighted the importance of IL-6 in clearance of EPEC in children (66, 67). In addition, a chronic Th17 response is strongly associated with the development of IBD (68).

Studies have suggested that TNF plays an important role in clearance of EPEC in children (67) and control of bacterial load in animal models of infection (69, 70). TNF is produced rapidly in response to local or systemic tissue damage and is a potent activator of macrophages and neutrophils. Failure to control TNF production can lead to severe tissue damage and organ-specific tissue pathology as a result of chronic activation of immune cells and inflammatory responses. In the intestine, increased TNF levels are highly associated with tissue pathology and disease, including Crohn's disease and IBD (71). Mice that are TNF deficient (TNFRp55$^{-/-}$) demonstrate increased tissue pathology during *C. rodentium* infection, including pronounced hyperplasia, increased T-cell infiltrate, and higher levels of mucosal IFN-γ and IL-12 production (69). However, TNFRp55$^{-/-}$ mice are not compromised in their ability to clear infection compared to wild-type C57BL/6 mice, and despite the compensatory increase in cytokine production, they experience significantly higher bacterial loads than wild-type mice (69). Overall, it is plausible that the increased bacterial burden may be a major contributing factor to increased tissue pathology in the absence of TNF and that TNF plays a major role in limiting bacterial replication.

TLRs have been strongly implicated in mediating inflammation during *C. rodentium* infection. For instance, TLR4, although not protective, mediates chemokine induction and subsequent neutrophil and macrophage recruitment and is partially responsible for tissue pathology during *C. rodentium* infection (72). Conversely, TLR2 is not critical in mediating proinflammatory responses associated with *C. rodentium*-induced colitis but helps maintain mucosal integrity during *C. rodentium* infection (65). MyD88 is a key signaling adaptor protein that is shared among TLRs (except TLR3), IL-18R, and IL-1R and is essential in the control of mucosal infection by a number of pathogens (73–76). MyD88 activation initiates NF-κB activation and, in the intestine, plays an important role in maintaining tissue homeostasis (77). During *C. rodentium* infection MyD88 protects against bacteremia and severe pathology in C57BL/6 mice (12), and MyD88-deficient mice experience elevated bacterial load in the colon and peripheral tissues, which correlates with a decrease in neutrophil infiltration into the colonic tissue. MyD88-deficient mice also experience severe colonic ulceration and bleeding, resulting in high levels of morbidity and mortality. This severe disease phenotype is due to impaired epithelial barrier function and defective cellular proliferation, followed by an inability to repair mucosal damage.

MyD88 association with the IL-1R but not the IL-18R affords protection against increased mortality and pathology during infection with *C. rodentium* (78). IL-1R-deficient mice infected with *C. rodentium* have disease similar to that seen in MyD88-deficient mice, but they do not experience higher bacterial loads or an inability to recruit neutrophils and mediate tissue damage repair. The mice are unable to induce production of

proinflammatory cytokines IL-6 and IFNγ during infection, suggesting that the bacteria are mediating severe tissue pathology rather than a dysregulated immune response and that IL-1R signaling regulates susceptibility to *C. rodentium* infection (78). The importance of IL-1R signaling is supported by work showing that mice deficient for the inflammasome components Nlrc4, Nlrp3, and caspase-1 that lead to IL-1β and IL-18 production were highly susceptible to *C. rodentium* and had exacerbated intestinal inflammation and increased bacterial load after day 10 of infection (79). Similar observations were made in IL-1β- and IL-18-deficient mice. Despite the fact that Nlrc4-deficient mice were highly susceptible to *C. rodentium* infection, only Nlrp3 and not Nlrc4 induced caspase-1 activation (79).

INHIBITION OF INFLAMMATORY SIGNALING BY A/E PATHOGENS

Many of the early studies on EHEC/EPEC-mediated inflammation focused on the proinflammatory response characterized by marked infiltration and transmigration of PMNs in the intestinal epithelium. However, some 10 years ago, researchers made the remarkable observation that EHEC and EPEC strains could inhibit NF-κB and MAPK activation as well as IκB degradation and the production of proinflammatory cytokines, including IL-8 and IL-6, early in the course of infection (80, 81). Furthermore, this inhibitory mechanism was dependent on the presence of a functional T3SS, revealing for the first time that EHEC/EPEC suppressed host inflammatory responses in a type III-dependent manner (80). This suggested that EHEC/EPEC had evolved specialized mechanisms to dampen the early inflammatory response during infection, possibly to persist for longer periods before the overall immune response would inevitably clear the bacteria.

The early discovery of the LEE pathogenicity island and its association with A/E lesion formation led to the intensive study of LEE-encoded effectors and their involvement in cytoskeletal reorganization and disruption of tight junctions in the epithelium (82). However, the activity of these effectors could not explain how A/E pathogens suppressed the production of inflammatory cytokines. Characterization of several non-

LEE-encoded effectors over the past 3 years has revealed that A/E pathogens inject multiple effector proteins into the host cell that specifically target innate immune factors. These effectors have highly specific and diverse functions that interfere with a range of innate signaling pathways, including NF-κB and MAPK activation (83–86) (Fig. 1).

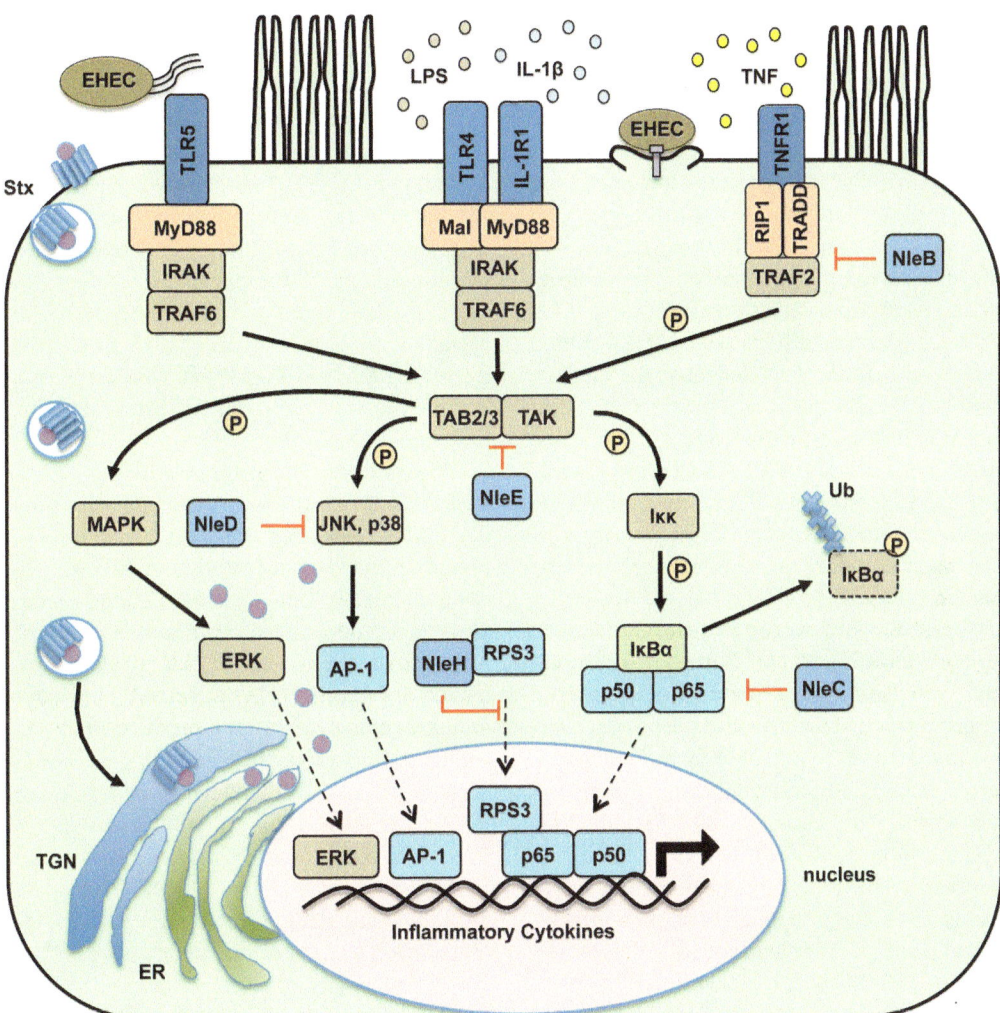

FIGURE 1 Inflammatory signaling pathways stimulated and inhibited by EHEC. EHEC products such as flagellin and LPS stimulate TLR signaling and the production of cytokines by epithelial cells during infection. At the same time, T3SS effector proteins injected by the LEE-encoded translocon inhibit inflammatory signaling at different points in the various pathways. The signaling factors and T3SS effector proteins, and their mechanisms of action, are described in detail in the main text. TGN, trans-Golgi network; ER, endoplasmic reticulum; Ub, ubiquitin; P, phosphorylation. doi:10.1128/microbiolspec.EHEC-0012-2013.f1

EFFECTORS THAT INHIBIT INFLAMMATORY SIGNALING

The Cysteine Methyltransferase, NleE

NleE was one of the first non-LEE-encoded effectors of EHEC implicated in the inhibition of NF-κB signaling (84, 85). This 27-kDa translocated effector is highly conserved across A/E pathogens and was first discovered as a type III secreted effector of *C. rodentium* (87, 88). In EHEC O157:H7 strain EDL933, NleE is located on the virulence-associated O-island (OI) 122, whereas in EPEC O127:H6 E2348/69, NleE is encoded on genomic pathogenicity island integrative element six (IE6) (89, 90). In both cases, NleE is encoded directly downstream of the type III effector NleB1. Homology at the amino acid level between NleE from EHEC and EPEC is high (99% identity) and also between EHEC/EPEC and *C. rodentium* (~85% identity). Furthermore, all *Shigella* species carry a strong homolog of NleE known as OspZ, which share about 74% identity with NleE from EHEC/EPEC (91). Overall, the high level of similarity between the NleE/OspZ effectors suggests that their function is conserved across the species.

Initial in vivo studies revealed no significant colonization defect for a *C. rodentium* Δ*nleE* mutant compared to the wild type strain, although the Δ*nleE* mutant was outcompeted by the wild type in a competitive infection (88). A subsequent study, however, argued that NleE was essential for full colonization in mice and that NleE contributed to disease pathology during *C. rodentium* infection (90). In addition, NleE is consistently associated with the virulence profile of O157 and non-O157 EHEC and EPEC strains (92, 93).

More recent in vitro studies have shown that NleE from EHEC, EPEC, rabbit specific EPEC (REPEC), and *C. rodentium* and full-length OspZ from *Shigella* species inhibit the activation of NF-κB in cultured epithelial cells (84, 85). Despite the translocation of numerous type III effectors into the host cell during infection, NleE is sufficient to block TNF-induced IκB degradation and nuclear translocation of activated NF-κB (p65) in epithelial cells when it is expressed ectopically (84, 85). Furthermore, NleE-dependent inhibition of NF-κB activation contributes significantly to the low levels of *IL8* expression and IL-8 production in EPEC-infected epithelial cells (84, 85). A concurrent study also demonstrated the ability of NleE to inhibit nuclear translocation of activated NF-κB and proinflammatory cytokine production in human dendritic cells (DCs) (94). IL-8, TNF, and IL-6 expression and production were suppressed in the presence of EPEC-delivered NleE in human monocyte DCs and Peyer's patch DCs. The follicle-associated epithelium is rich in DCs and forms the interface between the luminal contents of the intestinal tract and the gut-associated lymphoid tissue. Therefore it seems plausible that enteric pathogens such as EHEC, EPEC, and *Shigella* sp. would evolve a mechanism to evade immune detection at this junction.

The precise mechanism by which NleE inhibits NF-κB activation was described recently (95). NleE possesses a unique *S*-adenosyl-L-methionine-dependent methyltransferase activity that modifies a cysteine residue in the zinc finger domain of the signaling adaptor proteins TAB2 and TAB3 (TAK1-binding proteins 2 and 3) (95). TAB2/3 contain zinc finger domains, which bind K63-linked polyubiquitin chains on the target TNF receptor-associated proteins (TRAF) 2/6 following activation of TNF or TLR/IL-1R signaling. The binding of TAB2/3 to ubiquitinated TRAF allows TAK1 to form a complex with IKK and subsequently phosphorylate IKKβ, which leads to the degradation of IκB and activation of NF-κB. The modification by NleE abolishes ubiquitin-chain binding of the zinc finger domains of TAB2/3 and thereby disrupts NF-κB signaling in host cells. The activity of NleE depends on a conserved 6-amino-acid motif, [209]IDSYMK[214], within the C-terminal region that is essential for its ability to block NF-κB activation and modify TAB2/3 (85, 95).

The Metalloprotease Effectors, NleC and NleD

The non-LEE effectors NleC and NleD are potent inhibitors of two key innate signaling networks through their proteolytic activities. While NleC targets NF-κB Rel proteins for degradation, NleD cleaves the MAPK enzymes JNK and p38 (96–99). NleC was first identified as an effector secreted by the LEE-encoded T3SS of *C. rodentium* (87). The 37-kDa effector is conserved across A/E pathogens, encoded on OI 36 in EHEC O157:H7 EDL933 and on prophage four (PP4) in EPEC O127:H6 E2348/69 (89, 100). EHEC and EPEC NleC share 100% amino acid sequence similarity and are 95% similar to NleC from *C. rodentium*. NleD is a 30-kDa effector that is also conserved across A/E pathogens and is also encoded on OI 36 in EHEC O157:H7 EDL933 and on PP4 in EPEC O127:H6 E2348/69. EPEC and EHEC NleD share 99% amino acid similarity and 77% with NleD from *C. rodentium*.

Initial studies on NleC and NleD revealed that both effectors were translocated into host cells by the EPEC T3SS. However, when tested in calves or lambs, Δ*nleC* and Δ*nleD* deletion mutants of EHEC O157:H7 exhibited wild-type virulence (101). Similarly, neither *nleC* nor *nleD* mutants of *C. rodentium* were attenuated in mice (88), although mice infected with an *nleC* deletion mutant showed increased pathology, suggesting that NleC assists in reducing the severity of colitis during *C. rodentium* infection (99).

The study by Marchès et al. (2005) (101) first identified putative zinc metalloprotease motifs, HEXXH, within both NleC and NleD although no further research on the significance of the motif for protein function was conducted until recently. In fact, NleC degrades the NF-κB subunit, p65, and this function is dependent on the zinc metalloprotease motif, [183]HEIIH (96–99, 102). Degradation is direct, as recombinant His[6]-NleC cleaves both p65 and p50 in epithelial cell lysates (98). Furthermore, mutation of the histidines or the glutamate within the metalloprotease motif of NleC renders the protein inactive (96–99, 102), a phenomenon that had been previously observed for other zinc metalloproteases (103). Although two independent research groups concurrently identified specific cleavage sites of recombinant p65 by NleC, they were not consistent with each other (96, 102). Both groups identified the site within the N-terminal Rel homology domain of p65. However, whereas Baruch et al. (96) put the cleavage site between residue C38 and E39 by N-terminal sequencing, Yen et al. (102) proposed the cleavage site to be between residue P10 and A11. Either way, since the Rel homology domain of NF-κB proteins is required for binding to NF-κB consensus, dimerization, and nuclear localization (104), cleavage by NleC would render p65 unable to bind target genes and activate transcription of inflammatory cytokines.

To date, there is conflicting evidence on whether NleC degrades other NF-κB proteins in addition to p65. Whereas one study showed clear degradation of p50 with NleC expressed ectopically or delivered by the T3SS (98), another group stated NleC could not cleave p50 directly (102). Similarly, while IκB degradation was observed upon ectopic expression of NleC (97), another study found that reduced levels of IκB in EPEC infected cells could not be attributed directly to NleC (102). Yet another study suggested that NleC cleaves the acetyltransferase p300, a transcriptional coactivator of many host cell genes, including p65 (105, 106). NleC delivered by the T3SS bound to endogenous p300 and reduced host cell nuclear levels of p300 in a metalloprotease-dependent manner (106). Furthermore, overexpression of p300 in host cells resulted in a significant increase in IL-8 production during wild-type EPEC infection, suggesting that NleC-dependent cleavage of p300 assists suppression of IL-8 during infection (106). In addition, NleC was observed to target MAPK signaling by inhibiting phosphorylation of p38 during EPEC infection, although this was independent of the zinc metalloprotease motif (99).

Given the diversity and complexity of innate immune signaling pathways, it is perhaps not surprising that A/E pathogens have evolved mechanisms to target specific pathways such as NF-κB signaling at multiple levels. Although there is apparent redundancy in the functions of NleE and NleC, secreted IL-8 levels are significantly higher upon infection with a double *nleC/nleE* mutant than single *nleE* or *nleC* mutants, suggesting that NleC and NleE act synergistically to inhibit IL-8 production (96, 98, 99, 102, 106).

NleD-mediated proteolytic degradation of the MAPKs p38 and JNK also depends on the [141]HELLH motif of NleD (96). JNK cleavage by wild-type EPEC is detected as early as 30 min post infection with the kinase completely degraded by 2.5 h post infection. NleD directly cleaves JNK and p38 within a conserved activation loop present in both signaling kinases. ERK, however, is unaffected by the metalloprotease, which is likely explained by the fact that it does not contain the activation loop present in p38 and JNK. The specific cleavage site of JNK as determined by N-terminal sequencing occurs after residue P184 within a TPY motif (96). Phosphorylation of T183 and Y185 in this motif is required for JNK activation; therefore, NleD selectively cleaves the signaling kinase at its site of activation. As with NleC, mutation of the glutamate within the metalloprotease motif of NleD renders NleD inactive, and furthermore, although not sufficient, NleD contributes to the relief of IL-8 suppression in wild-type EPEC infection of cultured epithelial cells (96).

Overall, NleC and NleD are zinc-containing metalloproteases that provide yet another mechanism by which EHEC and EPEC control the host inflammatory response during infection. Although the degradation of p65, JNK, and p38 may appear extraneous given the efficiency of NleE alone to inhibit IL-8 production, NleC and NleD clearly amplify this modulatory effect, demonstrating that EHEC and EPEC are highly evolved bacteria that specifically target signaling networks at multiple levels.

The Glycosyl Transferase, NleB

The non-LEE-encoded effector B (NleB) was first identified as a secreted effector of *C. rodentium* and is conserved across all A/E pathogens (87). EHEC and EPEC carry two copies of *nleB* (*nleB1* and *nleB2*) whereas *C. rodentium* carries only one that is most similar to NleB1. NleB1 from EPEC and EHEC share a high level of amino acid identity (~98%) with each other and about 89% amino acid identity and ~96% similarity with NleB from *C. rodentium*. In EHEC O157:H7, the *nleB1* gene is located on genomic OI 122, which is consistently present in Stx-producing EHEC strains that cause outbreaks and severe disease (107). NleB2 is located on a separate pathogenicity island in both EHEC and EPEC strains and has not yet been implicated in disease epidemiology of either pathotype. NleB2 is highly related to NleB1 from both EHEC and EPEC strains and shares 60 to 62% amino acid identity and ~80% similarity.

Early work in vivo using a signature-tagged mutagenesis approach revealed that *nleB* is essential for full colonization of mice by *C. rodentium*, thereby establishing the effector as an important virulence determinant of A/E pathogens (88). Furthermore, a study on the transmissibility of *C. rodentium* showed that NleB contributes significantly to the fitness of the pathogen (90). Strong homologs of NleB (termed SseK1, SseK2, and SseK3) exist in a number of *Salmonella* spp., where their function and contribution to virulence are still under investigation (108).

Recent studies in cell signaling have shown that transient expression of NleB1 in cultured epithelial cells inhibits IκB degradation and NF-κB activation in response to TNF but not IL-1β stimulation (85, 96). This suggested that the point at which NleB1 interrupted NF-κB signaling was downstream of TNF-R1 but upstream of IKK and TAK1, where the TNF and IL-1 pathways converge. Despite compelling evidence for NleB-mediated inhibition of NF-κB activation from TNF-R1, subsequent work showed that the effector had no significant role in blocking production of the NF-κB-dependent cytokine, IL-8, during EPEC infection (J. S. Pearson, unpublished data).

A recent study revealed that NleB of *C. rodentium* is a member of the GT-8 family of glycosyltransferase proteins and exhibits GlcNAc transferase activity (109). NleB1 and NleB2 contain a catalytic motif (DxD) that is conserved across all A/E pathogens and the SseK proteins from *Salmonella* spp. The host protein GAPDH was proposed as the target for NleB binding and O-GlcNAc modification. Coincidently, the study proposed a role for GAPDH binding and promoting TRAF2 ubiquitylation and subsequent NF-κB activation, and thus concluded that GlcNAcylation of GAPDH by NleB hindered TRAF2-mediated NF-κB activation (109). In this instance, modification of GAPDH may explain previous observations that NleB could block TNF-induced IκB degradation and p65 nuclear translocation (84, 85); however, this study did not examine the cytokine production that would implicate NleB in dampening the inflammatory response during infection. This work did, however, show that a *C. rodentium* nleB mutant complemented with a catalytic mutant of NleB (NleB$_{AAA}$) was attenuated to similar levels as the nleB mutant, indicating that the DxD motif contributes to the virulence of A/E pathogens (109).

NleH and the Inhibition of NF-κB Signaling

NleH1 and NleH2 are non-LEE-encoded effectors of A/E pathogens that also contribute to suppression of NF-κB activation. The effectors share 84% amino acid identity and are located on separate genomic islands in both EHEC and EPEC strains. Studies to date have been varied in defining the mechanism by which the NleH effectors interfere with NF-κB signaling. Initial work showed that EHEC nleH1 and nleH2 bind to a recently discovered non-Rel NF-κB subunit, ribosomal protein S3 (RPS3), a K homology domain protein that regulates NF-κB dependent transcription (83, 110). NleH has significant sequence similarity

to the *Shigella* effector OspG, a Ser/Thr protein kinase that prevents ubiquitylation and subsequent degradation of phospho-IκBα and downstream activation of NF-κB (111). Alanine replacement of the conserved lysine residue at position 53 abolishes the kinase activity of OspG (83). NleH1 and NleH2 are also autophosphorylated Ser/Thr protein kinases, but their kinase activity is independent of their interaction with RPS3 (110). NleH1 but not NleH2 reduces the nuclear abundance of RPS3 with no effect on other NF-κB signaling factors, and in fact, the N-terminal region of NleH1 is sufficient to perform this function (83). The mechanism by which NleH1 executes this through inhibition of IKKβ-mediated phosphorylation of RPS3 at serine residue 209 was demonstrated during EHEC O157:H7 infection in vitro and in vivo (112). NleH2 does not inhibit RPS3 nuclear translocation, but it does increase expression from an AP-1-dependent luciferase reporter; hence it may affect a different signaling pathway (83). Another study suggested that NleH1 and NleH2 suppress TNF-induced IκBα degradation in cultured epithelial cells by interfering with phospho-IκBα ubiquitylation, which is dependent on conserved lysine residues, K159 and K169, in NleH1 and NleH2, respectively, that are implicated in kinase activity (86).

A number of studies conducted with animal models have tried to dissect the contribution of NleH to pathogenesis and inflammatory responses in the host; however, they have been somewhat inconsistent. *C. rodentium* carries only one copy of NleH and is functionally most similar to NleH1 of EHEC/EPEC (83). Initial work showed that an nleH mutant was attenuated early in infection (6 days post infection) in C57BL/6 mice, but not at later time points (10 days post infection) (113), plus the mutant strain was cleared more rapidly than the wild-type strain (114), suggesting the effector plays a role in persistence of the pathogen. As for inflammation, one study showed that wild-type *C. rodentium* infection in C57BL/6 mice induces higher transcription of TNF mRNA in mouse tissues than an *nleH* mutant at 14 days post infection (114), whereas a more recent study showed the opposite at 7 days post infection (115), suggesting an anti-inflammatory role for NleH in vivo earlier in infection. The former study also showed that NleH induced an increase in NF-κB activity in the colonic mucosa of C57BL/6 mice at later time points; however, there was no overall significant tissue damage or increase in T-cell infiltrate in the colons of mice infected with either wild type or an *nleH* mutant of *C. rodentium* over 14 days of infection (114). Another study suggested that KC (a functional homolog of human IL-8) secretion in streptomycin-treated C57BL/6 mice increases upon infection with an EPEC *nleH1nleH2* double mutant, suggesting that NleH contributes to the inhibition of inflammation in EPEC-infected mice (86). However, it is difficult to reconcile this result with the fact that EPEC colonization of mice does not lead to a productive infection (116).

Despite intense interest in the mechanism of action and function of NleH, neither NleH1 nor NleH2 inhibits NF-κB activity to the same extent as NleE or NleC in vitro (85). Given the possible additional role of NleH in cell death signaling (117), the mechanism of action of NleH and its contribution to EPEC/EHEC pathogenesis warrant further investigation.

Another Role for the Translocated Intimin Receptor (Tir)?

Two studies have identified a role for Tir in inhibiting host innate signaling mechanisms. Ruchaud-Sparagano et al. (2012) (118) showed that transient expression of Tir from EPEC in HeLa cells could inhibit TNF-induced NF-κB activation by binding and degrading the cytoplasmic TRAF2 in a proteasome-independent manner. Consequently, a slight yet significant increase in IL-8 production in HeLa cells infected with an EPEC *tir* deletion mutant was observed (118). A subsequent study showed that Tir shares sequence similarity with host cellular immunoreceptor tyrosine-based inhibition motifs (ITIMs) and inhibits TLR

signaling (119) these motifs are critical negative regulators of eukaryotic immunoreceptor signaling pathways (120, 121). Tir was required for the inhibition of proinflammatory cytokine (TNF and IL-6) expression during EPEC and *C. rodentium* infection in vitro and in vivo, respectively (119). However, the essential role of Tir in intimate adherence and hence effector translocation makes it difficult to attribute this property directly to Tir, particularly as the background strain used was JPN15, which has been cured of the plasmid that encodes bundle-forming pili and already adheres less efficiently to cells than wild-type EPEC. Likewise, while deletion of *tir* enhanced NF-κB, Erk, JNK, and p38 MAPK activation during EPEC infection of cultured epithelial cells (119), this was not distinguished from the role of Tir in adherence and the translocation of other effectors. Nevertheless, this study provided evidence that Tir binds directly to SHP-1 (119), a phosphatase that suppresses cellular immune responses by dephosphorylation of NF-κB and MAPKs (122). Downregulation or mutagenesis of SHP-1 increased TNF and IL-6 production in EPEC-infected cells, and recruitment of SHP-1 by Tir was essential for its inhibitory function. Finally, Tir also enhanced SHP-1 association with TRAF6 by an unknown mechanism, which in turn prevented TRAF6 ubiquitylation and subsequent signaling cascades.

Proinflammatory T3SS Effectors

Evidence for the proinflammatory activity of some T3SS effector proteins comes primarily from the study of genomic island deletion mutants. An EPEC mutant lacking genomic islands PP4 and IE6 (encoding seven effectors including NleE, NleC, and NleD) induced p65 nuclear translocation and pronounced IL-8 secretion independently of TNF stimulation (85, 98). Nuclear translocation of p65 induced by the PP4/IE6 double mutant was greater than for a T3SS mutant (*escN*), suggesting that other translocated effectors stimulated NF-κB activity (85). Recent work on the WxxxE effector, EspT, supports the idea that some effectors directly activate NF-κB signaling (123). EspT induced IL-8 and IL-1β secretion, which required the small GTPase Rac1 (123). A similar phenomenon was evident during *Salmonella* sp. infection where activation of Rac1 and Cdc42 by the WxxxE effector, SopE, also induced NF-κB activation. NF-κB activation by Rac1 occurred via NOD1 and RIPK2 signaling and was also applicable to the sensing of peptidoglycan by NOD1 (124). This study concluded that NOD1 senses microbial infection by monitoring the activation state of small GTPases, such as Rac1 and Cdc42. Because the WxxxE effectors have known roles in cytoskeletal rearrangement and bacterial invasion, this illustrates the point that the activity of an effector protein on a host cell protein may inadvertently stimulate an inflammatory response.

The pO157-Encoded SteC Protease from EHEC Impairs Neutrophil Function

In addition to the type III effectors, a number of other EHEC-related virulence factors are believed to counteract host inflammatory processes. StcE (secreted protease of C1-esterase inhibitor) is a type II secreted protease encoded by the pO157 virulence plasmid of EHEC strains that cleaves the protein backbone of mucin-type glycoproteins (125). Mucins are found on almost all hematopoietic cells, including neutrophils, and play critical roles in cellular interactions within the immune system. CD43 and CD45 are neutrophil-specific mucins (126, 127) specifically targeted by StcE for proteolysis, which leads to increased neutrophil adhesion, impaired migratory capacity, and an enhanced oxidative burst (128). The impaired neutrophil function may contribute to increased tissue damage and inflammation or, conversely, may act to dampen the inflammatory response due to a lack of neutrophil mobility. Indeed, neutrophils isolated from children with HUS are more adherent to the vascular endothelium (129).

CONCLUSIONS

Given that the gastrointestinal tract is the largest inflammatory organ in the mammalian body, it is not surprising that EHEC/EPEC and other enteric bacterial pathogens including *Salmonella* sp. and *Shigella* sp. have evolved numerous mechanisms for diverting or arresting inflammatory processes in order to persist and cause disease. Although there is a certain level of redundancy between the effectors and their functions, each one plays a contributing role in counteracting the inflammatory responses of the host. This diversity of effector protein function exemplifies how A/E pathogens have evolved to evade numerous host defense mechanisms, especially innate immunity. Current research efforts to determine effector function are revealing more novel enzymatic functions and targets for effectors and at the same time refining our understanding of host-pathogen interactions.

ACKNOWLEDGMENT

We declare no conflicts of interest with regard to the manuscript.

CITATION

Pearson JS, Hartland EL. 2014. The inflammatory response during enterohemorrhagic *Escherichia coli* infection. Microbiol Spectrum 2(4):EHEC-0012-2013.

REFERENCES

1. **Akira S, Uematsu S, Takeuchi O.** 2006. Pathogen recognition and innate immunity. *Cell* **124:**783–801.
2. **O'Neill LA, Golenbock D, Bowie AG.** 2013. The history of Toll-like receptors—redefining innate immunity. *Nat Rev Immunol* **13:**453–460.
3. **Latz E, Xiao TS, Stutz A.** 2013. Activation and regulation of the inflammasomes. *Nat Rev Immunol* **13:**397–411.
4. **Robins-Browne RM, Hartland EL.** 2002. *Escherichia coli* as a cause of diarrhea. *J Gastroenterol Hepatol* **17:**467–475.

5. **Kagnoff MF, Eckmann L.** 1997. Epithelial cells as sensors for microbial infection. *J Clin Invest* **100:**6–10.

6. **Badea L, Beatson SA, Kaparakis M, Ferrero RL, Hartland EL.** 2009. Secretion of flagellin by the LEE- encoded type III secretion system of enteropathogenic *Escherichia coli*. *BMC Microbiol* **9:**30.

7. **Miyamoto Y, Iimura M, Kaper JB, Torres AG, Kagnoff MF.** 2006. Role of Shiga toxin versus H7 flagellin in enterohaemorrhagic *Escherichia coli* signalling of human colon epithelium in vivo. *Cell Microbiol* **8:**869–879.

8. **Schuller S, Lucas M, Kaper JB, Giron JA, Phillips AD.** 2009. The ex vivo response of human intestinal mucosa to enteropathogenic *Escherichia coli* infection. *Cell Microbiol* **11:**521–530.

9. **Moon HW, Whipp SC, Argenzio RA, Levine MM, Giannella RA.** 1983. Attaching and effacing activities of rabbit and human enteropathogenic *Escherichia coli* in pig and rabbit intestines. *Infect Immun* **41:**1340–1351.

10. **Tzipori S, Gibson R, Montanaro J.** 1989. Nature and distribution of mucosal lesions associated with enteropathogenic and enterohemorrhagic *Escherichia coli* in piglets and the role of plasmid-mediated factors. *Infect Immun* **57:**1142–1150.

11. **Tzipori S, Robins-Browne RM, Gonis G, Hayes J, Withers M, McCartney E.** 1985. Enteropathogenic *Escherichia coli* enteritis: evaluation of the gnotobiotic piglet as a model of human infection. *Gut* **26:**570–578.

12. **Lebeis SL, Bommarius B, Parkos CA, Sherman MA, Kalman D.** 2007. TLR signaling mediated by MyD88 is required for a protective innate immune response by neutrophils to *Citrobacter rodentium*. *J Immunol* **179:**566–577.

13. **Slutsker L, Ries A, Greene KD, Wells JG, Hutwagner L, Griffin PM.** 1997. *Escherichia coli* O157:H7 diarrhea in the United States: clinical and epidemiologic features. *Ann Intern Med* **126:**505–513.

14. **Elliott E, Li Z, Bell C, Stiel D, Buret AG, Wallace J, Brzuszczak I, O'Loughlin EV.** 1994. Modulation of host response to *Escherichia coli* O157:H7 infection by anti-CD18 antibody in rabbits. *Gastroenterology* **106:**1554–1561.

15. **Amulic B, Cazalet C, Hayes GL, Metzler KD, Zychlinsky A.** 2012. Neutrophil function: from mechanisms to disease. *Ann Rev Immunol* **30:**459–489.

16. **Brinkmann V, Reichard U, Goosmann C, Fauler B, Uhlemann Y, Weiss DS, Weinrauch Y, Zychlinsky A.** 2004. Neutrophil extracellular traps kill bacteria. *Science* **303:**1532–1535.

17. **Phillipson M, Kubes P.** 2011. The neutrophil in vascular inflammation. *Nat Med* **17:**1381–1390.

18. **Sadik CD, Kim ND, Luster AD.** 2011. Neutrophils cascading their way to inflammation. *Trends Immunol* **32:**452–460.

19. **Ley K, Laudanna C, Cybulsky MI, Nourshargh S.** 2007. Getting to the site of inflammation: the leukocyte adhesion cascade updated. *Nat Rev Immunol* **7:**678–689.

20. **Dahan S, Busuttil V, Imbert V, Peyron J-F, Rampal P, Czerucka D.** 2002. Enterohemorrhagic *Escherichia coli* infection induces interleukin-8 production via activation of mitogen-activated protein kinases and the transcription factors NF-κB and AP-1 in T84 cells. *Infect Immun* **70:**2304–2310.

21. **Savkovic S, Koutsouris A, Hecht G.** 1997. Activation of NF-kappaB in intestinal epithelial cells by enteropathogenic *Escherichia coli*. *Am J Physiol* **273:**1160–1167.

22. **Jung HC, Eckmann L, Yang SK, Panja A, Fierer J, Morzycka-Wroblewska E, Kagnoff MF.** 1995. A distinct array of proinflammatory cytokines is expressed in human colon epithelial cells in response to bacterial invasion. *J Clin Invest* **95:**55–65.

23. **McCormick BA, Colgan SP, Delp-Archer C, Miller SI, Madara JL.** 1993. *Salmonella typhimurium* attachment to human intestinal epithelial monolayers: transcellular signalling to subepithelial neutrophils. *J Cell Biol* **123:**895–907.

24. **Philpott DJ, Yamaoka S, Israel A, Sansonetti PJ.** 2000. Invasive *Shigella flexneri* activates NF-κB through a lipopolysaccharide-dependent innate intracellular response and leads to IL-8 expression in epithelial cells. *J Immunol* **165:**903–914.

25. **Mukaida N, Mahe Y, Matsushima K.** 1990. Cooperative interaction of nuclear factor-kappa B- and cis-regulatory enhancer binding protein-like factor binding elements in activating the interleukin-8 gene by pro-inflammatory cytokines. *J Biol Chem* **265:**21128–21133.

26. **Li Q, Verma IM.** 2002. NF-κB regulation in the immune system. *Nat Rev Immunol* **2:**725–734.

27. **Ghosh S, May MJ, Kopp EB.** 1998. NF-κB and Rel proteins: evolutionarily conserved mediators of immune responses. *Annu Rev Immunol* **16:**225–260.

28. **Verma IM, Stevenson JK, Schwarz EM, Van Antwerp D, Miyamoto S.** 1995. Rel/NF-kappa B/I kappa B family: intimate tales of association and dissociation. *Genes Dev* **9:**2723–2735.

29. **May MJ, Ghosh S.** 1998. Signal transduction through NF-κB. *Immunol Today* **19:**80–88.

30. **Schulze-Luehrmann J, Ghosh S.** 2006. Antigen-receptor signaling to nuclear factor κB. *Immunity* **25:**701–715.

31. **Perkins N.** 2007. Integrating cell-signaling pathways with NF-κB and IKK function. *Nat Rev Mol Cell Biol* **8:**49–62.

32. **Mukaida N, Okamoto S, Ishikawa Y, Matsushima K.** 1994. Molecular mechanism of interleukin-8 gene expression. *J Leuk Biol* **56:** 554–558.

33. **Czerucka D, Dahan S, Mograbi B, Rossi B, Rampal P.** 2001. Implication of mitogen-activated protein kinases in T84 cell responses to enteropathogenic *Esherichia coli* infection. *Infect Immun* **69:**1298–1305.

34. **de Grado M, Rosenberger CM, Gauthier A, Vallance BA, Finlay BB.** 2001. Enteropathogenic *Escherichia coli* infection induces expression of the early growth response factor by activating mitogen-activated protein kinase cascades in epithelial cells. *Infect Immun* **69:**6217–6224.

35. **Davis RJ.** 1993. The mitogen-activated protein kinase signal transduction pathway. *J Biol Chem* **268:**14553–14556.

36. **Davis RJ.** 2000. Signal transduction by the JNK group of MAP kinases. *Cell* **103:**239–252.

37. **Johnson GL, Lapadat R.** 2002. Mitogen-activated protein kinase pathways mediated by ERK, JNK, and p38 protein kinases. *Science* **298:**1911–1912.

38. **Thorpe CM, Hurley BP, Lincicome LL, Jacewicz MS, Keusch GT, Acheson DWK.** 1999. Shiga toxins stimulate secretion of interleukin-8 from intestinal epithelial cells. *Infect Immun* **67:**5985–5993.

39. **Thorpe CM, Smith WE, Hurley BP, Acheson DWK.** 2001. Shiga toxins induce, superinduce, and stabilize a variety of C-X-C chemokine mRNAs in intestinal epithelial cells, resulting in increased chemokine expression. *Infect Immun* **69:**6140–6147.

40. **Yamasaki C, Natori Y, Zeng XT, Ohmura M, Yamasaki S, Takeda Y.** 1999. Induction of cytokines in a human colon epithelial cell line by Shiga toxin 1 (Stx1) and Stx2 but not by non-toxic mutant Stx1 which lacks N-glycosidase activity. *FEBS Lett* **442:**231–234.

41. **Berin MC, Darfeuille-Michaud A, Egan LJ, Miyamoto Y, Kagnoff MF.** 2002. Role of EHEC O157:H7 virulence factors in the activation of intestinal epithelial cell NF-*κ*B and MAP kinase pathways and the upregulated expression of interleukin 8. *Cell Microbiol* **4:**635–648.

42. **Zhou X, Girón JA, Torres AG, Crawford JA, Negrete E, Vogel SN, Kaper JB.** 2003. Flagellin of enteropathogenic *Escherichia coli* stimulates interleukin-8 production in T84 cells. *Infect Immun* **71:**2120–2129.

43. **Elliott E, Li Z, Bell C, Stiel D, Buret A, Wallace J, Brzuszczak I, O'Loughlin E.** 1994. Modulation of host response to *Escherichia coli* O157:H7 infection by anti-CD18 antibody in rabbits. *Gastroenterology* **106:**1554–1561.

44. **Li Z, Bell C, Buret AG, Robins-Browne RM, Stiel D, O'Loughlin EV.** 1993. The effect of enterohemorrhagic *Escherichia coli* O157:H7 on intestinal structure and solute transport in rabbits. *Gastroenterology* **104:**467–474.

45. **Gewirtz AT, Navas TA, Lyons S, Godowski PJ, Madara JL.** 2001. Cutting edge: bacterial flagellin activates basolaterally expressed TLR5 to induce epithelial proinflammatory gene expression. *J Immunol* **167:**1882–1885.

46. **McNamara BP, Koutsouris A, O'Connell CB, Nougayrede JP, Donnenberg MS, Hecht G.** 2001. Translocated EspF protein from enteropathogenic *Escherichia coli* disrupts host intestinal barrier function. *J Clin Invest* **107:**621–629.

47. **Muza-Moons MM, Koutsouris A, Hecht G.** 2003. Disruption of cell polarity by enteropathogenic *Escherichia coli* enables basolateral membrane proteins to migrate apically and to potentiate physiological consequences. *Infect Immun* **71:** 7069–7078.

48. **Kinnebrew MA, Buffie CG, Diehl GE, Zenewicz LA, Leiner I, Hohl TM, Flavell RA, Littman DR, Pamer EG.** 2012. Interleukin 23 production by intestinal CD103(+)CD11b(+) dendritic cells in response to bacterial flagellin enhances mucosal innate immune defense. *Immunity* **36:**276–287.

49. **Phillips A, Frankel G.** 2000. Intimin-mediated tissue specificity in enteropathogenic *Escherichia coli* interaction with human intestinal organ cultures. *J Infect Dis* **181:**1496–1500.

50. **Cashman SB, Morgan JG.** 2009. Transcriptional analysis of Toll-like receptors expression in M cells. *Mol Immunol* **47:**365–372.

51. **Hurley BP, Thorpe CM, Acheson DWK.** 2001. Neutrophil translocation across intestinal epithelial cells is enhanced by neutrophil transmigration. *Infect Immun* **69:**6148–6155.

52. **Te Loo DM, IIinsbergh VW, Heuvell LP, Monnens LA.** 2001. Detection of verotoxin bound to circulationg polymorphonuclear leukocytes of patients with hemolytic uremic syndrome. *J Am Soc Nephrol* **12:**800–806.

53. **Stearns-Kurosawa DJ, Oh S-Y, Cherla RP, Lee M-S, Tesh VL, Papin J, Henderson J, Kurosawa S.** 2013. Distinct renal pathology and a chemotactic phenotype after enterohemorrhagic *Escherichia coli* Shiga toxins in non-human primate models of hemolytic uremic syndrome. *Am J Pathol* **182:**1227–1238.

54. **Ledesma MA, Ochoa SA, Cruz A, Rocha-Ramirez LM, Mas-Oliva J, Eslava CA, Giron JA, Xicohtencatl-Cortes J.** 2010. The hemorrhagic coli pilus (HCP) of *Escherichia coli* O157:H7 is an inducer of proinflammatory cytokine secretion in intestinal epithelial cells. *PLoS One* **5:** e12127.

55. **Luperchio SA, Schauer DB.** 2001. Molecular pathogenesis of *Citrobacter rodentium* and transmissible murine colonic hyperplasia. *Microb Infect* **3**:333–340.

56. **Ma C, Wickham ME, Guttman JA, Deng W, Walker J, Madsen KL.** 2006. *Citrobacter rodentium* infection causes both mitochondrial dysfunction and intestinal epithelial barrier disruption in vivo: role of mitochondrial associated protein (Map). *Cell Microbiol* **8**:1669–1686.

57. **Maaser C, Housley MP, Iimura M, Smith JR, Vallance BA, Finlay BB, Schreiber JR, Varki NM, Kagnoff MF, Eckmann L.** 2004. Clearance of *Citrobacter rodentium* requires B cells but not secretory immunoglobulin A (IgA) or IgM antibodies. *Infect Immun* **72**:3315–3324.

58. **Simmons CP, Clare S, Ghaem-Maghami M, Uren TK, Rankin J, Huett A, Goldin R, Lewis DJ, MacDonald TT, Strugnell RA, Frankel G, Dougan G.** 2003. Central role for B lymphocytes and CD4+ T cells in immunity to infection by the attaching and effacing pathogen *Citrobacter rodentium*. *Infect Immun* **71**:5077–5086.

59. **Wang Y, Xiang GS, Kourouma F, Umar S.** 2006. *Citrobacter rodentium*-induced NF-κB activation in hyperproliferating colonic epithelia: role of p65 (Ser536) phosphorylation. *Br J Pharmacol* **148**:814–824.

60. **Khan MA, Bouzari S, Ma C, Rosenberger CM, Bergstrom KSB, Gibson DL, Steiner TS, Vallance BA.** 2008. Flagellin-dependent and -independent inflammatory responses. *Infect Immun* **76**:1410–1422.

61. **Wei OL, Hillard A, Kalman D, Sherman MA.** 2005. Mast cells limit systemic bacterial dissemenation but not colitis in response to *Citrobacter rodentium*. *Infect Immun* **73**:1978–1985.

62. **Higgins LM, Frankel G, Douce G, Dougan G, MacDonald TT.** 1999. *Citrobacter rodentium* infection in mice elicits a mucosal Th1 cytokine response and lesions similar to those in murine inflammatory bowel disease. *Infect Immun* **67**:3031–3039.

63. **Geddes K, Rubino SJ, Magalhaes JG, Streutker C, Le Bourhis L, Cho JH, Robertson SJ, Kim CJ, Kaul R, Philpott DJ, Girardin SE.** 2011. Identification of an innate T helper type 17 response to intestinal bacterial pathogens. *Nat Med* **17**:837–844.

64. **Zheng Y, Valdez PA, Danilenko DM, Hu Y, Sa SM, Gong Q, Abbas AR, Modrusan Z, Ghilardi N, de Sauvage FJ, Ouyang W.** 2008. Interleukin-22 mediates early host defense against attaching and effacing bacterial pathogens. *Nat Med* **14**:282–289.

65. **Gibson DL, Ma A, Rosenberger CM, Bergstrom KSB, Valdez Y, Huang JT, Khan MA, Vallance BA.** 2008. Toll-like receptor 2 plays a critical role in maintaining mucosal integrity during *Citrobacter rodentium*-induced colitis. *Cell Microbiol* **10**:388–403.

66. **Eckmann L.** 2006. Animal models of inflammatory bowel disease: lessons from enteric infections. *Ann New York Acad Sci* **1072**:28–38.

67. **Long KZ, Rosado JL, Santos JI, Haas M, Al Mamun A, DuPont HL, Nanthakumar NN, Estrada-Garcia T.** 2010. Associations between mucosal innate and adaptive immune responses and resolution of diarrheal pathogen infections. *Infect Immun* **78**:1221–1228.

68. **Sarra M, Pallone F, Macdonald TT, Monteleone G.** 2010. IL-23/IL-17 axis in IBD. *Inflamm Bowel Dis* **16**:1808–1813.

69. **Gonçalves NS, Ghaem-Maghami M, Monteleone G, Frankel G, Dougan G, Lewis DJ, Simmons CP, MacDonald TT.** 2001. Critical role for tumour necrosis factor alpha in controlling the number of lumenal pathogenic bacteria and immunopathology in infectious colitis. *Infect Immun* **69**:6651–6659.

70. **Ramirez K, Huerta R, Oswald E, Garcia-Tovar C, Hernandez JM, Navarro-Garcia F.** 2005. Role of EspA and intimin in expression of proinflammatory cytokines from enterocytes and lymphocytes by rabbit enteropathogenic *Escherichia coli*-infected rabbits. *Infect Immun* **73**:103–113.

71. **Kollias G, Douni E, Kassiotis G, Kontoyiannis D.** 1999. On the role of tumor necrosis factor and receptors in models of multiorgan failure, rheumatoid arthritis, multiple sclerosis and inflammatory bowel disease. *Immunol Rev* **169**:175–194.

72. **Khan MA, Ma C, Knodler LA, Valdez Y, Rosenberger CM, Deng W, Finlay BB, Vallance BA.** 2006. Toll-like receptor 4 contributes to colitis development but not to host defense during *Citrobacter rodentium* infection in mice. *Infect Immun* **74**:2522–2536.

73. **Hawn TR, Smith KD, Aderem A, Skerrett SJ.** 2006. Myeloid differentiation primary response gene (88)- and toll-like receptor 2-deficient mice are susceptible to infection with aerosolized *Legionella pneumophila*. *J Infect Dis* **193**:1693–1702.

74. **Skerrett SJ, Liggitt HD, Hajjar AM, Wilson CB.** 2004. Cutting edge: myeloid differentiation factor 88 is essential for pulmonary host defense against *Pseudomonas aeruginosa* but not *Staphylococcus aureus*. *J Immunol* **172**:3377–3381.

75. **Watson RO, Novik V, Hofreuter D, Lara-Tejero M, Galan JE.** 2007. A MyD88-deficient mouse model reveals a role for Nramp1 in *Campylobacter jejuni* infection. *Infect Immun* **75**:1994–2003.

76. **Weiss DS, Takeda K, Akira S, Zychlinsky A, Moreno E.** 2005. MyD88, but not Toll-like

receptors 4 and 2, is required for efficient clearance of *Brucella abortus*. *Infect Immun* **73**:5137–5143.

77. **Akira S, Takeda K.** 2004. Toll-like receptor signalling. *Nat Rev Immunol* **4**:499–511.

78. **Lebeis SL, Powell KR, Merlin D, Sherman MA, Kalman D.** 2009. Interleukin-1 receptor signaling protects mice from lethal intestinal damage caused by the attaching and effacing pathogen *Citrobacter rodentium*. *Infect Immun* **77**:604–614.

79. **Liu Z, Zaki MH, Vogel P, Gurung P, Finlay BB, Deng W, Lamkanfi M, Kanneganti TD.** 2012. Role of inflammasomes in host defense against *Citrobacter rodentium* infection. *J Biol Chem* **287**:16955–16964.

80. **Hauf N, Charkraborty T.** 2003. Suppression of NF-kappaB activation and proinflammatory cytokine expression by Shiga toxin-producing *Escherichia coli*. *J Immunol* **170**:2074–2082.

81. **Ruchaud-Sparagano M, Maresca M, Kenny B.** 2007. Enteropathogenic *Escherichia coli* (EPEC) inactivate innate immune responses prior to compromising epithelial barrier function. *Cell Microbiol* **9**:1909–1921.

82. **Wong ARC, Pearson JS, Bright MD, Munera D, Robinson KS, Lee SF, Frankel G, Hartland EL.** 2011. Enteropathogenic and enterohaemorrhagic *Escherichia coli*: even more subversive elements. *Mol Microbiol* **80**:1420–1438.

83. **Gao X, Wan F, Mateo K, Callegari E, Wang D, Deng W, Puente J, Li F, Chaussee MS, Finlay BB, Lenardo MJ, Hardwidge PR.** 2009. Bacterial effector binding to ribosomal protein S3 subverts NF-κB function. *PLoS Pathog* **5**:1–18.

84. **Nadler C, Baruch K, Kobi S, Mills E, Haviv G, Farago M, Alkalay I, Bartfeld S, Meyer T, Ben-Neriah Y, Rosenshine I.** 2010. The type III secretion effector NleE inhibits NF-κB activation. *PLoS Pathog* **6**:1–11.

85. **Newton HJ, Pearson JS, Badea L, Kelly M, Lucas M, Holloway G, Wagstaff KM, Dunstone MA, Sloan J, Whisstock J, Kaper JB, Robins-Browne RM, Jans DA, Frankel G, Philips AD, Coulson BS, Hartland EL.** 2010. The type III effectors NleE and NleB from enteropathogenic *E. coli* and OspZ from *Shigella* block nuclear translocation of NF-κB p65. *PLoS Pathog* **6**:1–16.

86. **Royan SV, Jones RM, Koutsouris A, Roxas JL, Falzari K, Weflen AW, Kim A, Bellmeyer A, Turner JR, Neish AS, Rhee K-J, Viswanathan VK, Hecht GA.** 2010. Enteropathogenic *E. coli* non-LEE encoded effectors NleH1 and NleH2 attenuate NF-κB activation. *Mol Microbiol* **78**:1232–1245.

87. **Deng W, Puente JL, Gruenheid S, Li Y, Vallance BA, Vazquez A, Barba J, Ibarra JA, O'Donnell P, Metalnikov P, Ashman K, Lee S, Goode D, Pawson T, Finlay BB.** 2004. Dissecting virulence:

systematic and functional analyses of a pathogenicity island. *Proc Natl Acad Sci USA* **101**:3597–3602.

88. **Kelly M, Hart E, Mundy R, Marchès O, Wiles S, Badea L, Luck S, Tauschek M, Frankel G, Robins-Browne RM, Hartland EL.** 2006. Essential role of the type III secretion system effector NleB in colonization of mice by *Citrobacter rodentium*. *Infect Immun* **74**:2328–2337.

89. **Iguchi A, Thomson NR, Ogura Y, Saunders D, Ooka T, Henderson IR, Harris D, Asadulghani M, Kurokawa K, Dean P, Kenny B, Quail MA, Thurston S, Dougan G, Hayashi T, Parkhill J, Frankel G.** 2009. Complete genome sequence and comparative genome analysis of enteropathogenic *Escherichia coli* O127:H6 strain E2348/69. *J Bacteriol* **191**:347.

90. **Wickham ME, Lupp C, Vazquez A, Marscarenhas M, Coburn B, Coombes BK, Karmali MA, Puente JL, Deng W, Finlay BB.** 2007. *Citrobacter rodentium* virulence in mice associates with bacterial load and the type III effector NleE. *Microb Infect* **9**:400–407.

91. **Zurawski DV, Mumy KL, Badea L, Prentice JA, Hartland EL, McCormick BA, Maurelli AT.** 2008. The NleE/OspZ family of effector proteins is required for polymorphonuclear transepithelial migration, a characteristic shared by enteropathogenic *Escherichia coli* and *Shigella flexneri* infections. *Infect Immun* **76**:369–379.

92. **Bugarel M, Martin A, Fach P, Beutin L.** 2011. Virulence gene profiling of enterohemorrhagic (EHEC) and enteropathogenic (EPEC) *Escherichia coli* strains: a basis for molecular risk assessment of typical and atypical EPEC strains. *BMC Microbiol* **11**:142–152.

93. **Buvens G, Piérard D.** 2012. Virulence profiling and disease association of verotoxin-producing *Escherichia coli* O157 and non-O157 isolates in Belgium. *Foodborne Path Dis* **9**:1–6.

94. **Vossenkämper A, Marches O, Fairclough PD, Warnes G, Stagg AJ, Lindsay JO, Evans PC, Luong IA, Croft NM, Naik S, Frankel G, MacDonald TT.** 2010. Inhibition of NF-κB signaling in human dentritic cells by the enteropathogenic *Escherichia coli* protein NleE. *J Immunol* **185**:4118–4127.

95. **Zhang L, Ding X, Cui J, Xu H, Chen J, Gong Y-N, Hu Z, Zhou Y, Ge J, Lu Q, Liu L, Chen S, Shao F.** 2011. Cysteine methylation disrupts ubiquitin-chain sensing in NF-κB activation. *Nature* **481**:204–208.

96. **Baruch K, Gur-Arie L, Nadler C, Koby S, Yerushalmi G, Ben-Neriah Y, Yogev O, Shaulian E, Guttman C, Zarivach R, Rosenshine I.** 2011. Metalloprotease type III effectors that specifically cleave JNK and NF-κB. *EMBO J* **30**:221–231.

97. **Mühlen S, Ruchaud-Sparagano M-H, Kenny B.** 2011. Proteasome-independent degradation of canonical NFκB complex components by the NleC protein of pathogenic *Escherichia coli*. *J Biol Chem* **286:**5100–5107.

98. **Pearson JS, Riedmaier P, Marches O, Frankel G, Hartland EL.** 2011. A type III effector protease NleC from enteropathogenic *Escherichia coli* targets NF-κB for degradation. *Mol Microbiol* **80:**219–230.

99. **Sham HP, Shames SR, Croxen MA, Ma C, Chan JM, Khan MA, Wickham M, Deng W, Finlay BB, Vallance BA.** 2011. Attaching and effacing bacterial effector NleC suppresses epithelial inflammatory responses by inhibiting NF-κB and p38 mitogen-activated protein kinase activation. *Infect Immun* **79:**3552–3562.

100. **Perna NT, Plunkett G, Burland V, Mau B, Glasner JD, Rose DJ, Mayhew GF, Evans PS, Gregor J, Kirkpatrick HA, Posfai G, Hackett J, Klink S, Boutin A, Shao Y, Miller L, Grotbeck EJ, Davis NW, Lim A, Dimalanta ET, Potamousis KD, Apodaca J, Anantharaman TS, Lin J, Yen G, Schwartz DC, Welch RA, Blattner FR.** 2001. Genome sequence of enterohaemorrhagic *Escherichia coli* O157:H7. *Nature* **409:**529–533.

101. **Marchès O, Wiles S, Dziva F, La Ragione RM, Schüller S, Best A, Phillips AD, Hartland EL, Woodward MJ, Stevens MP, Frankel G.** 2005. Characterization of two non-locus of enterocyte effacement-encoded type III-translocated effectors, NleC and NleD, in attaching and effacing pathogens. *Infect Immun* **73:**8411–8417.

102. **Yen H, Ooka T, Iguchi A, Hayashi T, Sugimoto N, Tobe T.** 2010. NleC, a type III secretion protease, compromises NF-kappaB activation by targeting p65/RelA. *PLoS Pathog* **6:**e1001231.

103. **Jongeneel CV, Bouvier J, Bairoch A.** 1989. A unique signature identifies a family of zinc-dependent metallopeptidases. *FEBS Lett.* **242:**211–214.

104. **Gilmore TD, Wolenski FS.** 2012. NF-κB: where did it come from and why? *Immunol Rev* **246:**14–35.

105. **Hayden MS, Ghosh S.** 2008. Shared principles in NF-κB signaling. *Cell* **132:**344–362.

106. **Shames SR, Bhavsar AP, Croxen MA, Law RJ, Mak SHC, Deng W, Li Y, Bidshari R, de Hoog CL, Foster LJ, Finlay BB.** 2011. The pathogenic *Escherichia coli* type III secreted protease NleC degrades the host acetyltransferase p300. *Cell Microbiol* **13:**1542–1557.

107. **Karmali MA, Mascarenhas M, Shen S, Ziebell K, Johnson S, Reid-Smith R, Isaac-Renton J, Clark C, Rahn K, Kaper JB.** 2003. Association of genomic O island 122 of *Escherichia coli* EDL 933 with verocytotoxin-producing *Escherichia coli* seropathotypes that are linked to epidemic and/or serious disease. *J Clin Microbiol* **41:**4930–4940.

108. **Brown NF, Coombes BK, Bishop JL, Wickham ME, Lowden MJ, Gal-Mor O, Goode DL, Boyle EC, Sanderson KL, Finlay BB.** 2011. *Salmonella* phage ST64B encodes a member of the SseK/NleB effector family. *PLoS One* **6:**e17824.

109. **Gao X, Wang X, Pham TH, Feuerbacher LA, Lubos M-L, Huang M, Olsen R, Mushegian A, Slawson C, Hardwidge PR.** 2013. NleB, a bacterial effector with glycosyltransferase activity, targets GADPH function to inhibit NF-κB activation. *Cell Host Microb* **13:**87–99.

110. **Wan F, Anderson D, Barnitz R, Snow A, Bidere N, Zheng L, Hegde V, Lam L, Staudt L, Levens D, Deutsch W, Lenardo M.** 2007. Ribosomal protein S3: a KH domain subunit in NF-κB complexes that mediates selective gene regulation. *Cell* **131:**927–939.

111. **Kim DW, Lenzen G, Page AL, Legrain P, Sansonetti PJ, Parsot C.** 2005. The *Shigella flexneri* effector OspG interferes with innate immune responses by tagretting ubiquitin-conjugating enzymes. *Proc Natl Acad Sci USA* **102:**14046–14051.

112. **Wan F, Weaver A, Gao X, Bern M, Hardwidge PR, Lenardo MJ.** 2011. IKKβ phosphorylation regulates RPS3 nuclear translocation and NF-κB function during infection with *Escherichia coli* strain O157:H7. *Nat Immunol* **12:**335–343.

113. **Garcia-Angulo VA, Deng W, Thomas NA, Finlay BB, Puente JL.** 2008. Regulation of expression and secretion of NleH, a new non-locus of enterocyte effacement-encoded effector in *Citrobacter rodentium*. *J Bacteriol* **190:**2388–2399.

114. **Hemrajani C, Marches O, Wiles S, Girard F, Dennis A, Dziva F, Best A, Phillips AD, Berger CN, Mousnier A, Crepin VF, Kruidenier L, Woodward MJ, Stevens MP, La Ragione RM, MacDonald TT, Frankel G.** 2008. Role of NleH, a type III secreted effector from attaching and effacing pathogens, in colonization of the bovine, ovine, and murine gut. *Infect Immun* **76:**4804–4813.

115. **Pham TH, Gao X, Tsai K, Olsen R, Wan F, Hardwidge PR.** 2012. Functional differences and interactions between the *E. coli* type III secretion system effectors NleH1 and NleH2. *Infect Immun* **80:**2133–2140.

116. **Mundy R, Girard F, FitzGerald AJ, Frankel G.** 2006. Comparison of colonization dynamics and pathology of mice infected with enteropathogenic *Escherichia coli*, enterohaemorrhagic *E. coli* and *Citrobacter rodentium*. *FEMS Microbiol Lett* **265:**126–32.

117. **Hemrajani C, Berger CN, Robinson KS, Marchès O, Mousnier A, Frankel G.** 2010. NleH effectors interact with Bax inhibitor-1 to block apoptosis during enteropathogenic *Escherichia coli* infection. *Proc Natl Acad Sci USA* **107:**3129–3134.

118. **Ruchaud-Sparagano M-H, Mühlen S, Dean P, Kenny B.** 2011. The enteropathogenic *E. coli* (EPEC) Tir effector inhibits NF-κB activity by targeting TNFα receptor-associated factors. *PLoS Pathog* **7:**1–14.

119. **Yan D, Wang X, Luo L, Cao X, Ge B.** 2012. Inhibition of TLR signaling by a bacterial protein containing immunoreceptor tyrosine-based inhibitory motifs. *Nat Immunol* **13:**1063–1071.

120. **Barrow AD, Trowsdale J.** 2006. You say ITAM and I say ITIM, let's call the whole thing off: the ambiguity of immunoreceptor signaling. *Eur J Immunol* **36:**1646–1653.

121. **Daeron M, Jaeger S, Du Pasquier L, Vivier E.** 2008. Immunoreceptor tyrosine-based inhibition motifs: a quest in the past and future. *Immunol Rev* **224:**11–43.

122. **Nandan D, Lo R, Reiner NE.** 1999. Activation of phosphotyrosine phosphatase activity attenuates mitogen-activated protein kinase signaling and inhibits c-FOS and nitric oxide synthase expression in macrophages infected with *Leishmania donovani*. *Infect Immun* **67:**4055–4063.

123. **Raymond B, Crepin VF, Collins JW, Frankel G.** 2011. The WxxxE effector EspT triggers expression of immune mediators in an Erk/JNK and NF-kappaB-dependent manner. *Cell Microbiol* **13:**1881–1893.

124. **Keestra AM, Winter MG, Auburger JJ, Frassle SP, Xavier MN, Winter SE, Kim A, Poon V, Ravesloot MM, Waldenmaier JF, Tsolis RM, Eigenheer RA, Baumler AJ.** 2013. Manipulation of small Rho GTPases is a pathogen-induced process detected by NOD1. *Nature* **496:**233–237.

125. **Lathem WW, Grys TE, Witowski SE, Torres AG, Kaper JB, Tarr PI, Welch RA.** 2002. StcE, a metalloprotease secreted by *Escherichia coli* O157:H7, specifically cleaves C1 esterase inhibitor. *Mol Microbiol.* **45:**277–288.

126. **Hermiston ML, Xu Z, Weiss A.** 2003. CD45: a critical regulator of signaling thresholds in immune cells. *Annu Rev Immunol* **21:**107–137.

127. **Ostberg J, Barth RK, Frelinger JG.** 1998. The Roman god Janus: a paradigm for the function of CD43. *Immunol Today* **19:**546–550.

128. **Szabady RL, Lokuta MA, Walters KB, Huttenlocher A, Welch RA.** 2009. Modulation of neutrophil function by a secreted mucinase of *Escherichia coli* O157:H7. *PLoS Pathog* **5:**e1000320.

129. **Forsyth KD, Simpson AC, Fitzpatrick MM, Barratt TM, Levinsky RJ.** 1989. Neutrophil-mediated endothelial injury in haemolytic uremic syndrome. *Lancet* **2:**411–414.

New Therapeutic Developments against Shiga Toxin-Producing Escherichia coli

17

ANGELA R. MELTON-CELSA[1] and ALISON D. O'BRIEN[1]

BACKGROUND

Shiga toxin (Stx)-producing *Escherichia coli* (STEC) colonizes the intestine and causes hemorrhagic colitis. STEC encodes a variety of colonization factors, but a significant subset of STEC, the enterohemorrhagic *E. coli* (EHEC) strains, have the locus of enterocyte effacement (LEE), the products of which allow the bacteria to intimately adhere to and form attaching and effacing lesions on intestinal tissue. The O157:H7 strains, which are responsible for the majority of large outbreaks due to STEC infection, are members of the EHEC group. All STEC strains make one or more Stxs; these pathogens may produce two immunologically distinct but highly similar Stxs, Stx1 and Stx2. These toxins are briefly described in the section on therapeutics targeted to the Stxs.

Some individuals infected with STEC manifest a serious sequela called hemolytic uremic syndrome (HUS), a thrombotic microangiopathy defined by the presence of hemolytic anemia, thrombocytopenia, and renal failure. The initial insult leading to the development of the thrombotic microangiopathy is damage by the Stx(s) to vascular endothelial cells that express the toxin

[1]Department of Microbiology and Immunology, Uniformed Services University of the Health Sciences, Bethesda, MD 20814.

Enterohemorrhagic Escherichia coli *and Other Shiga Toxin-Producing* E. coli
Edited by Vanessa Sperandio and Carolyn J. Hovde
© 2015 American Society for Microbiology, Washington, DC
doi:10.1128/microbiolspec.EHEC-0013-2013

receptor, globotriaosylceramide (Gb3). The Stx-mediated injury to endothelial cells initiates a cascade of events that lead to the activation of platelets and the formation of thrombi in the small vessels of the kidney and sometimes the central nervous system. Prevention of HUS is important as HUS can lead to death or long-term consequences such as hypertension and renal disease (1–3). Because STEC strains are intestinal bacterial pathogens, the inclination by physicians is to treat with antibiotics to eliminate the organisms from the gut. However, several studies show that treatment of STEC-infected patients with certain antibiotics may lead to an increase in HUS (antibiotic use is discussed further in the next section). Therefore, the focus for therapeutics against STEC and HUS has been to find (i) compounds that act at the level of the bacterium but do not cause an increase in Stx production; (ii) receptor mimics or other molecules that alter trafficking of the toxin or the cellular response to the toxin; (iii) antibodies directed against the Stxs; or (iv) therapies to prevent or treat the HUS disease process. The therapies discussed in this article are listed in Table 1.

THE DIFFICULTIES WITH ANTIBIOTIC USE FOR STEC INFECTIONS

In the United States, antibiotics are not recommended for treatment of STEC infections because of the increased risk for the development of HUS (4, 5). However, even though antibiotics are contraindicated for those with STEC infection, a recent study at FoodNet sites found that antibiotics are commonly used in those with proven O157 infection (6). The issue of the use of antibiotics to treat STEC infections is confounded by a number of factors: (i) some, but not all, antibiotics at sublethal doses increase the expression of Stx by inducing the lysogenic phage that encodes the toxin genes (see diagram in Fig. 1) (7); (ii) treatment of STEC with a lethal dose of antibiotics may cause release of a large bolus

of toxin at one time as the bacteria die—a result that would be bad for a patient; (iii) in some STEC strains, the Stxs are either chromosomally encoded or are associated with defective bacteriophages, and thus antibiotic treatment may not alter Stx expression; (iv) antibiotics may be used only in the most ill patients, a fact that may skew the apparent risk for HUS development; and (v) published studies on the risk of HUS associated with antibiotic treatment used different treatments administered at different stages in the disease process. In addition, although several studies demonstrate an increased risk for HUS after treatment with antibiotics (8, 9), others do not find an increased risk (10). Conversely, several studies indicate that the use of β-lactam antibiotics, particularly in the first 2 to 3 days of illness, and perhaps associated with age <13 years, is correlated with increased HUS risk (11). In contrast, many people were treated with fosfomycin during a large O157:H7 outbreak in Japan without an apparent increase in HUS, and perhaps even a protective effect if the antibiotic was given within the first 2 days of illness (12). However, the low overall HUS rate in the Sakai outbreak (13) suggests that the causative strain may (fortunately) have had reduced virulence compared with O157 strains from other outbreaks, and, as such, would make that outbreak a poor platform on which to make generalized decisions about treatment with antibiotics.

During the O104:H4 outbreak in Germany in 2011 caused by an unusual enteroaggregative *E. coli* (EAEC) strain that makes Stx2, antibiotics were used in many patients, some of whom were given up to three antibiotics at various times after disease onset (14). Because of the vastly different protocols used to treat patients during the outbreak in Germany, it is difficult to make generalized conclusions about antibiotic use based on that outbreak; however, overall there was not definitive evidence of a benefit due to antibiotic treatment. Although one surprising study asserted ciprofloxacin treatment reduced the HUS incidence in patients, the number of treated patients was small (n = 5), and in two of those patients HUS did develop (15). Furthermore, in several patients from the O104:H4 outbreak HUS was reported to develop after ciprofloxacin or metronidazole treatment (16, 17). We consider the use of ciprofloxacin to be especially dangerous because of the evidence that ciprofloxacin increases Stx expression in the EAEC O104:H4 strain (18) and STEC O157 strains. Finally, one unique aspect of the outbreak in Germany was the use of azithromycin to reduce shedding of O104:H4 in patients who appeared to be long-term carriers (19).

Animal studies also give contradictory information about the use of antibiotics for the treatment of STEC infections. In two studies fosfomycin reduced colonization and mortality from STEC infection in mice (20, 21), but another study did not show a protective effect by that antibiotic (22). In a mouse model in which the mice are starved for protein calories, the use of trimethoprim-sulfamethoxazole was detrimental when given between days 3 and 5 post infection, whereas the same antibiotic as well as ampicillin and fosfomycin were protective when given anywhere from days 1 to 5 after infection (23). Nevertheless, we should note that the use of ampicillin to treat an infection with an organism (*Shigella dysenteriae* or the Stx2+ EAEC O104:H4, respectively) resistant to that antibiotic appears to be detrimental to humans and mice (24, 25).

Overall, the current data do not support the use of antibiotics for the treatment of STEC infection in humans, and the majority of the evidence indicates that antibiotic treatment is potentially detrimental. The empirical use of antibiotics could be dangerous, since some of these treatments could increase the risk of HUS. Furthermore, since there is no clear demonstrable benefit to the use of antibiotics in STEC-infected patients, and because antibiotic use itself may pose possible risk to the patient (allergy or other risks), we believe antibiotics should not be used to treat STEC infections.

COMPOUNDS DIRECTED TOWARD STEC

Pyocin

A novel strategy to kill O157 strains based on a modified pyocin (a bactericidal protein, similar to bacteriophage tail fibers, made by *Pseudomonas aeruginosa* that usually targets other *P. aeruginosa* cells) that consists of the R2 tail fiber fused to a phage spike protein specific for O157 was found to kill O157 strains without increasing expression of Stx (26). Subsequent tests demonstrated that a slightly modified version of the pyocin, AvR2-V10.3, given 3 h after infection and once daily for the next 2 days to infant rabbits infected with O157:H7 strain EDL933 reduced colonization and prevented diarrhea at the highest dose of pyocin administered (27). When the modified pyocin was given to EDL933-infected infant rabbits that exhibited diarrhea, the amount of loose stool decreased relative to the control animals given buffer alone and the numbers of EDL933 organisms decreased in the intestines and stools. One problem with the pyocin approach is that it is specific for serogroup, so other pyocins would have to be developed for non-O157 STEC strains.

A Small Molecule and a Divalent Cation

Another alternative strategy to antibiotics directed toward the bacterium uses a small molecule, LED209, which inhibits the activity of the QseC sensor that is involved in regulating some of the proteins involved in EHEC adherence and, indirectly, Stx2 expression. Although LED209 inhibited the formation of attaching and effacing lesions by EHEC on HeLA cells, the compound was not effective in reducing colonization or disease in infant rabbits (28). The reason for LED209 failing to protect in the rabbits may be that the concentration of the compound was not high enough at the site of EHEC colonization in the gut.

Similar to LED209, treatment of STEC infection with the divalent cation zinc reduces

TABLE 1 Therapies directed against STEC, the Stx receptor, Stx function, or the cellular response to Stx

Target[a]	Therapeutic	Function	Model used to test function or efficacy	Reference(s)
STEC or STEC pathogenesis	Pyocin (AvR2-V10.3)	Kills O157	In vitro growth; infant rabbits	26, 27
	LED209	Reduces attaching and effacing lesions on HeLa cells but did not reduce colonization or disease in rabbits	HeLa cells; infant rabbits	28
	Zinc	Reduces stx transcription; reduces adherence	HeLa cells; rabbit ileal loops (using rabbit enteropathogenic E. coli transduced with phage carrying stx_1 or stx_2)	29
Receptor (Gb3) synthesis inhibitor	C-9	Receptor generation; glucosylceramide synthase	HRTEC; Stx2-intoxicated rats	36, 37
Receptor analogs (B)	SYNSORB Pk	Receptor analog	Phase I and II trials	39, 41
	STARFISH	Receptor analog	Vero cells; Stx1-intoxicated mice (does not protect Stx2-intoxicated mice)	42, 43
	Daisy	Receptor analog	Vero cells; Stx1- and Stx2-intoxicated mice; B2F1-infected, streptomycin-treated mice	43
	SUPER TWIG (1)6	Receptor analog	Vero cells; Stx1- and Stx2-intoxicated mice; O157-infected, protein-calorie-deficient mice	44
	Gb3 polymers	Receptor analog	Mice	45
	TVP (also known as Ac-PPP-tet)	Receptor analog, binds Stx2 but not Stx1	Vero cells; O157-infected, protein-calorie-deficient mice, rabbit ileal loops; baboon	46, 49, 50
	MMA-tet	Receptor analog	Vero cells; O157-infected, protein-calorie-deficient mice	51
	Probiotic that displays Galα1-4Galβ1-4Glc-	Receptor analog	Vero cells; streptomycin-treated mice infected with an Stx2 or Stx2dact producer	47, 52–54
	HuSAP	Binds Stx2 but not Stx1	Stx1- or Stx2-intoxicated mice (only protects Stx2-intoxicated animals)	57–59

Antitoxin antibodies (B, U, T, or E)	Polyclonal anti-Stx2	Neutralize Stx2	Gnotobiotic pigs	62
	caStx1 (human/mouse chimeric version of 13C4)	Block Stx1 binding (B subunit)	Vero cells; Stx1-intoxicated mice; phase I and II trials	64, 68, 70
	caStx2 (anti-A subunit, human/mouse chimeric of 11E10)	Alters intracellular trafficking and inhibits enzymatic function	Vero cells; streptomycin-treated mice; phase I and II trials	63, 68–70
	Anti-Stx1 (5-5B)	Neutralize Stx1 (B subunit)	Ramos cells	71
	Urtoxazumab (also called TMA-15; humanized version of VTm1.1)	Neutralize Stx2 (B subunit)	Human renal adenocarcinoma cells; Stx2-intoxicated mice; B2F1-infected streptomycin-treated mice, phase I trial	73–75
	Anti-Stx1 monoclonals	Neutralize Stx1 (most are anti-B subunit)	HeLa cells; Stx1-intoxicated mice	76
	Anti-Stx2 (5C12)	Neutralize Stx2 (A subunit)	HeLa cells; Stx2-intoxicated mice; B2F1-infected streptomycin-treated mice; O157-infected gnotobiotic piglets	77, 78
Toxin transport (U or T)	Exo2	Blocks transport at the level of the early endosome/trans-Golgi interface	Vero cells	82, 83
	Retro-2^{cycl}	Toxin transport	HeLa cells	85, 86
	Chloroquine	Toxin transport	HEp-2 cells	87
	Small molecules	Toxin transport	HeLa cells	88
	Eeyarestatin 1	Intracellular trafficking	HeLa cells	90
	Nitrobenzyl-thioinosine	Stx1 trafficking (retention in early endosomes)	Human renal cortical epithelial cells	91
	Manganese	Protects HeLa cells and BALB/c mice from Stx1-mediated lethality; blocks StxB1 trafficking; does not protect against Stx2	HeLa cells; Stx1-intoxicated BALB/c mice (does not protect Stx2-intoxicated mice)	92, 93
Toxin processing (N)	Furin inhibitor	Cleavage of A subunit to A_1 and A_2	HEp-2 cells	94
Toxin function (E)	Small molecule	Enzymatic activity	Cell-free reporter assay	89
Cellular or host response to toxin (ribotoxic stress response or apoptosis)	Imatinib	Ribotoxic stress response inhibitor	HCT-8 cells; infant rabbit	98
	Ouabain	Apoptosis inhibition	Rat proximal tubule cells (RPTC)	99
Complement factor C5	Eculizumab	Anticomplement factor C5	STEC-infected patients	101, 105, 106

[a]Step in Fig. 2 at which therapeutic acts.

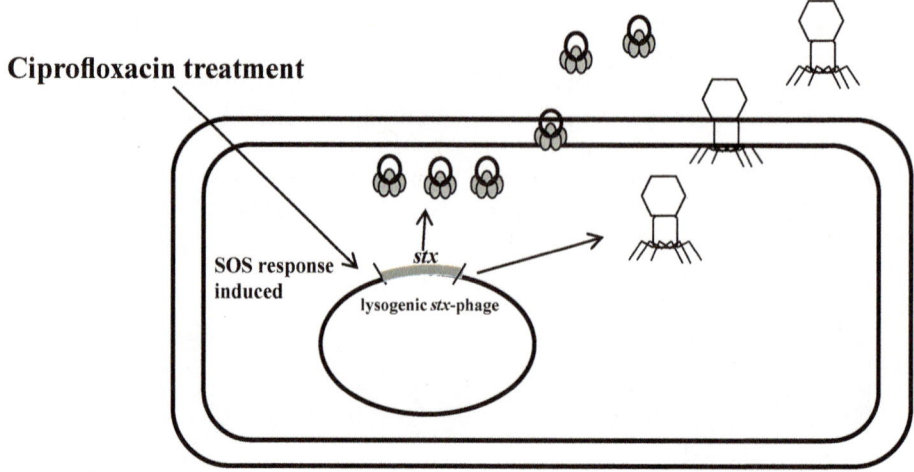

FIGURE 1 Induction of the *stx*-encoding phage by antibiotics such as ciprofloxacin. Subinhibitory concentrations of some antibiotics induce the lysogenic *stx* phage to enter the lytic cycle, and as consequence, *stx*-bacteriophage are made and released. Furthermore, 10- to 100-fold more toxin is made and released from the cell. doi:10.1128/microbiolspec.EHEC-0013-2013.f1

adherence to a HeLa cell monolayer, though through a different mechanism, perhaps due to decreased expression of *E. coli*-secreted proteins EspA and EspB (29). In addition to a reduction in adherence to cells, zinc treatment also appears to decrease expression of the *stx* genes. The potential therapeutic effect of zinc was tested in a rabbit ileal loop model, in which rabbit enteropathogenic *E. coli* strains transduced with a bacteriophage that encodes *stx₁* or *stx₂* were inoculated into the loop in the presence or absence of zinc. The presence of 1 mM of zinc reduced the amount of toxin found in the loops, decreased adherence, and lessened histological damage. The next avenue to explore for zinc is to determine if the cation could prevent disease or mortality in an oral infection model.

THERAPEUTICS THAT INTERFERE WITH TOXIN BINDING, UPTAKE, TRAFFICKING, OR FUNCTION

Stx1 and Stx2 are the key STEC virulence factors that lead to the development of HUS. Therefore, therapeutics that impede toxin binding, uptake, trafficking, or function are strong contenders for treatment of STEC infections. Stx1 and Stx2 share about 56% homology at the amino acid level but are immunologically distinct. Structurally, the toxins consist of five identical B subunits and a single A subunit. The B pentamer binds to the cellular receptor, Gb3. The A subunit contains the enzymatic function of the toxin and removes an adenine residue from the 28S rRNA, an action that destroys ribosome function. After binding to Gb3, the toxin receptor complex is taken up by clathrin-dependent and -independent mechanisms. The toxin then traffics in a retrograde direction from the endosome to the Golgi apparatus to the endoplasmic reticulum (ER). The A subunit of the toxin can be nicked within the Golgi body such that the enzymatic function (within A_1) is separated by a disulfide bond from the A_2 peptide that links A_1 to the B pentamer. The nicked toxin traffics next to the ER where the disulfide bond between A_1 and A_2 is reduced. The A_1 subunit then enters the cytoplasm and targets the ribosome. The damage to the ribosome halts protein synthesis and can lead to a ribotoxic stress response and apoptosis (see review in reference 30). A model of Stx trafficking is shown in Fig. 2.

Stx1 and Stx2 are the major toxin types produced by STEC that cause human disease. There are subtypes of both Stx1 and Stx2, however, and toxin nomenclature was recently updated so that the prototype Stxs are now designated Stx1a and Stx2a. The less specific Stx1 and Stx2 designations are used when the toxin subtype is unknown (31). We use the more general designations for the toxins elsewhere in this review for simplicity. Besides the prototypic Stx2a, two Stx2 subtypes are associated with HUS, Stx2c and Stx2d (32, 33). Stx2c and Stx2d have two amino acid differences in the B subunit as compared to Stx2a, and those changes are responsible for reduced cytotoxicity on Vero cells, though Stx2d is just as toxic to mice as

Stx2a (34). Stx2d has two amino acid differences from Stx2a and Stx2c in the A subunit, and those two changes contribute to the capacity of the toxin to exhibit increased toxicity when treated with elastase from intestinal mucus (35), a phenotype indicated with the designation Stx2dact.

Receptor Synthesis Inhibitor: C-9

One step at which to stop Stx is at the level of the toxin receptor. Various laboratories have tried to protect cells or animals from Stxs with compounds that prevent Gb3 synthesis or that bind to the toxin to prevent toxin-receptor interaction. For example, a molecule called C-9 was used in human renal tubular epithelial cells (HRTEC) to prevent the conversion of ceramide to glucosylceramide, an early step

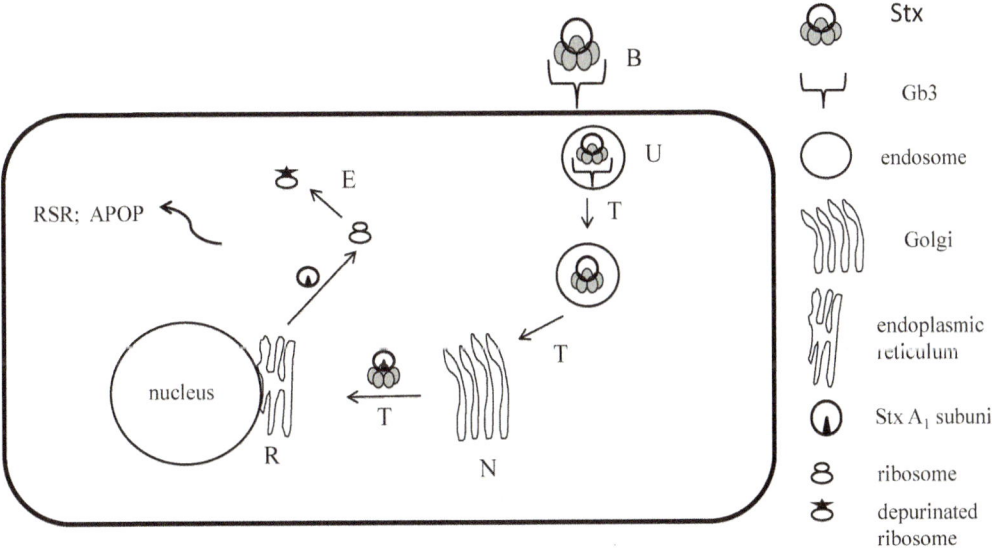

FIGURE 2 Cellular trafficking of Stx and points in the pathway where therapeutics function. Therapeutics can interfere with Stx action at several points as it traffics into and through the cell. The steps in toxin trafficking are briefly diagramed above and outlined here. The toxin first binds (B) to the receptor Gb3. The toxin/Gb3 complex is taken up (U) by both clathrin-dependent and -independent mechanisms, and then traffics (T) from the early endosome to the late endosome to the trans-Golgi apparatus. Within the Golgi the toxin A subunit is nicked (N), but the toxin remains intact due to a disulfide bond between the A_1 and A_2 subunits. The nicked toxin continues to traffic (T) along the retrograde pathway to the endoplasmic reticulum. The disulfide bond in the A subunit is reduced (R) within the endoplasmic reticulum and the A_1 subunit enters the cytoplasm where it exerts its enzymatic (E) attack and depurinates the ribosome. The action of the toxin within the cell can lead to a ribotoxic stress response (RSR) and apoptosis (APOP). doi:10.1128/microbiolspec.EHEC-0013-2013.f2

in Gb3 synthesis. HRTEC pretreated with C-9 for 24 h exhibited reduced levels of Gb3 and decreased sensitivity to Stx2 (36). In rats treated with C-9 for 2 days before intoxication and for 4 days post intoxication, about 50% protection from injection with bacterial supernatants that contained Stx2 was observed (37). Treated rats also showed lower rises in serum creatinine and urea and reduced renal tubular injury.

Receptor Analogs

The first drug tested in humans intended for the treatment of HUS was SYNSORB Pk, a silicon dioxide compound that contains the trisaccharide component of Gb3 (38, 39). The theory behind the use of SYNSORB Pk is that the compound would bind up free Stx(s) within the intestines of infected patients and prevent that toxin from binding to the functional receptor so that the toxin could not act either locally or systemically. SYNSORB Pk was shown to neutralize both Stx1 and Stx2 on human renal adenocarcinoma cells (40). Although SYNSORB Pk was tolerated well in the phase I trial (41), the double-blind placebo-controlled trial suggested no difference between treated and placebo groups (39). The reason for the lack of efficacy by SYNSORB Pk in the latter study may be that the patients were enrolled after HUS diagnosis, whereas the best time to neutralize toxin to prevent HUS is most likely before development of this serious sequela.

A number of other Stx receptor analogs in addition to SYNSORB Pk have been developed, such as STARFISH (42), Daisy (43), SUPER TWIG (1)6 (44), Gb3 polymers (45), and Ac-PPPtet (46), as well as a probiotic that displays an Stx binder on its surface (47). STARFISH is a five-"armed" molecule with two receptor mimics at the end of each arm; the compound appears to bind two B pentamers at the same time and neutralizes both Stx1 and Stx2 in vitro. However, STARFISH has a higher avidity for Stx1 than Stx2 and was only able to protect mice injected with

Stx1 but not Stx2 (43). Because of the lack of protective efficacy by STARFISH in the Stx2-injection model, a similar but modified divalent trisaccharide inhibitor called Daisy was synthesized (43). Daisy protects Vero cells and mice from both Stx1 and Stx2, and in a streptomycin-treated mouse model, prevented death due to infection with O91:H21 STEC strain B2F1 in 50% of the animals (43). The SUPER TWIG (1)6 identified by Nishikawa's group carries six trisaccharides and blocks the binding of both Stx1 and Stx2 to Vero cells, protects mice from Stx2 when administered together with the toxin, and prevents death of O157:H7-infected mice in a protein-calorie-deficient animal model when given intravenously after infection (44). However, one point to note is that SUPER TWIG (1)12, developed in the latter study, neutralized as well as SUPER TWIG (1)6 in vitro but did not protect in vivo, a finding that demonstrates the importance of testing potential therapeutics in an animal model system. To find a receptor analog that could be given orally, the Nishikawa group developed Gb3 polymers that bind to the Stxs with higher affinity than SUPER TWIG (1)6, and those polymers, when given by gavage twice daily in the protein-calorie-deficient mouse model on days 3 to 5 after infection, protect mice from death (45). The strong neutralization effect observed with SUPER TWIG and the Gb3 polymers is due to the fact that the B pentamer of the Stxs contains multiple Gb3-binding sites, so receptor mimics that display multiple copies of the Gb3 trisaccharide bind the toxin tightly. A similar approach as described above was used by another group to display the receptor sugar moiety on chitosan, and they similarly found that the analog neutralizes in vitro and in vivo (48).

According to Nishikawa, however, a drawback to the multiple trisaccharide display approach is the complexity of synthesis (46, 49); therefore, a tetravalent peptide library was screened to find molecules that bind the Stx2 pentamer (49). In that latter screen, tetravalent peptides were identified that form a complex with Stx2 and neutralize the toxin on Vero cells but do not prevent the toxin/peptide moiety from binding to the cell. Rather, the peptide-bound Stx2 was unable to reach the ER. Furthermore, O157-infected protein-calorie-deficient mice gavaged after infection with the peptides were protected from death; the protection appeared to be dose dependent and treatment had to be started by day 3 post infection. Further testing of the optimized tetrapeptide, Ac-PPP-tet (later renamed TVP), demonstrated that the compound inhibits Stx2-mediated fluid accumulation in rabbit ileal loops (46). However, the tetravalent peptide only protects when administered orally and not intravenously in the protein-calorie-deficient mouse model, so TVP efficacy was tested in the baboon model of intoxication (50). Baboons given Stx2 and TVP simultaneously were protected from death, renal injury, and thrombocytopenia, but not anemia. TVP also rescued 75% of animals that received the drug 24 h post intoxication and a supplemental dose on days 2, 3, and 4. Untreated but intoxicated baboons died by day 6. The authors of that study suggest that the advantage of TVP compared to other Stx binders is that the compound is cell-permeable. However, it is unclear how TVP would rescue intoxicated cells since the proposed mechanism for action is for the Stx2-TVP complex to traffic differently within the cell than Stx2 alone, and indeed, in vitro, cells are only protected if the TVP is added to cells at the same time as Stx2. In a recent study, the Nishikawa group identified another tetravalent peptide, MMA-tet, that inhibits both Stx1 and Stx2 on Vero cells when it is added at the same time as the toxins, and when administered orally, protects protein-calorie-deficient mice infected with STEC strain N-9 (51). The mechanism whereby MMA-tet protects cells and animals is unclear: MMA-tet forms a complex with the Stx1 B subunit but does not appear to alter binding or trafficking of the MMA-tet–StxB1 through the cell, at least as far as the ER. However, MMA-tet pretreatment of Vero cells did prevent protein synthesis inhibition by Stx1, a finding that may indicate that MMA-tet somehow prevents the toxin A_1 subunit from reaching the cytoplasm from the ER.

Finally, a novel modification of the receptor analog approach to bind up Stx specifically within the intestine was designed by Paton's group: the core trisaccharide from Gb3, Galα1-4Galβ1-4Glc-, was displayed on the lipopolysaccharide (LPS) structure of a commensal *E. coli* to create a probiotic (52). The constructed strain, CWG308:pJCP-Gb3, neutralizes Stx1, Stx2, Stx2c, and Stx2d on Vero cells. In addition, streptomycin-treated mice are protected from infection with either an Stx2- or Stx2d-producer when fed the Galα1-4Galβ1-4Glc-probiotic twice daily. What was not clear in that study was whether the mice would continue to do well once the probiotic was discontinued. However, the probiotic with a receptor-analog approach might be possible as long as the infecting strain did not persist for long-term colonization. Additionally, it may be necessary to monitor shedding of the infecting strain over time. CWG308:pJCP-Gb3 was found to be effective even when killed prior to administration; however, the formaldehyde-treated probiotic had to be administered three times daily for complete protection (53). To develop a probiotic that might be able to be used in humans, a K-12 strain that expressed the same Galα1-4Galβ1-4Glc- epitope on its LPS was created (54). In that strain, additional mutations were incorporated so that no antibiotic resistance markers were needed for plasmid maintenance. The K-12 probiotic that displays Galα1-4Galβ1-4Glc- neutralizes both in vitro and in vivo (54).

Human Serum Amyloid Component P (HuSAP)

Reports that there is an Stx2-neutralizing component in normal human serum that was not immunoglobulin were published in 1993 (55, 56). Later, HuSAP was identified as the protein from plasma that can neutralize Stx2, but not Stx1 (57). HuSAP administered intravenously protects BALB/c mice given Stx2 1 h later (58). Another group showed similar results, though they administered the HuSAP twice daily intraperitoneally and the Stx2 by the subcutaneous route (59). This group then challenged transgenic C57Bl/6 mice that express HuSAP with two 50% lethal doses of Stx2. The HuSAP-transgenic mice survived nearly twice as long as the control mice in that study. The reason that the HuSAP-injected BALB/c mice survived longer than the HuSAP-transgenic mice was hypothesized to be because the circulating levels of HuSAP were higher in the BALB/c mice injected with HuSAP than in the transgenic animals. If, however, the HuSAP-transgenic mice were injected with a combination of LPS and Stx2, the animals were not protected, even though the LPS did not prevent HuSAP-Stx2 interaction or the neutralization by HuSAP of Stx2 for Vero cells in vitro (60). Exactly how HuSAP binds Stx2 is not clear; Marcato et al. reported that the interaction requires both toxin A and B subunits and cannot be competed with Daisy (61). The latter finding suggests that HuSAP does not bind within the Gb3-binding sites on the B pentamer.

ANTIBODIES TO THE Stxs

Because the Stxs are the primary factors responsible for the development of HUS, neutralization of the toxins is a critical therapeutic approach for STEC infections. Polyclonal antiserum against Stx2 protects gnotobiotic piglets from infection by O157:H7 strain 86-24 (62). Furthermore, monoclonal antibodies to the toxins are protective in animal models, and two groups have taken such therapeutics into phase II safety trials, as detailed below.

Monoclonal antibodies specific for the Stx1 B subunit (13C4) and the Stx2 A subunit (11E10) were developed in the O'Brien laboratory in the mid to late 1980s (63, 64). Those antibodies neutralize the toxins on Vero cells (65, 66), and the Stx2 monoclonal antibody is

protective in mice (67). The epitopes for both antibodies have been mapped (65, 66); of note, although 11E10 neutralizes Stx2 in a cell-free protein synthesis assay, the antibody also alters the intracellular localization of the toxin such that the toxin-antibody complex does not reach the cytoplasm (65). Both 11E10 and 13C4 were transformed into human/mouse chimeras by genetic techniques. The "humanized" versions of anti-Stx1 and anti-Stx2 neutralize the toxins on Vero cells and in either a mouse intoxication model (Stx1) or the streptomycin-treated mouse model of infection with strain B2F1 (68). The humanized versions of the antibodies (cαStx1 for 13C4 and cαStx2 for 11E10) were evaluated in phase I (69, 70) and II trials. The preliminary results of the phase II trial indicate that the antibodies were safe and well tolerated in sick children infected with STEC.

Takeda's group developed monoclonal antibodies that neutralize Stx1 or Stx2 by preventing receptor binding (71, 72). The anti-Stx2 monoclonal was later humanized and named TMA-15. TMA-15 demonstrates Vero cell neutralization and protective efficacy in animal models (73, 74). Antibody TMA-15 was renamed urtoxazumab and evaluated in safety trials in healthy adults and STEC-infected children (75). Urtoxazumab was well tolerated in STEC-infected children, but no efficacy data are yet published.

Finally, Tzipori's group generated fully humanized neutralizing monoclonal anti-Stx1 and -Stx2 antibodies in transgenic mice (76, 77). The anti-Stx2 monoclonal antibody, 5C12, was developed further, and shown to protect piglets and streptomycin-treated mice from STEC strains that produce Stx2 or Stx2dact, respectively (77, 78). Similarly to 11E10, antibody 5C12 neutralizes Stx2 by altering its intracellular trafficking pattern and also has the capacity to prevent protein synthesis inhibition by Stx2 in a cell-free assay (79, 80). The Tzipori group also evaluated isotype variants and Fab and F(ab')$_2$ fragments of 5C12 in in vitro and in vivo models of efficacy and found that all of the isotype variants demonstrate in vitro and in vivo neutralization but the Fab and F(ab')$_2$ were only efficacious in the in vitro assay (81). One point of note from the latter study is that the IgG4 variant of 5C12 was the least protective of the isotype variants in vitro, but was as protective in vivo as the best in vitro neutralizer (the IgG3 variant). These latter results indicate once again that in vitro and in vivo neutralization data do not always correlate.

Nonantibody Inhibitors of Toxin Trafficking or the Cellular Response to Stx

Brefeldin A (BFA), a fungal toxin that disrupts the Golgi apparatus but also causes tabulation of early endosomes, is a known inhibitor of Stx transport; in fact, the protective effect of BFA against the Stxs helped define the pathway by which Stx is transported through the cell. Because BFA is a global inhibitor of protein transport within the cell, it is not included in Table 1. Another Golgi disruptor, Exo2, a small molecule that does not inhibit cholera toxin trafficking, prevents Stx from reaching the Golgi apparatus (82). Unfortunately, cell disruptors such as BFA and Exo2 are toxic to cells. Another drawback to molecules such as BFA and Exo2 is that treated cells recover over time, such that incubation of the cells with the inhibitor and the toxin for longer periods reduces the effectiveness of the therapeutic. Exo2 was recently derivatized to generate an inhibitor with lower toxicity that still interferes with Stx trafficking (83). Another small molecule, Retro-2, blocks Stx from reaching the *trans*-Golgi network (84, 85) and is effective against ricin, a toxin that traffics in the same manner as the Stxs. Recently, a derivative of Retro-2, Retro-2^{cycl}, was identified. Retro-2^{cycl} is 100-fold more active on HeLa cells than Retro-2 (86). However, the drawback to Retro-2^{cycl} and derivatives is that the cells needed to be pretreated with the drug to observe protection from the toxin (84, 86). Another compound known to protect cells from Stx, chloroquine, was also recently shown to permit the toxin to reach the ER

(87). Since chloroquine does not inhibit the enzymatic activity of Stx nor degrade the toxin in vitro, the drug likely acts by inhibiting the toxin from exiting the ER.

Further small-molecule screens have identified other compounds that inhibit Stx and ricin transport within Vero cells (88) or the enzymatic activity of both toxins (which have the same mode of action) (89). Another small molecule, eeyarestatin I, which interferes with intracellular trafficking among cellular compartments, was shown to cause a lag in protein synthesis inhibition in HeLa cells by Stx (90). Finally, nitrobenzylthioinosine was shown in 2002 to inhibit the trafficking of Stx1, trapping the toxin within the early endosome (91); however, no additional information on the use of this compound to protect cells or animals has been published.

A recent report suggests that manganese inhibits intracellular trafficking of the B subunit of Stx (92), a finding that may or may not extend to the holotoxin. However, manganese treatment increased the viability of HeLa cells incubated with Stx1 in that study and protected BALB/c mice injected with 500 ng of Stx1 and injected with daily doses of manganese (at least 10 mg/kg daily injections were required). Whether such levels of manganese would be reasonably achievable in a patient is not clear. In addition, manganese fails to protect HeLa cells from Stx2 intoxication (93), a finding that makes the potential use of manganese less likely as Stx2 is more commonly associated with HUS development than Stx1.

A COMPOUND THAT INTERFERES WITH TOXIN PROCESSING

Because the A subunit must be cleaved by a furin- or trypsin-like protease so that the A$_1$ or enzymatically active moiety of the toxin may be separated from the holotoxin, furin inhibitors were tested in a cell-based assay for the capacity to protect HEp-2 cells from Stx (94). Although one of the inhibitors showed moderate inhibition activity against Stx, the inhibition was overcome at higher concentrations of Stx. Furthermore, since the toxin can be cleaved by enzymes in intestinal mucus (35), as well as by intracellular proteases (95), and because Stx mutated at the trypsin-sensitive site can still be cleaved (96, 97), we suspect that it will be difficult to protect animals from Stx with protease inhibitors.

Therapies That Interfere with the Cellular Response to Stx

Inhibitors of mitogen-activated protein kinase pathways were used to assess potential protective efficacy against the Stx2-mediated ribotoxic stress response in HCT-8 cells or for oral gavage of Stx2 into infant rabbits (98). The authors observed modest protective capacity by imatinib in HCT-8 cells and against some aspects of Stx2 intoxication in the animals, specifically heterophil infiltration into the intestine. That study indicates that the infant rabbit model of Stx2 gavage may prove useful for the evaluation of therapeutics. However, such a model would be expensive, and determining effective therapeutic doses and the timing for those doses might prove challenging.

A recent paper suggests that an inhibitor of apoptosis, ouabain, protects rat renal proximal tubules from Stx-mediated cell death (99). Kidneys from ouabain-treated mice that were given about four 50% lethal doses of Stx2 showed reduced podocyte depletion as compared to phosphate-buffered saline–treated mice. However, no data were reported on whether the ouabain could protect the mice from Stx2-mediated lethality.

TREATMENT ONCE HUS IS DIAGNOSED

What do you do for patients who already have HUS? We leave discussion of specific clinical interventions such as dialysis and apheresis for medical experts. In terms of a therapeutic that may interfere in the etiology of HUS,

Lapeyraque and colleagues reported in 2011 the use of eculizumab (an antibody directed against complement protein C5 used to treat *atypical* HUS, a disease that may arise due to complement dysregulation) in three children with HUS due to STEC infection (100). Although the sick children exhibited improvements in platelet counts and reductions in lactate dehydrogenase, plasma exchange and hemodialysis were used concurrently. In addition, no children in the small study received just antibody infusion alone. Eculizumab was also used to treat many patients with HUS during the O104:H4 outbreak in Germany in 2011. Although there was no evidence of efficacy for eculizumab in treating HUS from the outbreak in Germany, the cohorts are difficult to compare because the patients received other concurrent treatments, such as plasma exchange, antibiotic therapy, immunoglobulin G immunoadsorption, and dialysis (14, 101–105). In addition, many of the HUS patients given eculizumab were quite ill and had neurological complications. Therefore, a randomized controlled trial is necessary to answer the question about the efficacy of eculizumab for the treatment of Stx-associated HUS.

CONCLUSION

Since the first outbreak of STEC infection in the United States in 1982, much research has focused on ways to neutralize the action of the Stxs. At this time, only SYNSORB Pk, urtoxazumab, and the Shiga monoclonal antibodies cαStx1 and cαStx2 have been tested in phase I and II trials intended to treat or prevent HUS. Efforts to prevent STEC infection, such as elimination from the food supply and proper food handling, are important as well as we try to stop the development of the potentially deadly HUS. Lastly, consideration should be given to the development of a vaccine against the Stxs, minimally for those in research or clinical labs who are exposed to STEC.

ACKNOWLEDGMENT

We declare no conflicts of interest with regard to the manuscript.

CITATION

Melton-Celsa AR, O'Brien AD. 2014. New therapeutic developments against Shiga toxin-producing *Escherichia coli*. Microbiol Spectrum 2(5):EHEC-0013-2013.

REFERENCES

1. **Rosales A, Hofer J, Zimmerhackl LB, Jungraithmayr TC, Riedl M, Giner T, Strasak A, Orth-Holler D, Wurzner R, Karch H.** 2012. Need for long-term follow-up in enterohemorrhagic *Escherichia coli*-associated hemolytic uremic syndrome due to late-emerging sequelae. *Clin Infect Dis* **54:**1413–1421.

2. **Palermo MS, Exeni RA, Fernandez GC.** 2009. Hemolytic uremic syndrome: pathogenesis and update of interventions. *Expert Rev Anti Infect Ther* **7:**697–707.

3. **Spinale JM, Ruebner RL, Copelovitch L, Kaplan BS.** 2013. Long-term outcomes of Shiga toxin hemolytic uremic syndrome. *Pediatr Nephrol* **28:**2097–2105.

4. **Thielman NM, Guerrant RL.** 2004. Clinical practice. Acute infectious diarrhea. *N Engl J Med* **350:**38–47.

5. **Molbak K, Mead PS, Griffin PM.** 2002. Antimicrobial therapy in patients with *Escherichia coli* O157:H7 infection. *JAMA* **288:**1014–1016.

6. **Nelson JM, Griffin PM, Jones TF, Smith KE, Scallan E.** 2011. Antimicrobial and antimotility agent use in persons with Shiga toxin-producing *Escherichia coli* O157 infection in FoodNet sites. *Clin Infect Dis* **52:**1130–1132.

7. **Karch H, Goroncy-Bermes P, Opferkuch W, Kroll H-P, O'Brien A.** 1985. Subinhibitory concentrations of antibiotics modulate amount of Shiga-like toxin produced by *Escherichia coli*, p 239–245. *In* Adam D, Hahn H, Opferkuch W (ed), *The Influence of Antibiotics on the Host-Parasite Relationship II*. SpringerVerlag, Berlin, Germany.

8. **Wong CS, Jelacic S, Habeeb RL, Watkins SL, Tarr PI.** 2000. The risk of the hemolytic-uremic syndrome after antibiotic treatment of *Escherichia coli* O157:H7 infections. *N Engl J Med* **342:**1930–1936.

9. **Slutsker L, Ries AA, Malone K, Wells JG, Greene KD, Griffin PM.** 1998. A nationwide case-control study of *Escherichia coli* O157:H7 infection in the United States. *J Infect Dis* **177:**962–966.

10. **Panos GZ, Betsi GI, Falagas ME.** 2006. Systematic review: are antibiotics detrimental or beneficial for the treatment of patients with *Escherichia coli* O157:H7 infection? *Aliment Pharmacol Ther* **24:**731–742.

11. **Smith KE, PWilke PR, Reiter PL, Hedican EB, Bender JB, Hedberg CW.** 2012. Antibiotic treatment of *Escherichia coli* O157 infection and the risk of hemolytic uremic syndrome, Minnesota. *Pediatr Infect Dis J* **31:**37–41.

12. **Ikeda K, Ida O, Kimot K, Takatorige T, Nakanishi N, Tatar K.** 1999. Effect of early fosfomycin treatment on prevention of hemolytic uremic syndrome accompanying *Escherichia coli* O157:H7 infection. *Clin Nephrol* **52:**357–362.

13. **Fukushima H, Hashizume T, Morita Y, Tanaka J, Azuma K, Mizumoto Y, Kaneno M, Matsuura M, Konma K, Kitani T.** 1999. Clinical experiences in Sakai City Hospital during the massive outbreak of enterohemorrhagic *Escherichia coli* O157 infections in Sakai City, 1996. *Pediatr Int* **41:**213–217.

14. **Menne J, Nitschke M, Stingele R, Abu-Tair M, Beneke J, Bramstedt J, Bremer JP, Brunkhorst R, Busch V, Dengler R, Deuschl G, Fellermann K, Fickenscher H, Gerigk C, Goettsche A, Greeve J, Hafer C, Hagenmuller F, Haller H, Herget-Rosenthal S, Hertenstein B, Hofmann C, Lang M, Kielstein JT, Klostermeier UC, Knobloch J, Kuehbacher M, Kunzendorf U, Lehnert H, Manns MP, Menne TF, Meyer TN, Michael C, Munte T, Neumann-Grutzeck C, Nuernberger J, Pavenstaedt H, Ramazan L, Renders L, Repenthin J, Ries W, Rohr A, LRump LC, Samuelsson O, Sayk F, Schmidt BM, Schnatter S, Schocklmann H, Schreiber S, von Seydewitz CU, Steinhoff J, Stracke S, Suerbaum S, van de Loo A, Vischedyk M, Weissenborn K, Wellhoner P, Wiesner M, Zeissig S, Buning J, Schiffer M, Kuehbacher T.** 2012. Validation of treatment strategies for enterohaemorrhagic *Escherichia coli* O104:H4 induced haemolytic uraemic syndrome: case-control study. *BMJ* **345:**e4565.

15. **Geerdes-Fenge HF, Lobermann M, Nurnberg M, Fritzsche C, Koball S, Henschel J, Hohn R, Schober HC, Mitzner S, Podbielski A, Reisinger EC.** 2013. Ciprofloxacin reduces the risk of hemolytic uremic syndrome in patients with *Escherichia coli* O104:H4-associated diarrhea. *Infection* **41:**669–673.

16. **Binks S, Regan K, Richenberg J, Chevassut T.** 2012. Microbes without frontiers: severe haemolytic-uraemic syndrome due to *E coli* O104:H4. *BMJ Case Rep* **2012.**

17. Colic E, Dieperink H, Titlestad K, Tepel M. 2011. Management of an acute outbreak of diarrhoea-associated haemolytic uraemic syndrome with early plasma exchange in adults from southern Denmark: an observational study. *Lancet* **378**:1089–1093.

18. Bielaszewska M, Idelevich EA, Zhang W, Bauwens A, Schaumburg F, Mellmann A, Peters G, Karch H. 2012. Effects of antibiotics on Shiga toxin 2 production and bacteriophage induction by epidemic *Escherichia coli* O104:H4 strain. *Antimicrob Agents Chemother* **56**:3277–3282.

19. Nitschke M, Sayk F, Hartel C, Roseland RT, Hauswaldt S, Steinhoff J, Fellermann K, Derad I, Wellhoner P, Buning J, Tiemer B, Katalinic A, Rupp J, Lehnert H, Solbach W, Knobloch JK. 2012. Association between azithromycin therapy and duration of bacterial shedding among patients with Shiga toxin-producing enteroaggregative *Escherichia coli* O104:H4. *JAMA* **307**:1046–1052.

20. Isogai E, Isogai H, Hayashi S, Kubota T, Kimura K, Fujii N, Ohtani T, Sato K. 2000. Effect of antibiotics, levofloxacin and fosfomycin, on a mouse model with *Escherichia coli* O157 infection. *Microbiol Immunol* **44**:89–95.

21. Zhang X, McDaniel AD, Wolf LE, Keusch GT, Waldor MK, Acheson DW. 2000. Quinolone antibiotics induce Shiga toxin-encoding bacteriophages, toxin production, and death in mice. *J Infect Dis* **181**:664–670.

22. Yoshimura K, Fujii J, Taniguchi H, Yoshida S. 1999. Chemotherapy for enterohemorrhagic *Escherichia coli* O157:H infection in a mouse model. *FEMS Immunol Med Microbiol* **26**:101–108.

23. Kurioka T, Yunou Y, Harada H, Kita E. 1999. Efficacy of antibiotic therapy for infection with Shiga-like toxin-producing *Escherichia coli* O157:H7 in mice with protein-calorie malnutrition. *Eur J Clin Microbiol Infect Dis* **18**:561–571.

24. Zangari T, Melton-Celsa AR, Panda A, Boisen N, Smith MA, Taratov I, De Tolla LJ, Nataro JP, O'Brien AD. 2013. Virulence of the Shiga toxin type 2-expressing *Escherichia coli* O104:H4 German outbreak isolate in two animal models. *Infect Immun* **81**:1562–1574.

25. Al-Qarawi S, Fontaine RE, Al-Qahtani MS. 1995. An outbreak of hemolytic uremic syndrome associated with antibiotic treatment of hospital inpatients for dysentery. *Emerg Infect Dis* **1**:138–140.

26. Scholl D, Cooley M, Williams SR, Gebhart D, Martin D, Bates A, Mandrell R. 2009. An engineered R-type pyocin is a highly specific and sensitive bactericidal agent for the food-borne pathogen *Escherichia coli* O157:H7. *Antimicrob Agents Chemother* **53**:3074–3080.

27. Ritchie JM, Greenwich JL, Davis BM, Bronson RT, Gebhart D, Williams SR, Martin D, Scholl D, Waldor MK. 2011. An *Escherichia coli* O157-specific engineered pyocin prevents and ameliorates infection by *E. coli* O157:H7 in an animal model of diarrheal disease. *Antimicrob Agents Chemother* **55**:5469–5474.

28. Rasko DA, Moreira CG, Li de R, Reading NC, Ritchie JM, Waldor MK, Williams N, Taussig R, Wei S, Roth M, Hughes DT, Huntley JF, Fina MW, Falck JR, Sperandio V. 2008. Targeting QseC signaling and virulence for antibiotic development. *Science* **321**:1078–1080.

29. Crane JK, Byrd IW, Boedeker EC. 2011. Virulence inhibition by zinc in Shiga-toxigenic *Escherichia coli*. *Infect Immun* **79**:1696–1705.

30. Tesh VL. 2012. The induction of apoptosis by Shiga toxins and ricin. *Curr Top Microbiol Immunol* **357**:137–178.

31. Scheut F, Teel LD, Beutin L, Pierard D, Buvens G, Karch H, Mellmann A, Caprioli A, Tozzoli R, Morabito S, Strockbine NA, Melton-Celsa AR, Sanchez M, Persson S, O'Brien AD. 2012. Multicenter evaluation of a sequence-based protocol for subtyping Shiga toxins and standardizing Stx nomenclature. *J Clin Microbiol* **50**:2951–2963.

32. Bielaszewska M, Friedrich AW, Aldick T, Schurk-Bulgrin R, Karch H. 2006. Shiga toxin activatable by intestinal mucus in *Escherichia coli* isolated from humans: predictor for a severe clinical outcome. *Clin Infect Dis* **43**:1160–1167.

33. Friedrich AW, Bielaszewska M, Zhang WL, Pulz M, Kuczius T, Ammon A, Karch H. 2002. *Escherichia coli* harboring Shiga toxin 2 gene variants: frequency and association with clinical symptoms. *J Infect Dis* **185**:74–84.

34. Lindgren SW, Samuel JE, Schmitt CK, O'Brien AD. 1994. The specific activities of Shiga-like toxin type II (SLT-II) and SLT-II-related toxins of enterohemorrhagic *Escherichia coli* differ when measured by Vero cell cytotoxicity but not by mouse lethality. *Infect Immun* **62**:623–631.

35. Melton-Celsa AR, Darnell SC, O'Brien AD. 1996. Activation of Shiga-like toxins by mouse and human intestinal mucus correlates with virulence of enterohemorrhagic *Escherichia coli* O91:H21 isolates in orally infected, streptomycin-treated mice. *Infect Immun* **64**:1569–1576.

36. Silberstein C, Copeland DP, Chiang W-L, Repetto HA, Ibarra C. 2008. A glucosylceramide synthase inhibitor prevents the cytotoxic effects of Shiga toxin-2 on human renal tubular epithelial cells. *J Epithel Biol Pharmacol* **1**:71–75.

37. Silberstein C, Lucero MS, Zotta E, Copeland DP, Lingyun L, Repetto HA, Ibarra C. 2011. A glucosylceramide synthase inhibitor protects rats

against the cytotoxic effects of Shiga toxin 2. *Pediatr Res* **69:**39–394.

38. **Armstrong GD, Fodor E, Vanmaele R.** 1991. Investigation of Shiga-like toxin binding to chemically synthesized oligosaccharide sequences. *J Infect Dis* **164:**1160–1167.

39. **Trachtman H, Cnaan A, Christen E, Gibbs K, Zhao S, Acheson DW, Weiss R, Kaskel FJ, Spitzer A, Hirschman GH.** 2003. Effect of an oral Shiga toxin-binding agent on diarrhea-associated hemolytic uremic syndrome in children: a randomized controlled trial. *JAMA* **290:**1337–1344.

40. **Takeda T, Yoshino K, Adachi E, Sato Y, Yamagata K.** 1999. In vitro assessment of a chemically synthesized Shiga toxin receptor analog attached to chromosorb P (Synsorb Pk) as a specific absorbing agent of Shiga toxin 1 and 2. *Microbiol Immunol* **43:**331–337.

41. **Armstrong GD, Rowe PC, Goodyer P, Orrbine E, Klassen TP, Wells G, MacKenzie A, Lior H, Blanchard C, Auclair F, et al.** 1995. A phase I study of chemically synthesized verotoxin (Shiga-like toxin) Pk-trisaccharide receptors attached to chromosorb for preventing hemolytic-uremic syndrome. *J Infect Dis* **171:**1042–1045.

42. **Kitov PI, Sadowska JM, Mulvey G, Armstrong GD, Ling H, Pannu NS, Read RJ, Bundle DR.** 2000. Shiga-like toxins are neutralized by tailored multivalent carbohydrate ligands. *Nature* **403:**66–672.

43. **Mulve GL, Marcato P, Kitov PI, Sadowska J, Bundle DR, Armstrong GD.** 2003. Assessment in mice of the therapeutic potential of tailored, multivalent Shiga toxin carbohydrate ligands. *J Infect Dis* **187:**640–649.

44. **Nishikawa K, Matsuoka K, Kita E, Okabe N, Mizuguchi M, Hino K, Miyazawa S, Yamasaki C, Aoki J, Takashima S, Yamakawa Y, Nishijima M, Terunuma D, Kuzuhara H, Natori Y.** 2002. A therapeutic agent with oriented carbohydrates for treatment of infections by Shiga toxin-producing *Escherichia coli* O157:H7. *Proc Natl Acad Sci USA* **99:**7669–7674.

45. **Watanabe M, Matsuoka K, Kita E, Igai K, Higashi N, Miyagawa A, Watanabe T, Yanoshita R, Samejima Y, Terunuma D, Natori Y, Nishikawa K.** 2004. Oral therapeutic agents with highly clustered globotriose for treatment of Shiga toxigenic *Escherichia coli* infections. *J Infect Dis* **189:**360–368.

46. **Watanabe-Takahashi M, Sato T, Dohi T, Noguchi N, Kano F, Murata M, Hamabata T, Natori Y, Nishikawa K.** 2010. An orally applicable Shiga toxin neutralizer functions in the intestine to inhibit the intracellular transport of the toxin. *Infect Immun* **78:**177–183.

47. **Paton AW, Morona R, Paton JC.** 2001. Neutralization of Shiga toxins Stx1, Stx2c, and Stx2e by recombinant bacteria expressing mimics of globotriose and globotetraose. *Infect Immun* **69:**1967–1970.

48. **Li X, Wu P, Cheng S, Lv X.** 2012. Synthesis and assessment of globotriose-chitosan conjugate, a novel inhibitor of Shiga toxins produced by *Escherichia coli*. *J Med Chem* **55:**2702–2710.

49. **Nishikawa K, Watanabe M, Kita E, Igai K, Omata K, Yaffe MB, Natori Y.** 2006. A multivalent peptide library approach identifies a novel Shiga toxin inhibitor that induces aberrant cellular transport of the toxin. *FASEB J* **20:**2597–2599.

50. **Stearns-Kurosawa DJ, Collins V, Freeman S, Debord D, Nishikawa K, Oh SY, Leibowitz CS, Kurosawa S.** 2011. Rescue from lethal Shiga toxin 2-induced renal failure with a cell-permeable peptide. *Pediatr Nephrol* **26:**2031–2039.

51. **Tsutsuki K, Watanabe-Takahashi M, Takenaka Y, Kita E, Nishikawa K.** 2013. Identification of a peptide-based neutralizer that potently inhibits both Shiga toxins 1 and 2 by targeting specific receptor-binding regions. *Infect Immun* **81:**2133–2138.

52. **Paton AW, Morona R, Paton JC.** 2000. A new biological agent for treatment of Shiga toxigenic *Escherichia coli* infections and dysentery in humans. *Nat Med* **6:**265–270.

53. **Paton JC, Rogers TJ, Morona R, Paton AW.** 2001. Oral administration of formaldehyde-killed recombinant bacteria expressing a mimic of the Shiga toxin receptor protects mice from fatal challenge with Shiga-toxigenic *Escherichia coli*. *Infect Immun* **69:**1389–1393.

54. **Pinyon RA, Paton JC, Paton AW, JBotten JA, Morona R.** 2004. Refinement of a therapeutic Shiga toxin-binding probiotic for human trials. *J Infect Dis* **189:**1547–1555.

55. **Bitzan M, Klemt M, Steffens R, Muller-Wiefel DE.** 1993. Differences in verotoxin neutralizing activity of therapeutic immunoglobulins and sera from healthy controls. *Infection* **21:**140–145.

56. **Bitzan M, Ludwig K, Klemt M, Konig H, Buren J, Muller-Wiefel DE.** 1993. The role of *Escherichia coli* O157 infections in the classical (enteropathic) haemolytic uraemic syndrome: results of a Central European, multicentre study. *Epidemiol Infect* **110:**183–196.

57. **Kimura T, Tani S, Matsumoto Yi Y, Takeda T.** 2001. Serum amyloid P component is the Shiga toxin 2-neutralizing factor in human blood. *J Biol Chem* **276:**41576–41579.

58. **Kimura T, Tani S, Motoki M, Matsumoto Y.** 2003. Role of Shiga toxin 2 (Stx2)-binding protein, human serum amyloid P component (HuSAP), in Shiga toxin-producing *Escherichia*

coli infections: assumption from in vitro and in vivo study using HuSAP and anti-Stx2 humanized monoclonal antibody TMA-15. *Biochem Biophys Res Commun* **305**:1057–1060.

59. **Armstrong GD, Mulvey GL, Marcato P, Griener TP, Kahan MC, Tennent GA, Sabin CA, Chart H, Pepys MB.** 2006. Human serum amyloid P component protects against *Escherichia coli* O157:H7 Shiga toxin 2 in vivo: therapeutic implications for hemolytic-uremic syndrome. *J Infect Dis* **193**:1120–1124.

60. **Griener TP, Strecker JG, Humphries RM, Mulvey GL, Fuentealba C, Hancock RE, Armstrong GD.** 2011. Lipopolysaccharide renders transgenic mice expressing human serum amyloid P component sensitive to Shiga toxin 2. *PLoS One* **6**:e21457.

61. **Marcato P, Vander Helm K, Mulvey GL, Armstrong GD.** 2003. Serum amyloid P component binding to Shiga toxin 2 requires both a subunit and B pentamer. *Infect Immun* **71**:607–6078.

62. **Donohue-Rolfe A, Kondova I, Mukherjee J, Chios K, Hutto D, Tzipori S.** 1999. Antibody-based protection of gnotobiotic piglets infected with *Escherichia coli* O157:H7 against systemic complications associated with Shiga toxin 2. *Infect Immun* **67**:3645–3648.

63. **Perera LP, Marques LR, O'Brien AD.** 1988. Isolation and characterization of monoclonal antibodies to Shiga-like toxin II of enterohemorrhagic *Escherichia coli* and use of the monoclonal antibodies in a colony enzyme-linked immunosorbent assay. *J Clin Microbiol* **26**:2127–2131.

64. **Strockbine NA, Marques LR, Holmes RK, O'Brien AD.** 1985. Characterization of monoclonal antibodies against Shiga-like toxin from *Escherichia coli*. *Infect Immun* **50**:69–700.

65. **Smith MJ, Melton-Celsa AR, Sinclair JF, Carvalho HM, Robinson CM, O'Brien AD.** 2009. Monoclonal antibody 11E10, which neutralizes shiga toxin type 2 (Stx2), recognizes three regions on the Stx2 A subunit, blocks the enzymatic action of the toxin in vitro, and alters the overall cellular distribution of the toxin. *Infect Immun* **77**:2730–2740.

66. **Smith MJ, Carvalho HM, Melton-Celsa AR, O'Brien AD.** 2006. The 13C4 monoclonal antibody that neutralizes Shiga toxin type 1 (Stx1) recognizes three regions on the Stx1 B subunit and prevents Stx1 from binding to its eukaryotic receptor globotriaosylceramide. *Infect Immun* **74**:6992–6998.

67. **Sauter KA, Melton-Celsa AR, Larkin K, Troxell ML, O'Brien AD, Magun BE.** 2008. Mouse model of hemolytic-uremic syndrome caused by endotoxin-free Shiga toxin 2 (Stx2) and protection from lethal outcome by anti-Stx2 antibody. *Infect Immun* **76**:4469–4478.

68. **Edwards AC, Melton-Celsa AR, Arbuthnott K, Stinson JR, Schmitt CK, Wong HC, O'Brien AD.** 1998. Vero cell neutralization and mouse protective efficacy of humanized monoclonal antibodies against *Escherichia coli* toxins Stx1 and Stx2, p 388–392. *In* Kaper JB, O'Brien AD (ed), *Escherichia coli O157:H7 and Other Shiga Toxin-Producing* E. coli *Strains.* ASM Press, Washington, DC.

69. **Dowling TC, Chavaillaz PA, Young DG, Melton-Celsa A, O'Brien A, Thuning-Roberson C, Edelman R, Tacket CO.** 2005. Phase 1 safety and pharmacokinetic study of chimeric murine-human monoclonal antibody c alpha Stx2 administered intravenously to healthy adult volunteers. *Antimicrob Agents Chemother* **49**:1808–1812.

70. **Bitzan M, Poole R, Mehran M, Sicard E, Brockus C, Thuning-Roberson C, Riviere M.** 2009. Safety and pharmacokinetics of chimeric anti-Shiga toxin 1 and anti-Shiga toxin 2 monoclonal antibodies in healthy volunteers. *Antimicrob Agents Chemother* **53**:3081–3087.

71. **Nakao H, Kataoka C, Kiyokawa N, Fujimoto J, Yamasaki S, Takeda T.** 2002. Monoclonal antibody to Shiga toxin 1, which blocks receptor binding and neutralizes cytotoxicity. *Microbiol Immunol* **46**:777–780.

72. **Nakao H, Kiyokawa N, Fujimoto J, Yamasaki S, Takeda T.** 1999. Monoclonal antibody to Shiga toxin 2 which blocks receptor binding and neutralizes cytotoxicity. *Infect Immun* **67**:5717–5722.

73. **Kimura T, Co MS, Vasquez M, Wei S, Xu H, Tani S, Sakai Y, Kawamura T, Matsumoto Y, Nakao H, Takeda T.** 2002. Development of humanized monoclonal antibody TMA-15 which neutralizes Shiga toxin 2. *Hybrid Hybridomics* **21**:161–168.

74. **Yamagami S, Motoki M, Kimura T, Izumi H, Takeda T, Katsuura Y, Matsumoto Y.** 2001. Efficacy of postinfection treatment with anti-Shiga toxin (Stx) 2 humanized monoclonal antibody TMA-15 in mice lethally challenged with Stx-producing *Escherichia coli*. *J Infect Dis* **184**:738–742.

75. **Lopez EL, Contrini MM, Glatstein E, Gonzalez Ayala S, Santoro R, Allende D, Ezcurra G, Teplitz E, Koyama T, Matsumoto Y, Sato H, Sakai K, Hoshide S, Komoriya K, Morita T, Harning R, Brookman S.** 2010. Safety and pharmacokinetics of urtoxazumab, a humanized monoclonal antibody, against Shiga-like toxin 2 in healthy adults and in pediatric patients infected with Shiga-like toxin-producing *Escherichia coli*. *Antimicrob Agents Chemother* **54**:239–243.

76. **Mukherjee J, Chios K, Fishwild D, Hudson D, O'Donnell S, Rich SM, Donohue-Rolfe A, Tzipori S.** 2002. Production and characterization of protective human antibodies against Shiga toxin 1. *Infect Immun* **70:**5896–5899.

77. **Mukherjee J, Chios K, Fishwild D, Hudson D, O'Donnell S, Rich SM, Donohue-Rolfe A, Tzipori S.** 2002. Human Stx2-specific monoclonal antibodies prevent systemic complications of *Escherichia coli* O157:H7 infection. *Infect Immun* **70:**612–619.

78. **Sheoran AS, Chapman S, Singh P, Donohue-Rolfe A, Tzipori S.** 2003. Stx2-specific human monoclonal antibodies protect mice against lethal infection with *Escherichia coli* expressing Stx2 variants. *Infect Immun* **71:**3125–3130.

79. **Krautz-Peterson G, Chapman-Bonofiglio S, Boisvert K, Feng H, Herman IM, Tzipori S, Sheoran AS.** 2008. Intracellular neutralization of Shiga toxin 2 by an a subunit-specific human monoclonal antibody. *Infect Immun* **76:**1931–1939.

80. **Jeong KI, Chapman-Bonofiglio S, Singh P, Lee J, Tzipori S, Sheoran AS.** 2010. In vitro and in vivo protective efficacies of antibodies that neutralize the RNA N-glycosidase activity of Shiga toxin 2. *BMC Immunol* **11:**16.

81. **Akiyoshi DE, Sheoran AS, Rich CM, Richard L, Chapman-Bonofiglio S, Tzipori S.** 2010. Evaluation of Fab and F(ab')2 fragments and isotype variants of a recombinant human monoclonal antibody against Shiga toxin 2. *Infect Immun* **78:**1376–1382.

82. **Spooner RA, Watson P, Smith DC, Boal F, Amessou M, Johannes L, Clarkson GJ, Lord JM, Stephens DJ, Roberts LM.** 2008. The secretion inhibitor Exo2 perturbs trafficking of Shiga toxin between endosomes and the trans-Golgi network. *Biochem J* **414:**471–484.

83. **Guetzoyan LJ, Spooner RA, Boal F, Stephens DJ, Lord JM, Roberts LM, Clarkson GJ.** 2010. Fine tuning Exo2, a small molecule inhibitor of secretion and retrograde trafficking pathways in mammalian cells. *Mol Biosyst* **6:**2030–2038.

84. **Stechmann B, Bai SK, Gobbo E, Lopez R, Merer G, Pinchard S, Panigai L, Tenza D, Raposo G, Beaumelle B, Sauvaire D, Gillet D, Johannes L, Barbier J.** 2010. Inhibition of retrograde transport protects mice from lethal ricin challenge. *Cell* **141:**231–242.

85. **Park JG, Kahn JN, Tumer NE, Pang YP.** 2012. Chemical structure of Retro-2, a compound that protects cells against ribosome-inactivating proteins. *Sci Rep* **2:**631.

86. **Noel R, Gupta N, Pons V, Goudet A, Garcia-Castillo MD, Michau A, Martinez J, Buisson DA, Johannes L, Gillet D, Barbier J, Cintrat JC.** 2013. *N*-methyl dihydroquinazolinones derivatives of Retro-2 with enhanced efficacy against Shiga toxin. *J Med Chem* **56:**3404–3413.

87. **Dyve Lingelem AB, Bergan J, Sandvig K.** 2012. Inhibitors of intravesicular acidification protect against Shiga toxin in a pH-independent manner. *Traffic* **13:**443–454.

88. **Saenz JB, Doggett TA, Haslam DB.** 2007. Identification and characterization of small molecules that inhibit intracellular toxin transport. *Infect Immun* **75:**4552–4561.

89. **Wahome PG, Bai Y, Neal LM, Robertus JD, Mantis NJ.** 2010. Identification of small-molecule inhibitors of ricin and Shiga toxin using a cell-based high-throughput screen. *Toxicon* **56:**313–323.

90. **Aletrari MO, McKibbin C, Williams H, Pawar V, Pietroni P, Lord JM, Flitsch SL, Whitehead R, Swanton E, High S, Spooner RA.** 2011. Eeyarestatin 1 interferes with both retrograde and anterograde intracellular trafficking pathways. *PLoS One* **6:**e22713.

91. **Sekino T, Kiyokawa N, Taguchi T, Ohmi K, Nakajima H, Suzuki T, Furukawa S, Nakao H, Takeda T, Fujimoto J.** 2002. Inhibition of Shiga toxin cytotoxicity in human renal cortical epithelial cells by nitrobenzylthioinosine. *J Infect Dis* **185:**785–796.

92. **Mukhopadhyay S, Linstedt AD.** 2012. Manganese blocks intracellular trafficking of Shiga toxin and protects against Shiga toxicosis. *Science* **335:** 332–335.

93. **Mukhopadhyay S, Redler B, Linstedt AD.** 2013. Shiga toxin-binding site for host cell receptor GPP130 reveals unexpected divergence in toxin-trafficking mechanisms. *Mol Biol Cell* **24:**2311–2318.

94. **Becker GL, Lu Y, Hardes K, Strehlow B, Levesque C, Lindberg I, Sandvig K, Bakowsky U, Day R, Garten W, Steinmetzer T.** 2012. Highly potent inhibitors of proprotein convertase furin as potential drugs for treatment of infectious diseases. *J Biol Chem* **287:**21992–22003.

95. **Garred O, van Deurs B, Sandvig K.** 1995. Furin-induced cleavage and activation of Shiga toxin. *J Biol Chem* **270:**10817–10821.

96. **Garred O, Dubinina E, Holm PK, Olsnes S, van Deurs B, Kozlov JV, Sandvig K.** 1995. Role of processing and intracellular transport for optimal toxicity of Shiga toxin and toxin mutants. *Exp Cell Res* **218:**39–49.

97. **Burgess BJ, Roberts LM.** 1993. Proteolytic cleavage at arginine residues within the hydrophilic disulphide loop of the *Escherichia coli* Shiga-like toxin I A subunit is not essential for cytotoxicity. *Mol Microbiol* **10:**171–179.

98. **Stone SM, Thorpe CM, Ahluwalia A, Rogers AB, Obata F, Vozenilek A, Kolling GL, Kan AV, Magun BE, Jandhyala DM.** 2012. Shiga toxin 2-induced intestinal pathology in infant rabbits is

A-subunit dependent and responsive to the tyrosine kinase and potential ZAK inhibitor imatinib. *Front Cell Infect Microbiol* **2:**135.

99. Burlaka I, Liu XL, Rebetz J, Arvidsson I, Yang L, Brismar H, Karpman D, Aperia A. 2013. Ouabain protects against Shiga toxin-triggered apoptosis by reversing the imbalance between Bax and Bcl-xL. *J Am Soc Nephrol* **24:**1413–1423.

100. Lapeyraque AL, Malina M, Fremeaux-Bacchi V, Boppel T, Kirschfink M, Oualha M, Proulx F, Clermont MJ, Le Deist F, Niaudet P, Schaefer F. 2011. Eculizumab in severe Shiga-toxin-associated HUS. *N Engl J Med* **364:**2561–2563.

101. Greinache A, Friesecke S, Abel P, Dressel A, Stracke S, Fiene M, Ernst F, Selleng K, Weissenborn K, Schmidt BM, Schiffer M, Felix SB, Lerch MM, Kielstein JT, Mayerle J. 2011. Treatment of severe neurological deficits with IgG depletion through immunoadsorption in patients with *Escherichia coli* O104:H4-associated haemolytic uraemic syndrome: a prospective trial. *Lancet* **378:**1166–1173.

102. Ullrich S, Bremer P, Neumann-Grutzeck C, Otto H, Ruther C, von Seydewitz CU, Meyer GP, Ahmadi-Simab K, Rother J, Hogan B, Schwenk W, Fischbach R, Caselitz J, Puttfarcken J, Huggett S, Tiedeken P, Pober J, Kirkiles-Smith NC, Hagenmuller F. 2013. Symptoms and clinical course of EHEC O104 infection in hospitalized patients: a prospective single center study. *PLoS One* **8:**e55278.

103. Loos S, Ahlenstiel T, Kranz B, Staude H, Pape L, Hartel C, Vester U, Buchtala L, Benz K, Hoppe B, Beringer O, Krause M, Muller D, Pohl M, Lemke J, Hillebrand G, Kreuzer M, Konig J, Wigger M, Konrad M, Haffner D, Oh J, Kemper MJ. 2012. An outbreak of Shiga toxin-producing *Escherichia coli* O104:H4 hemolytic uremic syndrome in Germany: presentation and short-term outcome in children. *Clin Infect Dis* **55:**753–759.

104. Hauswaldt S, Nitschke M, Sayk F, Solbach W, Knobloch JK. 2013. Lessons learned from uutbreaks of Shiga toxin producing *Escherichia coli. Curr Infect Dis Rep* **15:**9.

105. Kielstein JT, Beutel G, Fleig S, Steinhoff J, Meyer TN, Hafer C, Kuhlmann U, Bramstedt J, Panzer U, Vischedyk M, Busch V, Ries W, Mitzner S, Mees S, Stracke S, Nurnberger J, Gerke P, Wiesner M, Sucke B, Abu-Tair M, Kribben A, Klause N, Schindler R, Merkel F, Schnatter S, Dorresteijn EM, Samuelsson O, Brunkhorst R. 2012. Best supportive care and therapeutic plasma exchange with or without eculizumab in Shiga-toxin-producing *E. coli* O104:H4 induced haemolytic-uraemic syndrome: an analysis of the German STEC-HUS registry. *Nephrol Dial Transplant* **27:**3807–3815.

106. Trachtman H, Austin C, Lewinski M, Stahl RA. 2012. Renal and neurological involvement in typical Shiga toxin-associated HUS. *Nat Rev Nephrol* **8:**658–669.

HOST DETERMINANTS OF DISEASE
AND HOST RESPONSE

Risk Factors for Shiga Toxin-Producing *Escherichia coli-*Associated Human Diseases

18

MARTA RIVAS,[1] ISABEL CHINEN,[1] ELIZABETH MILIWEBSKY,[1] and MARCELO MASANA[2]

INTRODUCTION

Shiga toxin-producing *Escherichia coli* (STEC) strains emerged in the late 1970s or early 1980s as highly significant zoonotic threats to public health. In 1982, two outbreaks of severe bloody diarrhea, related to a previously rare serotype of *E. coli*, O157:H7, were reported in the United States (1).

At present, we know that STEC strains are an important cause of morbidity and mortality, with associated loss of life years and diminished health-related quality of life. The clinical manifestations of infection range from symptom-free carriage to nonbloody diarrhea, hemorrhagic colitis (HC), and hemolytic-uremic syndrome (HUS) (2). The linkage between STEC infection and the development of HUS was established by Karmali and colleagues in 1983 to 1985 (3, 4).

HUS is a systemic thrombotic microangiopathy caused by different etiologies and mechanisms, involving acute kidney failure that may result in death or end-stage renal disease (ESRD), a serious chronic condition that reduces life expectancy. Patients with ESRD are initially treated with peritoneal dialysis or

[1]Instituto Nacional de Enfermedades Infecciosas, ANLIS "Dr. C. G. Malbrán," (1281) Buenos Aires, Argentina; [2]Instituto Tecnología de Alimentos, Centro de Investigación de Agroindustria, Instituto Nacional de Tecnología Agropecuaria, (B1708WAB) Morón, Pcia. de Buenos Aires, Argentina.

Enterohemorrhagic Escherichia coli *and Other Shiga Toxin-Producing* E. coli
Edited by Vanessa Sperandio and Carolyn J. Hovde
© 2015 American Society for Microbiology, Washington, DC
doi:10.1128/microbiolspec.EHEC-0002-2013

hemodialysis and may later be eligible for kidney transplantation (5). The cascade leading from gastrointestinal infection to renal impairment is complex, the production of Shiga toxin 1, Shiga toxin 2, and/or their subtypes (Stx1a, Stx1c, Stx2a, Stx2b, Stx2c, Stx2d$_{activatable}$, and Stx2f) being the primary virulence trait responsible for human disease. However, a mosaic of different virulence traits, comprising several adhesins and other toxins that may play a role in pathogenesis, has also been described (2).

Cattle have been recognized as the main reservoir for STEC for more than 30 years; however, several different surveys have demonstrated that STEC strains occurred in the gastrointestinal tracts of other domestic animals such as sheep, goats, water buffalos, pigs, dogs, and cats (6). Humans usually become infected by eating undercooked beef products, but secondary sources, including leafy green vegetables, apple cider, and dairy products that have been contaminated with manure, are also vehicles for food-borne infection (7). Infections have also been caused by drinking or swimming in contaminated water, person-to-person transmission, or contact with infected animals (6, 8).

STEC isolates that cause human infections belong to a large number of O:H serotypes, and O157:H7 is the most prevalent serotype associated with large outbreaks and sporadic cases of HC and HUS in many countries (9).

In 2003, Karmali et al. (10) proposed classifying STEC serotypes into five seropathotypes (A to E) based on their reported frequencies in human illness and their known associations with outbreaks and severe outcomes, including HC and HUS. Seropathotype A (O157:H7 and O157:NM), considered the most virulent, is associated with the highest incidence in human disease and is often involved in outbreaks. Seropathotype B (O26:H11 and NM; O45:H2 and NM; O103:H2, H11, H25, and NM; O111:H8 and NM; O121:H19 and H7; and O145:NM) is associated with severe human disease, but at a lower frequency, and is uncommonly involved

with outbreaks. Seropathotypes C (O91:H21, O104:H21 and O113:H21, among others) and D have a low incidence in human illness and are rarely associated with outbreaks. Finally, seropathotype E is composed of many serotypes that have not been implicated in human diseases.

SURVEILLANCE AND DISEASE TRENDS

Surveillance practices vary considerably among countries, and therefore caution is required when comparing STEC incidence rates among countries.

In the United States, *E. coli* O157:H7 infection became nationally notifiable in 1995. Since 2000, all STEC infections that cause human illness are notifiable to the Nationally Notifiable Diseases Surveillance System (NNDSS) in the United States. In 2011, the Centers for Disease Control and Prevention (CDC) Emerging Infections Program analyzed the data gathered from the Foodborne Diseases Active Surveillance Network (FoodNet). A total of 521 laboratory-confirmed cases of STEC non-O157 and 463 of STEC O157 infections were identified, with incidence rates of 1.10 and 0.97 per 100,000 persons, respectively. Moreover, FoodNet ascertained 96 HUS cases, including 93 (97%) postdiarrheal HUS cases in 2010. The population under surveillance was 47,505,580, which represents 15.2% of the total U.S. population. According to CDC, illnesses caused by non-O157 STEC serogroups tended to be less severe than those caused by *E. coli* O157:H7 because they required less hospitalization (18% versus 43.4%), the death rate was lower (0.19% versus 0.43%), and HUS developed in fewer patients (1.7% versus 6.3%) (11).

In Canada, STEC infection has been classified as a notifiable disease since 1990. C-EnterNet is a national integrated enteric pathogen surveillance system that collects information on both cases and source of exposure in two sentinel sites, Ontario (since 2005) and British Columbia (since 2010).

In 2010, the incidence of illnesses caused by STEC was 2.2 and 2.9 per 100,000 persons in each site, respectively. The national incidence rate was 2.3 per 100,000 persons in 2008 (12).

In 2010, the European Centre for Disease Prevention and Control reported 3,710 confirmed cases of STEC infection, with an incidence of 0.96 per 100,000 persons. The annual Community Summary Report on the European Union gave an incidence of all STEC infections in Austria, Belgium, Finland, and Italy as 1 case per 100,000 or less; and in Germany, the United Kingdom, the Netherlands, Denmark, Sweden, and Ireland, as 1.2, 1.8, 2.9, 3.2, 3.6, and 4.4 cases per 100,000 persons, respectively (13).

In Australia, information on the incidence of STEC infections and HUS is obtained from the Australian NNDSS, and has been mandatory in all jurisdictions since 2000, except Queensland and Western Australia, where the incidence became notifiable in 2001. For the 11-year period from 2000 to 2010, the overall annual rate was 0.4 cases per 100,000 persons, and the annual rate of notification for HUS was 0.07 cases per 100,000 persons, while neighboring New Zealand reported a STEC infection rate of 3.3 cases per 100,000 per year (14).

In Latin America, STEC infections are endemic and contribute to the burden of acute diarrheal syndrome in children less than 5 years of age, being responsible for 2% of total cases of acute diarrhea, and in a few studies correspond to 20 to 30% of bloody diarrhea (8).

Important differences exist in the incidence of STEC infections and HUS in South America. A regional network for surveillance purposes is still nonexistent, and data are restricted to only a few countries. HUS is endemic in some countries of the southern cone region, and reporting is mandatory only in Argentina, Bolivia, Chile, and Paraguay.

In Brazil, STEC infections are important public health issues, at least in some regions, but in general, the incidence is relatively low (15). In Chile, a National Surveillance System

was established in 1999, and all clinical laboratories must report and send isolates to the Reference Laboratory. In Uruguay, reports of HUS are not mandatory, and only a few cases are recognized each year (16). In Paraguay, reporting of HUS has been mandatory since 2005, and the estimated annual incidence is 0.6 cases per 100,000 children under 5 years old (8).

In Argentina, STEC-associated illnesses are a serious public health concern. Data on human STEC infections are gathered through different strategies: (i) reporting of clinical HUS cases to the National Health Surveillance System (in Argentina the system is named Sistema Nacional de Vigilancia de la Salud [SNVS]); reports, which have been mandatory since 2000, must be immediate and individualized; (ii) the Sentinel Surveillance System through 25 HUS sentinel units; (iii) the laboratory-based surveillance system through the National Diarrheal and Foodborne Pathogens Network; and (iv) the Molecular Surveillance through the PulseNet of Latin America and Caribbean.

Postdiarrheal HUS is endemic, with the highest rate of pediatric cases globally. Over the last 10 years, approximately 400 HUS cases were reported annually. The incidence ranged from 10 to 17 cases per 100,000 children less than 5 years of age, and lethality was between 1 and 4% (Fig. 1).

Between 2004 and 2010, a total of 1,245 O157 and non-O157 STEC strains, isolated from HUS (597) and bloody (335) and non-bloody (167) diarrhea cases, healthy carriers (74), and unspecified pathologies (72), were confirmed by the laboratory-based surveillance system and the HUS sentinel units. Multiple serotypes were identified, but O157:H7 was the predominant (>70%), and O145:NM (13.6%) was the second most important serotype identified. Among the STEC O157 strains, the $stx_{2a}/stx_{2c}/eae/ehxA$ genotype prevailed (>80%). For the non-O157 strains, the genotypes were more diverse, but the full virulent $stx_{2a}/eae/ehxA$ genotype was prevalent (>60%). In Argentina, outbreaks are identified

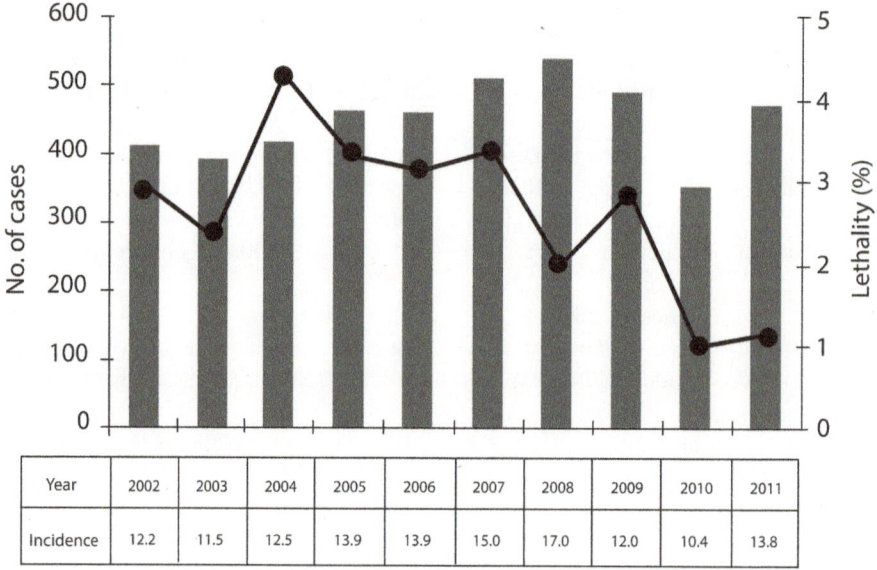

Year	2002	2003	2004	2005	2006	2007	2008	2009	2010	2011
Incidence	12.2	11.5	12.5	13.9	13.9	15.0	17.0	12.0	10.4	13.8

FIGURE 1 Number of HUS cases, incidence rates, and percentages of lethality in Argentina, 2002–2011. doi:10.1128/microbiolspec.EHEC-0002-2013.f1

through the surveillance system of HUS and STEC-associated diseases. The definition of an outbreak used for this analysis is two or more linked cases. Pulsed-field gel electrophoresis (PFGE) and phage typing are used to establish the clonal relatedness of the isolates. In the period 2004 to 2010, a total of 12 outbreaks of bloody diarrhea and HUS cases associated with O157 and non-O157 STEC strains occurred in kindergartens, families, and the community. The outbreak size ranged from 2 to 32 cases, and two patients with HUS died. Person-to-person transmission was the main route identified.

As a part of PulseNet Latin America and Caribbean, national databases were created for O157 and non-O157 *E. coli*, including strains isolated since 1988 from different sources. Among O157 strains, two patterns, named AREXHX01.011 and AREXHX01.022, are prevalent, representing around 13% of the database (Fig. 2a and 2b). Pattern 011 and other related patterns, with 95% similarity, are part of the hypervirulent clone described in different countries (Fig. 2c). Pattern 011, which has been prevalent in Argentina in the past 10 years, is identical to the most prevalent pattern in Sweden (named SMI-H) and to the most common type, named 047, in human infections in the United States (17). Among the non-O157 strains, the PFGE patterns are more diverse and two patterns, named AREXSX01.0006 (O145) and AREXPX01.0008 (O113), are prevalent.

COST OF STEC-ASSOCIATED DISEASES

Severe illnesses with long-term sequelae caused by STEC have a social and economic cost to the community and the health system. However, it is difficult to compare data from cost of illness studies among countries because of differences in definitions of costs, methodologies used, and income distribution.

From FoodNet data (2005–2008), non-O157 STEC strains are estimated to cause 168,698 illnesses each year, and *E. coli* O157:H7, 96,534 cases in the United States, with more than 3,600 hospitalizations and 30 deaths (18). Frenzen et al. (19) estimated the total annual cost of illness (COI) due

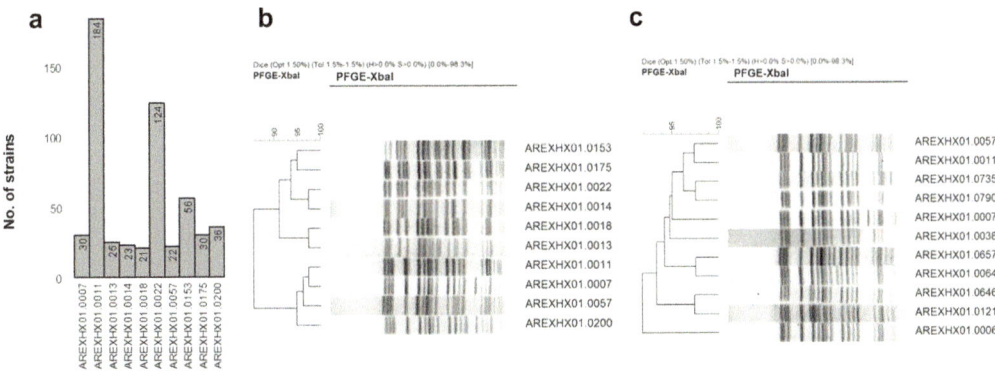

FIGURE 2 (a) Top ten XbaI-PFGE patterns associated with human STEC O157 strains in Argentina. (b) Dendrogram of the top ten XbaI-PFGE patterns. (c) Dendrogram of AREHXHX01.0011 pattern and other related patterns. doi:10.1128/microbiolspec.EHEC-0002-2013.f2

to STEC O157 at USD$405 million in 2003, Buzby and Roberts (20) updated this estimate to USD$459 million in 2007, and Scharff et al. (21) suggested that STEC O157 infections cost about USD$990 million in 2009 to U.S. residents.

In the Netherlands, Havelaar et al. (22) described the burden of disease associated with STEC O157 at the population level using the public health indicator "Disability-Adjusted Life Years" (DALYs), and showed that mortality due to HUS, ESRD, and dialysis due to ESRD constitute the main determinants. Tariq et al. (23) evaluated the societal impact of STEC O157 infection using a combination of DALYs and COI, including direct health care costs and indirect non-health care costs. Total annual COI due to STEC O157 infection for the Dutch society was estimated at €9.1 million. The authors concluded that, compared to other food-borne pathogens, STEC O157 infections result in a relatively low burden and low annual costs at the societal level, but the burden and costs per case are high.

In Australia, McPherson et al. (24) estimated the cost of STEC infections at approximately AUD$2.6 million each year.

In Argentina, Caletti et al. (25) evaluated the direct health care costs and indirect

non-health care costs of 231 HUS cases attending the Hospital Nacional de Pediatría "Dr. Juan P. Garrahan" in Buenos Aires City, in the period 1987 to 2003. The total annual cost for HUS cases was approximately USD$2 million.

RISK FACTORS

Some determinants of the pathogen and its reservoir, the host, and cultural and dietary behaviors have been described as risk factors for acquiring an STEC-associated disease.

Pathogen Factors

Several determinants have been described as risk factors that may play a role in the outcome of an STEC infection, such as the initial bacterial inoculum, the amount and type of Stx produced, and the serotype of the infecting strain; the ability of *E. coli* to horizontally acquire specific genetic elements known as pathogenicity islands (PAI), and *stx* genes from free bacteriophages in the environment and in the mammalian hosts; and the improved adaptation of the bacteria to human hosts (26).

Among STEC strains, *E. coli* O157:H7 has become a significant food-borne pathogen, exhibiting some characteristic features, such as low infectious dose (~100 to 500 organisms) and acid tolerance that certainly favors their transmission to humans by the food chain. In addition to STEC O157, only a restricted number of STEC serotypes (mainly those of seropathotype B) are associated with outbreaks and HUS. Moreover, Stx types differ in their biological activity and association with disease. There is evidence of a linkage of Stx2a (formerly Stx2) with a higher risk of severe human disease. The presence of both stx_{2a} and eae genes in an STEC isolate is considered to be a predictor of HUS (27). However, it has been shown that highly pathogenic strains producing Stx2d$_{activatable}$ are eae negative (28).

Besides Stx and lipopolysaccharide, other putative virulence factors, including adhesins, other toxins and proteases are required to develop disease. Extensive evidence suggests that a major pathogen determinant is the presence of specific PAIs. A number of PAIs, including the locus of enterocyte effacement (LEE), play a major role in enhancing the ability of some serotypes to cause severe human disease. In addition to the proteins encoded in the LEE, the type III secretion system also secretes other effectors encoded outside the LEE. At least three of these non-LEE-encoded effectors have been linked to non-O157 STEC strains that cause HUS (29).

As the population of STEC O157 strains increased in frequency and spread geographically, it has genetically diversified. Isolates of STEC O157 from clinical and bovine sources have been shown to be genotypically diverse by different methods, including PFGE, octomer-based genome scanning, and multilocus variable number of tandem repeats analysis. Studies of prophage and prophage remnants in STEC O157 strains have indicated that genotypic diversity is largely attributable to bacteriophage-related insertions, deletions, and duplications of variable sizes of DNA fragments (30).

After the description of the hypervirulent clade of O157 associated with the raw spinach outbreak in the United States in 2006 (30), several studies were performed in the northern hemisphere countries to assess the frequency of this clade in human disease and cattle. However, no data were available from the southern hemisphere countries.

As notable differences were observed in the prevalence and severity of human diseases caused by O157 isolates in Argentina (high) and Australia (low), Mellor et al. (31) compared human and bovine O157 isolates from both countries. The locus-specific polymorphism analysis genotyping revealed that lineage I/II (LI/II) *E. coli* O157 isolates were the most prevalent in Argentina (88%) and Australia (88%). Argentinean LI/II isolates were shown to belong to clades 4 (30%) and 8 (65%) while Australian LI/II isolates were identified as clades 6 (15%), 7 (80%), and 8 (2%). In Argentina, clade 8 isolates dominated in both cattle (50%) and humans (80%); meanwhile, in Australia clade 7 dominated in both cattle (70%) and humans (90%).

Host Factors

Several host factors influence the risk of acquiring STEC infection, including age, immunity, health status, the use of antibiotics and antimotility agents, stress, and genetic factors.

The highest age-specific frequency of HUS is in infants and young children. It declines with increasing age and increases again in the elderly, probably due to changes in the immunity status. Gastric acidity is an important initial host barrier to ingested pathogens. Its protective role against *E. coli* O157:H7 infection has been suggested because individuals with low gastric acidity are at a significantly higher risk for HUS developing than those with normal physiological gastric function.

The possible role of stress as a risk factor for severe disease, and host genetic factors that may influence host-pathogen interactions, including the innate immune response to infection and the nature of the toxin-cell interaction, have been described. The genes that regulate the gut colonization by *E. coli* O157:H7 may be modulated by hormone-like soluble factors produced by other bacteria in a process known as quorum sensing. The quorum-sensing pathway could be activated by host stress hormones such as epinephrine and norepinephrine (32).

Behavior and Cultural Factors

Since the emergence of STEC, case-control and population-based studies, varying in sizes and rigor, have been conducted to examine risk factors in associated infections.

In the 1990s, studies to evaluate risk factors were focused on sporadic cases of *E. coli* O157 infection. The consumption of undercooked hamburgers and meat, eating in restaurants or fast-food establishments, living or working on or visiting a cattle farm, drinking untreated surface water, swimming in contaminated water, contact with animal feces, and consumption of raw milk were the main risk factors identified.

In a literature review of several articles published in the past decade, it was observed that several studies were conducted to learn more about the risk factors of both non-O157 and STEC O157 infections. The main findings of some of these studies are summarized in Table 1.

Cattle Management Factors

The role of cattle as a source of human infection has been extensively studied and reviewed, mostly in reference to STEC O157 (42). Prevalence of STEC O157 in cattle feces and hides is highly variable (43), dependent on region, farm and cattle type, age, and season, among other factors. The degree of herd infection is uneven, with herds usually being colonized by a low number of predominant strains (44). In Argentina, a similar pattern was found at the abattoirs, where STEC O157 was detected in approximately 15% of arriving lots, with an average prevalence in feces of 4.1% (45).

In the animal production environment, the prevalence of other STEC serogroups, mainly those named "big six" (O26, O45, O103, O111, O121, O145), is less well known. Recent studies have shown that only a small fraction of strains from those O groups isolated from cattle carry Stx genes (46, 47).

Risk of infection could be increased in spring and summer, as most reports state that during warmer months there is a higher prevalence of STEC in cattle (48). The risk of colonization and shedding through the use of different cattle diets is a complex issue, thoroughly revised lately (49). Attention has been given to finishing diets with an increased ratio of grains that could enhance STEC O157 shedding. Inclusion of orange peel (50) or a dietary shift to forage has been proposed to decrease STEC O157 shedding. However, much less is known about the effect of diets on the ecology of other STEC serotypes.

The emergence of super-shedding bovines is a relevant risk for STEC O157 contamination of the beef supply chain. Super-shedders have been linked to the diversity of STEC O157 prevalence in cattle populations (51) and to the increasing spread of hide contamination in feedlot cattle (52). These studies and others (53) have shown that by controlling super-shedding bovines there would be a high impact on preventing STEC O157 infection in humans.

STEC strains are able to survive some months in the environment, feces slurries, and cattle manure (54–56). A recent and systematic review (57) concluded that there is a complex relationship among animal reservoirs, pathogens, and the environment leading to the contamination of fresh produce from the environment. Manure and fecal contamination of irrigation water were the most important media for STEC presence in fresh produce at preharvest. The importance of minimizing these sources of contamination has been highlighted by STEC outbreaks with fresh produce such as spinach (58).

Mapping studies have given epidemiological evidence of the significance of environmental contamination for STEC human

TABLE 1 Risk factors for STEC infections identified by case-control studies and population-based studies, 2000–2012

Year of study	Country	Study design and size	Risk factors identified	Conclusions	Advice to change behaviors and reduce health risks	Reference
2000–2001	France	Matched case-control study to evaluate risk factors for sporadic HUS cases in children. 105 cases; 196 controls.	Eating undercooked ground beef; contact with a person with diarrhea; drinking well water during the summer period	Adequate cooking of ground beef may reduce the incidence of STEC infection. Assiduous attention to hygienic measures to prevent the spread of STEC within families and child-care facilities has the potential to further reduce HUS episodes in childhood.	Proper cooking of ground beef (National Public Health Campaign, 2006). Revision of safety control measures at all levels of the ground beef food chain. Specific education program for professionals involved in the meat industry.	33
2001–2002	Argentina	Matched case-control study to evaluate risk factors for sporadic STEC infections in children enrolled in two sites, Mendoza (urban and semirural area) and Buenos Aires (urban area). 150 cases; 299 controls.	Eating undercooked beef at any place; contact with a young child with diarrhea; attending a day care center or kindergarten; living in or visiting a place with farm animals; contact with farm animals and cattle manure; having nonparental income	Meat-related dietary habits and animal exposures were linked to illness. Person-to-person spread was an important mode of transmission. Some risk factors were specific by location. Eating a wider variety of fruits and vegetables and washing hands after handling raw beef, especially with soap and water, were protective factors.	Apply effective safe practices at all stages of the food chain (industry, government, and consumers). Ensure that beef is well cooked at home and outside home. Establish an educational program to avoid risks of dietary habits and behaviors, and recommendations to protect people through hand washing.	34
2001–2003	Germany	Matched case-control study to evaluate risk factors for sporadic STEC infections in different age groups. 202 cases; 202 controls.	Contact with small ruminants and consuming raw milk (children <3 years); playing in a sandbox (children <3 years and aged 3–9 years); swimming in nonpublic swimming pools (children aged 3–9 years); consuming lamb meat and raw fermented spreadable sausages (cases ≥10 years or older)	Risk factors were age specific. In children, food-borne transmission played a lesser role in both acquiring STEC infection and developing HUS. Consumption of lamb meat and raw spreadable sausages was identified as risk factor for the first time.	Modify the upper pH limit (5.6) of raw spreadable sausage because STEC survives under acidic conditions (pH 4.0).	35

1997–2006	Finland	Population-based study applying a statistical model to distinguish between risk factors for occurrence and incidence of STEC diseases. 131 cases.	Increased occurrence and incidence: proportion of beef cattle to human population; proportion of population with higher education related to consumption habits of undercooked meat. Increased incidence: proportion of fresh water; number of cultivated farms; proportion of low-income households with children	Socioeconomic factors, like low-level income and education, were important for acquiring STEC infections. Ecological factors such as the relation between population and density of beef cattle were important for the incidence of the disease.	Applying good hygiene in animal and slaughter process and food retail. Consumer education. Proper food handling. Up-to-date national legislation and regulation.	36
2003–2007	Australia	Case-control study to evaluate risk factors in sporadic O157 and non-O157 STEC infections. 213 cases; 304 controls.	STEC O157: Consumption of hamburgers; eating at restaurants; having occupational exposure to red meat. Non-O157 STEC (O111, O26, O103, O113, O172): occupational exposure to animals; consumption of sliced processed chicken meat and sliced corned beef; eating out at a catered event; camping in the bush	Risk factors were serogroup specific. Hamburgers and ground beef have not been implicated in outbreaks of STEC and have not been previously considered a source of infection.	Design educational programs for people who live or work with animals or raw meat, as well as for those who enjoy camping. Recommend that consumers and food handlers cook hamburgers thoroughly.	37
2000–2009	United States	Population-based study to compare risk factors in patients infected with O157 and non-O157 STEC. 392 patients.	O157 and O103 STEC: eating pink hamburger; handling raw ground beef. Non-O157 STEC (O111, O103, O26): international travel in the 7 days prior to symptoms. Non-O157 STEC (O111, O103, O26, O45): consumption of untreated surface water	Some STEC serogroups such as O103 seem to have an epidemiological and exposure profile similar to O157, and they likely occupy a similar ecologic niche. Other serogroups are quite different (O45) and may not be able to be managed through identical measures to control O157.	Continue population-level monitoring of the epidemiology of STEC to determine longer-term trends and opportunities for control.	38

(Continued on next page)

TABLE 1 Risk factors for STEC infections identified by case-control studies and population-based studies, 2000-2012 *(Continued)*

Year of study	Country	Study design and size	Risk factors identified	Conclusions	Advice to change behaviors and reduce health risks	Reference
2002–2009	Argentina	Case-control study to identify risks factors for sporadic STEC infections in children aged up to 6 from the Central Eastern area. 63 cases; 374 controls	Eating food prepared outside home	Protective effects of a diet that includes vegetables.	Plan strategies for local prevention to diminish the incidence of HUS in the region under study.	39
2007	Argentina	Epidemiological survey to evaluate risk factors for STEC infections in different socioeconomic groups. 883 students aged 10–12 years from elementary public schools of an urban area of Buenos Aires Province.	Eating commercially prepared precooked or homemade hamburgers, exposure to water of swimming pools; no hand wash after going to the toilet or before eating food	Differences in the frequency of hamburger consumption were observed among children from different socioeconomic strata. Children from high and medium strata attending private swimming pools and children from low stratum attending public swimming pools were at risk.	Improve educational programs to enhance personal hygiene, adequate meat handling and cooking techniques, and maintenance of safe recreational water.	40
2003–2012	Argentina	Epidemiological survey to evaluate risk factors for sporadic HUS cases and STEC infections in urban and semirural area of Río Negro Province. 42 cases.	Eating undercooked ground beef, sausages, barbecue, and unpasteurized milk; contact with farm animals; poor hygiene in food and father with rural activities, and poor hygiene of work clothes	Food, contact with rural workers, and environment were risk factors for children.	Improve the surveillance system, considering the particular conditions of the region. Establish a health education program sustained over time.	41

infection in rural areas (59–61). In Argentina, Tanaro et al. (62) have shown a degree of contamination of surface waters in rural areas of a cattle-producing region of Entre Rios Province.

Armstrong et al. (63) suggested, as a probable cause for the emergence of STEC O157, the changes in modern livestock and food processing industries, characterized by the concentration, homogenization, and increased scale of operations. Within the abattoir, the hide-removal operation is a critical point for carcass contamination (53). In addition, lairage areas have been identified as key for STEC dissemination as most isolates from carcasses were traceable to the lairage environment rather than to the original feedlot (64).

Studies conducted in Argentina showed that 11 STEC strains (ten O157:H7 and one O178:H19) isolated in abattoirs had similar phage type-XbaI-PFGE pattern-*stx* genotype profile as those responsible for 19 HUS cases in the same period (45, 65). For STEC O157 it was possible to link 12% of reported human infections to the bovine reservoir (66). In the same study, it was estimated that, at slaughter, ca. 38,000 bovines would carry STEC O157 in their feces per each clinical case of HUS and bloody and nonbloody diarrhea cases reported.

Risk factors identified in cattle production and at the abattoir can be integrated as components of risk assessment models that allow estimates of the risk of STEC infection for the population from consumption of a specific meat product (67). They are also useful tools to evaluate the outcome of mitigation strategies in cattle production and the meat industry (67, 68). Risk assessment studies have also been conducted to estimate the risk of STEC in ground beef hamburgers in Argentina (69).

NEW SCENARIO AND LEARNED LESSONS FROM OUTBREAKS

The epidemiological profile of food-borne diseases has changed dramatically in recent decades. Some of the contributing factors for the emergence of outbreaks with different epidemiological characteristics and for the widespread epidemics are (i) the development of new food processing technologies and foods, (ii) the more centralized and rapid food distribution systems, (iii) the changes in consumer preferences and behaviors, and (iv) the considerable increase in the volume of food products traded internationally. Additionally, the enormous increase in global travel allows individuals to be infected in one country and to become ill on their return to their country of origin (70).

This epidemiological change was also influenced by the genetic variation and "relentless evolution" of the O157 pathogen population (71). Mellmann et al. (72) remarked that bacterial evolution is an ongoing process that undoubtedly will lead to the emergence of other successful pathogenic clones of *E. coli* in the future.

Since the emergence of STEC as a food-borne pathogen in 1982 (1), large outbreaks of gastrointestinal disease, involving numerous persons and associated with different sources and vehicles of transmission, have been described worldwide.

Each outbreak showed different aspects of complexity, from the detection of the source and the vehicles of transmission to the need for the development of control strategies to avoid the occurrence of new cases. These large outbreaks were of great concern and challenge for the public health system, compelling the food regulatory and health agencies and the food industry to establish improved guidelines to control and prevent new incidents.

Large outbreaks have been commonly notified in industrialized countries like the United States (73), Japan (74), Germany (75), Canada (76), among others. In these countries, the advanced epidemiological system has contributed to a better investigation and understanding of these events. Furthermore, small indoor outbreaks were described worldwide, mainly in families and child-care centers with a lower frequency (77, 78).

In this context, it is interesting to point out the lessons learned about risk factors, treatment, diagnosis, and advances in health and food regulations from some emblematic outbreaks (Table 2).

The large outbreak of *E. coli* O157:H7 infection that occurred in four western states of the United States in February 1993 was the first event with a great population and media impact. For the first time, people were massively informed about good practices for food handling regarding this pathogen. The outbreak investigation allowed us to know about the low infectious dose of this organism, its capacity to survive for a long time (more than 2 months) in frozen storage, and the temperature and time combinations (68°C, 15 sec) that ensure a safe hamburger cooking condition. These findings were used as strong arguments to enforce a zero tolerance policy for this microorganism in processed food and for the need for a pronounced decrease in the contamination of raw ground beef (73, 79, 80).

At present, the Sakai outbreak of *E. coli* O157 infection in July 1996 was the largest outbreak ever experienced, because of the number of people affected (74). As described by Fukushima et al. (81), lunch foods contaminated with *E. coli* O157 and supplied to elementary schools by a centralized distribution system were the cause of this massive outbreak. It is interesting to note the clinical experiences in the treatment received by the affected children. Almost all patients were treated with antibiotics, fosfomycin and lactobacilli, from the early days of the illness. An evaluation of the outcomes revealed that these treatments were more than satisfactory compared to previous reports. However, the favorable outcome could not be only attributed to the effectiveness of antibiotics. Other factors, like age and racial differences, might also be responsible.

The spinach outbreak in the United States in July 2006 highlighted the importance of fresh produce as a vehicle in STEC infections and also the role of international trade because the product was exported to Canada and Mexico. During the epidemiological survey, most of the patients (95%) reported consuming uncooked fresh spinach during the 10 days before the onset of the illness (58). Women (71%) were the most affected, probably reflecting that women are more likely to consume fresh vegetables. Moreover, the outbreak was related to the practice of consuming prewashed, bagged leafy greens, as the general public assumed food handling in farms and in processing facilities to be safe. The sequencing of the genome of the TW14359 strain, responsible of this outbreak, helped identify the genetic factors that enhanced the ability of this strain to cause such a high number of HUS cases (82).

Manning et al. (30) published a comparison among outbreaks with different severity of the reported illness. The 1993 outbreak in western North America and the large 1996 outbreak in Japan had low rates of hospitalization and HUS, whereas the 2006 North American spinach outbreak had high rates of both hospitalization (50%) and HUS (10%). Single nucleotide polymorphism genotyping revealed the genetic variability among pathogenic strains associated with clinical infection. Their results support the hypothesis that the clade 8 lineage has recently acquired novel factors that contribute to the enhanced virulence.

The massive outbreak of bloody diarrhea and HUS that occurred in 2011 in Germany and other 13 European countries, the United States, and Canada was a challenge for clinicians and microbiologists given its atypical presentation. The seriousness of the illness and the fatalities, coupled with the lack of a definitive source of the causative agent, created a highly negative impact on the population and gained the front page of newspapers around the world (83).

This unprecedented outbreak affected mainly adults (90%), predominantly women, and resulted in an unusually high number of HUS cases (n=855). The augmented adherence of the strain to the intestinal epithelium facilitating systemic absorption of Stx could explain the high progression to HUS. The outbreak provided important new insight into novel antibiotic strategies in the treatment of HUS in adults and for decolonization of long-term STEC carriers (84).

The characterization of the O104:H4 outbreak strain revealed an unusual combination of pathogenic features typical of enteroaggregative *E. coli* combined with the capacity to produce Stx. Additionally, isolates displayed an extended-spectrum β-lactamase phenotype, carrying plasmid-borne blaCTX-M-15 and blaTEM-1 genes (85). DNA sequencing data rapidly revealed that outbreak strain was a new hybrid of two types of pathogenic *E. coli*. Up to date, nine O104:H4 isolates have been sequenced (https://github.com/ehec-outbreak-crowdsourced/BGI-data-analysis/wiki; http://www.bgisequence.com/eu/index.php?cID=194). Different methodologies for detection and characterization of the O104:H4 strain in clinical and food samples were developed in a short time (85, 86). Possible mitigation options for safe consumption of raw vegetables were advised after a fast-track assessment of the consumer exposure to STEC through this type of food (87).

A restaurant cohort study and the trace-back and trace-forward data analysis of the Task Force EHEC contributed to the identification of fenugreek seeds as sources of transmission. This study also contributed to confirm the implications of international trade in food-borne outbreaks (88).

One of the most important lessons from the O104 outbreak is the successful cooperation among health and food networks and agencies. This type of work could in a short time produce valuable epidemiological and microbiological information, essential for developing public health measures to improve the management of future outbreak situations (75). This outbreak emphasized the importance of common alert and surveillance systems for the early detection of international outbreaks and for a better assessment of their spread.

CONCLUSIONS

Differences in the frequency and the severity of STEC-associated human diseases are observed from country to country. The more reliable information is provided by developed countries as better enteric pathogen surveillance systems are in place. Because of the severity and the long-term sequelae of STEC-associated illnesses, they have a high social and economic cost for both the affected families and the health system. In Argentina, because of the number of HUS cases reported each year, those social and economic costs are particularly significant.

The risks for acquiring an STEC infection are associated with several determinants of the pathogen and its reservoir and with biological and cultural factors of the host. The best knowledge about risk factors was obtained from case-control and population-based studies. Main risk factors identified in earlier studies were dietary behaviors related to beef consumption, but at present they include a wider range of foods, such as fresh produce or sprouts. Other risky behaviors identified have been connected to environmental sources, as living in, working in, or visiting rural areas; swimming and camping in recreational areas; and being in contact with farm animals. Risk factors for STEC infection have also been identified in cattle management and at the abattoir, such as the effect of finishing diets, the existence of super-shedders in a herd, the cross-contamination in lairage areas, and the hide-removal step at the abattoir, among the most important ones. Another important risk factor is person-to-person transmission, especially for young children. In Argentina, beef is a traditional component of diet, with an average consumption of ca. 62.5 kg per capita per year. This high rate of consumption, and particularly some meat-related dietary habits, could be risk factors for STEC illness in our population.

A new risk scenario has emerged in the last decades due to the bacterial evolution that gave rise to the emergence of hypervirulent

TABLE 2 Features of major food-borne outbreaks associated with Shiga toxin-producing *E. coli*

Outbreak	No. of cases	No. of hosp.	No. of HUS	No. of deaths	Associated pathogen	Impact	Vehicle	Risk factors	Lessons learned
United States, 4 western states 1992–1993	501[a]	151[a]	45[a]	3[a]	STEC O157:H7 (stx_{1a}/stx_{2a})	First large outbreak associated with beef product. Different restaurants of the same food chain involved. High number of people affected. Population and media impact.	Burger	Centralized production and distribution system. Inadequate cooking.	Importance of: The low infectious dose. The survival in frozen storage. The safe cooking conditions. PFGE as subtyping technique for molecular surveillance. The mandatory notification of *E. coli* O157 infections. Good hygienic practices in food handling. Updating beef-related regulations.
Japan, Sakai City 1996	12,680[b]	398[c]	121[b]	3[b]	STEC O157:H7 (stx_{1a}/stx_{2a})	Massive and widespread outbreak. Children in elementary schools in different districts affected.	Radish sprouts, among others.	Centralized distribution of school lunch food. Raw vegetables.	Treatment of bloody diarrhea with fosfomycin and lactobacilli from the early days of the illness. Advances in food regulations. Food monitoring programs. Specific sanitization control measures in catering facilities. Hygiene guidelines for cooking practices at home.
United States, 26 states 2006	205[d]	95[d]	29[d]	2[d]	STEC O157:H7 (stx_{2a})	Large outbreak with high number of hospitalizations and HUS cases.	Fresh spinach.	Consumption of prewashed, bagged fresh produce. Emergence of hypervirulent *E. coli* O157 strain. International trade.	Recognition of fresh produce as vehicles of transmission. Genome sequence and new virulence determinants recognized. Advances in the knowledge of hypervirulent clades.

| Germany and other 13 European countries, United States, and Canada 2011 | 3,842[e] | ND[f] | 855[e] | 53[e] | STEC O104:H4 (stx_{2a}) | Large outbreak with: High proportion of adults affected, mainly women. Unusually high number of HUS in adults. Patients with severe neurological symptoms followed by death. | Fenugreek Sprouts/seeds. | Eating salads. Emergence of hypervirulent hybrid EAEC-STEC O104 strain, with enhanced colonization and long-term shedding. | New methodologies for STEC O104 detection in food and humans. Sequencing by new-generation technologies. New therapeutic approaches for adults: azithromycin treatment for HUS and decolonization of long-term STEC carriers. Cooperation and teamwork of networks. New vision: Emergence of a new *E. coli.* pathotype. Any STEC could be potentially pathogenic for humans. Open-source data release, and prompt crowd-sourced analyses. |

[a] Data from reference 79.
[b] Data from reference 81.
[c] Data from reference 74.
[d] Data from reference 58.
[e] Data from reference 84.
[f] ND, no data.

O157 clones with a worldwide distribution, and other STEC strains with unusual combinations of pathogenic features, such as the O104:H4 strain. The epidemiological changes were also influenced by the increase in centralized food production and distribution systems and the growth in the volume of international trade of food ingredients.

The learned lessons from large and emblematic outbreaks could be summarized as (i) the advances in the knowledge of virulence determinants of new pathogenic strains; (ii) the recognition of new vehicles of infection; (iii) the development of new methodologies for STEC detection in foods and humans; (iv) the improvement of food regulations and hygiene guidelines; (v) the new therapeutic approaches in the treatment of STEC-infected patients, especially HUS in adults; (vi) the establishment of continuous educational programs for food consumers; and (vii) the enhanced cooperation and teamwork of regional and international networks.

ACKNOWLEDGMENT

We declare no conflicts of interest with regard to the manuscript.

CITATION

Rivas M, Chinen I, Miliwebsky E, Masana M. 2014. Risk factors for Shiga toxin-producing *Escherichia coli*-associated human diseases. Microbiol Spectrum 2(5):EHEC-0002-2013.

REFERENCES

1. **Riley LW, Remis RS, Helgerson SD, McGee HB, Wells JG, Davis BR, Hebert RJ, Olcott ES, Johnson LM, Hargrett NT, Blake PA, Cohen ML.** 1983. Hemorrhagic colitis associated with a rare *Escherichia coli* O157:H7 serotype. *N Engl J Med* **308:**681–685.

2. **Gyles CL.** 2007. Shiga toxin-producing *Escherichia coli*: an overview. *J Anim Sci* **85:**E45–E62.

3. **Karmali MA, Steele BT, Petric M, Lim C.** 1983. Sporadic cases of hemolytic uremic syndrome associated with fecal cytotoxin and cytotoxin producing *Escherichia coli*. *Lancet* **i:**619–620.

4. **Karmali MA, Petric M, Lim C, Fleming PC, Arbus GS, Lior H.** 1985. The association between hemolytic uremic syndrome and infection by verotoxin-producing *Escherichia coli*. *J Infect Dis* **151:**775–782.

5. **Palermo MS, Exeni RA, Fernández GC.** 2009. Hemolytic uremic syndrome: pathogenesis and update of interventions. *Expert Rev Anti Infect Ther* **7:**697–707.

6. **Caprioli A, Morabito S, Brugère H, Oswald E.** 2005. Enterohaemorrhagic *Escherichia coli*: emerging issues on virulence and modes of transmission. *Vet Res* **36:**289–311.

7. **Erickson MC, Doyle MP.** 2007. Food as vehicle for transmission of Shiga toxin-producing *Escherichia coli*. *J Food Prot* **70:**2426–2449.

8. **Guth BEC, Prado V, Rivas M.** 2010. Shiga toxin-producing *Escherichia coli*, p 65–83. *In* Torres AG (ed), *Pathogenic* Escherichia coli *in Latin America*. Betham Science Publishers Ltd., Sharjah, United Arab Emirates.

9. **Mora A, Blanco M, Blanco JE, Dahbi G, López C, Justel P, Alonso MP, Echeita A, Bernárdez MI, González EA, Blanco J.** 2007. Serotypes, virulence genes and intimin types of Shiga toxin (verocytotoxin)-producing *Escherichia coli* isolates from minced beef in Lugo (Spain) from 1995 through 2003. *BMC Microbiol* **1:**7–13.

10. **Karmali MA, Mascarenhas M, Shen S, Ziebell K, Johnson S, Reid-Smith R, Isaac-Renton J, Clark C, Rahn K, Kaper JB.** 2003. Association of genomic O island 122 of *Escherichia coli* EDL 933 with verocytotoxin-producing *Escherichia coli* seropathotypes that are linked to epidemic and/or serious disease. *J Clin Microbiol* **41:**4930–4940.

11. **Centers for Disease Control and Prevention.** 2012. *Foodborne Diseases Active Surveillance Network (FoodNet): FoodNet Surveillance Report for 2011 (Final Report)*. Atlanta, Georgia: U.S. Department of Health and Human Services, Centers for Disease Control and Prevention, Atlanta, GA. http://www.cdc.gov/foodnet/PDFs/2011_annual_report_508c.pdf. Accessed 25 June 2014.

12. **Government of Canada, National Integrated Enteric Pathogen SurveillanceSystem (C-EnterNet).** 2012. *2011 Short Report*. Guelph, ON: Public Health Agency of Canada, Guelph, ON. http://www.phac-aspc.gc.ca/c-enternet/publications-eng.php#a2. Accessed 17 April 2013.

13. **European Centre for Disease Prevention and Control.** 2013. *Annual epidemiological report 2012. Reporting on 2010 surveillance data and 2011 epidemic intelligence data*. ECDC, Stockholm, Sweden. http://ecdc.europa.eu/en/Publications/Annual-Epidemiological-Report-2012.pdf. Accessed 17 April 2013.

14. **Vally H, Hall G, Dyda A, Raupach J, Knope K, Combs B, Desmarchelier P.** 2012. Epidemiology of Shiga toxin-producing *Escherichia coli* in Australia, 2000–2010. *BMC Public Health* **12:**63–74.

15. **Guth BEC, Picheth CF, Gomes TAT.** 2010. *Escherichia coli* situation in Brazil, p 162–178. *In* Torres AG (ed), *Pathogenic* Escherichia coli *in Latin America.* Betham Science Publishers Ltd., Sharjah, United Arab Emirates.

16. **Varela G, Gómez-Duarte OG, Ochoa T.** 2010. Diarrheigenic *Escherichia coli* in children from Uruguay, Colombia and Perú, p 209–222. *In* Torres AG (ed), *Pathogenic* Escherichia coli *in Latin America.* Betham Science Publishers Ltd., Sharjah, United Arab Emirates.

17. **Löfdahl S.** 2008. How global is VTEC?, p 65–67. *Epidemiology and Transmission of VTEC and Other Pathogenic* Escherichia coli. Swedish Institute for Infectious Disease Control, Stockholm, Sweden.

18. **Scallan E, Hoekstra RM, Angulo FJ, Tauxe RV, Widdowson MA, Roy SL, Jones JL, Griffin PM.** 2011. Foodborne illness acquired in the United States—major pathogens. *Emerg Infect Dis* **17:** 7–15.

19. **Frenzen PD, Drake A, Angulo FJ, the Emerging Infections Program Foodnet Working Group.** 2005. Economic cost of illness due to *Escherichia coli* O157 infections in the United States. *J Food Prot* **68:**2623–2630.

20. **Buzby JC, Roberts T.** 2009. The economics of enteric infections: human foodborne disease costs. *Gastroenterology* **136:**1851–1862.

21. **Scharff RL, McDowell J, Medeiros L.** 2009. Economic cost of foodborne illness in Ohio. *J Food Prot* **72:**128–136.

22. **Havelaar AH, van Duynhoven YTHP, Nauta MJ, Bouwknegt M, Heuvelink AE, de Wit GA, Nieuwenhuizen MGM, van de Kar NCAJ.** 2004. Disease burden in The Netherlands due to infections with Shiga toxin-producing *Escherichia coli* O157. *Epidemiol Infect* **132:**467–484.

23. **Tariq L, Haagsma J, Havelaar A.** 2011. Cost of illness and disease burden in The Netherlands due to infections with Shiga toxin–producing *Escherichia coli* O157. *J Food Prot* **74:**545–552.

24. **McPherson M, Kirk MD, Raupach J, Combs B, Butler JRG.** 2011. Economic costs of Shiga toxin-producing *Escherichia coli* infection in Australia. *Foodborne Pathog Dis* **8:**55–62.

25. **Caletti MG, Petetta D, Jaitt M, Casaliba S, Gimenez A.** 2006. Evaluación de costos directos e indirectos del tratamiento del Síndrome Urémico Hemolítico en sus distintas etapas evolutivas. *Medicina (Buenos Aires)* **66**(Suppl):22–26.

26. **Beutin L.** 2006. Emerging enterohaemorrhagic *Escherichia coli*, causes and effects of the rise of a human pathogen. *J Vet Med B Infect Dis Vet Public Health* **53:**299–305.

27. **Boerlin P, McEwen SA, Boerlin-Petzold F, Wilson JB, Johnson RP, Gyles CL.** 1999. Associations between virulence factors of Shiga toxin-producing *Escherichia coli* and disease in humans. *J Clin Microbiol* **37:**497–503.

28. **Bielaszewska M, Friedrich AW, Aldick T, Schurk-Bulgrin R, Karch H.** 2006. Shiga toxin activatable by intestinal mucus in *Escherichia coli* isolated from humans: predictor for a severe clinical outcome. *Clin Infect Dis* **43:**1160–1167.

29. **Coombes BK, Wickham ME, Mascarenhas M, Gruenheid S, Finlay BB, Karmali MA.** 2008. Molecular analysis as an aid to assess the public health risk of non-O157 Shiga toxin-producing *Escherichia coli* strains. *Appl Environ Microbiol* **74:**2153–2160.

30. **Manning SD, Motiwala AS, Springman AC, Qi W, Lacher DW, Ouellette LM, Mladonicky JM, Somsel P, Rudrik JT, Dietrich SE, Zhang W, Swaminathan B, Alland D, Whittam TS.** 2008. Variation in virulence among clades of *Escherichia coli* O157:H7 associated with disease outbreaks. *Proc Natl Acad Sci USA* **105:**4868–4873.

31. **Mellor GE, Sim EM, Barlow RS, D'Astek BA, Galli L, Chinen I, Rivas M, Gobius KS.** 2012. Phylogenetically related Argentinean and Australian *Escherichia coli* O157 isolates are distinguished by virulence clades and alternative Shiga toxin 1 and 2 prophages. *Appl Environ Microbiol* **78:**4724–4731.

32. **Clarke MB, Sperandio V.** 2005. Events at the host-microbial interface of the gastrointestinal tract III. Cell-to-cell signalling among microbial flora, host, and pathogens: there is a whole lot of talking going on. *Am J Physiol Gastrointest Liver Physiol* **288:**G1105–G1109.

33. **Vaillant V, Espié E, de Valk H, Durr U, Barataud D, Bouvet P, Grimont F, Desenclos JC.** 2009. Undercooked ground beef and person-to-person transmission as major risk factors for sporadic hemolytic uremic syndrome related to Shiga-toxin producing *Escherichia coli* infections in children in France. *Pediatr Infect Dis J* **28:**650–653.

34. **Rivas M, Sosa-Estani S, Rangel J, Caletti MG, Vallés P, Roldán CD, Balbi L, Marsano de Mollar MC, Amoedo D, Miliwebsky E, Chinen I, Hoekstra RM, Mead P, Griffin PM.** 2008. Risk factors for sporadic Shiga toxin-producing *Escherichia coli* infections in children, Argentina. *Emerg Infect Dis* **14:**763–771.

35. **Werber D, Behnke SC, Fruth A, Merle R, Menzler S, Glaser S, Kreienbrock L, Prager R,**

Tschape H, Roggentin P. 2007. Shiga toxin-producing *Escherichia coli* infection in Germany: different risk factors for different age groups. *Am J Epidemiol* **165**:425–434.

36. Jalava K, Ollgren J, Eklund M, Sitonen A, Kuusi M. 2011. Agricultural, socioeconomic and environmental variables as risks for human verotoxigenic *Escherichia coli* (VTEC) infection in Finland. *BMC Infect Dis* **11**:275.

37. McPherson M, Lalor K, Combs B, Raupach J, Stafford R, Kirk MD. 2009. Serogroup-specific risk factors for Shiga toxin–producing *Escherichia coli* infection in Australia. *Clin Infect Dis* **49**:249–256.

38. Hadler JL, Clogher P, Hurd S, Phan Q, Mandour M, Bemis K, Marcus R. 2011. Ten-year trends and risk factors for non-O157 Shiga toxin-producing *Escherichia coli* found through Shiga toxin testing, Connecticut, 2000–2009. *Clin Infect Dis* **53**:269–276.

39. Rivero MA, Passucci JA, Rodriguez EM, Signorini ML, Tarabla HD, Parma AE. 2011. Factors associated with sporadic verotoxigenic *Escherichia coli* infection in children with diarrhea from the Central Eastern Area of Argentina. *Foodborne Pathog Dis* **8**:901–906.

40. Bentancor AB, Ameal LA, Calvino MF, Martinez MC, Miccio L, Osvaldo J, Degregorio OJ. 2012. Risk factors for Shiga toxin-producing *Escherichia coli* infections in preadolescent schoolchildren in Buenos Aires, Argentina. *J Infect Dev Ctries* **6**:378–386.

41. Di Pietro S, Stafforini G, Cifone N, Alvarez ML, Arellano O, Manzini S, Haritchabalet K, Nóbile M, Chinen I, Miliwebsky E, Rivas M, Larrieu E. 2013. Epidemiología del síndrome urémico hemolítico, Viedma, Provincia de Río Negro 2003–2012. *Rev Arg Zoonosis* **III**:11–15.

42. Ferens WA, Hovde CJ. 2011. *Escherichia coli* O157:H7: animal reservoir and sources of human infection. *Foodborne Pathog Dis* **8**:465–487.

43. Rhoades JR, Duffy G, Koutsoumanis K. 2009. Prevalence and concentration of verocytotoxigenic *Escherichia coli*, *Salmonella enterica* and *Listeria monocytogenes* in the beef production chain: a review. *Food Microbiol* **4**:357–376.

44. Renter DG, Sargeant JM, Oberst RD, Samadpour M. 2003. Diversity, frequency, and persistence of *Escherichia coli* O157 strains from range cattle environments. *Appl Environ Microbiol* **69**:542–547.

45. Masana MO, Leotta GA, Del Castillo LL, D'Astek BA, Palladino PM, Galli L, Vilacoba E, Carbonari C, Rodríguez HR, Rivas M. 2010. Prevalence, characterization, and genotypic analysis of *Escherichia coli* O157:H7/NM from selected beef exporting abattoirs of Argentina. *J Food Prot* **73**:649–656.

46. Kalchayanand N, Arthur TM, Bosilevac JM, Wells JE, Wheeler TL. 2013. Chromogenic agar medium for detection and isolation of *Escherichia coli* serogroups O26, O45, O103, O121 and O145 from fresh beef and cattle feces. *J Food Prot* **76**:192–199.

47. Thomas KM, McCann MS, Collery MM, Logan A, Whyte P, McDowell DA, Duffy G. 2012. Tracking verotoxigenic *Escherichia coli* O157, O26, O111, O103 and O145 in Irish cattle. *Int J Food Microbiol* **153**:288–296.

48. Barkocy-Gallagher GA, Arthur TM, Rivera-Betancourt M, Nou X, Shackelford SD, Wheeler TL, Koohmaraie M. 2003. Seasonal prevalence of Shiga toxin-producing *Escherichia coli* including O157 and non-O157 serotypes, and *Salmonella* in commercial beef processing plants. *J Food Prot* **66**:1978–1986.

49. Callaway TR, Carr MA, Edrington TS, Anderson RC, Nisbet DJ. 2009. Diet, *Escherichia coli* O157:H7, and cattle: a review after 10 years. *Curr Issues Mol Biol* **11**:67–79.

50. Callaway TR, Carroll JA, Arthington JD, Edrington TS, Rossman ML, Carr MA, Krueger NA, Ricke SC, Crandall P, Nisbet DJ. 2011. *Escherichia coli* O157:H7 populations in ruminants can be reduced by orange peel product feeding. *J Food Prot* **74**:1917–1921.

51. Matthews L, McKendrick IJ, Ternent H, Gunn GJ, Synge B, Woolhouse ME. 2006. Super-shedding cattle and the transmission dynamics of *Escherichia coli* O157. *Epidemiol Infect* **134**:131–142.

52. Arthur TM, Keen JE, Bosilevac JM, Brichta-Harhay DM, Kalchayanand N, Shackelford SD, Wheeler TL, Nou X, Koohmaraie M. 2009. Longitudinal study of *Escherichia coli* O157:H7 in a beef cattle feedlot and role of high-level shedders in hide contamination. *Appl Environ Microbiol* **75**:6515–6523.

53. Arthur TM, Brichta-Harhay DM, Bosilevac JM, Kalchayanand N, Shackelford SD, Wheeler TL, Koohmaraie M. 2010. Super shedding of *Escherichia coli* O157:H7 by cattle and the impact on beef contamination. *Meat Sci* **86**:32–37.

54. Fukushima H, Hoshina K, Gomyoda M. 1999. Long-term survival of Shiga toxin-producing *Escherichia coli* O26, O111, and O157 in bovine feces. *Appl Environ Microbiol* **65**:5177–5181.

55. Fremaux B, Prigent-Combaret C, Delignette-Muller ML, Dothal M, Vernozy-Rozand C. 2007. Persistence of Shiga toxin-producing *Escherichia coli* O26 in cow slurry. *Lett Appl Microbiol* **45**:55–61.

56. Fremaux B, Delignette-Muller ML, Prigent-Combaret C, Gleizal A, Vernozy-Rozand C. 2007. Growth and survival of non-O157:H7 Shiga

toxin-producing *Escherichia coli* in cow manure. *J Appl Microbiol* **102:**89–99.

57. **Park S, Szonyi B, Gautam R, Nightingale K, Anciso J, Ivanek R.** 2012. Risk factors for microbial contamination in fruits and vegetables at the preharvest level: a systematic review. *J Food Prot* **11:**2055–2081.

58. **Centers for Disease Control and Prevention.** 2006. Ongoing multistate outbreak of *Escherichia coli* serotype O157:H7 infections associated with consumption of fresh spinach—United States, September 2006. *Morbid Mortal Wkly Rep* **55:** 1045–1046.

59. **Frank C, Kapfhammer S, Werber D, Stark K, Held L.** 2008. Cattle density and Shiga toxin-producing *Escherichia coli* infection in Germany: increased risk for most but not all serotypes. *Vector Borne Zoonotic Dis* **8:**635–643.

60. **Strachan NJC, Dunn GM, Locking ME, Reid TMS, Ogden ID.** 2006. *Escherichia coli* O157: burger bug or environmental pathogen? *Int J Food Microbiol* **112:**129–137.

61. **Kistemann T, Zimmer S, Vagsholm I, Andersson Y.** 2004. GIS-supported investigation of human EHEC and cattle VTEC O157 infections in Sweden: geographical distribution, spatial variation and possible risk factors. *Epidemiol Infect* **132:**495–505.

62. **Tanaro JD, Leotta GA, Lound LH, Galli L, Piaggio MC, Carbonari CC, Araujo S, Rivas M.** 2010. *Escherichia coli* O157 in bovine feces and surface water streams in a beef cattle farm of Argentina. *Foodborne Pathog Dis* **7:**475–477.

63. **Armstrong GL, Hollingsworth J, Morris JG Jr.** 1996. Emerging foodborne pathogens: *Escherichia coli* O157:H7 as a model of entry of a new pathogen into the food supply of the developed world. *Epidemiol Rev* **18:**29–51.

64. **Arthur TM, Bosilevac JM, Brichta-Harhay DM, Kalchayanand N, King DA, Shackelford SD, Wheeler TL, Koohmaraie M.** 2008. Source tracking of *Escherichia coli* O157:H7 and *Salmonella* contamination in the lairage environment of commercial U.S. beef processing plants and identification of an effective intervention. *J Food Prot* **71:**1752–1760.

65. **Masana MO, D'Astek BA, Palladino PM, Galli L, Del Castillo LL, Carbonari C, Leotta GA, Vilacoba E, Irino K, Rivas M.** 2011. Genotypic characterization of non-O157 Shiga toxin-producing *Escherichia coli* in beef abattoirs of Argentina. *J Food Prot* **74:**2008–2017.

66. **D'Astek BA, Del Castillo LL, Miliwebsky E, Carbonari C, Palladino PM, Deza N, Chinen I, Manfredi E, Leotta GA, Masana MO, Rivas M.** 2012. Subtyping of *Escherichia coli* O157:H7 strains isolated from human infections and healthy cattle in Argentina. *Foodborne Pathog Dis* **9:**457–464.

67. **Smith BA, Fazil A, Lammerding AM.** 2013. A risk assessment model for *Escherichia coli* O157:H7 in ground beef and beef cuts in Canada: evaluating the effects of interventions. *Food Control* **29:**364–381.

68. **Hurd HS, Malladi S.** 2012. An outcomes model to evaluate risks and benefits of *Escherichia coli* vaccination in beef cattle. *Foodborne Pathog Dis* **10:**952–961.

69. **Signorini M, Tarabla H.** 2009. Quantitative risk assessment for verocytotoxigenic *Escherichia coli* in ground beef hamburgers in Argentina. *Int J Food Microbiol* **132:**153–161.

70. **World Health Organization.** 1997. *Prevention and control of enterohaemorrhagic Escherichia coli (EHEC) infections*. Report of a WHO Consultation. World Health Organization, Geneva, Switzerland.

71. **Robins-Browne RM.** 2005. The relentless evolution of pathogenic *Escherichia coli*. *Clin Infect Dis* **41:**793–794.

72. **Mellmann A, Bielaszewska M, Zimmerhackl LB, Prager R, Harmsen D, Tschäpe H, Karch H.** 2005. Enterohemorrhagic *Escherichia coli* in human infection: in vivo evolution of a bacterial pathogen. *Clin Infect Dis* **41:**785–792.

73. **Centers for Disease Control and Prevention.** 1993. Multistate outbreak of *Escherichia coli* O157: H7 infections from hamburgers—western United States, 1992–1993. *Morbid Mortal Wkly Rep* **42:**258–263.

74. **Michino H, Araki K, Minami S, Takaya S, Sakai N, Miyazaki M, Ono A, Yanagawa H.** 1999. Massive outbreak of *Escherichia coli* O157:H7 infection in schoolchildren in Sakai City, Japan, associated with consumption of white radish sprouts. *Am J Epidemiol* **150:**787–796.

75. **STEC Workshop Reporting Group.** 2012. Experiences from the Shiga toxin-producing *Escherichia coli* O104:H4 outbreak in Germany and research needs in the field, Berlin, 28–29 November 2011. *Euro Surveill* **17**(7):pii=20091. http://www.eurosurveillance.org/ViewArticle. aspx?ArticleId=20091. Accessed 24 April 2013.

76. **Matsell DG, White CT.** 2009. An outbreak of diarrhea-associated childhood hemolytic uremic syndrome: the Walkerton epidemic. *Kidney Int* **75:**S35–S37.

77. **Snedeker KG, Shaw DJ, Locking ME, Prescott RJ.** 2009. Primary and secondary cases in *Escherichia coli* O157 outbreaks: a statistical analysis. *BMC Infect Dis* **9:**144.

78. **Miliwebsky E, Deza N, Carbonari C, D'Astek B, Baschkier A, Zolezzi G, Manfredi E, Chinen I, Rivas M.** *Outbreaks of Shiga toxin-producing Escherichia coli in Argentina, 2007–2011*, abstr. P-017, p 107. Abstr. VTEC 2012, Amsterdam, The Netherlands, 6–9 May 2012.

79. **Bell BP, Goldoft M, Griffin PM, Davis MA, Gordon DC, Tarr PI, Bartleson CA, Lewis JH, Barrett TJ, Wells JG, Baron R, Kobayashi JA.** 1994. A multistate outbreak of *Escherichia coli* O157:H7-associated bloody diarrhea and hemolytic uremic syndrome from hamburgers. The Washington experience. *JAMA* **272:**1349–1353.

80. **Tuttle J, Gomez T, Doyle MP, Wells JG, Zhao T, Tauxe RV, Griffin PM.** 1999. Lessons from a large outbreak of *Escherichia coli* O157:H7 infections: insights into the infectious dose and method of widespread contamination of hamburger patties. *Epidemiol Infect* **122:**185–192.

81. **Fukushima I, Hashizume T, Morita Y, Tanaka J, Azuma K, Mizumoto Y, Kaneno M, Matsuura M, Konma K, Kitani T.** 1999. Clinical experiences in Sakai City Hospital during the massive outbreak of enterohemorrhagic *Escherichia coli* O157 infections in Sakai City, 1996. *Pediatr Int* **41:**213–217.

82. **Kulasekara BR, Jacobs M, Zhou Y, Wu Z, Sims E, Saenphimmachak C, Rohmer L, Ritchie JM, Radey M, McKevitt M, Freeman TL, Hayden H, Haugen E, Gillett W, Fong C, Chang J, Beskhlebnaya V, Waldor MK, Samadpour M, Whittam TS, Kaul R, Brittnacher M, Miller SI.** 2009. Analysis of the genome of the *Escherichia coli* O157:H7 2006 spinach-associated outbreak isolate indicates candidate genes that may enhance virulence. *Infect Immun* **77:**3713–3721.

83. **Rubino S, Cappuccinelli P, Kelvin DJ.** 2011. *Escherichia coli* (STEC) serotype O104 outbreak causing haemolytic syndrome (HUS) in Germany and France. *J Infect Dev Ctries* **5:**437–440.

84. **Hauswaldt S, Nitschke M, Sayk F, Solbach W, Knobloch JKM.** 2013. Lessons learned from outbreaks of Shiga toxin-producing *Escherichia coli. Curr Infect Dis Rep* **15:**4–9.

85. **Bielaszewska M, Mellmann A, Zhang W, Köck R, Fruth A, Bauwens A, Peters G, Karch H.** 2011. Characterisation of the *Escherichia coli* strain associated with an outbreak of haemolytic uraemic syndrome in Germany, 2011: a microbiological study. *Lancet Infect Dis* **11:**671–676.

86. **Scheutz F, Møller Nielsen E, Frimodt-Møller J, Boisen N, Morabito S, Tozzoli R, Nataro JP, Caprioli A.** 2011. Characteristics of the enteroaggregative Shiga toxin/verotoxin-producing *Escherichia coli* O104:H4 strain causing the outbreak of haemolytic uraemic syndrome in Germany, May to June 2011. *Euro Surveill* **16**(24):pii=19889. http://www.eurosurveillance.org/ViewArticle.aspx?ArticleId=19889.

87. **European Food Safety Authority (EFSA).** 2011. Scientific Report of EFSA: Urgent advice on the public health risk of Shiga toxin-producing *Escherichia coli* in fresh vegetables. *EFSA J* **9:**2274–2324.

88. **Weiser AA, Gross S, Schielke A, Wigger JF, Ernert A, Adolphs J, Fetsch A, Müller-Graf C, Käsbohrer A, Mosbach-Schulz O, Appel B, Greiner M.** 2013. Trace-back and trace-forward tools developed *ad hoc* and used during the STEC O104:H4 outbreak 2011 in Germany and generic concepts for future outbreak situations. *Foodborne Pathog Dis* **10:**263–269.

Enterohemorrhagic *Escherichia coli* Pathogenesis and the Host Response

19

DIANA KARPMAN[1] and ANNE-LIE STÅHL[1]

INTRODUCTION

Bacterial exotoxins may cause damage to host cells by defined mechanisms. Depending on the presence of the globotriaosylceramide (Gb3) receptor, Shiga toxin may bind to cells and induce the ribotoxic stress response and apoptosis (1, 2). The toxin can also induce a proinflammatory response in cells, an effect that may be dissociated from ribosome inactivation and can even occur in cells lacking protein synthesis machinery. Bacterial lipopolysaccharide (LPS) induces a host response by binding to Toll-like receptor 4 (TLR4) and activating specific intracellular pathways. The activation of proinflammatory pathways, if excessive, promotes damage to the host. This article addresses enterohemorrhagic *Escherichia coli* (EHEC) pathogenesis and the host response, examining the innate and adaptive immune responses to the bacteria and virulence factors and how they affect the process of colonization, transfer and transport of virulence factors in the circulation, activation of thrombosis and inflammation, and specific end-organ damage to the kidney and the brain.

[1]Department of Pediatrics, Clinical Sciences, Lund University, Lund, Sweden.
Enterohemorrhagic Escherichia coli *and Other Shiga Toxin-Producing* E. coli
Edited by Vanessa Sperandio and Carolyn J. Hovde
© 2015 American Society for Microbiology, Washington, DC
doi:10.1128/microbiolspec.EHEC-0009-2013

EHEC COLONIZATION AND THE HOST RESPONSE IN THE INTESTINE

During intestinal colonization EHEC strains encounter chemical, mechanical, and biological barriers (3). Chemical encounters include saliva containing mucins and enzymes, acid stress in the stomach, bile secretion in the small intestine, and antimicrobial peptides throughout the intestine. The mechanical barrier consists of the mucus layer, and the biological barrier includes the intestinal microflora. These encounters, and the innate and acquired immune response to the pathogen, attempt to eliminate the bacteria. EHEC strains must overcome these barriers to colonize. After ingestion, strains survive stomach acidity by expressing acid-resistant systems (4–6). Interestingly, the response to acid stress is not only a survival strategy but was also found to activate certain EHEC properties associated with enhanced motility and cell adhesion, but not to affect the expression of Shiga toxin (Stx) (4, 6). From the stomach the bacteria pass to the small intestine where contact with bile may further promote migration (6).

Initial binding is assumed to occur at the follicle-associated epithelium of Peyer's patches and villi of the terminal ileum (7, 8). Bacteria may be taken up by intestinal M cells and transferred to underlying macrophages where they can survive and produce Stx (9). In the small and large intestine the bacteria come in contact with short-chain fatty acids, acetate, propionate, and butyrate, secreted by the intestinal flora as fermentation products of dietary carbohydrates (10). Low concentrations of butyrate were shown to upregulate EHEC virulence genes involved in motility and formation of attaching and effacing (A/E) lesions (11). This effect of butyrate was abrogated by deletion of the bacterial *lrp* gene encoding the leucine-responsive regulatory protein. High concentrations of short-chain fatty acids trigger expression of the bacterial *iha* gene conferring colonic adherence (12). Furthermore, butyrate treatment of colon cells in vitro resulted in enhanced expression of the

Stx receptor and thus increased susceptibility to the toxin (13).

Short-chain fatty acids also affect the expression of antimicrobial peptides, cathelicidin and defensins (14), thus creating a hostile environment for bacterial colonization, as shown by using the A/E-forming pathogen *Citrobacter rodentium* (15). In a mouse model inoculated with EHEC, cathelicidin protected wild-type mice from EHEC colonization, infection, and subsequent renal injury in comparison to knockout mice (3). The effect occurred at the intestinal level, as cathelicidin-sufficient mice were not protected from the effects of injected Stx. In addition to enhanced bacterial survival, cathelicidin-deficient mice had a thinner colonic mucus layer and thus a defective intestinal mechanical barrier. Although antimicrobial peptides protect the host from bacterial colonization, colonic pathogens may actually downregulate antimicrobial peptides, as demonstrated in shigellosis (16), to the advantage of the bacteria.

The interaction with the intestinal commensal microflora also activates communication between bacteria known as quorum sensing and between bacteria and host hormones. The interaction of bacteria with host catecholamines promotes virulence by activating bacterial motility (flagellar synthesis), formation of A/E lesions, and increased expression of Stx (17, 18). Presumably an increase in catecholamines in the circulation and the local intestinal environment could occur during hemorrhagic colitis.

The bacteria release Stx into the intestinal lumen and onto enterocytes. Toxin was demonstrated inside intestinal cells from a patient with EHEC infection, indicating that it can be taken up by these cells (19). These cells may express the Gb3 receptor to which the toxin binds (13, 20). The toxin may, however, undergo endocytosis by macropinocytosis (19). Furthermore, the toxin binds to intestinal Paneth cells (21). Alternatively, excess permeability of the mucosal barrier may allow Stx transport from the lumen in a paracellular fashion (22) although this remains a speculation. Thus the precise manner by which Stx binds to intestinal cells and is internalized or transported to the endothelium in the in vivo setting, is as yet, unclear. Stx leads to apoptosis of epithelial cells in vitro in human (23–25) and mouse intestinal cells (26). Apoptosis was demonstrated in the human intestine affected by EHEC infection, in a rabbit in vivo model (27), and in the murine EHEC model. The major virulence factor associated with this effect was determined to be Stx (28). The apoptosis-inducing effect initiates an inflammatory response and leukocyte influx of primarily phagocytes. Neutrophil migration toward the intestinal lumen occurs simultaneously and enhances Stx translocation from the lumen via enterocytes in vitro (29, 30). High leukocyte counts were demonstrated in feces of patients with *E. coli* O157:H7 infection (31), suggesting that a similar process may occur in vivo.

The intestinal inflammatory response is a paramount feature of host resistance to infection. The initial interaction between the intestinal mucosa and bacterial virulence factors may promote bacterial clearance by inducing an appropriate degree of inflammation. Invasive enteropathogenic strains induce an excessive release of chemokines from intestinal cell lines in comparison to noninvasive strains, including *E. coli* O157:H7 (32). All the same, in vitro studies have demonstrated release of interleukin-8 (IL-8) by T84 intestinal cells stimulated with EHEC, thus promoting inflammatory influx. Stx alone can also induce secretion of IL-8 and other C-X-C chemokines in the gut (33–35). Stx-induced IL-8 expression was associated with induction of mitogen-activated protein (MAP) kinase pathways Jun N-terminal protein kinase and stress-activated protein kinase and p38 in intestinal epithelial cells (23). Thus the apparently disparate proinflammatory and apoptotic pathways may converge via induction of host stress-activated MAP kinases. Stx induces the ribotoxic stress response, thereby inhibiting protein synthesis, but may, via eIF4E phosphorylation, promote translation of inflammatory mediators, so that both processes

occur simultaneously (36). Stx also induced the expression of tumor necrosis factor alpha (TNF-α) and IL-6 from murine peritoneal macrophages (37). In addition to Stx, long-polar fimbriae expressed by EHEC were recently shown to induce a proinflammatory response in T84 cells in a study showing that the NF-κB pathway was activated (38).

EHEC could, however, suppress the intestinal epithelial cytokine response to Stx (39). Likewise, EHEC, and Stx in particular, were shown to inhibit gamma interferon-mediated epithelial cell activation (40); both these effects could mitigate the host response and thus potentially promote bacterial colonization.

The importance of LPS in EHEC infection was demonstrated in a murine model using C3H/HeJ LPS-hyporesponsive mice in comparison to wild-type C3H/HeN mice. C3H/HeN mice developed earlier and simultaneous systemic and neurologic symptoms whereas C3H/HeJ mice exhibited a biphasic course of disease, first developing systemic symptoms and later severe neurologic symptoms (41). The discrepancy between LPS-responders and nonresponders was assumed to be related to the initial intestinal inflammatory response. LPS may induce a mucosal immune response during the initial phase of disease, which could facilitate bacterial clearance. A reduced initial response, due to lack of response to LPS, would promote more severe disease as clearance of bacteria from the intestine would be delayed, enabling bacterial proliferation and extended toxin release intestinally and systemically, thus explaining the biphasic prolonged course of disease in C3H/HeJ mice.

Pathogen-associated molecular patterns, or PAMPs, are specific pathogen-associated molecules recognized by cell receptors, such as TLRs, transmembrane receptors within the innate immune system. Stimulation of TLRs triggers a downstream cellular signal that results in the production and release of cytokines and chemokines. TLRs recruit intracellular adaptor proteins. MyD88 is an adaptor molecule common for all TLRs except for TLR3. LPS binds to the MD2-TLR4 receptor complex and initiates a signal cascade via MyD88-dependent or MyD88-independent pathways (42). TLR signaling depends on four adaptor proteins—MyD88, TIRAP (also known as MAL), TRIF (also known as TICAM1), and TRAM (also known as TICAM2)—that recruit downstream signaling components (43).

The importance of TLR4, TRIF, and MyD88 for the pathogenesis of EHEC infection was demonstrated in wild-type and knockout mice infected with *E. coli* O157:H7 (both Stx2-producing and non-producing) (44). Only Stx2-producing mice developed symptoms, and the most severe symptoms and pathology were demonstrated in MyD88-deficient mice. These mice also had the highest bacterial burden, suggesting that the immune response at the intestinal mucosa was essential for bacterial elimination. Even TLR4-knockout mice exhibited more severe disease than wild-type mice did. In contrast, an in vitro study showed that Stx uses TLR4 on intestinal cells for cellular uptake and transport (45); thus lack of TLR4 should have been protective in vivo, which was not the case.

THE ACQUIRED IMMUNE RESPONSE TO EHEC INFECTION

An acquired immune response develops after EHEC infection. In developing countries, enteropathogenic *E. coli* (EPEC) is a similar pathogen capable of causing diarrhea. In similarity to EHEC, it has a type-3 secretion system and can thus form A/E lesions on the intestinal epithelium, but EPEC does not produce Stx. Antibodies against antigens common to both EPEC and EHEC strains, such as *E. coli*-secreted proteins A and B as well as intimin, have been found in human serum, saliva, colostrum, and breast-milk (46–52).

EHEC-infected patients were also found to have serum and saliva antibodies against the infecting strain's LPS (53, 54) although the antibody levels decreased over time (55). Anti-LPS antibodies were also detected in umbilical cord sera of uninfected women in areas of EPEC endemicity (56). Patients may likewise develop serum antibodies against Stx2 and Stx1 (57), and a lesser antibody response to Stx and LPS was even detected in asymptomatic household contacts (58).

To what degree anti-Stx and anti-LPS antibodies exert a protective effect against infection is unclear. However, antibodies against common EPEC and EHEC antigens may have a protective effect and thus explain the low prevalence of EHEC infections in areas where EPEC is endemic (59). Colostrum from Brazilian mothers in areas of EPEC endemicity was shown to inhibit adherence of EHEC strains to HEp-2 cells (52). The protective effect of cross-immunity between EPEC and EHEC was demonstrated in a mouse model in which mice were first inoculated with EPEC, followed by inoculation with EHEC (60). Mice prechallenged with EPEC developed antibodies against common antigens and were protected from symptoms and pathology (intestinal and renal) caused by EHEC infection.

SHIGA TOXIN AND LPS INTERACTIONS WITH BLOOD CELLS

Neutrophils

Neutrophils are a prominent component of the acute inflammatory response, and levels of circulating neutrophils rise during an infectious process. In HUS, neutrophil counts are usually elevated at presentation (61), and the degree of neutrophilia at the onset of HUS is predictive of outcome (62). Neutrophils in HUS are activated and degranulated, thus releasing proteases (63, 64), as demonstrated by the presence of high levels of neutrophil elastase in patient sera (65, 66) and a higher capacity to adhere to cultured human endothelial cells and degrade fibronectin (67). Degranulation and activation of neutrophils correlated with poor prognosis (68, 69). Moreover, patient neutrophils were also delayed in their apoptotic program with an increased life span (63, 70).

Augmented levels of circulating apoptotic and necrotic leukocytes were found in patients with *E. coli* O104:H4-associated HUS (71). Plasma levels of leukocyte-derived microparticles were elevated as was binding of platelet microparticles to leukocytes. Inhibition of complement had only a moderate impact on microparticle levels while it increased the amount of apoptotic and necrotic leukocytes in the circulation (71).

Stx2 may promote neutrophilia in mice by triggering the release of cells of myeloid lineage from the bone marrow or by accelerated proliferation of polymorphonuclear leukocyte (PMN) progenitors (72). Whether Stx binds to neutrophils, or not, is debatable because the cells lack the Gb3 receptor but may bind to TLR4, see note added in proof (73–75). Studies on circulating neutrophils taken during the acute phase of HUS detected Stx bound to the surface of neutrophils (76–79) and to platelet-neutrophil aggregates (78) although the toxin apparently does not exert a cytotoxic effect on these cells.

In vitro studies have shown that Stx2 activates neutrophils, predominantly those in complex with platelets (78). Moreover, Stx may transfer from the neutrophil surface to endothelial cells, implying that neutrophils could serve as a carrier for the toxin in the circulation. In mice, Stx2 was shown to impair neutrophil migration (72). In addition, StcE, a metalloprotease released from *E. coli* O157:H7, increased the neutrophil oxidative burst and cell adhesion leading to impaired migration (80). Taken together, activation and degranulation of neutrophils in HUS and impaired migration leading to increased tissue destruction at sites of neutrophil infiltration can be explained by the interaction between neutrophils and Stx as well as StcE.

Monocytes

Monocytes are a critical effector component in the innate immune response by presenting antigens and producing cytokines, thus regulating innate and adaptive immune responses. In HUS, monocytes are differentiated toward an inflammatory phenotype with an increased population of cells with reduced CD14 and enhanced CD16 membrane expression (81). In addition, monocytes from patients with HUS exhibited reduced circulatory expression of function-related proteins such as CD11b, CD64, CD62L, and CX_3CR, possibly due to monocyte infiltration of tissue lesions (81, 82).

Patients with HUS were found to have Stx2 on monocytes and platelet-monocyte aggregates (78). Monocytes express small amounts of the Gb3 receptor that is slightly different from the Gb3 lipoforms present on endothelial cells (83). The number of Gb3 receptors on monocytes can be increased by LPS binding, thus leading to activation of the cells and enhanced binding of Stx (83). Stx is not cytotoxic for monocytes but triggers a variety of proinflammatory events, including in vitro synthesis and secretion of proinflammatory cytokines and chemokines such as IL-6, IL-8, TNF-α, and IL-1β (83).

Monocytes may indirectly contribute to the thrombotic process occurring during HUS. Stimulation of macrophage-like cells of the monocytic cell line THP-1 with Stx2 induced the release of macrophage-derived chemokine, RANTES, and IL-8, an effect that was further enhanced in the presence of LPS, leading to platelet activation and aggregation (84). Incubation of monocytes with Stx1 or Stx2 induced expression of tissue factor on their surfaces (78, 85), which was further enhanced upon coincubation with LPS and when monocytes were in complex with platelets (78), and thus presumably involved in the prothrombotic process occurring during HUS.

Monocyte-derived microparticles expressing tissue factor were detected in patients with HUS (78). Stx2 induced the release of tissue factor-expressing microparticles from monocytes in vitro. Monocyte-derived microparticles can deliver and transfer tissue factor to platelets (86) and neutrophils (87) and thus induce a prothrombotic surface. Taken together, Stx can bind to and activate mono-cytes and induce the formation of platelet-monocyte aggregates and the release of tissue factor-expressing microparticles with prothrombotic properties. However, the trans-fer of Stx from monocytes to endothelial cells probably does not occur in vivo (88), as both cells express receptors with similar affinities (83).

Platelets

Low platelet counts and formation of renal thrombi are characteristic of HUS, suggesting that platelet activation is involved in the path-ogenesis. Thrombocytopenia is assumed to be related to consumption of platelets in microthrombi and may be caused by the direct effects of Stx on platelets or by Stx-induced endothelial cell damage leading to second-ary platelet activation and the formation of microthrombi (89, 90). During the acute phase of HUS platelets are degranulated (91), with reduced intracellular levels of β-thromboglobulin and impaired aggregating response (92). Platelet-derived microparticles are increased, indicating platelet activation (78, 93), and plasma from patients with HUS induced aggregation of normal platelets (94). Mice inoculated with *E. coli* O157:H7 mimicked the human disease and developed thrombocy-topenia, which was also demonstrated in mice injected with Stx2 and LPS (44, 95).

Platelets bind Stx through Gb3 and an al-ternative glycosphingolipid receptor, termed band 0.03 (96, 97). Gb3 expression on resting platelets is very low (97), and the distribution in humans is quite heterogeneous (96). Stx is assumed to bind primarily to activated platelets (78, 97), and Stx circulates bound to platelets, in addition to leukocytes, during HUS (98). Upon binding to platelets, Stx is rapidly internalized, leading to further acti-vation, aggregation, structural changes en-hancing the surface area, and increased fibrinogen-binding capacity (99).

In addition to their role in thrombosis and hemostasis, platelets are involved in inflamma-tion and can influence both innate and adaptive immunity (100). Activated platelets interact with both neutrophils (101) and monocytes (102) and release chemokines such as platelet factor-4, macrophage inflammatory protein (MIP), RANTES, IL-8, β-thromboglobulin, and monocyte chemoattractant protein (MCP) from their α-granules, which potentiate the inflam-matory process (103).

Both human and murine platelets express TLR4 and other TLR receptors (98, 104, 105). Platelet TLRs mediate LPS-induced throm-bocytopenia and TNF-α production by leukocytes (105–109). Recent evidence sug-gests that platelets may bridge the innate and adaptive immune system by expressing immunostimulatory proteins such as CD40L and thereby stimulate CD8+ T-cell induction and adaptive immunity (110). Resting platelets must be primed (with LPS or other activators) before interaction with Stx (78, 111). LPS binds to platelets via a receptor complex of TLR4 and CD62 and activates them as shown by the expression of CD40L, activated GPIIb/IIIa receptor, and fibrinogen binding (98). GPIIb/IIIa and CD40L both play a central role in thrombotic diseases. In addition, CD40L can interact with CD40 on endothelial cells, trig-gering an inflammatory response leading to local or systemic release of MCP-1, VCAM, and ICAM (112–114).

LPS INTERACTION WITH BLOOD CELLS

Recognition of LPS by innate immune cells is vital for host defense against gram-negative bacteria. LPS, the major component of the outer membrane of gram-negative bacteria, circulates bound to platelets during acute HUS (98). Platelets from mice inoculated intraperitoneally with LPS showed surface-bound LPS and exhibited increased CD40L expression, suggesting that LPS activates platelets in the circulation in the murine model (98). Patients develop an antibody re-sponse to strain-specific LPS (115). The con-centration of LPS-binding protein, which binds LPS in plasma and transfers it to cell surfaces, is increased in the plasma of patients with HUS compared to those with uncomplicated EHEC diarrhea, suggesting that the acute-phase response to LPS is associated with disease severity (116).

Neutrophils and monocytes express TLR4 and respond to LPS stimulation by releasing proinflammatory cytokines (117). At the onset of HUS, neutrophils exhibited higher levels of TLR4 mRNA and TLR4 protein expression (118). TLR4 expression was correlated to increased circulating TNF-α levels. No differences were noted in TLR4 receptor expression on monocytes at the onset of HUS (118), indicating different regulation of TLR4 expression on neutrophils and monocytes.

TISSUE FACTOR

During the acute phase of HUS, patients have high plasma levels of tissue factor and tissue factor pathway inhibitor (119) and circulating platelet-leukocyte aggregates expressing tissue factor and tissue factor-bearing platelet microparticles (78). In vitro studies showed that Stx, in cooperation with LPS, induced aggregate formation between platelets and leukocytes, leading to release of platelet-derived microparticles with surface-bound tissue factor (78).

Tissue factor-positive microparticles may contribute to thrombosis. Tissue factor is a transmembrane glycoprotein receptor (120) for coagulation factor VII (121). By acting as a cofactor for factor VII, tissue factor promotes proteolysis and activation of factor VIIa, followed by formation of the prothrombinase complex (122) and conversion of prothrombin into thrombin, resulting in thrombus formation and further platelet activation. Expression of tissue factor on platelets is debated. Platelets contain tissue factor mRNA (123), which can be spliced into mature mRNA upon platelet activation, leading to minimal protein expression and procoagulant activity (124, 125). However, platelets seem to acquire

most of their tissue factor through interaction with tissue factor-bearing microparticles from monocytes (126, 127).

THE THROMBOTIC PROCESS IN HUS

The prothrombotic condition in HUS has been primarily ascribed to damage of the microvascular endothelium. When vascular injury occurs, platelets are recruited to the damaged site in a multistep process that involves the interaction of specific platelet cell-surface receptors with subendothelial matrix proteins such as von Willebrand factor (VWF), collagen, and fibronectin. The first event in this process is binding of platelet glycoprotein Ibα receptor (GPIbα) to the A1 domain of VWF after which a conformational change in the platelet integrin receptor GPIIb/IIIa occurs, allowing binding to both VWF and fibrinogen, inducing aggregation. After initial tethering steps, platelets become activated and firmly adhere to the vessel wall and form a clot. Adherent and activated platelets release potent platelet agonists such as thrombin, ADP, and thromboxane A2 from their intracellular granules, promoting further platelet activation and aggregation and resulting in rapid growth of the thrombus.

In HUS, platelets are activated (128) and degranulated (91, 92), and platelet-derived factors such as β-thromboglobulin and platelet factor-4 are elevated (129) as is VWF, which may be secreted from both platelets and endothelial cells (130, 131). VWF mediates platelet adhesion to activated endothelial cells in response to Stx (132). In addition, Stx delayed the cleavage of VWF-platelet strings by the metalloprotease ADAMTS13 on activated endothelial cells (133), thus potentiating thrombus growth in the presence of larger VWF multimers. Functional blockade of receptors or adhesive proteins, such GPIIb/IIIa, P-selectin, or VWF, was associated with a marked reduction of thrombi on endothelial cells (132). Both Stx and LPS activate platelets, especially under high shear stress, and costimulation with both factors simultaneously induced an additive effect on the formation of platelet-leukocyte aggregates expressing tissue factor (78, 97–99). In addition, platelets may be activated indirectly by additional factors such as chemokines and cytokines released by Stx-stimulated monocytes (84) or endothelial cells (134). Thus, platelet activation and thrombus formation may be caused directly by Stx and/or LPS or by the release of cytokines, indicating that platelet activation and inflammation are correlated events.

In parallel with the recruitment of platelets, the blood coagulation cascade is activated at the site of vessel injury (135). In HUS, there is no consumption of coagulation factors, but elevated levels of prothrombin fragment 1 + 2 (136–138), tissue plasminogen activator, tissue plasminogen activator inhibitor-1 (139), and D-dimers have been found, even before HUS develops, indicating enhanced thrombotic capacity and impaired fibrinolysis.

THE SYSTEMIC AND RENAL HOST RESPONSE

The precise mechanism by which Stx and other EHEC virulence factors, such as LPS, reach the kidney is, as yet, unknown. Although Stx and O157LPS have been shown to circulate bound to blood cells, as described above, the manner by which they transfer to resident target organ cells in the kidney, brain, or other organs has not been elucidated in the in vivo setting. However, the toxin was detected in glomeruli and tubuli of pediatric and geriatric patients with HUS (140, 141), indicating that it reaches the kidney. The extensive renal injury associated with renal cell apoptosis occurring during HUS (142) activates a variety of host responses, including the influx of leukocytes and the release of cytokines, both of which could enhance the tissue damage.

Leukocytosis was associated with the development of HUS (143, 144) and with a poor outcome (145), as described above. Renal influx of neutrophils was associated with increased mortality (61, 146). The functionality

of neutrophils correlated to patient renal dysfunction (68). Mice that were coinjected with Stx2 and LPS developed neutrophilia and monocytosis with markers of leukocyte activation (95). Neutrophil and macrophage accumulation was demonstrated in the murine kidneys (147, 148). Likewise, kidneys from rabbits injected with Stx2 showed PMN infiltrates correlating with levels of IL-8 (149). In vitro studies showed that stimulation of glomerular endothelial cells with Stx2 enhanced leukocyte adherence and migration under perfusion, effects mediated by MCP-1, IL-8, and fractalkine (CX3CL1) secreted from the cells (150, 151) and enhanced by TNF-α (152). Furthermore, Stx borne on leukocytes migrated through endothelial cells and induced their release of IL-8 and MCP-1 (153).

In addition to leukocytes, platelets have also been demonstrated in the glomerular microthrombi characteristic of human thrombotic microangiopathy (reviewed in reference 154), in the glomeruli of primates treated with Stx (155) and of mice injected with Stx2 and LPS (95). Low platelet counts during HUS are correlated to poor recovery of renal function (62, 156). Stx and LPS induce human microvascular endothelial cells to secrete factors that activate platelets (134). Although platelets are activated and secrete microparticles during HUS, and the role of platelets in the renal thrombotic events is evident, their contribution to the proinflammatory events occurring in the kidney has not been thoroughly investigated. For example, degranulation of platelet alpha granules upon activation will release microbicidal proteins, CC-chemokines, and CXC-chemokines (103, 154). The binding of specific ligands and release of potent platelet components could theoretically promote the inflammatory state in the kidney.

Proinflammatory and prothrombotic responses have been documented in HUS. Inflammatory and prothrombotic mediators, including cytokines, chemokines, soluble adhesion molecules, growth factors, cytokine receptors, tissue factor, and acute-phase response proteins, are elevated in patients with

EHEC-associated infection and HUS (116, 119, 138, 157–174) and could be correlated, in certain studies, to the progression of renal damage (reviewed in reference 89). Increased levels of chemoattractants, such as IL-8 (69), granulocyte colony-stimulating factor (159), and MCP-1 (157), could explain leukocyte influx. Stx and TNF-α act in synergy to cause cytotoxic effects on human endothelial cells (175–177). Elevated TNF-α may enhance the expression of the Stx receptor Gb3 on endothelial cells (178). Enhanced Gb3 expression was also demonstrated on cultured endothelial cells stimulated with LPS and IL-1 (175, 177, 179, 180), thus sensitizing the cells to Stx.

Animal models have further addressed the importance of inflammatory and chemotactic pathways for the pathogenesis of HUS. Gnotobiotic mice inoculated with *E. coli* O157:H7 exhibited TNF-α, IL-1, and IL-6 in the kidney (181). Certain mice were also treated with TNF-α, which led to enhanced kidney damage. TNF inhibition, with the protease inhibitor Nafamostat mesilate, reduced target-organ damage. Stx induced TNF synthesis in the mouse kidney while increasing renal sensitivity to the toxic effects of TNF (182). Interestingly, TNF had a protective effect when administered to mice before Stx1 (183) but exacerbated disease when given after Stx1. These results are in contrast to a previous study using neutralizing anti-TNF-α antibody before administration of Stx1, as well as TNF-α knockout mice, which could not demonstrate a role for TNF-α in Stx-induced toxicity (184). In vitro results indicated that Stx induced increased secretion of TNF-α from human renal proximal tubule cells (185) and that Stx2-induced TNF-α expression was diminished by blocking the p38 pathway (186).

HUS patients exhibited high levels of TNF-α in the circulation (163, 166, 168, 170) and in the urine, which did not correlate to levels in the blood, indicating local synthesis in the urinary tract (164). TNF-α is a proinflammatory cytokine mediating a cytotoxic effect on tumor cells, inflammation, and microvascular coagulation. The high urinary levels occurring

during HUS (164) may contribute to inflammation, thrombosis, and end-organ damage.

A recent study using baboons treated with Stx exhibited high renal mRNA levels of IL-8, MCP-1, and MIP-1α in contrast to a modest effect on TNF-α. These baboons had elevated urinary IL-6, IL-8, MCP-1, and VEGF (155). Similarly, a previous study in baboons showed that Stx infusion led to urinary secretion of TNF-α and IL-6 (187). Stimulation of proximal tubular cells with Stx increased expression of IL-6 mRNA and protein (185). A similar effect was noted in stimulated glomerular endothelial cells, particularly when cells were costimulated with LPS (188). Patients with HUS exhibit very high IL-6 levels in serum and urine (164). Elevated glomerular and tubular levels of IL-6 may promote renal injury. In addition to its role in immune regulation and inflammation, IL-6 may respond to glomerular injury by promoting mesangial proliferation (189).

The chemokine receptor CCR1 was shown to be involved in neutrophil and monocyte infiltrates in the kidney and in host survival after Stx2 treatment of CCR1-deficient mice (190). A slower increase in plasma TNF-α and IL-6 was also noted in the $CCR1^{-/-}$ mice compared to wild-type mice. Similarly, injection of Stx2 and LPS in a mouse model markedly increased renal expression of the chemoattractants CXCL1 and CXCL2, affecting neutrophil influx (147). In the same mouse model, macrophage influx was also demonstrated and associated with expression of monocyte chemoattractants: MCP-1/CCL2, MIP-1/CCL3, and RANTES (CCL5) (148). Taken together, the studies described provide evidence for chemokine- and cytokine-mediated inflammation that underscores the importance of the inflammatory response in the development of HUS.

The CXCR4/CXCR7/stromal cell-derived factor 1 (SDF-1) pathway, involved in renal homeostasis, was activated by Stx2 (174), thus enhancing endothelial activation and renal damage. The results were confirmed by using human microvascular endothelial cells in vitro and showed that low concentrations of Stx, which minimally affect protein synthesis, activated the CXCR4/CXCR7/SDF-1 pathway (134, 174). The importance of this finding was established in patients showing elevated plasma levels of SDF-1 and thus suggested an effect on the glomerular vasculature.

Tissue factor is a potent prothrombotic mediator and was found to be elevated in the plasma of HUS patients (119, 173), on circulating platelet-derived microparticles (78), and demonstrated in the kidney (191). Stx induced tissue factor activity in glomerular endothelial cells and proximal tubular cells, an effect enhanced in the presence of TNF-α (192, 193).

THE HOST RESPONSE IN THE BRAIN

The central nervous system (CNS) is also a target organ during HUS, and as many as 48 to- 66% of patients developed severe neurological manifestations during the *E. coli* O104:H4 epidemic in Germany is the spring of 2011 (194, 195). Shiga toxin may bind to the Gb3 receptor on neurons and endothelial cells in the human CNS (196). The induction of apoptosis (197, 198) and the inflammatory response may promote CNS injury. In a study conducted on pediatric patients with HUS and encephalopathy, versus patients with HUS without CNS symptoms or with EHEC-associated colitis without HUS, serum inflammatory mediators IL-6, soluble TNF receptor 1, and tissue inhibitor of metalloproteinase-1 were correlated to the presence of encephalopathy (199). Increased protein in the cerebrospinal fluid (200) also indicates an enhanced inflammatory state.

EHEC strains were used in animal models to reproduce neurological affection (41, 201) showing endothelial and neuronal damage, and Stx was detected in the brain cortex and spinal cord (201). The blood-brain barrier is generally affected by damage to cerebral capillary endothelial cells, perivascular pericytes, and possibly even astrocytes. Astrogliosis is triggered by multiple inflammatory mediators (202). In EHEC-infected mice astrocytes were activated in the medulla oblongata and spinal cord (203). EHEC-infected mice exhibited TNF-α in the brain, and intraperitoneal treatment with TNF-α given before and after EHEC inoculation worsened the neurological symptoms. The TNF inhibitor, Nafamostat mesilate, modified these responses and decreased cytokines in the brain, suggesting a role for TNF-α in the neurological manifestations (181). Furthermore, intravenous injection of Stx2 in rabbits induced brain edema that could be reversed by anti-inflammatory steroid treatment (204).

In vitro experiments using human brain endothelial cells have shown that TNF-α and IL-1β enhanced Stx toxicity (205). TNF increases the Gb3 receptor on human brain endothelium (206). The TNF effect was mitigated by inhibition of p38 MAP kinase (207). Similarly, TNF-α amplified the inflammatory response to Stx1 and LPS-treated rat astrocytes, enhancing PMN chemotaxis and cytotoxicity (208). Inflammatory mediators released from the stimulated astrocytes affected endothelial permeability and increased binding of PMNs and platelets to the endothelial cells (209). Thus in vivo and in vitro evidence points to a considerable inflammatory response in the CNS affecting the endothelium and astrocytes as well as the blood-brain barrier.

THE EFFECTS OF COMPLEMENT ACTIVATION

Complement activation via the alternative pathway occurs during EHEC-associated HUS. This has been documented by hypocomplementemia (low C3) and elevated plasma levels of degradation products such as complement factors Bb, C3a, and soluble C5b-9 during the acute phase of disease; this did not correlate with the severity of renal injury or occurrence of later complications and decreased upon recovery (210–213). However, circulating C3a and soluble C5b-9 may activate platelets (214, 215). One patient was also shown to have C3 deposits on circulating platelet-monocyte aggregates (213). Patient platelet- and monocyte-derived microparticles were shown to be coated with C3 and C9 deposits during the acute phase of HUS but not after recovery (213). These findings suggest that the extensive endothelial and blood cell activation occurring during HUS will lead to secondary complement activation. A direct effect of Stx and/or O157LPS was demonstrated when these were incubated in vitro with whole blood, leading to the formation of leukocyte-platelet aggregates and the release of platelet- and monocyte-derived microparticles, both with C3 and C9 deposits (213).

Complement could also be activated by dysfunctional inhibition. Stx was shown to activate the alternative pathway in vitro and to bind to cell-binding domains of the major soluble complement inhibitor, factor H, thus compromising the inhibitory effect and promoting complement activation (216). These studies were performed with very high concentrations of Stx2 and may thus not reflect the in vivo situation. Also, as the toxin does not circulate in free form, but is rather cell-bound in the circulation, the interaction with factor H may not occur in vivo (90). Factor H prevents complement activation via the alternative pathway on cell surfaces. Factor H binding to proximal tubular cells was impaired when cells were exposed to protein overload, resembling the situation during renal injury (217). Thus complement activation may also occur as a nonspecific consequence of tubular protein overload.

One study addressed activation of the mannan-binding lectin (MBL) pathway in EHEC-induced infection and HUS. MBL deficiency may predispose to infection, but no correlation was found between EHEC infection and MBL deficiency (218). The main function of the complement system is to dispose of foreign cells, such as bacterial pathogens, by opsonization and cytolysis (219). Complement is active in the colon (220), and its primary effect there is to promote bacterial clearance. Thus a certain degree of complement activation is protective at the mucosal surface. However, prolonged and amplified activation will promote platelet activation and endothelial damage in HUS (219, 221).

SUMMARY

The host responds to EHEC infections and, more specifically, to Shiga toxin and EHEC-derived LPS by activation of a large variety of cells, including blood cells, intestinal and renal epithelial cells, renal and cerebral endothelial cells, and astrocytes. Multiple chemokines and cytokines, as well as stress hormones and antimicrobial peptides, are secreted. They contribute to the inflammation and interact with bacteria or virulence factors in an attempt to clear the infection. The inflammatory responses may, however, have adverse effects on the host, thus worsening the outcome. In certain instances the bacterial virulence factors may hijack host responses to their own advantage. The release of tissue factor and thrombin promotes thrombosis, and the complement system is activated with secondary effects on the endothelium, tubular cells, and platelets. The inflammation and thrombosis arising explain some of the features of HUS. It is thus essential to understand not only the cellular signaling pathways used by EHEC virulence factors but also the pattern of inflammatory and thrombotic responses arising from the infection.

Treatments aimed at reducing injury to host organs should primarily attempt to neutralize Shiga toxin. However, as patients may manifest profound organ damage, treatments aimed at reducing the bacterial load, the inflammatory and thrombotic response, and the cell injury may be beneficial if instituted early in the course of infection. The specific pathways leading to host cell damage should therefore be determined, so that future treatments can effectively abrogate the inflammatory and thrombotic complications of this infection.

ACKNOWLEDGMENTS

Diana Karpman's research is supported by grants from The Swedish Research Council (K2013-64X-14008), The Torsten Söderberg Foundation, Crown Princess Lovisa's Society for Child Care, and The Konung Gustaf V:s 80-årsfond.

We declare no conflicts of interest with regard to the manuscript.

CITATION

Karpman D, Ståhl A. 2014. Enterohemorrhagic *Escherichia coli* pathogenesis and the host response. Microbiol Spectrum 2(5): EHEC-0009-2013.

NOTE ADDED IN PROOF

Shiga toxin may bind to TLR4 on neutrophils. (**Brigotti M, Carnicelli D, Arfilli V, Tamassia N, Borsetti F, Fabbri E, Tazzari PL, Ricci F, Pagliaro P, Spisni E, Cassatella MA. 2013.** Identification of TLR4 as the receptor that recognizes Shiga toxins in human neutrophils. *J Immunol* 191:4748–4758.)

REFERENCES

1. **Tesh VL.** 2012. The induction of apoptosis by Shiga toxins and ricin. *Curr Top Microbiol Immunol* 357:137–178.
2. **Jandhyala DM, Thorpe CM, Magun B.** 2012. Ricin and Shiga toxins: effects on host cell signal transduction. *Curr Top Microbiol Immunol* 357: 41–65.
3. **Chromek M, Arvidsson I, Karpman D.** 2012. The antimicrobial peptide cathelicidin protects mice from *Escherichia coli* O157:H7-mediated disease. *PLoS One* 7:e46476.
4. **House B, Kus JV, Prayitno N, Mair R, Que L, Chingcuanco F, Gannon V, Cvitkovitch DG, Barnett Foster D.** 2009. Acid-stress-induced changes in enterohaemorrhagic *Escherichia coli* O157 : H7 virulence. *Microbiology* 155:2907–2918.
5. **Foster JW.** 2004. *Escherichia coli* acid resistance: tales of an amateur acidophile. *Nat Rev Microbiol* 2:898–907.
6. **Barnett Foster D.** 2013. Modulation of the enterohemorrhagic *E. coli* virulence program through the human gastrointestinal tract. *Virulence* 4:315–323.
7. **Phillips AD, Navabpour S, Hicks S, Dougan G, Wallis T, Frankel G.** 2000. Enterohaemorrhagic *Escherichia coli* O157:H7 target Peyer's patches in humans and cause attaching/effacing lesions in both human and bovine intestine. *Gut* 47:377–381.
8. **Chong Y, Fitzhenry R, Heuschkel R, Torrente F, Frankel G, Phillips AD.** 2007. Human intestinal tissue tropism in *Escherichia coli* O157:H7— initial colonization of terminal ileum and Peyer's patches and minimal colonic adhesion ex vivo. *Microbiology* 153:794–802.
9. **Etienne-Mesmin L, Chassaing B, Sauvanet P, Denizot J, Blanquet-Diot S, Darfeuille-Michaud A, Pradel N, Livrelli V.** 2011. Interactions with M cells and macrophages as key steps in the pathogenesis of enterohemorrhagic *Escherichia coli* infections. *PLoS One* 6:e23594.
10. **Miller TL, Wolin MJ.** 1996. Pathways of acetate, propionate, and butyrate formation by the human fecal microbial flora. *Appl Environ Microbiol* 62: 1589–1592.
11. **Nakanishi N, Tashiro K, Kuhara S, Hayashi T, Sugimoto N, Tobe T.** 2009. Regulation of virulence by butyrate sensing in enterohaemorrhagic *Escherichia coli*. *Microbiology* 155:521–530.
12. **Herold S, Paton JC, Srimanote P, Paton AW.** 2009. Differential effects of short-chain fatty acids and iron on expression of iha in Shiga-toxigenic *Escherichia coli*. *Microbiology* 155:3554–3563.
13. **Jacewicz MS, Acheson DW, Mobassaleh M, Donohue-Rolfe A, Balasubramanian KA, Keusch GT.** 1995. Maturational regulation of globotriaosylceramide, the Shiga-like toxin 1 receptor, in cultured human gut epithelial cells. *J Clin Invest* 96:1328–1335.
14. **Schauber J, Svanholm C, Termen S, Iffland K, Menzel T, Scheppach W, Melcher R, Agerberth B, Luhrs H, Gudmundsson GH.** 2003. Expression of the cathelicidin LL-37 is modulated by short chain fatty acids in colonocytes: relevance of signalling pathways. *Gut* 52:735–741.
15. **Iimura M, Gallo RL, Hase K, Miyamoto Y, Eckmann L, Kagnoff MF.** 2005. Cathelicidin mediates innate intestinal defense against colonization with epithelial adherent bacterial pathogens. *J Immunol* 174:4901–4907.
16. **Islam D, Bandholtz L, Nilsson J, Wigzell H, Christensson B, Agerberth B, Gudmundsson G.** 2001. Downregulation of bactericidal peptides in enteric infections: a novel immune escape mechanism with bacterial DNA as a potential regulator. *Nat Med* 7:180–185.
17. **Pacheco AR, Sperandio V.** 2009. Inter-kingdom signaling: chemical language between bacteria and host. *Curr Opin Microbiol* 12:192–198.
18. **Hughes DT, Clarke MB, Yamamoto K, Rasko DA, Sperandio V.** 2009. The QseC adrenergic

signaling cascade in enterohemorrhagic *E. coli* (EHEC). *PLoS Pathog* **5:**e1000553.

19. **Malyukova I, Murray KF, Zhu C, Boedeker E, Kane A, Patterson K, Peterson JR, Donowitz M, Kovbasnjuk O.** 2009. Macropinocytosis in Shiga toxin 1 uptake by human intestinal epithelial cells and transcellular transcytosis. *Am J Physiol Gastrointest Liver Physiol* **296:**G78–G92.

20. **Zumbrun SD, Hanson L, Sinclair JF, Freedy J, Melton-Celsa AR, Rodriguez-Canales J, Hanson JC, O'Brien AD.** 2010. Human intestinal tissue and cultured colonic cells contain globotriaosylceramide synthase mRNA and the alternate Shiga toxin receptor globotetraosylceramide. *Infect Immun* **78:**4488–4499.

21. **Schüller S, Heuschkel R, Torrente F, Kaper JB, Phillips AD.** 2007. Shiga toxin binding in normal and inflamed human intestinal mucosa. *Microbes Infect* **9:**35–39.

22. **Bell CJ, Elliott EJ, Wallace JL, Redmond DM, Payne J, Li Z, O'Loughlin EV.** 2000. Do eicosanoids cause colonic dysfunction in experimental *E. coli* O157:H7 (EHEC) infection? *Gut* **46:**806–812.

23. **Smith WE, Kane AV, Campbell ST, Acheson DW, Cochran BH, Thorpe CM.** 2003. Shiga toxin 1 triggers a ribotoxic stress response leading to p38 and JNK activation and induction of apoptosis in intestinal epithelial cells. *Infect Immun* **71:**1497–1504.

24. **Schüller S, Frankel G, Phillips AD.** 2004. Interaction of Shiga toxin from *Escherichia coli* with human intestinal epithelial cell lines and explants: Stx2 induces epithelial damage in organ culture. *Cell Microbiol* **6:**289–301.

25. **Barnett Foster D, Abul-Milh M, Huesca M, Lingwood CA.** 2000. Enterohemorrhagic *Escherichia coli* induces apoptosis which augments bacterial binding and phosphatidylethanolamine exposure on the plasma membrane outer leaflet. *Infect Immun* **68:**3108–3115.

26. **Kashiwamura M, Kurohane K, Tanikawa T, Deguchi A, Miyamoto D, Imai Y.** 2009. Shiga toxin kills epithelial cells isolated from distal but not proximal part of mouse colon. *Biol Pharm Bull* **32:**1614–1617.

27. **Keenan KP, Sharpnack DD, Collins H, Formal SB, O'Brien AD.** 1986. Morphologic evaluation of the effects of Shiga toxin and *E. coli* Shiga-like toxin on the rabbit intestine. *Am J Pathol* **125:**69–80.

28. **Békássy ZD, Calderon Toledo C, Leoj G, Kristoffersson A, Leopold SR, Perez MT, Karpman D.** 2011. Intestinal damage in enterohemorrhagic *Escherichia coli* infection. *Pediatr Nephrol* **26:**2059–2071.

29. **Hurley BP, Jacewicz M, Thorpe CM, Lincicome LL, King AJ, Keusch GT, Acheson DW.** 1999. Shiga toxins 1 and 2 translocate differently across polarized intestinal epithelial cells. *Infect Immun* **67:**6670–6677.

30. **Hurley BP, Thorpe CM, Acheson DW.** 2001. Shiga toxin translocation across intestinal epithelial cells is enhanced by neutrophil transmigration. *Infect Immun* **69:**6148–6155.

31. **Slutsker L, Ries AA, Greene KD, Wells JG, Hutwagner L, Griffin PM.** 1997. *Escherichia coli* O157:H7 diarrhea in the United States: clinical and epidemiologic features. *Ann Intern Med* **126:** 505–513.

32. **Jung HC, Eckmann L, Yang SK, Panja A, Fierer J, Morzycka-Wroblewska E, Kagnoff MF.** 1995. A distinct array of proinflammatory cytokines is expressed in human colon epithelial cells in response to bacterial invasion. *J Clin Invest* **95:**55–65.

33. **Thorpe CM, Hurley BP, Lincicome LL, Jacewicz MS, Keusch GT, Acheson DW.** 1999. Shiga toxins stimulate secretion of interleukin-8 from intestinal epithelial cells. *Infect Immun* **67:**5985–5993.

34. **Thorpe CM, Smith WE, Hurley BP, Acheson DW.** 2001. Shiga toxins induce, superinduce, and stabilize a variety of C-X-C chemokine mRNAs in intestinal epithelial cells, resulting in increased chemokine expression. *Infect Immun* **69:**6140–6147.

35. **Yamasaki C, Natori Y, Zeng XT, Ohmura M, Yamasaki S, Takeda Y, Natori Y.** 1999. Induction of cytokines in a human colon epithelial cell line by Shiga toxin 1 (Stx1) and Stx2 but not by nontoxic mutant Stx1 which lacks N-glycosidase activity. *FEBS Lett* **442:**231–234.

36. **Colpoys WE, Cochran BH, Carducci TM, Thorpe CM.** 2005. Shiga toxins activate translational regulation pathways in intestinal epithelial cells. *Cell Signal* **17:**891–899.

37. **Tesh VL, Ramegowda B, Samuel JE.** 1994. Purified Shiga-like toxins induce expression of proinflammatory cytokines from murine peritoneal macrophages. *Infect Immun* **62:**5085–5094.

38. **Farfan MJ, Cantero L, Vergara A, Vidal R, Torres AG.** 2013. The long polar fimbriae of STEC O157:H7 induce expression of pro-inflammatory markers by intestinal epithelial cells. *Vet Immunol Immunopathol* **152:**126–131.

39. **Bellmeyer A, Cotton C, Kanteti R, Koutsouris A, Viswanathan VK, Hecht G.** 2009. Enterohemorrhagic *Escherichia coli* suppresses inflammatory response to cytokines and its own toxin. *Am J Physiol Gastrointest Liver Physiol* **297:**G576–G581.

40. **Ho NK, Ossa JC, Silphaduang U, Johnson R, Johnson-Henry KC, Sherman PM.** 2012. Enterohemorrhagic *Escherichia coli* O157:H7 Shiga toxins inhibit gamma interferon-mediated cellular activation. *Infect Immun* **80:**2307–2315.

41. **Karpman D, Connell H, Svensson M, Scheutz F, Alm P, Svanborg C.** 1997. The role of lipopoly-saccharide and Shiga-like toxin in a mouse model of *Escherichia coli* O157:H7 infection. *J Infect Dis* **175:**611–620.

42. **Takeda K, Akira S.** 2004. TLR signaling pathways. *Semin Immunol* **16:**3–9.

43. **Moresco EM, LaVine D, Beutler B.** 2011. Toll-like receptors. *Curr Biol* **21:**R488–R493.

44. **Calderon Toledo C, Rogers TJ, Svensson M, Tati R, Fischer H, Svanborg C, Karpman D.** 2008. Shiga toxin-mediated disease in MyD88-deficient mice infected with *Escherichia coli* O157:H7. *Am J Pathol* **173:**1428–1439.

45. **Torgersen ML, Engedal N, Pedersen AM, Husebye H, Espevik T, Sandvig K.** 2011. Toll-like receptor 4 facilitates binding of Shiga toxin to colon carcinoma and primary umbilical vein endothelial cells. *FEMS Immunol Med Microbiol* **61:**63–75.

46. **Carbonare CB, Carbonare SB, Carneiro-Sampaio MM.** 2003. Early acquisition of serum and saliva antibodies reactive to enteropathogenic *Escherichia coli* virulence-associated proteins by infants living in an endemic area. *Pediatr Allergy Immunol* **14:**222–228.

47. **Parissi-Crivelli A, Parissi-Crivelli JM, Giron JA.** 2000. Recognition of enteropathogenic *Escherichia coli* virulence determinants by human colostrum and serum antibodies. *J Clin Microbiol* **38:**2696–2700.

48. **Karpman D, Békássy ZD, Sjögren AC, Dubois MS, Karmali MA, Mascarenhas M, Jarvis KG, Gansheroff LJ, O'Brien AD, Arbus GS, Kaper JB.** 2002. Antibodies to intimin and *Escherichia coli* secreted proteins A and B in patients with enterohemorrhagic *Escherichia coli* infections. *Pediatr Nephrol* **17:**201–211.

49. **Sjögren AC, Kaper JB, Caprioli A, Karpman D.** 2004. Enzyme-linked immunosorbent assay for detection of Shiga toxin-producing *Escherichia coli* infection by antibodies to *Escherichia coli* secreted protein B in children with hemolytic uremic syndrome. *Eur J Clin Microbiol Infect Dis* **23:**208–211.

50. **Noguera-Obenza M, Ochoa TJ, Gomez HF, Guerrero ML, Herrera-Insua I, Morrow AL, Ruiz-Palacios G, Pickering LK, Guzman CA, Cleary TG.** 2003. Human milk secretory antibodies against attaching and effacing *Escherichia coli* antigens. *Emerg Infect Dis* **9:**545–551.

51. **Loureiro I, Frankel G, Adu-Bobie J, Dougan G, Trabulsi LR, Carneiro-Sampaio MM.** 1998. Human colostrum contains IgA antibodies reactive to enteropathogenic *Escherichia coli* virulence-associated proteins: intimin, BfpA, EspA, and EspB. *J Pediatr Gastroenterol Nutr* **27:**166–171.

52. **Palmeira P, Carbonare SB, Amaral JA, Tino-De-Franco M, Carneiro-Sampaio MM.** 2005. Colostrum from healthy Brazilian women inhibits adhesion and contains IgA antibodies reactive with Shiga toxin-producing *Escherichia coli. Eur J Pediatr* **164:**37–43.

53. **Bitzan M, Moebius E, Ludwig K, Müller-Wiefel DE, Heesemann J, Karch H.** 1991. High incidence of serum antibodies to *Escherichia coli* O157 lipopolysaccharide in children with hemolytic-uremic syndrome. *J Pediatr* **119:**380–385.

54. **Ludwig K, Grabhorn E, Bitzan M, Bobrowski C, Kemper MJ, Sobottka I, Laufs R, Karch H, Müller-Wiefel DE.** 2002. Saliva IgM and IgA are a sensitive indicator of the humoral immune response to *Escherichia coli* O157 lipopolysaccharide in children with enteropathic hemolytic uremic syndrome. *Pediatr Res* **52:**307–313.

55. **Ludwig K, Bitzan M, Bobrowski C, Müller-Wiefel DE.** 2002. *Escherichia coli* O157 fails to induce a long-lasting lipopolysaccharide-specific, measurable humoral immune response in children with hemolytic-uremic syndrome. *J Infect Dis* **186:**566–569.

56. **Palmeira P, Yu Ito L, Arslanian C, Carneiro-Sampaio MM.** 2007. Passive immunity acquisition of maternal anti-enterohemorrhagic *Escherichia coli* (EHEC) O157:H7 IgG antibodies by the newborn. *Eur J Pediatr* **166:**413–419.

57. **Ludwig K, Karmali MA, Sarkim V, Bobrowski C, Petric M, Karch H, Muller-Wiefel DE.** 2001. Antibody response to Shiga toxins Stx2 and Stx1 in children with enteropathic hemolytic-uremic syndrome. *J Clin Microbiol* **39:**2272–2279.

58. **Ludwig K, Sarkim V, Bitzan M, Karmali MA, Bobrowski C, Ruder H, Laufs R, Sobottka I, Petric M, Karch H, Muller-Wiefel DE.** 2002. Shiga toxin-producing *Escherichia coli* infection and antibodies against Stx2 and Stx1 in household contacts of children with enteropathic hemolytic-uremic syndrome. *J Clin Microbiol* **40:**1773–1782.

59. **Martinez MB, Taddei CR, Ruiz-Tagle A, Trabulsi LR, Giron JA.** 1999. Antibody response of children with enteropathogenic *Escherichia coli* infection to the bundle-forming pilus and locus of enterocyte effacement-encoded virulence determinants. *J Infect Dis* **179:**269–274.

60. **Calderon Toledo C, Arvidsson I, Karpman D.** 2011. Cross-reactive protection against entero-hemorrhagic *Escherichia coli* infection by enteropathogenic E. coli in a mouse model. *Infect Immun* **79:**2224–2233.

61. **Walters MD, Matthei IU, Kay R, Dillon MJ, Barratt TM.** 1989. The polymorphonuclear leucocyte count in childhood haemolytic uraemic syndrome. *Pediatr Nephrol* **3**:130–134.

62. **Robson WL, Fick GH, Wilson PC.** 1988. Prognostic factors in typical postdiarrhea hemolytic-uremic syndrome. *Child Nephrol Urol* **9**:203–207.

63. **Fernandez GC, Gomez SA, Rubel CJ, Bentancor LV, Barrionuevo P, Alduncin M, Grimoldi I, Exeni R, Isturiz MA, Palermo MS.** 2005. Impaired neutrophils in children with the typical form of hemolytic uremic syndrome. *Pediatr Nephrol* **20**:1306–1314.

64. **Milford D, Taylor CM, Rafaat F, Halloran E, Dawes J.** 1989. Neutrophil elastases and haemolytic uraemic syndrome. *Lancet* **2**:1153.

65. **Fitzpatrick MM, Shah V, Filler G, Dillon MJ, Barratt TM.** 1992. Neutrophil activation in the haemolytic uraemic syndrome: free and complexed elastase in plasma. *Pediatr Nephrol* **6**:50–53.

66. **Hughes DA, Smith GC, Davidson JE, Murphy AV, Beattie TJ.** 1996. The neutrophil oxidative burst in diarrhoea-associated haemolytic uraemic syndrome. *Pediatr Nephrol* **10**:445–447.

67. **Forsyth KD, Simpson AC, Fitzpatrick MM, Barratt TM, Levinsky RJ.** 1989. Neutrophil-mediated endothelial injury in haemolytic uraemic syndrome. *Lancet* **2**:411–414.

68. **Fernandez GC, Gomez SA, Ramos MV, Bentancor LV, Fernandez-Brando RJ, Landoni VI, Lopez L, Ramirez F, Diaz M, Alduncin M, Grimoldi I, Exeni R, Isturiz MA, Palermo MS.** 2007. The functional state of neutrophils correlates with the severity of renal dysfunction in children with hemolytic uremic syndrome. *Pediatr Res* **61**:123–128.

69. **Fitzpatrick MM, Shah V, Trompeter RS, Dillon MJ, Barratt TM.** 1992. Interleukin-8 and polymorphoneutrophil leucocyte activation in hemolytic uremic syndrome of childhood. *Kidney Int* **42**:951–956.

70. **Liu J, He T, He Y, Zhang Z, Akahoshi T, Kondo H, Zhong S.** 2002. Prolongation of functional lifespan of neutrophils by recombinant verotoxin 2. *Chin Med J (Engl)* **115**:900–903.

71. **Ge S, Hertel B, Emden SH, Beneke J, Menne J, Haller H, von Vietinghoff S.** 2012. Microparticle generation and leucocyte death in Shiga toxin-mediated HUS. *Nephrol Dial Transplant* **27**:2768–2775.

72. **Fernandez GC, Lopez MF, Gomez SA, Ramos MV, Bentancor LV, Fernandez-Brando RJ, Landoni VI, Dran GI, Meiss R, Isturiz MA, Palermo MS.** 2006. Relevance of neutrophils in the murine model of haemolytic uraemic syndrome: mechanisms involved in Shiga toxin type 2-induced neutrophilia. *Clin Exp Immunol* **146**:76–84.

73. **Fukuda MN, Dell A, Oates JE, Wu P, Klock JC, Fukuda M.** 1985. Structures of glycosphingolipids isolated from human granulocytes. The presence of a series of linear poly-N-acetyllactosaminylceramide and its significance in glycolipids of whole blood cells. *J Biol Chem* **260**:1067–1082.

74. **Arfilli V, Carnicelli D, Rocchi L, Ricci F, Pagliaro P, Tazzari PL, Brigotti M.** 2010. Shiga toxin 1 and ricin A chain bind to human polymorphonuclear leucocytes through a common receptor. *Biochem J* **432**:173–180.

75. **Geelen JM, van der Velden TJ, Te Loo DM, Boerman OC, van den Heuvel LP, Monnens LA.** 2007. Lack of specific binding of Shiga-like toxin (verocytotoxin) and non-specific interaction of Shiga-like toxin 2 antibody with human polymorphonuclear leucocytes. *Nephrol Dial Transplant* **22**:749–755.

76. **Te Loo DM, van Hinsbergh VW, van den Heuvel LP, Monnens LA.** 2001. Detection of verocytotoxin bound to circulating polymorphonuclear leukocytes of patients with hemolytic uremic syndrome. *J Am Soc Nephrol* **12**:800–806.

77. **Tazzari PL, Ricci F, Carnicelli D, Caprioli A, Tozzi AE, Rizzoni G, Conte R, Brigotti M.** 2004. Flow cytometry detection of Shiga toxins in the blood from children with hemolytic uremic syndrome. *Cytometry B Clin Cytom* **61**:40–44.

78. **Ståhl AL, Sartz L, Nelsson A, Békássy ZD, Karpman D.** 2009. Shiga toxin and lipopolysaccharide induce platelet-leukocyte aggregates and tissue factor release, a thrombotic mechanism in hemolytic uremic syndrome. *PLoS One* **4**:e6990.

79. **Brigotti M, Tazzari PL, Ravanelli E, Carnicelli D, Rocchi L, Arfilli V, Scavia G, Minelli F, Ricci F, Pagliaro P, Ferretti AV, Pecoraro C, Paglialonga F, Edefonti A, Procaccino MA, Tozzi AE, Caprioli A.** 2011. Clinical relevance of Shiga toxin concentrations in the blood of patients with hemolytic uremic syndrome. *Pediatr Infect Dis J* **30**:486–490.

80. **Szabady RL, Lokuta MA, Walters KB, Huttenlocher A, Welch RA.** 2009. Modulation of neutrophil function by a secreted mucinase of *Escherichia coli* O157:H7. *PLoS Pathog* **5**:e1000320.

81. **Fernandez GC, Ramos MV, Gomez SA, Dran GI, Exeni R, Alduncin M, Grimoldi I, Vallejo G, Elias-Costa C, Isturiz MA, Palermo MS.** 2005. Differential expression of function-related antigens on blood monocytes in children with hemolytic uremic syndrome. *J Leukoc Biol* **78**:853–861.

82. **Ramos MV, Fernandez GC, Patey N, Schierloh P, Exeni R, Grimoldi I, Vallejo G, Elias-Costa C,**

Del Carmen Sasiain M, Trachtman H, Combadiere C, Proulx F, Palermo MS. 2007. Involvement of the fractalkine pathway in the pathogenesis of childhood hemolytic uremic syndrome. *Blood* **109**:2438–2445.

83. van Setten PA, Monnens LA, Verstraten RG, van den Heuvel LP, van Hinsbergh VW. 1996. Effects of verocytotoxin-1 on nonadherent human monocytes: binding characteristics, protein synthesis, and induction of cytokine release. *Blood* **88**:174–183.

84. Guessous F, Marcinkiewicz M, Polanowska-Grabowska R, Keepers TR, Obrig T, Gear AR. 2005. Shiga toxin 2 and lipopolysaccharide cause monocytic THP-1 cells to release factors which activate platelet function. *Thromb Haemost* **94**:1019–1027.

85. Murata K, Higuchi T, Takada K, Oida K, Horie S, Ishii H. 2006. Verotoxin-1 stimulation of macrophage-like THP-1 cells up-regulates tissue factor expression through activation of c-Yes tyrosine kinase: possible signal transduction in tissue factor up-regulation. *Biochim Biophys Acta* **1762**:835–843.

86. Del Conde I, Shrimpton CN, Thiagarajan P, Lopez JA. 2005. Tissue-factor-bearing microvesicles arise from lipid rafts and fuse with activated platelets to initiate coagulation. *Blood* **106**:1604–1611.

87. Egorina EM, Sovershaev MA, Olsen JO, Osterud B. 2008. Granulocytes do not express but acquire monocyte-derived tissue factor in whole blood: evidence for a direct transfer. *Blood* **111**:1208–1216.

88. Geelen JM, van der Velden TJ, van den Heuvel LP, Monnens LA. 2007. Interactions of Shiga-like toxin with human peripheral blood monocytes. *Pediatr Nephrol* **22**:1181–1187.

89. Proulx F, Seidman EG, Karpman D. 2001. Pathogenesis of Shiga toxin-associated hemolytic uremic syndrome. *Pediatr Res* **50**:163–171.

90. Zoja C, Buelli S, Morigi M. 2010. Shiga toxin-associated hemolytic uremic syndrome: pathophysiology of endothelial dysfunction. *Pediatr Nephrol* **25**:2231–2240.

91. Fong JS, Kaplan BS. 1982. Impairment of platelet aggregation in hemolytic uremic syndrome: evidence for platelet "exhaustion." *Blood* **60**:564–570.

92. Sassetti B, Vizcarguenaga MI, Zanaro NL, Silva MV, Kordich L, Florentini L, Diaz M, Vitacco M, Sanchez Avalos JC. 1999. Hemolytic uremic syndrome in children: platelet aggregation and membrane glycoproteins. *J Pediatr Hematol Oncol* **21**:123–128.

93. Galli M, Grassi A, Barbui T. 1996. Platelet-derived microvesicles in thrombotic thrombocytopenic purpura and hemolytic uremic syndrome. *Thromb Haemost* **75**:427–431.

94. Walters MD, Levin M, Smith C, Nokes TJ, Hardisty RM, Dillon MJ, Barratt TM. 1988. Intravascular platelet activation in the hemolytic uremic syndrome. *Kidney Int* **33**:107–115.

95. Keepers TR, Psotka MA, Gross LK, Obrig TG. 2006. A murine model of HUS: Shiga toxin with lipopolysaccharide mimics the renal damage and physiologic response of human disease. *J Am Soc Nephrol* **17**:3404–3414.

96. Cooling LL, Walker KE, Gille T, Koerner TA. 1998. Shiga toxin binds human platelets via globotriaosylceramide (Pk antigen) and a novel platelet glycosphingolipid. *Infect Immun* **66**:4355–4366.

97. Ghosh SA, Polanowska-Grabowska RK, Fujii J, Obrig T, Gear AR. 2004. Shiga toxin binds to activated platelets. *J Thromb Haemost* **2**:499–506.

98. Ståhl AL, Svensson M, Morgelin M, Svanborg C, Tarr PI, Mooney JC, Watkins SL, Johnson R, Karpman D. 2006. Lipopolysaccharide from enterohemorrhagic *Escherichia coli* binds to platelets through TLR4 and CD62 and is detected on circulating platelets in patients with hemolytic uremic syndrome. *Blood* **108**:167–176.

99. Karpman D, Papadopoulou D, Nilsson K, Sjögren AC, Mikaelsson C, Lethagen S. 2001. Platelet activation by Shiga toxin and circulatory factors as a pathogenetic mechanism in the hemolytic uremic syndrome. *Blood* **97**:3100–3108.

100. Semple JW, Italiano JE Jr, Freedman J. 2011. Platelets and the immune continuum. *Nat Rev Immunol* **11**:264–274.

101. Moore KL, Patel KD, Bruehl RE, Li F, Johnson DA, Lichenstein HS, Cummings RD, Bainton DF, McEver RP. 1995. P-selectin glycoprotein ligand-1 mediates rolling of human neutrophils on P-selectin. *J Cell Biol* **128**:661–671.

102. Michelson AD, Barnard MR, Krueger LA, Valeri CR, Furman MI. 2001. Circulating monocyte-platelet aggregates are a more sensitive marker of in vivo platelet activation than platelet surface P-selectin: studies in baboons, human coronary intervention, and human acute myocardial infarction. *Circulation* **104**:1533–1537.

103. Gear AR, Camerini D. 2003. Platelet chemokines and chemokine receptors: linking hemostasis, inflammation, and host defense. *Microcirculation* **10**:335–350.

104. Shiraki R, Inoue N, Kawasaki S, Takei A, Kadotani M, Ohnishi Y, Ejiri J, Kobayashi S, Hirata K, Kawashima S, Yokoyama M. 2004. Expression of Toll-like receptors on human platelets. *Thromb Res* **113**:379–385.

105. Andonegui G, Kerfoot SM, McNagny K, Ebbert KV, Patel KD, Kubes P. 2005. Platelets express

functional Toll-like receptor-4. *Blood* **106**:2417–2423.

106. **Aslam R, Speck ER, Kim M, Crow AR, Bang KW, Nestel FP, Ni H, Lazarus AH, Freedman J, Semple JW.** 2006. Platelet Toll-like receptor expression modulates lipopolysaccharide-induced thrombocytopenia and tumor necrosis factor-alpha production in vivo. *Blood* **107**:637–641.

107. **Cicala C, Santacroce C, Itoh H, Douglas GJ, Page CP.** 1997. A study on rat platelet responsiveness following intravenous endotoxin administration. *Life Sci* **60**:PL31–PL38.

108. **Jayachandran M, Brunn GJ, Karnicki K, Miller RS, Owen WG, Miller VM.** 2007. In vivo effects of lipopolysaccharide and TLR4 on platelet production and activity: implications for thrombotic risk. *J Appl Physiol* **102**:429–433.

109. **Scott T, Owens MD.** 2008. Thrombocytes respond to lipopolysaccharide through Toll-like receptor-4, and MAP kinase and NF-kappaB pathways leading to expression of interleukin-6 and cyclooxygenase-2 with production of prostaglandin E2. *Mol Immunol* **45**:1001–1008.

110. **Elzey BD, Tian J, Jensen RJ, Swanson AK, Lees JR, Lentz SR, Stein CS, Nieswandt B, Wang Y, Davidson BL, Ratliff TL.** 2003. Platelet-mediated modulation of adaptive immunity. A communication link between innate and adaptive immune compartments. *Immunity* **19**:9–19.

111. **Viisoreanu D, Polanowska-Grabowska R, Suttitanamongkol S, Obrig TG, Gear AR.** 2000. Human platelet aggregation is not altered by Shiga toxins 1 or 2. *Thromb Res* **98**:403–410.

112. **Gawaz M, Neumann FJ, Dickfeld T, Koch W, Laugwitz KL, Adelsberger H, Langenbrink K, Page S, Neumeier D, Schomig A, Brand K.** 1998. Activated platelets induce monocyte chemotactic protein-1 secretion and surface expression of intercellular adhesion molecule-1 on endothelial cells. *Circulation* **98**:1164–1171.

113. **Henn V, Slupsky JR, Grafe M, Anagnostopoulos I, Forster R, Muller-Berghaus G, Kroczek RA.** 1998. CD40 ligand on activated platelets triggers an inflammatory reaction of endothelial cells. *Nature* **391**:591–594.

114. **Hollenbaugh D, Mischel-Petty N, Edwards CP, Simon JC, Denfeld RW, Kiener PA, Aruffo A.** 1995. Expression of functional CD40 by vascular endothelial cells. *J Exp Med* **182**:33–40.

115. **Caprioli A, Luzzi I, Rosmini F, Resti C, Edefonti A, Perfumo F, Farina C, Goglio A, Gianviti A, Rizzoni G.** 1994. Community-wide outbreak of hemolytic-uremic syndrome associated with non-O157 verocytotoxin-producing *Escherichia coli*. *J Infect Dis* **169**:208–211.

116. **Proulx F, Seidman E, Mariscalco MM, Lee K, Caroll S.** 1999. Increased circulating levels of lipopolysaccharide binding protein in children with *Escherichia coli* O157:H7 hemorrhagic colitis and hemolytic uremic syndrome. *Clin Diagn Lab Immunol* **6**:773.

117. **Jerala R.** 2007. Structural biology of the LPS recognition. *Int J Med Microbiol* **297**:353–363.

118. **Valles PG, Melechuck S, Gonzalez A, Manucha W, Bocanegra V, Valles R.** 2012. Toll-like receptor 4 expression on circulating leucocytes in hemolytic uremic syndrome. *Pediatr Nephrol* **27**:407–415.

119. **Kamitsuji H, Nonami K, Murakami T, Ishikawa N, Nakayama A, Umeki Y.** 2000. Elevated tissue factor circulating levels in children with hemolytic uremic syndrome caused by verotoxin-producing *E. coli*. *Clin Nephrol* **53**:319–324.

120. **Edgington TS, Mackman N, Brand K, Ruf W.** 1991. The structural biology of expression and function of tissue factor. *Thromb Haemost* **66**:67–79.

121. **Rao LV, Rapaport SI, Bajaj SP.** 1986. Activation of human factor VII in the initiation of tissue factor-dependent coagulation. *Blood* **68**:685–691.

122. **Monroe DM, Hoffman M.** 2006. What does it take to make the perfect clot? *Arterioscler Thromb Vasc Biol* **26**:41–48.

123. **Camera M, Frigerio M, Toschi V, Brambilla M, Rossi F, Cottell DC, Maderna P, Parolari A, Bonzi R, De Vincenti O, Tremoli E.** 2003. Platelet activation induces cell-surface immunoreactive tissue factor expression, which is modulated differently by antiplatelet drugs. *Arterioscler Thromb Vasc Biol* **23**:1690–1696.

124. **Schwertz H, Tolley ND, Foulks JM, Denis MM, Risenmay BW, Buerke M, Tilley RE, Rondina MT, Harris EM, Kraiss LW, Mackman N, Zimmerman GA, Weyrich AS.** 2006. Signal-dependent splicing of tissue factor pre-mRNA modulates the thrombogenicity of human platelets. *J Exp Med* **203**:2433–2440.

125. **Panes O, Matus V, Saez CG, Quiroga T, Pereira J, Mezzano D.** 2007. Human platelets synthesize and express functional tissue factor. *Blood* **109**:5242–5250.

126. **del Conde I, Nabi F, Tonda R, Thiagarajan P, Lopez JA, Kleiman NS.** 2005. Effect of P-selectin on phosphatidylserine exposure and surface-dependent thrombin generation on monocytes. *Arterioscler Thromb Vasc Biol* **25**:1065–1070.

127. **Østerud B, Olsen JO.** 2013. Human platelets do not express tissue factor. *Thromb Res* **132**:112–115.

128. **Bolande RP, Kaplan BS.** 1985. Experimental studies on the hemolytic-uremic syndrome. *Nephron* **39**:228–236.

129. **Appiani AC, Edefonti A, Bettinelli A, Cossu MM, Paracchini ML, Rossi E.** 1982. The relationship between plasma levels of the factor VIII

complex and platelet release products (beta-thromboglobulin and platelet factor 4) in children with the hemolytic-uremic syndrome. *Clin Nephrol* **17**:195–199.

130. van de Kar NC, van Hinsbergh VW, Brommer EJ, Monnens LA. 1994. The fibrinolytic system in the hemolytic uremic syndrome: in vivo and in vitro studies. *Pediatr Res* **36**:257–264.

131. Tsai HM, Chandler WL, Sarode R, Hoffman R, Jelacic S, Habeeb RL, Watkins SL, Wong CS, Williams GD, Tarr PI. 2001. von Willebrand factor and von Willebrand factor-cleaving metal-loprotease activity in *Escherichia coli* O157:H7-associated hemolytic uremic syndrome. *Pediatr Res* **49**:653–659.

132. Morigi M, Galbusera M, Binda E, Imberti B, Gastoldi S, Remuzzi A, Zoja C, Remuzzi G. 2001. Verotoxin-1-induced up-regulation of adhesive molecules renders microvascular endothelial cells thrombogenic at high shear stress. *Blood* **98**:1828–1835.

133. Nolasco LH, Turner NA, Bernardo A, Tao Z, Cleary TG, Dong JF, Moake JL. 2005. Hemo-lytic uremic syndrome-associated Shiga toxins promote endothelial-cell secretion and impair ADAMTS13 cleavage of unusually large von Willebrand factor multimers. *Blood* **106**:4199–4209.

134. Guessous F, Marcinkiewicz M, Polanowska-Grabowska R, Kongkhum S, Heatherly D, Obrig T, Gear AR. 2005. Shiga toxin 2 and lipopoly-saccharide induce human microvascular endo-thelial cells to release chemokines and factors that stimulate platelet function. *Infect Immun* **73**:8306–8316.

135. Furie B, Furie BC. 2008. Mechanisms of thrombus formation. *N Engl J Med* **359**:938–949.

136. Nevard CH, Jurd KM, Lane DA, Philippou H, Haycock GB, Hunt BJ. 1997. Activation of coag-ulation and fibrinolysis in childhood diarrhoea-associated haemolytic uraemic syndrome. *Thromb Haemost* **78**:1450–1455.

137. Van Geet C, Proesmans W, Arnout J, Vermylen J, Declerck PJ. 1998. Activation of both coagu-lation and fibrinolysis in childhood hemolytic uremic syndrome. *Kidney Int* **54**:1324–1330.

138. Chandler WL, Jelacic S, Boster DR, Ciol MA, Williams GD, Watkins SL, Igarashi T, Tarr PI. 2002. Prothrombotic coagulation abnormalities preceding the hemolytic-uremic syndrome. *N Engl J Med* **346**:23–32.

139. Bergstein JM, Riley M, Bang NU. 1992. Role of plasminogen-activator inhibitor type 1 in the pathogenesis and outcome of the hemolytic ure-mic syndrome. *N Engl J Med* **327**:755–759.

140. Chaisri U, Nagata M, Kurazono H, Horie H, Tongtawe P, Hayashi H, Watanabe T, Tapchaisri P, Chongsa-nguan M, Chaicumpa W. 2001. Lo-calization of Shiga toxins of enterohaemorrhagic *Escherichia coli* in kidneys of paediatric and geri-atric patients with fatal haemolytic uraemic syn-drome. *Microb Pathog* **31**:59–67.

141. Uchida H, Kiyokawa N, Horie H, Fujimoto J, Takeda T. 1999. The detection of Shiga toxins in the kidney of a patient with hemolytic uremic syndrome. *Pediatr Res* **45**:133–137.

142. Karpman D, Håkansson A, Perez MT, Isaksson C, Carlemalm E, Caprioli A, Svanborg C. 1998. Apoptosis of renal cortical cells in the hemolytic-uremic syndrome: in vivo and in vitro studies. *Infect Immun* **66**:636–644.

143. Buteau C, Proulx F, Chaibou M, Raymond D, Clermont MJ, Mariscalco MM, Lebel MH, Seidman E. 2000. Leukocytosis in children with *Escherichia coli* O157:H7 enteritis developing the hemolytic-uremic syndrome. *Pediatr Infect Dis J* **19**:642–647.

144. Salzman MB, Ettenger RB, Cherry JD. 1991. Leukocytosis in hemolytic-uremic syndrome. *Pediatr Infect Dis J* **10**:470–471.

145. Coad NA, Marshall T, Rowe B, Taylor CM. 1991. Changes in the postenteropathic form of the he-molytic uremic syndrome in children. *Clin Nephrol* **35**:10–16.

146. Inward CD, Howie AJ, Fitzpatrick MM, Rafaat F, Milford DV, Taylor CM. 1997. Renal histo-pathology in fatal cases of diarrhoea-associated haemolytic uraemic syndrome. British Associa-tion for Paediatric Nephrology. *Pediatr Nephrol* **11**:556–559.

147. Roche JK, Keepers TR, Gross LK, Seaner RM, Obrig TG. 2007. CXCL1/KC and CXCL2/MIP-2 are critical effectors and potential targets for therapy of *Escherichia coli* O157:H7-associated renal inflammation. *Am J Pathol* **170**:526–537.

148. Keepers TR, Gross LK, Obrig TG. 2007. Mono-cyte chemoattractant protein 1, macrophage in-flammatory protein 1 alpha, and RANTES recruit macrophages to the kidney in a mouse model of hemolytic-uremic syndrome. *Infect Immun* **75**:1229–1236.

149. Garcia A, Marini RP, Catalfamo JL, Knox KA, Schauer DB, Rogers AB, Fox JG. 2008. Intra-venous Shiga toxin 2 promotes enteritis and renal injury characterized by polymorphonuclear leu-kocyte infiltration and thrombosis in Dutch Belt-ed rabbits. *Microbes Infect* **10**:650–656.

150. Zoja C, Angioletti S, Donadelli R, Zanchi C, Tomasoni S, Binda E, Imberti B, te Loo M, Monnens L, Remuzzi G, Morigi M. 2002. Shiga toxin-2 triggers endothelial leukocyte adhesion and transmigration via NF-kappaB dependent up-regulation of IL-8 and MCP-1. *Kidney Int* **62**:846–856.

151. Zanchi C, Zoja C, Morigi M, Valsecchi F, Liu XY, Rottoli D, Locatelli M, Buelli S, Pezzotta A, Mapelli P, Geelen J, Remuzzi G, Hawiger J. 2008. Fractalkine and CX3CR1 mediate leukocyte capture by endothelium in response to Shiga toxin. *J Immunol* **181**:1460–1469.

152. Morigi M, Micheletti G, Figliuzzi M, Imberti B, Karmali MA, Remuzzi A, Remuzzi G, Zoja C. 1995. Verotoxin-1 promotes leukocyte adhesion to cultured endothelial cells under physiologic flow conditions. *Blood* **86**:4553–4558.

153. Brigotti M, Caprioli A, Tozzi AE, Tazzari PL, Ricci F, Conte R, Carnicelli D, Procaccino MA, Minelli F, Ferretti AV, Paglialonga F, Edefonti A, Rizzoni G. 2006. Shiga toxins present in the gut and in the polymorphonuclear leukocytes circulating in the blood of children with hemolytic-uremic syndrome. *J Clin Microbiol* **44**:313–317.

154. Karpman D, Manea M, Vaziri-Sani F, Ståhl AL, Kristoffersson AC. 2006. Platelet activation in hemolytic uremic syndrome. *Semin Thromb Hemost* **32**:128–145.

155. Stearns-Kurosawa DJ, Oh SY, Cherla RP, Lee MS, Tesh VL, Papin J, Henderson J, Kurosawa S. 2013. Distinct renal pathology and a chemotactic phenotype after enterohemorrhagic *Escherichia coli* Shiga toxins in non-human primate models of hemolytic uremic syndrome. *Am J Pathol* **182**:1227–1238.

156. Lopez EL, Devoto S, Fayad A, Canepa C, Morrow AL, Cleary TG. 1992. Association between severity of gastrointestinal prodrome and long-term prognosis in classic hemolytic-uremic syndrome. *J Pediatr* **120**:210–215.

157. van Setten PA, van Hinsbergh VW, van den Heuvel LP, Preyers F, Dijkman HB, Assmann KJ, van der Velden TJ, Monnens LA. 1998. Monocyte chemoattractant protein-1 and Interleukin-8 levels in urine and serum of patents with hemolytic uremic syndrome. *Pediatr Res* **43**:759–767.

158. Decaluwe H, Harrison LM, Mariscalco MM, Gendrel D, Bohuon C, Tesh VL, Proulx F. 2006. Procalcitonin in children with Escherichia coli O157:H7 associated hemolytic uremic syndrome. *Pediatr Res* **59**:579–583.

159. Proulx F, Toledano B, Phan V, Clermont MJ, Mariscalco MM, Seidman EG. 2002. Circulating granulocyte colony-stimulating factor, C-X-C, and C-C chemokines in children with *Escherichia coli* O157:H7 associated hemolytic uremic syndrome. *Pediatr Res* **52**:928–934.

160. Masri C, Proulx F, Toledano B, Clermont MJ, Mariscalco MM, Seidman EG, Carcillo J. 2000. Soluble Fas and soluble Fas-ligand in children with *Escherichia coli* O157:H7-associated hemo-lytic uremic syndrome. *Am J Kidney Dis* **36**:687–694.

161. Proulx F, Turgeon JP, Litalien C, Mariscalco MM, Robitaille P, Seidman E. 1998. Inflammatory mediators in *Escherichia coli* O157:H7 hemorrhagic colitis and hemolytic-uremic syndrome. *Pediatr Infect Dis J* **17**:899–904.

162. Litalien C, Proulx F, Mariscalco MM, Robitaille P, Turgeon JP, Orrbine E, Rowe PC, McLaine PN, Seidman E. 1999. Circulating inflammatory cytokine levels in hemolytic uremic syndrome. *Pediatr Nephrol* **13**:840–845.

163. van de Kar NC, Sauerwein RW, Demacker PN, Grau GE, van Hinsbergh VW, Monnens LA. 1995. Plasma cytokine levels in hemolytic uremic syndrome. *Nephron* **71**:309–313.

164. Karpman D, Andreasson A, Thysell H, Kaplan BS, Svanborg C. 1995. Cytokines in childhood hemolytic uremic syndrome and thrombotic thrombocytopenic purpura. *Pediatr Nephrol* **9**: 694–699.

165. Proulx F, Litalien C, Turgeon JP, Mariscalco MM, Seidman E. 2000. Circulating levels of transforming growth factor-beta1 and lymphokines among children with hemolytic uremic syndrome. *Am J Kidney Dis* **35**:29–34.

166. Inward CD, Varagunam M, Adu D, Milford DV, Taylor CM. 1997. Cytokines in haemolytic uraemic syndrome associated with verocytotoxin-producing *Escherichia coli* infection. *Arch Dis Child* **77**:145–147.

167. Inward CD, Pall AA, Adu D, Milford DV, Taylor CM. 1995. Soluble circulating cell adhesion molecules in haemolytic uraemic syndrome. *Pediatr Nephrol* **9**:574–578.

168. Murata A, Shimazu T, Yamamoto T, Taenaka N, Nagayama K, Honda T, Sugimoto H, Monden M, Matsuura N, Okada S. 1998. Profiles of circulating inflammatory- and anti-inflammatory cytokines in patients with hemolytic uremic syndrome due to *E. coli* O157 infection. *Cytokine* **10**:544–548.

169. Yamamoto T, Nagayama K, Satomura K, Honda T, Okada S. 2000. Increased serum IL-10 and endothelin levels in hemolytic uremic syndrome caused by *Escherichia coli* O157. *Nephron* **84**:326–332.

170. Lopez EL, Contrini MM, Devoto S, de Rosa MF, Grana MG, Genero MH, Canepa C, Gomez HF, Cleary TG. 1995. Tumor necrosis factor concentrations in hemolytic uremic syndrome patients and children with bloody diarrhea in Argentina. *Pediatr Infect Dis J* **14**:594–598.

171. Nevard CH, Blann AD, Jurd KM, Haycock GB, Hunt BJ. 1999. Markers of endothelial cell activation and injury in childhood haemolytic uraemic syndrome. *Pediatr Nephrol* **13**:487–492.

172. **Caletti MG, Balestracci A, Roy AH.** 2010. Levels of urinary transforming growth factor beta-1 in children with D+ hemolytic uremic syndrome. *Pediatr Nephrol* **25:**1177–1180.

173. **Bhowmik D.** 2001. Elevated tissue factor levels in children with hemolytic uremic syndrome. *Clin Nephrol* **55:**262.

174. **Petruzziello-Pellegrini TN, Yuen DA, Page AV, Patel S, Soltyk AM, Matouk CC, Wong DK, Turgeon PJ, Fish JE, Ho JJ, Steer BM, Khajoee V, Tigdi J, Lee WL, Motto DG, Advani A, Gilbert RE, Karumanchi SA, Robinson LA, Tarr PI, Liles WC, Brunton JL, Marsden PA.** 2012. The CXCR4/CXCR7/SDF-1 pathway contributes to the pathogenesis of Shiga toxin-associated hemolytic uremic syndrome in humans and mice. *J Clin Invest* **122:**759–776.

175. **Louise CB, Obrig TG.** 1991. Shiga toxin-associated hemolytic-uremic syndrome: combined cytotoxic effects of Shiga toxin, interleukin-1 beta, and tumor necrosis factor alpha on human vascular endothelial cells in vitro. *Infect Immun* **59:**4173–4179.

176. **Keusch GT, Acheson DW, Aaldering L, Erban J, Jacewicz MS.** 1996. Comparison of the effects of Shiga-like toxin 1 on cytokine- and butyrate-treated human umbilical and saphenous vein endothelial cells. *J Infect Dis* **173:**1164–1170.

177. **van Setten PA, van Hinsbergh VW, van der Velden TJ, van de Kar NC, Vermeer M, Mahan JD, Assmann KJ, van den Heuvel LP, Monnens LA.** 1997. Effects of TNF alpha on verocytotoxin cytotoxicity in purified human glomerular microvascular endothelial cells. *Kidney Int* **51:**1245–1256.

178. **van de Kar NC, Monnens LA, Karmali MA, van Hinsbergh VW.** 1992. Tumor necrosis factor and interleukin-1 induce expression of the verocytotoxin receptor globotriaosylceramide on human endothelial cells: implications for the pathogenesis of the hemolytic uremic syndrome. *Blood* **80:**2755–2764.

179. **Louise CB, Obrig TG.** 1992. Shiga toxin-associated hemolytic uremic syndrome: combined cytotoxic effects of shiga toxin and lipopolysaccharide (endotoxin) on human vascular endothelial cells in vitro. *Infect Immun* **60:**1536–1543.

180. **Kaye SA, Louise CB, Boyd B, Lingwood CA, Obrig TG.** 1993. Shiga toxin-associated hemolytic uremic syndrome: interleukin-1 beta enhancement of Shiga toxin cytotoxicity toward human vascular endothelial cells in vitro. *Infect Immun* **61:**3886–3891.

181. **Isogai E, Isogai H, Kimura K, Hayashi S, Kubota T, Fujii N, Takeshi K.** 1998. Role of tumor necrosis factor alpha in gnotobiotic mice infected with an *Escherichia coli* O157:H7 strain. *Infect Immun* **66:**197–202.

182. **Harel Y, Silva M, Giroir B, Weinberg A, Cleary TB, Beutler B.** 1993. A reporter transgene indicates renal-specific induction of tumor necrosis factor (TNF) by Shiga-like toxin. Possible involvement of TNF in hemolytic uremic syndrome. *J Clin Invest* **92:**2110–2116.

183. **Lentz EK, Cherla RP, Jaspers V, Weeks BR, Tesh VL.** 2010. Role of tumor necrosis factor alpha in disease using a mouse model of Shiga toxin-mediated renal damage. *Infect Immun* **78:**3689–3699.

184. **Wolski VM, Soltyk AM, Brunton JL.** 2002. Tumour necrosis factor alpha is not an essential component of verotoxin 1-induced toxicity in mice. *Microb Pathog* **32:**263–271.

185. **Hughes AK, Stricklett PK, Kohan DE.** 1998. Shiga toxin-1 regulation of cytokine production by human proximal tubule cells. *Kidney Int* **54:**1093–1106.

186. **Nakamura A, Johns EJ, Imaizumi A, Yanagawa Y, Kohsaka T.** 2001. Activation of beta(2)-adrenoceptor prevents shiga toxin 2-induced TNF-alpha gene transcription. *J Am Soc Nephrol* **12:**2288–2299.

187. **Taylor FB Jr, Tesh VL, DeBault L, Li A, Chang AC, Kosanke SD, Pysher TJ, Siegler RL.** 1999. Characterization of the baboon responses to Shiga-like toxin: descriptive study of a new primate model of toxic responses to Stx-1. *Am J Pathol* **154:**1285–1299.

188. **Hughes AK, Stricklett PK, Kohan DE.** 2001. Shiga toxin-1 regulation of cytokine production by human glomerular epithelial cells. *Nephron* **88:**14–23.

189. **Akira S, Hirano T, Taga T, Kishimoto T.** 1990. Biology of multifunctional cytokines: IL 6 and related molecules (IL 1 and TNF). *FASEB J* **4:**2860–2867.

190. **Ramos MV, Auvynet C, Poupel L, Rodero M, Mejias MP, Panek CA, Vanzulli S, Combadiere C, Palermo M.** 2012. Chemokine receptor CCR1 disruption limits renal damage in a murine model of hemolytic uremic syndrome. *Am J Pathol* **180:**1040–1048.

191. **Karpman D, Sartz L, Johnson S.** 2010. Pathophysiology of typical hemolytic uremic syndrome. *Semin Thromb Hemost* **36:**575–585.

192. **Nestoridi E, Kushak RI, Duguerre D, Grabowski EF, Ingelfinger JR.** 2005. Up-regulation of tissue factor activity on human proximal tubular epithelial cells in response to Shiga toxin. *Kidney Int* **67:**2254–2266.

193. **Nestoridi E, Tsukurov O, Kushak RI, Ingelfinger JR, Grabowski EF.** 2005. Shiga toxin enhances functional tissue factor on human glomerular endothelial cells: implications for the pathophysiology of hemolytic uremic syndrome. *J Thromb Haemost* **3:**752–762.

194. **Braune SA, Wichmann D, von Heinz MC, Nierhaus A, Becker H, Meyer TN, Meyer GP, Muller-Schulz M, Fricke J, de Weerth A, Hoepker WW, Fiehler J, Magnus T, Gerloff C, Panzer U, Stahl RA, Wegscheider K, Kluge S.** 2013. Clinical features of critically ill patients with Shiga toxin-induced hemolytic uremic syndrome. *Crit Care Med* 41:1702–1710.

195. **Magnus T, Rother J, Simova O, Meier-Cillien M, Repenthin J, Moller F, Gbadamosi J, Panzer U, Wengenroth M, Hagel C, Kluge S, Stahl RK, Wegscheider K, Urban P, Eckert B, Glatzel M, Fiehler J, Gerloff C.** 2012. The neurological syndrome in adults during the 2011 northern German *E. coli* serotype O104:H4 outbreak. *Brain* 135:1850–1859.

196. **Obata F, Tohyama K, Bonev AD, Kolling GL, Keepers TR, Gross LK, Nelson MT, Sato S, Obrig TG.** 2008. Shiga toxin 2 affects the central nervous system through receptor globotriaosylceramide localized to neurons. *J Infect Dis* 198:1398–1406.

197. **Ergonul Z, Hughes AK, Kohan DE.** 2003. Induction of apoptosis of human brain microvascular endothelial cells by shiga toxin 1. *J Infect Dis* 187:154–158.

198. **Fujii J, Wood K, Matsuda F, Carneiro-Filho BA, Schlegel KH, Yutsudo T, Binnington-Boyd B, Lingwood CA, Obata F, Kim KS, Yoshida S, Obrig T.** 2008. Shiga toxin 2 causes apoptosis in human brain microvascular endothelial cells via C/EBP homologous protein. *Infect Immun* 76:3679–3689.

199. **Shiraishi M, Ichiyama T, Matsushige T, Iwaki T, Iyoda K, Fukuda K, Makata H, Matsubara T, Furukawa S.** 2008. Soluble tumor necrosis factor receptor 1 and tissue inhibitor of metalloproteinase-1 in hemolytic uremic syndrome with encephalopathy. *J Neuroimmunol* 196:147–152.

200. **Sheth KJ, Swick HM, Haworth N.** 1986. Neurological involvement in hemolytic-uremic syndrome. *Ann Neurol* 19:90–93.

201. **Fujii J, Kita T, Yoshida S, Takeda T, Kobayashi H, Tanaka N, Ohsato K, Mizuguchi Y.** 1994. Direct evidence of neuron impairment by oral infection with verotoxin-producing *Escherichia coli* O157:H- in mitomycin-treated mice. *Infect Immun* 62:3447–3453.

202. **Sofroniew MV, Vinters HV.** 2010. Astrocytes: biology and pathology. *Acta Neuropathol* 119:7–35.

203. **Amran MY, Fujii J, Suzuki SO, Kolling GL, Villanueva SY, Kainuma M, Kobayashi H, Kameyama H, Yoshida S.** 2013. Investigation of encephalopathy caused by Shiga toxin 2c-producing *Escherichia coli* infection in mice. *PLoS One* 8:e58959.

204. **Fujii J, Kinoshita Y, Matsukawa A, Villanueva SY, Yutsudo T, Yoshida S.** 2009. Successful steroid pulse therapy for brain lesion caused by Shiga toxin 2 in rabbits. *Microb Pathog* 46:179–184.

205. **Ramegowda B, Samuel JE, Tesh VL.** 1999. Interaction of Shiga toxins with human brain microvascular endothelial cells: cytokines as sensitizing agents. *J Infect Dis* 180:1205–1213.

206. **Eisenhauer PB, Chaturvedi P, Fine RE, Ritchie AJ, Pober JS, Cleary TG, Newburg DS.** 2001. Tumor necrosis factor alpha increases human cerebral endothelial cell Gb3 and sensitivity to Shiga toxin. *Infect Immun* 69:1889–1894.

207. **Stricklett PK, Hughes AK, Kohan DE.** 2005. Inhibition of p38 mitogen-activated protein kinase ameliorates cytokine up-regulated shigatoxin-1 toxicity in human brain microvascular endothelial cells. *J Infect Dis* 191:461–471.

208. **Landoni VI, de Campos-Nebel M, Schierloh P, Calatayud C, Fernandez GC, Ramos MV, Rearte B, Palermo MS, Isturiz MA.** 2010. Shiga toxin 1-induced inflammatory response in lipopolysaccharide-sensitized astrocytes is mediated by endogenous tumor necrosis factor alpha. *Infect Immun* 78:1193–1201.

209. **Landoni VI, Schierloh P, de Campos Nebel M, Fernandez GC, Calatayud C, Lapponi MJ, Isturiz MA.** 2012. Shiga toxin 1 induces on lipopolysaccharide-treated astrocytes the release of tumor necrosis factor-alpha that alter brain-like endothelium integrity. *PLoS Pathog* 8:e1002632.

210. **Monnens L, Molenaar J, Lambert PH, Proesmans W, van Munster P.** 1980. The complement system in hemolytic-uremic syndrome in childhood. *Clin Nephrol* 13:168–171.

211. **Robson WL, Leung AK, Fick GH, McKenna AI.** 1992. Hypocomplementemia and leukocytosis in diarrhea-associated hemolytic uremic syndrome. *Nephron* 62:296–299.

212. **Thurman JM, Marians R, Emlen W, Wood S, Smith C, Akana H, Holers VM, Lesser M, Kline M, Hoffman C, Christen E, Trachtman H.** 2009. Alternative pathway of complement in children with diarrhea-associated hemolytic uremic syndrome. *Clin J Am Soc Nephrol* 4:1920–1924.

213. **Ståhl AL, Sartz L, Karpman D.** 2011. Complement activation on platelet-leukocyte complexes and microparticles in enterohemorrhagic *Escherichia coli*-induced hemolytic uremic syndrome. *Blood* 117:5503–5513.

214. **Polley MJ, Nachman R.** 1978. The human complement system in thrombin-mediated platelet function. *J Exp Med* 147:1713–1726.

215. **Polley MJ, Nachman RL.** 1983. Human platelet activation by C3a and C3a des-arg. *J Exp Med* 158:603–615.

216. **Orth D, Khan AB, Naim A, Grif K, Brockmeyer J, Karch H, Joannidis M, Clark SJ, Day AJ, Fidanzi S, Stoiber H, Dierich MP, Zimmerhackl LB, Wurzner R.** 2009. Shiga toxin activates complement and binds factor H: evidence for an active role of complement in hemolytic uremic syndrome. *J Immunol* **182:**6394–6400.

217. **Buelli S, Abbate M, Morigi M, Moioli D, Zanchi C, Noris M, Zoja C, Pusey CD, Zipfel PF, Remuzzi G.** 2009. Protein load impairs factor H binding promoting complement-dependent dysfunction of proximal tubular cells. *Kidney Int* **75:**1050–1059.

218. **Proulx F, Wagner E, Toledano B, Decaluwe H, Seidman EG, Rivard GE.** 2003. Mannan-binding lectin in children with *Escherichia coli* O157:H7 haemmorrhagic colitis and haemolytic uraemic syndrome. *Clin Exp Immunol* **133:**360–363.

219. **Karpman D, Tati R.** 2013. Complement activation in thrombotic microangiopathy. *Hamostaseologie* **33:**96–104.

220. **Ueki T, Mizuno M, Uesu T, Kiso T, Nasu J, Inaba T, Kihara Y, Matsuoka Y, Okada H, Fujita T, Tsuji T.** 1996. Distribution of activated complement, C3b, and its degraded fragments, iC3b/C3dg, in the colonic mucosa of ulcerative colitis (UC). *Clin Exp Immunol* **104:**286–292.

221. **Noris M, Mescia F, Remuzzi G.** 2012. STEC-HUS, atypical HUS and TTP are all diseases of complement activation. *Nat Rev Nephrol* **8:**622–633.

The Interplay between the Microbiota and Enterohemorrhagic Escherichia coli

20

REED PIFER[1,2] and VANESSA SPERANDIO[1,2]

INTRODUCTION

The human gastrointestinal (GI) tract harbors trillions of bacterial cells belonging to more than 1,000 species (1), and the number of bacterial cells within the GI tract is 10 times higher than the number of human cells within our bodies (2). The GI microbiota plays essential roles in human nutrition, physiology, development, immunity, and behavior, with disruption of the structure and balance of this community leading to dysbiosis and disease (3–5). This fundamental association between host and bacteria relies on chemical signaling and nutrient availability and exchange. It is also clear that this important balance between host and microbiota can be severely disrupted by environmental stimuli. Two of the most common insults on the microbiota that induce dysbiosis are antibiotic treatment and infectious diseases. Both insults can lead to several disease states ranging from autism, to inflammatory bowel disease, to inflammatory bowel syndrome (IBS). It is also noteworthy that stress exacerbates these syndromes (3).

Broad-spectrum antibiotics alter the microbiome by reducing diversity and shifting community composition (6). Although most of the microbiota return after treatment is ceased, some members of this community are lost indefinitely

[1]Department of Microbiology, The University of Texas Southwestern Medical Center, Dallas, TX 75390; [2]Department of Biochemistry, The University of Texas Southwestern Medical Center, Dallas, TX 75390.

Enterohemorrhagic Escherichia coli *and Other Shiga Toxin-Producing* E. coli
Edited by Vanessa Sperandio and Carolyn J. Hovde
© 2015 American Society for Microbiology, Washington, DC
doi:10.1128/microbiolspec.EHEC-0015-2013

(6). These community shifts cause changes in the metabolic profiles of the intestine, decreasing concentrations of amino acids and short-chain fatty acids (SCFAs) and increasing oligosaccharide levels, suggesting that microbial fermentation of carbohydrates, a fundamental feature of the microbial flora, was disrupted (7–11). Depletion of SCFAs through antibiotic treatment has critical implications in intestinal health and immunity. SCFAs are rapidly absorbed in the colon and provide a preferred energy source to enterocytes, regulating cell proliferation, differentiation, and apoptosis (12, 13), as well as many aspects of intestinal immunity (12, 13). Butyrate is the SCFA with the strongest effect on cell cycle and plays an anti-inflammatory role in the gut. These anti-inflammatory properties are beneficial because they prevent overinflammation and carcinogenesis (14). Butyrate acts by inducing prostaglandin E_2 (15). It is also worth mentioning that in Crohn's disease (an inflammatory bowel disease pattern) there are decreased levels of prostaglandin E_2 and butyrate (16), again suggesting that disruption of the metabolome by dysbiosis has important consequences in GI tract health. Another fundamental impact of dysbyosis in the metabolome is disruption of carbohydrate fermentation. Primary fermenters such as *Bacteroides* spp. are the gateway for the entrance of carbohydrates in the network of syntrophic links in the microbiota, with *Bacteriodes* spp., a prominent glycophagic species, degrading complex carbohydrates into monosaccharides that can be consumed by other members of the gut microbiota (17).

The second important insult that causes dysbiosis is infection by an invading pathogen. It is known that invading enteric pathogens such as *Salmonella* spp. and *Citrobacter rodentium* cause inflammation within the gut that in turn diminishes the overall numbers of bacteria in the microbiota, acting as a competition advantage to the pathogen (18, 19). Additionally, infection with *C. rodentium* also causes significant changes in the structure of the microbial community, decreasing the number of anaerobes and increasing the numbers of G*ammaproteobacteria* (18). *C. rodentium* is a murine pathogen that models the enteric infection of the human pathogen enterohemorrhagic *Escherichia coli* (EHEC). The most compelling evidence for the involvement of the microbiota in disease states comes from IBS, with the chief support coming from data on postinfection IBS, whereby IBS ensues following an episode of bacteriological gastroenteritis (3). The highest supported incidence of postinfection IBS is related to the Walkerton outbreak, when postinfection IBS developed in the majority of individuals after an EHEC infection (20).

MICROBIOTA/EHEC QUORUM-SENSING SIGNALING INTERACTIONS WITHIN THE INTESTINE

EHEC colonizes the human colon, where it has to interact and successfully compete with the microbiota to find a colonization niche. The first appreciation of EHEC-microbiota interactions on EHEC pathogenesis came from the observation that EHEC employs quorum-sensing (QS) signaling to regulate expression of its virulence genes (21). QS is a bacterial cell-to-cell signaling mechanism through which bacterial cells assess the density of their population. These bacteria secrete hormone-like compounds, usually referred to as autoinducers. When these autoinducers reach a threshold concentration, they interact with transcriptional regulators to drive bacterial gene expression (22). Because EHEC infection requires a low infectious dose, estimated to be between 50 and 100 CFU, it was deemed unlikely that EHEC was responding to self-produced signals upon infection of the host, and it was proposed that EHEC was sensing autoinducers produced by the GI microbiota (21). The identity of this QS signal was initially a matter of debate. QS was first described in the regulation of bioluminescence in *Vibrio fischeri* and *Vibrio harveyi* (23, 24). The luciferase operon in *V. fischeri* is regulated by two proteins, LuxI, which is responsible for the production of the acyl-homoserine-lactone (AHL) autoinducer, and LuxR, which is activated by this autoinducer to increase transcription of the luciferase operon (25, 26). Since this initial description, homologs of LuxR-LuxI have been identified in other gram-negative bacteria, and in all of these LuxR-LuxI systems, the bacteria produce an AHL, which binds to the LuxR protein and regulates the transcription of several genes involved in a variety of phenotypes (27–29). *E. coli* has a LuxR homolog, SdiA (30), but does not have a *luxI* gene and does not produce AHLs (31, 32). Because of these features it was considered unlikely for many years that EHEC had any functional QS systems. However, the discovery of the autoinducer-2

(AI-2) system, present in both Gram-positive and Gram-negative bacteria (including EHEC), suggested that other QS systems, evolved to promote interspecies communication within bacterial populations, may be employed by EHEC to regulate its virulence repertoire in the GI tract (21, 33).

AI-2 was originally identified in *V. harveyi* as an inducer of cell density-dependent bioluminescence. AI-2 is synthesized from *S*-adenosylmethionine in three enzymatic steps, the last of which is catalyzed by LuxS producing 4,5-dihydroxy-2,3-pentanedione that spontaneously cyclizes to form (2*S*,4*S*)-2-methyl-2,3,3,4-tetrahydroxytetrahydrofuran (AI-2) (Fig. 1) (34). LuxS is highly conserved, with homologs found in a wide variety of pathogenic and commensal bacteria within

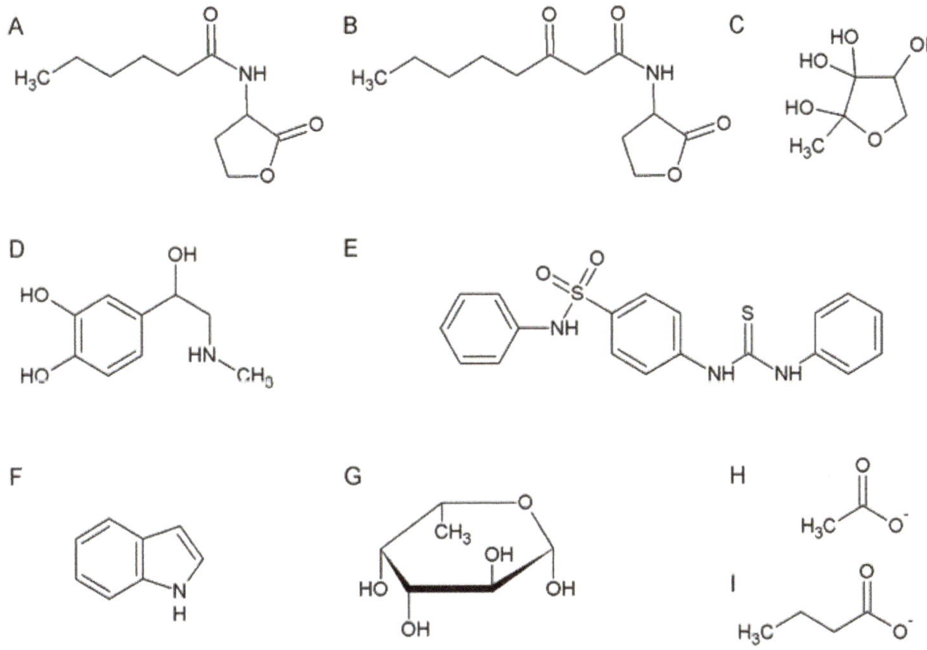

FIGURE 1 Structure of known quorum-sensing ligands. *N*-hexanoyl-l-homoserine lactone (C6-HSL) (A) and *N*-(3-Oxo-octanoyl)-l-homoserine lactone (3-oxo-C8-HSL) (B) stabilize SdiA, which can suppress T3S. (2*S*,4*S*)-2-methyl-2,3,3,4-tetrahydroxytetrahydrofuran (AI-2) (C) appears to have a surprisingly modest effect on virulence. QseC responds to host-derived epinephrine (D) and is antagonized by synthetic LED209 (E), perhaps yielding clues to the identity of AI-3. Indole (F) is a tryptophan-derived metabolite that influences motility and type III secretion. Mucin degradation releases fucose (G), which activates the FusKR two-component system to downregulate T3S. SCFAs, including acetate (H) and butyrate (I), induce motility of EHEC, while only butyrate induces T3S via Lrp activity. doi:10.1128/microbiolspec.EHEC-0015-2013.f1

Bacteroidetes, Firmicutes, and *Proteobacteria.* Human stool samples and *in vitro* propagated commensal bacteria from healthy individuals are capable of inducing luminescence from *V. harveyi,* suggesting that AI-2 is produced within the GI tract (35). EHEC encodes a LuxS homolog, and preconditioned media from this strain are capable of inducing luminescence from *V. harveyi* in a *luxS*-dependent manner (36), suggesting that EHEC is capable of producing AI-2. Supernatants of EHEC grown to late exponential phase contain a signal capable of inducing transcription of many EHEC virulence genes, including the locus of enterocyte effacement (LEE) that encodes a type III secretion system (T3SS) essential for EHEC to attach to and efface enterocytes and promote disease (37, 38), the flagellum regulon, and expression of Shiga toxin (35, 39). Purified and synthetic AI-2 was unable to govern LEE and flagella expression (40), suggesting the existence of an additional class of autoinducer (AI-3). The identity of AI-3 is not yet known, but it is methanol soluble (35), likely tyrosine derived (36), and shares signaling mechanisms with host-derived epinephrine and norepinephrine (Fig. 1). AI-3 signal is produced by a variety of enterobacterial species, including pathogenic strains and normal flora such as *Enterobacter cloacae* (36), and is also present in the stool of humans (35). The reliance of a functional *luxS* gene in the presence of AI-3 in EHEC was not due to LuxS being involved in the synthesis of AI-3, but to a metabolic shift that occurs in a *luxS* mutant (LuxS is involved in the central methyl cycle in bacterial cells) that leads to decreased AI-3 production (36).

Interkingdom Chemical Signaling in EHEC-Host Interactions

The AI-3 QS signaling system intersects with the host adrenergic signaling system (epinephrine/norepinephrine [NE] hormones) (40). In fact, this intersection occurs at a biochemical level, with the same bacterial receptor, the membrane-bound histidine sensor

kinase (HK) QseC, sensing the bacterial signal AI-3 and the host signals epinephrine and/or NE (41). Both epinephrine and NE are present in the GI tract. NE is synthesized within the adrenergic neurons within the enteric nervous system (42). Epinephrine is synthesized in the central nervous system and in the adrenal medulla; it acts in a systemic manner after being released into the bloodstream, thereby reaching the intestine (43). Both hormones modulate intestinal smooth muscle contraction, submucosal blood flow, and chloride and potassium secretion in the intestine (44). Epinephrine and NE are recognized by adrenergic receptors in mammalian cells; Freddolino et al. (45) reported that the ligand-binding sites for epinephrine and NE in a human adrenergic receptor are similar. Extensive evidence indicates that both epinephrine and NE are recognized by the same receptors and play important biological roles in the human GI tract.

The AI-3/epinephrine/NE interkingdom signaling cascade activates expression of virulence genes in EHEC (39, 40, 46, 47). The host hormones epinephrine/NE are specifically sensed by two HKs: QseC and QseE, which are the first bacterial adrenergic receptors identified (46, 48). QseE is downstream of QseC in this signaling cascade, given that transcription of *qseE* is activated through QseC (49). In addition to sensing these host hormones, QseC also senses the bacterial signal AI-3 (46). The QseC sensor is the first example of a receptor for both a bacterial and a host signal, integrating bacterial-host signaling at the biochemical level. QseE, however, does not sense AI-3, thereby discriminating between host- and bacteria-derived signals (49). Upon sensing their respective signals, QseC and QseE activate virulence gene expression and pathogenesis *in vitro* and *in vivo* in EHEC (46, 48, 50). QseC and QseB constitute a cognate two-component system, and the *qseBC* genes are cotranscribed in an operon that is also autoregulated (51). QseC transfers its phosphate to three response regulators (RRs): its cognate RR QseB and the noncognate QseF and KdpE RRs (52). QseE only transfers its phosphate to its cognate RR QseF (53). The concerted action of these RRs activates the EHEC virulence repertoire (Fig. 2).

QseC facilitates both phosphorylation and dephosphorylation of QseB. Unphosphorylated QseB binds upstream of the *flhDC* promoter (FlhDC operons are the master regulators of the flagella regulon) and inhibits transcription of these master activators of motility, whereas phosphorylated QseB induces expression (52, 54). QseF indirectly stimulates the expression of *espFU/tccP* (49), a critical non-LEE-encoded effector required for actin recruitment at the site of pedestal formation. Activation of QseC induces Shiga toxin expression (50) and this is

dependent on QseF (52). KdpE functions as an activator of *LEE1* (*LEE1* encodes the Ler activator of all LEE genes [55]) expression (52) by direct binding upstream of the distal *LEE1* promoter (56) (Fig. 2). An important note is that the QseC signaling cascade can be inhibited by the antivirulence drug LED209, preventing activation of the EHEC virulence repertoire (50).

Nutrient Signaling in EHEC-Microbiota Interactions and Virulence Regulation

The mammalian GI tract harbors trillions of indigenous bacteria whose coexistence relies on the ability of each member to use one or a

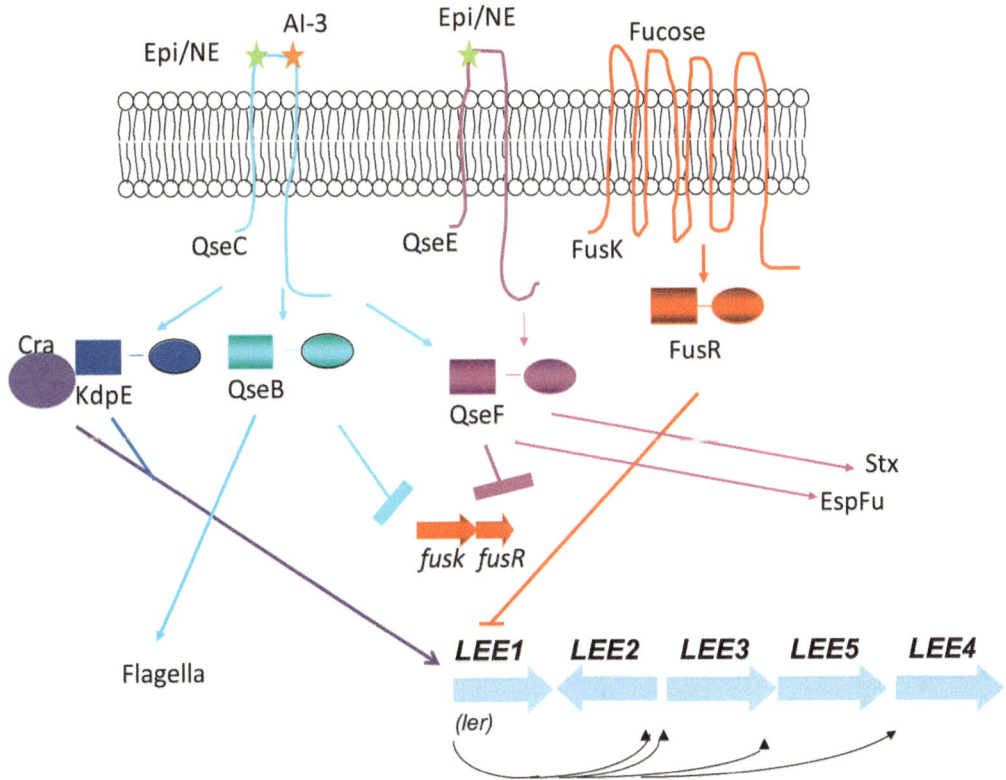

FIGURE 2 The QseC signaling cascade. QseC senses AI-3 epinephrine and norepinephrine and phosphorylates QseB, QseF, and KdpE. QseE senses epinephrine and phosphorylates QseF. QseB activates flagella expression. QseF indirectly promotes Shiga toxin and EspFu expression. KdpE together with Cra activate LEE gene expression. Both QseBC and QseEF repress *fusKR* expression. FusK senses fucose and phosphorylates FusR that represses the LEE. doi:10.1128/microbiolspec.EHEC-0015-2013.f2

few limiting resources (1). Invading pathogens have to compete with the microbiota for these resources to establish colonization. These pathogens tend to be aggressive and greedy in search of a colonization niche, and achieve this purpose by precisely coordinating expression of an arsenal of virulence genes. Linking carbon and nitrogen metabolism to the precise coordination of virulence expression is a key step in the adaptation of pathogens toward recognizing suitable sites for colonization, and a means to tip the scale in the tug-of-war between pathogens and the microbiota.

EHEC is no stranger to this "struggle" for nutrients. In addition to purpose-built signaling molecules like the autoinducers, the interplay between the nutrient requirements of normal flora and EHEC is important in determining virulence. Glucose and other monomeric dietary sugars are ideal carbon sources for EHEC, but these molecules are scarce in the lower GI tract. The proximal small intestine is the site of host absorption of simple sugars, whereas the distal small intestine houses a vast population of commensal bacteria scavenging for free sugars. As a result, the colonic home of EHEC is a gluconeogenic environment. Njoroge and colleagues uncovered the importance of glucose availability in regulating T3SS by EHEC (56). Stemming from the observation that high-glucose growth media suppressed type III secretion (T3S) while low-glucose conditions induced LEE expression, the authors uncovered a role for the catabolite repressor/activator protein (Cra) in *ler* regulation. Indeed, a *cra*-deficient mutant of EHEC exhibited diminished attaching and effacing (A/E) lesion formation, *LEE1* promoter activity, transcript levels, and EspA secretion under low-glucose conditions. However, under glucose-rich conditions no effect was seen. A Cra-binding site was identified upstream of the distal *LEE1* promoter, and binding to this site in an electrophoretic mobility shift assay could be prevented by the inclusion of the glycolytic intermediates fructose-1-phosphate or fructose-1,6-bisphosphate.

Cra was found to directly interact with the RR KdpE, previously found to positively regulate *LEE1* (52). Interestingly, *KdpE*-dependent LEE regulation was also found to be in effect only in low-glucose conditions, and KdpE binding *in vitro* to the *LEE1* promoter was diminished by phosphorylation (56). Presumably, under high-glucose conditions, KdpE is phosphorylated by its cognate sensor kinase, KdpD, which is activated by the glucose-sensitive IIA^{Ntr} phosphotransfer system (57). These data suggest that Cra and KdpE act in concert to induce T3S by inducing *ler* expression and that this control is only active under glucose-limiting conditions, such as those found within the colon.

While a gradient of diet-derived glucose may provide an indicator of progress through the length of the GI tract, spatial regulation of virulence factor expression likely requires EHEC to distinguish luminal from host membrane proximal environments (Fig. 3). A thick layer of goblet cell-derived mucus partitions the bacteria-laden GI lumen from the host enterocytes. The mucous layer is composed of a dense matrix of cross-linked mucin proteins heavily decorated with O-linked glycans. Within the colon, Muc2 represents the dominant species of mucin linked to glycans composed of GalNAc, NANA, GlcNAc, galactose, fucose (Fig. 1), and mannose monosaccharides, listed in terms of decreasing relative abundance (58). Members of the normal flora of the colon express mucolytic enzymes to harvest carbon from this barrier and from dietary polysaccharides. *Bacteroides thetaiotaomicron*, a well-studied commensal, is capable of metabolizing pectins, starches, fructan, alpha-glucans, and a number of glycans originating from host tissues (59). In contrast, *E. coli* is an organism adapted for exploitation of monosaccharides and tricarboxylic acid cycle intermediates liberated from complex carbohydrates by these commensals (61, 62). It is not surprising that mucin-derived carbon-source sensing has been adapted by EHEC to regulate virulence mechanisms.

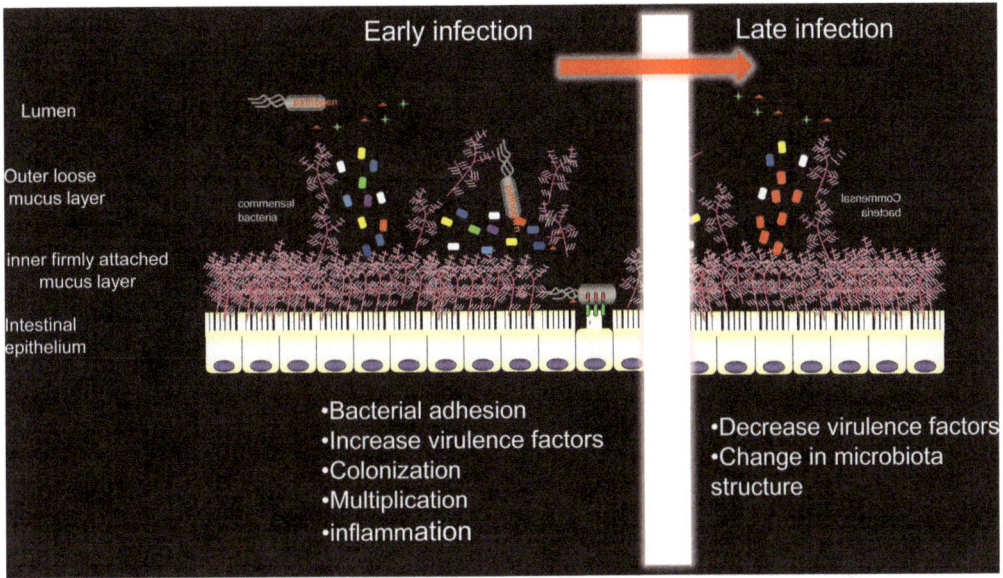

FIGURE 3 EHEC gastrointestinal colonization. doi:10.1128/microbiolspec.EHEC-0015-2013.f3

Pacheco et al. describe a novel two-component system that enables regulation of virulence gene expression and carbon-source choice by EHEC upon sensing fucose (62). The *fusKR* operon is within the genomic O-island 20, found only in O55:H7 descendant strains and within *C. rodentium*. FusK is a transmembrane HK that autophosphorylates at a histidine residue and transfers the phospho-group to an aspartate of FusR to regulate DNA binding. The authors observe that the FusKR two-component system functions as a repressor of T3S. Knockout of either *fusK* or *fusR* results in increased A/E lesion formation in a cell culture infection model, increased LEE transcript levels, and higher levels of EspB in culture supernatants compared to wild type. Given that FusK shares sequence homology with UhpB, a glucose-6-phosphate sensor kinase, the authors explored whether FusK was responsive to sugar monomers and found that fucose, but not other sugars, was sufficient to induce autophosphorylation of the kinase. Expression of *LEE1* is diminished in wild-type EHEC grown with fucose as the sole carbon source when compared with glucose; however, no difference is seen in *fusK*-deficient EHEC. Similarly, mucin was found to diminish *LEE1* expression when EHEC was grown in the presence of *B. thetaiotaomicron*, which expresses fucosidases capable of liberating fucose from mucin. The effect was not seen with cocultures grown in the presence of fucose rather than mucin. Altogether, these data suggest that fucose serves as a signal to downregulate T3S via FusKR when EHEC is in the lumen.

As the mucous barrier limits the approach of mucolytic commensals toward the epithelial surface, fucose liberation occurs in the colonic lumen, but not in close contact with the intestinal epithelium. LEE expression within the lumen is not appropriate for EHEC; therefore, a luminal signal to decrease T3S would serve to ensure efficient resource expenditure. While in the lumen, fucose-FusK signaling satisfies this need. However, upon reaching the epithelium, FusKR would be detrimental to EHEC as it would inhibit establishment of pedestals. EHEC has used the interkingdom signaling systems of QseBC and

QseEF to alleviate this issue while at the epithelial surface. QseB directly binds to and inhibits transcription of the *fusKR* operon, while QseF likely induces expression of a *fusKR* repressor (62). Additionally, EHEC also uses ethanolamine, a vast source of nitrogen within the intestine that is present within membranes, to regulate LEE and Shiga toxin expression (63).

Intriguingly, FusKR was also found to downregulate a putative fucose transporter protein encoded near *fusKR*. Deletion of this major facilitator superfamily protein locus, *z0461*, dramatically diminishes the expression of genes involved in fucose utilization and delays growth in media containing fucose as the sole carbon source. On the surface, this observation represents a paradox in that EHEC senses fucose, a prime carbon source, and yet diminishes its own capacity to use fucose. This can perhaps be explained by observations concerning carbon-source preference in physiological settings of bovine small intestine contents (64) and a murine infection model (65). Though fucose utilization contributes to EHEC growth within bovine small intestinal contents, deficiencies within this pathway are not as dramatic as those seen in galactose or mannose utilization pathways (64). Analysis of mouse colonization of mutants of EDL933 or MG1655 demonstrates that fucose is used by both virulent and avirulent strains of *E. coli*, whereas galactose and mannose are primarily the forte of EHEC and dispensable for MG1655 (65). Therefore, downregulation of fucose utilization systems may be a mechanism of EHEC to avoid carbon-source overlap with commensal *E. coli* and thus avoid direct competition. This possibility is further corroborated by the work of Kamada et al. (66); these authors observed that expression of virulence genes by the EHEC surrogate murine model *C. rodentium* is activated at the onset of infection, and at later time points, *C. rodentium* infection backfires, triggering a bloom of Gammaproteobacteria that effectively compete with *C. rodentium* for carbon sources. This bloom of *Gammaproteobacteria* upon *C. rodentium* infection has also been previously reported by Lupp et al. (18), further suggesting that competition for carbon sources with the microbiota plays an important role in EHEC clearance from the GI tract.

Fermentation of starches within the colon is primarily mediated by *Firmicutes*, such as *Faecalibacterium prausnitzii* and *Eubacterium rectale*, and also *Bacteroides* spp. to produce SCFAs. The importance of the microbiota to SCFA production is clear from the very low level of these metabolites in germfree animals (67). Acetate, propionate, and butyrate (Fig. 1) dominate the SCFA population and contribute broadly to host physiology (68). SCFAs are present in the colon in millimolar concentrations and exist in much lower concentrations in the upper GI tract. Thus, they represent an excellent indicator of arrival within the large intestine in much the same way that abundant glucose is indicative of localization within the small intestine. As such, EHEC uses these molecules as a cue to govern expression of virulence genes.

Butyrate, but not acetate or propionate, is capable of inducing T3S from the Sakai strain of EHEC (69); 20 mM butyrate increases adhesion to Caco-2 cells 10-fold over control and facilitates the formation of microcolonies (70). Leucine is able to induce *LEE4* and *LEE5* protein expression similarly to butyrate (69). Both leucine- and butyrate-driven expression of the LEE is dependent on expression of the leucine-responsive protein (Lrp) transcription factor. Given the similarities in the structure of leucine and butyrate and the observation that the M124R leucine-insensitive mutation to Lrp eliminates butyrate-induced LEE expression, it is likely that butyrate is directly sensed by Lrp. In addition to requiring Lrp, butyrate-induced T3S requires Ler and the *ler* activator PchA. In fact, the promoter of *pchA* cannot be substituted without abolishing the butyrate effect, implying a cascade of events: butyrate binds Lrp, which then activates transcription of *pchA*, which in turn activates transcription of the *LEE1*-encoded *ler*.

Acetate, propionate, and butyrate increase motility of the Sakai strain (70). Synthesis of the flagellin subunit, FliC, is increased, and the frequency of flagellated bacteria increases upon exposure to these SCFAs. Two mechanisms of induction are in play: activation of the class 1 *flhDC* promoter by butyrate and induction of class 2 *flgN* by acetate, propionate, and butyrate. Butyrate-induced *flhDC* expression requires *lrp*, whereas *lrp* is dispensable for FlgN accumulation.

SCFAs heavily influence the colonic epithelium by serving as an energy source for host use, promoting nutrient absorption, and regulating cell differentiation. Zumbrun and colleagues recently demonstrated that butyrate increases globotriaosylceramide expression on human colonic epithelial cells (71). Mice fed a high-fiber diet, which increases the concentration of butyrate within the intestinal lumen, express higher globotriaosylceramide levels within the intestinal epithelium and in kidneys. When challenged with EHEC, mice fed a high-fiber diet experience a higher pathogenic burden, significant weight loss, and decreased survival compared to animals fed a low-fiber diet. Interestingly, a high-fiber diet diminishes the frequency of *Escherichia* spp. within the gut while increasing total flora levels. These results suggest that a high-fiber diet, via increased butyrate levels, may increase susceptibility of animals to Shiga toxin released from EHEC by promoting the expression of the receptor necessary for toxicity. Additionally, a high-fiber diet may reduce competition faced by EHEC during infection by reducing levels of normal *E. coli* flora while simultaneously increasing the prevalence of other species that may benefit EHEC.

Another important aspect of the role of SCFAs and the outcome of EHEC infections stems from the observation that probiotic strains of *Bifidobacterium longum* that produce high levels of acetate effectively prevent Shiga toxin translocation through the intestinal epithelium. The authors propose that acetate produced by these probiotic bacteria improves intestinal defense and protects the host against lethal infection (72).

The interplay of microbiota and host diet likely creates a multidimensional gradient of metabolites within the GI tract of mammals. EHEC adapted to colonization of the lower intestinal tract uses these gradients to regulate expression of complex, metabolically expensive virulence systems such as the LEE-encoded T3SS to maximize fitness within an animal host. Under such a model, localization within the colon is indicated by increased concentrations of fermentation products, such as butyrate, and increasingly gluconeogenic conditions, which promote expression of the LEE and, thus, attachment as is appropriate. A luminal-epithelial axis is created within the colon by the presence of luminal sugars liberated from carbohydrates, such as fucose derived from mucin, that suppress LEE expression until the epithelium is within reach.

MICROBIOTA COMPOSITION AND SUSCEPTIBILITY TO EHEC INFECTIONS

Within the scope of the complex associations between EHEC and different members of the GI microbiota, a burning question is to how much microbiota differential composition contributes to host resistance to EHEC infections? EHEC infections can vary in their degree of severity, ranging from watery diarrhea, to severe bloody diarrhea, to hemolytic-uremic syndrome. It has been reported that antibiotic treatment that alters the microbial composition of the GI microbiota may increase susceptibility to GI infection by *C. rodentium* (73). It has also been reported that microbiota transplantation from susceptible mice to mice resistant to *C. rodentium* infection also increased susceptibility of resistant mice to this pathogen. In the opposite experiment, transplantation of microbiota from resistant mice to susceptible mice was protective (74). Additionally, the combination of certain microbiota members with different metabolites can result in different outcomes on EHEC virulence gene expression. Fucose released from the mucus by

B. thetaiotaomicron decreases LEE gene expression (62), while under gluconeogenic conditions in the absence of fucose, *B. thetaiotaomicron* promotes LEE gene expression (56). It has also been reported that expression of Shiga toxin is decreased in the presence of *B. thetaiotaomicron* in Leedle and Hesplee medium (75).

DIFFERENTIAL SIGNALING SYSTEMS GOVERNING HOST ASSOCIATIONS

The main environmental reservoir of EHEC is ruminants, and it is estimated that 70 to 80% of cattle herds in the United States are colonized with EHEC (76–81). EHEC is an example of a bacterium that behaves as a commensal or a pathogen, depending on its host. EHEC is a commensal in the GI tract of adult cattle but is a human pathogen (82). EHEC colonizes the large intestine of humans, forming A/E lesion, thought to be largely responsible for promoting disease (82). The genes for A/E lesion formation are encoded within the LEE (82). The LEE and A/E lesion

formation are also necessary for EHEC colonization of the recto-anal junction (RAJ) of cattle, facilitating shedding of this pathogen in the environment (83). In addition to the LEE, EHEC uses the glutamate decarboxylase (*gad*) acid resistance system to survive passage through the acidic stomachs of these animals to reach its site of colonization, the RAJ (84). SdiA is a regulator that senses AHL QS signaling molecules and aids EHEC survival and colonization of the bovine GI tract. AHLs are prominent within cattle rumen but absent in the other sections of the GI tract. Through SdiA, transcription of the LEE genes is decreased by rumen AHLs, while transcription of the *gad* acid-resistant system is increased. Expression of the LEE in the rumen would be an unnecessary energy burden for EHEC in this GI compartment. However, in preparation for the acidic distal stomachs, the EHEC *gad* is activated in the rumen. SdiA-AHL signaling aids EHEC in gauging these environments and modulates gene expression toward adaptation to a commensal lifestyle in cattle. Consequently, an *sdiA* mutant is deficient for cattle colonization (85, 86) (Fig. 4).

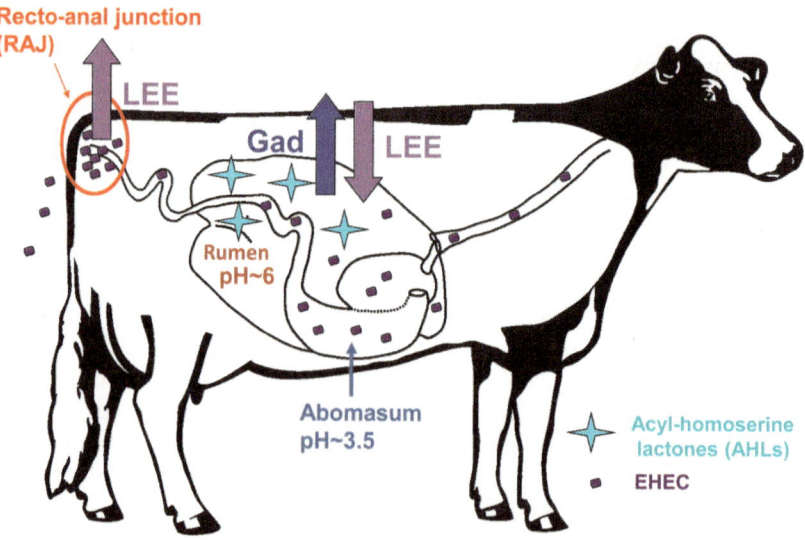

FIGURE 4 EHEC cattle colonization. Within the rumen EHEC senses AHLs through SdiA to decrease LEE expression and increase *gad* expression. Within the RAJ, in the absence of AHLs, LEE expression is promoted. doi:10.1128/microbiolspec.EHEC-0015-2013.f4

AHLs are synthesized from *S*-adenosyl-methionine and an acylated acyl carrier protein by AHL synthases, such as LuxI of *V. fischeri* or RhlI of *Pseudomonas aeruginosa* (87). AHLs are sensed by a class of unstable transcription factors exemplified by LuxR of *V. fischeri*. LuxR homologs consist of an N-terminal autoinducer binding domain and a C-terminal helix-turn-helix motif. AHL binding promotes LuxR folding and dimerization to allow for binding to a target sequence within a promoter to regulate gene expression. Acyl chain length and structure provide specificity for AHL sensing by different species. *E. coli* does not express any known AHL synthase but does encode for a *luxR* homolog, *sdiA*.

Initial studies aimed at understanding the role of *sdiA* in virulence of EHEC observed that overexpression of SdiA diminished production of *LEE4* and *LEE5* encoded proteins, as well as decreasing flagellin expression and soft agar motility (88). Like other LuxR homologs, SdiA tends to fold poorly in the absence of exogenous AHLs and accumulates in inclusion bodies. Overexpression of SdiA likely results in a small, transient population of soluble molecules that is capable of affecting transcription. Indeed, transcriptional analysis of *sdiA*-deficient EHEC reveals that the glutamate-dependent acid resistance (*gad*) genes are activated by SdiA independently of exogenous AHL activity, although AHLs enhance this activation (85). In contrast, SdiA regulation of the LEE T3SS is AHL dependent. SdiA-deficient EHEC has no change in LEE expression relative to wild type in the absence of AHL. Exogenous oxo-C6-homoserine lactone (Fig. 1) diminishes *LEE1* transcription and EspA secretion in an SdiA-dependent manner. SdiA stabilized by exogenous AHLs directly binds to the *LEE1* promoter to repress *ler* transcription and diminishes T3S through this action (85).

Nuclear magnetic resonance structural analysis of SdiA of *E. coli* has demonstrated that the protein is capable of fruitfully interacting with at least three additional AHLs: C8-HSL, oxo-C8-HSL (Fig. 1), and C6-HSL

(89). This, combined with the lack of endogenous AHLs from *E. coli*, implies that SdiA likely functions as a sensor for an array of foreign AHLs found within the surroundings of *E. coli*. In support of this notion, AHLs have been found within the rumen contents of cattle fed both grain and forage diets (86, 90, 91). However, chemical analysis of lower GI tract contents from cattle and nonruminant animals has not revealed the presence of AHLs, which may be due to inadequate sensitivity of the methods used, alkaline instability of AHLs, or potentially the existence of exotic homoserine lactones or non-AHL small molecules that may influence SdiA function (40).

These reports highlight how different chemical signaling systems can be employed by bacteria to adapt to either pathogenic or commensal lifestyles in different hosts and that this signaling system aids this human pathogen to adapt to a commensal lifestyle in cattle, its main reservoir.

CONCLUDING REMARKS

The increasing knowledge of the essential roles of the microbiota in human health is opening many avenues of research to understand differential host susceptibility to infectious diseases. These studies are fundamental for enteric pathogens, which inhabit a complex and dynamic environment. It is fascinating to envision how such few EHEC organisms (circa 50 CFU) efficiently manage to establish themselves in the host and cause disease. It is becoming clear that EHEC is very crafty at reading many cues provided by both the host and the microbiota and rapidly adapting its virulence program toward successful host infection. We are also at the tip of an iceberg in determining how different microbiota enterotypes may determine the severity of EHEC disease. One should also take into consideration that differences in diets and antibiotic regimens, which cause shifts in the composition of the GI microbial flora, may also influence the outcome of EHEC disease.

ACKNOWLEDGMENT

We declare no conflicts of interest with regard to the manuscript.

CITATION

Pifer R, Sperandio V. 2014. The interplay between the microbiota and enterohemorrhagic *Escherichia coli*. Microbiol Spectrum 2(5): EHEC-0015-2013.

REFERENCES

1. **Gill SR, Pop M, Deboy RT, Eckburg PB, Turnbaugh PJ, Samuel BS, Gordon JI, Relman DA, Fraser-Liggett CM, Nelson KE.** 2006. Metagenomic analysis of the human distal gut microbiome. *Science* **312:**1355–1359.

2. **Hooper LV, Gordon JI.** 2001. Commensal host-bacterial relationships in the gut. *Science* **292:**1115–1118.

3. **Grenham S, Clarke G, Cryan JF, Dinan TG.** 2011. Brain-gut-microbe communication in health and disease. *Front Physiol* **2:**94.

4. **Gordon JI, Klaenhammer TR.** 2011. A rendezvous with our microbes. *Proc Natl Acad Sci USA* **108 Suppl 1:**4513–4515.

5. **Gonzalez A, Stombaugh J, Lozupone C, Turnbaugh PJ, Gordon JI, Knight R.** 2011. The mind-body-microbial continuum. *Dialogues Clin Neurosci* **13:**55–62.

6. **Dethlefsen L, Relman DA.** 2011. Incomplete recovery and individualized responses of the human distal gut microbiota to repeated antibiotic perturbation. *Proc Natl Acad Sci USA* **108 Suppl 1:**4554–4561.

7. **Romick-Rosendale LE, Goodpaster AM, Hanwright PJ, Patel NB, Wheeler ET, Chona DL, Kennedy MA.** 2009. NMR-based metabonomics analysis of mouse urine and fecal extracts following oral treatment with the broad-spectrum antibiotic enrofloxacin (Baytril). *Magn Reson Chem* **47 Suppl 1:**S36–S46.

8. **Yap IK, Li JV, Saric J, Martin FP, Davies H, Wang Y, Wilson ID, Nicholson JK, Utzinger J, Marchesi JR, Holmes E.** 2008. Metabonomic and microbiological analysis of the dynamic effect of vancomycin-induced gut microbiota modification in the mouse. *J Proteome Res* **7:**3718–3728.

9. **Martin FP, Wang Y, Sprenger N, Yap IK, Lundstedt T, Lek P, Rezzi S, Ramadan Z, van Bladeren P, Fay LB, Kochhar S, Lindon JC, Holmes E, Nicholson JK.** 2008. Probiotic modulation of symbiotic gut microbial-host metabolic interactions in a humanized microbiome mouse model. *Mol Syst Biol* **4:**157.

10. **Woodmansey EJ, McMurdo ME, Macfarlane GT, Macfarlane S.** 2004. Comparison of compositions and metabolic activities of fecal microbiotas in young adults and in antibiotic-treated and non-antibiotic-treated elderly subjects. *Appl Environ Microbiol* **70:**6113–6122.

11. **Hoverstad T, Carlstedt-Duke B, Lingaas E, Midtvedt T, Norin KE, Saxerholt H, Steinbakk M.** 1986. Influence of ampicillin, clindamycin, and metronidazole on faecal excretion of short-chain fatty acids in healthy subjects. *Scand J Gastroenterol* **21:**621–626.

12. **Millard AL, Mertes PM, Ittelet D, Villard F, Jeannesson P, Bernard J.** 2002. Butyrate affects differentiation, maturation and function of human monocyte-derived dendritic cells and macrophages. *Clin Exp Immunol* **130:**245–255.

13. **Hossain Z, Konishi M, Hosokawa M, Takahashi K.** 2006. Effect of polyunsaturated fatty acid-enriched phosphatidylcholine and phosphatidylserine on butyrate-induced growth inhibition, differentiation and apoptosis in Caco-2 cells. *Cell Biochem Funct* **24:**159–165.

14. **Hamer HM, Jonkers D, Venema K, Vanhoutvin S, Troost FJ, Brummer RJ.** 2008. Review article: the role of butyrate on colonic function. *Aliment Pharmacol Ther* **27:**104–119.

15. **Usami M, Kishimoto K, Ohata A, Miyoshi M, Aoyama M, Fueda Y, Kotani J.** 2008. Butyrate and trichostatin A attenuate nuclear factor kappaB activation and tumor necrosis factor alpha secretion and increase prostaglandin E2 secretion in human peripheral blood mononuclear cells. *Nutr Res* **28:**321–328.

16. **Jansson J, Willing B, Lucio M, Fekete A, Dicksved J, Halfvarson J, Tysk C, Schmitt-Kopplin P.** 2009. Metabolomics reveals metabolic biomarkers of Crohn's disease. *PLoS One* **4:**e6386.

17. **Fischbach MA, Sonnenburg JL.** 2011. Eating for two: how metabolism establishes interspecies interactions in the gut. *Cell Host Microbe* **10:**336–347.

18. **Lupp C, Robertson ML, Wickham ME, Sekirov I, Champion OL, Gaynor EC, Finlay BB.** 2007. Host-mediated inflammation disrupts the intestinal microbiota and promotes the overgrowth of *Enterobacteriaceae*. *Cell Host Microbe* **2:**204.

19. **Stecher B, Robbiani R, Walker AW, Westendorf AM, Barthel M, Kremer M, Chaffron S, Macpherson AJ, Buer J, Parkhill J, Dougan G, von Mering C, Hardt WD.** 2007. *Salmonella enterica* serovar Typhimurium exploits inflammation to compete with the intestinal microbiota. *PLoS Biol* **5:**2177–2189.

20. **Dunlop SP, Jenkins D, Neal KR, Spiller RC.** 2003. Relative importance of enterochromaffin cell hyperplasia, anxiety, and depression in postinfectious IBS. *Gastroenterology* **125:**1651–1659.

21. **Sperandio V, Mellies JL, Nguyen W, Shin S, Kaper JB.** 1999. Quorum sensing controls expression of the type III secretion gene transcription and protein secretion in enterohemorrhagic and enteropathogenic *Escherichia coli*. *Proc Natl Acad Sci USA* **96:**15196–15201.

22. **Fuqua WC, Winans SC, Greenberg EP.** 1994. Quorum sensing in bacteria: the LuxR-LuxI family of cell density-responsive transcriptional regulators. *J Bacteriol* **176:**269–275.

23. **Nealson KH, Platt T, Hastings JW.** 1970. Cellular control of the synthesis and activity of the bacterial luminescent system. *J Bacteriol* **104:**313–322.

24. **Nealson KH, Hastings JW.** 1979. Bacterial bioluminescence: its control and ecological significance. *Microbiol Rev* **43:**496–518.

25. **Engebrecht J, Nealson K, Silverman M.** 1983. Bacterial bioluminescence: isolation and genetic analysis of functions from *Vibrio fischeri*. *Cell* **32:**773–781.

26. **Engebrecht J, Silverman M.** 1984. Identification of genes and gene products necessary for bacterial bioluminescence. *Proc Natl Acad Sci USA* **81:**4154–4158.

27. **Parsek MR, Greenberg EP.** 2000. Acyl-homoserine lactone quorum sensing in gram-negative bacteria: a signaling mechanism involved in associations with higher organisms. *Proc Natl Acad Sci USA* **97:**8789–8793.

28. **Davies DG, Parsek MR, Pearson JP, Iglewski BH, Costerton JW, Greenberg EP.** 1998. The involvement of cell-to-cell signals in the development of a bacterial biofilm. *Science* **280:**295–298.

29. **de Kievit TR, Iglewski BH.** 2000. Bacterial quorum sensing in pathogenic relationships. *Infect Immun* **68:**4839–4849.

30. **Wang XD, de Boer PA, Rothfield LI.** 1991. A factor that positively regulates cell division by activating transcription of the major cluster of essential cell division genes of *Escherichia coli*. *Embo J* **10:**3363–3372.

31. **Swift S, Lynch MJ, Fish L, Kirke DF, Tomas JM, Stewart GS, Williams P.** 1999. Quorum sensing-dependent regulation and blockade of exoprotease production in *Aeromonas hydrophila*. *Infect Immun* **67:**5192–5199.

32. **Michael B, Smith JN, Swift S, Heffron F, Ahmer BM.** 2001. SdiA of *Salmonella enterica* is a LuxR homolog that detects mixed microbial communities. *J Bacteriol* **183:**5733–5742.

33. **Surett MG, Bassler BL.** 1998. Quorum sensing in *Escherichia coli* and *Salmonella typhimurium*. *Proc Natl Acad Sci USA* **95:**7046–7050.

34. **Schauder S, Shokat K, Surette MG, Bassler BL.** 2001. The LuxS family of bacterial autoinducers: biosynthesis of a novel quorum-sensing signal molecule. *MolMicrobiol* **41:**463–476.

35. **Sperandio V, Torres AG, Jarvis B, Nataro JP, Kaper JB.** 2003. Bacteria-host communication: the language of hormones. *Proc Natl Acad Sci USA* **100:**8951–8956.

36. **Walters M, Sircili MP, Sperandio V.** 2006. AI-3 synthesis is not dependent on luxS in *Escherichia coli*. *JBacteriol* **188:**5668–5681.

37. **McDaniel TK, Jarvis KG, Donnenberg MS, Kaper JB.** 1995. A genetic locus of enterocyte effacement conserved among diverse enterobacterial pathogens. *Proc Natl AcadSci USA* **92:**1664–1668.

38. **Jarvis KG, Giron JA, Jerse AE, McDaniel TK, Donnenberg MS, Kaper JB.** 1995. Enteropathogenic *Escherichia coli* contains a putative type III secretion system necessary for the export of proteins involved in attaching and effacing lesion formation. *Proc Natl Acad Sci USA* **92:**7996–8000.

39. **Sperandio V, Torres AG, Giron JA, Kaper JB.** 2001. Quorum sensing is a global regulatory mechanism in enterohemorrhagic *Escherichia coli* O157:H7. *J Bacteriol* **183:**5187–5197.

40. **Swearingen MC, Sabag-Daigle A, Ahmer BM.** 2013. Are these acyl-homoserine lactones within mammalian intestines? *J Bacteriol* **195:**173–179.

41. **Clarke MB, Hughes DT, Zhu C, Boedeker EC, Sperandio V.** 2006. The QseC sensor kinase: a bacterial adrenergic receptor. *Proc Natl Acad Sci USA* **103:**10420–10425.

42. **Furness JB.** 2000. Types of neurons in the enteric nervous system. *J Auton Nerv Syst* **81:**87–96.

43. **Purves D, Fitzpatrick D, Williams SM, McNamara JO, Augustine GJ, Katz LC, LaMantia A.** 2001. *Neuroscience*, 2nd ed. Sinauer Associates, Inc., Sunderland, MA.

44. **Horger S, Schultheiss G, Diener M.** 1998. Segment-specific effects of epinephrine on ion transport in the colon of the rat. *Am J Physiol* **275:**G1367–G1376.

45. **Fredollino PL, Kalani MY, Vaidihi N, Floriano WB, Hall SE, Trabanino RJ, Kam VW, Goddard WA.** 2004. Predicted 3D structure for the human beta 2 adrenergic receptor and its binding site for agonists and antagonists. *Proc Natl Acad Sci USA* **101:**2736–2741.

46. **Clarke MB, Hughes DT, Zhu C, Boedeker EC, Sperandio V.** 2006. The QseC sensor kinase: a bacterial adrenergic receptor. *Proc Natl Acad Sci USA* **103:**10420–10425.

47. **Walters M, Sircili MP, Sperandio V.** 2006. AI-3 synthesis is not dependent on luxS in *Escherichia coli. J Bacteriol* **188:**5668–5681.

48. **Reading NC, Rasko DA, Torres AG, Sperandio V.** 2009. The two-component system QseEF and the membrane protein QseG link adrenergic and stress sensing to bacterial pathogenesis. *Proc Natl Acad Sci USA* **106:**5889–5894.

49. **Reading NC, Torres AG, Kendall MM, Hughes DT, Yamamoto K, Sperandio V.** 2007. A novel two-component signaling system that activates transcription of an enterohemorrhagic *Escherichia coli* effector involved in remodeling of host actin. *J Bacteriol* **189:**2468–2476.

50. **Rasko DA, Moreira CG, Li de R, Reading NC, Ritchie JM, Waldor MK, Williams N, Taussig R, Wei S, Roth M, Hughes DT, Huntley JF, Fina MW, Falck JR, Sperandio V.** 2008. Targeting QseC signaling and virulence for antibiotic development. *Science* **321:**1078–1080.

51. **Clarke MB, Sperandio V.** 2005. Transcriptional autoregulation by quorum sensing *E. coli* regulators B and C (QseBC) in enterohemorrhagic *E. coli* (EHEC). *Mol Microbiol* **58:**441–455.

52. **Hughes DT, Clarke MB, Yamamoto K, Rasko DA, Sperandio V.** 2009. The QseC adrenergic signaling cascade in enterohemorrhagic *E. coli* (EHEC). *PLoS Pathog* **5:**e1000553.

53. **Yamamoto K, Hirao K, Oshima T, Aiba H, Utsumi R, Ishihama A.** 2005. Functional characterization in vitro of all two-component signal transduction systems from *Escherichia coli. J Biol Chem* **280:**1448–1456.

54. **Clarke MB, Sperandio V.** 2005. Transcriptional regulation of flhDC by QseBC and sigma (FliA) in enterohaemorrhagic *Escherichia coli. Mol Microbiol* **57:**1734–1749.

55. **Mellies JL, Elliott SJ, Sperandio V, Donnenberg MS, Kaper JB.** 1999. The Per regulon of enteropathogenic *Escherichia coli*: identification of a regulatory cascade and a novel transcriptional activator, the locus of enterocyte effacement (LEE)-encoded regulator (Ler). *Mol Microbiol* **33:**296–306.

56. **Njoroge JW, Nguyen Y, Curtis MM, Moreira CG, Sperandio V.** 2012. Virulence meets metabolism: Cra and KdpE gene regulation in enterohemorrhagic *Escherichia coli. MBio* **3:**e00280-00212.

57. **Luttmann D, Heermann R, Zimmer B, Hillmann A, Rampp IS, Jung K, Gorke B.** 2009. Stimulation of the potassium sensor KdpD kinase activity by interaction with the phosphotransferase protein IIA(Ntr) in *Escherichia coli. Mol Microbiol* **72:**978–994.

58. **Herrmann A, Davies JR, Lindell G, Martensson S, Packer NH, Swallow DM, Carlstedt I.** 1999. Studies on the "insoluble" glycoprotein complex from human colon. Identification of reduction-insensitive MUC2 oligomers and C-terminal cleavage. *J BiolChem* **274:**15828–15836.

59. **Martens EC, Lowe EC, Chiang H, Pudlo NA, Wu M, McNulty NP, Abbott DW, Henrissat B, Gilbert HJ, Bolam DN, Gordon JI.** 2011. Recognition and degradation of plant cell wall polysaccharides by two human gut symbionts. *PLoS Biol* **9:**e1001221.

60. **Chang DE, Smalley DJ, Tucker DL, Leatham MP, Norris WE, Stevenson SJ, Anderson AB, Grissom JE, Laux DC, Cohen PS, Conway T.** 2004. Carbon nutrition of *Escherichia coli* in the mouse intestine. *Proc Natl Acad Sci USA* **101:** 7427–7432.

61. **Miranda RL, Conway T, Leatham MP, Chang DE, Norris WE, Allen JH, Stevenson SJ, Laux DC, Cohen PS.** 2004. Glycolytic and gluconeogenic growth of *Escherichia coli* O157:H7 (EDL933) and *E. coli* K-12 (MG1655) in the mouse intestine. *Infect Immun* **72:**1666–1676.

62. **Pacheco AR, Curtis MM, Ritchie JM, Munera D, Waldor MK, Moreira CG, Sperandio V.** 2012. Fucose sensing regulates bacterial intestinal colonization. *Nature* **492:**113–117.

63. **Kendall MM, Gruber CC, Parker CT, Sperandio V.** 2012. Ethanolamine controls expression of genes encoding components involved in interkingdom signaling and virulence in enterohemorrhagic *Escherichia coli* O157:H7. *MBio* **3:** e00050-12.

64. **Bertin L, Grilli S, Spagni A, Fava F.** 2013. Innovative two-stage anaerobic process for effective codigestion of cheese whey and cattle manure. *Bioresour Technol* **128:**779–783.

65. **Fabich AJ, Jones SA, Chowdhury FZ, Cernosek A, Anderson A, Smalley D, McHargue JW, Hightower GA, Smith JT, Autieri SM, Leatham MP, Lins JJ, Allen RL, Laux DC, Cohen PS, Conway T.** 2008. Comparison of carbon nutrition for pathogenic and commensal *Escherichia coli* strains in the mouse intestine. *Infect Immun* **76:**1143–1152.

66. **Kamada N, Kim YG, Sham HP, Vallance BA, Puente JL, Martens EC, Nunez G.** 2012. Regulated virulence controls the ability of a pathogen to compete with the gut microbiota. *Science* **336:**1325–1329.

67. **Hoverstad T, Midtvedt T.** 1986. Short-chain fatty acids in germfree mice and rats. *J Nutr* **116:**1772–1776.

68. **Topping DL, Clifton PM.** 2001. Short-chain fatty acids and human colonic function: roles of resistant starch and nonstarch polysaccharides. *PhysiolRev* **81:**1031–1064.

69. **Nakanishi N, Tashiro K, Kuhara S, Hayashi T, Sugimoto N, Tobe T.** 2009. Regulation of virulence by butyrate sensing in enterohaemorrhagic *Escherichia coli*. *Microbiology* **155:**521–530.

70. **Tobe T, Nakanishi N, Sugimoto N.** 2011. Activation of motility by sensing short-chain fatty acids via two steps in a flagellar gene regulatory cascade in enterohemorrhagic *Escherichia coli*. *Infect Immun* **79:**1016–1024.

71. **Zumbrun SD, Melton-Celsa AR, Smith MA, Gilbreath JJ, Merrell DS, O'Brien AD.** 2013. Dietary choice affects Shiga toxin-producing *Escherichia coli* (STEC) O157:H7 colonization and disease. *Proc Natl Acad Sci USA* **110:**E2126–E2133.

72. **Fukuda S, Toh H, Hase K, Oshima K, Nakanishi Y, Yoshimura K, Tobe T, Clarke JM, Topping DL, Suzuki T, Taylor TD, Itoh K, Kikuchi J, Morita H, Hattori M, Ohno H.** 2011. Bifidobacteria can protect from enteropathogenic infection through production of acetate. *Nature* **469:**543–547.

73. **Wlodarska M, Willing B, Keeney KM, Menendez A, Bergstrom KS, Gill N, Russell SL, Vallance BA, Finlay BB.** 2011. Antibiotic treatment alters the colonic mucus layer and predisposes the host to exacerbated *Citrobacter rodentium*-induced colitis. *Infect Immun* **79:**1536–1545.

74. **Willing BP, Vacharaksa A, Croxen M, Thanachayanont T, Finlay BB.** 2011. Altering host resistance to infections through microbial transplantation. *PLoS One* **6:**e26988.

75. **de Sablet T, Chassard C, Bernalier-Donadille A, Vareille M, Gobert AP, Martin C.** 2009. Human microbiota-secreted factors inhibit shiga toxin synthesis by enterohemorrhagic *Escherichia coli* O157:H7. *Infect Immun* **77:**783–790.

76. **Brown CA, Harmon BG, Zhao T, Doyle MP.** 1997. Experimental *Escherichia coli* O157:H7 carriage in calves. *Appl Environ Microbiol* **63:**27–32.

77. **Cray WC Jr, Moon HW.** 1995. Experimental infection of calves and adult cattle with *Escherichia coli* O157:H7. *Appl Environ Microbiol* **61:**1586–1590.

78. **Dean-Nystrom EA, Bosworth BT, Cray WC Jr, Moon HW.** 1997. Pathogenicity of I O157:H7 in the intestines of neonatal calves. *Infect Immun* **65:**1842–1848.

79. **Woodward MJ, Gavier-Widen D, McLaren IM, Wray C, Sozmen M, Pearson GR.** 1999. Infection of gnotobiotic calves with *Escherichia coli* O157: H7 strain A84. *Vet Rec* **144:**466–470.

80. **Wray C, McLaren IM, Randall LP, Pearson GR.** 2000. Natural and experimental infection of normal cattle with *Escherichia coli* O157. *Vet Rec* **147:**65–68.

81. **Nguyen Y, Sperandio V.** 2012. Enterohemorrhagic *E. coli* (EHEC) pathogenesis. *Front Cell Infect Microbiol* **2:**90.

82. **Kaper JB, Nataro JP, Mobley HL.** 2004. Pathogenic *Escherichia coli*. *Nat Rev Microbiol* **2:**123–140.

83. **Sheng H, Lim JY, Knecht HJ, Li J, Hovde CJ.** 2006. Role of *Escherichia coli* O157:H7 virulence factors in colonization at the bovine terminal rectal mucosa. *Infect Immun* **74:**4685–4693.

84. **Price SB, Wright JC, DeGraves FJ, Castanie-Cornet MP, Foster JW.** 2004. Acid resistance systems required for survival of *Escherichia coli* O157:H7 in the bovine gastrointestinal tract and in apple cider are different. *Appl Environ Microbiol* **70:**4792–4799.

85. **Hughes DT, Terekhova DA, Liou L, Hovde CJ, Sahl JW, Patankar AV, Gonzalez JE, Edrington TS, Rasko DA, Sperandio V.** 2010. Chemical sensing in mammalian host-bacterial commensal associations. *Proc Natl Acad Sci USA* **107:**9831–9836.

86. **Sheng H, Nguyen Y, Hovde CJ, Sperandio V.** 2013. SdiA aids enterohemorrhagic *Escherichi coli* carriage by cattle fed a forage or grain diet. *Infect Immun* **81:**3472–3478.

87. **Parsek MR, Val DL, Hanzelka BL, Cronan JE Jr, Greenberg EP.** 1999. Acyl homoserine-lactone quorum-sensing signal generation. *Proc Natl AcadSci USA* **96:**4360–4365.

88. **Kanamaru K, Kanamaru K, Tatsuno I, Tobe T, Sasakawa C.** 2000. SdiA, an *Escherichia coli* homologue of quorum-sensing regulators, controls the expression of virulence factors in enterohaemorrhagic *Escherichia coli* O157:H7. *Mol Microbiol* **38:**805–816.

89. **Yao Y, Martinez-Yamout MA, Dickerson TJ, ABrogan AP, Wright PE, Dyson HJ.** 2006. Structure of the *Escherichia coli* quorum sensing protein SdiA: activation of the folding switch by acyl homoserine lactones. *J Mol Biol* **355:**262–273.

90. **Edrington TS, Farrow RL, Sperandio V, Hughes DT, Lawrence TE, Callaway TR, Anderson RC, anNisbet DJ.** 2009. Acyl-homoserine-lactone autoinducer in the gastrointestinal [corrected] tract of feedlot cattle and correlation to season, *E. coli* O157:H7 prevalence, and diet. *Curr Microbiol* **58:**227–232.

91. **Erickson DL, Nsereko VL, Morgavi DP, Selinger LB, Rode LM, Beauchemin KA.** 2002. Evidence of quorum sensing in the rumen ecosystem: detection of *N*-acyl homoserine lactone autoinducers in ruminal contents. *Can J Microbiol* **48:**374–378.

PREVENTION AND CONTROL STRATEGIES

"Preharvest" Food Safety for *Escherichia coli* O157 and Other Pathogenic Shiga Toxin-Producing Strains

21

THOMAS E. BESSER,[1] CARRIE E. SCHMIDT,[1]
DEVENDRA H. SHAH,[1] and SMRITI SHRINGI[1]

INTRODUCTION

Upton Sinclair's novel, *The Jungle*, which described horrific conditions in historical Chicago meat packing plants, engendered numerous reforms and regulations of the industry, including the Pure Food and Drug Act and the Meat Inspection Act of 1906, which in turn led to vast improvements in the sanitary conditions under which meat and meat products were handled. The massive and highly publicized 1993 outbreak of *Escherichia coli* O157 associated with Jack in the Box had a similar broad impact for the microbiological safety of food, including the classification of this pathogen as an "adulterant" in ground beef, and led to the implementation of the formal Pathogen Reduction and Hazard Analysis and Critical Control Point Program for this bacterium and other food-borne agents in meat processing plants. These changes were credited with significant reduction in the incidence of human infection with *E. coli* O157 in the United States over the subsequent several years; however, this trend did not continue, and in recent years the incidence of disease due to *E. coli* O157 has remained stubbornly stable. Incidence of disease caused by non-O157 Shiga toxin-producing *E. coli* (STEC) has paradoxically steadily increased, although this trend is undoubtedly due in part to increased use of more efficient diagnostic procedures.

[1]Veterinary Microbiology and Pathology, Washington State University, Pullman, WA 99164.

Enterohemorrhagic Escherichia coli *and Other Shiga Toxin-Producing* E. coli
Edited by Vanessa Sperandio and Carolyn J. Hovde
© 2015 American Society for Microbiology, Washington, DC
doi:10.1128/microbiolspec.EHEC-0021-2013

The continued occurrence of disease outbreaks from *E. coli* O157 and other pathogenic STEC strains linked to ground beef indicates the limitations of postprocessing interventions to completely eliminate risk of human exposure through contaminated meat and meat products. New evidence of disease linked to other sources, including contaminated produce, water, and other environmental exposures including direct animal contacts, indicates that this group of pathogens has a more complex ecology than may have been previously recognized. This article addresses some of the data supporting this complexity to explain why human disease incidence is not declining, discusses the implications of the different genetic lineages of *E. coli* O157 on sources and severity of human infection, and reviews the benefits and limitations of control measures directed toward reducing the prevalence and shedding level of *E. coli* O157 and other pathogenic STEC strains by cattle, otherwise known as preharvest food safety in cattle production.

Twenty Years after the "Jack in the Box" Outbreak, Why Is *E. coli* O157 Still a Problem?

Why has the incidence of human infection with *E. coli* O157 and other STEC pathogens remained stubbornly steady despite the implementation of stringent regulations and large investments in improved equipment and processing methods in meat packaging plants? One important factor is seasonal variation, or the marked increase in the numbers of cattle shedding *E. coli* O157 in their feces accompanied by increased contamination of hair coats (hides) during summer months. This seasonal variation results in increased contamination pressure, potentially overwhelming the control measures that are otherwise effective in preventing meat contamination during the rest of the year. The effects of higher contamination of cattle that overwhelm the control measures could be mitigated, at least in part, by adding a final decontamination step such as gamma irradiation for meat products

of beef origin. However, in the absence of such a highly effective decontamination step, further reductions in meat-borne exposures to *E. coli* O157 may require interventions that reduce the degree of contamination of cattle sent to slaughter. Over the years, it has become clear that apart from ground beef, there are numerous vehicles for *E. coli* O157 that can result in human exposure, including fresh produce, drinking and recreational water, direct contacts with animals, and other environmental sources and reservoirs. This complex ecology of *E. coli* O157 likely contributes to seasonal infection pressure on cattle as well, and needs to be addressed in order to develop highly effective methods to reduce cattle infections with *E. coli* O157 and other pathogenic STEC strains.

How Do Foods of Bovine Origin Become Contaminated with *E. coli* O157?

There is a strong correlation between *E. coli* O157 prevalence in the feces and on the hair coats of cattle entering slaughter plants and carcass contamination during processing (1). Recent studies have begun to characterize the level of pathogen reduction in cattle feces that may be necessary to significantly reduce the hide and carcass contamination during processing. Woerner et al. showed that fecal pen prevalence exceeding 20% was associated with hide contamination prevalence of 80% or more (2). Similarly, Arthur et al. determined that slaughter cattle from feedlot pens with more than 20% positive fecal pats had both higher hide contamination rates (25.5%) and higher carcass contamination at pre-evisceration (14.3%), post-evisceration (2.9%), and post-final intervention (0.7%) stages (3). Comparative figures for slaughter cattle from feedlot pens with <20% positive fecal pat samples were lower hide contamination (5%) and carcass contamination 6.3%, 0% and 0% at pre-evisceration, post-evisceration, and post-final intervention stages, respectively (3). Overall, these data suggest that 20% fecal pat prevalence may be a functional threshold or marker for predicting groups of feedlot cattle having increased risk of hide or carcass contamination. Management practices that consistently result in fecal pat prevalence of less than 20% may therefore be required to accomplish further progress in preharvest food safety.

Can Live Cattle Be Managed to Reduce or Prevent *E. coli* O157 Infection?

Heavily contaminated cattle entering meat processing plants can apparently overwhelm the best sanitary procedures in practice; therefore, preharvest interventions in cattle rearing, management and husbandry, transport, and lairage that can effectively reduce the frequency of cattle infection with *E. coli* O157 offer the potential to reduce human exposures. In the last 2 decades, the development of preharvest interventions has remained a major focus of the food safety research in the United States. The early emphasis of preharvest food safety research was based on the hypothesis that the emergence of *E. coli* O157 disease in humans resulted from relatively recent changes in cattle management practices that favored this pathogen. Examples of such management practices included increased grain components in cattle feeds (4, 5), the use of antimicrobial drugs and other growth-promoting feed additives (6–11), increased intensity of cattle production, rearing larger herds and increased confinement (12–16), and the adoption of new methods of manure handling and disposal on farms (17–19). Unfortunately, however, each of these attractive hypotheses has since either been refuted or shown to have only minor influence on cattle infection with *E. coli* O157, as described in several comprehensive recent reviews (20–23). The hypothesis that *E. coli* O157 infection of cattle results from high-grain diets and that feeding hay to the cattle would eliminate the problem merits particular note; while it has not been supported by subsequent research (5, 24), it is still frequently cited as if true in the news media responses to each new *E. coli* O157 disease outbreak, demonstrating a clear disconnect between scientific data and the popular support for an idea.

Unfortunately, with the exception of cattle vaccination against *E. coli* O157, the efforts to identify cattle management practices that consistently result in significant reductions in the frequency of cattle infection with *E. coli* O157 have largely failed (reviewed in references 20–23).

VACCINATION OF CATTLE AGAINST *E. COLI* O157: A RAY OF LIGHT

Although certain interventions [for example, probiotics (20, 22)] show some promise for preharvest food safety against *E. coli* O157, vaccines have been the most effective interventions documented to date. Currently, two commercial vaccines against *E. coli* O157 in cattle have been developed and are available in at least some locations: a type III secretion system (T3SS) protein-based (Bioniche Life Sciences Inc., Belleville, Ontario, Canada) and a siderophore receptor and porin (SRP) protein-based (Epitopix, LLC, Wilmar, Minnesota) vaccine.

Cattle Vaccine Mechanisms

These vaccines target different mechanisms to induce immunity against *E. coli* O157 in cattle; T3SS proteins play important roles in bacterial adherence to the bovine intestinal epithelium, whereas SRP proteins are important for iron acquisition and survival of bacteria within the host. The products of T3SS genes such as *eae* and *tir* (intimin and Tir), encoded within the locus of enterocyte effacement (LEE), play key roles in the colonization of bovine intestines by *E. coli* O157 (25–30). Translocation of Tir and other effector proteins into host cells requires the T3SS-secreted EspA protein, which forms filaments connecting the bacteria to the host cell surface, as well as EspB and EspD, which are thought to form a membrane pore [reviewed by Frankel et al. (31) and Caron et al. (32)]. The T3SS protein-based vaccine strategy results in induction of mucosal antibodies capable of blocking adherence and subsequent colonization of the bovine intestinal mucosa by *E. coli* O157. Under low-iron conditions, bacteria produce a high-affinity iron transport system (e.g., SRP proteins) to bring the required nutrient inside the bacterial cell (33). The SRP protein-based vaccine results in induction of antibodies that bind to SRP located on the outer membrane of the bacterial cell, subsequently blocking iron transport into the cell, compromising the bacterial cell iron acquisition. Blocking iron transport by anti-SRP antibodies renders the bacteria at a selective disadvantage in a mixed microbial environment, resulting in reduced colonization. These approaches were recognized over a decade ago, resulting in a number of subsequent vaccine trials using purified T3SS or SRP protein-based vaccines. Although vaccines targeting T3SS proteins and SRP function via two entirely different mechanisms, recent meta-analysis studies suggest that both vaccines are efficacious at reducing the proportion of culture-positive animals (34–37).

Efficacy of Vaccination

Although the effectiveness of current vaccines in terms of reduced carcass contamination and ultimately reduced human illnesses is unknown, if 20% fecal prevalence is considered as a functional threshold marker for significantly reduced hide and carcass contamination, then the current vaccine efficacy would have to effectively reduce pen prevalence to <20%. Ideally, the precise efficacy of each vaccine can be calculated; however, significant variation in the efficacy of current vaccines is reported in different trials, and recent meta-analyses of multiple vaccine studies suggest that the efficacy of current vaccines is largely uncertain (36, 37). Consequently, Vogstad et al. simulated the uncertainty about vaccine efficacy using a log-normal distribution and estimated that the mean efficacy of current T3SS protein-based vaccine is approximately 58% (36). Using this vaccine efficacy, the authors developed a stochastic simulation

model to compare distributions of *E. coli* O157 fecal shedding prevalence between cattle vaccinated with T3SS protein vaccine and non-vaccinated cattle.

The model outputs included distributions of fecal pen prevalence of *E. coli* O157 among vaccinated and nonvaccinated summer-fed cattle and nonvaccinated winter-fed cattle. One of the outcomes of this model was a reduction in the percentage of high-prevalence pens among immunized cattle fed in the summer. In this model, approximately 58% of pens of nonvaccinated, summer-fed cattle showed fecal prevalence of >20%. In contrast, when summer-fed cattle were vaccinated with the T3SS protein-based vaccine, the percentage of pens with >20% fecal prevalence was reduced to approximately 30%. These results suggest that vaccination as an intervention in cattle prior to slaughter may roughly halve the number of pens with fecal prevalence of >20%, a significant improvement but still leaving 30% of pens with fecal prevalence >20%. As already discussed in this article, according to Arthur et al. (3) and Woener et al. (2), pens with >20% fecal prevalence contribute significantly to hide and carcass contamination. In turn, hide and carcass contamination can compromise apparent vaccine efficacy due to cross-contamination of hides during transport to harvest (38) or cross-contamination of carcasses during processing (39). On the basis of a postulated threshold effect involving vaccine-induced reductions in shedding density (reductions in the numbers of animals with fecal shedding exceeding 10^3 CFU/g *E. coli* O157, also known as super-shedders), Matthews et al. recently proposed that cattle vaccination would in fact produce substantially greater reductions in human disease caused by *E. coli* O157 than predicted based solely on effects on cattle shedding prevalence (40). Overall, it is still questionable whether the current vaccines would provide sufficient efficacy to accomplish the goal of controlling or reducing postharvest *E. coli* O157 contamination of cattle-derived food products.

Is It Practical To Vaccinate Cattle?

Recently, Withee et al. (41) combined quantitative risk assessment and marginal economic analysis to estimate the cost-benefit ratio of the "hypothetical O157:H7 vaccine" to prevent human food-borne illness. These authors determined that vaccinating the entire U.S. herd would be an effective intervention for preventing *E. coli* O157 illness in humans; however, the true efficiency of vaccination will primarily depend on three factors: (i) overall efficacy of the vaccine, (ii) herd coverage of immunity, and (iii) the cost of vaccine per unit. For example, the authors estimated that if the vaccine efficacy and coverage for herd immunity were assumed at 100% and the vaccine cost was assumed to be $3.00 per unit, then vaccination will optimally prevent approximately 21,000 human illnesses each year (41), or one-third to one-fifth of the annual burden of disease as estimated by the CDC (42, 43). This level of control would require vaccinating 22 million cattle intended for slaughter each year at a total cost of $66 million. In this scenario, the total benefits expected to accrue as a result of preventing 21,000 human illnesses would be $131 million (21,000 forgone cases times $6,256 per case). In contrast, if the vaccine efficacy was assumed at 50% (close to the estimated efficacy of current vaccines) and required herd coverage for immunity was assumed at 100%, then a $4.00 per unit cost of vaccination will optimally produce approximately 5,000 forgone illnesses (41). Therefore, even the moderate efficacy of current vaccines is predicted to prevent several thousand food-borne illnesses each year; however, there is still clearly significant room for the improvement of the efficacy of current vaccines and vaccination strategies.

Possible Future Directions for Vaccine Development

Given that two current vaccines provide protection by completely unrelated mechanisms, it is possible that simultaneous vaccination with both currently available products could have synergistic effects and result in significantly improved efficacy; however, no published studies in the literature address this possibility. Alternatively, new vaccines may be developed with improved efficacy. Dziva et al. (27) showed that in addition to the genes encoded on LEE-T3SS, *E. coli* O157 colonization in cattle is mediated by numerous other cell surface structures, including fimbriae, outer membrane proteins, O antigens, and other bacterial proteins. These authors have identified a novel fimbrial locus (z2199–z2206; ecs2114–ecs2107/locus 8) required for intestinal colonization in calves, and demonstrated that a deletion mutant is rapidly outcompeted by the parent strain in coinfection studies (27). For another example, Torres et al. (44) described two chromosomal operons (*lpf1* and *lpf2*) in *E. coli* O157 closely related to the long polar fimbrial (*lpf*) operon of *Salmonella enterica* serovar Typhimurium that have been associated with the appearance of long fimbriae that enhance colonization in animal models (reviewed in reference 45). Finally, in studies that used bovine terminal rectal primary epithelial cells and bovine intestinal tissue explants, the H7 flagellum acted as an adhesin to bovine intestinal epithelium and contributed to initiation of intestinal colonization (46, 47). A following study showed that immunization of cattle with H7 flagellin reduced colonization rates and delayed peak bacterial shedding following subsequent oral challenge with *E. coli* O157 (48). Based on these data, incorporation of one or more of these antigens, perhaps in combination with antigens used in the currently available vaccines, may further enhance vaccine efficacy.

CATTLE INFECTION WITH *E. COLI* O157 AS AN ECOLOGICAL PROBLEM

As microbiological methods were developed to efficiently detect *E. coli* O157 and other pathogenic STEC strains in cattle feces and environmental samples, and as more epidemiological studies in cattle herds were completed, several observations with profound implications for preharvest food safety were made. These included (i) the ubiquitous presence of *E. coli* O157 and other pathogenic STEC strains on cattle farms during the summer (49) but its relative absence during the winter, (ii) the similarity in prevalence of infection among cattle raised under drastically different management conditions ranging from dispersed distribution of animals on pastures to housing in a highly concentrated fashion in feedlots, (iii) the transient nature of STEC colonization of individual animals, typically lasting one to a few weeks, (iv) the sporadic occurrence of herd outbreaks of high prevalence *E. coli* O157 fecal shedding that present all the hallmarks of food- or waterborne transmission, and (v) the detection of *E. coli* O157 fecal shedding in a very wide range of other mammalian and avian species. Basically, these observations are inconsistent with the widely held idea that cattle are the central sustaining reservoir for *E. coli* O157 and instead support the idea that cattle are just one more mammalian host periodically infected with this agent following oral exposures, albeit a host with particular significance for human exposure due to its use for producing human foods. The following sections of this article explore what is known of the ecology of this agent.

"Reservoirs" of *E. coli* O157 on Cattle Farms

The study of reservoirs is complex, and a variety of reservoir models exist for different pathogens. Much work has gone into identifying the reservoir for *E. coli* O157 to formulate strategies for controlling this pathogen on farms in pursuit of preharvest food safety. The clearest type of reservoir is a *biological reservoir*, a site or host where the agent can always be found and serves as a source of the infection for target populations. Complex reservoirs may include maintenance host populations that persistently harbor the infectious

agent, as well as nonmaintenance (incidental or amplifying) host populations that do not harbor the microorganism indefinitely, but aid in the dissemination and amplification of the pathogen. Haydon et al. explain that the number of maintenance host populations is generally limited, whereas the number of nonmaintenance host populations may be unlimited (50). These definitions may be useful in considering the role(s) such populations may play in the seasonal occurrence of *E. coli* O157 on farms and in understanding how these populations may serve as targets for preharvest control of these bacteria.

Cattle as Reservoirs

Many human outbreaks with *E. coli* O157 have been associated with the consumption of contaminated foods of bovine origin or with direct contact with cattle or farms where infected cattle are raised (51, 52). Cattle are the sole animal host known to demonstrate site-specific intestinal colonization with this agent, at the recto-anal junction (RAJ). RAJ colonization among cattle has been observed on several dairy and beef farms without resulting in a detectable illness in these animals (53). Nearly all cattle herds, including both beef and dairy types, may be colonized. As discussed previously in this article, fecal shedding is associated with hide contamination, which has been demonstrated as a main source of meat contamination at slaughter (1, 54); thus research has been directed to the identification of preslaughter interventions that can decrease RAJ colonization and fecal shedding of these bacteria. Vaccination (discussed above) of cattle may be promising to accomplish this goal; however, identifying ways to reduce or eliminate the source of cattle infection is equally important.

While cattle are likely an important part of the reservoir for *E. coli* O157 on farms, several pieces of evidence have raised questions on whether cattle are truly a maintenance population for this pathogen. First, cattle typically shed *E. coli* O157 only transiently during summer months, and levels and prevalence of cattle shedding cease or decrease drastically during winter months (55, 56). Second, a single strain of *E. coli* O157 frequently predominates on individual farms over periods of multiple years, despite essentially disappearing from cattle populations each winter. This tendency is particularly interesting on large feedlots that go through multiple animal population turnovers annually, with incoming cattle originating from many diverse sources (57). These data suggest that farms may contain other noncattle maintenance hosts (reservoirs) of *E. coli* O157 and also cast doubt on whether the cattle themselves make up the true maintenance host population. Recently, it has been experimentally demonstrated that the seasonal differences in *E. coli* O157 shedding by cattle are not due to intrinsic factors within the animals. Cattle given identical challenge doses of *E. coli* O157 shed the agent in similar amounts and for similar durations, regardless of the season of exposure (58); this is the outcome predicted of an amplifying host population, where the source is the key factor in duration and level of bacterial colonization in cattle. If cattle are simply an amplifying host population, it seems clear that identification of the true maintenance reservoir(s) of *E. coli* O157 is critical to the development of truly preventive systems for management of *E. coli* O157 on cattle farms.

Survivability of *E. coli* O157 in the Environment

E. coli O157 is surprisingly persistent in environmental sites, documented to survive in ovine manure for 21 months (19). The environments (bedding materials and water) of experimentally infected steers maintain detectable viable *E. coli* O157 for at least 14 weeks after inoculation of the cattle (59). Interestingly, in this study, *E. coli* O157 was cultured from the bedding and water even during weeks when it was not possible to recover *E. coli* O157 from cattle fecal samples (59). In a longitudinal, year-long study of

naturally occurring *E. coli* O157 infection of cattle on two feedlots in southern Alberta, *E. coli* O157 was cultured from only 0.8% of the fecal pats, but 12% of the water troughs sampled were found positive. *E. coli* O157 was also cultured from 1.7% of feed bunk feed samples but not from fresh total mixed rations (60). Culture-positive water troughs occurred seasonally: 35% of water troughs sampled during the summer on one feedlot were culture-positive for *E. coli* O157, compared to 0% sampled during the winter (60). This seasonal variation clearly parallels the seasonality of cattle infection on farms and also raises the question of whether the water contamination is the source of, or results from, the cattle infection.

Water as a Reservoir

As described above, water is one of the most commonly contaminated materials on cattle farms. In culture-positive water troughs *E. coli* O157 is consistently detected more frequently in sediments than in the water column (59, 61). Water trough sediment consists of feed and fecal material admixed with numerous bacteria and protozoa, with rare metazoan species (nematodes and rotifers). Viable *E. coli* O157 in the sediment layers of water troughs can persist for greater than 245 days (61). One hypothesis is that ambient temperature during the summer is more permissible for growth of bacteria and, therefore, may result in increased bacterial populations in water troughs during the summer compared to the winter season (62). While increased ambient temperature likely plays a role in proliferation of bacteria, there may be other factors that influence seasonal variation and overall survival of these bacteria in water troughs. For example, mean coliform counts were significantly higher in water troughs that were cleaned at least every 2 months compared to those that were cleaned less frequently (63). Additionally, use of chlorinated or hyperchlorinated water in trough microcosms failed to eliminate *E. coli* O157 (61). These data suggest that there are likely additional factors other than ambient temperature that contribute to the survival and proliferation of *E. coli* O157 in water troughs.

Role of Protozoa

While the relationship between *E. coli* O157 and protozoa has not yet been clarified, LeJeune et al. demonstrated a significant increase in the quantity of free-living protozoa within water trough sediment in the winter compared to the summer (63). Other studies have demonstrated grazing of bacteria by protozoa collected from soil, lakes, and streams. While many bactivorous protozoa will feed on any available food, preferential grazing for bacteria also occurs. For example, *E. coli* O157 containing Stx2a-encoding bacteriophage are relatively resistant to grazing by *Tetrahymena* sp. (64, 65). Survival of *E. coli* O157 within the food vacuoles and excretory vacuoles of protozoa isolated from dairy lagoon wastewater suggests that protozoa may be vehicles for dissemination of the bacterium to crops (66). Many free-living protozoa form cysts under stressful conditions such as temperature or salinity changes and food deprivation, and these cysts can persist in the environment for decades. While several bacterial genera including *Legionella, Mycobacterium, and Listeria* spp. have been shown to survive within such protozoan cysts (67–70), research is needed to determine whether this may also be true for *E. coli* O157.

Role of Environmental Invertebrates

Apart from protozoa, invertebrate organisms such as nematodes and rotifers that have the potential for harboring *E. coli* O157 also inhabit water troughs and soils on cattle farms. Research has shown that *E. coli* O157 can amplify and persist for 5 days or more within one such free-living nematode, *Caenorhabditis elegans* (71). The association of *C. elegans* with *Salmonella* spp. has been more thoroughly investigated; when *C. elegans* is exposed to

Salmonella serovar Newport, these bacteria can be detected in nematode progeny for at least the subsequent two generations (71). Another bactivorous, free-living nematode, *Diploscapter* spp., has been demonstrated to migrate rapidly toward colonies of *E. coli* O157 and to shed viable bacterial cells for at least a day after exposure (72). Free-living nematodes protect themselves in scarcity of food or harsh environmental conditions by forming arrested-development larvae (dauer) stages, however, it has not yet been determined whether dauer stages can harbor foodborne pathogens and subsequently act as a source of contamination or as a reservoir for these pathogens.

Role of Flies

Many different families of flies are present on cattle farms. Flies mostly multiply during the spring and are in constant contact with cattle and feed during summer and early autumn months. Many flies lay their eggs in cattle feces, which hatch into larvae (maggots) that feed on manure before maturing into pupae within a week (73). Pupae contain a hard durable shell that allows them to survive under harsh conditions; most flies survive in this stage over the winter (74, 75). Adult flies that emerge from pupae typically survive for only a few weeks. Because of flies' close interactions with cattle on farms, some investigators have studied flies as a component of the reservoir of *E. coli* O157. These bacteria can be cultured from adult houseflies found in feed bunks and cattle feed storage sheds during summer months (76). In an *E. coli* O157 outbreak at a nursery school in Japan, the strains of *E. coli* O157 isolated from patients matched those detected in houseflies collected from within the school (77), and the possibility that the flies were acting as mechanical vectors able to disseminate bacteria to food and eating utensils was considered. Subsequent research suggested that flies may be more than just mechanical vectors. After oral infection of adult houseflies with *E. coli* O157, bacteria were identified in the alimentary canals of 30% of these flies up to 3 days postinfection. Orally infected flies with actively proliferating *E. coli* O157 on their mouthparts demonstrate cellular lesions similar to the attaching and effacing lesions seen in the colonized RAJ of cattle (78).

Role of Birds

E. coli O157 has been cultured from wild birds on cattle farms in many investigations. Birds, much like flies, may be seen as a general nuisance on farms and may act to contaminate cattle feeds and water sources, as well as disseminate bacteria within and between farms. A surveillance study determined that 3% of European starlings and 4% of the cattle study population were culture-positive for *E. coli* O157. In addition, these birds frequently visited the same farms on daily feeding forays but returned nightly to share a communal roost with birds that visited other farms, providing a potential method for pathogen dissemination (79). Poultry are readily experimentally colonized with *E. coli* O157 (80), but contamination of poultry products is very rare and human infection with *E. coli* O157 resulting from contaminated poultry has rarely been documented.

Role of Mammals

E. coli O157 fecal shedding has been detected in many different domestic animal species, including dogs, cats, horses, and sheep. Colonization of the ovine RAJ has been demonstrated but seems to occur less efficiently (81). Colonization in wildlife including feral swine, deer, raccoons, opossums, and rats has also been reported. Deer have been frequently documented to shed *E. coli* O157, and human infections have been traced to contaminated venison (82, 83). Swine are readily experimentally colonized with *E. coli* O157, but the prevalence of natural infection is very low (84, 85). In contrast, feral swine have been demonstrated to shed *E. coli* O157 and were suggested to play a role in dissemination of this agent to fresh produce that resulted in a large human outbreak of disease (86).

The Need for a Better Understanding of the Ecology and Reservoir Structure of *E. coli* O157

As mentioned previously, the reservoir for *E. coli* O157 is very complex. Based on the descriptions by Haydon et al. of complex reservoirs (50), there is likely one or more maintenance host populations that could include role(s) for organisms such as protozoa, invertebrates, or flies on cattle farms. Presence of a maintenance host population outside cattle is suggested by the fact that although swine and poultry, like cattle, are readily colonized with E. coli O157 in experimental settings, contamination of pork or poultry meats with this agent is relatively rare (87, 88). One possible explanation for this low prevalence may be that swine and poultry are typically reared in confinement in the United States, which may shield them from exposure to environmental sources of *E. coli* O157 infection. If so, this suggests that management systems to reduce cattle exposure to environmental sources of *E. coli* O157 may be required to reduce their prevalence of infection.

It is also possible that the bacteria can survive without hosts in soil or water environments during the winter, amplifying each spring (as ambient temperatures increase) to levels that are infectious to cattle. Several vertebrates, including birds, cattle, and other mammals, likely act at least as nonmaintenance host populations that aid in dissemination and amplification of these bacteria, especially during the summer months. More research leading to a better understanding of the complex reservoirs of *E. coli* O157 may lead to improved targeting of these bacteria and improved preharvest control on cattle farms along with better strategies to reduce environmental and non-beef-product-related exposures contributing to human infection.

E. COLI O157 GENOTYPES, HUMAN DISEASE, AND PREHARVEST FOOD SAFETY

Various genotyping methods including multilocus enzyme electrophoresis (89, 90), octamer-based genome scanning (91, 92), whole-genome PCR scanning (93), pulsed-field gel electrophoresis(94), Shiga toxin-associated bacteriophage insertion, typing (95), lineage-specific polymorphism assay, (96), comparative genomic hybridization, (97, 98), optical mapping (99), and single nucleotide polymorphism typing (100, 101) have been used to decipher the population structure of *E. coli* O157 (102). These studies revealed that bacteriophages play an important role in establishing the genetic diversity among *E. coli* O157 isolates and that certain specific genetic lineages of *E. coli* O157 are associated with most human disease. These strongly disease-associated genotypes have been termed clinical genotypes whereas other lineages, less frequently isolated from humans with illness compatible with *E. coli* O157 infection, have been termed bovine-biased genotypes (91, 92, 96, 103–107). In general, the various genotyping methods are concordant in their identification of clinical genotypes of *E. coli* O157 (108, 109). Populations of *E. coli* O157 in different geographical regions differ significantly in the relative frequency of particular genotypes in different countries, and generally clinical genotypes are more frequent in cattle populations in countries with higher incidences of hemolytic-uremic syndrome, a severe form of illness associated with *E. coli* O157 infection (110–115). On the other hand, at least some genotypes isolated from clinical illness in humans are not represented in cattle, indicating the presence of non-cattle-associated reservoirs or sources of human infection (101).

Given the similar prevalence of cattle infection with clinical and bovine-biased lineages in the United States, it seems likely that people in this country are similarly exposed to both clinical and bovine-biased genotypes of *E. coli* O157 via ground beef, other cattle-origin meats, and cattle environments. Therefore, the preponderance of human disease associated with clinical genotypes in the United States may simply be the result of relatively higher virulence of clinical genotype strains. This possibility has two important implications for preharvest food safety: First, the virulence differences among *E. coli* O157 genotypes suggest the possibility or likelihood that these genotypes may respond differently to preharvest food safety interventions due to other intrinsic biological differences associated with their genotypes, and second, that in evaluating the efficacy of preharvest food safety interventions it is important to demonstrate specific reductions of clinical genotypes, rather than assuming that any prevalence or shedding reductions include clinical genotypes. Recent studies have shown that different lineages of *E. coli* O157 may differ in their ability to persist on cattle farms through various seasons, cattle diets, and animal husbandry practices. Vanaja et al. (116) demonstrated that certain cattle-associated genotypes expressed gene repertoires expected to improve their resistance to adverse environmental conditions in comparison to genotypes more commonly associated with clinical disease. Some genotypes of *E. coli* O157 are more resistant to stress factors such as heat and starvation compared to other genotypes (117). It is similarly possible that different lineages of *E. coli* O157 may respond differently to preharvest control measures such as vaccines, probiotics, bacteriophage treatments, or animal husbandry interventions. Therefore, further studies are required to (i) specifically target bacterial genetic factors that are responsible for the differential response of different lineages of *E. coli* O157 to various preharvest control measures, and (ii) to confirm that any preharvest control measures put into practice are effective against clinical genotypes. These studies will aid in identifying tools to improve the current preharvest food safety measures or formulate new better ways to reduce prevalence and shedding of *E. coli* O157 on cattle farms with consistent and reliable results.

CONCLUSION

Preharvest food safety for *E. coli* O157 is the term used for management systems that reduce the prevalence and/or magnitude of shedding of this agent by cattle populations to reduce the risk of contamination of cattle-derived food products and subsequent human exposures. Decades of research have provided a better understanding of the epidemiology and ecology of *E. coli* O157 on cattle farms, but only limited progress on preharvest food safety goals has been made. Although several interventions (certain feed ingredients, probiotics, and vaccines) have been identified with statistically significant impacts on cattle shedding of *E. coli* O157, the impact of these potential interventions remains insufficient due to their limited efficacy, practical difficulties with their implementation, or inconsistency in their results, leading to limited uptakes by producers. To date, the promise of the preharvest food safety approach to reducing human infection with *E. coli* O157 has not been fulfilled. A more holistic approach, with complex ecology and genetics of this bacterium in mind, is needed toward identifying true maintenance host populations and developing strategies to control *E. coli* O157 and other pathogenic STEC strains in these maintenance host populations.

ACKNOWLEDGMENT

We acknowledge support provided by the Agriculture and Food Research Initiative of the USDA National Institute of Agriculture grants 2009-04248 and 2010-04487.

We declare no conflicts of interest with regard to the manuscript.

CITATION

Besser TE, Schmidt CE, Shah DH, Shringi S. 2014. "Preharvest" food safety for *Escherichia coli* O157 and other pathogenic Shiga toxin-producing strains. Microbiol Spectrum 2(5): EHEC-0021-2013.

REFERENCES

1. **Elder RO, Keen JE, Siragusa GR, Barkocy-Gallagher GA, Koohmaraie M, Laegreid WW.** 2000. Correlation of enterohemorrhagic *Escherichia coli* O157 prevalence in feces, hides, and carcasses of beef cattle during processing. *Proc Natl Acad Sci USA* **97:**2999–3003.

2. **Woerner DR, Ransom JR, Sofos JN, Dewell GA, Smith GC, Salman MD, Belk KE.** 2006. Determining the prevalence of *Escherichia coli* O157 in cattle and beef from the feedlot to the cooler. *J Food Prot* **69:**2824–2827.

3. **Arthur TM, Keen JE, Bosilevac JM, Brichta-Harhay DM, Kalchayanand N, Shackelford SD, Wheeler TL, Nou X, Koohmaraie M.** 2009. Longitudinal study of *Escherichia coli* O157:H7 in a beef cattle feedlot and role of high-level shedders in hide contamination. *Appl Environ Microbiol* **75:**6515–6523.

4. **Grauke LJ, Wynia SA, Sheng HQ, Yoon JW, Williams CJ, Hunt CW, Hovde CJ.** 2003. Acid resistance of *Escherichia coli* O157:H7 from the gastrointestinal tract of cattle fed hay or grain. *Vet Microbiol* **95:**211–225.

5. **Hovde CJ, Austin PR, Cloud KA, Williams CJ, Hunt CW.** 1999. Effect of cattle diet on *Escherichia coli* O157:H7 acid resistance. *Appl Environ Microbiol* **65:**3233–3235.

6. **Jacob ME, Fox JT, Narayanan SK, Drouillard JS, Renter DG, Nagaraja TG.** 2008. Effects of feeding wet corn distillers grains with solubles with or without monensin and tylosin on the prevalence and antimicrobial susceptibilities of fecal foodborne pathogenic and commensal bacteria in feedlot cattle. *J Anim Sci* **86:**1182–1190.

7. **McAllister TA, Bach SJ, Stanford K, Callaway TR.** 2006. Shedding of *Escherichia coli* O157:H7 by cattle fed diets containing monensin or tylosin. *J Food Prot* **69:**2075–2083.

8. **Swyers KL, Carlson BA, Nightingale KK, Belk KE, Archibeque SL.** 2011. Naturally colonized beef cattle populations fed combinations of yeast culture and an ionophore in finishing diets containing dried distiller's grains with solubles had similar fecal shedding of *Escherichia coli* O157:H7. *J Food Prot* **74:**912–918.

9. **Edrington TS, Callaway TR, Bischoff KM, Genovese KJ, Elder RO, Anderson RC, Nisbet DJ.** 2003. Effect of feeding the ionophores monensin and laidlomycin propionate and the antimicrobial bambermycin to sheep experimentally infected with *E. coli* O157:H7 and Salmonella typhimurium. *J Anim Sci* **81:**553–560.

10. **Reinstein S, Fox JT, Shi X, Alam MJ, Renter DG, Nagaraja TG.** 2009. Prevalence of *Escherichia coli* O157:H7 in organically and naturally

raised beef cattle. *Appl Environ Microbiol* **75:**5421–5423.

11. **LeJeune JT, Christie NP.** 2004. Microbiological quality of ground beef from conventionally-reared cattle and "raised without antibiotics" label claims. *J Food Prot* **67:**1433–1437.

12. **Renter DG, Checkley SL, Campbell J, King R.** 2004. Shiga toxin-producing *Escherichia coli* in the feces of Alberta feedlot cattle. *Can J Vet Res* **68:**150–153.

13. **Renter DG, Sargeant JM, Hungerford LL.** 2004. Distribution of *Escherichia coli* O157:H7 within and among cattle operations in pasture-based agricultural areas. *Am J Vet Res* **65:**1367–1376.

14. **Renter DG, Sargeant JM, Oberst RD, Samadpour M.** 2003. Diversity, frequency, and persistence of *Escherichia coli* O157 strains from range cattle environments. *Appl Environ Microbiol* **69:**542–547.

15. **Hancock DD, Besser TE, Kinsel ML, Tarr PI, Rice DH, Paros MG.** 1994. The prevalence of *Escherichia coli* O157. H7 in dairy and beef cattle in Washington State. *Epidemiol Infect* **113:**199–207.

16. **Hutchison ML, Walters LD, Avery SM, Munro F, Moore A.** 2005. Analyses of livestock production, waste storage, and pathogen levels and prevalences in farm manures. *Appl Environ Microbiol* **71:**1231–1236.

17. **Herriott DE, Hancock DD, Ebel ED, Carpenter LV, Rice DH, Besser TE.** 1998. Association of herd management factors with colonization of dairy cattle by Shiga toxin-positive *Escherichia coli* O157. *J Food Prot* **61:**802–807.

18. **Ravva SV, Sarreal CZ, Duffy B, Stanker LH.** 2006. Survival of *Escherichia coli* O157:H7 in wastewater from dairy lagoons. *J Appl Microbiol* **101:**891–902.

19. **Kudva IT, Blanch K, Hovde CJ.** 1998. Analysis of *Escherichia coli* O157:H7 survival in ovine or bovine manure and manure slurry. *Appl Environ Microbiol* **64:**3166–3174.

20. **Berry ED, Wells JE.** 2010. *Escherichia coli* O157:H7: recent advances in research on occurrence, transmission, and control in cattle and the production environment. *Adv Food Nutri Res* **60:**67–117.

21. **Food Safety and Inspection Service.** 2010. *Preharvest Management Controls and Intervention options for Reducing* Escherichia coli *O157:H7 Shedding in Cattle.* U.S. Department of Agriculture, Food Safety and Inspection Service, Washington, DC http://www.fsis.usda.gov/shared/PDF/Reducing_Ecoli_Shedding_In_Cattle_0510.pdf?redirecthttp=true.

22. **LeJeune JT, Wetzel AN.** 2007. Preharvest control of *Escherichia coli* O157 in cattle. *J Anim Sci* **85:**E73–E80.

23. **Jacob ME, Callaway TR, Nagaraja TG.** 2009. Dietary interactions and interventions affecting *Escherichia coli* O157 colonization and shedding in cattle. *Foodborne Pathog Dis* **6:**785–792.

24. **Callaway TR, Carr MA, Edrington TS, Anderson RC, Nisbet DJ.** 2009. Diet, *Escherichia coli* O157:H7, and cattle: a review after 10 years. *Curr Issues Mol Biol* **11:**67–79.

25. **Dean-Nystrom EA, Bosworth BT, Moon HW, O'Brien AD.** 1998. *Escherichia coli* O157:H7 requires intimin for enteropathogenicity in calves. *Infect Immun* **66:**4560–4563.

26. **Cornick NA, Booher SL, Moon HW.** 2002. Intimin facilitates colonization by *Escherichia coli* O157:H7 in adult ruminants. *Infect Immun* **70:**2704–2707.

27. **Dziva F, van Diemen PM, Stevens MP, Smith AJ, Wallis TS.** 2004. Identification of *Escherichia coli* O157:H7 genes influencing colonization of the bovine gastrointestinal tract using signature-tagged mutagenesis. *Microbiology* **150:**3631–3645.

28. **Naylor SW, Roe AJ, Nart P, Spears K, Smith DG, Low JC, Gally DL.** 2005. *Escherichia coli* O157:H7 forms attaching and effacing lesions at the terminal rectum of cattle and colonization requires the LEE4 operon. *Microbiology* **151:**2773–2781.

29. **Stevens MP, Roe AJ, Vlisidou I, van Diemen PM, La Ragione RM, Best A, Woodward MJ, Gally DL, Wallis TS.** 2004. Mutation of toxB and a truncated version of the efa-1 gene in *Escherichia coli* O157:H7 influences the expression and secretion of locus of enterocyte effacement-encoded proteins but not intestinal colonization in calves or sheep. *Infect Immun* **72:**5402–5411.

30. **Sheng H, Lim JY, Knecht HJ, Li J, Hovde CJ.** 2006. Role of *Escherichia coli* O157:H7 virulence factors in colonization at the bovine terminal rectal mucosa. *Infect Immun* **74:**4685–4693.

31. **Frankel G, Phillips AD, Rosenshine I, Dougan G, Kaper JB, Knutton S.** 1998. Enteropathogenic and enterohaemorrhagic *Escherichia coli*: more subversive elements. *Mol Microbiol* **30:**911–921.

32. **Caron E, Crepin VF, Simpson N, Knutton S, Garmendia J, Frankel G.** 2006. Subversion of actin dynamics by EPEC and EHEC. *Curr Opin Microbiol* **9:**40–45.

33. **Neilands JB.** 1995. Siderophores: structure and function of microbial iron transport compounds. *J Biol Chem* **270:**26723–26726.

34. **Varela NP, Dick P, Wilson J.** 2013. Assessing the existing information on the efficacy of bovine vaccination against *Escherichia coli* O157:H7—a systematic review and meta-analysis. *Zoonoses Public Health* **60:**253–268.

35. **Vogstad AR, Moxley RA, Erickson GE, Klopfenstein TJ, Smith DR.** 2013. Assessment of heterogeneity of efficacy of a three-dose regimen

of a type III secreted protein vaccine for reducing STEC O157 in feces of feedlot cattle. *Foodborne Pathog Dis* **10:**678–683.

36. **Vogstad AR, Moxley RA, Erickson GE, Klopfenstein TJ, Smith DR.** 2013. Stochastic simulation model comparing distributions of STEC O157 faecal shedding prevalence between cattle vaccinated with type Iii secreted protein vaccines and non-vaccinated cattle. *Zoonoses Public Health* **61:**283–289.

37. **Snedeker KG, Campbell M, Sargeant JM.** 2012. A systematic review of vaccinations to reduce the shedding of *Escherichia coli* O157 in the faeces of domestic ruminants. *Zoonoses Public Health* **59:**126–138.

38. **Arthur TM, Bosilevac JM, Brichta-Harhay DM, Guerini MN, Kalchayanand N, Shackelford SD, Wheeler TL, Koohmaraie M.** 2007. Transportation and lairage environment effects on prevalence, numbers, and diversity of *Escherichia coli* O157:H7 on hides and carcasses of beef cattle at processing. *J Food Prot* **70:**280–286.

39. **Jordan D, McEwen SA, Lammerding AM, McNab WB, Wilson JB.** 1999. A simulation model for studying the role of pre-slaughter factors on the exposure of beef carcasses to human microbial hazards. *Prev Vet Med* **41:**37–54.

40. **Matthews L, Reeve R, Gally DL, Low JC, Woolhouse ME, McAteer SP, Locking ME, Chase-Topping ME, Haydon DT, Allison LJ, Hanson MF, Gunn GJ, Reid SW.** 2013. Predicting the public health benefit of vaccinating cattle against *Escherichia coli* O157. *Proc Natl Acad Sci U S A* **110:**16265–16270.

41. **Withee J, Williams M, Disney T, Schlosser W, Bauer N, Ebel E.** 2009. Streamlined analysis for evaluating the use of preharvest interventions intended to prevent *Escherichia coli* O157:H7 illness in humans. *Foodborne Pathog Dis* **6:**817–825.

42. **Mead PS, Slutsker L, Dietz V, McCaig LF, Bresee JS, Shapiro C, Griffin PM, Tauxe RV.** 1999. Food-related illness and death in the United States. *Emerg Infect Dis* **5:**607–625.

43. **Scallan E, Hoekstra RM, Angulo FJ, Tauxe RV, Widdowson MA, Roy SL, Jones JL, Griffin PM.** 2011. Foodborne illness acquired in the United States—major pathogens. *Emerg Infect Dis* **17:**7–15.

44. **Torres AG, Kanack KJ, Tutt CB, Popov V, Kaper JB.** 2004. Characterization of the second long polar (LP) fimbriae of *Escherichia coli* O157:H7 and distribution of LP fimbriae in other pathogenic E. coli strains. *FEMS Microbiol Lett* **238:**333–344.

45. **Lloyd SJ, Ritchie JM, Torres AG.** 2012. Fimbriation and curliation in *Escherichia coli* O157:H7: a paradigm of intestinal and environmental colonization. *Gut Microbes* **3:**272–276.

46. **Mahajan A, Currie CG, Mackie S, Tree J, McAteer S, McKendrick I, McNeilly TN, Roe A, La Ragione RM, Woodward MJ, Gally DL, Smith DG.** 2009. An investigation of the expression and adhesin function of H7 flagella in the interaction of *Escherichia coli* O157 : H7 with bovine intestinal epithelium. *Cell Microbiol* **11:**121–137.

47. **Erdem AL, Avelino F, Xicohtencatl-Cortes J, Giron JA.** 2007. Host protein binding and adhesive properties of H6 and H7 flagella of attaching and effacing *Escherichia coli. J Bacteriol* **189:**7426–7435.

48. **McNeilly TN, Naylor SW, Mahajan A, Mitchell MC, McAteer S, Deane D, Smith DG, Low JC, Gally DL, Huntley JF.** 2008. *Escherichia coli* O157:H7 colonization in cattle following systemic and mu cosal immunization with purified H7 flagellin. *Infect Immun* **76:**2594–2602.

49. **Barkocy-Gallagher GA, Arthur TM, Rivera-Betancourt M, Nou X, Shackelford SD, Wheeler TL, Koohmaraie M.** 2003. Seasonal prevalence of Shiga toxin-producing *Escherichia coli*, including O157:H7 and non-O157 serotypes, and *Salmonella* in commercial beef processing plants. *J Food Prot* **66:**1978–1986.

50. **Haydon DT, Cleaveland S, Taylor LH, Laurenson MK.** 2002. Identifying reservoirs of infection: a conceptual and practical challenge. *Emerg Infect Dis* **8:**1468–1473.

51. **Armstrong GL, Hollingsworth J, Morris JG Jr.** 1996. Emerging foodborne pathogens: *Escherichia coli* O157:H7 as a model of entry of a new pathogen into the food supply of the developed world. *Epidemiol Rev* **18:**29–51.

52. **Sargeant JM, Amezcua MR, Rajic A, Waddell L.** 2007. Pre-harvest interventions to reduce the shedding of *E. coli* O157 in the faeces of weaned domestic ruminants: a systematic review. *Zoonoses Public Health* **54:**260–277.

53. **Lim JY, Li J, Sheng H, Besser TE, Potter K, Hovde CJ.** 2007. *Escherichia coli* O157:H7 colonization at the rectoanal junction of long-duration culture-positive cattle. *Appl Environ Microbiol* **73:**1380–1382.

54. **Jacob ME, Renter DG, Nagaraja TG.** 2010. Animal- and truckload-level associations between *Escherichia coli* O157:H7 in feces and on hides at harvest and contamination of preevisceration beef carcasses. *J Food Prot* **73:**1030–1037.

55. **Hancock DD, Besser TE, Rice DH, Herriott DE, Tarr PI.** 1997. A longitudinal study of *Escherichia coli* O157 in fourteen cattle herds. *Epidemiol Infect* **118:**193–195.

56. **Hancock DD, Rice DH, Thomas LA, Dargatz DA, Besser TE.** 1997. Epidemiology of *Escherichia coli* O157:H7 in feedlot cattle. *J. Food Protect* **60:**462–465.

57. **LeJeune JT, Besser TE, Rice DH, Berg JL, Stilborn RP, Hancock DD.** 2004. Longitudinal study of fecal shedding of *Escherichia coli* O157:H7 in feedlot cattle: predominance and persistence of specific clonal types despite massive cattle population turnover. *Appl Environ Microbiol* **70:**377–384.

58. **Hovde CJ, Sheng H, Baker K, Deobald C, Davis MA, Minnich SA, Besser TE.** 2012. *Experimental evaluation of the basis of seasonal vbariation in bovine shedding of STEC O157, abstr P206.* 8th International Symposium on Shiga Toxin-Producing *Escherichia coli* Infections, Amsterdam, The Netherlands.

59. **Davis MA, Cloud-Hansen KA, Carpenter J, Hovde CJ.** 2005. *Escherichia coli* O157:H7 in environments of culture-positive cattle. *Appl Environ Microbiol* **71:**6816–6822.

60. **Van Donkersgoed J, Berg J, Potter A, Hancock D, Besser T, Rice D, LeJeune J, Klashinsky S.** 2001. Environmental sources and transmission of *Escherichia coli* O157 in feedlot cattle. *Can Vet J* **42:**714–720.

61. **LeJeune JT, Besser TE, Hancock DD.** 2001. Cattle water troughs as reservoirs of *Escherichia coli* O157. *Appl Environ Microbiol* **67:**3053–3057.

62. **Gautam R, Bani-Yaghoub M, Neill WH, Dopfer D, Kaspar C, Ivanek R.** 2011. Modeling the effect of seasonal variation in ambient temperature on the transmission dynamics of a pathogen with a free-living stage: example of *Escherichia coli* O157: H7 in a dairy herd. *Prev Vet Med* **102:**10–21.

63. **LeJeune JT, Besser TE, Merrill NL, Rice DH, Hancock DD.** 2001. Livestock drinking water microbiology and the factors influencing the quality of drinking water offered to cattle. *J Dairy Sci* **84:**1856–1862.

64. **Steinberg KM, Levin BR.** 2007. Grazing protozoa and the evolution of the *Escherichia coli* O157:H7 Shiga toxin-encoding prophage. *Proc Biol Sci* **274:**1921–1929.

65. **Lainhart W, Stolfa G, Koudelka GB.** 2009. Shiga toxin as a bacterial defense against a eukaryotic predator, *Tetrahymena thermophila. J Bacteriol* **191:**5116–5122.

66. **Ravva SV, Sarreal CZ, Mandrell RE.** 2010. Identification of protozoa in dairy lagoon wastewater that consume *Escherichia coli* O157:H7 preferentially. *PLoS One* **5:**e15671.

67. **Steinert M, Birkness K, White E, Fields B, Quinn F.** 1998. *Mycobacterium avium* bacilli grow saprozoically in coculture with *Acanthamoeba polyphaga* and survive within cyst walls. *Appl Environ Microbiol* **64:**2256–2261.

68. **Kilvington S, Price J.** 1990. Survival of *Legionella pneumophila* within cysts of *Acanthamoeba polyphaga* following chlorine exposure. *J Appl Bacteriol* **68:**519–525.

69. **Pushkareva VI, Ermolaeva SA.** 2010. *Listeria monocytogenes* virulence factor Listeriolysin O favors bacterial growth in co-culture with the ciliate *Tetrahymena pyriformis*, causes protozoan encystment and promotes bacterial survival inside cysts. *BMC Microbiol* **10:**26.

70. **El-Etr SH, Margolis JJ, Monack D, Robison RA, Cohen M, Moore E, Rasley A.** 2009. *Francisella tularensis* type A strains cause the rapid encystment of *Acanthamoeba castellanii* and survive in amoebal cysts for three weeks postinfection. *Appl Environ Microbiol* **75:**7488–7500.

71. **Kenney SJ, Anderson GL, Williams PL, Millner PD, Beuchat LR.** 2005. Persistence of *Escherichia coli* O157:H7, *Salmonella* Newport, and *Salmonella* Poona in the gut of a free-living nematode, *Caenorhabditis elegans*, and transmission to progeny and uninfected nematodes. *Int J Food Microbiol* **101:**227–236.

72. **Gibbs DS, Anderson GL, Beuchat LR, Carta LK, Williams PL.** 2005. Potential role of *Diploscapter* sp. strain LKC25, a bacterivorous nematode from soil, as a vector of food-borne pathogenic bacteria to preharvest fruits and vegetables. *Appl Environ Microbiol* **71:**2433–2437.

73. **Smith T.** 1908. The housefly as an agent in the dissemination of infectious disease. *Am J Public Hygiene* **18:**312–324.

74. **West LS.** 1951. *The Housefly, Its Natural History, Medical Importance, and Control.* Comstock Pub Co, Ithaca, NY.

75. **DeBartolo A.** 1986. Buzz off! the housefly has made a pest of himself for 25 million years. *Chicago Tribune.* http://articles.chicagotribune.com/1986-06-05/features/8602090713_1_maggots-labor-day-picnic-flies

76. **Alam MJ, Zurek L.** 2004. Association of *Escherichia coli* O157:H7 with houseflies on a cattle farm. *Appl Environ Microbiol* **70:**7578–7580.

77. **Wada A.** 1997. [Molecular analysis of enterohemorrhagic *Escherichia coli* O157:H7 isolates in Japan 1996 using pulsed-field gel electrophoresis]. *Nippon Rinsho* **55:**665–670 (In Japanese).

78. **Kobayashi M, Sasaki T, Saito N, Tamura K, Suzuki K, Watanabe H, Agui N.** 1999. Houseflies: not simple mechanical vectors of enterohemorrhagic *Escherichia coli* O157:H7. *Am J Trop Med Hyg* **61:**625–629.

79. **LeJeune JT, Homan J, Linz G, Pearl DL.** 2008. Role of the European starling in the transmission of E. coli O157 on dairy farms, p 31–34. *In* Timm RM, Madon MB (ed), *Proceedings of the 23rd Vertebrate Pest Conference*, University of California, Davis, CA.

80. **Stavric S, Buchanan B, Gleeson TM.** 1993. Intestinal colonization of young chicks with *Escherichia coli* O157:H7 and other verotoxin-producing serotypes. *J Appl Bacteriol* **74:**557–563.

81. **Best A, Clifford D, Crudgington B, Cooley WA, Nunez A, Carter B, Weyer U, Woodward MJ, La Ragione RM.** 2009. Intermittent *Escherichia coli* O157:H7 colonisation at the terminal rectum mucosa of conventionally-reared lambs. *Vet Res* **40**:9.

82. **Keene WE, Sazie E, Kok J, Rice DH, Hancock DD, Balan VK, Zhao T, Doyle MP.** 1997. An outbreak of *Escherichia coli* O157:H7 infections traced to jerky made from deer meat. *JAMA* **277**:1229–1231.

83. **Rabatsky-Ehr T, Dingman D, Marcus R, Howard R, Kinney A, Mshar P.** 2002. Deer meat as the source for a sporadic case of *Escherichia coli* O157:H7 infection, Connecticut. *Emerg Infect Dis* **8**:525–527.

84. **Chapman PA, Siddons CA, Gerdan Malo AT, Harkin MA.** 1997. A 1-year study of *Escherichia coli* O157 in cattle, sheep, pigs and poultry. *Epidemiol Infect* **119**:245–250.

85. **Booher SL, Cornick NA, Moon HW.** 2002. Persistence of *Escherichia coli* O157:H7 in experimentally infected swine. *Vet Microbiol* **89**:69–81.

86. **Jay MT, Cooley M, Carychao D, Wiscomb GW, Sweitzer RA, Crawford-Miksza L, Farrar JA, Lau DK, O'Connell J, Millington A, Asmundson RV, Atwill ER, Mandrell RE.** 2007. *Escherichia coli* O157:H7 in feral swine near spinach fields and cattle, central California coast. *Emerg Infect Dis* **13**:1908–1911.

87. **Levine P, Rose B, Green S, Ransom G, Hill W.** 2001. Pathogen testing of ready-to-eat meat and poultry products collected at federally inspected establishments in the United States, 1990 to 1999. *J Food Prot* **64**:1188–1193.

88. **Tutenel AV, Pierard D, Van Hoof J, Cornelis M, De Zutter L.** 2003. Isolation and molecular characterization of *Escherichia coli* O157 isolated from cattle, pigs and chickens at slaughter. *Int J Food Microbiol* **84**:63–69.

89. **Whittam TS, Wolfe ML, Wachsmuth IK, Orskov F, Orskov I, Wilson RA.** 1993. Clonal relationships among *Escherichia coli* strains that cause hemorrhagic colitis and infantile diarrhea. *Infect Immun* **61**:1619–1629.

90. **Feng P, Lampel KA, Karch H, Whittam TS.** 1998. Genotypic and phenotypic changes in the emergence of *Escherichia coli* O157:H7. *J Infect Dis* **177**:1750–1753.

91. **Kim J, Nietfeldt J, Benson AK.** 1999. Octamer-based genome scanning distinguishes a unique subpopulation of *Escherichia coli* O157:H7 strains in cattle. *Proc Natl Acad Sci USA* **96**:13288–13293.

92. **Kim J, Nietfeldt J, Ju J, Wise J, Fegan N, Desmarchelier P, Benson AK.** 2001. Ancestral divergence, genome diversification, and phylogeographic variation in subpopulations of sorbitol-negative, beta-glucuronidase-negative enterohemorrhagic *Escherichia coli* O157. *J Bacteriol* **183**:6885–6897.

93. **Ohnishi M, Terajima J, Kurokawa K, Nakayama K, Murata T, Tamura K, Ogura Y, Watanabe H, Hayashi T.** 2002. Genomic diversity of enterohemorrhagic *Escherichia coli* O157 revealed by whole genome PCR scanning. *Proc Natl Acad Sci USA* **99**:17043–17048.

94. **Kudva IT, Evans PS, Perna NT, Barrett TJ, DeCastro GJ, Ausubel FM, Blattner FR, Calderwood SB.** 2002. Polymorphic amplified typing sequences provide a novel approach to *Escherichia coli* O157:H7 strain typing. *J Clin Microbiol* **40**:1152–1159.

95. **Shaikh N, Tarr PI.** 2003. *Escherichia coli* O157:H7 Shiga toxin-encoding bacteriophages: integrations, excisions, truncations, and evolutionary implications. *J Bacteriol* **185**:3596–3605.

96. **Yang Z, Kovar J, Kim J, Nietfeldt J, Smith DR, Moxley RA, Olson ME, Fey PD, Benson AK.** 2004. Identification of common subpopulations of non-sorbitol-fermenting, beta-glucuronidase-negative *Escherichia coli* O157:H7 from bovine production environments and human clinical samples. *Appl Environ Microbiol* **70**:6846–6854.

97. **Wick LM, Qi W, Lacher DW, Whittam TS.** 2005. Evolution of genomic content in the stepwise emergence of *Escherichia coli* O157:H7. *J Bacteriol* **187**:1783–1791.

98. **Zhang Y, Laing C, Steele M, Ziebell K, Johnson R, Benson AK, Taboada E, Gannon VP.** 2007. Genome evolution in major *Escherichia coli* O157:H7 lineages. *BMC Genomics* **8**:121.

99. **Kotewicz ML, Mammel MK, LeClerc JE, Cebula TA.** 2008. Optical mapping and 454 sequencing of *Escherichia coli* O157:H7 isolates linked to the US 2006 spinach-associated outbreak. *Microbiology* **154**:3518–3528.

100. **Manning SD, Motiwala AS, Springman AC, Qi W, Lacher DW, Ouellette LM, Mladonicky JM, Somsel P, Rudrik JT, Dietrich SE, Zhang W, Swaminathan B, Alland D, Whittam TS.** 2008. Variation in virulence among clades of *Escherichia coli* O157:H7 associated with disease outbreaks. *Proc Natl Acad Sci USA* **105**:4868–4873.

101. **Bono JL, Smith TP, Keen JE, Harhay GP, McDaneld TG, Mandrell RE, Jung WK, Besser TE, Gerner-Smidt P, Bielaszewska M, Karch H, Clawson ML.** 2012. Phylogeny of Shiga toxin-producing *Escherichia coli* O157 isolated from cattle and clinically ill humans. *Mol Biol Evol* **29**:2047–2062.

102. **Karama M, Gyles CL.** 2010. Methods for genotyping verotoxin-producing *Escherichia coli*. *Zoonoses Public Health* **57**:447–462.

103. **Chapman PA, Siddons CA.** 1994. A comparison of strains of *Escherichia coli* O157 from humans and cattle in Sheffield, United Kingdom. *J Infect Dis* **170:**251–253.

104. **Besser TE, Shaikh N, Holt NJ, Tarr PI, Konkel ME, Malik-Kale P, Walsh CW, Whittam TS, Bono JL.** 2007. Greater diversity of Shiga toxin-encoding bacteriophage insertion sites among *Escherichia coli* O157:H7 isolates from cattle than in those from humans. *Appl Environ Microbiol* **73:**671–679.

105. **Lejeune JT, Abedon ST, Takemura K, Christie NP, Sreevatsan S.** 2004. Human *Escherichia coli* O157:H7 genetic marker in isolates of bovine origin. *Emerg Infect Dis* **10:**1482–1485.

106. **Bono JL, Keen JE, Clawson ML, Durso LM, Heaton MP, Laegreid WW.** 2007. Association of *Escherichia coli* O157:H7 tir polymorphisms with human infection. *BMC Infect Dis* **7:**98.

107. **Clawson ML, Keen JE, Smith TP, Durso LM, McDaneld TG, Mandrell RE, Davis MA, Bono JL.** 2009. Phylogenetic classification of *Escherichia coli* O157:H7 strains of human and bovine origin using a novel set of nucleotide polymorphisms. *Genome Biol* **10:**R56.

108. **Whitworth J, Zhang Y, Bono J, Pleydell E, French N, Besser T.** 2010. Diverse genetic markers concordantly identify bovine origin *Escherichia coli* O157 genotypes underrepresented in human disease. *Appl Environ Microbiol* **76:**361–365.

109. **Laing CR, Buchanan C, Taboada EN, Zhang Y, Karmali MA, Thomas JE, Gannon VP.** 2009. In silico genomic analyses reveal three distinct lineages of *Escherichia coli* O157:H7, one of which is associated with hyper-virulence. *BMC Genomics* **10:**287.

110. **Whitworth JH, Fegan N, Keller J, Gobius KS, Bono JL, Call DR, Hancock DD, Besser TE.** 2008. International comparison of clinical, bovine, and environmental *Escherichia coli* O157 isolates on the basis of Shiga toxin-encoding bacteriophage insertion site genotypes. *Appl Environ Microbiol* **74:**7447–7450.

111. **Leotta GA, Miliwebsky ES, Chinen I, Espinosa EM, Azzopardi K, Tennant SM, Robins-Browne RM, Rivas M.** 2008. Characterisation of Shiga toxin-producing *Escherichia coli* O157 strains isolated from humans in Argentina, Australia and New Zealand. *BMC Microbiol* **8:**46.

112. **Mellor GE, Sim EM, Barlow RS, D'Astek BA, Galli L, Chinen I, Rivas M, Gobius KS.** 2012. Phylogenetically related Argentinean and Australian *Escherichia coli* O157 isolates are distinguished by virulence clades and alternative Shiga toxin 1 and 2 prophages. *Appl Environ Microbiol* **78:**4724–4731.

113. **Mellor GE, Besser TE, Davis MA, Beavis B, Jung W, Smith HV, Jennison AV, Doyle CJ, Chandry PS, Gobius KS, Fegan N.** 2013. Multilocus genotype analysis of *Escherichia coli* O157 isolates from Australia and the United States provides evidence of geographic divergence. *Appl Environ Microbiol* **79:**5050–5058.

114. **Franz E, van Hoek AH, van der Wal FJ, de Boer A, Zwartkruis-Nahuis A, van der Zwaluw K, Aarts HJ, Heuvelink AE.** 2012. Genetic features differentiating bovine, food, and human isolates of shiga toxin-producing *Escherichia coli* O157 in The Netherlands. *J Clin Microbiol* **50:**772–780.

115. **Lee K, French NP, Hara-Kudo Y, Iyoda S, Kobayashi H, Sugita-Konishi Y, Tsubone H, Kumagai S.** 2011. Multivariate analyses revealed distinctive features differentiating human and cattle isolates of Shiga toxin-producing *Escherichia coli* O157 in Japan. *J Clin Microbiol* **49:**1495–1500.

116. **Vanaja SK, Springman AC, Besser TE, Whittam TS, Manning SD.** 2010. Differential expression of virulence and stress fitness genes between *Escherichia coli* O157:H7 strains with clinical or bovine-biased genotypes. *Appl Environ Microbiol* **76:**60–68.

117. **Lee K, French NP, Jones G, Hara-Kudo Y, Iyoda S, Kobayashi H, Sugita-Konishi Y, Tsubone H, Kumagai S.** 2012. Variation in stress resistance patterns among stx genotypes and genetic lineages of shiga toxin-producing *Escherichia coli* O157. *Appl Environ Microbiol* **78:**3361–3368.

Peri- and Postharvest Factors in the Control of Shiga Toxin-Producing *Escherichia coli* in Beef

22

RODNEY A. MOXLEY[1] and GARY R. ACUFF[2]

ATTRIBUTION OF BEEF AS A SOURCE OF STEC-ASSOCIATED ILLNESS

Based on outbreak data acquired from 1998 to 2008 by the U.S. Centers for Disease Control and Prevention (CDC), known food-borne etiological agents were estimated to have caused 4,589 outbreaks, 9,638,301 cases, 57,462 hospitalizations, and 1,451 deaths per year in the United States (1). From these data, it was estimated that 482,199 cases (5.0%), 2,650 hospitalizations (0.03%), and 51 deaths (0.0005%) were attributable to bacteria consumed from beef. Of the 4,589 outbreaks, 103 (2.2%) and 3 (0.065%) were further attributable to beef-acquired Shiga toxin-producing *Escherichia coli* (STEC) O157 and non-O157 strains. Of the outbreaks caused by STEC O157 and non-O157 STEC strains for all food commodities combined, 103 of 186 (55.3%) and 3 of 6 (50%), respectively, were attributable to beef. On the basis of data acquired from the CDC from 2000 to 2008, of 9,388,075 cases of domestically acquired food-borne illness in the United States caused by 31 major pathogens, 63,153 (0.67%) were due to STEC O157 and 112,752 (1.20%) were due to non-O157 STEC infection (2). In the same study, of 35,796 hospitalizations, 2,138 (5.97%) were due to STEC O157 and 271 (0.76%) were due to non-O157 STEC; of 861 deaths, 20 (2.32%) were due to STEC O157 and 0 were due to non-O157 STEC infection.

[1]School of Veterinary Medicine and Biomedical Sciences, University of Nebraska-Lincoln, Lincoln, NE 68583-0905; [2]Department of Animal Science, Texas A&M University, College Station, TX 77843-2471.

Enterohemorrhagic Escherichia coli *and Other Shiga Toxin-Producing* E. coli
Edited by Vanessa Sperandio and Carolyn J. Hovde
© 2015 American Society for Microbiology, Washington, DC
doi:10.1128/microbiolspec.EHEC-0017-2013

PERIHARVEST STEC PREVALENCE

Ruminant Intestines and Feces Are the Major Reservoir of STEC

STEC strains are part of the normal intestinal flora of healthy cattle and sheep and, consequently, are shed in their feces (3–5). *E. coli* O157:H7 is the prototype of STEC; however, it is only one of the more than 435 serotypes and 120 O serogroups of STEC known to colonize the intestines of cattle (3, 4). More than 470 different STEC serotypes have been isolated from humans, with most of these serotypes identified in cattle, beef, or both, and more than 100 were associated with human disease (4, 6). In North America, cattle are a major reservoir of STEC; however, in countries such as Australia, sheep are of greater significance (4). Although many different serotypes of STEC are carried and subclinically shed by healthy cattle of all ages, a subset naturally causes diarrheal disease in young calves (7). Experimentally, young calves and, especially, neonates develop more extensive intestinal colonization with STEC than do older animals. This increased susceptibility may be due to lack of immunity and to a reduced rate of epithelial cell turnover compared to that of the older animal (8, 9). Animals that shed higher concentrations of STEC in their feces pose a greater risk for transmission to other animals around them, and also a greater risk of meat contamination during the slaughtering process (10, 11). Colonization of the recto-anal junction in cattle of different ages with *E. coli* O157:H7 has been associated with higher-level shedding and increased risk of transmission, the so-called "super-shedder" state (12).

Fecal, Hide, and Carcass Prevalence Is Correlated, and Hides Are the Major Vehicles of Contamination of Carcass Surfaces

The exposed surface of the hide and hair of cattle accumulates dust, dirt, and fecal material. Since the 1970s the hide has been recognized as the primary source of bacterial contamination of carcass surfaces, contamination that occurs during the process of hide removal at slaughter (13). Many studies support this hypothesis and have led to the conclusion that the hide is the major source of carcass contamination by *E. coli* O157:H7 and non-O157 STEC (14–18). Carcass surfaces coming in contact with droplets aerosolized from hides during removal were shown to contain higher counts of aerobic bacteria and *Enterobacteriaceae* and also a higher prevalence of pathogens (19). *E. coli* O157:H7 has been isolated from up to 23% of air samples from hide removal areas, in contrast to 0% for air samples from evisceration and fabrication areas at different plants (19). Aerosolization of droplets from hides occurs especially through the use of equipment and processes that remove the hide with considerable force, such as mechanical and hydraulic hide pullers.

Fecal, hide, and carcass prevalence of *E. coli* O157 is directly correlated, and most contamination is thought to occur from animals within the same lots (16, 20). In one study, the frequency of *E. coli* O157:H7 or O157:NM in feces and on hides within groups of fed cattle from single sources (lots) presented for slaughter at meat processing plants in the midwestern United States was determined, as was the frequency of carcass contamination during processing from cattle within the same lots (16). In that study, *E. coli* O157 prevalence was 28% in feces and 11% on hides. Carcass samples were taken at three points during processing: preevisceration, postevisceration before antimicrobial intervention, and postprocessing after carcasses entered the cooler. Of 30 lots sampled, 87% had at least one *E. coli* O157-positive preevisceration sample, 57% of lots were positive postevisceration, and 17% had positive postprocessing samples. Prevalence of *E. coli* O157 in the three postprocessing samples was 43%, 18%, and 2%, respectively. Reduction in carcass prevalence from preevisceration to postprocessing suggested that sanitary procedures were effective within the processing plants.

Most *E. coli* O157 strains that contaminate carcasses originate from animals within the same lot going to slaughter. Evidence for this was provided in a study that compared pulsed-field gel electrophoresis (PFGE) patterns of isolates from feces, hides, and carcasses (20). Approximately 68% of *E. coli* O157 isolates from carcasses had the same PFGE pattern as those from feces and hides. On individual carcasses, isolates recovered before evisceration matched 65.3% of those recovered after evisceration. Also, on individual carcasses, 66.7% of the isolates recovered in the cooler matched those recovered before evisceration. PFGE genotyping confirmed that the majority of *E. coli* O157 found on the carcass is the result of preevisceration contamination. Hence, the data indicated the need to apply additional in-plant intervention strategies aimed at preventing direct contamination of the carcasses early in processing.

In a study in which beef carcass sponge samples were collected at four large processing plants in the United States, 53.9% of preevisceration samples and 8.3% of postprocessing samples were positive for non-O157 STEC (21). Altogether, 361 non-O157 STEC isolates were recovered, belonging to 41 different O serogroups. O serogroups that previously had been associated with human disease accounted for 49% of the isolates. The significant decrease in prevalence of STEC detected from preevisceration to postprocessing was attributed to the various interventions in place, viz., steam vacuum, hot water, organic acids, and steam pasteurization.

In another study of fed beef cattle harvested at three midwestern beef processing plants, the prevalence of *E. coli* O157:H7 in samples was 5.9% in feces, 60.6% on hides, 26.7% on dehided carcass surfaces prior to the preevisceration wash, and 1.2% on carcasses sampled at chilling (postintervention) at concentrations of approximately <3.0 cells per 100 cm^2 (15). Somewhat different results were found for the prevalence of *E. coli* O157:H7

and non-O157 STEC in feces and hides. The highest *E. coli* O157:H7 prevalence in feces was detected in the summer, and the highest on hides was from spring through fall. In contrast, non-O157 STEC prevalence in feces was lower in the summer and higher in the spring and fall and also peaked in the fall on hides. The efficiency in the recovery of non-O157 STEC could have influenced these results. The prevalence of Shiga toxin gene (*stx*)-positive bacteria as detected by PCR was, in descending order, 96.6% on carcasses prior to evisceration, 92.0% on hides, 34.3% in feces, and 16.2% on carcasses after interventions. The approximate concentration of non-O157 STEC and *stx*-positive cells on postintervention carcasses was ≥3.0 cells per 100 cm^2 for only 4% of carcasses. Pathogen prevalence on hides may reflect several sources of contamination, such as soils, feces from other animals, and possibly lairage. However, these results further confirmed that hides were the major source of contamination for beef carcasses and that postharvest interventions used by the beef industry were effective.

The hide-level prevalence of STEC has been shown to increase as a result of commingling cattle, e.g., in pens, sales barns, trucks, and lairage at the abattoir, in several studies (22–24). The effect of commingling on hide-level prevalence during lairage at the abattoir was demonstrated experimentally with nonpathogenic bacteria carrying antibiotic-resistant markers (viz., *E. coli* K-12 and *Pseudomonas fluorescens*) (22). At the abattoir, the initial prevalence of animals positive for the hide marker (11.1%) inoculated at unloading increased to 100% on hides before skinning and to 88.8% on skinned carcasses. In addition, another marker inoculated on environmental surfaces in lairage pens, races, and the stunning box was detected on 83.3% of hides before skinning and 88.8% of skinned carcasses. These results demonstrated that both the livestock market process and the unloading-to-skinning process at abattoirs can facilitate the extensive spread of microbial contamination on hides, not just within, but also between, batches of animals.

The significant role of cross-contamination of hides at the abattoir has also been demonstrated by characterization of the phage and verocytotoxin (Shiga toxin) types of isolates (24). The majority of cattle (84%) were found to have subtypes of STEC O157 on their hides that had not been found previously in any animal from the farm of origin, strongly suggesting that contamination occurred once animals had left the farm of origin. Several variables and factors were found to be strongly associated with cross-contamination of cattle hides at the univariate level: commercial transport to slaughter; transport with other animals; use of a crush (restraining crate used when reading ear tags); line automation; and increasing slaughterhouse throughput.

Studies on the effects of transportation on hide contamination with STEC O157 have yielded variable results. Although transport stress may possibly lead to immunosuppression (25), and transport and fasting are believed to increase fecal shedding of STEC O157 (23, 26, 27), other studies have found that transport of cattle had no influence on shedding (28–30). Stanford et al. (31) concluded that transportation did not affect prevalence of hide contamination with *E. coli* O157, and the feedlot pen prevalence had a greater effect on hide contamination at the slaughter plant than transportation factors, including temperature-humidity index, loading density, and duration of transport.

POSTHARVEST STEC PREVALENCE

The United States Department of Agriculture (USDA), Food Safety and Inspection Service (FSIS) in October 1994 declared *E. coli* O157:H7 an adulterant in raw ground beef, and began a sampling program to test for this organism in raw ground beef prepared in federally inspected plants and retail stores. In January 1999, the FSIS expanded that declaration to include nonintact beef, such as

mechanically tenderized or reconstructed products (32). Based on STEC isolates submitted to the CDC from 1983 to 2002, six O groups comprising 13 serotypes were identified as the cause of 71% of non-O157 STEC disease in the United States (6). The six O serogroups, O26, O111, O103, O121, O45, and O145, were responsible for 22, 16, 12, 8, 7, and 5% of cases, respectively. Serotypes included O26:H11 or nonmotile (NM); O45:H2 or NM; O103:H2, H11, H25, or NM; O111:H8 or NM; O121:H19 or H7; and O145:NM. On September 20, 2011, the USDA-FSIS declared O26, O45, O103, O111, O121, and O145 adulterants in certain raw beef products (33).

In response to the need for uniform detection methods for STEC in meat products, the USDA-FSIS published preferred methods in the Microbiology Laboratory Guidebook (MLG). The most recent MLG specifies different protocols for O157:H7 and non-O157 STEC that involve screening of samples by real-time PCR or lateral flow device (in the case of O157:H7), and if positive, culture procedures that involve selective enrichment broth, immunomagnetic separation, chromogenic agar plating, agglutination testing, and subsequent confirmation steps. The past 2 decades of research have resulted in the development of effective reagents and methodologies for detection, isolation, and identification of STEC O157:H7. In contrast, immunological reagents have only recently become commercially available for non-O157 STEC, and although DNA- and culture-based protocols have been developed, investigators have reported difficulty in their detection (34, 35). Part of the problem is the lack of biochemical differences between non-O157 STEC and other *E. coli* strains, in contrast to the relatively unique clone of O157:H7 that was responsible for most outbreaks of illness worldwide. In addition, six times as many serogroups are being targeted, and within some of these serogroups the organisms are biochemically diverse.

According to the most recent FSIS MLG for detection of *E. coli* O157:H7 from meat, confirmation of the sample as positive requires cultural isolation of the organism, biochemical identification of the isolate as *E. coli*, and on this isolate, serological or genetic (PCR) detection of O157, and detection of at least one of the following: Shiga toxin production, *stx*, or the H7 gene. Similarly, according to the most recent MLG for detection of non-O157 STEC, confirmation of a sample as positive requires cultural isolation of the organism, biochemical identification of the isolate as *E. coli*, and on this isolate, serological (agglutination test) and genetic (PCR) detection of O26, O45, O103, O111, O121, or O145, and genetic (PCR) detection of both the *stx* and intimin (*eae*) genes.

Bosilevac and Koohmaraie conducted a large-scale study to determine the prevalence and virulence gene characteristics of non-O157 STEC in commercial ground beef samples in the United States (36). A total of 4,133 samples were cultured; of these samples, *stx* was detected in 1,006 (24.3%) and STEC in 300 (7.3%). A total of 338 unique STEC isolates that belonged to 99 different serotypes were obtained from these samples; the most frequent serotype identified was O113:H21. Only six isolates qualified as FSIS-defined adulterants; four were O103:H2, one was O26:H11, and one was O26:H21. Only four other isolates that were deemed pathogenic STEC were detected, based on a PCR screen for a number of virulence genes, and these isolates came from enrichments that were negative for the intimin gene (*eae*). The FSIS MLG for non-O157 STEC involves classifying a sample as negative and not subject to further testing if it screens negative by real-time PCR for *eae*. Of the six isolates that classified as adulterants, a number of them (although O group, *stx* and *eae* positive) lacked other virulence factors associated with severe disease. The authors noted that, "narrowly focusing on only the described top six STEC serogroups poses the problem of identifying numerous isolates of little pathogenic concern while missing other significant pathogenic STEC serogroups, especially since nearly one-third of pathogenic STEC strains are not within the top six serogroups." Clearly, improved methodologies are needed for the accurate testing of meat samples for non-O157 STEC.

PERI- AND POSTHARVEST INTERVENTIONS

USDA-FSIS Directive 7120.1 (http://www.fsis.usda.gov/OPPDE/rdad/FSISDirectives/7120.1.pdf) provides a list of "Safe and Suitable Ingredients Used in the Production of Meat, Poultry, and Egg Products." Table 1 includes those antimicrobials approved for use in beef, including hides, carcasses, primals, subprimals, cuts, ground beef, sausages, cooked product, ready-to-eat, and other products. Most of these antimicrobials are chemicals, but they also include biologicals (e.g., bacteriophage for use on hides, and food-grade bacteria such as *Lactobacillus* sp. and other genera). Although approved for use, efficacy of these antimicrobials is not a requirement for inclusion in this list. Many studies testing the efficacy of different antimicrobials against STEC on hides, carcasses, parts, and products have been published; however, they would need to be approved for use by the USDA-FSIS before they could be implemented in plants. Several hide and carcass interventions discussed below were developed to reduce STEC contamination of beef carcasses and processed beef. Reductions in the prevalence of *E. coli* O157:H7 on hides are directly correlated with lower carcass prevalence rates (14, 37).

Hide Interventions

It has been surmised for some time that contamination on the hide of cattle was the primary source of carcass contamination with enteric pathogens; therefore, numerous reports investigate the possibility of cleaning the hide before removal. Byrne et al. (38) reported in a study conducted in Ireland that washing cattle with a power-hose for 3 min would remove all visible fecal contamination and

TABLE 1 USDA-FSIS table of safe and suitable antimicrobials for beef[a]

Substance	Product
Acetic acid	Dried and fermented sausages
Aqueous mixture of peroxyacetic acid, hydrogen peroxide, acetic acid, sulfuric acid (optional), and 1-hydroxyethylidine-1, 1-diphosphonic acid (HEDP)	Use in process water used for washing, rinsing, or cooling whole or cut meat including carcasses, parts, trim, and organs
Aqueous solution of sodium diacetate, lactic acid, pectin, and acetic acid	Cooked meat products
Aqueous solution of hydrochloric acid, phosphoric acid, and lactic acid	Raw and ready-to-eat (RTE) poultry carcasses, parts, trim, and organs, and beef products
Aqueous solution of sodium octanoate or octanoic acid and either glycerin and/or propylene glycol and/or a polysorbate surface active agent	Fresh meat primals and subprimals and cuts
Aqueous solution of sulfuric acid and sodium sulfate	In form of spray, wash or dip on surface of meat and poultry products
Blend of citric acid and sorbic acid	To reduce microbial load of purge trapped inside soaker pads in packages of raw whole muscle cuts of meat
Blend of lactic acid, citric acid, and potassium hydroxide	Beef carcasses, heads, and organs including unskinned livers, tongues, tails, primal cuts, subprimal cuts, and trimmings
Blend of salt, sodium acetate, lemon extract, and grapefruit extract	Ground beef, cooked, cured, comminuted sausages (e.g., bologna), and RTE whole muscle meat products; beef steaks
Blend of salt, lactic acid, sodium diacetate, and mono- and diglycerides	Various nonstandardized RTE meat products and standardized meat poultry products that permit use of any safe and suitable antimicrobial agent
Mixture of hops beta acids, egg white lysozyme, and cultured skim milk	In dressing used in refrigerated meat salads
Mixture of maltodextrin, cultured dextrose, sodium diacetate, egg white lysozyme, and nisin preparation	In salads, sauces, and dressings to which fully cooked meat will be added
Acidified sodium chlorite	Meat carcasses, parts and organs; processed, comminuted, or formed meat products (including RTE)
Ammonium hydroxide	Beef carcasses (in hot boxes and holding coolers) and boneless beef trimmings
Anhydrous ammonia	Ground beef
Bacteriophage preparation (*E. coli* O157:H7 targeted)	On hides of live animals in holding pens prior to slaughter
Bacteriophage preparation (*E. coli* O157:H7 targeted)	Red meat parts and trim prior to grinding
Calcium hypochlorite	Red meat carcasses down to a quarter of a carcass; in water used in meat processing; beef primals
Chlorine dioxide	Red meat carcasses down to a quarter of a carcass; in water used in meat processing; beef primals
Chlorine gas	Beef trimmings prior to grinding and beef subprimals; bologna in edible casing; fully cooked meat products in impermeable and permeable prestuck casings; separated beef heads and associated offal products (e.g., hearts, livers, tails, tongues); in brine to cool fully cooked RTE meat products: sausages and similar products in natural casings
Citric acid	In meat products (e.g., beef injected with cultured substrates) and RTE meat products (e.g., hot dogs and

(Continued on next page)

TABLE 1 (Continued)

Substance	Product
	luncheon meat). Cultured substrates are not intended for use in infant formula or foods
Cultured substrates that are produced by the fermentation of natural food sources such as dairy sources, fruit- and vegetable-based sources, and others; substrate is fermented to organic acids by individual microorganisms including *Streptococcus thermophilus*, *Bacillus coagulans*, *Lactobacillus acidophilus*, and others	In enhanced meat and poultry products (e.g., beef injected with a solution) and RTE meat products (e.g., hot dogs)
Cultured sugar (derived from corn, cane, or beets)	In enhanced meat products (e.g., beef injected with a solution) and RTE meat products (e.g., hot dogs)
Cultured sugar and vinegar (derived from corn, cane, or beets)	For use in water applied to beef hides, carcasses, heads, trim, parts, and organs
1,3-dibromo-5,5-dimethylhydantoin (DBDMH)	In casings and on cooked (RTE) meat products
Egg white lysozyme	Red meat carcasses down to a quarter of a carcass
Electrolytically generated hypochlorous acid	In water used in meat processing Beef primals Meat carcasses, parts, trim, and organs
An aqueous solution of citric and hydrochloric acids	To adjust the acidity in various meat products
A blend of citric acid, hydrochloric acid, and phosphoric acid	In casings and on cooked (RTE) meat products
Hops beta acids	In water or ice used for processing meat products; in water or ice, used as either spray or dip, for meat (hides on or off)
Hypobromous acid	Livestock carcasses prior to fabrication (i.e., pre- and postchill), offal, and variety meats; beef subprimals and trimmings; beef heads and tongues
Lactic acid	RTE cooked sausages (e.g., frankfurters, bologna, etc.) and cooked, cured whole muscle products; nonstandardized comminuted meat products (e.g., beef patties), ground beef, and raw whole muscle beef cuts
Lactic acid bacteria mixture consisting of *Lactobacillus acidophilus* (NP35, NP51), *Lactobacillus lactis* (NP7), and *Pediococcus acidilactici* (NP3)	Beef carcasses and parts
Lactoferrin	Fresh cuts of meat, nonstandardized RTE comminuted meat products and standardized RTE comminuted meat products that permit the use of any safe and suitable antimicrobial agent
Lauramide arginine ethyl ester (LAE), silicon dioxide, and refined sea salt	Fresh cuts of meat, nonstandardized RTE comminuted meat products and standardized RTE comminuted meat products that permit the use of any safe and suitable antimicrobial agent
LAE dissolved at specified concentrations in either propylene glycol, glycerin, or water to which may be added a polysorbate surface active agent	RTE meat products; ground beef
LAE	Cooked, RTE meat products containing sauces; meat soups; in casings and on cooked (RTE) meat
Nisin preparation	Frankfurters and other similar cooked meat sausages
Blend of encapsulated nisin preparation, rosemary extract and salt	Cooked (RTE) meat sausages and cured meat products
Blend of nisin preparation, rosemary extract, salt, maltodextrin, and cultured dextrose	Cooked (RTE) meat sausages and cured meat products

(Continued on next page)

TABLE 1 USDA-FSIS table of safe and suitable antimicrobials for beef[a] *(Continued)*

Substance	Product
Blend of nisin preparation, rosemary extract, salt, and sodium diacetate	As part of a carcass wash applied prechill
Organic acids (i.e., lactic, acetic, and citric acid)	All meat products
Ozone	Meat carcasses, parts, trim, and organs
Peroxyacetic acid, octanoic acid, acetic acid, hydrogen peroxide, peroxyoctanoic acid, and HEDP	Process water for washing, rinsing, cooling, or otherwise for processing meat carcasses, parts, trim, and organs
Mixture of peroxyacetic acid, hydrogen peroxide, acetic acid, and HEDP	Process water for washing, rinsing, cooling, or otherwise for processing meat carcasses, parts, trim, and organs
Combination of two aqueous mixtures (FCN 323 and FCN 880) of peroxyacetic (peracetic) acid, hydrogen peroxide, acetic acid, and stabilizer HEDP	Red meat carcasses, parts, and trim
Aqueous mixture of peroxyacetic acid, hydrogen peroxide, acetic acid, HEDP, and sulfuric acid	Water or ice for washing, rinsing, cooling, or processing whole or cut meat including carcasses, parts, trim, and organs
Mixture of peroxyacetic acid, hydrogen peroxide, acetic acid, and HEDP	In process water or ice for washing, rinsing, storing, or cooling of processed and preformed meat products
Aqueous mixture of peroxyacetic acid, hydrogen peroxide, acetic acid, and HEDP	In process water used for washing, rinsing, cooling or otherwise processing meat carcasses, parts, trim, and organs
Aqueous mixture of peroxyacetic acid, hydrogen peroxide, acetic acid, and HEDP	In process water or ice used for washing, rinsing, cooling or processing whole or cut meat including parts, trim, and organs
Aqueous mixture of peroxyacetic acid, hydrogen peroxide, HEDP, and optionally sulfuric acid	Red meat carcasses, parts, trim, and organs
Aqueous mixture of peroxyacetic acid, hydrogen peroxide, HEDP, dipicolinic acid, and sulfuric acid	Use as a spray, rinse, dip, chiller water, or scald water for raw meat carcasses, parts, trim, and organs
Mixture of peroxyacetic acid, hydrogen peroxide, acetic acid, HEDP, and water	Various meat products which permit addition of antimicrobial agents, e.g., hot dogs
Potassium diacetate	Various RTE meat products, e.g., hot dogs
Solution of water, lactic acid, propionic acid, and acidic calcium sulfate	Raw comminuted beef
Solution of water, acidic calcium sulfate, and lactic acid	Raw whole muscle beef cuts and cooked roast beef and similar cooked beef products (e.g., corned beef, pastrami, etc.)
Solution of water, acidic calcium sulfate, lactic acid, and sodium phosphate	Beef jerky
Solution of water, acidic calcium sulfate, lactic acid, and sodium	Meat sausages including those with standards of identity which permit the use of antimicrobial agents
Skim milk or dextrose cultured with *Propionibacterium freudenreichii* subsp. Shermanii	RTE meat products that permit the use of any safe and suitable antimicrobial agent
Sodium benzoate	Nonstandardized and standardized comminuted meat and poultry products which permit ingredients of this type
Sodium citrate buffered with citric acid	RTE meat products that permit the use of any safe and suitable antimicrobial agent
Sodium diacetate, sodium propionate, and sodium benzoate	Red meat carcasses down to quarter of carcass; water used in meat processing; beef primals

(Continued on next page)

TABLE 1 (Continued)

Substance	Product
Sodium hypochlorite	Component of marinades used for raw meat products; raw beef carcasses, subprimals, and trimmings; RTE meat products
Sodium metasilicate	RTE meat, where antimicrobials are permitted
Sodium propionate/propionic acid	RTE meat and poultry, where antimicrobials are permitted

[a]For a complete listing, which includes other meats, poultry and eggs, and antimicrobials targeting other specific organisms, please refer to USDA-FSIS Directive 7120.1 Safe and Suitable Ingredients Used in the Production of Meat, Poultry, and Egg Products, http://www.fsis.usda.gov/OPPDE/rdad/FSISDirectives/7120.1.pdf.

reduce the presence of *E. coli* O157:H7 inoculated onto the hide. However, it was reported that the wash did not significantly reduce the numbers of *E. coli* O157:H7 transferred from the hide to the carcass during slaughter/dressing procedures. However, Nou et al. (39) demonstrated that prevalence of *E. coli* O157:H7 could be significantly reduced on carcasses by removing bacterial contamination on the hide before removal.

Several studies evaluated the effectiveness of incorporating sanitizers into hide washes to reduce the potential spread of pathogens from the hide to carcass surfaces. Mies et al. (40) investigated the implementation of a commercial cattle hide wash system that evaluated water washes, 0.5% lactic acid (LA), and 50 ppm chlorine and reported that bacterial numbers actually increased after the treatments. Bosilevac et al. (41) reported that ozonated and electrolyzed oxidizing water treatments reduced *Enterobacteriaceae* counts (EBC) by 3.4 and 4.3 log_{10} CFU per 100 cm^2, respectively. Small et al. (42) compared hide decontamination treatments and found that the greatest bacterial reduction (2.3 log_{10} CFU per cm^2) could be attained by clipping the hair from the hide and then singeing with a handheld blowtorch.

A number of studies investigated the efficacy of different hide sanitization treatments, some of which have been commercialized, for reducing *E. coli* O157:H7 numbers. Baird et al. (43) inoculated beef hide sections with bovine fecal slurries and treated them with various potential wash sanitizers. The authors reported

that the greatest reductions in coliform counts on the hide sections resulted from treatment with 1% cetylpyridinium chloride (CPC), 2% LA, and 3% hydrogen peroxide (4.5, 4.1, and 3.9 log_{10} CFU per 100 cm^2, respectively). In an investigation of several different potential hide sanitizer treatments, Carlson et al. (44) reported that 2.4% potassium cyanate, 6.2% sodium sulfide, and 1.5% sodium hydroxide followed by high-pressure washing with 0.02% chlorinated water caused the greatest reductions in numbers of *E. coli* O157:H7, achieving reductions ranging from 4.8 to 5.1 log_{10} CFU per cm^2.

Bosilevac et al. (37) tested a water wash plus CPC treatment as a hide intervention when applied to cattle in the holding pens of a commercial processing plant. Cattle were washed with water the day before harvest, and before stunning, were sprayed with 1% CPC. Hides and carcasses after hide removal but before evisceration were sampled to determine aerobic plate counts (APC), EBC, and *E. coli* O157 prevalence. The prevalence of *E. coli* O157 on hides was reduced by 18% (from 56 to 34%) and that on carcasses prior to evisceration by 20% (from 23 to 3%). On preevisceration carcasses, APC were decreased by approximately 77,000 CFU per 100 cm^2 and EBC by approximately 1,150 CFU per 100 cm^2. It was concluded that this treatment has great potential and deserves further evaluation.

One of the more novel approaches to prevent contamination on the hide from reaching the carcass surface during slaughter/dressing was published by Antic et al. (45). The Serbian investigators coated cattle hides with a solution of food-grade resin (shellac) in ethanol, hypothesizing that the immobilization of bacteria on the hide would reduce transmission of bacteria to the carcass. The shellac-based treatment was reported to have successfully reduced hide-level *E. coli* O157:H7 prevalence by 3.7 $\log_{10}$ CFU per cm^2.

Another novel approach to reduce STEC contamination on the hide is the application of bacteriophage (46). One product has received approval in the United States for use on the hides of live cattle in the holding pens 1 to 4 h before slaughter and hide removal, and specifically targets *E. coli* O157:H7.

Chemical Dehairing

Chemical dehairing is a process patented by Bowling and Clayton (47) that involves treatment of the hide with a sodium sulfide solution, followed by a hydrogen peroxide solution and water washing. Treatment with sodium sulfide solution dissolves and removes hair and extraneous matter from the skin surface, and hydrogen peroxide neutralizes the pH. Additional steps, e.g., water rinse prior to the sodium sulfide or additional neutralization steps, may be involved, depending on the protocol. The first studies evaluating chemical dehairing as a hide intervention were conducted by Schnell et al. (48). This study involved 10 grain-fed steers or heifers to be dehaired and 10 controls that were slaughtered and dressed without dehairing. Excised hide samples from conventional and dehaired carcasses were analyzed for APC, total coliform counts (TCC), and *E. coli* biotype I counts. Dehairing reduced the amount of visible contamination on beef carcasses, but dehaired cattle had significantly higher TCC (*P*<0.05) and no significant difference in APC or *E. coli* counts from that of conventionally slaughtered cattle. The authors hypothesized that the lack of reduction in APC and *E. coli* could have been the result of aerosol, human, and equipment contamination in the facility.

Castillo et al. (49) found that a chemical dehairing process significantly reduced APC, TCC, and *E. coli*, as well as *E. coli* O157:H7 and *Salmonella enterica* serovar Typhimurium on artificially inoculated hide pieces. Pieces of hide (4 cm^2) were contaminated with bovine feces containing both rifampicin-resistant *E. coli* O157:H7 and serovar Typhimurium to yield approximately 5.0 $\log_{10}$ CFU/cm^2 of each pathogen, or with noninoculated feces, which produced an approximate final APC of 6.0 $\log_{10}$ CFU/cm^2 and a coliform and *E. coli* count of 5.0 $\log_{10}$ CFU/cm^2. Counts of pathogens, APC, coliforms, and *E. coli* were conducted before and after chemical dehairing. Chemical dehairing significantly reduced serovar Typhimurium and *E. coli* O157:H7 populations from 5.1 to 5.3 $\log_{10}$ CFU/cm^2 to <0.5 $\log_{10}$ CFU/cm^2, and reduced APC, coliforms, and *E. coli* counts by 3.4, 3.9, and >4.3 $\log_{10}$ CFU/cm^2, respectively. The authors concluded that since the hide is a major source of fecal contamination of beef carcass surfaces, chemical dehairing may be beneficial in reducing overall contamination of carcasses.

Nou et al. (39) tested the efficacy of chemical dehairing on reducing the prevalence of *E. coli* O157:H7 and other bacteria on the surfaces of preeviscerated beef carcasses from which hides had been removed. Hides were sampled immediately after stunning, before exsanguination or any antimicrobial intervention, to confirm that bacterial loads were not significantly different between the control and treatment groups. Carcasses were sampled immediately after hide removal and before evisceration. Total APC and EBC on hides in both control and treatment groups were not significantly different. Preevisceration carcasses processed after chemical dehairing had approximately 2 logs lower APC and EBC compared with those processed by conventional procedures (*P*<0.0001). In addition, the prevalence of *E. coli* O157:H7 was lower on chemically dehaired (1%) than control (conventionally processed) preevisceration carcasses (50%, *P*<0.05). The data indicated that chemical dehairing of cattle hides is an effective intervention to reduce the incidence of hide-to-carcass contamination with *E. coli* O157:H7 and potentially other pathogens. However, although chemical dehairing was found to be an effective hide intervention, the industry has not considered it feasible to implement (18).

Alternatives to Chemical Dehairing and CPC

Hide interventions proven to significantly reduce carcass contamination in processing facilities include chemical dehairing, CPC washing, and a 65°C sodium hydroxide wash followed by a water rinse (50). Because CPC is not yet approved for use in beef processing plants, Bosilevac et al. (50) tested other chemicals and antimicrobial compounds on hides that are approved for use in beef processing plants, albeit for carcass and boneless beef trim decontamination. In vitro experiments were conducted on cattle hides to evaluate 4% trisodium phosphate, 4% phosphoric acid, 1.6% sodium hydroxide, and 4% chlorofoam (chlorinated alkaline detergent containing 1,200 ppm free chlorine at pH 7.0) as washes. A rinse step, consisting of water or acidified chlorine, was used following all wash treatments. These wash treatments reduced hide coliform counts by 1.5 to 2.5 $\log_{10}$ CFU per 100 cm^2, and acidified (pH 7.0) chlorine rinses (200 or 500 ppm) further reduced coliforms by 10 to 100 CFU per 100 cm^2. Removal of excess liquid by vacuuming treated areas reduced the microbial load by approximately 10 CFU per 100 cm^2. An online hide-wash cabinet that delivered a sodium hydroxide wash and a chlorinated (1 ppm) water rinse reduced APC and EBC on hides by 2.1 and 3.4 $\log_{10}$ CFU per 100 cm^2, respectively, and reduced the prevalence of *E. coli* O157 on hides from 44 to 17%. This hide washing procedure further resulted in a reduction of APC and EBC on carcasses before evisceration by 0.8 $\log_{10}$ CFU/100 cm^2, and *E. coli* O157 prevalence from 17 to 2%. These results provided further evidence that

decontamination of hides also significantly reduces contamination of carcasses during processing.

The efficacy of hypobromous acid (HOBr) as a hide intervention was studied (51). Hides after removal from carcasses at a beef processing plant were sprayed with 220 or 500 ppm of HOBr. HOBr at a concentration of 220 ppm significantly reduced bacterial counts (APCs, total coliforms, and *E. coli*) on hides by 2.2 log CFU/100 cm^2 (51). HOBr at a concentration of 500 ppm significantly reduced these bacterial counts by 3.3, 3.7, and 3.8 log CFU per 100 cm^2, respectively, demonstrating a dose effect with HOBr. It was concluded that a HOBr wash would reduce the pathogen prevalence and concentrations of spoilage bacteria on hides and decrease the risk of carcass contamination.

CARCASS, PRIMAL, SUBPRIMAL, AND TRIM SURFACES

Water Rinsing

Empey and Scott (52) reported that washing carcasses with cold water reduces their bacterial populations (13). However, Bell (53) found that cold water carcass washing was ineffective in removing microbial contamination and tended to bring about a posterior to anterior redistribution of microbial contamination, resulting in increased counts at forequarter sites. Patterson (54) found that beef carcasses treated with a steam and hot water spray (80 to 96°C) for 2 min had significantly reduced bacterial numbers compared to untreated carcasses.

Hot water treatments of beef carcasses to reduce *E. coli* O157:H7 and other bacteria applied through a model carcass spray cabinet were tested by Castillo et al. (55). Paired hot carcass surface regions with varying fat characteristics were inoculated with bovine feces containing 10^6 CFU bacteria per g. Carcass surfaces then were exposed to a warm water wash followed or not by a hot (95°C) water spray, which raised the carcass surface temperature to 82 to 85°C in 1 to 2 sec. The effect of time between the application of feces and water treatment was also evaluated. Warm water wash followed by hot water spray provided mean log reductions per cm^2 of 3.7, 2.9, 3.3, and 3.3 of *E. coli* O157:H7, APC, coliform, and thermotolerant coliform counts, respectively. Carcass surface region, but not an increase in time (30 min) before treatment, affected the efficacy of hot water treatments. This study also resulted in the conclusion that coliform counts may be used to verify the efficacy of hot water interventions used as critical control points in a hazard analysis critical control point (HACCP) system.

Steam and Steam Vacuuming

The use of steam-vacuum systems instead of knife trimming for physical removal of small areas of fecal contamination from beef carcasses was approved by the FSIS in 1996 (56). Steam-vacuum systems deliver 82 to 88°C water via spray nozzles at the carcass surface while the vacuum removes any loose material. Commercial steam-vacuum systems have been reported to reduce total bacterial populations and populations of *E. coli* O157:H7 on beef carcass surfaces by 3.0 and 5.5 log_{10} CFU/cm^2, respectively (57, 58). Steam vacuuming has been implemented in most beef processing plants in the United States at various stages in the slaughter or dressing process (18). Knife trimming and steam vacuuming of visible contamination in localized areas of the carcass surface have been reported to be useful for pathogen reduction. In addition, application of steam vacuuming is commonly applied to areas of the carcass surface believed to be "hot spots" (e.g., hide removal pattern lines). These techniques, however, cannot be used efficiently for the entire carcass (59) and are intended for spot treatment only. Although the technology has been reported to be successful in reducing carcass contamination (57, 58, 60), a report by Castillo et al. (61) questioned overall effectiveness.

Castillo et al. (61) evaluated a steam-vacuum system designed to be a spot-cleaning method for removal of fecal contamination on the surface of carcasses and subsequent reduction of *E. coli* contamination of hot beef carcasses. The efficacy of accompanying treatments, which included hot (95°C) or warm (55°C) water, 2% LA spray, or combinations of both methods, was also assessed. In this study, 0.025 g of bovine fecal material was used as a vehicle to deliver contaminating bacteria to a 5-cm^2 area on three specific regions of hot carcass surfaces, viz., the outside round, brisket, and clod. These regions were removed from the rest of the carcass before inoculation, but the "hot" carcass temperature was maintained by placing the meat in insulated containers. It was reported that all treatments significantly reduced the numbers of APC, EBC, total coliforms, thermotolerant coliforms, and *E. coli* on beef carcass surfaces. However, steam vacuuming alone resulted in significantly smaller reductions than those obtained by a combination of steam vacuuming with any subsequent sanitizing treatment. Steam vacuuming reduced the number of different indicator organisms tested by ca. 3.0 log cycles; however, it was also observed that the treatment spread the bacterial contamination to areas of the carcass surface adjacent to the contaminated sites. This relocated contamination was most effectively reduced by treating the area with a combination of hot water followed by LA.

The application of steam to carcass surfaces was shown by Dorsa et al. (57) to be effective for carcass decontamination, and these authors, with Frigoscandia, subsequently developed commercial cabinets for application of steam, calling the treatment "steam pasteurization." Phebus et al. (62) designed an experimental steam pasteurization chamber for laboratory testing and reported reductions of *E. coli* O157:H7 and certain other bacterial pathogens by 3.4 to 3.7 log cycles on hot beef carcass surfaces. However, steam pasteurization alone showed no greater reductions than other treatments such as knife trimming or steam vacuuming. Nutsch et al. (63) conducted evaluations of commercial steam pasteurization application in a beef processing plant. Carcasses were treated with a preliminary water wash, followed by passing through air blowers to eliminate excessive humidity that would favor steam condensation. The carcasses were then passed, treated with steam within a chamber, followed by transfer to another section of the cabinet where cold water was applied. After treatment, it was reported that APCs were reduced on carcasses from initial counts of 2.1 to 2.2 $\log_{10}$ CFU/cm^2 to 0.6 to 0.8 $\log_{10}$ CFU/cm^2. Counts of *E. coli* were also reduced from original counts of 0.6 to 1.5 $\log_{10}$ CFU/cm^2 to undetectable levels after 6 or 8 sec of steam treatment.

Organic Acids and Miscellaneous Sanitizers

Various studies evaluated the efficacy of organic acids for sanitizing whole carcass sides (13, 60, 64), and over time LA became the most commonly used organic acid for carcass decontamination in commercial practice. Many processors implemented LA washes on preevisceration carcasses and final carcass sprays before carcasses entered the cooler. Although the most common application of LA is currently on hot carcass surfaces, Castillo et al. (65) reported that the treatment is also effective, although to a lesser degree, on chilled carcass surfaces. Kotula and Thelappurate (66) reported that APC and *E. coli* counts increased more rapidly on untreated steaks than on steaks treated with acetic acid or LA.

A concern related to spraying beef carcasses with organic acid is the reported resistance of *E. coli* O157:H7 to low pH (67–69). However, several studies indicated that lactic or acetic acid sprays, when applied at 55°C, can effectively reduce levels of *Salmonella* sp. and *E. coli* O157:H7 (65, 70, 71). Successful reduction of bacteria on meat surfaces by using organic acids or other sanitizers requires that the sanitizing solution be allowed

to contact the bacterial cells. If bacteria are hidden in small knife cuts or under tissue and the organic acid cannot contact the cell, the desired antibacterial effect is unlikely.

Pittman et al. (72) tested the efficacy of LA as an initial and secondary subprimal intervention. Sections of chilled beef subprimals (beef round peeled knuckle and beef brisket flats) having 100 cm^2 of exposed lean surface were inoculated with *E. coli* O157:H7, non-O157 STEC, or nonpathogenic (biotype I) *E. coli*, the last as surrogates for *E. coli* O157:H7. After 30 min at 4°C to allow for bacterial attachment, sections were sprayed with LA in a custom-built spray cabinet. Treatments were applied at 1 of 16 combinations of two LA concentrations (2.0 or 5.0%), two LA temperatures (22 or 48°C), two pressures (1.03 or 4.83 bar), and two flow rates (0.22 or 6.22 liters per min). Sections were allowed to drip for 10 sec and were then vacuum packaged, sealed, and stored at 4°C until bacteria were enumerated or given a second LA treatment 24 h later. The initial application of LA spray reduced *E. coli* surrogate, *E. coli* O157:H7, and non-O157 STEC counts from 6.0 CFU per cm^2 to 3.6, 4.4, and 4.4 log CFU per cm^2, respectively. The second application further reduced *E. coli* surrogate, *E. coli* O157:H7, and non-O157 STEC counts to 2.6, 3.2, and 3.6 log CFU per cm^2, respectively. LA sprays were effective as both the initial and secondary treatments on beef subprimals in reducing pathogenic and nonpathogenic *E. coli*, as well as naturally occurring microflora.

Fouladkhah et al. (73) tested whether the six non-O157 STEC serogroups currently classified as adulterants in beef by the USDA-FSIS (O26, O45, O103, O111, O121, and O145) could be reduced on beef trimming pieces by LA treatments previously shown to be effective against *E. coli* O157:H7. Beef trimming samples weighing approximately 100 g were inoculated with approximately 1,000 CFU per cm^2 via micropipette and, after time was allowed for bacterial attachment (10 min at 4°C), were immersed for 30 sec in 5% LA solutions at 25 or 55°C. Treatments resulted in reductions of 0.5 to 0.9 (25°C LA) and 1.0 to 1.4 (55°C LA) $\log_{10}$ CFU per cm^2 ($P<0.05$) for *E. coli* O157:H7 and non-O157 STEC. It was concluded that the LA treatment used against *E. coli* O157:H7 on beef trimmings should also be effective against the six non-O157 STEC serogroups.

Kalchayanand et al. (74) inoculated prerigor beef flanks to determine if antimicrobial interventions currently used by the meat industry have a similar effect in reducing non-O157 STEC serogroups O26, O45, O103, O111, O121, and O145 compared to *E. coli* O157:H7. The surfaces of the beef flanks were inoculated with a high (5 ×10^4 CFU per cm^2) or low (5 × 10^1 CFU per cm^2) concentration of bacteria and given 15 min at room temperature to allow for bacterial attachment. After inoculation, flanks were subjected to a 15-sec spray treatment with one of the following FSIS-approved treatments using a model spray cabinet with three oscillating spray nozzles at 60 cycles per min: acidified sodium chlorite (ASC) (1,000 ppm), peroxyacetic acid (200 ppm), LA (4%), or hot water (85°C). Surviving bacterial concentrations were determined within 10 min after treatment or after 48 h storage at 2 to 4°C. Against both high- and low-level inoculation, hot water, LA, peroxyacetic acid, and ASC ranked in this order as the most to least effective. While hot water and LA (reductions of 3.2 to 4.2 and 1.6 to 2.7 log CFU per cm^2, respectively, for nonchilled specimens) were effective against all STEC strains ($P \leq 0.05$), peroxyacetic acid was not effective against O111, and ASC was not effective against O26, O111, and O145. Bacterial reductions with ASC were increased by storage at 4°C for 48 h; hence, it was concluded that this compound might be a long-acting microbial inhibitor and suitable as a prepackaged meat intervention. Storage at 4°C provided little additional reduction of pathogen levels with the other treatments. The levels of reduction of non-O157 STEC achieved by these antimicrobial interventions were comparable to that of *E. coli* O157:H7; however, that low levels of pathogens were still detectable indicated that none of the treatments would result in their total elimination.

In an attempt to identify effective and inexpensive antimicrobial interventions that could be used in very small meat plants (i.e., <10 employees and generating average annual revenue of $2.5 million or less), the relative effectiveness of eight antimicrobial compounds (acetic acid, citric acid, LA, peroxyacetic acid, ASC, chlorine dioxide, sodium hypochlorite, and aqueous ozone) was tested (75). These compounds were applied to beef plate piece surfaces that had been inoculated with fecal slurry containing a pathogen cocktail of STEC O157:H7, serovar Typhimurium, *Campylobacter coli*, and *Campylobacter jejuni* with small, handheld spraying equipment. The relative antimicrobial effectiveness from greatest to least was as follows: organic acids, peroxyacetic acid, chlorinated compounds, and aqueous ozone. A 2% LA rinse provided 3.5- to 6.4-log CFU/cm^2 reductions across all four bacterial populations studied.

The reduction of *E. coli* O157:H7 and serovar Typhimurium on beef carcass surfaces through application of ASC solutions was investigated and reported by Castillo et al. (61). When phosphoric acid was used to acidify sodium chlorite, the resulting ASC solution reduced populations of both pathogens by 3.8 to 3.9 log cycles. However, when ASC solutions were prepared through acidification with citric acid, reductions were reported to range from 4.5 to 4.6 log cycles.

Carvacrol is a monoterpenoid phenol present as an essential oil in *Origanum vulgare* (oregano) and several other plants and has been shown to have activity against *E. coli* O157:H7 in edible apple films (76). A study was conducted to determine whether carvacrol has activity against *E. coli* O157 on cattle hides and beef carcass cuts (77). In this study, carvacrol was sprayed onto hides and beef carcass cuts at concentrations of 0, 10, 20, and 30 mg per ml. These surfaces were then inoculated with *E. coli* O157 at a concentration of 5 to 6 $\log_{10}$ CFU per cm^2, with 10 min allowed for bacterial attachment. After this time, the hide and carcass cut surfaces were swabbed and cultured, and surviving *E. coli* O157 concentrations were determined. *E. coli* O157 concentrations were reduced on carcass cuts and hides treated with carvacrol at 30 mg per ml by approximately 1.4 and 1.6 $\log_{10}$ CFU per cm^2, respectively ($P<0.05$). It was concluded that carvacrol has the potential to control *E. coli* O157 on bovine hide and carcass cuts, but should be further studied.

Bacteriophage

A bacteriophage product has been approved for use on red meat trim and parts intended to be ground (47). This product was reported to eliminate 95 to 100% of *E. coli* O157:H7 when sprayed on the surfaces of beef (47).

Combination Treatments

Reports have indicated that a combination of treatments, also known as a multiple hurdle treatment (14, 55, 62), may be required during processing to reduce pathogen contamination. Numerous carcass decontamination methods have been investigated alone and in combination for their ability to reduce pathogens on meat; however, results and conclusions are varied and often contradictory. For example, Gill and Landers (78) reported that spraying beef carcasses with 2% LA, steam-vacuuming, or trimming was ineffective, and that only steam or hot water treatments substantially reduced bacterial contamination. The same authors indicated that a 200-ppm peroxyacetic acid carcass spray was likely also ineffective, but efficacy was difficult to determine due to a subsequent steam treatment.

Gill and Badoni (79) evaluated the effects of 0.02% peroxyacetic acid, 0.16% ASC, 2% LA, and 4% LA on chilled beef surfaces and determined that peroxyacetic acid and ASC produced a negligible effect on coliforms or *E. coli*, and both treatments were less effective than 4% LA. The authors surmised that evaluation of antimicrobial treatments may produce inconsistent results due to different types of meat surfaces to which they are applied during investigations, or that the results may be a factor of the surface microflora composition as influenced by prior antimicrobial treatments. King et al. (80) reported that peroxyacetic acid concentrations up to 600 ppm were ineffective for antimicrobial treatment of chilled inoculated beef carcass surfaces.

Elramady et al. (81), using in vitro broth culture and cattle hide model experiments, studied the efficacy of chitosan acetate (CA), sodium dodecyl sulfate (SDS), and LA as individual treatments and in combination against *E. coli* O157:H7. CA is a naturally occurring substance with demonstrated antibacterial properties, and SDS is a food additive that has been shown to enhance the antibacterial effects of organic acids, hence the reason for being tested in this study. CA as a treatment consisted of 1% chitosan in 1% acetic acid solution, and the SDS treatment consisted of a 1% or 2% solution. LA was applied as a 1% solution, and the CA-SDS treatment consisted of 1% chitosan in 1% acetic acid mixed with 1% SDS. LA-SDS treatments were tested in two different concentrations, viz., 1% LA mixed with 1% SDS, and 1% LA mixed with 2% SDS. In the in vitro broth experiments, all treatments resulted in a significant reduction in survival of *E. coli* O157:H7. However, only 1% LA plus 1% SDS and 1% LA plus 2% SDS treatments resulted in significant reductions of *E. coli* O157:H7 on hides. These treatments resulted in 4.6 and 4.7 $\log_{10}$ CFU per cm^2 reductions, respectively, compared to phosphate buffer control treated hides, which had 6.0 $\log_{10}$ CFU per cm^2 surviving cells. The antibacterial efficacy of 1% LA was significantly enhanced when combined with 1% SDS. A low-concentration LA-SDS combination treatment as a wash may potentially reduce the risk of *E. coli* O157:H7 contamination on hides.

Signorini and Tarabla (82) used a stochastic simulation model to assess the effects of measures implemented in the agri-food chain to reduce the contamination of ground beef with STEC. A published risk assessment model developed in Argentina was used as baseline scenario. Control measures assessed were based on a reduction in herd prevalence of infection due to vaccination, reduction in opportunity for cross-contamination in the slaughterhouses by the introduction of an online hide-wash cabinet, and control of storage temperature in slaughterhouses, retail stores, and home. Additionally, the increase of feedlot production was modeled. Simulations suggested that the greatest potential impact was associated with hide-wash cabinet and vaccination, measures aimed to reduce the STEC prevalence and concentration in the cattle hides at the beginning of the food chain. Control of storage temperature was not effective if cross-contamination of the carcasses with the pathogen was not prevented or reduced. An increased production (fattening) of cattle in feedlots may raise the risk of STEC infection and its sequelae. This information can be used as a basis for measures of risk management.

In some cases, the combination of treatments has not afforded any greater efficacy than the component treatments applied individually. For example, Bosilevac et al. (unpublished data) reported that *E. coli* O157 prevalence was reduced by 81% and 35% by hot water and LA treatments, respectively, but only 79% by both treatments combined.

Numerous options are available for decontaminating meat surfaces; however, none of the approved interventions are capable of eliminating the presence of pathogens. Frequently, processors in the meat industry use several redundant intervention technologies in an attempt to decrease risk of pathogen contamination, but a guarantee of complete absence of bacterial contamination is not possible (15, 16). Proper end-user handling of meat products is required for assurance of safety. Research continues to seek novel interventions in an attempt to assist meat processors minimize or eliminate enteric origin pathogens, as well as unknown and emerging food-borne pathogens.

HACCP SYSTEMS AND VALIDATION

All U.S. establishments producing raw beef products are required to use HACCP to control contamination with food-borne pathogens such as the STEC strains that have been declared adulterants by the USDA-FSIA. To meet these requirements, slaughter establishments have implemented a variety of carcass decontamination procedures such as critical control points (CCPs) that are essential to the safety of beef. In the last quarter of 2002, the USDA-FSIS issued a notice reminding slaughter establishments that all CCPs must be validated to ensure they can successfully prevent, eliminate, or reduce STEC O157:H7. The regulatory agency indicated that until establishments have collected data to demonstrate that CCPs function properly under actual in-plant conditions, the effectiveness of the CCP would be considered theoretical and not validated. The FSIS also noted that many establishments have not validated CCPs based on actual in-plant conditions.

Microbiological testing can play a unique role in verification and validation activities. Detection of food-borne pathogens is not considered to be an effective tool for monitoring CCPs within a slaughter or processing HACCP plan. Pathogens are often absent from carcass surfaces and, when present, their uneven distribution makes it difficult to obtain a truly representative sample. In contrast, microbiological testing can be applied within a HACCP plan to validate and verify the effectiveness of carcass decontamination procedures.

Because of the difficulty in consistently finding and documenting reductions of levels of enteric pathogens on carcass surfaces, an ideal solution can be the use of nonpathogenic surrogate bacteria that are capable of indicating the probable reduction of pathogens. Surrogate bacteria are required to have very similar growth and resistance characteristics to the pathogens of concern. After known amount of the surrogate is inoculated to a carcass surface, the effectiveness of a CCP can be validated by comparing surviving levels of the surrogate on the carcass surface following the processing step.

Although verification and validation of HACCP systems may initially seem intimidating, careful thought and planning can make the process logical, reasonable, and extremely helpful. Assistance is available through many tools, such as rapid microbiological tests, extensive publication of research results in the scientific literature, and numerous HACCP experts. The human tendency is to find a single tool that works and use it to excess; however, successful verification and validation will most likely be attained through the utilization of as many of the tools as possible. Regular challenging of the validity of a HACCP system through verification will only serve to strengthen confidence in the ability of the process to control hazards.

ACKNOWLEDGMENTS

This project was supported by Agriculture and Food Research Initiative Competitive Grant no. 2012-68003-30155 from the USDA National Institute of Food and Agriculture.

We declare no conflicts of interest with regard to the manuscript.

CITATION

Moxley RA, Acuff GR. 2014. Peri- and postharvest factors in the control of Shiga toxin-producing *Escherichia coli* in beef. Microbiol Spectrum 2(4):EHEC-0017-2013.

REFERENCES

1. **Painter JA, Hoekstra RM, Ayers T, Tauxe RV, Braden CR, Angulo FJ, Griffin PM.** 2013. Attribution of foodborne illnesses, hospitalizations, and deaths to food commodities by using outbreak data, United States, 1998–2008. *Emerg Infect Dis* **19:**407–415.

2. **Scallan E, Hoekstra RM, Angulo FJ, Tauxe RV, Widdowson MA, Roy SL, Jones JL, Griffin PM.** 2011. Foodborne illness acquired in the United States—major pathogens. *Emerg Infect Dis* **17:**7–15.

3. **Blanco M, Blanco JE, Mora A, Dahbi G, Alonso MP, González, Bernárdez MI, Blanco J.** 2004.

4. **Gyles CL.** 2007. Shiga toxin-producing *Escherichia coli*: an overview. *J Anim Sci* **85:**E45–E62.

5. **Moxley RA.** 2004. *Escherichia coli* O157:H7: an update on intestinal colonization and virulence mechanisms. *Anim Health Res Rev* **5:**15–33.

6. **Brooks JT, Sowers EG, Wells JG, Greene KD, Griffin PM, Hoekstra RM, Strockbine NA.** 2005. Non-O157 Shiga toxin-producing *Escherichia coli* infections in the United States, 1983–2002. *J Infect Dis* **192:**1422–1429.

7. **Moxley RA, Smith DR.** 2010. Attaching-effacing *Escherichia coli* infections in cattle. *Vet Clin North Am Food Anim Pract* **26:**29–56.

8. **Cray WC, Moon HW.** 1995. Experimental infection of calves and adult cattle with *Escherichia coli* O157:H7. *Appl Environ Microbiol* **61:**1586–1590.

9. **Dean-Nystrom EA, Bosworth BT, Cray WC, Moon HW.** 1997. Pathogenicity of *Escherichia coli* O157:H7 in the intestines of neonatal calves. *Infect Immun* **65:**1842–1848.

10. **Matthews L, McKendrick IJ, Ternent H, Gunn GJ, Synge B, Woolhouse MEJ.** 2006. Supershedding cattle and the transmission dynamics of *Escherichia coli* O157. *Epidemiol Infect* **134:**131–142.

11. **Omisakin F, MacRae M, Ogden ID, Strachan NJC.** 2003. Concentration and prevalence of *Escherichia coli* O157 in cattle feces at slaughter. *Appl Environ Microbiol* **69:**2444–2447.

12. **Cobbold RN, Hancock DD, Rice DH, Berg J, Stilborn R, Hovde CJ, Besser TE.** 2007. Rectoanal junction colonization of feedlot cattle by *Escherichia coli* O157:H7 and its association with supershedders and excretion dynamics. *Appl Environ Microbiol* **73:**1563–1568.

13. **Dickson JS, Anderson ME.** 1992. Microbiological decontamination of food animal carcasses by washing and sanitizing systems: a review. *J Food Prot* **55:**133–140.

14. **Arthur TM, Bosilevac JM, Nou X, Shackelford SD, Wheeler TL, Kent MP, Jaroni D, Pauling B, Allen DM, Koohmaraie M.** 2004. *Escherichia coli* O157 prevalence and enumeration of aerobic bacteria, *Enterobacteriaceae*, and *Escherichia coli* O157 at various steps in commercial beef processing plants. *J Food Prot* **67:**658–665.

15. **Barkocy-Gallagher GA, Arthur TM, Rivera-Betancourt M, Nou X, Shackelford SD, Wheeler TL, Koohmaraie M.** 2003. Seasonal prevalence of Shiga toxin-producing *Escherichia coli*, including O157:H7 and non-O157 serotypes, and *Salmonella*

Serotypes, virulence genes, and intimin types of Shiga toxin (verotoxin)-producing *Escherichia coli* isolates from cattle in Spain and identification of a new intimin variant gene (*eae-xi*). *J Clin Microbiol* **42:**645–651.

in commercial beef processing plants. *J Food Prot* 66:1978–1986.

16. **Elder RO, Keen JE, Siragusa GR, Barkocy-Gallagher GA, Koohmaraie M, Laegreid WW.** 2000. Correlation of enterohemorrhagic *Escherichia coli* O157 prevalence in feces, hides, and carcasses of beef cattle during processing. *Proc Natl Acad Sci USA* 97:2999–3003.

17. **Keen JE, Elder RO.** 2002. Isolation of Shiga-toxigenic *Escherichia coli* O157 from hide surfaces and the oral cavity of finished beef feedlot cattle. *J Am Vet Med Assoc* 220:756–763.

18. **Koohmaraie M, Arthur TM, Bosilevac JM, Guerini M, Shackelford SD, Wheeler TM.** 2005. Post-harvest interventions to reduce/eliminate pathogens in beef. *Meat Sci* 71:79–91.

19. **Schmidt JW, Arthur TM, Bosilevac JM, Kalchayanand N, Wheeler TL.** 2012. Detection of *Escherichia coli* O157:H7 and *Salmonella enterica* in air and droplets at three U.S. commercial beef processing plants. *J Food Prot* 75:2213–2218.

20. **Barkocy-Gallagher GA, Arthur TM, Siragusa GR, Keen JE, Elder RO, Laegreid WW, Koohmaraie M.** 2001. Genotypic analyses of *Escherichia coli* O157:H7 and O157 nonmotile isolates recovered from beef cattle and carcasses at processing plants in the Midwestern states of the United States. *Appl Environ Microbiol* 67:3810–3818.

21. **Arthur TM, Barkocy-Gallagher GA, Rivera-Betancourt M, Koohmaraie M.** 2002. Prevalence and characterization of non-O157 Shiga toxin-producing *Escherichia coli* on carcasses in commercial beef cattle processing plants. *Appl Environ Microbiol* 68:4847–4852.

22. **Collis VJ, Reid CA, Hutchison ML, Davies MH, Wheeler KPA, Small A, Buncic S.** 2004. Spread of marker bacteria from the hides of cattle in a simulated livestock market and at an abattoir. *J Food Prot* 67:2397–2402.

23. **Dewell GA, Simpson CA, Dewell RD, Hyatt DR, Belk KE, Scanga JA, Morley PS, Grandin T, Smith GC, Dargatz DA, Wagner BA, Salman MD.** 2008. Impact of transportation and lairage on hide contamination with *Escherichia coli* O157 in finished beef cattle. *J Food Prot* 71:1114–1118.

24. **Mather AE, Reid SW, McEwen SA, Ternent HE, Reid-Smith RJ, Boerlin P, Taylor DJ, Steele WB, Gunn GJ, Mellor DJ.** 2008. Factors associated with cross-contamination of hides of Scottish cattle by *Escherichia coli* O157. *Appl Environ Microbiol* 74:6313–6319.

25. **Stanger KJ, Ketheesan N, Parker AJ, Colman CJ, Lazzaroni SM, Fitzpatrick LA.** 2005. The effect of transportation on the immune status of *Bos indicus* steers. *J Anim Sci* 83:2632–2636.

26. **Arthur TM, Bosilevac JM, Brichta-Harhay DM, Guerini MN, Kalchayanand N, Shackelford SD, Wheeler TL, Koohmaraie M.** 2007. Transportation and lairage environment effects on prevalence, numbers and diversity of *Escherichia coli* O157:H7 on hides and carcasses of beef cattle at processing. *J Food Prot* 70:280–286.

27. **Bach SJ, McAllister TA, Mears GJ, Schwartzkopf-Genswein KS.** 2004. Long-haul transport and lack of preconditioning increases fecal shedding of *Escherichia coli* O157:H7 by calves. *J Food Prot* 67:672–678.

28. **Barham AR, Barman BL, Johnson AK, Allen DM, Blanton JR Jr, Miller MF.** 2002. Effects on the transportation of beef cattle from the feedyard to the packing plant on prevalence levels of *Escherichia coli* O157 and *Salmonella* spp. *J Food Prot* 65:280–283.

29. **Fegan N, Higgs G, Duffy LL, Barlow RS.** 2009. The effects of transport and lairage on counts of *Escherichia coli* O157 in the feces and on the hides of individual cattle. *Foodborne Pathog Dis* 6:1113–1120.

30. **Minihan D, Whyte P, O'Mahony M, Clegg T, Collins JD.** 2003. An investigation on the effect of transport and lairage on the faecal shedding prevalence of *Escherichia coli* O157 in cattle. *J Med Vet Ser B* 50:378–382.

31. **Stanford K, Bryan M, Peters J, González LA, Stephens TP, Schwartzkopf-Genswein.** 2011. Effects of long- or short-haul transportation of slaughter heifers and cattle liner microclimate on hide contamination with *Escherichia coli* O157. *J Food Prot* 74:1605–1610.

32. **Federal Register.** 1999. Beef products contaminated with *Escherichia coli* O157:H7. U.S. Department of Agriculture, Food Safety and Inspection Service, Code of Federal Regulations. January 19, 1999. *Fed Regist* 64:2803–2805.

33. **Federal Register.** 2011. Shiga toxin-producing *Escherichia coli* in certain raw beef products. U.S. Department of Agriculture, Food Safety and Inspection Service, Code of Federal Regulations. September 20, 2011. *Fed Regist* 76:58157–58165.

34. **Brusa V, Aliverti V, Aliverti F, Ortega EE, de la Torre JH, Linares LH, Sanz ME, Etcheverria AI, Padola NL, Galli L, Peral Garcia P, Copes J, Leotta GA.** 2013. Shiga toxin-producing *Escherichia coli* in beef retail markets from Argentina. *Front Cell Infect Microbiol* 2:1–6.

35. **Kalchayanand N, Arthur TM, Bosilevac JM, Wells JE, Wheeler TL.** 2013. Chromogenic agar medium for detection and isolation of *Escherichia coli* serogroups O26, O45, O103, O111, O121, and O145 from fresh beef and cattle feces. *J Food Prot* 76:192–199.

36. **Bosilevac JM, Koohmaraie M.** 2011. Prevalence and characterization of non-O157 Shiga toxin-producing *Escherichia coli* isolates from commercial ground beef in the United States. *Appl Environ Microbiol* **77:**2103–2112.

37. **Bosilevac JM, Arthur TM, Wheeler TL, Shackelford SD, Rossman M, Reagan JO, Koohmaraie M.** 2004. Prevalence of *Escherichia coli* O157 and levels of aerobic bacteria and *Enterobacteriaceae* are reduced when hides are washed and treated with cetylpyridinium chloride at a commercial beef processing plant. *J Food Prot* **67:**646–650.

38. **Byrne CM, Bolton DJ, Sheridan JJ, McDowell DA, Blair IS.** 2000. The effects of preslaughter washing on the reduction of *Escherichia coli* O157:H7 transfer from cattle hides to carcasses during slaughter. *Lett Appl Microbiol* **30:**142–145.

39. **Nou X, Rivera-Betancourt M, Bosilevac JM, Wheeler TL, Shackelford SD, Gwartney BL, Reagan JO, Koohmaraie M.** 2003. Effect of chemical dehairing on the prevalence of *Escherichia coli* O157:H7 and the levels of aerobic bacteria and *Enterobacteriaceae* on carcasses in a commercial beef processing plant. *J Food Prot* **66:**2005–2009.

40. **Mies PD, Covington BR, Harris KB, Lucia LM, Acuff GR, Savell JW.** 2004. Decontamination of cattle hides prior to slaughter using washes with and without antimicrobial agents. *J Food Prot* **67:**579–582.

41. **Bosilevac JM, Shackelford SD, Brichta DM, Koohmaraie M.** 2005. Efficacy of ozonated and electrolyzed oxidative waters to decontaminate hides of cattle before slaughter. *J Food Prot* **68:**1393–1398.

42. **Small A, Wells-Burr B, Buncic S.** 2005. An evaluation of selected methods for the decontamination of cattle hides prior to skinning. *Meat Sci* **69:**263–268.

43. **Baird BE, Lucia LM, Acuff GR, Harris KB, Savell JW.** 2006. Beef hide antimicrobial interventions as a means of reducing bacterial contamination. *Meat Sci* **73:**245–248.

44. **Carlson BA, Ruby J, Smith GC, Sofos JN, Bellinger GR, Warren-Serna W, Centrella B, Bowling RA, Belk KE.** 2008. Comparison of antimicrobial efficacy of multiple beef hide decontamination strategies to reduce levels of *Escherichia coli* O157:H7 and *Salmonella*. *J Food Prot* **71:**2223–2227.

45. **Antic D, Blagojevic B, Ducic M, Mitrovic R, Nastasijevic I, Buncic S.** 2010. Treatment of cattle hides with shellac-in-ethanol solution to reduce bacterial transferability—a preliminary study. *Meat Sci* **85:**77–81.

46. **Sillankorva SM, Oliveira H, Azeredo J.** 2012. Bacteriophages and their role in food safety. *Int J Microbiol* **2012:**863945.

47. **Bowling RA, Clayton RP.** 1992. *Method for dehairing animals.* U.S. patent 5,149,295.

48. **Schnell TD, Sofos JN, Littlefield VG, Morgan JB, Gorman BM, Clayton RP, Smith GC.** 1995. Effects of postexanguination dehairing on the microbial load and visual cleanliness of beef carcasses. *J Food Prot* **58:**1297–1302.

49. **Castillo A, Dickson JS, Clayton RP, Lucia LM, Acuff GR.** 1998. Chemical dehairing of bovine skin to reduce pathogenic bacteria and bacteria of fecal origin. *J Food Prot* **61:**623–625.

50. **Bosilevac JM, Nou X, Osborn MS, Allen DM, Koohmaraie M.** 2005. Development and evaluation of an on-line hide decontamination procedure for use in a commercial beef processing plant. *J Food Prot* **68:**265–272.

51. **Schmidt JW, Wang R, Kalchayanand N, Wheeler TL, Koohmaraie.** 2012. Efficacy of hypobromous acid as a hide-on carcass antimicrobial intervention. *J Food Prot* **75:**955–958.

52. **Empey WA, Scott WJ.** 1939. Investigation of chilled beef. Part 1. Microbial contamination acquired in the meat works. *Council for Sci. and Indus. Res. Bull. No. 126*, Melbourne, Australia.

53. **Bell RG.** 1997. Distribution and sources of microbial contamination on beef carcasses. *J Appl Microbiol* **82:**292–300.

54. **Patterson JT.** 1969. Hygiene in meat processing plants. *Rec Agric Res Minist Agric NI* **18:**85–87.

55. **Castillo A, Lucia LM, Goodson KJ, Savell JW, Acuff GR.** 1998. Use of hot water for beef carcass decontamination. *J Food Prot* **61:**19–25.

56. **Federal Register.** 1996. Notice of policy change; achieving the zero tolerance performance standard for beef carcasses by knife trimming and vacuuming with hot water or steam; use of acceptable carcass interventions for reducing carcass contamination without prior agency approval. U.S. Department of Agriculture, Food Safety and Inspection Service. *Fed Regist* **61:**15024–15027.

57. **Dorsa WJ, Cutter CN, Siragusa GR.** 1996. Effectiveness of a steam-vacuum sanitizer for reducing *Escherichia coli* O157:H7 inoculated to beef carcass surface tissue. *Lett Appl Microbiol* **23:**61–63.

58. **Dorsa WJ, Cutter CN, Siragusa GR, Koohmaraie M.** 1996. Microbial decontamination of beef and sheep carcasses by steam, hot water spray washes, and a steam-vacuum sanitizer. *J Food Prot* **59:**127–135.

59. **Dorsa WJ, Cutter CN, Siragusa GR.** 1997. Effects of acetic acid, lactic acid and trisodium phosphate

on the microflora of refrigerated beef carcass surface tissue inoculated with *Escherichia coli* O157:H7, *Listeria innocua,* and *Clostridium sporogenes. J Food Prot* **60:**619–624.

60. **Dorsa WJ.** 1997. New and established carcass decontamination procedures commonly used in the beef-processing industry. *J Food Prot* **60:**1146–1151.

61. **Castillo A, Lucia LM, Goodson KJ, Savell JW, Acuff GR.** 1999. Decontamination of beef carcass surface tissue by steam vacuuming alone and combined with hot water and lactic acid sprays. *J Food Prot* **62:**146–151.

62. **Phebus RK, Nutsch AL, Schafer DE, Wilson RC, Riemann MJ, Leising JD, Kastner CL, Wolf JR, Prasai RK.** 1997. Comparison of steam pasteurization and other methods for reduction of pathogens on surfaces of freshly slaughtered beef. *J Food Prot* **60:**476–484.

63. **Nutsch AL, Phebus RK, Riemann MJ, Schafer DE, Boyer JE, Wilson CR, Leising JD, Kastner CL.** 1997. Evaluation of a steam pasteurization process in a commercial beef processing facility. *J Food Prot* **60:**485–492.

64. **Siragusa GR.** 1995. The effectiveness of carcass decontamination systems for controlling the presence of pathogens on the surfaces of meat animal carcasses. *J Food Saf* **15:**229–238.

65. **Castillo A, Lucia LM, Roberson DB, Stevenson TH, Mercado I, Acuff GR.** 2001. Lactic acid sprays reduce bacterial pathogens on cold beef carcass surfaces and in subsequently produced ground beef. *J Food Prot* **64:**58–62.

66. **Kotula KL, Thelappurate R.** 1994. Microbiological and sensory attributes of retail cuts of beef treated with acetic and lactic solutions. *J Food Prot* **57:**665–670.

67. **Cheng CM, Kaspar CW.** 1998. Growth and processing conditions affecting acid tolerance in *Escherichia coli* O157:H7. *Food Microbiol* **15:**157–166.

68. **Padhye NV, Doyle MP.** 1992. *Escherichia coli* O157:H7: epidemiology, pathogenesis, and methods for detection in foods. *J Food Prot* **55:**555–565.

69. **Wang G, Doyle MP.** 1998. Heat shock response enhances acid tolerance of *Escherichia coli* O157:H7. *Lett Appl Microbiol* **26:**31–34.

70. **Castillo A, Lucia LM, Goodson KJ, Savell JW, Acuff GR.** 1998. Comparison of water wash, trimming, and combined hot water and lactic acid treatments for reducing bacteria of fecal origin on beef carcasses. *J Food Prot* **61:**823–828.

71. **Hardin MD, Acuff GR, Lucia LM, Oman JS, Savell JW.** 1995. Comparison of methods for contamination removal from beef carcass surfaces. *J Food Prot* **58:**368–374.

72. **Pittman CI, Geornaras I, Woerner DR, Nightingale KK, Sofos JN, Goodridge L, Belk KE.** 2012. Evaluation of lactic acid as an initial and secondary subprimal intervention for *Escherichia coli* O157:H7, non-O157 Shiga toxin-producing *E. coli,* and a nonpathogenic *E. coli* surrogate for *E. coli* O157:H7. *J Food Prot* **75:**1701–1708.

73. **Fouladkhah A, Geornaras I, Yang H, Belk KE, Nightingale KK, Woerner DR, Smith GC, Sofos JN.** 2012. Sensitivity of Shiga toxin-producing *Escherichia coli,* multidrug-resistant *Salmonella,* and antibiotic-susceptible *Salmonella* to lactic acid on inoculated beef trimmings. *J Food Prot* **75:**1751–1758.

74. **Kalchayanand N, Arthur TM, Bosilevac JM, Schmidt JW, Wang R, Shackelford SD, Wheeler TL.** 2012. Evaluation of commonly used antimicrobial interventions for fresh beef inoculated with Shiga toxin-producing *Escherichia coli* serotypes O26, O45, O103, O111, O121, O145, and O157:H7. *J Food Prot* **75:**1207–1212.

75. **Yoder SF, Henning WR, Mills EW, Doores S, Ostiguy N, Cutter CN.** 2012. Investigation of chemical rinses suitable for very small meat plants to reduce pathogens on beef surfaces. *J Food Prot* **75:**14–21.

76. **Du WX, Olsen CW, Avena-Bustillos RJ, McHugh TH, Levin CE, Friedman M.** 2008. Storage stability and antibacterial activity against *Escherichia coli* O157:H7 of carvacrol in edible apple films made by two different casting methods. *J Agric. Food Chem* **56:**3082–3088.

77. **McDonnell MJ, Rivas L, Burgess CM, Fanning S, Duffy G.** 2012. Evaluation of carvacrol for the control of *Escherichia coli* O157 on cattle hide and carcass cuts. *Foodborne Pathog Dis* **9:**1049–1052.

78. **Gill CO, Landers C.** 2003. Microbiological effects of carcass decontaminating treatments at four beef packing plants. *Meat Sci* **65:**1005–1011.

79. **Gill CO, Badoni M.** 2004. Effects of peroxyacetic acid, acidified sodium chlorite or lactic acid solutions on the microflora of chilled beef carcasses. *Int J Food Microbiol* **91:**43–50.

80. **King DA, Lucia LM, Castillo A, Acuff GR, Harris KB, Savell JW.** 2005. Evaluation of peroxyacetic acid as a post-chilling intervention for control of *Escherichia coli* O157:H7 and *Salmonella* Typhimurium on beef carcass surfaces. *Meat Sci* **69:**401–407.

81. **Elramady MG, Aly SS, Rossitto PV, Crook JA, Cullor JS.** 2013. Synergistic effects of lactic acid and sodium dodecyl sulfate to decontaminate *Escherichia coli* O157:H7 on cattle hide sections. *Foodborne Pathog Dis* **10:**661–663.

82. **Signorini ML, Tarabla HD.** 2010. Interventions to reduce verocytotoxigenic *Escherichia coli* in ground beef in Argentina. *Prev Vet Med* **94:**36–42.

Veterinary Public Health Approach to Managing Pathogenic Verocytotoxigenic *Escherichia coli* in the Agri-Food Chain

23

GERALDINE DUFFY[1] and EVONNE McCABE[1]

INTRODUCTION

It is well documented that animals and, in particular, ruminants can carry a range of potentially harmful pathogens, including verocytotoxigenic *Escherichia coli* (VTEC), in their gastrointestinal tract. VTEC can reportedly survive for several months in the environment, in feces, and in soil, which allows for the recycling of VTEC among food animals and wildlife and prolonged environmental contamination. VTEC contamination of fresh produce may arise from irrigation water, manure or compost applied to soil as a fertilizer, and feces of wildlife or farmed animals. Table 1 summarizes some of the diverse VTEC outbreaks over the past few years (2006–2013). Of significant interest is that apart from the newly emerging vehicles of infection, serogroups other than O157 are increasingly causing outbreaks, many of which have severe outcomes with cases of hemolytic-uremic syndrome (HUS) and fatalities. While outbreaks are important and gain notoriety, the contribution of sporadic cases of human VTEC infection cannot be ignored. The data show that although foods of animal origin such as meat and dairy products and fresh produce such as salads and vegetables are well recognized important vectors of infection, there have also been VTEC outbreaks related to direct contact with fecal matter through recreational activities including visiting petting zoos, attending agricultural fairs,

[1]Teagasc Food Research Centre, Ashtown, Dublin 15, Ireland.

Enterohemorrhagic Escherichia coli *and Other Shiga Toxin-Producing* E. coli
Edited by Vanessa Sperandio and Carolyn J. Hovde
© 2015 American Society for Microbiology, Washington, DC
doi:10.1128/microbiolspec.EHEC-0023-2013

TABLE 1 Selected VTEC outbreaks (2006–2013) highlighting different vehicles of infection and serogroups

Country	Reference(s)	Year	Serotype	No. of cases	No. of HUS cases	No. of deaths	Likely source or mode of transmission
United Kingdom	Launders et al., 2013 (142)	2013	O157	19	0	0	Watercress
Denmark	Soborg et al., 2013 (143)	2012	O157	13	8	0	Ground beef
United States	Slayton et al., 2013 (144)	2012	O157	58	3	0	Romaine lettuce
Germany	Karch et al., 2012 (10)	2011	O104	4,321	852	32	Sprouts
United States	McCollum et al., 2010 (104)	2010	O157	41	1	0	Cheese Gouda
United States	Taylor et al., 2013 (145)	2010	O145	33	3	0	Romaine lettuce
Netherlands	Greenland et al., 2009 (146)	2009	O157	20	0	0	Steak tartare
United States	Neil et al., 2012 (147)	2009	O157	72	10	0	Cookie dough
United States	CDC, 2010 (148)	2008	O157	99	1	0	Ground beef
Ireland	O'Sullivan et al., 2008 (89)	2008	O157	148	NA[a]	NA	Private wells drinking water
Belgium	De Schrijver et al., 2008 (107)	2008	O26 & O145	12	3	0	Homemade ice cream, fecal material from farm
Netherlands/ Iceland	Friesema et al., 2008 (149)	2007	O157	50	0	0	Lettuce
Denmark	Ethelberg et al., 2009 (150)	2007	O26	20	0	0	Fermented beef sausage
United States	Wendel et al., 2009 (151); Jay et al., 2007 (69)	2006	O157	199	31	3	Fresh spinach and feral pigs

[a]Data not available.

and swimming in contaminated water (1). With so many routes of potential transmission, it is clear that the management of this pathogen requires a multidisciplinary approach with cooperation among the disciplines of human medicine, animal and veterinary sciences, and environmental and food scientists. A veterinary public health approach to managing VTEC focuses on collation of data on prevalence of VTEC in different sources, the importance of each source as a reservoir of human pathogen VTEC, and risk factors for transmission to humans.

This article provides an overview of the risks for transmission of human pathogenic VTEC from the various routes of transmission, including animals, the farm environment, and production of milk and meat, and highlights general management approaches to reduce the risk of human exposure from such sources.

HUMAN PATHOGENIC VTEC

VTEC strains are a heterogeneous group of *E. coli* serogroups, and not all will cause

human illness. The diversity of serogroups and virulence factors among human-disease-causing VTEC strains poses considerable challenges in assessing the risk posed by VTEC recovered from the agri-food chain. The dominant pathogenic VTEC serotype remains *E. coli* O157:H7 (2, 3) and has the strongest association with HUS worldwide (4). However, a common pattern being observed in the European Union and the United States (5) is the increasing number of reported cases now attributed to non-O157 serotypes. Approximately half of all confirmed VTEC cases in the European Union are now associated with serogroup O157 (6), and in the United States in the period 2000 to 2010, O157 cases decreased from 2.17 to 0.95/100,000 whereas non-O157 cases increased from 0.12 to 0.95/100,000 (5). Of the non-O157 cases, six VTEC serogroups are most commonly isolated from patients: O26, O103, O145, O111, O21, and O45. However, it should be noted that not all strains belonging to these serogroups will cause severe illness and that other non-O157 VTEC serogroups also cause illness. Pathogenic VTEC strains are categorized as enterohemorrhagic *E. coli*, and usually in these strains, a large outer membrane protein (94–97 kDa) called intimin mediates the intimate contact between the bacterium and the enterocyte cytoplasmic membrane (attachment) and the destruction of the enterocyte microvilli (effacement). The genetic determinants for this (*eae, tir, esc,* and *sep* genes) are grouped together on the chromosome, forming a pathogenicity island called LEE, for locus of enterocyte effacement (7). The *eae* gene encodes for intimin, which is responsible for the intimate attachment, and at least five different forms (α, β, γ, δ, ε) have been reported for VTEC strains (8). The *tir* gene codes for a type III-secreted translocated intimin receptor (Tir protein). The *esp* genes code for type III-secreted translocated Esp proteins. There are, however, reports of LEE-negative *E. coli* O157 clones causing illness in humans (9), indicating that such strains express alternative adherence factors that allow them to colonize the intestinal tract.

In 2011, an outbreak strain that contained a combination of *vt* genes and virulence factors normally seen in enteroaggregative *E. coli* (EAEC) emerged as the principal protagonist in a major outbreak in Germany. The outbreak resulted in 4,321 confirmed cases, including 852 cases of HUS, with 54 deaths reported in 14 European Union member states, the United States, and Canada (10). This highly unusual hybrid organism combined the *vt* genes and the adhesion mechanisms of an EAEC (11) instead of the Tir/intimin and type III secretion and the system usually seen in enterohemorrhagic *E. coli*. Moreover, the strain also demonstrated extended spectrum beta-lactamase phenotype, thus further underlining the enhanced virulence of the strain and the severity of the outbreak, aided by the specific combination of enhanced adhesion, survival fitness, *vt2* production, and antibiotic resistance as a result of the high genome plasticity of this *E. coli* pathogen. This outbreak clone was a clear example of gene acquisition by means of lateral gene transfer that resulted in an accumulation of synergistic virulence factors. This illustrates that pathotypes of *E. coli* can have common characteristics that overlap and that they do and have the potential to evolve rather than stand as a fixed archetype (12).

Although no single marker or combination of markers defines "pathogenic" VTEC associated with human disease (13), the presence of *vt2-* and *eae*-positive strains is associated with a high risk of more serious illness and an increased risk of HUS (14), but other virulence gene combinations and/or serotypes may also be associated with serious disease in humans, including HUS. Patient-associated factors such as age, immune status, and the antibiotic therapy administered may also influence the likelihood and severity of disease.

The concept of seropathotype has evolved, which ranks the potential for a particular VTEC to cause serious human illness on the basis of both serotype and the presence of particular virulence genes. A recent European Food Safety Authority Scientific Opinion (15) proposed that any isolates of VTEC serogroups O157, O26, O103, O145, O111, or O104 that also have *eae* (intimin) or *aaiC* (secreted protein of EAEC) plus *aggR* (plasmid-encoded regulator) genes should be considered as presenting a potentially high risk for diarrhea and HUS. For any other VTEC serogroups with the same combination of virulence genes, the potential risk is regarded as high for diarrhea, but currently unknown for HUS. While for any serogroups not having this gene combination, the currently available data do not allow any inference regarding potential risks. This concept allows food business operators in Europe to make an assessment of the risk posed by a VTEC isolate recovered from a food, animal, or environmental sample. Nonetheless, to support stakeholders and food business operators, more information is urgently needed on what constitutes a human virulent VTEC, and this will remain an active area of research, and rapidly evolving techniques such as whole genome sequencing will progress this area.

CARRIAGE OF VTEC IN ANIMALS

VTEC strains have been isolated from a variety of animal carriers. including cattle, sheep, horses, deer, goats, dogs, geese, pigs, and wild birds. Recent studies on prevalence of VTEC in ruminant animals and meats are reviewed in Table 2. Moreover, feces from cattle, sheep, goats, deer, and rabbits have all been linked to cases of infection (16–20). Transmission to humans can occur as a result of direct contact with VTEC-contaminated fecal material, from handling or petting animals, or by exposure to fecally contaminated mud or vegetation during recreational activities.

Bovine Carriage of VTEC

Cattle are recognized as a principal reservoir for VTEC (21). VTEC is generally a transient member of the intestinal microflora and only rarely causes disease in young, weakened calves. Although cattle have been shown to harbor this pathogen on occasion in their rumen, VTEC is found more frequently in the distal portion of the bovine gastrointestinal tract, with the rectal-anal junction identified as the predominant colonization site (22, 23). Rhoades et al. extensively reviewed the prevalence of VTEC in the beef chain and the fecal prevalence of *E. coli* O157 in cattle and showed it varied from 0 to 48.8% (24–27). In the United States, it is reported that most bovine animals have been exposed to *E. coli* O157 by the time of weaning (28), and the reported prevalence in cattle ranges from 10 to 28% (29). In the European Union, studies have shown that VTEC prevalence in feces ranged in individual animals from 2.1 to 13.5%, in herds from 6.1 to 16.2%, and in slaughter batches from 13 to 20.2%, and for O157 this ranged from 0.3 to 2.3%, 1.5 to 13.7%, and 5.5 to 20.2%, respectively (6).

Shedding of VTEC by cattle is generally intermittent, with herd members remaining negative for months, with only a proportion sporadically becoming positive for a few weeks at a time (30, 31). Carriage rates are higher in calves than in adult cattle, and while detected all year-round, carriage rates are subject to seasonal effects, with higher rates reported in spring and summer. Other factors thought to influence carriage rates are the ages of animals and farming and husbandry practices (32).

The number of VTEC (CFU g^{-1}) pathogens being shed in the feces of individual animals is considered important in the context of hide, environmental, and subsequent carcass contamination. The typical pattern of shedding in a herd is sporadic, with intense periods of shedding interspersed with periods of non-shedding (33). Ogden et al. have also reported that concentrations of *E. coli* O157 being shed in the feces of positive cattle were highest during summer months (34). The phenomenon of "super-shedding" animals (those shedding >10^4 CFU/g feces) is thought to be a significant contributor in the dissemination of O157 VTEC within and between herds and within abattoirs (35–37). However, quantitative data are few relative to prevalence data as the routine detection methods generally employed in surveys are designed to yield data only on presence of the pathogen and not on the numbers present. As a result, data on concentrations of VTEC and on occurrence of super-shedders are limited. Thomas et al. (38) examined feces and hide for concentrations of six VTEC serogroups and showed that the vast majority of samples had counts below the limit of detection of the count method and samples with detectable counts ranged from 60 to 100 CFU/cm^2 on hide and 100 to 1300 CFU g^{-1} in feces. An abattoir study in the United Kingdom found that 70% of *E. coli* O157-positive animals shed <100 CFU g^{-1} of feces, but in some individuals concentrations could be as high as 10^6 CFU g^{-1} of feces (39). These authors also showed that 9% of the animals shedding *E. coli* O157:H7 at slaughter produced over 96% of the total O157:H7 fecal load for the group. It has also been hypothesized that high-level carriage of these microorganisms is a consequence of intestinal colonization whereas low levels

TABLE 2 Selected VTEC prevalence studies (2007–2013) showing pathogenic VTEC prevalence in ruminant animals (cattle and sheep)

Sample type (n = sample no.)	Serotype(s)	Country	Prevalence (%)	Reference
Cattle				
Carcass (300)	O157	Spain	14.7%	Ramoneda et al., 2013 (152)
Feces (1,145)[a]	O157	United States	19.7%,	Dargatz et al., 2013 (153)
Feces (1145)	O45, O103, O121, O145, O26, O111	United States	13.8%, 9.9%, 9.3%, 5.5%, 1.1%, 0.5%	Dargatz et al., 2013
Feces (250)	O157	Mexico	5.2%	Narvaez-Bravo et al., 2013 (154)
Hides (250)	O157	Mexico	11.7%	Narvaez-Bravo et al., 2013
Carcass (250)	O157	Mexico	0.8%	Narvaez-Bravo et al., 2013
Feces (301)	O157	Ireland	2.7%	Thomas et al., 2012 (38)
Feces (402)	O26, O103, O145	Ireland	1.5%, 8.5%, 0.7%	Thomas et al., 2012
Hide (301)	O157	Ireland	18.9%	Thomas et al., 2012
Hide (402)	O26, O103, O145	Ireland	6%, 27.1%, 2.5%	Thomas et al., 2012
Carcass (301)	O157	Ireland	0.7%	Thomas et al., 2012
Carcass (402)	O26, O103, O145	Ireland	0.5%, 5.5%, 0.7%	Thomas et al., 2012
Feces (399)	O26, O103, O111 O145	Belgium	6%, 6%, 6%, 6%	Joris et al., 2011 (47)
Carcass (474)	O157	Denmark	3.4%	Breum et al., 2010 (155)
Carcass (1,622)	O157	Argentina	2.6%	Masana et al 2010 (156)
Feces (1,622)	O157	Argentina	4.1%	Masana et al., 2010
Ear (446)	O157	Sweden	12%	Boqvist et al., 2009 (157)
Feces (1,758)	O157	Sweden	3.4%	Boqvist et al., 2009
Sheep				
Fleece (500)	O157	Ireland	0.8%	Thomas et al., 2013 (61)
Fleece (500)	O26, O103, O145	Ireland	1.0%, 16.8%, 0.2%	Thomas et al., 2013
Carcass (500)	O157	Ireland	0.6%	Thomas et al., 2013
Carcass (500)	O26, O103, O145	Ireland	0.4%, 13.6%, 0	Thomas et al., 2013
Feces (17,550)[b]	O103	Sweden	0.7%	Sekse et al., 2013 (58)
Feces (492)	O157	Sweden	1.8%	Soderlund et al., 2012 (51)
Ear (105)	O157	Sweden	1.9%	Soderlund et al., 2012
Feces (1,082)	O157	Scotland	3.4%	Evans et al., 2011 (56)
Feces (1,082)	O26, O103, O145	Scotland	3.4%, 5.2%, 2.3%	Evans et al., 2011
Feces (491)[c]	O26	Norway	17.9%	Sekse et al., 2011 (52)
Feces (533)	O157	Italy	7.1%	Franco et al., 2009 (54)
Fleece (400)	O157	Ireland	5.75%	Lenahan et al., 2007 (60)
Carcass (400)	O157	Ireland	1.5%	Lenahan et al., 2007

[a]Individual fecal swabs were collected from cattle approximately 60 days after their arrival in the feedlot and were pooled for evaluation (153).
[b]The investigation of fecal samples from 585 sheep flocks resulted in 1,222 *E. coli* O103 isolates that were analyzed for the presence of *eae* and *stx* genes; the study documented a low prevalence (0.7%) of isolates potentially pathogenic to humans (58).
[c]This number represents the number of flocks that were tested in the study using fecal samples to determine the prevalence; in total, 491 flocks were tested and *E. coli* O26 was detected in 17.9% of the flocks (52).

within individual animals may be a result of environmental exposure with no significant colonization (32, 40). Horizontal transmission between animals may be facilitated by contaminated water and feed, with water troughs potentially playing a role in the ecology of VTEC on the farm (41–46).

Recent studies have examined the human virulence potential of different bovine VTEC isolates, and an interesting picture is seen for O157 versus non-O157 VTEC serogroups. Whereas >90% of *E. coli* O157 isolates recovered from the beef chain in Ireland possessed *vt* and *eae* in combination, <10% of non-O157 VTEC serogroups fell into this category (38). In a Belgian study on O26, O103, O145, and O111 in cattle feces, about 6% of samples were positive and about 50% of isolates had key human virulence genes (47, 48). A further study showed that for *E. coli* O157 genotypes associated with human illness a minor sub-population was in the bovine reservoir. (47, 48) This shows that while VTEC isolates in beef play a role in human illness, a risk assessment of their virulence potential is essential.

Carriage of VTEC in Sheep and Goats

The presence of naturally occurring VTEC has been widely reported in sheep, and studies indicate that sheep feces and sheep meat are important reservoirs of human pathogenic VTEC (49–55). Seasonal prevalence of VTEC in sheep has been reported (54, 56), with incidences peaking in summer months in a similar trend to that in cattle and human infections (45). Much of the data on VTEC colonization of small ruminants relate to *E. coli* O157. When colonized, small ruminants generally show no clinical symptoms of illness and reinfection occurs frequently, although in young unweaned lambs or kids, scouring or diarrhea may occur. Some animals, particularly those that are persistently colonized, can excrete exceptionally high numbers of *E. coli* O157 (>10,000 CFU/g) in their feces (54, 57). There is some evidence of higher shedding of O157 in adult sheep and hoggets than in lambs (56).

   E. coli O103 is one of the most common non-O157 VTEC serotypes isolated from human cases in Europe, and a severe human outbreak was reported in Norway in 2006 that was caused by *E. coli* O103:H25 (58). This was due to fermented sausage mainly consisting of mutton, which was shown to be the food source of the outbreak (59). However, only eae⁺vt⁻ *E .coli* O103:H25 was ever isolated from food and sheep, but it was proposed that the infection was caused by VTEC and that recovered isolates probably had lost their *vt* genes during laboratory cultivation. In studies carried out in Ireland, the profile of *E. coli* O157:H7 recovered from sheep showed that 29/33 (87%) of isolates contained *vt2*, *eae*, and *hlyA* genes, whereas only five isolates (15%) contained the *vtx*1 gene (60, 61). Thomas et al. showed recovery of pathogenic O26, O103,

and O145 (with *eae* and *vt*) at level <1% on sheep fleece and carcases (61). For O26, isolates with *vt* and *eae* represented about 50% of all isolates of the serogroups recovered, while for O103 it was only about 0.01% of all isolates. As for bovine isolates, these data highlight the need to risk assess any recovered VTEC regardless of serogroup for key virulence genes (61).

Carriage of VTEC in Pigs

Although pigs are also known carriers of O157 VTEC, prevalence rates are relatively low, ranging from 0.4 to 2.1% (62–65). In pigs, VTEC can cause edema disease and may also be involved in postweaning diarrhea syndromes. In general, it is VTEC serogroups O138, O139, and O141 that are implicated in illness in pigs, and these have not been widely linked to human illness. The presence of *E. coli* O157 in pigs has been reported in the United States (62), Norway, and the Netherlands (66, 67). In a study carried out in Ireland, the prevalence of *E. coli* O157:H7 in pigs was found to be very low, and only four isolates were recovered from 1,710 pigs examined. However, three of these four contained the genes *vtx2*, *eae*, and *hlyA*, indicating their potential to cause illness in humans (68). While pork meat and products have generally not been implicated in human illness, in California in 2006, feral swine were implicated in contamination of agricultural fields and surface waterways with *E. coli* O157:H7, which caused a large outbreak linked to bagged spinach. However, *E. coli* O157 isolates were also recovered from cattle on a nearby ranch that had a similar subtype to the swine isolates as determined by using multilocus variable tandem repeat analysis and pulsed-field gel electrophoresis (PFGE), suggesting that not only had swine-to-swine transmission taken place but, in addition, that interspecies transmission between cattle and swine had occurred, facilitated through a common source of exposure such as water or soil (69). More data on the human virulence assessment of non-O157 serogroups isolated from pigs would provide important information on the likely future importance of their role as a vector of human VTEC infection.

ANIMAL WASTE AS A SOURCE OF VTEC FOR HUMANS

As animal waste can carry VTEC, there is a risk of direct infection of humans from contact with livestock and/or waste in the farm environment. A number of cases of human infection have been associated with visits to farms, agricultural fairs, and petting farms. Visitors to farms are likely to be at a higher risk of infection than farming families who may possess a degree of acquired immunity (70). Awareness by farmers and farming families of the potential dangers of VTEC and the steps required to protect themselves, and any visitors, from VTEC infection is essential. Full consideration of hygiene management is needed at open farms/fairs to reduce risks for visitors. Washing hands after contact with animals, especially before eating and drinking, is a simple control measure; therefore, open farms/fairs should be adequately equipped with hand-washing facilities, and provide designated dining areas that are segregated from animal holding areas. Farmers should be aware that farmyard surfaces such as gates and stiles pose a direct risk as *E. coli* O157:H7 has the potential to persist for long periods on these surfaces (70). In addition, adequate guidance and supervision before and during farm visits are a key factor in reducing the risk to children under 5 years of age who come into direct contact with animals.

Survival of VTEC in Animal Waste

Animal waste in the environment can arise from feces directly excreted into the environment by grazing or wild animals and from animal waste collected and stored as slurry or manure. Manures are feces that may have undergone some period of storage, and slurries are mixtures that include manure, urine, and leftover feed that are held in a tank or pit, generally under anaerobic conditions. Such waste is a valuable source of soil nutrients and is generally spread back to the land as fertilizer. If such waste is contaminated with VTEC, it poses a reservoir of contamination for animals, water supplies, crops, and fresh produce.

In experiments on the survival of *E. coli* O157:H7 in feces spread outdoors on grass under ambient weather conditions, the pathogen could be recovered directly from feces on the grass for 50 days, and when feces were no longer visible on the grass, enrichment techniques showed the organism was recoverable from the underlying soil for a further 49 days (71). A similar study in the United Kingdom (72) reported the survival of *E. coli* O157 in cattle feces for >50 days but reported much shorter survival times in cattle slurry in which it fell to undetectable levels within 10 days. However, McGee et al. reported that *E. coli* O157:H7 was recoverable from bovine slurry for up to 3 months (73). Semenov et al. showed the influence of aerobic and anaerobic conditions on the survival of *E. coli* O157:H7 in cattle manure and slurry (74). The data showed that *E. coli* O157:H7 survived significantly longer under anaerobic than under aerobic conditions, with survival ranging from approximately 2 weeks for aerobic manure and slurry to more than 6 months for anaerobic manure at 16°C. The importance of changes in microbial community and chemical composition of manure and slurry was highlighted as affecting the survival of *E. coli* O157:H7 in different oxygen conditions.

In further field experiments to determine the survival times of pathogens in livestock manures during storage and following land application, VTEC survived for less than 1 month in solid manure heaps where temperatures greater than 55°C were obtained. Following spreading of manure to land, *E. coli* O157 generally survived in the soil for up to 1 month after application to both sandy arable and clay loam grassland soils (75).

When left on the ground surface, pathogens such as VTEC are exposed to the full force of fluctuating environmental conditions that has been shown to increase the inactivation rate as compared to the immediate incorporation into the soil (75, 76). It has been found that *E. coli* O157 can survive on surface vegetation for 6 weeks but persisted when injected subsurface (77). VTEC may then be transported over land or below the surface flow to surface and ground waters. Less work has been focused on survival of non-O157 VTEC; however, a study that investigated the survival of *E. coli* O26, O111, and O157 in bovine feces (78) stored at 5, 15, and 25°C demonstrated that all three pathogens survived at 5 and 25°C for 1 to 4 weeks and at 15°C for 1 to 8 weeks when inoculated at a low concentration. At high concentrations, O26, O111, and O157 survived at 25°C for 3 to 12 weeks, at 15°C for 1 to 18 weeks, and at 5°C for 2 to 14 weeks, respectively.

In a study by Bolton et al. soil was inoculated with O26:H11, O113:H4, ONT:H4, O2:H27, O116:H28, O6:H8, ONT:H27, O119:H5, O145:H27, O20:H19, O174:H21, O168:H8, O136:H2, and O86:H21; the samples were stored at 10°C for up to 201 days and showed *D* values (time required to achieve a 90% or 1 log reduction) ranging from 50.26 to 75.60 days in sandy loam and 31.60 to 48.25 days in clay-based soils (79).

Control in Animal Waste

Such long-term survival of *E. coli* O157:H7 in manure and slurry emphasizes the need for appropriate farm waste management to curtail environmental spread of this bacterium and the risk of human exposure.

General measures to control the spread of pathogens including VTEC on the farm are outlined in farm quality assurance and good agricultural practices schemes (80). General measures to control the spread and persistence of VTEC on the farm include good management and hygiene, clean and dry bedding, appropriate stocking rates, well-ventilated housing with good floor drainage, and practicing a closed herd policy (81, 82).

Manure can be subjected to active treatment processes including composting, heat drying, and digestion. In general, these treatments are according to legislative guidance, which are primarily devised to achieve efficient and sustainable nutrient use. But any treatment will also have implications for the pathogen/VTEC load in the final output. The regulations generally specify the storage requirements and spreading restrictions to prevent diffuse and point-source contamination of adjacent surface and ground waters.

Composting is a useful way of managing solid animal waste. Under controlled conditions of aeration, moisture, particle size, and carbon-nitrogen ratio of the combustible material, composting temperatures of 55 to 65°C can be reached, which should be sufficient to inactivate VTEC. Lung et al. showed that *E. coli* O157:H7 inoculated at a level of log_{10} 7.00 CFU g^{-1} was nondetectable after 72 h in composted manure (83). Research by Shepherd et al. showed that compost heaps covered with finished compost maintained temperatures under the physical covering that were around 7 to 15.5°C higher than in an uncovered heap, resulting in a faster reduction of *E. coli* O157:H7 than was observed in heaps covered with fresh straw or left uncovered (84). This validated recommendations of the U.S. Environmental Protection Agency for covering fresh compost (84).

Treatments particularly suited to slurries include anaerobic digestion typically at 30 to 35°C with 12 days of retention for pig slurry or 20 days for cattle and poultry slurry. The addition of lime (quick lime or slaked lime to raise the pH to 12 for at least 2 h) should also result in significant pathogen reduction (85).

VTEC in Water

Surface water and unprotected groundwater may become contaminated with VTEC from livestock effluent or fecal contamination from humans, livestock, wild animals, and domestic pets. Water for both drinking and recreational swimming has been implicated in a number of VTEC outbreaks (86). The susceptibility of drinking water from private water supplies in rural areas to VTEC contamination has been highlighted, with a private water supply being considered as one that is not provided by a water company and includes wells, boreholes, springs, streams, rivers, lakes, or ponds. Such private water supplies are common in both Ireland and Scotland, where rates of human VTEC illness are high, and in both countries private water can be considered a potentially important risk factor for VTEC human infection (87–89). A large *E. coli* O157 outbreak involving a private group water scheme was reported in a rural area of Ireland, involving 18 cases of infection, including 2 HUS cases (46). A study by Tanaro et al. showed that on a beef farm in Argentina, non-O157 isolates with identical PFGE profiles were recovered from cattle rectal swabs, and water streams on the farm highlighted the circulation of VTEC strains among the animals, the environment, and the water course (90). Other studies have shown that the vulnerability of groundwater supplies is influenced by soil types, with groundwater being most at risk where the subsoils are absent or thin and in areas of karstic limestone, where surface streams sink underground at swallow holes (91, 92).

Studies have showed that *E. coli* O157:H7 survived for 13 weeks in lake water held at 15° C (93), while other studies have reported survival in farm water for up to 14 days at temperatures of <15°C (94), 70 days at 5°C, and 40 days at 21°C (95). In addition to temperature, the indigenous microflora (93), scarcity of nutrients (96), and protozoan predations also influence survival. Recent passage of VTEC through the bovine gastrointestinal tract may enhance survival of O157 in water (97).

Contaminated water troughs and livestock drinking water supplies have a role in the transmission of VTEC among animals (33), and large numbers of animals can be infected over a short period (42). VTEC can survive in water trough sediments for at least 4 months and appears to multiply there, especially in warm weather. On many farms, troughs are seldom cleaned so that thick sediments accumulate and remain a long-term potential source for continual recycling of VTEC among a herd. Farmers should clean water troughs frequently to prevent the accumulation of sediments (42). Water trough design and location are also important factors in reducing the possibility of direct fecal contamination. Water troughs should be positioned away from feed troughs/feed passageways, as contamination of water with feed can provide a nutrient-rich substrate for bacterial growth and survival at the bottom of the trough (42). Water troughs should not be located in shaded areas (42), as direct sunlight has a bactericidal effect.

VTEC IN FOODS OF ANIMAL ORIGIN

Food of animal origin such as milk and dairy products may be contaminated with VTEC during milking, and the meat carcass may be contaminated during slaughter and dressing operations.

VTEC in Milk

Although the public health risks associated with consumption of raw milk and raw milk products have been well documented, such products continue to be consumed and promoted by pro-raw milk advocates because of perceived organoleptic and health benefits over pasteurized products (98, 99). However, a number of serious outbreaks of *E. coli* O157 have been associated with the consumption of raw milk (100, 101), raw milk cheese (102), and unpasteurized Gouda cheese (103, 104). Other outbreaks were linked to butter made from raw milk and to commercial ice creams, where cross-contamination was identified as a possible source (105). Allerberger et al. reported two cases of HUS due to *E. coli* O26 linked to the consumption of raw cow's milk (106). However, pasteurized products have also been implicated in outbreaks, and *E. coli* O145 and O26 caused VTEC infection with five HUS cases among consumers of ice cream produced and sold in September 2007 at a farm in Belgium. The ice cream was made from pasteurized milk and most likely was contaminated by one of the food handlers (107). An outbreak of *E. coli* O111 associated with a correctional facility dairy in Colorado showed inmates employed at the dairy might have acquired VTEC O111 infection on the job or transported contaminated clothing or other items into the main correctional facility and kitchen, thereby exposing other inmates (108).

Contamination of raw milk can occur either during milking or after milking from on-farm environmental contamination. A study by Murphy et al. examined milk from 97 Irish dairy cattle farms and isolated *E. coli* O157:H7 from 12% of samples on these farms (109). A subsequent study by Murphy et al. examined Irish raw milk specifically destined to be used in raw milk cheese and detected VTEC in raw milk from sheep and goats in addition to cows (110). Data collated from European Union member states in 2011 (111) showed 5.3% (3/57) of raw milk samples in Germany contained VTEC and 2.6% (1/39) raw milk samples in Belgium contained *E. coli* O157. Among five reporting member states, 1.8% (36/2,045) of all dairy samples were positive for VTEC and 0.1% for *E. coli* O157. Between 2008 and 2011, an Italian study that monitored for the prevalence of *E. coli* O157:H7 in raw milk (60,907 samples) sold by self-service vending machines (*n* = 1,239) in seven Italian regions found just 24/60,907 (0.04%) of samples were positive for *E. coli* O157:H7.

VTEC may reportedly behave differently in raw and pasteurized milk and in derived products. Growth of VTEC may be slower in raw milk and products because of the presence of lactoperoxidase enzymes, natural inhibitors for many pathogenic bacteria in raw milk, and may also be due to the higher numbers of competitive microflora in such products (112). The behavior and the potential for survival/growth of VTEC in dairy products such as cheese depend on factors such as the moisture content, pH, and competitive microflora. Thus the additional hurdles imposed during the hard cheese manufacturing process, including low water activity and pH that develop during the curing process, and the changes in the competing microflora, reduce the survival and growth potential of the pathogen. Peng et al. showed survival of non-O157 following the production and ripening of semi-hard raw milk cheese (113). Miszczycha et al. examined the growth and survival of four VTEC serotypes (O157:H7, O26:H11, O103:H2, and O145:H28) in raw-milk cheeses manufactured and ripened according to five technological schemes: blue-type cheese, uncooked pressed cheese with long ripening and short ripening steps, cooked cheese, and lactic cheese (114). VTEC grew 2 to 3 $\log_{10}$ CFU g^{-1} in the blue-type cheese and the two uncooked pressed cheeses during the first 24 h of cheese making, but levels then progressively decreased in cheeses that were ripened for more than 6 months. In the cooked and lactic cheeses, VTEC did not grow but was detectable at the end of ripening and storage. Interestingly, a serotype effect was observed in all cheeses, with O157:H7 growing slower and less persistently than the other serotypes, indicating that the risk posed by non-O157 serotypes may be higher and the design of safety measures should take this into account.

Several studies reported the ability of *E. coli* O157:H7 to survive and to grow during storage of fermented dairy products. *E. coli* O157:H7 inoculated into commercial products could be recovered for up to 12 days from yogurt, 28 days from sour cream, and 32 to 35 days from buttermilk (115, 116).

VTEC and Meat

Since the first confirmed case in 1982, beef-associated VTEC O157 outbreaks were widely reported (3, 117–119). Between 2007 and 2009, of the 57 verified food-borne outbreaks of VTEC in the European Union, 8 were linked to the consumption of bovine meat or bovine meat products. During the years 2010 and 2011, of the 16 outbreaks reported with strong evidence, 2 were linked to the consumption of bovine meat or bovine meat products (6).

The hide and fleece of animals presented for slaughter are recognized as important sources of VTEC contamination for the carcass. Reported prevalence of *E. coli* O157 on hide varies from 4.7% (120), 17.6 % (38), to 75.7% for feedlot cattle (121). On sheep fleece the reported prevalence of *E. coli* O157 was 5.75 (60) and <1 % (61). Studies looking at an extended range of VTEC serogroups (O26, O111, O145, O103) recovered <1% on bovine hide and on sheep fleece recovered O26 at 1%, with neither O103, O145, nor O111 recovered (38, 61).

Transmission of *E. coli* O157:H7 and other VTEC serotypes can occur rapidly in groups of animals, with contamination of the bovine pens and hides occurring in less than 24 h (122). Thus conditions in transport from farm to factory and in lairage have a significant impact on the level of VTEC contamination on hide or fleece of animals presented for slaughter. The risk of introducing VTEC to a slaughter batch is increased by mixing animals from different farms (123, 124). The length of time animals are in transit to the abattoir and held in lairage also increases the risk of having VTEC-positive hide samples at slaughter, compared with cattle transported a shorter distance (125, 126). *E. coli* O157 has been shown to survive on the hides of live cattle for approximately 9 days. Efforts to reduce the level of hide or fleece soiling are thus warranted for control of VTEC. Farmers can employ suitable animal husbandry facilities and practices, such as bedding quality, stocking density, and feeding regimen, that give clean animals for slaughter although a clean animal does not guarantee absence of VTEC. Livestock transporters can also play an important role in the cleanliness and dryness of animals on arrival at the abattoir by ensuring that trucks are thoroughly washed and disinfected between loads and by not overloading the animals.

For the abattoir, knowledge from the farm of the VTEC prevalence in the slaughter batch could allow risk management of animals coming in for slaughter, including logistic slaughter, but currently there is insufficient information on VTEC at the farm level to implement this approach (6, 127).

The risk of fecal contamination on the carcass at slaughter can be reduced by specific procedures, including "rodding" (a technique used to separate the esophagus from the trachea and diaphragm). Bagging and tying of the bung can also help prevent contamination of the carcass. Removal of hides should be carried out in a manner that avoids contact between the hide and the carcass. This can be achieved by a number of measures, including the use of hide pulling equipment and using clean equipment (immersion of knives in water at 82°C) for the dehiding operation.

The reported prevalence of *E. coli* O157 on prechill carcasses (postevisceration) is low (≤3%) (27). There is a low prevalence (<0.5%) of other clinically significant VTEC serogroups (O145, O111, O103, and O26) on prechill beef and sheep carcasses (38, 61). The similarity of PFGE profiles for VTEC isolates in a study by Thomas et al. indicated the origin of VTEC on a carcass could be its own hide or the hides or fleece of other animals being slaughtered on the same day (38, 61). VTEC that may also be present in the air of the slaughter house, particularly at the hide removal area, may be an underestimated source of carcass contamination (128). The use of carcass antimicrobial interventions is commonplace in the U.S. beef industry (129) and include the use of organic acids, acidified sodium chlorite, trisodium phosphate, activated lactoferrin, chlorine, and chlorine dioxide and hot water. Studies by Kalchayanand et al. indicated that the reductions in non-O157 VTEC by the commonly used antimicrobial interventions were at least as great as for O157 (130). Such antimicrobial interventions for beef carcasses are not currently in use in the European Union beef industry sector. This is mainly due to customer preferences and associated costs.

The survival and concentrations of *E. coli* O157:H7 and other VTEC serogroups on fresh meat during distribution can be affected by the storage temperatures (131, 132), the packaging environment (133), and the competitive microflora (134). The processing of beef cuts into ground beef can lead to transfer of pathogen from the surface of beef into the center of the product, and beef-processing equipment and knives or needles used to cut into or inject whole muscle (in tenderizing beef) can play a role in this spread of contamination (135–137). Studies on thermal inactivation of different single serotypes of VTEC O26:H11, O45:H2, O103:H2, O104:H4, O111:H⁻, O121:H19, O145:NM, and O157:H7 in wafers of ground beef showed that cooking times and temperatures effective for inactivating a serotype O157:H7 strain of *E. coli* in ground beef were equally effective against the seven non-O157:H7 strains investigated (135).

VTEC can reportedly survive in ready-to-eat, dry, and semidry fermented meats, posing a particular risk in these commodities (138). This has led to a recommendation that processing protocols achieve a $\log_{10}$ 5.0 CFU g^{-1} reduction in numbers of *E. coli* O157:H7. Such reductions may be achieved by additional thermal processing, and a number of published studies outline heating steps that can be introduced into the manufacturing process after fermentation to achieve the required decline in pathogen numbers (139). In the absence of a thermal processing step, an extended fermentation or maturation period may prove effective in limiting pathogen numbers.

CONCLUSION

It is clear that there are multiple routes for transmission of human pathogenic VTEC along the farm-to-fork chain. Emerging challenges are new vehicles of infection and a changing profile of the VTEC serogroups that are causing human illness. A lack of clarity on what constitutes a human pathogenic VTEC poses additional challenges in assessing and managing public health risk. Whole genome sequencing approaches are now helping identify the genetic differences between human pathogenic and nonpathogenic VTEC, but more research is needed into the interactions between virulence factors and the human or animal host. Much of the knowledge to date has been obtained from in vitro observations, which do not necessarily reflect what is happening in vivo within a specific host. It is also likely that the microbiota plays a crucial part in disease dynamics and in host-pathogen and host-commensal and commensal-pathogen interactions. For example, signals from the commensal flora may affect expression of virulence determinants in pathogens (140). Therefore, the challenges that lie ahead are to improve the understanding of the ecology of the pathogen at sites of colonization and in key contamination in the agri-food environment that will underpin strategies for the prevention of transmission to target this diverse group of pathogens.

In addition, among members of a serotype there may be significant strain differences that warrant further research and that need to be considered for effective management practices. Data are now building on the phenotypic characteristics and behavior of non-O157 VTEC in the food chain that indicate that they behave in a similar way to O157 and that current general interventions implemented at pre- or postharvest stages of the food chain will yield similar reductions regardless of serogroup (130, 135, 141). However, there are some recent studies that indicate that this may not always be case, i.e., in semi-hard raw-milk cheese, non-O157 VTEC persisted longer than O157 (114). There is an urgent need for robust studies with more serogroups and more strains to validate survival kinetics of diverse VTEC strains and to assess the impact of food control measures to ensure that differences are true serogroup differences and not related to an interstrain impact. There is a need for better epidemiological linkage of strains from animal, farm environment, and foods and humans and for source attribution of VTEC, in particular, for emerging serogroups.

This is challenging, but with so many potential vehicles and routes of infection, some risk ranking at a country or regional level of the pathways and products, based on the relevant regional epidemiological and practices, would be useful in effectively directing and focusing veterinary public health risk management approaches and resources.

ACKNOWLEDGMENT

We declare no conflicts of interest with regard to the manuscript.

CITATION

Duffy G, McCabe E. 2014. Veterinary public health approach to managing pathogenic verocytotoxigenic *Escherichia coli* in the agri-food chain. Microbiol Spectrum 2(5):EHEC-0023-2013.

REFERENCES

1. **Fairbrother JM, Nadeau E.** 2006. *Escherichia coli*: on-farm contamination of animals. *Rev Sci Tech* **25:**555–569.
2. **Adak GK, Long SM, O'Brien SJ.** 2002. Trends in indigenous foodborne disease and deaths, England and Wales: 1992 to 2000. *Gut* **51:**832–841.
3. **Riley LW, Remis RS, Helgerson SD, McGee HB, Wells JG, Davis BR, Hebert RJ, Olcott ES, Johnson LM, Hargrett NT, Blake PA, Cohen ML.** 1983. Hemorrhagic colitis associated with a rare *Escherichia coli* serotype. *N Engl J Med* **308:**681–685.
4. **Tarr PI, Gordon CA, Chandler WL.** 2005. Shiga-toxin-producing *Escherichia coli* and haemolytic uraemic syndrome. *Lancet* **365:**1073–1086.
5. **Centers for Disease Control and Prevention.** 2012. *Foodborne Diseases Active Surveillance Network (FoodNet): FoodNet Surveillance Report for 2011 (Final Report)*. U.S. Department of Health and Human Services, Centers for Disease Control and Prevention, Atlanta, GA.
6. **European Food Safety Authority, European Centre for Disease Prevention and Control.** 2013. The European Union Summary Report on Trends and Sources of Zoonoses, Zoonotic Agents and Food-borne Outbreaks in 2011. *EFSA Journal* **11(4):**3129.
7. **Kaper JB.** 1998. Enterohemorrhagic *Escherichia coli*. *Curr Opin Microbiol* **1:**103–108.
8. **Law D.** 2000. Virulence factors of *Escherichia coli* O157 and other Shiga toxin-producing E. coli. *J Appl Microbiol* **88:**729–745.
9. **Paton AW, Paton JC.** 1999. Direct detection of Shiga toxigenic *Escherichia coli* strains belonging to serogroups O111, O157, and O113 by multiplex PCR. *J Clin Microbiol* **37:**3362–3365.
10. **Karch H, Denamur E, Dobrindt U, Finlay BB, Hengge R, Johannes L, Ron EZ, Tonjum T, Sansonetti PJ, Vicente M.** 2012. The enemy within us: lessons from the 2011 European *Escherichia coli* O104:H4 outbreak. *EMBO Mol Med* **4:**841–848.
11. **Bielaszewska M, Mellmann A, Zhang W, Kock R, Fruth A, Bauwens A, Peters G, Karch H.** 2011. Characterisation of the *Escherichia coli* strain associated with an outbreak of haemolytic uraemic syndrome in Germany, 2011: a microbiological study. *Lancet Infect Dis* **11:**671–676.
12. **Rasko DA, Webster DR, Sahl JW, Bashir A, Boisen N, Scheutz F, Paxinos EE, Sebra R, Chin CS, Iliopoulos D, Klammer A, Peluso P, Lee L, Kislyuk AO, Bullard J, Kasarskis A, Wang S, Eid J, Rank D, Redman JC, Steyert SR, Frimodt-Moller J, Struve C, Petersen AM, Krogfelt KA, Nataro JP, Schadt EE, Waldor MK.** 2011. Origins of the *E. coli* strain causing an outbreak of hemolytic-uremic syndrome in Germany. *N Engl J Med* **365:**709–717.
13. **Beutin L, Martin A.** 2012. Outbreak of Shiga toxin-producing *Escherichia coli* (STEC) O104:H4 infection in Germany causes a paradigm shift with regard to human pathogenicity of STEC strains. *J Food Prot* **75:**408–418.
14. **Ethelberg S, Olsen KE, Scheutz F, Jensen C, Schiellerup P, Enberg J, Petersen AM, Olesen B, Gerner-Smidt P, Molbak K.** 2004. Virulence factors for hemolytic uremic syndrome, Denmark. *Emerg Infect Dis* **10:**842–847.
15. **European Food Safety Authority, EFSA Panel on Biological Hazards (BIOHAZ).** 2013. Scientific Opinion on VTEC-seropathotype and scientific criteria regarding pathogenicity assessment. *EFSA Journal* **11(4):**3138.
16. **Bettelheim KA.** 2003. Non-O157 verotoxin-producing *Escherichia coli*: a problem, paradox, and paradigm. *Exp Biol Med (Maywood)* **228:**333–344.
17. **Beutin L, Geier D, Steinruck H, Zimmermann S, Scheutz F.** 1993. Prevalence and some properties of verotoxin (Shiga-like toxin)-producing *Escherichia coli* in seven different species of healthy domestic animals. *J Clin Microbiol* **31:**2483–2488.

18. Blanco M, Blanco JE, Mora A, Dahbi G, Alonso MP, Gonzalez EA, Bernardez MI, Blanco J. 2004. Serotypes, virulence genes, and intimin types of Shiga toxin (verotoxin)-producing *Escherichia coli* isolates from cattle in Spain and identification of a new intimin variant gene (*eae-xi*). *J Clin Microbiol* **42**:645–651.

19. Sachdeva S, Ahmad G, Malhotra P, Mukherjee P, Chauhan VS. 2004. Comparison of immunogenicities of recombinant *Plasmodium vivax* merozoite surface protein 1 19- and 42-kiloDalton fragments expressed in *Escherichia coli*. *Infect Immun* **72**:5775–5782.

20. Scaife HR, Cowan D, Finney J, Kinghorn-Perry SF, Crook B. 2006. Wild rabbits (*Oryctolagus cuniculus*) as potential carriers of verocytotoxin-producing *Escherichia coli*. *Vet Rec* **159**:175–178.

21. Caprioli A, Morabito S, Brugere H, Oswald E. 2005. Enterohaemorrhagic *Escherichia coli*: emerging issues on virulence and modes of transmission. *Vet Res* **36**:289–311.

22. Naylor SW, Low JC, Besser TE, Mahajan A, Gunn GJ, Pearce MC, McKendrick IJ, Smith DG, Gally DL. 2003. Lymphoid follicle-dense mucosa at the terminal rectum is the principal site of colonization of enterohemorrhagic *Escherichia coli* O157:H7 in the bovine host. *Infect Immun* **71**:1505–1512.

23. Arthur TM, Brichta-Harhay DM, Bosilevac JM, Kalchayanand N, Shackelford SD, Wheeler TL, Koohmaraie M. 2010. Super shedding of *Escherichia coli* O157:H7 by cattle and the impact on beef carcass contamination. *Meat Sci* **86**:32–37.

24. Hussein HS, Bollinger LM. 2005. Prevalence of Shiga toxin-producing *Escherichia coli* in beef. *Meat Sci* **71**:676–689.

25. Hussein HS, Bollinger LM. 2005. Prevalence of Shiga toxin-producing *Escherichia coli* in beef cattle. *J Food Prot* **68**:2224–2241.

26. Naylor SW, Roe AJ, Nart P, Spears K, Smith DG, Low JC, Gally DL. 2005. *Escherichia coli* O157:H7 forms attaching and effacing lesions at the terminal rectum of cattle and colonization requires the LEE4 operon. *Microbiology* **151**:2773–2781.

27. Rhoades JR, Duffy G, Koutsoumanis K. 2009. Prevalence and concentration of verocytotoxigenic *Escherichia coli*, *Salmonella enterica* and *Listeria monocytogenes* in the beef production chain: a review. *Food Microbiol* **26**:357–376.

28. Laegreid WW, Elder RO, Keen JE. 1999. Prevalence of *Escherichia coli* O157:H7 in range beef calves at weaning. *Epidemiol Infect* **123**:291–298.

29. Karmali MA, Gannon V, Sargeant JM. 2010. Verocytotoxin-producing *Escherichia coli* (VTEC). *Vet Microbiol* **140**:360–370.

30. Blanco J, Blanco M, Blanco JE, Mora A, Gonzalez EA, Bernardez MI, Alonso MP, Coira A, Rodriguez A, Rey J, Alonso JM, Usera MA. 2003. Verotoxin-producing *Escherichia coli* in Spain: prevalence, serotypes, and virulence genes of O157:H7 and non-O157 VTEC in ruminants, raw beef products, and humans. *Exp Biol Med (Maywood)* **228**:345–351.

31. Monaghan A, Byrne B, Fanning S, Sweeney T, McDowell D, Bolton DJ. 2011. Serotypes and virulence profiles of non-O157 Shiga toxin-producing *Escherichia coli* isolates from bovine farms. *Appl Environ Microbiol* **77**:8662–8668.

32. Naylor SW, Gally DL, Low JC. 2005. Enterohaemorrhagic *E. coli* in veterinary medicine. *Int J Med Microbiol* **295**:419–441.

33. Hancock DD, Besser TE, Rice DH, Ebel ED, Herriott DE, Carpenter LV. 1998. Multiple sources of *Escherichia coli* O157 in feedlots and dairy farms in the northwestern USA. *Prev Vet Med* **35**:11–19.

34. Ogden ID, MacRae M, Strachan NJ. 2004. Is the prevalence and shedding concentrations of *E. coli* O157 in beef cattle in Scotland seasonal? *FEMS Microbiol Lett* **233**:297–300.

35. Matthews L, McKendrick IJ, Ternent H, Gunn GJ, Synge B, Woolhouse ME. 2006. Super-shedding cattle and the transmission dynamics of *Escherichia coli* O157. *Epidemiol Infect* **134**:131–142.

36. Matthews L, Low JC, Gally DL, Pearce MC, Mellor DJ, Heesterbeek JA, Chase-Topping M, Naylor SW, Shaw DJ, Reid SW, Gunn GJ, Woolhouse ME. 2006. Heterogeneous shedding of *Escherichia coli* O157 in cattle and its implications for control. *Proc Natl Acad Sci USA* **103**:547–552.

37. Duffy G, Cummins E, Nally P, OB S, Butler F. 2006. A review of quantitative microbial risk assessment in the management of *Escherichia coli* O157:H7 on beef. *Meat Sci* **74**:76–88.

38. Thomas KM, McCann MS, Collery MM, Logan A, Whyte P, McDowell DA, Duffy G. 2012. Tracking verocytotoxigenic *Escherichia coli* O157, O26, O111, O103 and O145 in Irish cattle. *Int J Food Microbiol* **153**:288–296.

39. Omisakin F, MacRae M, Ogden ID, Strachan NJ. 2003. Concentration and prevalence of *Escherichia coli* O157 in cattle feces at slaughter. *Appl Environ Microbiol* **69**:2444–2447.

40. Low JC, McKendrick IJ, McKechnie C, Fenlon D, Naylor SW, Currie C, Smith DG, Allison L, Gally DL. 2005. Rectal carriage of enterohemorrhagic *Escherichia coli* O157 in slaughtered cattle. *Appl Environ Microbiol* **71**:93–97.

41. LeJeune JT, Besser TE, Hancock DD. 2001. Cattle water troughs as reservoirs of *Escherichia coli* O157. *Appl Environ Microbiol* **67**:3053–3057.

42. **LeJeune JT, Besser TE, Merrill NL, Rice DH, Hancock DD.** 2001. Livestock drinking water microbiology and the factors influencing the quality of drinking water offered to cattle. *J Dairy Sci* **84:**1856–1862.

43. **Lejeune JT, Besser TE, Rice DH, Hancock DD.** 2001. Methods for the isolation of water-borne *Escherichia coli* O157. *Lett Appl Microbiol* **32:**316–320.

44. **Sargeant JM, Sanderson MW, Smith RA, Griffin DD.** 2003. *Escherichia coli* O157 in feedlot cattle feces and water in four major feeder-cattle states in the USA. *Prev Vet Med* **61:**127–135.

45. **Carroll AM, Gibson A, McNamara EB.** 2005. Laboratory-based surveillance of human verocytotoxigenic *Escherichia coli* infection in the Republic of Ireland, 2002–2004. *J Med Microbiol* **54:**1163–1169.

46. **Mannix M, O'Connell N, McNamara E, Fitzgerald A, Prendiville T, Norris T, Greally T, Fitzgerald R, Whyte D, Barron D, Monaghan R, Whelan E, Carroll A, Curtin A, Collins C, Quinn J, O'Dea F, O'Riordan M, Buckley J, McCarthy J, Mc Keown P.** 2005. Large *E. coli* O157 outbreak in Ireland, October-November 2005. *Euro Surveill* **10:**E051222 051223.

47. **Joris MA, Pierard D, De Zutter L.** 2011. Occurrence and virulence patterns of *E. coli* O26, O103, O111 and O145 in slaughter cattle. *Vet Microbiol* **151:**418–421.

48. **Franz E, van Hoek AH, van der Wal FJ, de Boer A, Zwartkruis-Nahuis A, van der Zwaluw K, Aarts HJ, Heuvelink AE.** 2012. Genetic features differentiating bovine, food, and human isolates of shiga toxin-producing *Escherichia coli* O157 in The Netherlands. *J Clin Microbiol* **50:**772–780.

49. **Blanco M, Blanco JE, Mora A, Rey J, Alonso JM, Hermoso M, Hermoso J, Alonso MP, Dahbi G, Gonzalez EA, Bernardez MI, Blanco J.** 2003. Serotypes, virulence genes, and intimin types of Shiga toxin (verotoxin)-producing *Escherichia coli* isolates from healthy sheep in Spain. *J Clin Microbiol* **41:**1351–1356.

50. **Rey J, Blanco JE, Blanco M, Mora A, Dahbi G, Alonso JM, Hermoso M, Hermoso J, Alonso MP, Usera MA, Gonzalez EA, Bernardez MI, Blanco J.** 2003. Serotypes, phage types and virulence genes of shiga-producing *Escherichia coli* isolated from sheep in Spain. *Vet Microbiol* **94:**47–56.

51. **Soderlund R, Hedenstrom I, Nilsson A, Eriksson E, Aspan A.** 2012. Genetically similar strains of *Escherichia coli* O157:H7 isolated from sheep, cattle and human patients. *BMC Vet Res* **8:**200.

52. **Sekse C, Sunde M, Lindstedt BA, Hopp P, Bruheim T, Cudjoe KS, Kvitle B, Urdahl AM.** 2011. Potentially human-pathogenic *Escherichia*

coli O26 in Norwegian sheep flocks. *Appl Environ Microbiol* **77:**4949–4958.

53. **Sanchez S, Martinez R, Garcia A, Blanco J, Blanco JE, Blanco M, Dahbi G, Lopez C, Mora A, Rey J, Alonso JM.** 2009. Longitudinal study of Shiga toxin-producing *Escherichia coli* shedding in sheep feces: persistence of specific clones in sheep flocks. *Appl Environ Microbiol* **75:**1769–1773.

54. **Franco A, Lovari S, Cordaro G, Di Matteo P, Sorbara L, Iurescia M, Donati V, Buccella C, Battisti A.** 2009. Prevalence and concentration of verotoxigenic *Escherichia coli* O157:H7 in adult sheep at slaughter from Italy. *Zoonoses Public Health* **56:**215–220.

55. **Brandal LT, Sekse C, Lindstedt BA, Sunde M, Lobersli I, Urdahl AM, Kapperud G.** 2012. Norwegian sheep are an important reservoir for human-pathogenic *Escherichia coli* O26:H11. *Appl Environ Microbiol* **78:**4083–4091.

56. **Evans J, Knight H, McKendrick IJ, Stevenson H, Varo Barbudo A, Gunn GJ, Low JC.** 2011. Prevalence of *Escherichia coli* O157:H7 and serogroups O26, O103, O111 and O145 in sheep presented for slaughter in Scotland. *J Med Microbiol* **60:**653–660.

57. **Ogden ID, MacRae M, Strachan NJ.** 2005. Concentration and prevalence of *Escherichia coli* O157 in sheep faeces at pasture in Scotland. *J Appl Microbiol* **98:**646–651.

58. **Sekse C, Sunde M, Hopp P, Bruheim T, Cudjoe KS, Kvitle B, Urdahl AM.** 2013. Occurrence of potentially human-pathogenic *Escherichia coli* O103 in Norwegian sheep. *Appl Environ Microbiol* **79:**7502–7509.

59. **Schimmer B, Nygard K, Eriksen HM, Lassen J, Lindstedt BA, Brandal LT, Kapperud G, Aavitsland P.** 2008. Outbreak of haemolytic uraemic syndrome in Norway caused by stx2-positive *Escherichia coli* O103:H25 traced to cured mutton sausages. *BMC Infect Dis* **8:**41.

60. **Lenahan M, O'Brien S, Kinsella K, Sweeney T, Sheridan JJ.** 2007. Prevalence and molecular characterization of *Escherichia coli* O157:H7 on Irish lamb carcasses, fleece and in faeces samples. *J Appl Microbiol* **103:**2401–2409.

61. **Thomas KM, McCann MS, Collery MM, Moschonas G, Whyte P, McDowell DA, Duffy G.** 2013. Transfer of verocytotoxigenic *Escherichia coli* O157, O26, O111, O103 and O145 from fleece to carcass during sheep slaughter in an Irish export abattoir. *Food Microbiol* **34:**38–45.

62. **Feder I, Wallace FM, Gray JT, Fratamico P, Fedorka-Cray PJ, Pearce RA, Call JE, Perrine R, Luchansky JB.** 2003. Isolation of *Escherichia coli* O157:H7 from intact colon fecal samples of swine. *Emerg Infect Dis* **9:**380–383.

63. **Chapman PA, Cerdan Malo AT, Ellin M, Ashton R, Harkin.** 2001. *Escherichia coli* O157 in cattle and sheep at slaughter, on beef and lamb carcasses and in raw beef and lamb products in South Yorkshire, UK. *Int J Food Microbiol* **64:**139–150.

64. **Callaway TR, Anderson RC, Tellez G, Rosario C, Nava GM, Eslava C, Blanco MA, Quiroz MA, Olguin A, Herradora M, Edrington TS, Genovese KJ, Harvey RB, Nisbet DJ.** 2004. Prevalence of *Escherichia coli* O157 in cattle and swine in central Mexico. *J Food Prot* **67:**2274–2276.

65. **Chapman PA, Siddons CA, Gerdan Malo AT, Harkin MA.** 1997. A 1-year study of *Escherichia coli* O157 in cattle, sheep, pigs and poultry. *Epidemiol Infect* **119:**245–250.

66. **Eriksson E, Nerbrink E, Borch E, Aspan A, Gunnarsson A.** 2003. Verocytotoxin-producing *Escherichia coli* O157:H7 in the Swedish pig population. *Vet Rec* **152:**712–717.

67. **Heuvelink AE, Zwartkruis-Nahuis JT, van den Biggelaar FL, van Leeuwen WJ, de Boer E.** 1999. Isolation and characterization of verocytotoxin-producing *Escherichia coli* O157 from slaughter pigs and poultry. *Int J Food Microbiol* **52:**67–75.

68. **Lenahan M, O'Brien SB, Byrne C, Ryan M, Kennedy CA, McNamara EB, Fanning S, Sheridan JJ, Sweeney T.** 2009. Molecular characterization of Irish *E. coli* O157:H7 isolates of human, bovine, ovine and porcine origin. *J Appl Microbiol* **107:**1340–1349.

69. **Jay MT, Cooley M, Carychao D, Wiscomb GW, Sweitzer RA, Crawford-Miksza L, Farrar JA, Lau DK, O'Connell J, Millington A, Asmundson RV, Atwill ER, Mandrell RE.** 2007. *Escherichia coli* O157:H7 in feral swine near spinach fields and cattle, central California coast. *Emerg Infect Dis* **13:**1908–1911.

70. **Williams AP, Avery LM, Killham K, Jones DL.** 2005. Persistence of *Escherichia coli* O157 on farm surfaces under different environmental conditions. *J Appl Microbiol* **98:**1075–1083.

71. **Bolton DJ, Byrne CM, Sheridan JJ, McDowell DA, Blair IS.** 1999. The survival characteristics of a non-toxigenic strain of *Escherichia coli* O157:H7. *J Appl Microbiol* **86:**407–411.

72. **Maule A.** 2000. Survival of verocytotoxigenic *Escherichia coli* O157 in soil, water and on surfaces. *Symp Ser Soc Appl Microbiol*71S–78S.

73. **McGee P, Bolton DJ, Sheridan JJ, Earley B, Leonard N.** 2001. The survival of *Escherichia coli* O157:H7 in slurry from cattle fed different diets. *Lett Appl Microbiol* **32:**152–155.

74. **Semenov AV, van Overbeek L, Termorshuizen AJ, van Bruggen AH.** 2011. Influence of aerobic and anaerobic conditions on survival of *Escherichia coli* O157:H7 and *Salmonella enterica* serovar Typhimurium in Luria-Bertani broth, farmyard manure and slurry. *J Environ Manage* **92:**780–787.

75. **Nicholson FA, Groves SJ, Chambers BJ.** 2005. Pathogen survival during livestock manure storage and following land application. *Bioresour Technol* **96:**135–143.

76. **Hutchison ML, Walters LD, Moore A, Crookes KM, Avery SM.** 2004. Effect of length of time before incorporation on survival of pathogenic bacteria present in livestock wastes applied to agricultural soil. *Appl Environ Microbiol* **70:**5111–5118.

77. **Avery SM, Moore A, Hutchison ML.** 2004. Fate of *Escherichia coli* originating from livestock faeces deposited directly onto pasture. *Lett Appl Microbiol* **38:**355–359.

78. **Fukushima H, Hoshina K, Gomyoda M.** 1999. Long-term survival of shiga toxin-producing *Escherichia coli* O26, O111, and O157 in bovine feces. *Appl Environ Microbiol* **65:**5177–5181.

79. **Bolton DJ, Monaghan A, Byrne B, Fanning S, Sweeney T, McDowell DA.** 2011. Incidence and survival of non-O157 verocytotoxigenic *Escherichia coli* in soil. *J Appl Microbiol* **111:**484–490.

80. **EC No 853/2004.** 2004. Regulation (EC) No 853/2004 of the European Parliament and of the Council of 29 April 2004, laying down specific hygiene rules for food of animal origin. *Official Journal of the European Union L 139 of 30 April 2004.*

81. **Lyons NA, Smith RP, Rushton J.** 2013. Cost-effectiveness of farm interventions for reducing the prevalence of VTEC O157 on UK dairy farms. *Epidemiol Infect* **141:**1905–1919.

82. **Ellis-Iversen J, Smith RP, Van Winden S, Paiba GA, Watson E, Snow LC, Cook AJ.** 2008. Farm practices to control *E. coli* O157 in young cattle—a randomised controlled trial. *Vet Res* **39:**3.

83. **Lung AJ, Lin CM, Kim JM, Marshall MR, Nordstedt R, Thompson NP, Wei CI.** 2001. Destruction of *Escherichia coli* O157:H7 and *Salmonella enteritidis* in cow manure composting. *J Food Prot* **64:**1309–1314.

84. **Shepherd MW Jr, Kim J, Jiang X, Doyle MP, Erickson MC.** 2011. Evaluation of physical coverings used to control *Escherichia coli* O157:H7 at the compost heap surface. *Appl Environ Microbiol* **77:**5044–5049.

85. **Bujoczek G, Reiners RS, Olaszkiewicz JA.** 2001. Abiotic factors affecting inactivation of pathogens in sludge. *Water Sci Technol* **44:**79–84.

86. **Locking M, Allison L, Rae L, Pollock K, Hanson M.** 2006. VTEC infections and livestock-related exposures in Scotland, 2004. *Euro Surveill* **11:**E060223 060224.

87. **Money P, Kelly AF, Gould SW, Denholm-Price J, Threlfall EJ, Fielder MD.** 2010. Cattle, weather and water: mapping *Escherichia coli* O157:H7 infections in humans in England and Scotland. *Environ Microbiol* **12:**2633–2644.

88. **Health Protection Surveillance Centre (HPSC).** 2005. *Report of the HPSC Sub-Committee on Verotoxigenic E. coli.* http://www.hpsc.ie/A-Z/ Gastroenteric/VTEC/Guidance/Reportofthe HPSCSub-CommitteeonVerotoxigenicEcoli/.

89. **O'Sullivan MB, Garvey P, O'Riordan M, Coughlan H, McKeown P, Brennan A, McNamara E.** 2008. Increase in VTEC cases in the south of Ireland: link to private wells? *Euro Surveill* **13.**

90. **Tanaro JD, Galli L, Lound LH, Leotta GA, Piaggio MC, Carbonari CC, Irino K, Rivas M.** 2012. Non-O157:H7 Shiga toxin-producing *Escherichia coli* in bovine rectums and surface water streams on a beef cattle farm in Argentina. *Foodborne Pathog Dis* **9:**878–884.

91. **Department of the Environment and Local Government, Environmental Protection Agency, Geological Survey of Ireland (DELG, EPA, GSI).** 1999. *Groundwater Protection Schemes Report.* GSI, Dublin, Ireland.

92. **Food Safety Authority of Ireland.** 2010. Report of the Scientific Committee of the Food Safety Authority of Ireland. *The Prevention of Verocytotoxigenic Escherichia coli (VTEC) Infection: A Shared Responsibility*, 2nd Ed. FSAI, Dublin, Ireland.

93. **Wang G, Doyle MP.** 1998. Heat shock response enhances acid tolerance of *Escherichia coli* O157:H7. *Lett Appl Microbiol* **26:**31–34.

94. **McGee P, Bolton DJ, Sheridan JJ, Earley B, Kelly G, Leonard N.** 2002. Survival of *Escherichia coli* O157:H7 in farm water: its role as a vector in the transmission of the organism within herds. *J Appl Microbiol* **93:**706–713.

95. **Rice EW, Johnson CH, Wild DK, Reasoner DJ.** 1992. Survival of *Escherichia coli* O157:H7 in drinking water associated with a waterborne disease outbreak of hemorrhagic colitis. *Lett Appl Microbiol* **15:**38–40.

96. **Fremaux B, Prigent-Combaret C, Delignette-Muller ML, Mallen B, Dothal M, Gleizal A, Vernozy-Rozand C.** 2008. Persistence of Shiga toxin-producing *Escherichia coli* O26 in various manure-amended soil types. *J Appl Microbiol* **104:**296–304.

97. **Scott L, McGee P, Sheridan JJ, Earley B, Leonard N.** 2006. A comparison of the survival in feces and water of *Escherichia coli* O157:H7 grown under laboratory conditions or obtained from cattle feces. *J Food Prot* **69:**6–11.

98. **Macdonald LE, Brett J, Kelton D, Majowicz SE, Snedeker K, Sargeant JM.** 2011. A systematic review and meta-analysis of the effects of pasteurization on milk vitamins, and evidence for raw milk consumption and other health-related outcomes. *J Food Prot* **74:**1814–1832.

99. **Jay-Russell MT.** 2010. Raw (unpasteurized) milk: are health-conscious consumers making an unhealthy choice? *Clin Infect Dis* **51:**1418–1419.

100. **Denny J, Bhat M, Eckmann K.** 2008. Outbreak of *Escherichia coli* O157:H7 associated with raw milk consumption in the Pacific Northwest. *Foodborne Pathog Dis* **5:**321–328.

101. **Guh A, Phan Q, Randall N, Purviance K, Milardo E, Kinney S, Mshar P, Kasacek W, Cartter M.** 2010. Outbreak of *Escherichia coli* O157 associated with raw milk, Connecticut, 2008. *Clin Infect Dis* **51:**1411–1417.

102. **Gaulin C, Levac E, Ramsay D, Dion R, Ismail J, Gingras S, Lacroix C.** 2012. *Escherichia coli* O157:H7 outbreak linked to raw milk cheese in Quebec, Canada: use of exact probability calculation and casecase study approaches to foodborne outbreak investigation. *J Food Prot* **75:**812–818.

103. **Honish L, Predy G, Hislop N, Chui L, Kowalewska-Grochowska K, Trottier L, Kreplin C, Zazulak I.** 2005. An outbreak of *E. coli* O157:H7 hemorrhagic colitis associated with unpasteurized gouda cheese. *Can J Public Health* **96:**182–184.

104. **McCollum JT, Williams NJ, Beam SW, Cosgrove S, Ettestad PJ, Ghosh TS, Kimura AC, Nguyen L, Stroika SG, Vogt RL, Watkins AK, Weiss JR, Williams IT, Cronquist AB.** 2012. Multistate outbreak of *Escherichia coli* O157:H7 infections associated with in-store sampling of an aged raw-milk Gouda cheese, 2010. *J Food Prot* **75:**1759–1765.

105. **Rangel JM, Sparling PH, Crowe C, Griffin PM, Swerdlow DL.** 2005. Epidemiology of *Escherichia coli* O157:H7 outbreaks, United States, 1982–2002. *Emerg Infect Dis* **11:**603–609.

106. **Allerberger F, Friedrich AW, Grif K, Dierich MP, Dornbusch HJ, Mache CJ, Nachbaur E, Freilinger M, Rieck P, Wagner M, Caprioli A, Karch H, Zimmerhackl LB.** 2003. Hemolytic-uremic syndrome associated with enterohemorrhagic *Escherichia coli* O26:H infection and consumption of unpasteurized cow's milk. *Int J Infect Dis* **7:**42–45.

107. **De Schrijver K, Buvens G, Posse B, Van den Branden D, Oosterlynck O, De Zutter L, Eilers K, Pierard D, Dierick K, Van Damme-Lombaerts R, Lauwers C, Jacobs R.** 2008. Outbreak of verocytotoxin-producing *E. coli* O145 and O26 infections associated with the consumption of ice cream produced at a farm, Belgium, 2007. *Euro Surveill* **13.**

108. **Centers for Disease Control and Prevention.** 2012. Outbreak of Shiga toxin-producing *Escherichia coli* O111 infections associated with a

correctional facility dairy—Colorado, 2010. *Morb Mortal Wkly Rep* **61:**149–152.

109. **Murphy M, Carroll A, Whyte P, O'Mahony M, Anderson W, McNamara E, Fanning S.** 2005. Prevalence and characterization of *Escherichia coli* O26 and O111 in retail minced beef in Ireland. *Foodborne Pathog Dis* **2:**357–360.

110. **Murphy M, Buckley JF, Whyte P, O'Mahony M, Anderson W, Wall PG, Fanning S.** 2007. Surveillance of dairy production holdings supplying raw milk to the farmhouse cheese sector for *Escherichia coli* O157, O26 and O111. *Zoonoses Public Health* **54:**358–365.

111. **European Food Safety Authority, European Centre for Disease Prevention and Control.** 2012. The European Union Summary Report on Trends and Sources of Zoonoses, Zoonotic Agents and Food-borne Outbreaks in 2010. *EFSA Journal* **10(3):**2597.

112. **Schvartzman MS, Maffre A, Tenenhaus-Aziza F, Sanaa M, Butler F, Jordan K.** 2011. Modelling the fate of *Listeria monocytogenes* during manufacture and ripening of smeared cheese made with pasteurised or raw milk. *Int J Food Microbiol* **145 Suppl 1:**S31–S38.

113. **Peng S, Hoffmann W, Bockelmann W, Hummerjohann J, Stephan R, Hammer P.** 2013. Fate of Shiga toxin-producing and generic *Escherichia coli* during production and ripening of semihard raw milk cheese. *J Dairy Sci* **96:**815–823.

114. **Miszczycha SD, Perrin F, Ganet S, Jamet E, Tenenhaus-Aziza F, Montel MC, Thevenot-Sergentet D.** 2013. Behavior of different Shiga toxin-producing *Escherichia coli* serotypes in various experimentally contaminated raw-milk cheeses. *Appl Environ Microbiol* **79:**150–158.

115. **Dineen SS, Takeuchi K, Soudah JE, Boor KJ.** 1998. Persistence of *Escherichia coli* O157:H7 in dairy fermentation systems. *J Food Prot* **61:**1602–1608.

116. **McIngvale SC, Chen XQ, McKillip JL, Drake MA.** 2000. Survival of *Escherichia coli* O157:H7 in buttermilk as affected by contamination point and storage temperature. *J Food Prot* **63:**441–444.

117. **Currie A, MacDonald J, Ellis A, Siushansian J, Chui L, Charlebois M, Peermohamed M, Everett D, Fehr M, Ng LK.** 2007. Outbreak of *Escherichia coli* O157:H7 infections associated with consumption of beef donair. *J Food Prot* **70:**1483–1488.

118. **Ethelberg S, Lisby M, Vestergaard LS, Enemark HL, Olsen KE, Stensvold CR, Nielsen HV, Porsbo LJ, Plesner AM, Molbak K.** 2009. A foodborne outbreak of *Cryptosporidium hominis* infection. *Epidemiol Infect* **137:**348–356.

119. **King LA, Mailles A, Mariani-Kurkdjian P, Vernozy-Rozand C, Montet MP, Grimont F, Pihier N, Devalk H, Perret F, Bingen E, Espie E, Vaillant V.** 2009. Community-wide outbreak of *Escherichia coli* O157:H7 associated with consumption of frozen beef burgers. *Epidemiol Infect* **137:**889–896.

120. **Elder RO, Keen JE, Siragusa GR, Barkocy-Gallagher GA, Koohmaraie M, Laegreid WW.** 2000. Correlation of enterohemorrhagic *Escherichia coli* O157 prevalence in feces, hides, and carcasses of beef cattle during processing. *Proc Natl Acad Sci USA* **97:**2999–3003.

121. **Arthur TM, Bosilevac JM, Nou X, Shackelford SD, Wheeler TL, Kent MP, Jaroni D, Pauling B, Allen DM, Koohmaraie M.** 2004. *Escherichia coli* O157 prevalence and enumeration of aerobic bacteria, Enterobacteriaceae, and *Escherichia coli* O157 at various steps in commercial beef processing plants. *J Food Prot* **67:**658–665.

122. **McGee P, Scott L, Sheridan JJ, Earley B, Leonard N.** 2004. Horizontal transmission of *Escherichia coli* O157:H7 during cattle housing. *J Food Prot* **67:**2651–2656.

123. **Arthur TM, Bosilevac JM, Brichta-Harhay DM, Kalchayanand N, King DA, Shackelford SD, Wheeler TL, Koohmaraie M.** 2008. Source tracking of *Escherichia coli* O157:H7 and *Salmonella* contamination in the lairage environment at commercial U.S. beef processing plants and identification of an effective intervention. *J Food Prot* **71:**1752–1760.

124. **Arthur TM, Bosilevac JM, Nou X, Shackelford SD, Wheeler TL, Koohmaraie M.** 2007. Comparison of the molecular genotypes of *Escherichia coli* O157:H7 from the hides of beef cattle in different regions of North America. *J Food Prot* **70:**1622–1626.

125. **Dewell GA, Simpson CA, Dewell RD, Hyatt DR, Belk KE, Scanga JA, Morley PS, Grandin T, Smith GC, Dargatz DA, Wagner BA, Salman MD.** 2008. Impact of transportation and lairage on hide contamination with *Escherichia coli* O157 in finished beef cattle. *J Food Prot* **71:**1114–1118.

126. **Jacob ME, Renter DG, Nagaraja TG.** 2010. Animal- and truckload-level associations between *Escherichia coli* O157:H7 in feces and on hides at harvest and contamination of preevisceration beef carcasses. *J Food Prot* **73:**1030–1037.

127. **European Food Safety Authority, EFSA Panel on Biological Hazards.** 2013. Scientific Opinion on the public health hazards to be covered by inspection of meat (bovine animals). *EFSA Journal* **11(6):**3266.

128. **Schmidt JW, Arthur TM, Bosilevac JM, Kalchayanand N, Wheeler TL.** 2012. Detection of *Escherichia coli* O157:H7 and Salmonella

enterica in air and droplets at three U.S. commercial beef processing plants. *J Food Prot* **75:** 2213–2218.

129. **Koohmaraie M, Arthur TM, Bosilevac JM, Guerini M, Shackelford SD, Wheeler TL.** 2005. Post-harvest interventions to reduce/eliminate pathogens in beef. *Meat Sci* **71:**79–91.

130. **Kalchayanand N, Arthur TM, Bosilevac JM, Schmidt JW, Wang R, Shackelford SD, Wheeler TL.** 2012. Evaluation of commonly used antimicrobial interventions for fresh beef inoculated with Shiga toxin-producing *Escherichia coli* serotypes O26, O45, O103, O111, O121, O145, and O157:H7. *J Food Prot* **75:**1207–1212.

131. **Jacob R, Porto-Fett AC, Call JE, Luchansky JB.** 2009. Fate of surface-inoculated *Escherichia coli* O157:H7, *Listeria monocytogenes*, and *Salmonella typhimurium* on kippered beef during extended storage at refrigeration and abusive temperatures. *J Food Prot* **72:**403–407.

132. **Adler JM, Geornaras I, Belk KE, Smith GC, Sofos JN.** 2012. Thermal inactivation of *Escherichia coli* O157:H7 inoculated at different depths of non-intact blade-tenderized beef steaks. *J Food Sci* **77:**M108–M114.

133. **Kudra LL, Sebranek JG, Dickson JS, Mendonca AF, Larson EM, Jackson-Davis AL, Lu Z.** 2011. Effects of vacuum or modified atmosphere packaging in combination with irradiation for control of *Escherichia coli* O157:H7 in ground beef patties. *J Food Prot* **74:**2018–2023.

134. **Vold L, Holck A, Wasteson Y, Nissen H.** 2000. High levels of background flora inhibits growth of *Escherichia coli* O157:H7 in ground beef. *Int J Food Microbiol* **56:**219–225.

135. **Luchansky JB, Porto-Fett AC, Shoyer BA, Phillips J, Eblen D, Evans P, Bauer N.** 2013. Thermal inactivation of a single strain each of serotype O26:H11, O45:H2, O103:H2, O104:H4, O111:H(-), O121:H19, O145:NM, and O157:H7 cells of Shiga toxin-producing *Escherichia coli* in wafers of ground beef. *J Food Prot* **76:**1434–1437.

136. **Flores RA, Tamplin ML.** 2002. Distribution patterns of *Escherichia coli* O157:H7 in ground beef produced by a laboratory-scale grinder. *J Food Prot* **65:**1894–1902.

137. **Flores RA.** 2004. Distribution of *Escherichia coli* O157:H7 in beef processed in a table-top bowl cutter. *J Food Prot* **67:**246–251.

138. **Riordan DC, Duffy G, Sheridan J, Eblen BS, Whiting RC, Blair IS, McDowell DA.** 1998. Survival of *Escherichia coli* O157:H7 during the manufacture of pepperoni. *J Food Prot* **61:** 146–151.

139. **Riordan DC, Duffy G, Sheridan JJ, Whiting RC, Blair IS, McDowell DA.** 2000. Effects of acid adaptation, product pH, and heating on survival of *Escherichia coli* O157:H7 in pepperoni. *Appl Environ Microbiol* **66:**1726–1729.

140. **Lupp C, Robertson ML, Wickham ME, Sekirov I, Champion OL, Gaynor EC, Finlay BB.** 2007. Host-mediated inflammation disrupts the intestinal microbiota and promotes the overgrowth of Enterobacteriaceae. *Cell Host Microbe* **2:**119–129.

141. **Vasan A, Leong WM, Ingham SC, Ingham BH.** 2013. Thermal tolerance characteristics of non-O157 Shiga toxigenic strains of *Escherichia coli* (STEC) in a beef broth model system are similar to those of O157:H7 STEC. *J Food Prot* **76:**1120–1128.

142. **Launders N, Byrne L, Adams N, Glen K, Jenkins C, Tubin-Delic D, Locking M, Williams C, Morgan D.** 2013. Outbreak of Shiga toxin-producing *E. coli* O157 associated with consumption of watercress, United Kingdom, August to September 2013. *Euro Surveill* **18.**

143. **Soborg B, Lassen SG, Muller L, Jensen T, Ethelberg S, Molbak K, Scheutz F.** 2013. A verocytotoxin-producing *E. coli* outbreak with a surprisingly high risk of haemolytic uraemic syndrome, Denmark, September-October 2012. *Euro Surveill* **18.**

144. **Slayton RB, Turabelidze G, Bennett SD, Schwensohn CA, Yaffee AQ, Khan F, Butler C, Trees E, Ayers TL, Davis ML, Laufer AS, Gladbach S, Williams I, Gieraltowski LB.** 2013. Outbreak of Shiga toxin-producing *Escherichia coli* (STEC) O157:H7 associated with romaine lettuce consumption, 2011. *PLoS One* **8:**e55300.

145. **Taylor EV, Nguyen TA, Machesky KD, Koch E, Sotir MJ, Bohm SR, Folster JP, Bokanyi R, Kupper A, Bidol SA, Emanuel A, Arends KD, Johnson SA, Dunn J, Stroika S, Patel MK, Williams I.** 2013. Multistate outbreak of *Escherichia coli* O145 infections associated with romaine lettuce consumption, 2010. *J Food Prot* **76:**939–944.

146. **Greenland K, de Jager C, Heuvelink A, van der Zwaluw K, Heck M, Notermans D, van Pelt W, Friesema I.** 2009. Nationwide outbreak of STEC O157 infection in the Netherlands, December 2008-January 2009: continuous risk of consuming raw beef products. *Euro Surveill* **14. pii: 19129.**

147. **Neil KP, Biggerstaff G, MacDonald JK, Trees E, Medus C, Musser KA, Stroika SG, Zink D, Sotir MJ.** 2012. A novel vehicle for transmission of *Escherichia coli* O157:H7 to humans: multistate outbreak of *E. coli* O157:H7 infections associated with consumption of ready-to-bake commercial prepackaged cookie dough—United States, 2009. *Clin Infect Dis* **54:**511–518.

148. **Centers for Disease Control and Prevention.** 2010. Two multistate outbreaks of Shiga toxin-producing *Escherichia coli* infections linked to

beef from a single slaughter facility—United States, 2008. *Morb Mortal Wkly Rep* **59:**557–560.

149. **Friesema I, Sigmundsdottir G, van der Zwaluw K, Heuvelink A, Schimmer B, de Jager C, Rump B, Briem H, Hardardottir H, Atladottir A, Gudmundsdottir E, van Pelt W.** 2008. An international outbreak of Shiga toxin-producing *Escherichia coli* O157 infection due to lettuce, September-October 2007. *Euro Surveill* **13. pii: 19065.**

150. **Ethelberg S, Smith B, Torpdahl M, Lisby M, Boel J, Jensen T, Nielsen EM, Molbak K.** 2009. Outbreak of non-O157 Shiga toxin-producing *Escherichia coli* infection from consumption of beef sausage. *Clin Infect Dis* **48:**e78–e81.

151. **Wendel AM, Johnson DH, Sharapov U, Grant J, Archer JR, Monson T, Koschmann C, Davis JP.** 2009. Multistate outbreak of *Escherichia coli* O157:H7 infection associated with consumption of packaged spinach, August-September 2006: the Wisconsin investigation. *Clin Infect Dis* **48:**1079–1086.

152. **Ramoneda M, Foncuberta M, Simon M, Sabate S, Ferrer MD, Herrera S, Landa B, Muste N, Marti R, Trabado V, Carbonell O, Vila M, Espelt M, Ramirez B, Duran J.** 2013. Prevalence of verotoxigenic *Escherichia coli* O157 (VTEC O157) and compliance with microbiological safety standards in bovine carcasses from an industrial beef slaughter plant. *Lett Appl Microbiol* **56:**408–413.

153. **Dargatz DA, Bai J, Lubbers BV, Kopral CA, An B, Anderson GA.** 2013. Prevalence of *Escherichia coli* O-types and Shiga toxin genes in fecal samples from feedlot cattle. *Foodborne Pathog Dis* **10:**392–396.

154. **Narvaez-Bravo C, Miller MF, Jackson T, Jackson S, Rodas-Gonzalez A, Pond K, Echeverry A, Brashears MM.** 2013. *Salmonella* and *Escherichia coli* O157:H7 prevalence in cattle and on carcasses in a vertically integrated feedlot and harvest plant in Mexico. *J Food Prot* **76:**786–795.

155. **Breum SO, Boel J.** 2010. Prevalence of *Escherichia coli* O157 and verocytotoxin producing *E. coli* (VTEC) on Danish beef carcasses. *Int J Food Microbiol* **141:**90–96.

156. **Masana MO, Leotta GA, Del Castillo LL, D'Astek BA, Palladino PM, Galli L, Vilacoba E, Carbonari C, Rodriguez HR, Rivas M.** 2010. Prevalence, characterization, and genotypic analysis of *Escherichia coli* O157:H7/NM from selected beef exporting abattoirs of Argentina. *J Food Prot* **73:**649–656.

157. **Boqvist S, Aspan A, Eriksson E.** 2009. Prevalence of verotoxigenic *Escherichia coli* O157:H7 in fecal and ear samples from slaughtered cattle in Sweden. *J Food Prot* **72:**1709–1712.

Clinical Studies of *Escherichia coli* O157:H7 Conjugate Vaccines in Adults and Young Children

24

SHOUSUN CHEN SZU[1] and AMINA AHMED[2]

INTRODUCTION

Shiga toxin (Stx)-producing *Escherichia coli* (STEC) is a food-borne pathogen that can lead to complications such as hemorrhagic colitis and hemolytic-uremic syndrome (HUS), serious sequelae. In the United States, the most common *E. coli* serotype causing outbreaks is O157:H7, although non-O157 serotypes also cause the same disease, but in much fewer cases. The highest incidence rate is among children of preschool age (1, 2).

Prevention of *E. coli* O157 infection has been difficult because of the broad spectrum of contaminated sources, ranging from food such as beef, milk, produce, and fruits, to nonfood origins such as pool water and petting zoo animals (3, 4). Chemical or antimicrobial interventions are difficult to apply and have shown limited effectiveness (5). Since cattle are the major animal reservoir for *E. coli* O157, cattle vaccines to reduce carriage of *E. coli* O157 also were studied to a great extent and reached moderate success (6, 7).

An ideal *E. coli* O157 vaccine for humans should be safe and immunogenic in children and elicit bactericidal antibody that kills the inoculum upon contact. The infectious dose found in *E. coli* O157 outbreaks was usually low, around

[1]Eunice Kennedy Shriver National Institute of Child Health and Human Development, National Institutes of Health, Bethesda, MD 20892; [2]Levine Children's Specialty Center—Pediatric Infectious Disease, Carolina Medical Centers, Charlotte, NC 28203.

Enterohemorrhagic Escherichia coli *and Other Shiga Toxin-Producing* E. coli
Edited by Vanessa Sperandio and Carolyn J. Hovde
© 2015 American Society for Microbiology, Washington, DC
doi:10.1128/microbiolspec.EHEC-0016-2013

10^2 CFU (8). High levels of serum lipopoly-saccharide (LPS) antibodies are detected following environmental exposure or symptomatic infection with *E. coli* O157 (9, 10). Although the protective role of these antibodies is unknown, evidence suggests that antibodies to the LPS of other enteric pathogens confer immunity by lysing the pathogens in the intestine (11). Convalescence from shigellosis, for example, provides LPS-specific resistance to infection, and vaccination with a *Shigella sonnei* LPS-based conjugate is efficacious in preventing infection (12). In clinical trials of a *Salmonella enterica* serovar Paratyphi A LPS-based conjugate vaccine, high levels of immunoglobulin G (IgG) anti-LPS with bactericidal activities were also induced in adults and preschool children (13). Because of the similarities between *E. coli* O157 and these gram-negative enteropathogens, vaccine-induced LPS antibodies to *E. coli* O157 may confer protection (14, 15).

Based on this postulation, our goal is to develop vaccines that elicit serum bactericidal IgG in young children. Several investigational polysaccharide conjugate vaccines were prepared at the National Institutes of Health. The vaccines were composed of detoxified LPS for *E. coli* O157, covalently linked to the carrier protein rEPA, a recombinant exoprotein of *Pseudomonas aeruginosa*. Here we review the clinical studies of O157-rEPA investigational vaccines conducted in adults and children 2 to 5 years old (16–19).

One of the major virulence factors in enterohemorrhagic *E. coli* (EHEC) is the Stx secreted by both the O157 and non-O157 serotypes. Therefore, an ideal human vaccine is one that elicits neutralizing antibody against Stx. Passive immunization with neutralizing antibody remains the only effective therapy for many other toxin-mediated diseases (20). However, up until the present, the only data supporting the development of Stx-based vaccines for both active and passive immunization are in preclinical stage. The Stx-based vaccine is reviewed briefly in this article.

METHODS AND RESULTS

Investigational LPS-Based Vaccines

The O-antigen of *E. coli* O157 consists of a linear copolymer of the tetrasaccharide repeating unit, α-D-GalpNAc(1-2)-α-D-PerpNAc-(1-3)-α-L-Fucp-(1-4)-β-D-Glcp-(1-3), where PerpNAc represents perosamine, 4-amino-4,6-dideoxy-D-mannose (17). The O-antigen can be extracted from LPS and detoxified by acetic acid (yielding O-specific polysaccharide [OSP]) or by anhydrous hydrazine to remove the O-acyl-linked lipid chains (yielding DeALPS) (13). Both OSP and DeALPS are not immunogenic without conjugation to a carrier protein. Three investigational vaccines had been prepared by covalently linking to the carrier protein rEPA and designated as O157 OSP-rEPA$_1$, OSP-rEPA$_2$, and DeALPS-rEPA. The carrier protein rEPA was chosen because it has been demonstrated to be clinically safe and served effectively in several polysaccharide conjugates (21). It also has the advantage of not overloading the already crowded routine vaccines with additional doses of diphtheria or tetanus toxoids.

The conjugates passed preclinical immunogenicity tests in mice and followed the safety requirements of the Code of Federal Regulations and were approved by the U.S. Food and Drug Administration as investigational vaccines. Each 0.5-ml dose contained 25 µg of polysaccharide and an approximately equal amount of rEPA. This dosage followed established studies on conjugate vaccines for noncapsular polysaccharides and is slightly higher than the licensed *Haemophilus influenzae* type b, pnemococcal, or meningococcal conjugate vaccines (12, 13).

Clinical Studies

Both the phase I and phase II clinical studies of O157 conjugate vaccines were conducted at the Carolina Medical Centers, Charlotte, NC. Briefly, in phase I, 87 healthy adults were injected once with one of the three conjugates (18). After safety and immunogenicity were demonstrated, the conjugate that elicited the highest antibody was chosen for the phase II study where 49 children, 2 to 5 years old, were recruited and divided into two groups receiving one or two doses of the vaccine (19).

The *E. coli* O157 conjugate vaccines were safe for all ages. There were no fever cases except for one child who had a temperature of 38.2°C 72 h after the second dose was administered. The local reactions were all mild and subsided within 24 h. In phase I, serum assays, including lactate dehydrogenase or alkaline phosphatase, serum bilirubin, and indirect bilirubin, were performed 1 week after injection to evaluate liver function. Six (7%) had asymptomatic elevations (up to 35% above the normal range) in one or more serum assays that returned to normal within 5 weeks. There were no significant differences in serum transaminase levels between pre- and postvaccination in children in the trial.

Antibody Responses

The serum anti-LPS IgG responses and bactericidal titers in vaccinees were used as the end-point markers for vaccine evaluation. The responses were examined before and 1, 4, and 26 weeks after immunization for adults and 1, 6, 10, and 26 weeks after the first injection for children.

All adults had low levels of preimmune anti-LPS IgG (measured in enzyme-linked immunosorbent assay [ELISA] units), and this level was approximately two times higher than those detected in young children (vide infra) (10, 18). The higher background level in adults could be a result of longer environmental exposure to cross-reactive organisms containing perosamine residue in their LPS, such as *Citrobacter* species, *Yersinia enterocolitica*, *Salmonella urbana*, *Pseudomonas maltophilia*, and *Brucella melitensis* (22–28).

All three conjugates elicited a significant rise of anti-LPS IgG in just 1 week after the injection, with 82% having greater than a 4-fold rise (Fig. 1). The escalation of antibodies shortly after immunization is important since the vaccine could be considered as a useful control measure during outbreaks before the source of contamination is identified and to block primary or secondary transmissions (29). The antibody levels continued to rise 4 weeks after the injection, and the geometric mean (GM) in the recipients of OSP-rEPA was slightly higher than that induced by the conjugate prepared with hydrazine-treated LPS, DeALPS-rEPA. At 6 months, the levels of anti-LPS IgG waned to ~33 ELISA units (EU) for all three conjugates, with 97% of volunteers continuing to have greater than a 10-fold rise than the pre-injection levels.

To a lesser degree than the IgG response, O157-rEPA conjugates also induced increases in serum anti-LPS IgM and IgA levels. Interestingly, there is no correlation between the serum IgG and IgM or IgA antibody titers.

The highest incidence of HUS caused by *E. coli* O157 infection occurred in children under 6 years of age (30–32). We chose this target age group, children 2 to 5 years old, to study the safety and immunogenicity of our O157-rEPA conjugate. Children had very low anti-LPS IgG preinjection levels; however, within 1 week of injection, 81% responded with a greater than 4-fold rise in their serum anti-LPS IgG levels (Fig. 2). The antibody

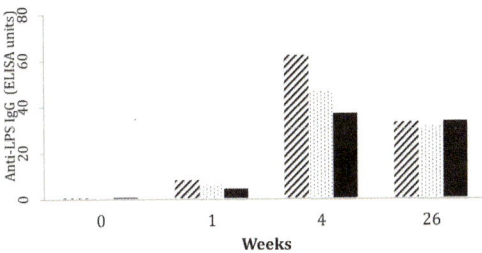

FIGURE 1 Serum anti-Vi IgG response in healthy adults receiving one injection of OSP-rEPA (striped bars), DeALPS-rEPA$_1$ (dotted bars), or DeALPS-rEPA$_2$ (solid bars); *n* = 29 in each group.
doi:10.1128/microbiolspec.EHEC-0016-2013.f1

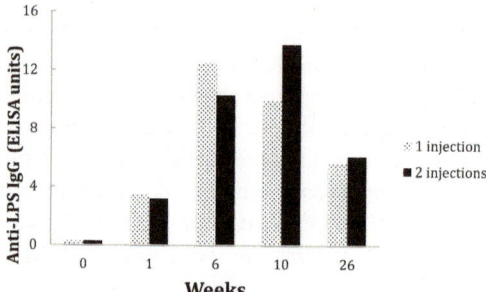

FIGURE 2 Serum anti-Vi IgG response in children 2 to 5 years old receiving one injection (dotted bars) or a booster dose 6 weeks later (solid bars); $n = 25$ in each group. No statistical difference between the groups at all periods.
doi:10.1128/microbiolspec.EHEC-0016-2013.f2

levels continued to rise; 6 weeks after the first injection, the GM reached 11.36 EU, with 98% having >10-fold increase compared with the preinjection levels (one child had a 6-fold rise). At all time intervals, the postinjection GM of anti-LPS IgG is significantly higher than the GM of the preinjection level.

Children who received a second injection of O157-rEPA at week 6 had an increase of antibodies measured 4 weeks later. However, at 26 weeks there was no difference in anti-LPS IgG levels between the groups receiving one or two injections. The lack of a booster response in this age group was also observed in other polysaccharide conjugate vaccines, such as *Salmonella* serovar Paratyphi A and *Shigella flexneri* type 2a (12, 13). At 26 weeks, all children in the study except one continued to have >4-fold higher anti-LPS IgG than their preinjection levels (Fig. 2).

The serum bactericidal assay has been a reliable and reproducible functional bioassay for gram-negative organisms such as *Salmonella* serovar Typhimurium and *Neisseria meningitidis* group C and serves as a good surrogate for protection (33, 34). The assay is mediated by antibody- and complement-induced lysis of the bacterial cells, mimicking the killing of the inoculums in vivo. In Table 1 we list the bactericidal titers and corresponding levels of anti-LPS IgG and IgM in representative serum samples from children before and 26 weeks after the first injection. After the sera were treated with 2-mercapto-ethanol to inactivate IgM function, we observed that there was a direct correlation between the level of IgG anti-LPS and the bactericidal titers ($R^2 = 0.78$).

There is a possibility for the O157 LPS-based vaccines to protect against other pathogens that have cross-reactive LPS (22, 24). For instance, the LPS of *Vibrio cholerae* O:1 constitutes a monosaccharide repeat of perosamine, coinciding with one of the four sugars in *E. coli* O157 O-antigen. It has been reported that LPS antibodies against *V. cholerae* O:1 cross-react with *E. coli* O157 (22). We also observed some low-level cross-reaction in sera from children injected with O157-rEPA with *V. cholerae* O:1 serotype Inaba, but not with serotype Ogawa. Interestingly, there is no correlation between the anti-LPS titers to *E. coli* O157 and those to *V. cholerae* ($R^2 < 0.2$) (19).

TABLE 1 Reciprocal bactericidal activity of serum LPS antibodies elicited in 2- to 5-year-old children injected with *E. coli* O157-rEPA conjugates[a]

Volunteer no.	IgG titer (IgG ELISA units)	IgM titer (IgM ELISA units)	Bactericidal titer[b]	
			Whole serum	Serum treated with 2-ME[c]
ECO 161	100	100	1:1280	1:640
023	33.63	6.81	1:2560	1:1,280
033	24.44	7.82	1:320	1:160
035	17.21	7.72	1:320	1:320
041	14.23	3.76	1:160	1:160
043	11.28	3.12	1:320	1:80
047	12.98	18.94	1:320	1:80
048	21.87	12.25	1:160	1:160
054	18.55	16.00	1:320	1:160
055	21.60	4.61	1:160	1:80

[a]Sera collected from children 42 days after the first injection of the conjugate vaccine.
[b]Titers are expressed as the inverse of dilution giving 50% killing. The anti-LPS IgG titers are calculated using a reference serum randomly assigned 100 EU for IgG (ECO 161). Similarly, a separate reference serum assigned 100 EU for IgM (ECO 110). Correlation coefficient for IgG vs. serum, $R^2 = 0.75$; for IgG vs. serum treated with 2-mercaptoethanol (2-ME), $R^2 = 0.78$. All bactericidal titers compared with IgM, no correlation ($R^2 < 0.10$).
[c]Sera treated with 50 mM 2-ME. Reference serum (ECO 161) was from the phase I study in adults injected with the OSP-rEPA.

The most essential virulence factor of STEC is Stx, in particular type 2. Its role as both a prophylactic and a therapeutic antigen has also been observed in animal models and in epidemiology findings (9, 35). The major hindrance in development of an Stx2-based vaccine is the difficulty in producing a sufficient amount for vaccine preparation. In a proof-of-principle test, we conjugated OSP with the nontoxic recombinant B subunit of Stx1 (Stx1B) by two methods, by direct attachment or by linker adipic dihydrazide. Weaning mice injected with either conjugate elicited bactericidal antibodies to *E. coli* O157 and neutralizing antibodies to holotoxin Stx1 in vitro (Table 2). However, there was no observed cross-neutralization against Stx2 (36). Mutants with various promoters for Stx2B fragments have been constructed and offered future potential in this approach (unpublished data).

DISCUSSION AND FUTURE DEVELOPMENT

The simple thesis of this review is to demonstrate the possibility that, similar to polio, typhoid fever, and cholera, parenterally administered vaccines can protect against orally transmitted diseases such as infections with *E. coli* O157. The outermost carbohydrate moiety of *E. coli* O157 is the O-antigen of LPS,

TABLE 2 Neutralization titers of Stx1 in sera from mice injected with *E. coli* O157 OSP conjugated with Stx1B

Immunogen	n	Titer to Stx1[a] GM (25–75%)	Titer to Stx2
E. coli O157 LPS	5	<10	<10
OSP-AH-Stx1B[b]	10	8,040 (6,400–15,250)	<10
OSP-Stx1B[c]	10	14,400 (12,250–18,600)	<10

[a]Neutralization of Stx1 or Stx2 was measured by using HeLa cell monolayers incubated with dilutions of 100 pg of toxin/ml of serum. Sera from mice immunized with saline or LPS alone showed titers of <10. The titers are the highest serum dilutions to yield 50% neutralization. 14,440 vs. 8,040, P<0.03.
[b]O-polysaccharide conjugated with Shiga toxin 1B with a linker.
[c]O-polysaccharide conjugate with Shiga toxin 1B without a linker.

and conjugates synthesized with O-antigen were shown to be safe and immunogenic and elicited bactericidal antibodies in young children. *Shigella* sp. and *E. coli* are closely related in genetics and pathogenicity (14, 15). Evidence from *S. sonnei* and *S. flexneri* type 2b efficacy trials showed that OSP conjugate vaccines, based on the same construct, are efficacious against similar disease even during an outbreak (29).

We have noticed an age-dependent anti-LPS IgG response between pre- and postinjection sera. The higher background level of anti-LPS in adults compared to that in children was also noticed by others (10). These preexisting LPS antibodies could come from prolonged exposure to other gram-negative organisms containing cross-reactive LPS, and may in turn explain the age-related incidence rate of *E. coli* O157 (1, 2, 30, 31, 32). Adults also responded with about 4-fold higher anti-LPS IgG than children at all time intervals. However, at 26 weeks, the level of antibodies in children remained about 12 times higher than the adults' preimmune level, implying that the children after vaccination had elevated immunity to *E. coli* O157.

There are advantages to using LPS as the vaccine source: the O-specific antigen is stable, its purity and chemical composition can be identified unambiguously, the polysaccharide can be produced in large quantity and is suitable for vaccine production, the detoxification procedures are well established, the residual endotoxicity level can be validated to meet the requirements of the regulatory guidelines, and, most of all, serum LPS antibodies elicited by the conjugates demonstrated bactericidal activity against *E. coli* O157 and can be adopted as a functional bioassay for potency test.

However, there are limitations of LPS-based vaccines. For example, the induced LPS antibodies do not neutralize Stx, the major virulence factor of EHEC. This shortcoming limits its usefulness in prophylaxis and treatment during an outbreak, especially for non-O157 outbreaks. In one attempt to compensate for this shortcoming, we conjugated OSP with the B subunit of Stx1. Mice injected with OSP-Stx1B elicited bactericidal antibodies with high neutralization titers against Stx1 (36). Since Stx type 2 is the most common type found in EHEC outbreaks, an ideal vaccine would include Stx2 B-subunit or nontoxic recombinant Stx2 mutants as part of the vaccine component (37–39). Another obstacle that human vaccine development faces, either LPS- or Stx-based vaccine, is the planning of an efficacy trial to demonstrate the effectiveness. Most EHEC cases occur in outbreaks at no particular locations or regions, and because of the unpredictable nature of the disease epidemiology, designing an efficacy trial for future human vaccines bears inherent difficulty.

LPS that enables gram-negative organisms to escape complement fixation was considered as one of the virulence factors for *E. coli* O157 (40). There are other virulence and attachment factors such as adhesin intimin, Tir, and EspA proteins, and some showed various degrees of protection against *E. coli* O157 in animal models (41–44). Attempts to include these protein antigens as chimeric recombinant proteins in transgenic plants have also reached some preclinical success (43, 44). In the future development of *E. coli* O157 human vaccine, including such virulence factors either as the carrier protein for an OSP conjugate or as separate components in a combined formulation, could be beneficial. Concurrent immunization with multiple antigens may generate synergistic protection, broaden the coverage to the non-O157 STEC serotypes, and facilitate both preventive and therapeutic treatments (45).

Although plasmaphoresis is a common emergency intervention for patients with HUS, the plasma used was not enriched with specific Stx neutralization antibodies. Other nonspecific measures aimed at lowering systemic Stx levels in patients include immunoadsorption, IgG replacement activated charcoal absorption, or kidney dialysis, and their effectiveness remains controversial (46, 47). Several reviews

showed monoclonal antitoxin with humanized, chimeric, or human monoclonal antibodies produced in transgenic mice was successful in animal models and offered high prospect (48–50). A recent Stx challenge study showed that administration of Stx2A or Stx2B human monoclonal antibodies could significantly reduce the Stx accumulation in kidney, accompanied by a short-term elevation of Stx-antibody complex in liver during clearance (51). With these plausible results, clinical demonstration of passive immunization with these or similar Stx-neutralizing antibodies is much anticipated.

ACKNOWLEDGMENTS

Research performed in this review was funded by the Intramural Research of Eunice Kennedy Shriver National Institute of Child Health and Human Development, National Institutes of Health, and by Carolinas Medical Center, Charlotte, NC. The authors have no conflict of interests or financial obligations with the subject matter or materials discussed in the article. No writing assistance was used in the production of this article.

CITATION

Szu SC, Ahmed A. 2014. Clinical studies of *Escherichia coli* O157:H7 conjugate vaccines in adults and young children. Microbiol Spectrum 2(4):EHEC-0016-2013.

REFERENCES

1. **Centers for Disease Control and Prevention.** 2006. Ongoing multistate outbreak of *Escherichia coli* serotype O157:H7 infections associated with consumption of fresh spinach—United States, September 2006. *Morb Mortal Wkly Rep* **55:**1045–1046.
2. **Page AV, Liles WC.** 2013. Enterohemorrhagic *Escherichia coli* infections and hemolytic uremic syndrome. *Med Clin N Am* **97:**681–695.
3. **Kassenborg HD, Hedberg CW, Hoekstra M, Evans MC, Chin AE, Marcus R, Vugia DJ, Smith K, Ahuja SD, Slutsker L, Griffin PM, Emerging Infections Program FoodNet**

Working Group. 2004. Farm visits and undercooked hamburgers as major risk factors for sporadic *Escherichia coli* O157:H7 infection: data from a case-control study in 5 FoodNet sites. *Clin Infect Dis* **38** Suppl 3:S271–S278.

4. **Goode B, O'Reilly C, Dunn J, Fullerton K, Smith S, Ghneim G, Keen J, Durso L, Davies M, Montgomery S.** 2009. Outbreak of *Escherichia coli* O157:H7 infections after petting zoo visits, North Carolina State Fair, October-November 2004. *Arch Pediatr Adolesc Med* **163:**42–48.

5. **Durak MZ, Churey JJ, Worobo RW.** 2012. Efficacy of UV, acidified sodium hypochlorite, and mild heat for decontamination of surface and infiltrated *Escherichia coli* O157:H7 on green onions and baby spinach. *J Food Prot* **75:**1198–1206.

6. **Sheng H, Lim JY, Knecht HJ, Li J, Hovde CJ.** 2006. Role of *Escherichia coli* O157:H7 virulence factors in colonization at the bovine terminal rectal mucosa. *Infect Immun* **74:**4685–4693.

7. **Varela NP, Dick P, Wilson J.** 2012. Assessing the existing information on the efficacy of bovine vaccination against *Escherichia coli* O157:H7—a systematic review and meta-analysis. *Zoonoses Public Health* **60:**253–268.

8. **Teunis P, Takumi K, Shinagawa K.** 2004. Dose response for infection by *Escherichia coli* O157:H7 from outbreak data. *Risk Anal* **24:**401–407.

9. **Reymond D, Johnson RP, Karmali MA, Petric M, Winkler M, Johnson S, Rahn K, Renwick S, Wilson J, Clarke RC, Spika J.** 1996. Neutralizing antibodies to *Escherichia coli* Vero cytotoxin 1 and antibodies to O157 lipopolysaccharide in healthy farm family members and urban residents. *J Clin Microbiol* **34:**2053–2057.

10. **Navarro A, Eslava C, Hernandez U, Navarro-Henze JL, Aviles M, Garcia-de la Torre G, Cravioto A.** 2003. Antibody responses to *Escherichia coli* O157 and other lipopolysaccharides in healthy children and adults. *Clin Diagn Lab Immunol.* **10:**797–801.

11. **Robbins JB, Schneerson R, Szu SC.** 1995. Perspective: hypothesis: serum IgG antibody is sufficient to confer protection against infectious diseases by inactivating the inoculum (Review). *J Infect Dis* **171:**1387–1398.

12. **Passwell JH, Ashkenzi S, Banet-Levi Y, Ramon-Saraf R, Farzam N, Lerner-Geva L, Even-Nir H, Yerushalmi B, Chu C, Shiloach J, Robbins JB, Schneerson R, Israeli Shigella Study Group.** 2010. Age-related efficacy of *Shigella* O-specific polysaccharide conjugates in 1-4-year-old Israeli children. *Vaccine* **28:**2231–2235.

13. **Konadu EY, Lin FY, Hó VA, Thuy NT, Van Bay P, Thanh TC, Khiem HB, Trach DD, Karpas AB, Li J, Bryla DA, Robbins JB, Szu SC.** 2000. Phase 1 and phase 2 studies of *Salmonella enterica*

serovar Paratyphi A O-specific polysaccharide-tetanus toxoid conjugates in adults, teenagers, and 2- to 4-year-old children in Vietnam. *Infect Immun* **68:**1529–1534.

14. **Hayashi T, Makino K, Ohnishi M, Kurokawa K, Ishii K, Yokoyama K, Han CG, Ohtsubo E, Nakayama K, Murata T, Tanaka M, Tobe T, Iida T, Takami H, Honda T, Sasakawa C, Ogasawara N, Yasunaga T, Kuhara S, Shiba T, Hattori M, Shinagawa H.** 2001. Complete genome sequence of enterohemorrhagic *Escherichia coli* O157:H7 and genomic comparison with a laboratory strain K-12. *DNA Res* **8:**11–22. (Erratum, **8:**96.)

15. **Zhang Y, Lin K.** 2012. A phylogenomic analysis of *Escherichia coli/Shigella* group: implications of genomic features associated with pathogenicity and ecological adaptation. *BMC Evol Biol* **12:**174.

16. **Konadu E, Parke JC Jr, Donohue-Rolfe A, Calderwood SB, Robbins JB, Szu SC.** 1998. Synthesis and immunologic properties of O-specific polysaccharide-protein conjugate vaccines for prevention and treatment of infections with *Escherichia coli* O157 and other causes of hemolytic-uremic syndrome, p 419–424. *In* Kaper JB, O'Brien AD (ed), *Escherichia coli O157:H7 and Other Shiga Toxin-Producing E. coli Strains.* ASM Press, Washington, DC.

17. **Konadu E, Robbins JB, Shiloach J, Bryla DA, Szu SC.** 1994. Preparation, characterization, and immunological properties in mice of *Escherichia coli* O157 O-specific polysaccharide-protein conjugate vaccines. *Infect Immun* **62:**5048–5054.

18. **Konadu EY, Parke JC Jr, Tran HT, Bryla DA, Robbins JB, Szu SC.** 1998. Investigational vaccine for *Escherichia coli* O157: phase 1 study of O157 O-specific polysaccharide-*Pseudomonas aeruginosa* recombinant exoprotein A conjugates in adults. *J Infect Dis* **177:**383–387.

19. **Ahmed A, Li J, Shiloach Y, Robbins JB, Szu SC.** 2006. Safety and immunogenicity of *Escherichia coli* O157 O-specific polysaccharide conjugate vaccine in 2-5-year-old children. *J Infect Dis* **193:**515–521.

20. **Chow SK, Casadevall A.** 2012. Monoclonal antibodies and toxins—a perspective on function and isotype. *Toxins (Basel)* **4:**430–454.

21. **Szu SC, Taylor DN, Trofa AC, Clements JD, Shiloach J, Sadoff JC, Bryla DA, Robbins JB.** 1994. Laboratory and preliminary clinical characterization of Vi capsular polysaccharide-protein conjugate vaccines. *Infect Immun* **62:**4440–4444.

22. **Chart H, Rowe B.** 1993. Antibody cross-reactions with lipopolysaccharide from *E. coli* O157 after cholera vaccination. *Lancet* **341:**1282.

23. **Nichiuchi Y, Doe M, Hotta H, Kobayashi K.** 2000. Structure and serologic properties of

O-specific polysaccharide from *Citrobacter freundii* possessing cross-reactivity with *Escherichia coli* O157:H7. *FEMS Immunol Med Microbiol* **28**:163–171.

24. **Chart H, Cheasty T, Cope D, Gross RJ, Rowe B.** 1991. The serological relationship between *Yersinia enterocolitica* O9 and *Escherichia coli* O157 using sera from patients with yersiniosis and haemolytic uraemic syndrome. *Epidemiol Infect* **107**:349–356.

25. **DiFabio JL, Perry MB, Bundle DR.** 1987. Analysis of the lipopolysaccharide of *Pseudomonas maltophilia* 555. *Biochem Cell Biol* **65**:968–977.

26. **Samuel G, Hogbin JP, Wang L, Reeves PR.** 2004. Relationships of the *Escherichia coli* O157, O111, and O55 O-antigen gene clusters with those of *Salmonella enterica* and *Citrobacter freundii*, which express identical O antigens. *J Bacteriol* **186**:6536–6543.

27. **Bettelheim KA, Evangelidis H, Pearce JL, Sowers E, Strockbine NA.** 1993. Isolation of a *Citrobacter freundii* strain which carries the *Escherichia coli* O157 antigen. *J Clin Microbiol* **31**:760–761.

28. **Vinogradov E, Conlan JW, Perry MB.** 2000. Serological cross-reaction between the lipopolysaccharide O-polysaccharide antigens of *Escherichia coli* O157:H7 and strains of *Citrobacter freundii* and *Citrobacter sedlakii*. *FEMS Microbiol Lett* **190**:157–161.

29. **Cohen D, Ashkenazi S, Green MS, Gdalevich M, Robin G, Slepon R, Yavzori M, Orr N, Block C, Ashkenazi I, Shemer J, Taylor DN, Hale TL, Sadoff JC, Pavliakova D, Schneerson R, Robbins JB.** 1997. Double-blind vaccine-controlled randomised efficacy trial of an investigational *Shigella sonnei* conjugate vaccine in young adults. *Lancet* **349**:155–159.

30. **Espié E, Grimont F, Mariani-Kurkdjian P, Bouvet P, Haeghebaert S, Filliol I, Loirat C, Decludt B, Minh NN, Vaillant V, de Valk H.** 2008. Surveillance of hemolytic uremic syndrome in children less than 15 years of age, a system to monitor O157 and non-O157 Shiga toxin-producing *Escherichia coli* infections in France, 1996–2006. *Pediatr Infect Dis J* **27**:595–601.

31. **Eklund M, Nuorti JP, Ruutu P, Siitonen A.** 2005. Shigatoxigenic *Escherichia coli* (STEC) infections in Finland during 1998–2002: a population-based surveillance study. *Epidemiol Infect* **133**:845–852.

32. **Proctor ME, Davis JP.** 2000. Escherichia coli O157:H7 infections in Wisconsin, 1992–1999. *WMJ* **99**:32–37.

33. **Gill CJ, Ram S, Welsch JA, Detora L, Anemona A.** 2011. Correlation between serum bactericidal activity against *Neisseria meningitidis* serogroups A, C, W-135 and Y measured using human versus rabbit serum as the complement source. *Vaccine* **30**:29–34.

34. **Watson DC, Robbins JB, Szu SC.** 1992. Protection of mice against *Salmonella typhimurium* with an O-specific polysaccharide-protein conjugate vaccine. *Infect Immun* **60**:4679–4686.

35. **Mohawk KL, Melton-Celsa AR, Robinson CM, O'Brien AD.** 2010. Neutralizing antibodies to Shiga toxin type 2 (Stx2) reduce colonization of mice by Stx2-expressing *Escherichia coli* O157:H7. *Vaccine* **28**:4777–4785.

36. **Konadu E, Donohue-Rolfe A, Calderwood SB, Pozsgay V, Shiloach J, Robbins JB, Szu SC.** 1999. Syntheses and immunologic properties of *Escherichia coli* O157 O-specific polysaccharide and Shiga toxin 1 B subunit conjugates in mice. *Infect Immun* **67**:6191–6193.

37. **Marcato P, Griener TP, Mulvey GL, Armstrong GD.** 2005. Recombinant Shiga toxin B-subunit-keyhole limpet hemocyanin conjugate vaccine protects mice from Shigatoxemia. *Infect Immun* **73**:6523–6529.

38. **Perera LP, Samuel JE, Holmes RK, O'Brien AD.** 1991. Identification of three amino acid residues in the B subunit of Shiga toxin and Shiga-like toxin type II that are essential for holotoxin activity. *J Bacteriol* **173**:1151–1160.

39. **Suhan ML, Hovde CJ.** 1998. Disruption of an internal membrane-spanning region in Shiga toxin 1 reduces cytotoxicity. *Infect Immun* **66**:5252–5259.

40. **Miyashita A, Iyoda S, Ishii K, Hamamoto H, Sekimizu K, Kaito C.** 2012. Lipopolysaccharide O-antigen of enterohemorrhagic *Escherichia coli* O157:H7 is required for killing both insects and mammals. *FEMS Microbiol Lett* **333**:59–68.

41. **Bergan J, Dyve Lingelem AB, Simm R, Skotland T, Sandvig K.** 2012. Shiga toxins. *Toxicon* **60**:1085–1107.

42. **Melton-Celsa A, Mohawk K, Teel L, O'Brien A.** 2012. Pathogenesis of Shiga-toxin producing *Escherichia coli*. *Curr Top Microbiol Immunol* **357**:67–103.

43. **Amani J, Mousavi SL, Rafati S, Salmanian AH.** 2011. Immunogenicity of a plant-derived edible chimeric EspA, Intimin and Tir of *Escherichia coli* O157:H7 in mice. *Plant Sci* **180**:620–627.

44. **Judge NA, Mason HS, O'Brien AD.** 2004. Plant cell-based intimin vaccine given orally to mice primed with intimin reduces time of *Escherichia coli* O157:H7 shedding in feces. *Infect Immun* **72**:168–175.

45. **Zangari T, Melton-Celsa AR, Panda A, Boisen N, Smith MA, Taratov I, De Tolla LJ, Nataro JP, O'Brien AD.** 2013. Virulence of the Shiga toxin type 2-expressing *Escherichia coli* O104:H4

German outbreak isolate in two animal models. *Infect Immun* **81:**1562–1574.

46. Menne J, Nitschke M, Stingele R, Abu-Tair M, Beneke J, Bramstedt J, Bremer JP, Brunkhorst R, Busch V, Dengler R, Deuschl G, Fellermann K, Fickenscher H, Gerigk C, Goettsche A, Greeve J, Hafer C, Hagenmüller F, Haller H, Herget-Rosenthal S, Hertenstein B, Hofmann C, Lang M, Kielstein JT, Klostermeier UC, Knobloch J, Kuehbacher M, Kunzendorf U, Lehnert H, Manns MP, Menne TF, Meyer TN, Michael C, Münte T, Neumann-Grutzeck C, Nuernberger J, Pavenstaedt H, Ramazan L, Renders L, Repenthin J, Ries W, Rohr A, Rump LC, Samuelsson O, Sayk F, Schmidt BM, Schnatter S, Schöcklmann H, Schreiber S, von Seydewitz CU, Steinhoff J, Stracke S, Suerbaum S, van de Loo A, Vischedyk M, Weissenborn K, Wellhöner P, Wiesner M, Zeissig S, Büning J, Schiffer M, Kuehbacher T; EHEC-HUS consortium. 2012. Validation of treatment strategies for enterohaemorrhagic *Escherichia coli* O104:H4 induced haemolytic uraemic syndrome: case-control study. *BMJ* **345:** e4565.

47. Kielstein JT, Beutel G, Fleig S, Steinhoff J, Meyer TN, Hafer C, Kuhlmann U, Bramstedt J, Panzer U, Vischedyk M, Busch V, Ries W, Mitzner S, Mees S, Stracke S, Nürnberger J, Gerke P, Wiesner M, Sucke B, Abu-Tair M, Kribben A, Klause N, Schindler R, Merkel F, Schnatter S, Dorresteijn EM, Samuelsson O, Brunkhorst R; Collaborators of the DGfN STEC-HUS registry. 2012. Best supportive care and therapeutic plasma exchange with or without eculizumab in Shiga-toxin-producing *E. coli* O104:H4 induced haemolytic-uraemic syndrome: an analysis of the German STEC-HUS registry. *Nephrol Dial Transplant* **27:**3807–3815.

48. Tzipori S, Sheoran A, Akiyoshi D, Donohue-Rolfe A, Trachtman H. 2004. Antibody therapy in the management of Shiga toxin-induced hemolytic uremic syndrome. *Clin Microbiol Rev* **17:** 926–941.

49. Sheoran AS, Chapman-Bonofiglio S, Harvey BR, Mukherjee J, Georgiou G, Donohue-Rolfe A, Tzipori S. 2005. Human antibody against Shiga toxin 2 administered to piglets after the onset of diarrhea due to *Escherichia coli* O157:H7 prevents fatal systemic complications. *Infect Immun* **73:**4607–4613.

50. He X, McMahon S, Skinner C, Merrill P, Scotcher MC, Stanker LH. 2013. Development and characterization of monoclonal antibodies against Shiga toxin 2 and their application for toxin detection in milk. *J Immunol Methods* **389:**18–28.

51. Sheoran A, Jeong KI, Mukherjee J, Wiffin A, Singh P, Tzipori S. 2012. Biodistribution and elimination kinetics of systemic Stx2 by the Stx2A and Stx2B subunit-specific human monoclonal antibodies in mice. *BMC Immunol* **13:**27.

25

Vaccination of Cattle against *Escherichia coli* O157:H7

DAVID R. SMITH[1]

INTRODUCTION

Human infection with Shiga toxin-producing *Escherichia coli* O157:H7 (STEC O157) is relatively rare, but the consequences can be serious, especially in the very young and the elderly. Outcomes associated with STEC O157 infection include hemorrhagic colitis, renal failure, and death (1–5). In 2012, the overall laboratory-confirmed annual incidence of STEC O157 in the United States was 1.1 cases per 100,000 population (6). However, the incidence in children less than 5 years of age was 4.7 cases per 100,000 population (6).

Infection from STEC O157 occurs directly or indirectly via fecal-oral transmission (7). People are exposed to STEC O157 through a variety of sources, including direct contact with human or animal feces and indirect contact via contaminated food, water, or soil (8). The primary route of transmission of STEC O157 is contaminated food (9, 10); however, large outbreaks have been associated with contamination of municipal water supplies (11–14). Important environmental hazards for human exposure to STEC include daycare facilities, nursing homes, children playing with a sick friend, swimming pools, contaminated food and water, and direct exposure to animal environments such as farms, petting zoos, or livestock exhibitions (9, 10, 15). Approximately one-third

[1]College of Veterinary Medicine, Mississippi State University, Mississippi State, MS 39762-6100.
Enterohemorrhagic Escherichia coli *and Other Shiga Toxin-Producing* E. coli
Edited by Vanessa Sperandio and Carolyn J. Hovde
© 2015 American Society for Microbiology, Washington, DC
doi:10.1128/microbiolspec.EHEC-0006-2013

of human infections are attributed to consumption of ground or nonintact beef (16). Some of the earliest and most notorious outbreaks of STEC O157 infection were associated with the consumption of undercooked ground beef sandwiches, resulting in the infection being commonly known as "hamburger disease" (2, 17–19).

STEC has been recovered from many animal species, but ruminants are particularly prone to colonization (20). Of ruminants, cattle populations are widely recognized as an important reservoir of STEC O157 for human exposure in the United States (8, 21).

A variety of vehicles, other than food, have been important in the fecal-oral transmission of STEC strains to humans, including fomites such as dust (22) and water (11–14) and vectors such as flies (23–27). Other animals, besides cattle, have caused important STEC outbreaks in humans because they served as vehicles for fruit or vegetable crop contamination. For example, a large STEC O157 outbreak in the United States and Canada was due to consumption of spinach that was contaminated in the field by feces from feral pigs that had contact with cattle pastures (28). In Oregon, deer were the source of feces that contaminated strawberries with STEC O157, resulting in one death and at least 14 illnesses (29).

Circumstantial evidence supports the contention that cattle are the primary reservoir for human exposure to STEC in North America. First, there is strong correlation between seasonal variability in incidence of human STEC O157 illness, prevalence of ground beef contamination with STEC O157, and prevalence of STEC O157 shedding by cattle in feedlots, all greater in summer months than winter months (30). This relationship may indicate that STEC O157, originating in or on cattle, contaminates ground beef to eventually become the source for subsequent human infection (30). In addition, there is a correlation between the prevalence of carriage of STEC O157 in feces or on hides of live cattle entering the abattoir and subsequent rates of carcass contamination (31, 32). Finally, since 1998 in the United States, human incidence of STEC O157 has decreased (6), largely because of interventions taken in abattoirs to reduce the flow of STEC O157 from live cattle into the beef supply (33, 34). The decrease in incidence since 1998 is greater than the proportion of illnesses attributable to contaminated beef, suggesting that decreasing the bacterial flow from beef prevented secondary cases of person-to-person STEC O157 infection. Unfortunately, the incidence of STEC O157 infection has not changed meaningfully or statistically compared to the average annual incidence during 2006–2008, suggesting that additional actions, for example, at the preharvest level, are necessary to further reduce rates of STEC O157 illness (6).

PREHARVEST ECOLOGY OF STEC O157

Cattle are colonized by STEC O157 primarily at the terminal rectum (35, 36). Colonization by STEC O157 requires attachment to intestinal epithelium and induces attaching and effacing lesions. Following STEC O157 infection in cattle of all ages, inflammation and innate and adaptive immune responses occur (37), supporting the contention that STEC O157 is a bovine pathogen (37, 38). However, this latter point remains controversial because infection does not result in clinically observable signs of illness in adult cattle (39, 40). In any case, not all cattle shedding STEC in their feces are currently colonized; some may be shedding ingested organisms that are simply passing through the intestinal tract (41, 42). The duration of infection in cattle is variable but short-lived, approximating a month (41, 43–45). In field settings, reinfection is common (44).

Prevalence of STEC O157 carriage by feedlot cattle varies widely within and across seasons and is affected by both incidence and duration of shedding (44, 46, 47). The probability of cattle carrying STEC depends on both gut and environmental conditions that change over time. As with all *E. coli* strains, conditions of the bovine gut that favor STEC O157 may increase colonization and duration of shedding. Factors of the environment that favor STEC O157 survival or opportunities for fecal-oral transmission increase the incidence of exposure. This is because pathogenic and commensal *E. coli* strains have two principal habitats: a primary habitat in the lower intestine of warm-blooded animals and a secondary habitat in water, sediment, and soil (48). The suitability of the primary habitat is influenced by factors such as physical characteristics (e.g., pH); the host's diet, immune system, and physiological state; and interactions with other microorganisms in the same region. The suitability of the secondary habitat is also complex and dependent on physical factors, climatic and meteorological factors, nutrients, and interactions with other microorganisms within the ecosystem. In contrast to the primary habitat, which is uniformly warm, approximately 37°C, and nutrient rich, the secondary habitat may have extremes in temperatures and is typically nutrient deficient (48). Environmental conditions that favor survival and fecal-oral transmission have been associated with greater rates of exposure and shedding in feedlot cattle (46, 47).

Transmission heterogeneity, or super-spreading, is the phenomenon of a minority of infected individuals being responsible for transmitting the majority of new infections (49, 50). At a given point in time, STEC O157-infected cattle shed the organism at varying concentrations in feces (42, 51, 52). Therefore, some cattle may contribute vastly more STEC organisms into the environment, and possibly to other cattle, than others. Cattle that shed STEC at greater than 10^3 or 10^4 CFU/g of feces, or cattle that are culture-positive for prolonged periods, have variously been defined by the term super-shedder (42, 51). It has been proposed that super-shedding status is indicative of cattle colonized by STEC rather than cattle experiencing simple pass-through of organisms (42). Because of

the greater number of organisms being shed, cattle designated as super-shedders may have an important effect on environmental contamination and subsequent transmission within cattle production settings (53) or in lairage (54). The relevance of super-shedding to STEC O157 control is not clear. Super-shedding of STEC O157 in feces does not appear to be a persistent state, and we do not yet understand if super-shedding is a characteristic of certain cattle or merely a stage of infection that cattle transition through following infection. It has been observed that detection of super-shedding cattle is temporally correlated with periods of high prevalence, and super-shedder cattle appear to be a subset of fecal-culture-positive individuals within the population (42, 55). Super-shedding may not be necessary or sufficient for STEC O157 transmission, even in closed (all-in, all-out type) feeding systems (56). Rather than super-shedding cattle driving transmission of STEC to other cattle, super-shedding may be an outcome of environmental conditions that favor ingestion of the organism (47). When those conditions favor new host infections, then some cattle may become colonized and transiently shed large numbers of organisms, and because of favorable conditions for transmission, the duration of detectable shedding may be prolonged (44).

To reduce the prevalence of STEC O157 carriage by cattle, efforts have been attempted to make either the primary or secondary habitat less favorable to STEC O157 survival or growth (57–59). To date, efforts to make the cattle environment less hospitable to STEC O157, for example, by scraping pen surfaces or cleaning water tanks, have not effectively reduced STEC O157 carriage by cattle (60–62). However, several strategies for modifying the gut environment, including the use of vaccines; chemicals, such as sodium chlorate or antibiotics; and competing microorganisms, such as some strains of *Lactobacillus*, have effectively reduced the probability of cattle shedding STEC O157 in feces (63–66).

VACCINATION OF CATTLE AGAINST STEC O157

The objective of immunizing cattle against STEC O157 is to make the gut unfavorable for colonization, thereby reducing duration of carriage and minimizing shedding of the pathogen into the cattle environment (58). In theory, the benefit of vaccination within discrete populations (e.g., pens or herds of cattle) is reduced fecal-oral transmission within cattle environments, less contamination of cattle hides, and fewer pathogens carried into the abattoir at harvest. For vaccination to be useful as a preharvest intervention, the benefits must not be undone during subsequent management practices, such as transportation to the abattoir (67) or during holding in lairage (32, 68, 69). Preharvest interventions such as vaccination are not likely to be adopted widely by cattle producers until they are sufficiently valued in the marketplace to offset the cost of implementation.

Some candidate vaccines against STEC O157 have been tested in animal challenge studies or under field conditions of natural exposure. These vaccines either have undefined antigen targets in the form of bacterial extracts or are directed against specific antigens that function to enable bacterial colonization or survival. Unfortunately, because of serotype specificity, vaccines targeting STEC O157 may offer poor cross-protection against other STEC strains (70).

In randomized controlled studies, the strength of effect of a vaccine is often expressed as vaccine efficacy, a form of attributable fraction that measures the percentage of cases prevented by vaccination (71). Vaccine efficacy is calculated as 1 minus relative risk (72). In this case, relative risk is the probability of vaccinated cattle to carry STEC O157 divided by the probability of nonvaccinated cattle to carry the organism. The odds ratio is the statistical measure of association often reported from vaccine field studies because logistic regression is a commonly used method to analyze the data. Regardless of whether the comparison uses odds (i.e., odds ratio) or probability (i.e., relative risk), a value of 1 indicates no difference from the treatment. The further the value is from 1, toward 0 or infinity, the larger the measure of association. If the study is not a case-control study design, then odds ratio can be converted to relative risk after adjustment for marginal probabilities for disease and exposure (73). In studies with measures of fecal concentration, the measure of association may be expressed as the change in concentration due to vaccine treatment, which is often described as a logarithmic (base 10) reduction (74) and sometimes reported as a percentage (e.g., a decrease from 10,000 CFU/g of feces to 1,000 CFU/g of feces is a decrease of 1 $\log_{(10)}$ in CFU/g of feces and may be expressed as a 90% reduction in shedding concentration).

Vaccine Challenge Studies

STEC O157 colonizes bovine intestinal epithelial cells by a type III secreted protein (TTSP) system. Components of the TTSP system include:

- Intimin, an outer membrane bacterial receptor
- Translocated intimin receptor (Tir), a receptor injected into the host epithelial cell membrane
- EspA, an injection filament for delivering Tir to the host cell membrane
- EspB/EspD, which form a pore in the host cell membrane (7, 40, 75, 76)

The H7 flagellin is also believed to function in STEC O157:H7 colonization (77–79). For some STEC non-O157 serotypes, the enterohemorrhagic *E. coli* factor for adherence (*efa*-1) is important for colonization of bovine intestines, and STEC O157 carries a truncated form of the gene (80).

Vaccines targeting various STEC O15-specific antigens have been tested in animal challenge studies. Several studies have demonstrated immune response against the antigens but variable results regarding protection against STEC O157 infection. Suckling pigs whose dams were vaccinated with an intimin vaccine were protected from colonization or microscopic evidence of intestinal damage following oral challenge with 10^6 CFU of a Shiga toxin-negative strain of EHEC O157:H7 (81). Calves vaccinated with EspA developed antigen-specific antibody titers but failed to be protected against colonization with STEC O157 following challenge (82). Similarly, subunit vaccines targeting polypeptides of intimin or *efa*-1 elicited humoral responses in 2-week-old calves following intramuscular priming and intranasal booster doses, but the vaccine products failed to prevent shedding after STEC O157 or STEC O26 challenge (80). In the same study, a formalin-inactivated STEC O157 bacterin administered intramuscularly with subsequent intranasal booster doses also failed to reduce shedding in challenged calves (80). Two-month-old calves vaccinated intramuscularly with H7 flagellin had reduced rates of colonization and delayed peak bacterial shedding following oral challenge with STEC O157, but the calves did not show a reduction in total bacterial shedding (83). However, a vaccine prepared with intimin, EspA, and Tir did reduce STEC O157 colonization and bacterial counts in calves orally inoculated with STEC O157 (84). Also, lambs that had been vaccinated with intimin, EspA, and EspB shed fewer bacteria in feces than placebo-treated controls did following an oral challenge with STEC O157 (85). Six- to 8-month-old calves injected intramuscularly with a vaccine product containing intimin and EspB proteins developed an antibody response against the proteins and shed fewer STEC O157 bacteria in the first 13 days post challenge (86). Calves vaccinated with a bacterial supernatant with TTSP had reduced probability, magnitude, and duration of shedding of STEC O157 following challenge (87). In a follow-up study, calves receiving the same vaccine product were 21% less likely to shed STEC O157 in the feces and shed at a 1.4 $\log_{(10)}$ lower fecal concentration 3 to 6 days after experimental challenge with 10^9 CFU of STEC

O157 (74). Calves injected twice subcutaneously with an inactivated, whole-cell envelope vaccine (STEC O157 bacterial ghosts) demonstrated an antibody response and shed fewer STEC O157 post challenge (88). Vaccination of pregnant cows with intimin, EspA, EspB, and Shiga toxin 2 within 2 months of calving produced elevated serum and colostral antibodies against intimin and EspB and a moderate increase in EspA antibodies (89). Calves fed the dam's colostrum had significantly increased serum immunoglobulin G titers against intimin and EspB, but not EspA (89).

Siderophore receptor and porin (SRP) vaccines are targeted against bacterial cell membrane proteins used by gram-negative bacteria for iron transport in conditions of low iron supply (90). By limiting its uptake of iron, STEC O157 is placed at a competitive disadvantage relative to other gut microbiota (91). In a study of beef calves orally inoculated with STEC O157, the SRP vaccine reduced fecal prevalence and bacterial concentration to a level that approached statistical significance (90).

Vaccine Field Studies

The outcomes of experimental challenge studies may not predict the efficacy of a STEC O157 vaccine as it is used under field conditions because factors affecting rates of transmission, sources of pathogens, and dose-loads of exposure are complex and temporally dynamic in cattle production settings (44, 46, 47). Only a few STEC O157 vaccine products have been evaluated for efficacy in the conditions of natural STEC O157 exposure within cattle production systems. An uncharacterized bacterial extract did not reduce STEC O157 carriage in feedlot cattle (92). Another uncharacterized STEC O157 vaccine, administered to pregnant beef cattle during the last trimester of gestation, significantly increased antibody titers in the dam and subsequently the calf, but the study had insufficient power to evaluate efficacy at preventing shedding of

STEC O157 by the calves (93). Calves suckling cows that had been vaccinated against SRP antigens had significantly greater antibody titers against STEC O157 SRP at branding (i.e., 30 to 60 days of age), but neither the passively acquired antibodies nor active immunization significantly prevented STEC O157 shedding by the calves at feedlot entry (94).

Two vaccine products, one targeting TTSP, the other SRP, have been tested extensively in dry-lot beef feedlots under conditions typical of the Central Plains regions of the United States and Canada. These products were the subject of several systematic reviews and meta-analyses that found sufficient evidence to conclude that both vaccines effectively reduce the probability of feedlot cattle to shed STEC O157 in feces (63, 95). One meta-analysis of fecal shedding found the overall odds ratios (and 95% confidence intervals) for detecting STEC O157 in the feces of vaccinated cattle relative to nonvaccinated cattle to be 0.38 (0.29–0.51) and 0.42 (0.20–0.61) for TTSP and SRP vaccines, respectively (63). Given the overall fecal shedding prevalence of 15% observed in the TTSP studies (63), the odds ratio of 0.38 converts to a relative risk of 0.42 and vaccine efficacy of 0.58 (96). Another meta-analysis looked at all outcomes and reported that two doses of TTSP vaccine had odds ratios of 0.49 (0.40–0.60) for preharvest outcomes and 0.45 (0.34–0.60) for preharvest and at-harvest outcomes combined (95).

Details from individual studies provide additional information about the efficacy of STEC O157 vaccine products, although some details, such as antigen concentrations, have not always been reported. Using steers screened to be negative for STEC O157 carriage before the study start, researchers found that steers vaccinated twice with either 2 or 3 ml of SRP vaccine were 14 and 47% less likely than placebo-treated steers to have STEC O157 detected in either feces or recto-anal mucosa swab samples, respectively (97). Feedlot cattle receiving a 2-ml, two-dose SRP vaccine regimen did not differ from controls in STEC O157 carriage over the

postvaccination period except for the last day of the study (91). In a trial testing a 2-ml, three-dose SRP vaccine regimen against placebo-treated cattle, the vaccine was 85% effective in reducing the probability of detecting STEC O157 in feces and reduced STEC O157 concentration 1.7 logs compared to controls 56 days after the last dose of vaccine (91). In a vaccine trial conducted in a commercial feedlot, the SRP vaccine demonstrated 53% vaccine efficacy in reducing STEC O157 prevalence and 73% efficacy in reducing the prevalence of high shedders, defined as cattle shedding >10^4 CFU/g of feces (98). In that study, pens of cattle receiving vaccination had significantly reduced feed efficiency and rate of gain, which may represent an additional cost of the intervention (98).

Vaccinating feedlot cattle with a TTSP vaccine product failed to be efficacious in a large initial vaccine field trial (99). However, the vaccine product was reformulated and efficacy improved (99). Vaccine efficacy of a three-dose regimen of TTSP vaccine to reduce the probability of feedlot cattle shedding STEC O157 has ranged from 43 to 73% in several randomized controlled trials (87, 100–102). In addition, the vaccine was 92% and 98% effective in reducing the probability of colonization of the terminal rectum when two- (103) or three-dose (104) regimens, respectively, were used. Two doses of the same vaccine product significantly reduced carriage of STEC O157 by feedlot cattle (103, 105, 106), and it appears that two doses of vaccine may be sufficient to induce an effective immune response (95). However, three doses of vaccine were more effective than two doses in trials with direct comparisons (100, 107). This vaccine does not appear to affect growth performance (104, 107) or carcass quality (104, 106, 107).

The duration of immunity after vaccination is unknown because the evaluation period in feedlot studies has been relatively short, typically with postvaccination observation periods of between 60 and 100 days (63, 108, 109). Increasing or decreasing immunity would be evident as a statistical interaction between vaccine treatment and time elapsing since vaccination on the probability of cattle carrying the organism. This interaction has not been reported. Even though vaccine efficacy appears to persist sufficiently long enough for cattle on finishing diets, duration of immunity remains an important unmet area of investigation for beef and dairy young-stock and breeding cattle (109).

Cattle are typically managed as groups (e.g., pens or herds of cattle), which are fed and housed together. Similarly, cattle management practices such as vaccination are usually applied to the group, partly for ease of management and to provide protection to the group rather than simply the individual. The ability of groups to resist infection, or to limit the extent of infection within the group, is termed herd immunity (110). Herd immunity is a function of individual resistance to infection and the dynamics of transmission within the group (110, 111). Individuals lacking immunity may be protected from infection because of group-level factors; for example, the majority of individuals with immunity change the likelihood of exposure to those without (110).

The probability of cattle carrying STEC O157 in the gut or on their hides is affected by group-level factors. For example, the distribution of fecal prevalence of STEC O157 within pens of feedlot cattle tends to be greater or lesser than expected by binomial distribution around the mean (46), suggesting that, at a given point in time, cattle within pens behave similarly with respect to STEC O157 shedding (i.e., most cattle shedding or most not). Factors explaining the probability of cattle shedding the agent or having evidence of oral exposure are associated with characteristics of the pen environment that either favor survival of the organism (e.g., warm or wet) or increase opportunities for ingestion (e.g., mud or dust), indicating that sometimes the pen environment favors fecal-oral transmission and sometimes it does not (44, 46, 47, 112). Therefore, it is important to

evaluate group-level effects of vaccinating cattle against STEC O157. There is evidence that fecal-oral transmission of STEC O157 is reduced within pens of vaccinated cattle. Herd immunity was demonstrated in a longitudinal STEC O157 vaccine study as nonvaccinated cattle housed with vaccinated cattle were less likely to shed STEC O157 compared to cattle penned in the same feedyard where none of the cattle received vaccine (107). Vaccinated cattle housed together in large commercial feedyard pens were less likely to have oral exposure to STEC O157 compared to nonvaccinated cattle housed together in pens in the same feedyards, based on culturing ropes hung on feedbunk rails for cattle to chew (103). Culture of STEC O157 from ropes is correlated to fecal shedding prevalence (112), and more directly measures opportunities for oral exposure (113).

The value of considering the effects of group-level vaccination when designing a STEC O157 cattle vaccination program was demonstrated by the greater efficacy in reducing hide contamination when all cattle in a region of a feedyard were vaccinated compared to the efficacy observed when vaccinated and unvaccinated cattle were commingled within pens (106). Efficacy against hide contamination is important because the hides of cattle are the primary source of STEC O157 carcass contamination (32, 69, 114, 115). It was hypothesized that vaccination of all cattle within a region of a feedyard, or the entire feedyard, would result in a greater reduction in the load of organisms deposited by cattle into the environment and less subsequent contamination of hides than when vaccinated cattle are commingled in pens of nonvaccinated cattle (106). This finding illustrates that the goal of a cattle vaccination program against STEC O157 is to reduce environmental pathogen load to minimize ingestion of the organism or hide contamination, and this may be accomplished most effectively by administering the vaccine to all cattle within a production system (106).

Whatever efficacy a vaccine may have before harvest, it can be undone by events occurring during subsequent stages of the food system, such as cross-contamination of cattle hides with STEC O157 during transportation or while cattle are in lairage (32, 67, 116). However, the efficacy of preharvest interventions has persisted into the abattoir. In a randomized clinical trial to test a STEC O157 cattle vaccine, there was a significant increase in the prevalence of hide contamination between the time immediately before loading at the feedyard versus just before hide removal in the abattoir. However, vaccination treatments had equal efficacy for reducing hide contamination in the feedyard and at the abattoir. The preservation of vaccine efficacy into the abattoir may have been the result of efforts to load cattle by treatment groups into clean trucks for transportation to the abattoir (106). Therefore, to preserve vaccine efficacy, it may be necessary to devise methods for cattle handling so that preharvest benefits are retained post harvest.

Modeling STEC O157 Vaccine Usefulness

Ultimately, the reasons for vaccinating cattle against STEC O157 are to (i) benefit public health by preventing human STEC O157 infection and (ii) reduce costs to the beef industry due to recalls, lost product value, and liability. There is value in preventing human illness from direct contact with cattle or their environments, but this is a less common source of human illness compared to infections acquired through contaminated food, including beef, milk, and vegetable crops (9, 10). The primary value of vaccinating live cattle is the benefit to the postharvest sectors of the food system and the consumers of food products. An intervention is not likely to be used if the costs of the intervention exceed the benefits to the food industry or public health. Mathematical models provide a conceptual framework for understanding pathogen transmission dynamics. Models can help identify knowledge gaps, give insight into new research questions, and predict the usefulness of intervention strategies (117).

From a public health policy perspective, one might compare the cost of human illness to the cost of a preharvest intervention. If the marginal costs of vaccinating cattle were equivalent to the marginal benefit to public health, then as the cost of a vaccine intervention increased, fewer cattle would be vaccinated, and as a result, fewer human illnesses would be prevented. Similarly, the number of cattle that must receive an intervention to prevent a single human illness increases as the effectiveness of the product decreases (16). From a beef industry perspective, preharvest interventions might be valued on the basis of how cattle carrying STEC O157 into the abattoir affect subsequent food safety costs. For example, because an important source of STEC O157 carcass contamination is the hide (32), and fecal shedding prevalence above 20% has been associated with higher prevalence levels of hide contamination (118), postharvest sectors of the beef industry might benefit from preharvest interventions that supply cattle at harvest with less hide contamination and reduced, less variable, fecal shedding prevalence that does not overwhelm subsequent postharvest interventions.

Quantitative or qualitative models have been used to investigate the value of vaccinating cattle and other methods of intervention. Many models predict benefit to both public health and the beef industry from vaccinating cattle against STEC O157. For example, a model simulating ground beef contamination in Argentina predicted that vaccinating cattle and online hide washing would have the greatest impacts on reducing STEC O157 prevalence and concentration in ground beef product and the resulting numbers of human infections, hemolytic-uremic syndrome, and STEC O157-associated mortalities per ground beef meal (119). A stochastic simulation model based on U.S. beef production systems and risk for infection through consumption of ground beef also concluded that vaccination of cattle would have a strong impact on decreasing the number of human STEC O157 illnesses, the number of contaminated beef production lots, the likelihood of STEC O157 detection by regulatory testing, and the probability of outbreaks due to ground beef servings from the same lot (120). A simulation model was used to investigate infection transmission in pastured cattle systems. The modelers concluded that vaccine efficacy of 60% would be particularly effective in reducing levels of infection in a herd (121). Stochastic simulation of the distribution of pen-level fecal shedding prevalence in U.S. commercial beef feedyards predicted that vaccination of summer-fed cattle with a 58% effective product would eliminate pens of highest prevalence, resulting in a prevalence distribution similar to what is typically observed in winter-fed cattle. This model showed that a major effect of vaccination is reduced variability in shedding prevalence (122). The opinions of experts were used in a best-worse scaling evaluation to gain consensus on the effectiveness and practicality of on-farm methods to reduce human exposure to STEC O157 (123). Intervention methods were evaluated for effectiveness and practicality. By this process, vaccination of cattle was considered the most effective, and hand washing the most practical, method to reduce human exposure to STEC (123).

CONCLUSION

Ideally, preharvest interventions against STEC O157 should be

- Efficacious—cattle are less likely to carry the organism because of the intervention
- Useful—able to be practically applied within the beef production system
- Economical—add sufficient value to the product to offset the cost of the intervention

A number of STEC O157 antigens are being investigated as potential vaccine targets. Some vaccine products have demonstrated efficacy to reduce the prevalence of cattle carrying

STEC O157 by making the gut environment unfavorable to colonization. However, in conditions of natural exposure, efficacy afforded by vaccination depends on how the products are used to control environmental transmission within groups of cattle or throughout the production system (106). Preharvest benefits from vaccination may be nullified unless steps are taken to prevent cross-contamination of cattle or beef product throughout the food system (68). Although cattle vaccines against STEC O157 have gained either full or preliminary regulatory approval in Canada and the United States, it is not yet clear if they will be widely adopted by cattle feeders because there is not yet an economic signal to indicate that cattle vaccinated against STEC O157 are valued over other cattle.

ACKNOWLEDGMENT

I declare no conflicts of interest with regard to the manuscript.

CITATION

Smith DR. 2014. Vaccination of cattle against *Escherichia coli* O157:H7. Microbiol Spectrum 2(4):EHEC-0006-2013.

REFERENCES

1. **Scallan E, Hoekstra RM, Angulo FJ, Tauxe RV, Widdowson MA, Roy SL, Jones JL, Griffin PM.** 2011. Foodborne illness acquired in the United States—major pathogens. *Emerg Infect Dis* **17:**7–15.
2. **Riley LW, Remis RS, Helgerson SD, McGee HB, Wells JG, Davis BR, Hebert RJ, Olcott ES, Johnson LM, Hargrett NT, Blake PA, Cohen ML.** 1983. Hemorrhagic colitis associated with a rare *Escherichia coli* serotype. *N Engl J Med* **308:**681–685.
3. **Wells JG, Davis BR, Wachsmuth IK, Riley LW, Remis RS, Sokolow R, Morris GK.** 1983. Laboratory investigation of hemorrhagic colitis outbreaks associated with a rare *Escherichia coli* serotype. *J Clin Microbiol* **18:**512–520.
4. **Karmali MA, Petric M, Lim C, Fleming PC, Steele BT.** 1983. *Escherichia coli* cytotoxin, haemolytic-uraemic syndrome, and haemorrhagic colitis. *Lancet* **ii:**1299–1300.

5. **Karmali MA, Steele BT, Petric M, Lim C.** 1983. Sporadic cases of haemolytic-uraemic syndrome associated with faecal cytotoxin and cytotoxin-producing *Escherichia coli* in stools. *Lancet* **i:**619–620.

6. **Centers for Disease Control and Prevention.** 2013. Incidence and trends of infection with pathogens transmitted commonly through food—Foodborne Diseases Active Surveillance Network, 10 U.S. sites, 1996–2012. *Morb Mortal Wkly Rep* **62:**283–287.

7. **Nataro JP, Kaper JB.** 1998. Diarrheagenic *Escherichia coli*. *Clin Microbiol Rev* **11:**142–201.

8. **Sargeant JM, Smith DR.** 2003. The epidemiology of *Escherichia coli* O157:H7, p 131–141. *In* Torrence ME, Isaacson RE (ed), *Microbial Food Safety in Animal Agriculture: Current Topics.* Iowa State University Press, Ames, IA.

9. **Rangel JM, Sparling PH, Crowe C, Griffin PM, Swerdlow DL.** 2005. Epidemiology of *Escherichia coli* O157:H7 outbreaks, United States, 1982–2002. *Emerg Infect Dis* **11:**603–609.

10. **Sparling PH.** 1998. *Escherichia coli* O157:H7 outbreaks in the United States, 1982–1996. *J Am Vet Med Assoc* **213:**1733–1733.

11. **Swerdlow DL, Woodruff BA, Brady RC, Griffin PM, Tippen S, Donnell HD Jr, Geldreich E, Payne BJ, Meyer A Jr, Wells JG.** 1992. A waterborne outbreak in Missouri of *Escherichia coli* O157:H7 associated with bloody diarrhea and death. *Ann Intern Med* **117:**812–819.

12. **Kondro W.** 2000. *E. coli* outbreak deaths spark judicial inquiry in Canada. *Lancet* **355:**2058.

13. **Kondro W.** 2000. Canada reacts to water contamination. *Lancet* **355:**2228.

14. **Centers for Disease Control and Prevention.** 1999. Outbreak of *Escherichia coli* O157:H7 and *Campylobacter* among attendees of the Washington County Fair–New York, 1999. *Morb Mortal Wkly Rep* **48:**803–805.

15. **Feng P.** 1995. *Escherichia coli* serotype O157:H7: novel vehicles of infection and emergence of phenotypic variants. *Emerg Infect Dis* **1:**47–52.

16. **Withee J, Williams M, Schlosser W, Bauer N, Ebel E.** 2009. Streamlined analysis for evaluating the use of preharvest interventions intended to prevent *Escherichia coli* O157:H7 illness in humans. *Foodborne Pathog Dis* **6:**817–825.

17. **Kassenborg HD, Hedberg CW, Hoekstra M, Evans MC, Chin AE, Marcus R, Vugia DJ, Smith K, Ahuja SD, Slutsker L, Griffin PM.** 2004. Farm visits and undercooked hamburgers as major risk factors for sporadic *Escherichia coli* O157:H7 infection: data from a case-control study in 5 FoodNet sites. *Clin Infect Dis* **38** Suppl 3:S271–S278.

18. **Ryan CA, Tauxe RV, Hosek GW, Wells JG, Stoesz PA, McFadden HW Jr, Smith PW, Wright GF, Blake PA.** 1986. *Escherichia coli* O157:H7 diarrhea in a nursing home: clinical, epidemiological, and pathological findings. *J Infect Dis* **154:**631–638.

19. **Slutsker L, Ries AA, Maloney K, Wells JG, Greene KD, Griffin PM.** 1998. A nationwide case-control study of *Escherichia coli* O157:H7 infection in the United States. *J Infect Dis* **177:**962–966.

20. **Beutin L, Geier D, Steinruck H, Zimmermann S, Scheutz F.** 1993. Prevalence and some properties of verotoxin (Shiga-like toxin)-producing *Escherichia coli* in seven different species of healthy domestic animals. *J Clin Microbiol* **31:**2483–2488.

21. **Karmali MA, Gannon V, Sargeant JM.** 2010. Verocytotoxin-producing *Escherichia coli* (VTEC). *Vet Microbiol* **140:**360–370.

22. **Varma JK, Greene KD, Reller ME, DeLong SM, Trottier J, Nowicki SF, DiOrio M, Koch EM, Bannerman TL, York ST, Lambert-Fair MA, Wells JG, Mead PS.** 2003. An outbreak of *Escherichia coli* O157 infection following exposure to a contaminated building. *JAMA* **290:**2709–2712.

23. **Alam MJ, Zurek L.** 2004. Association of *Escherichia coli* O157:H7 with houseflies on a cattle farm. *Appl Environ Microbiol* **70:**7578–7580.

24. **Hancock DD, Besser TE, Rice DH, Ebel ED, Herriott DE, Carpenter LV.** 1998. Multiple sources of *Escherichia coli* O157 in feedlots and dairy farms in the northwestern USA. *Prev Vet Med* **35:**11–19.

25. **Janisiewicz WJ, Conway WS, Brown MW, Sapers GM, Fratamico P, Buchanan RL.** 1999. Fate of *Escherichia coli* O157:H7 on fresh-cut apple tissue and its potential for transmission by fruit flies. *Appl Environ Microbiol* **65:**1–5.

26. **Kobayashi M, Sasaki T, Saito N.** 1999. Houseflies: not simple mechanical vectors of enterohemorrhagic *Escherichia coli* O157:H7. *Am J Trop Med Hyg* **61:**625–629.

27. **Moriya K, Fujibayashi T, Yoshihara T, Matsuda A, Sumi N, Umezaki N, Kurahashi H, Agui N, Wada A, Watanabe H.** 1999. Verotoxin-producing *Escherichia coli* O157:H7 carried by the housefly in Japan. *Med Vet Entomol* **13:**214–216.

28. **Jay MT, Cooley M, Carychao D, Wiscomb GW, Sweitzer RA, Crawford-Miksza L, Farrar JA, Lau DK, O'Connell J, Millington A, Asmundson RV, Atwill ER, Mandrell RE.** 2007. *Escherichia coli* O157:H7 in feral swine near spinach fields and cattle, central California coast. *Emerg Infect Dis* **13:**1908–1911.

29. **Oregon Health Authority.** 2012. Strawberries, deer and other investigations. *CD Summary* **61**(13). https://public.health.oregon.gov/DiseasesConditions/CommunicableDisease/CDSummaryNewsletter/Documents/2012/ohd6113.pdf

30. **Williams MS, Withee JL, Ebel ED, Bauer NE, Scholosser WD, Disney WT, Smith DR, Moxley RA.** 2010. Determining relationships between the seasonal occurrence of *Escherichia coli* O157:H7 in live cattle, ground beef, and humans. *Foodborne Pathog Dis* **7:**1–8.

31. **Elder RO, Keen JE, Siragusa GR, Barkocy-Gallagher GA, Koohmaraie M, Laegreid WW.** 2000. Correlation of enterohemorrhagic *Escherichia coli* O157 prevalence in feces, hides, and carcasses of beef cattle during processing. *Proc Natl Acad Sci USA* **97:**2999–3003.

32. **Arthur TM, Bosilevac JM, Nou X, Shackelford SD, Wheeler TL, Kent MP, Jaroni D, Pauling B, Allen DM, Koohmaraie M.** 2004. *Escherichia coli* O157 prevalence and enumeration of aerobic bacteria, *Enterobacteriaceae*, and *Escherichia coli* O157 at various steps in commercial beef processing plants. *J Food Prot* **67:**658–665.

33. **Brichta-Harhay DM, Guerini MN, Arthur TM, Bosilevac JM, Kalchayanand N, Shackelford SD, Wheeler TL, Koohmaraie M.** 2008. *Salmonella* and *Escherichia coli* O157:H7 contamination on hides and carcasses of cull cattle presented for slaughter in the United States: an evaluation of prevalence and bacterial loads by immunomagnetic separation and direct plating methods. *Appl Environ Microbiol* **74:**6289–6297.

34. **Centers for Disease Prevention and Control.** 2006. Preliminary FoodNet data on the incidence of infection with pathogens transmitted commonly through food—10 States, United States, 2005. *Morbid Mortal Wkly Rep* **55:**392–395.

35. **Naylor SW, Low JC, Besser TE, Mahajan A, Gunn GJ, Pearce MC, McKendrick IJ, Smith DG, Gally DL.** 2003. Lymphoid follicle-dense mucosa at the terminal rectum is the principal site of colonization of enterohemorrhagic *Escherichia coli* O157:H7 in the bovine host. *Infect Immun* **71:**1505–1512.

36. **Grauke LJ, Kudva IT, Yoon JW, Hunt CW, Williams CJ, Hovde CJ.** 2002. Gastrointestinal tract location of *Escherichia coli* O157:H7 in ruminants. *Appl Environ Microbiol* **68:**2269–2277.

37. **Moxley RA, Smith DR.** 2010. Attaching-effacing *Escherichia coli* infections in cattle. *Vet Clin North Am Food Anim Pract* **26:**29–56.

38. **Phillips AD, Navabpour S, Hicks S, Dougan G, Wallis T, Frankel G.** 2000. Enterohaemorrhagic *Escherichia coli* O157:H7 target Peyer's patches in humans and cause attaching/effacing lesions in both human and bovine intestine. *Gut* **47:**377–381.

39. **Baehler AA, Moxley RA.** 2000. *Escherichia coli* O157:H7 induces attaching-effacing lesions in large intestinal mucosal explants from adult cattle. *FEMS Microbiol Lett* **185:**239–242.

40. **Moxley RA.** 2004. *Escherichia coli* O157:H7: an update on intestinal colonization and virulence mechanisms. *Anim Health Res Rev* **5:**15–33.

41. **Rice DH, Sheng HQ, Wynia SA, Hovde CJ.** 2003. Rectoanal mucosal swab culture is more sensitive than fecal culture and distinguishes *Escherichia coli* O157:H7-colonized cattle and those transiently shedding the same organism. *J Clin Microbiol* **41:**4924–4929.

42. **Naylor SW, Gally DL, Low JC.** 2005. Enterohaemorrhagic *E. coli* in veterinary medicine. *Int J Med Microbiol* **295:**419–441.

43. **Besser TE, Hancock DD, Pritchett LC, McRae EM, Rice DH, Tarr PI.** 1997. Duration of detection of fecal excretion of *Escherichia coli* O157:H7 in cattle. *J Infect Dis* **175:**726–729.

44. **Khaitsa ML, Smith DR, Stoner JA, Parkhurst AM, Hinkley S, Klopfenstein TJ, Moxley RA.** 2003. Incidence, duration, and prevalence of *Escherichia coli* O157:H7 fecal shedding by feedlot cattle during the finishing period. *J Food Prot* **66:**1972–1977.

45. **Sanderson MW, Besser TE, Gay JM, Gay CC, Hancock DD.** 1999. Fecal *Escherichia coli* O157:H7 shedding patterns of orally inoculated calves. *Vet Microbiol* **69:**199–205.

46. **Smith DR, Blackford MP, Younts SM, Moxley RA, Gray JT, Hungerford LL, Milton CT, Klopfenstein TJ.** 2001. Ecological relationships between the prevalence of cattle shedding *Escherichia coli* O157:H7 and characteristics of the cattle or conditions of the feedlot pen. *J Food Prot* **64:**1899–1903.

47. **Smith DR, Moxley RA, Clowser SL, Folmer JD, Hinkley S, Erickson GE, Klopfenstein TJ.** 2005. Use of rope devices to describe and explain the feedlot ecology of *Escherichia coli* O157:H7 by time and place. *Foodborne Pathog Dis* **2:**50–60.

48. **Savageau MA.** 1983. *Escherichia coli* habitats, cell types, and molecular mechanisms of gene control. *The American Naturalist* **122:**732–744.

49. **Galvani AP, May RM.** 2005. Epidemiology: dimensions of superspreading. *Nature* **438:**293–295.

50. **Lloyd-Smith JO, Schreiber SJ, Kopp PE, Getz WM.** 2005. Superspreading and the effect of individual variation on disease emergence. *Nature* **438:**355–359.

51. **Chase-Topping ME, McKendrick IJ, Pearce MC, MacDonald P, Matthews L, Halliday J, Allison L, Fenlon D, Low JC, Gunn G, Woolhouse MEJ.** 2007. Risk factors for the presence of high-level shedders of *Escherichia coli* O157 on Scottish farms. *J Clin Microbiol* **45:**1594–1603.

52. **Chase-Topping M, Gally D, Low C, Matthews L, Woolhouse M.** 2008. Super-shedding and the link between human infection and livestock

carriage of *Escherichia coli* O157. *Nat Rev Microbiol* **6**:904–912.

53. **Matthews L, Low JC, Gally DL, Pearce MC, Mellor DJ, Heesterbeek JA, Chase-Topping M, Naylor SW, Shaw DJ, Reid SW, Gunn GJ, Woolhouse ME.** 2006. Heterogeneous shedding of *Escherichia coli* O157 in cattle and its implications for control. *Proc Natl Acad Sci USA* **103**:547–552.

54. **Arthur TM, Brichta-Harhay DM, Bosilevac JM, Kalchayanand N, Shackelford SD, Wheeler TL, Koohmaraie M.** 2010. Super shedding of *Escherichia coli* O157:H7 by cattle and the impact on beef carcass contamination. *Meat Sci* **86**:32–37.

55. **Cobbold RN, Hancock DD, Rice DH, Berg J, Stilborn R, Hovde CJ, Besser TE.** 2007. Rectoanal junction colonization of feedlot cattle by *Escherichia coli* O157:H7 and its association with supershedders and excretion dynamics. *Appl Environ Microbiol* **73**:1563–1568.

56. **Williams ML, Pearl DL, Bishop KE, Lejeune JT.** 2013. Use of multiple-locus variable-number tandem repeat analysis to evaluate *Escherichia coli* O157 subtype distribution and transmission dynamics following natural exposure on a closed beef feedlot facility. *Foodborne Pathog Dis* **10**:827–834.

57. **Callaway TR, Carr MA, Edrington TS, Anderson RC, Nisbet DJ.** 2009. Diet, *Escherichia coli* O157:H7, and cattle: a review after 10 years. *Curr Issues Mol Biol* **11**:67–79.

58. **Smith DR, Vogstad AR.** 2012. Vaccination as a method of *E. coli* O157:H7 reduction in feedlot cattle, p 133–142. *In* Callaway TR, Edrington TS (ed), *On Farm Strategies to Control Foodborne Pathogens.* Nova Science Publishers Inc, Hauppauge, NY.

59. **Berry ED, Wells JE.** 2010. *Escherichia coli* O157:H7: recent advances in research on occurrence, transmission, and control in cattle and the production envrionment. *Adv Food Nutr Res* **60**:67–118.

60. **Smith DR, Klopfenstein T, Moxley RA, Milton CT, Hungerford LL, Gray JT.** 2002. An evaluation of three methods to clean feedlot water tanks. *The Bovine Practitioner* **36**:1–4.

61. **Folmer J, Macken C, Moxley R, Smith D, Brashears M, Hinkley S, Erickson G, Klopfenstein T.** 2003. Intervention strategies for reduction of *E. coli* O157:H7 in feedlot steers. *Nebraska Beef Cattle Report* **MP 80-A**:22–23.

62. **LeJeune JT, Besser TE, Rice DH, Berg JL, Stilborn RP, Hancock DD.** 2004. Longitudinal study of fecal shedding of *Escherichia coli* O157:H7 in feedlot cattle: predominance and persistence of specific clonal types despite massive cattle population turnover. *Appl Environ Microbiol* **70**:377–384.

63. **Snedeker KG, Campbell M, Sargeant JM.** 2012. A systematic review of vaccinations to reduce the shedding of *Escherichia coli* O157 in the faeces of domestic ruminants. *Zoonoses Public Health* **59**:126–138.

64. **Brashears MM, Galyean ML, Loneragan GH, Mann JE, Killinger-Mann K.** 2003. Prevalence of *Escherichia coli* O157:H7 and performance by beef feedlot cattle given *Lactobacillus* direct-fed microbials. *J Food Prot* **66**:748–754.

65. **Peterson RE, Klopfenstein TJ, Erickson GE, Folmer J, Hinkley S, Moxley RA, Smith DR.** 2007. Effect of *Lactobacillus acidophilus* strain NP51 on *Escherichia coli* O157:H7 fecal shedding and finishing performance in beef feedlot cattle. *J Food Prot* **70**:287–291.

66. **Loneragan GH, Brashears MM.** 2005. Preharvest interventions to reduce carriage of *E. coli* O157 by harvest-ready feedlot cattle. *Meat Sci* **71**:72–78.

67. **Miller MF, Loneragan GH, Harris DD, Adams KD, Brooks JC, Brashears MM.** 2008. Environmental dust exposure as a factor contributing to an increase in *Escherichia coli* O157:H7 and *Salmonella* populations on cattle hides in feedyards. *J Food Prot* **71**:2078–2081.

68. **Arthur TM, Bosilevac JM, Brichta-Harhay DM, Guerini MN, Kalchayanand N, Shackelford SD, Wheeler TL, Koohmaraie M.** 2007. Transportation and lairage environment effects on prevalence, numbers, and diversity of *Escherichia coli* O157:H7 on hides and carcasses of beef cattle at processing. *J Food Prot* **70**:280–286.

69. **Arthur TM, Bosilevac JM, Brichta-Harhay DM, Kalchayanand N, King DA, Shackelford SD, Wheeler TL, Koohmaraie M.** 2008. Source tracking of *Escherichia coli* O157:H7 and *Salmonella* contamination in the lairage environment at commercial U.S. beef processing plants and identification of an effective intervention. *J Food Prot* **71**:1752–1760.

70. **Asper DJ, Sekirov I, Finlay BB, Rogan D, Potter AA.** 2007. Cross reactivity of enterohemorrhagic Escherichia coli O157:H7-specific sera with non-O157 serotypes. *Vaccine* **25**:8262–8269.

71. **Dohoo IR, Martin SW, Stryhn H.** 2003. Measures of association, p 121–138. *In* Dohoo IR, Martin SW, Stryhn H (ed), *Veterinary Epidemiologic Research.* AVC, Inc, Charlottetown, PEI, Canada.

72. **Halloran ME.** 1998. Concepts of infectious disease epidemiology, p 529–554. *In* Rothman KJ, Greenland S (ed), *Modern Epidemiology*, 2nd ed. Lippincott-Raven, Philadelphia, PA.

73. **Beaudeau F, Fourichon C.** 1998. Estimating relative risk of disease from outputs of logistic regression when the disease is not rare. *Prev Vet Med* **36**:243–256.

74. **Allen KJ, Rogan D, Finlay BB, Potter AA, Asper DJ.** 2011. Vaccination with type III secreted proteins leads to decreased shedding in calves after experimental infection with *Escherichia coli* O157. *Can J Vet Res* **75:**98–105.

75. **Garmendia J, Frankel G, Creprin VF.** 2005. Enteropathogenic and enterohemorrhagic *Escherichia coli* infections: translocation, translocation, translocation. *Infect Immun* **73:**2573–2585.

76. **Goosney DL, Knoechel DG, Finlay BB.** 1999. Enteropathogenic *E. coli*, *Salmonella*, and *Shigella*: masters of host cell cytoskeletal exploitation. *Emerg Infect Dis* **5:**216–223.

77. **Mahajan A, Currie CG, Mackie S, Tree J, McAteer S, McKendrick I, McNeilly TN, Roe A, La Ragione RM, Woodward MJ, Gally DL, Smith DGE.** 2009. An investigation of the expression and adhesin function of H7 flagella in the interaction of *Escherichia coli* O157:H7 with bovine intestinal epithelium. *Cell Microbiol* **11:**121–137.

78. **Erdem AL, Avelino F, Xicohtencatl-Cortes J, Giron JA.** 2007. Host protein binding and adhesive properties of H6 and H7 flagella of attaching and effacing *Escherichia coli*. *J Bacteriol* **189:**7426–7435.

79. **Bretschneider G, Berberov EM, Moxley RA.** 2007. Reduced intestinal colonization of adult beef cattle by *Escherichia coli* O157:H7 *tir* deletion and nalidixix-acid-resistant mutants lacking flagellar expression. *Vet Microbiol* **125:**381–386.

80. **van Diemen PM, Dziva F, bu-Median A, Wallis TS, van den BH, Dougan G, Chanter N, Frankel G, Stevens MP.** 2007. Subunit vaccines based on intimin and Efa-1 polypeptides induce humoral immunity in cattle but do not protect against intestinal colonisation by enterohaemorrhagic *Escherichia coli* O157:H7 or O26:H–. *Vet Immunol Immunopathol* **116:**47–58.

81. **Dean-Nystrom EA, Gansheroff LJ, Mills M, Moon HW, O'Brien AD.** 2002. Vaccination of pregnant dams with intimin (O157) protects suckling piglets from *Escherichia coli* O157:H7 infection. *Infect Immun* **70:**2414–2418.

82. **Dziva F, Vlisidou I, Creprin VF, Wallis TS, Frankel G, Stevens MP.** 2007. Vaccination of calves with EspA, a key colonisation factor of *Escherichia coli* O157:H7, induces antigen-specific humoral responses but does not confer protection against intestinal colonisation. *Vet Microbiol* **123:**254–261.

83. **McNeilly TN, Naylor SW, Mahajan A, Mitchell MC, McAteer S, Deane D, Smith DGE, Low JC, Gally DL, Huntley JF.** 2008. *Escherichia coli* O157:H7 colonization in cattle following systemic and mucosal immunization with purified H7 flagellin. *Infect Immun* **76:**2594–2602.

84. **McNeilly TN, Mitchell MC, Rosser T, McAteer S, Low JC, Smith DGE, Huntley JF, Mahajan A, Gally DL.** 2010. Immunization of cattle with a combination of purified intimin-531, EspA and Tir significantly reduces shedding of *Escherichia coli* O157:H7 following oral challenge. *Vaccine* **28:**1422–1428.

85. **Yekta MA, Goddeeris BM, Vanrompay D, Cox E.** 2011. Immunization of sheep with a combination of intimin [gamma]; EspA and EspB decreases *Escherichia coli* O157:H7 shedding. *Vet Immunol Immunopathol* **140:**42–46.

86. **Vilte DA, Larzabal M, Garbaccio S, Gammella M, Rabinovitz BC, Elizondo AM, Cantet RJC, Delgado F, Meikle V, Cataldi A, Mercado EC.** 2011. Reduced faecal shedding of *Escherichia coli* O157:H7 in cattle following systemic vaccination with gamma-intimin C280 and EspB proteins. *Vaccine* **29:**3962–3968.

87. **Potter AA, Klashinsky S, Li Y, Frey E, Townsend H, Rogan D, Erickson G, Hinkley S, Klopfenstein T, Moxley RA, Smith DR, Finlay BB.** 2004. Decreased shedding of *Escherichia coli* O157:H7 by cattle following vaccination with type III secreted proteins. *Vaccine* **22:**362–369.

88. **Vilte DA, Larzabal M, Mayr UB, Garbaccio S, Gammella M, Rabinovitz BC, Delgado F, Meikle V, Cantet RJ, Lubitz P, Lubitz W, Cataldi A, Mercado EC.** 2012. A systemic vaccine based on *Escherichia coli* O157:H7 bacterial ghosts (BGs) reduces the excretion of *E. coli* O157:H7 in calves. *Vet Immunol Immunopathol* **146:**169–176.

89. **Rabinovitz BC, Gerhardt E, Tironi Farinati C, Abdala A, Galarza R, Vilte DA, Ibarra C, Cataldi A, Mercado EC.** 2012. Vaccination of pregnant cows with EspA, EspB, gamma-intimin, and Shiga toxin 2 proteins from *Escherichia coli* O157:H7 induces high levels of specific colostral antibodies that are transferred to newborn calves. *J Dairy Sci* **95:**3318–3326.

90. **Thornton AB, Thomson DU, Loneragan GH, Fox JT, Burkhardt DT, Emery DA, Nagaraja TG.** 2009. Effects of a siderophore receptor and porin proteins-based vaccination on fecal shedding of *Escherichia coli* O157:H7 in experimentally inoculated cattle. *J Food Prot* **72:**866–869.

91. **Thomson DU, Loneragan GH, Thornton AB, Lechtenberg KF, Emery DA, Burkhardt DT, Nagaraja TG.** 2009. Use of a siderophore receptor and porin proteins-based vaccine to control the burden of *Escherichia coli* O157:H7 in feedlot cattle. *Foodborne Pathog Dis* **6:**871–877.

92. **Woerner DR, Ransom JR, Sofos JN, Scanga JA, Smith GC, Belk KE.** 2006. Preharvest processes for microbial control in cattle. *Food Prot Trends* **26:**393–400.

93. Standley T, Paterson J, Skinner K, Rainey B, Roberts A, Geary T, Smith G, White R. 2008. The use of an experimental vaccine in gestating beef cows to reduce the shedding of *Escherichia coli* O157:H7 in the newborn calf. *The Professional Animal Scientist* **24**:4.

94. Wileman BW, Thomson DU, Olson KC, Jaeger JR, Pacheco LA, Bolte J, Burkhardt DT, Emery DA, Straub D. 2011. *Escherichia coli* O157:H7 shedding in vaccinated beef calves born to cows vaccinated prepartum with *Escherichia coli* O157:H7 SRP vaccine. *J Food Prot* **74**:1599–1604.

95. Varela NP, Dick P, Wilson J. 2012. Assessing the existing information on the efficacy of bovine vaccination against *Escherichia coli* O157:H7—a systematic review and meta-analysis. *Zoonoses Public Health* **60**:253–268.

96. Vogstad AR, Moxley RA, Erickson GE, Klopfenstein TJ, Smith DR. 2014. Stochastic simulation model comparing distributions of STEC O157 faecal shedding prevalence between cattle vaccinated with type III secreted protein vaccines and non-vaccinated cattle. *Zoonoses Public Health* **61**:283–289.

97. Fox JT, Thomson DU, Drouillard JS, Thornton AB, Burkhardt DT, Emery DA, Nagaraja TG. 2009. Efficacy of *Escherichia coli* O157:H7 siderophore receptor/porin proteins-based vaccine in feedlot cattle naturally shedding *E. coli* O157. *Foodborne Pathog Dis* **6**:893–899.

98. Cull CA, Paddock ZD, Nagaraja TG, Bello NM, Babcock AH, Renter DG. 2012. Efficacy of a vaccine and a direct-fed microbial against fecal shedding of *Escherichia coli* O157:H7 in a randomized pen-level field trial of commercial feedlot cattle. *Vaccine* **30**:6210–6215.

99. Van Donkersgoed J, Hancock D, Rogan D, Potter AA. 2005. *Escherichia coli* O157:H7 vaccine field trial in 9 feedlots in Alberta and Saskatchewan. *Can Vet J* **46**:724–728.

100. Moxley RA, Smith DR, Luebbe M, Erickson GE, Klopfenstein TJ, Rogan D. 2009. *Escherichia coli* O157:H7 vaccine dose-effect in feedlot cattle. *Foodborne Pathog Dis* **6**:879–884.

101. Peterson RE, Klopfenstein TJ, Moxley RA, Erickson GE, Hinkley S, Rogan D, Smith DR. 2007. Efficacy of dose regimen and observation of herd immunity from a vaccine against *Escherichia coli* O157:H7 for feedlot cattle. *J Food Prot* **70**:2561–2567.

102. Rich AR, Jepson AN, Luebbe M, Klopfenstein TJ, Smith DR, Moxley RA. 2010. Vaccination to reduce the prevalence of *Escherichia coli* O157:H7 in feedlot cattle fed wet distillers grains plus solubles. *Nebraska Beef Cattle Report* **MP93**:94–95.

103. Smith DR, Moxley RA, Peterson RE, Klopfenstein TJ, Erickson GE, Bretschneider G, Berberov EM, Clowser S. 2009. A two-dose regimen of a vaccine against type III secreted proteins reduced *Escherichia coli* O157:H7 colonization of the terminal rectum in beef cattle in commercial feedlots. *Foodborne Pathog Dis* **6**:155–161.

104. Peterson RE, Klopfenstein TJ, Moxley RA, Erickson GE, Hinkley S, Bretschneider G, Berberov EM, Rogan D, Smith DR. 2007. Effect of a vaccine product containing type III secreted proteins on the probability of *Escherichia coli* O157:H7 fecal shedding and mucosal colonization in feedlot cattle. *J Food Prot* **70**:2568–2577.

105. Smith DR, Moxley RA, Peterson RE, Klopfenstein T, Erickson GE, Clowser SL. 2008. A two-dose regimen of a vaccine against *Escherichia coli* O157:H7 type III secreted proteins reduced environmental transmission of the agent in a large-scale commercial beef feedlot clinical trial. *Foodborne Pathog Dis* **5**:589–598.

106. Smith DR, Moxley RA, Klopfenstein TJ, Erickson GE. 2009. A randomized longitudinal trial to test the effect of regional vaccination within a cattle feedyard on *Escherichia coli* O157:H7 rectal colonization, fecal shedding, and hide contamination. *Foodborne Pathog Dis* **6**:885–892.

107. Peterson RE, Klopfenstein TJ, Moxley RA, Erickson GE, Hinkley S, Rogan D, Smith DR. 2007. Efficacy of dose regimen and observation of herd immunity from a vaccine against *Escherichia coli* O157:H7 for feedlot cattle. *J Food Prot* **70**:2561–2567.

108. O'Connor AM, Sargeant JM, Gardner IA, Dickson JS, Torrence ME, Dewey CE, Dohoo IR, Evans RB, Gray JT, Greiner M, Keefe G, Lefebvre SL, Morley PS, Ramirez A, Sischo W, Smith DR, Snedeker K, Sofos J, Ward MP, Wills R. 2010. The REFLECT statement: methods and processes of creating reporting guidelines for randomized controlled trials for livestock and food safety by modifying the CONSORT statement. *Zoonoses Public Health* **57**:95–104.

109. Vogstad AR, Moxley RA, Erickson GE, Klopfenstein TJ, Smith DR. 2013. Assessment of heterogeneity of efficacy of a three-dose regimen of a type III secreted protein vaccine for reducing STEC O157 in feces of feedlot cattle. *Foodborne Pathog Dis* **10**:678–683.

110. Martin SW, Meek AH, Willeberg P. 1987. Descriptive epidemiology, p 79–120. *In* Martin SW, Meek AH, Willeberg P, *Veterinary Epidemiology: Principles and Methods*, 1st ed. Iowa State University Press, Ames. IA.

111. Fine PEM. 1993. Herd immunity: history, theory, practice. *Epidemiol Rev* **15**:265–302.

112. Smith DR, Gray JT, Moxley RA, Younts-Dahl SM, Blackford MP, Hinkley S, Hungerford LL, Milton CT, Klopfenstein TJ. 2004. A diagnostic strategy to determine the Shiga toxin-producing

Escherichia coli O157 status of pens of feedlot cattle. *Epidemiol Infect* **132**:297–302.

113. **Irwin KE, Smith DR, Gray JT, Klopfenstein TJ.** 2002. Behavior of cattle towards devices to detect food-safety pathogens in feedlot pens. *Bovine Pract* **36**:5–9.

114. **Arthur TM, Nou X, Bosilevac JM, Wheeler T, Koohmaraie M.** 2011. Survival of *Escherichia coli* O157:H7 on cattle hides. *Appl Environ Microbiol* **77**:3002–3008.

115. **Koohmaraie M, Arthur TM, Bosilevac JM, Guerini M, Shackelford SD, Wheeler TL.** 2005. Post-harvest interventions to reduce/eliminate pathogens in beef. *Meat Sci* **71**:79–91.

116. **Reicks AL, Brashears MM, Adams KD, Brooks JC, Blanton JR, Miller MF.** 2007. Impact of transportation of feedlot cattle to the harvest facility on the prevalence of *Escherichia coli* O157:H7, *Salmonella*, and total aerobic microorganisms on hides. *J Food Prot* **70**:17–21.

117. **Lanzas C, Lu Z, Grohn YT.** 2011. Mathematical modeling of the transmission and control of foodborne pathogens and antimicrobial resistance at preharvest. *Foodborne Pathog Dis* **8**:1–10.

118. **Arthur TM, Keen J, Bosworth BT, Brichta-Harhay DM, Kalchayanand N, Shackelford SD, Wheeler TL, Nou X, Koohmaraie M.** 2009. Longitudinal study of *Escherichia coli* O157-H7 in a beef cattle feedlot and role of high-level shedders in hide contamination. *Appl Environ Microbiol* **7**:6515–6523.

119. **Signorini ML, Tarabla HD.** 2010. Interventions to reduce verocytotoxigenic *Escherichia coli* in ground beef in Argentina: a simulation study. *Prev Vet Med* **94**:36–42.

120. **Hurd HS, Malladi S.** 2012. An outcomes model to evaluate risks and benefits of Escherichia coli vaccination in beef cattle. *Foodborne Pathog Dis* **9**:952–961.

121. **Wood JC, McKendrick IJ, Gettinby G.** 2006. A simulation model for the study of the within-animal infection dynamics of *E. coli* O157. *Prev Vet Med* **74**:180–193.

122. **Vogstad AR.** 2012. *Modeling the efficacy and effectiveness of Escherichia coli O157:H7 pre-harvest interventions.* MS thesis. University of Nebraska-Lincoln, Lincoln, NE.

123. **Cross P, Rigby D, Edwards-Jones G.** 2012. Eliciting expert opinion on the effectiveness and practicality of interventions in the farm and rural environment to reduce human exposure to *Escherichia coli* O157. *Epidemiol Infect* **140**:643–654.

ESCHERICHIA COLI O104:H4

Escherichia coli O104:H4 Pathogenesis: An Enteroaggregative E. coli/Shiga Toxin-Producing E. coli Explosive Cocktail of High Virulence

26

FERNANDO NAVARRO-GARCIA[1]

INTRODUCTION

In May 2011, an outbreak caused by *Escherichia coli* of serotype O104:H4 spread throughout Germany (1). The next month, France also reported a cluster of *E. coli* O104:H4 infections (2). A total of 46 deaths, 782 cases of hemolytic-uremic syndrome (HUS), and 3,128 cases of acute gastroenteritis were officially attributed to this new clone of enterohemorrhagic *E. coli* (EHEC) (last update from European Centre for Disease Prevention and Control, 27 July 2011). Most or all victims (although diagnosed in different countries in Europe) became infected in Germany or France. The phenotypic and genotypic characterization of the *E. coli* O104:H4 indicated that the isolates from the French and German outbreaks were common to both incidents. Fenugreek seeds imported from Egypt, from which sprouts were grown, were implicated as a common source. However, there is still much uncertainty about whether this is truly the common cause of the infections, as tests on the seeds did not allow the detection of any *E. coli* isolate of serotype O104:H4.

This large outbreak was caused by an unusual EHEC strain that is most similar to an enteroaggregative *E. coli* (EAEC) of serotype O104:H4. A significant difference, however, is the presence of a prophage encoding the Shiga toxin (Stx), which is characteristic of EHEC strains (3–5). This combination of

[1]Department of Cell Biology, Centro de Investigación y de Estudios Avanzados del IPN, México DF, Mexico.

Enterohemorrhagic Escherichia coli *and Other Shiga Toxin-Producing* E. coli
Edited by Vanessa Sperandio and Carolyn J. Hovde
© 2015 American Society for Microbiology, Washington, DC
doi:10.1128/microbiolspec.EHEC-0008-2013

genomic features, associating characteristics from both EAEC and EHEC, represents a new pathotype (Fig. 1). Because typical EAEC strains are isolated primarily from humans, the origin of this outbreak may not be zoonotic. That was recently confirmed by two surveys in Germany and France. No evidence of the Stx-producing *E. coli* O104:H4 outbreak strain or EAEC was found in cattle feces in northern Germany, the hot spot of the 2011 HUS outbreak area (6). Similarly, French cattle were not a reservoir of the highly virulent enteroaggregative Stx-producing *E. coli* of serotype O104:H4 (7). Recent identification from sporadic cases of HUS in France of EHEC clones similar to the one responsible of the outbreaks (8) suggests that the EHEC O104:H4 pathogen has become endemically established in Europe and very likely in the human population.

One burning question is what makes this outbreak EHEC O104:H4 strain so dangerous? One explanation is that this strain (in short, an EAEC with a phage coding for Stx type 2) is a better colonizer of the gut (Fig. 2). The enhanced adherence of this strain to intestinal epithelial cells might facilitate systemic absorption of Stx and could explain the high frequency of progression to HUS. It is believed that EAEC of serotype O104:H4 is by itself an emerging serovar that has acquired an original set of virulence factors (Fig. 1).

EAEC strains of serotype O104:H4 contain a large set of virulence-associated genes regulated by the AggR transcription factor (Fig. 1). They include the pAA plasmid genes encoding the aggregative adherence fimbriae (AAF), which anchor the bacterium to the intestinal mucosa (the aggregative, so-called stacked-brick, adherence pattern on intestinal

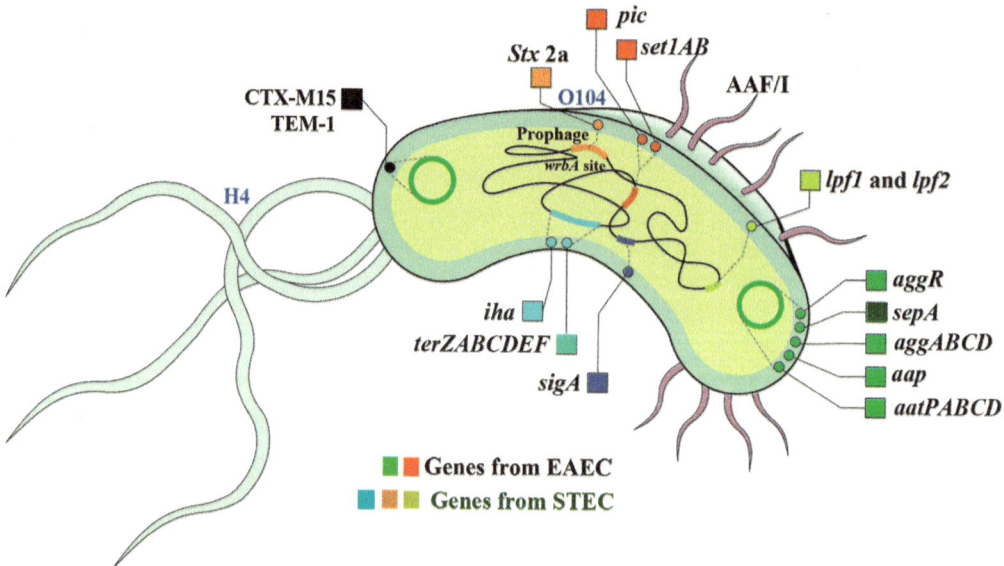

FIGURE 1 Hybrid characteristics of *E. coli* O104:H4 outbreak strain (EAEC/STEC). Schematic representation of the genes harbored by *E. coli* O104:H4; the main genes from EAEC or STEC are highlighted: *stx2* (coding for Stx 2), *pic*, *sigA*, and *sepA* (coding for the SPATE proteins); Pic, protein involved in intestinal colonization; SigA, a homolog of Pet, with cytotoxic activity; SepA, a colonization factor of *Shigella*), *set1AB* (coding for ShET1, a holotoxin AB5), *iha* (coding for Iha, a STEC adhesin that is an IrgA homolog), *aggR, aggABCD, aap, aatPABCD* (genes from EAEC plasmids coding for transcription regulator, AAF/I, dispersin, and dispersin transporter, respectively), *lpf1-2* (coding for Lpf of STEC), *terZABCDEF* (coding for a cluster for Tellurite resistance), CTX-M15 and TEM-1 (antibiotic resistance genes). SigA and SepA are SPATEs detected mainly in *Shigella* sp. doi:10.1128/microbiolspec.EHEC-0008-2013.f1

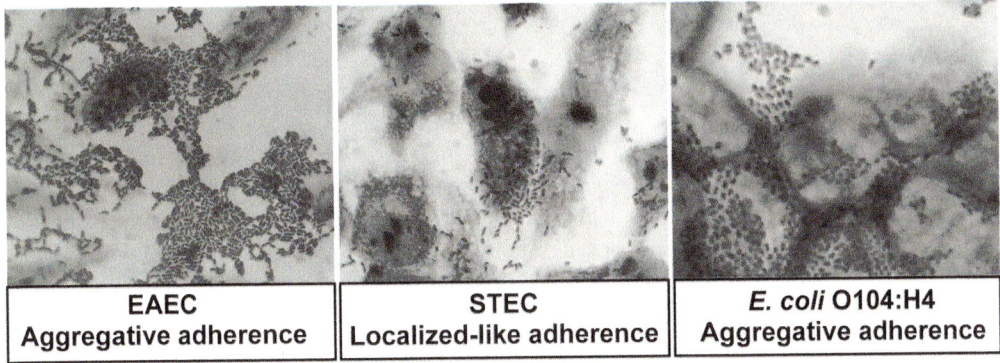

EAEC Aggregative adherence	STEC Localized-like adherence	*E. coli* O104:H4 Aggregative adherence

FIGURE 2 Adherence patterns of EAEC, STEC, and *E. coli* O104:H4 outbreak strain to epithelial cells. Subconfluent epithelial cell cultures are infected with the different bacterial strains. Cells are fixed and stained with Giemsa stain. From Scalesky et al. 1999. *Infect Immun* 67:3410; Paton et al. 2001. *Infect Immun* 69:6999; and Martina Bielaszewska. http://ecdc.europa.eu/en/press/events/Documents/22-231111-Breakthroughs-in-molecular-epidemiology-Bielaszewska.pdf. doi:10.1128/microbiolspec.EHEC-0008-2013.f2

epithelial cells) and induce inflammation, as well as a protein-coat secretion system (Aat) that secretes the protein dispersin (9). A switch of the virulence plasmid (pAA) together with the type of AAF could be an additional explanation for the higher virulence of this outbreak strain (Fig. 2). Indeed, the outbreak EHEC O104:H4 strain is similar to EAEC O104:H4 strain 55989, isolated in the late 1990s from a patient with persistent diarrhea in the Central African Republic, and to EHEC O104:H4 strain HUSEC041 that was associated in 2001 with very few HUS cases in Germany. Interestingly, EHEC O104:H4 strain HUSEC041 carries the plasmid encoding AAF/III (also present in EAEC strain 55989 in a different size). In contrast, outbreak EHEC O104:H4 isolates acquired a new plasmid, encoding AAF/I, and lost the AAF/III-encoding plasmid (3, 4). AAFs are encoded in high-molecular-weight plasmids (designated pAA), and their biogenesis employs the chaperone-usher secretion pathway (10–12). These fimbriae are members of the Dr superfamily of adhesins that includes strains of uropathogenic (UPEC) and diarrheagenic *E. coli* with a genetic organization consisting of chaperone, usher, and pilin subunits (13, 14). AAF/I are flexible, bundle-forming fimbriae (15), and the genes responsible for their

biogenesis are located in two regions of the plasmid: region 1, containing the genes that encode the pilin, chaperonin, and usher, and region 2, containing the regulator-encoding gene (designated *aggR*) (10, 15, 16).

EAEC strains of serotype O104:H4 also produce a variable number of serine protease autotransporters of *Enterobacteriaceae* (SPATEs) implicated in mucosal damage and colonization. The type V secretion pathway enables a family of proteins to reach the surface with a very limited number of accessory secretion factors because most information necessary to the translocation process is contained within the secreted protein itself. These proteins, which can carry out their own transport to the outer membrane, are autotransporter proteins. The current isolate diverges from common EAEC isolates in the number and nature of their SPATE proteases (Fig. 1), namely Pic, SigA, and SepA (3–5). Rasko and colleagues speculate that the combined activity of these SPATEs, together with other EAEC virulence factors, accounts for the increased uptake of Stx into the systemic circulation, resulting in the high rates of HUS. The *pic* gene has a unique characteristic among the autotransporter proteins since there are two oppositely oriented genes in tandem within the *pic* (*she*) gene, *set1B* and *set1A* (17), which encode the two subunits of the ShET1 toxin (18); this gene is also present in *Shigella flexneri* and UPEC. Pic is a mucinase (19) that has recently been shown to promote mucous secretion in the gut (20), and is responsible for the mucoid diarrhea that is a classic symptom of *Shigella* sp. and EAEC infection. Interestingly, EC55989 contains three copies of this gene, two on the chromosome (intact, EC55989_4682 and EC55989_3279) and one on the EAEC plasmid (55989p, truncated), all of which are conserved in the German outbreak strain. In addition, a fourth *pic* gene is present in the EAEC plasmid of the outbreak strain (but missing from EC55989) that seems to be intact. The outbreak strain also encodes SigA, a SPATE that cleaves the cytoskeletal protein spectrin, inducing rounding and exfoliation of enterocytes (21). The third SPATE, SepA, is associated with increased *S. flexneri* virulence (22), but its function is unknown.

As mentioned before, EAEC strains of serotype O104:H4, as a hybrid clone, also produce Stx2 and an adhesion-siderophore called Iha; both proteins are found with high prevalence in Stx-producing *E. coli* (STEC), including EHEC (Fig. 1). The ability of STEC to cause severe disease in humans is mainly associated with the production of Stx; two distinct groups, Stx1 and Stx2, with similar biological activity but different immunogenicity are well known (23). Members of the Stx1 group are antigenically similar, whereas those of the Stx2 group are heterogeneous and comprise several variants or subtypes (24). An interesting finding that highlights the relevance of Stx2 is that Stx2 has been epidemiologically more associated with severe disease in humans than Stx1 (25). Moreover, the *stx2d* activatable subtype has been associated with high virulence and the ability to cause HUS (26). On the other hand, Iha was first described as an adhesin in an EHEC O157:H7 strain and was named IrgA homolog adhesin, based on its homology to the IrgA enterobactin siderophore receptor of *Vibrio cholerae* (27) and its ability to confer epithelial cell adherence capability to a nonadherent K-12

strain when expressed from a multicopy plasmid (28). Recently, Iha was determined to have a dual-function urovirulence factor for *E. coli* clonal group A strains and other pathogenic *E. coli* strains. However, it still needs to be determined if the siderophore receptor activity, the adhesion phenotype, or both are important for the enhanced in vivo persistence of Iha demonstrated within the urinary tract (29).

E. COLI O104:H4 OUTBREAK STRAIN AS LOCUS OF ENTEROCYTE EFFACEMENT (LEE)-NEGATIVE STEC

The *E. coli* O104:H4 outbreak strain has been compared to typical EHEC outbreaks; however, EHEC only refers to a clinical condition, and the virulence factors of EHEC and STEC can be different. STEC strains isolated from humans with specific clinical signs are called EHEC (30). In humans, some strains cause severe inflammation of a section of the large intestine accompanied by hemorrhage of the intestinal mucosa and severe diarrhea (hemorrhagic colitis) or HUS, which can lead to kidney failure and even death. Thus, the STEC strains that cause these clinical pictures are designated as EHEC (31). EHEC strains are considered to be a subset of STEC, but our knowledge of STEC comes from outbreak investigations and studies of *E. coli* O157 (EHEC) infection, which was first identified as a pathogen in 1982. The non-O157 STEC serogroups are not nearly as well understood, partly because outbreaks caused by them are rarely identified. As a whole, the non-O157 serogroup is less likely than *E. coli* O157 to cause severe illness; however, some non-O157 STEC serogroups can cause the most severe manifestations of STEC illness (32).

Pathogenic STEC strains require additional virulence factors enabling adherence, for example, factors that permit colonization of intestinal epithelial cells (Fig. 2). EHEC strains contain an arrangement of virulence genes that contribute to their pathogenesis, such as the LEE pathogenicity island (PAI), the virulence plasmid pO157, and *stx*1 and/or *stx*2. LEE carries all genes necessary for the formation of the attaching and effacing (A/E) lesion (33). The A/E histopathology is characterized by effacement of the brush border microvilli, intimate bacterial adherence to the enterocyte apical plasma membrane, and the accumulation of polymerized actin beneath the attached bacteria (34). All A/E pathogens carry the LEE PAI (35) that encodes gene regulators (36, 37), the adhesin called intimin (38), a type III secretion system (T3SS) (35), chaperones (39, 40), and several secreted proteins, including the translocated intimin receptor called Tir (41). Upon contact with epithelial cells, EHEC injects a variety of effectors into the cells to modulate cellular functions involved in the host defense response, the dynamics of the cytoskeleton, and the maintenance of tight junctions. A major target of virulence factors is the cellular signaling cascade involved in the construction and modulation of the cytoskeleton and microfilaments (42). While the bacteria remain mostly extracellular in the lumen of the gut, the T3SS effectors of A/E pathogens access and manipulate the intracellular environment of host cells. The effectors subvert various host cell processes, which enable the bacteria to colonize, multiply, and contribute to the disease. Thus, two key factors encoded by LEE include the adhesin intimin (*eae*), which binds to Tir (*tir*), and both are essential for the intimate attachment of the bacteria with the cytoplasmic membrane of the host cell (33).

LEE appears to confer enhanced virulence, since LEE-positive STEC strains are much more commonly associated with HUS and outbreaks than LEE-negative STEC strains (43). EHEC has controversial definitions, most of them indicating that EHEC is a STEC harboring the *eae* gene, or a STEC implicated in the illness, or a STEC that has the same clinical, epidemiological, and pathogenic characteristics. These definitions are features of the EHEC O157 serogroup, but most of the

STEC strains have been designated as non-O157 strains. Persons with non-O157 STEC infection usually have less severe illness, and non-O157 STEC strains include many serogroups (~400) with varying virulence, some typically causing only mild diarrhea and others causing HUS and death; non-O157 isolates predominantly have Stx1. Additionally, there are regional variations in prevalence; however, six serogroups of STEC have been associated with 70% of the food safety illnesses in the United States (O26, O45, O103, O111, O121, and O145) and have been termed the "Big Six." U.S. Department of Agriculture's Food Safety and Inspection Service considers these six most frequent serogroups adulterants in ground beef, materials intended for ground beef production, and other nonintact beef products. The Food Safety and Inspection Service plans to test for these adulterants using a PCR protocol that initially targets the detection of Stx genes, *stx1* and/or *stx2*, and the intimin (*eae*) gene, and then tests for the six O groups. Besides *stx* genes and adherence factors, increasing evidence shows that differences in virulence between pathogenic and nonpathogenic bacterial strains can be attributed in part to virulence genes located in PAIs (43). PAIs usually contain blocks of virulence genes and are greater than 10 kb (44). Several PAIs have been identified and characterized in STEC. For instance, the chromosomal LEE PAI was identified in *E. coli* O157:H7 strain EDL933 (45). However, some LEE-positive STEC serotypes have never been associated with disease, and some LEE-negative STEC serotypes can cause HUS and outbreaks, indicating that virulence factors other than those in LEE may contribute to pathogenesis of STEC (43). Thus, in addition to the distribution of different PAIs (O island [OI]-122, OI-43/48, OI-57, high-pathogenicity island) and their virulence genes in STEC, the association of the PAIs and individual virulence genes with STEC seropathotypes linked to severe diseases and outbreaks was evaluated. Most OI-122, OI-43/48, and OI-57 virulence genes (*pagC, sen, nleB, efa-1, efa-2, terC, ureC, iha, aidA-1, nleG2-3, nleG6-2,* and *nleG5-2*) were highly prevalent in *eae*-positive STEC, but they were largely absent in *eae*-negative STEC, with the exception of *pagC* and *iha* (46). Phylogenetic analysis revealed that *iha* genes from *eae*-positive STEC had high similarity (99.6%), whereas they had lower sequence similarity (91.1 to 93.6%) with *iha* genes from *eae*-negative STEC, indicating that *iha* from *eae*-positive and *eae*-negative STEC may have evolved independently or have different origins (46). Indeed, it has been reported (47) that *iha* was carried by a 33,014-bp PAI in STEC serotype O91:H– strains (*eae*-negative). In addition, *iha* was found in pO113 plasmid of STEC serotype O113:H21 (*eae*-negative) (48). Furthermore, some of these factors can participate in bacterial regulation; it has been reported that lysogeny with Stx2-encoding bacteriophages represses T3SS in EHEC. Deletion of Stx2 phages from EHEC strains increased the level of T3SS whereas lysogeny decreased T3SS. A model is proposed in which Stx2-encoding bacteriophages regulate T3SS to coordinate epithelial cell colonization that is promoted by Stx and secreted effector proteins (49).

Interestingly, *E. coli* O104:H4 represents a pathogenic STEC paradigm shift, since it lacks the classic EHEC markers (the LEE PAI and the pO157) but possesses Stx2a and EAEC virulence factors, including an AAF (Fig. 1 and 2). Thus, the 2011 *E. coli* O104:H4 outbreak of hemorrhagic diarrhea in Germany is an example of the explosive cocktail of high virulence and resistance that can emerge in this species. In other words, *E. coli* O104:H4 is an *eae*-negative STEC (LEE-negative STEC) that does not possess a classical EHEC plasmid (pO157). To be pathogenic, a strain must have the necessary properties to cause disease in humans. These properties are called virulence factors. In the case of *E. coli* O104:H4 strains, their STEC-EAEC hybrid characteristics might contribute to make this outbreak EHEC O104:H4 strain so dangerous. Thus, EAEC strains of serotype O104:H4 contain a large set of virulence-associated genes regulated by the AggR transcription factor and other virulence factors encoded in EAEC plasmid. These include, among other factors, the pAA plasmid genes encoding the AAF, which anchor the bacterium to the intestinal mucosa, and the aggregative adherence pattern on intestinal epithelial cells (Fig. 2), also called stacked-brick pattern.

Stx2

The primary virulence factor in systemic host responses produced by clinical isolates of STEC is Stx2, but some isolates produce both Stx1 and Stx2, or rarely, only Stx1 (50, 51). Genes for Stx are located on a bacteriophage (a virus of bacteria) that is associated with all pathogenic STEC strains (Fig. 3). Numerous other factors produced by STEC are believed to act locally in the intestine rather than systemically as do the Stxs. Stx includes two major immunologically distinct forms (Stx1 and Stx2), with minor variants of Stx2 (Stx2a to h). In contrast to these genotypic differences, Stxs share many properties, including molecular structure, enzymatic activity, receptor specificity, and intracellular trafficking. All Stxs possess an AB5 structure with an enzymatically active A subunit of approximately 32 kDa in noncovalent association with five identical B subunits, with each B subunit being approximately 7.7 kDa in size (52). X-ray crystallographic analyses of Stxs have shown that the pentameric B subunits form a ring with the carboxy terminus of the A subunit interdigitated within the central pore. The A subunit has *N*-glycosidase activity, and the B subunit binds to a membrane glycolipid, globotriaosylceramide (Gb3) (Fig. 4). The association of *E. coli* Stxs with diarrhea-associated HUS was established in 1985 (33). In addition, Stx1 and Stx2 do not target exactly the same tissues and organs although both bind Gb3 and are capable of causing diarrhea-associated HUS (53, 54). Injection of animals with Stx1 or Stx2 results in preferential damage to organs including kidney and lung. Stx1 appears to target the lung whereas Stx2 prefers the kidney (55).

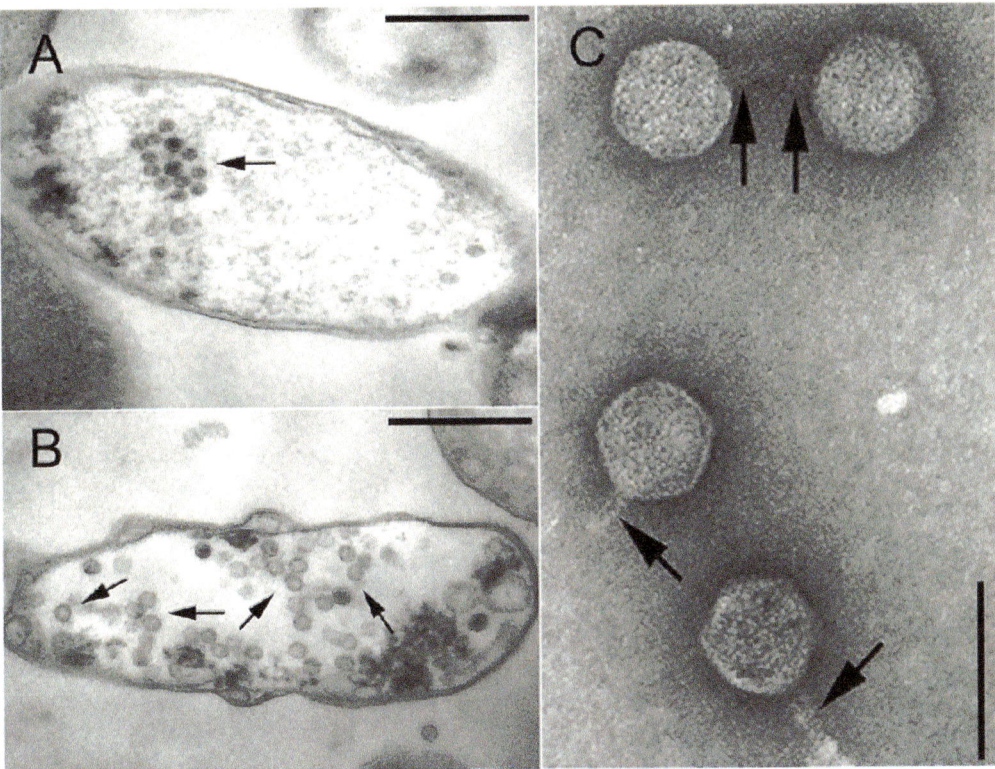

FIGURE 3 Transmission electron microscopy (TEM) of Stx2 phage (P13374) induction from lysogenic strain *E. coli* K-12 strain TPE2364 (C600) infected with phage lysates of *E. coli* O104:H4 strain CB13374. (A, B) Ultrathin sections of two bacterial cells (TPE2364) with maturating virion particles within the cytoplasm indicated by arrows (bars, 500 nm). (C) TEM of CsCl-purified, negatively stained phage (P13374) particles released by strain TPE2364 (bar, 100 nm). Short tails (arrows) and a hexagonal head are shown. From Beutin et al. 2012. *J Virol* 86:10444. doi:10.1128/microbiolspec.EHEC-0008-2013.f3

The Stx receptor is a major determinant and of central importance to diarrhea-associated HUS kidney disease (56). Gb3 is synthesized in the Golgi apparatus of select eukaryotic cells and is transported to the plasma membrane where it resides in the outer leaflet with its trisaccharide moiety facing outward and the hydrocarbon ceramide (C-16 to C-24) moiety noncovalently arranged within the plasma membrane. The binding subunit of Stx specifically recognizes the terminal alpha-1,4-di-galactose of the trisaccharide. A molecule of Stx contains five binding (B) subunits, each capable of binding one or more molecules of Gb3, resulting in cooperative high-affinity binding of Stx to cells. The importance of Gb3 in Stx action is evident from cell culture and animal studies where the absence of Gb3 eliminates the response to Stx (57). Gb3 may also be referred to as CD77 or the Pk blood antigen. Structure/function studies suggest that each toxin molecule may express 10 to 15 Gb3-binding sites per B-subunit pentamer (58), explaining the high affinity (dissociation constant $[K_d] \approx 10^{-9}$ M) of toxin binding. All Stxs, with the exception of one Stx2 variant called Stx2e, bind Gb3; Stx2e shows preferential binding to the glycolipid globotetraosylceramide. Structural differences in the toxin receptor also contribute to toxin susceptibility. Gb3 is heterogeneous, displaying variability in fatty acid chain length,

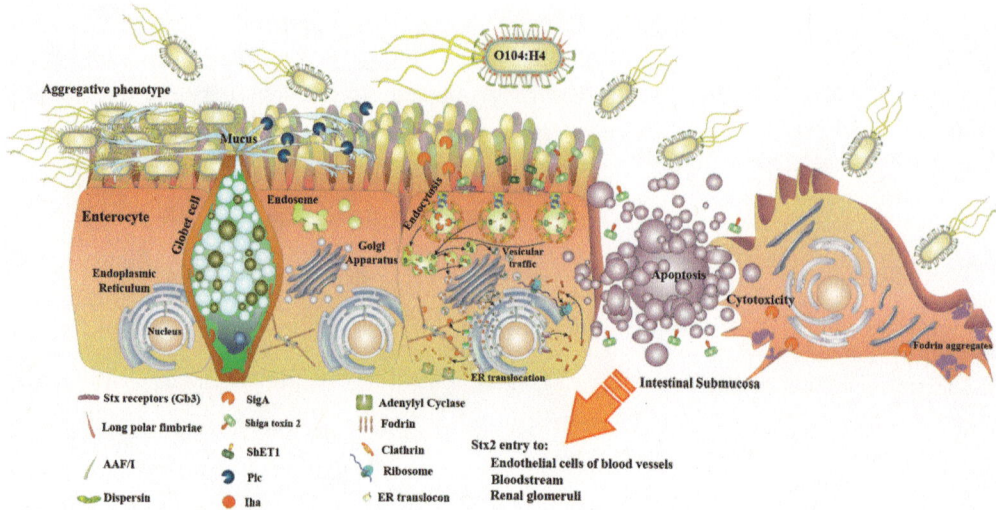

FIGURE 4 Schematic representation of *E. coli* O104:H4 virulence factors and their targets on the mucosal epithelium. The targets and virulence factors are extrapolated from their known function in other pathogens, and the action mechanism for ShET1 is hypothetical. doi:10.1128/microbiolspec.EHEC-0008-2013.f4

degree of bond saturation, and hydroxylation. Expression of Gb3 isoforms with long-chain, unsaturated fatty acids was associated with increased toxin sensitivity (59). It has been demonstrated that long-chain, unsaturated Gb3 isoforms were more likely to induce negative membrane curvature, leading to Gb3 clustering and formation of tubular invaginations (60). Finally, Gb3 association with lipid rafts was necessary for efficient intracellular transport of Stx1 B subunits (61). Polyunsaturated fatty acid incorporation into cell membranes, known to disrupt lipid rafts, protects cells from Stx intoxication (62). Thus, much or perhaps most of the Gb3 in a plasma membrane may not be reactive with Stx. This cryptic Gb3 may be a function of direct interaction of the Gb3 ceramide moiety with other membrane components, including cholesterol, other glycolipids, fatty acids, or proteins (63, 64). This possible mechanism helps explain why a tissue often appears to bind much less Stx compared to the total amount of Gb3 that is extractable from the tissue.

Most diarrhea-associated HUS is related to young children although individuals of all ages can develop this syndrome. A simple explanation is that more Gb3 is expressed in the kidneys of young people. However, this has proven not to be the case as there is more Gb3 extractable from kidneys of adults versus children (65, 66). An alternative explanation is that pediatric kidney expresses more Gb3 on cells that are more involved in biological responses leading to diarrhea-associated HUS. It is interesting that only pediatric kidney expressed Gb3 in glomeruli (66). Another concept, as mentioned above, is that the plasma membrane microenvironment nearby Gb3 dictates the biological response to Stx. Membrane Gb3 expression is a critical determinant of toxin sensitivity; cells expressing low Gb3 levels are sensitized to toxicity by increased membrane expression of toxin receptors, whereas cells selected for loss of Gb3 expression are resistant to Stxs (59, 67).

To be an effective protein synthesis inhibitor, Stx must reach the cytoplasm to access ribosomes (Fig. 4). Stx uses a retrograde transport pathway to reach the endoplasmic reticulum (ER), an intracellular compartment rich in membrane-associated ribosomes and containing the cellular machinery necessary for protein translocation into the cytoplasm.

Following binding and cross-linking of Gb3, Stx is internalized by clathrin-dependent or clathrin-independent mechanisms (60, 68). Following binding to Gb3 on target eukaryotic cells, the Stx-receptor complex is internalized and locates within endosomes. Rather than moving to lysosomes for degradation, the complex is transported in a retrograde manner through the Golgi apparatus to the ER (69). The resulting A1 and A2 fragments are linked through a disulfide bond; the latter fragment maintains association with the B pentamer. In the reductive environment of the ER, the disulfide bond is broken, and the 27-kDa A1 fragment translocates across the ER membrane, a process termed retrotranslocation. By this process, mammalian cells may recognize Stx delivered to the ER as misfolded proteins and activate the ER stress response, and the A1 fragment uses the ER-associated protein degradation pathway to reach ribosomes. To avoid the proteasome, A1 fragments contain few (Stx/Stx1) or no (Stx2) lysine residues (52), which may minimize ubiquitination, and A1 fragments associate with ribosomal proteins, which may inhibit proteasomal transport (70). Once the enzymatically active A1 subunit is released into the cytoplasm, it inactivates the eukaryotic ribosome by removal of a single adenine base from 28S rRNA within the large (60S) ribosomal subunit (71), acting as a highly specific *N*-glycosidase. This is an irreversible process that renders the ribosome defective for interaction with eukaryotic peptide elongation factor for binding of aminoacyl-tRNA and elongation of nascent peptide (72), resulting in inhibition of protein synthesis.

Stx can exert its effects on eukaryotic cells by one of three known mechanisms: (i) Inactivation of ribosomes and inhibition of cytoplasmic protein synthesis can result in cell death (73). (ii) Stx-dependent generation of the depurinated 28S rRNA in ribosomes initiates a signal-transduction response known as the ribotoxic stress response that leads to activation of cytokines, chemokines, or other factors that result in numerous events including apoptosis (Fig. 4) of the affected cell (74). The signaling pathways activated in the ribotoxic stress response include the mitogen-activated protein kinases (MAPK), such as p38MAPK (74, 75). Remarkably, these activities appear to be due to other than inhibition of protein synthesis. (iii) Receptor binding of Stx holotoxin or its B subunit alone can initiate a cytoplasmic signal-transduction cascade different from the ribotoxic stress response-activated pathway (76). Conserved features of Stx-induced apoptosis include routing active toxin to the ER where prolonged signaling through the ribotoxic and ER stress responses may activate apoptosis; alterations in the balanced expression of pro- and anti-apoptotic Bcl-2 proteins; a rapid activation of caspase 8, which, in turn, activates both caspase-dependent and mitochondrial-dependent apoptotic signaling pathways; and lack of signaling through Fas and tumor necrosis factor receptors to trigger apoptosis. Complement activation products have been detected in the serum and plasma of patients with HUS, and an in vitro study showed that Stx2 not only damages the kidney directly but also indirectly via complement in two ways. First, it activates complement, and second, it delays the functions of its control protein factor H on the cell surface, both known to damage the kidney (52).

Thus, Stx can exert different responses in a cell type-specific manner. The final result of these events can be cell death (apoptosis, necrosis) or inflammatory responses in cells that remain viable, and perhaps other intermediate responses that can be due to the Stx type, the host cell type, and/or the bacterial genetic background. STEC is known to cause hemorrhagic enterocolitis and HUS. Stx plays a role in the occurrence of blood in the feces and in HUS by its action on the endothelial cells of blood vessels in the intestinal submucosa and in the renal glomeruli (Fig. 4). Epidemiologically, Stx2 seems to be more important than Stx1 in development of HUS, which was also seen in the *E. coli* O104:H4 outbreak strain. The action of Stx is not limited to inhibition of protein synthesis. Stx induces macrophages to express tumor necrosis factor alpha (tumor necrosis factor-alpha), interleukin (IL)-1 beta, and IL-6 in vitro. These cytokines and lipopolysaccharide (LPS) are reported to increase the susceptibility of cells to Stx. A variety of cells such as tubular epithelial cells may be targets for Stx-mediated apoptosis. Apoptosis is considered to contribute to the pathogenesis of HUS caused by STEC (77). HUS is primarily due to the production and translocation of Stxs across the intestinal epithelial barrier (52). HUS may manifest after STEC is no longer detectable in the stool. Following toxin-mediated damage to colonic blood vessels, Stxs may enter the bloodstream although free toxins have not been detected in the circulation of patients with HUS. The risk of progression to extraintestinal complications is increased in patients infected with STEC expressing Stx2, either alone or in combination with Stx1 or Stx2c (78). The kidneys and central nervous system are most frequently damaged by Stxs. HUS is a constellation of hematological, renal, and neurological complications that develops in 10 to 15% of patients with hemorrhagic colitis. Complications include thrombocytopenia and hemolytic anemia with schistocytes (fragmented erythrocytes) present in blood smears. The characteristics of acute renal failure, which may follow STEC infection, include oliguria or anuria, swollen glomerular endothelial cells detached from the basement membrane, intraglomerular fibrin deposition, and thrombotic microangiopathy. Mesangiolysis or mesangial hyperplasia has been described in some HUS cases. Renal tubular injury may be present but is not a consistent finding late in the course of HUS (79). Approximately 66% of patients with HUS require dialysis. Central nervous system involvement may present as lethargy, irritability, seizures, paresis, and coma. Long-term sequelae include renal insufficiency, hypertension, hyperactivity and distractability, and insulin-dependent diabetes mellitus. Mortality from HUS is 3 to 5%. There is variability in signs and symptoms following ingestion of STEC. For example, patients may present with acute renal failure in the absence of bloody diarrhea (79).

Other events that may influence the pathogenesis of STEC are the phage origin of Stx and Stx variability (Fig. 3). Stx phages display extensive genetic mosaicism; however, genes encoding the Stx A and B subunits are generally located downstream of the antiterminator Q and the P′R promoter. As a consequence of this orientation, toxin genes are late genes optimally expressed upon induction of the lytic cycle. STEC may possess cryptic lambdoid prophages that serve as sources for recombination events, yielding novel toxin-converting phages, and Stx phages expressing new tail assemblies may expand the host range of toxin-producing organisms (80). Lysogenic conversion to the toxigenic phenotype may occur if recipient bacteria display phage receptors and possess integration sites within the genome (Fig. 3). Thus, Stx phages are responsible for the dissemination of *stx* genes in *E. coli* and other enteric bacteria. Following induction, Stx phages can infect other bacteria in vivo and in vitro. Stx phages may be considered to represent highly mobile genetic elements that play an important role in the expression of Stx, in horizontal gene transfer, and hence in genome diversification (81). On the other hand, purified Stx alone is capable of producing systemic complications, including HUS, in animal models of disease. Stx2a is more potent than Stx1. Epidemiologic studies suggest that Stx2 subtypes also differ in potency. Indeed, by examining protein synthesis inhibition using Vero monkey kidney cells and inhibition of metabolic activity using primary human renal proximal tubule epithelial cells, it was found that Stx2a, Stx2d, and elastase-cleaved Stx2d were at least 25 times more potent than Stx2b and Stx2c; in vivo this potency was also assessed in mice. Stx2b and Stx2c had potencies similar to that of Stx1, whereas Stx2a, Stx2d, and elastase-cleaved Stx2d were 40 to 400 times more potent than Stx1 (82). Furthermore, by using the classification for seropathotypes (A to E), it has been found that the *stx*(2) variant was mainly associated with strains of seropathotype A, whereas most of the strains of seropathotype C possessed the *stx*(2-vhb) variant, which was frequently associated with *stx*(2), *stx*(2-vha), or *stx*(2c). Levels of *stx*(2) and *stx*(2)-related mRNA were higher in strains belonging to seropathotype A and in those strains of seropathotype C that express the *stx*(2) variant than in the remaining strains of seropathotype C. The *stx*(2-vhb) genes were the least expressed (83). Another complicating factor in diarrhea-associated HUS is that antibiotics are not recommended in the earlier phases, i.e., before the appearance of bloody diarrhea because STEC bacteria respond to some antibiotics by producing excess Stx (84, 85).

Iha

The chromosomally encoded adherence-conferring protein Iha, a homolog of *V. cholerae* IrgA (28), is one of a range of novel adhesins identified among STEC strains (Fig. 4). Iha was first characterized in *E. coli* O157:H7, but it is distributed widely among LEE-negative and LEE-positive STEC strains and UPEC (28, 86). In a study in which the difference did not reach statistical significance, it was reported that an *iha* deletion mutant of O157:H7 STEC was impaired in adherence to HeLa cells. But there was a highly significant increase in adherence to both HeLa and MDBK cells when *iha* was expressed from a plasmid in a nonpiliated recombinant *E. coli* host (28). The potential importance of this adhesin lies in the high prevalence (91%) of *iha* in STEC belonging to different seropathotypes and the presence of multiple *iha* copies in some strains (28, 87).

In the UPEC strain CFT073, Iha functions as a urovirulence factor by contributing to colonization in a mouse urinary tract infection model (88). Additionally, Iha from UPEC strain UCB34 functions as a catecholate siderophore receptor in *E. coli* K-12. The capacity of Iha to transport siderophores is TonB-dependent, whereby the protein complex TonB/ExbB/ExbD provides the energy required for active transport (29, 89). Furthermore, *iha* expression is regulated by the protein Fur, a ferric uptake regulator (90). In the presence of iron, dimerized Fur binds to *iha* promoter regions through a sequence-specific protein-DNA interaction and represses *iha* transcription. Under iron-limiting conditions, Fur is unable to interact with the DNA, and as a consequence, *iha* transcription is derepressed (29, 90). Iron is an important environmental cue, and iron-limiting conditions induce virulence-related genes in a number of pathogens (91, 92). Free iron levels are typically low at mucosal surfaces, due to binding by host lactoferrin, and therefore the induction of iron-scavenging mechanisms is an important bacterial in vivo survival strategy (92). Recently, it was found that bacterial growth under conditions simulating colonic, but not ileal, short-chain fatty acid (SCFA) concentration increases *iha* expression in three tested STEC strains, with the strongest expression detected in LEE-negative STEC O113:H21 strain 98NK2, as expression of *iha* in O157:H7 STEC strain 98NK2 is subject to Fur-mediated iron repression. However, exogenous iron did not repress *iha* expression in the presence of colonic SCFAs in either 98NK2 or O157:H7 strain EDL933. Moreover, exposure to the iron chelator 2,2'-dipyridyl caused no further enhancement of *iha* expression over that induced by colonic SCFAs. These findings indicate that SCFAs regulate *iha* expression in STEC independently of iron. Increased expression of *iha* under colonic but not ileal SCFA conditions possibly may contribute to preferential colonization of the human colon by STEC (93).

Thus, Iha is a potentially important accessory virulence factor for several pathotypes of *E. coli*, including STEC, UPEC, and avian-pathogenic strains (28, 86, 92, 94–96), functioning as an adhesin and, at least for UPEC, as a catecholate siderophore receptor (29). This latter function is consistent with the fact that *iha* is induced under iron-limiting conditions and is subject to Fur-mediated re-pression in iron-replete environments in both UPEC and O157:H7 STEC strains (29, 90) and in the hypervirulent LEE-negative STEC strain 98NK2.

Long Polar Fimbriae (Lpf)

E. coli O157:H7 possesses two Lpf operons, *lpf1* and *lpf2*, both of which contain genes closely related to the Lpf of *Salmonella enterica* serovar Typhimurium (97). Expression of *lpf1* and *lpf2* is induced during the late exponen-tial-phase growth in tissue culture media at pH 6.5 and 37°C or under iron-restricted conditions (98), and has been found to influ-ence *E. coli* O157:H7 adherence to cultured epithelial monolayers (99, 100). The first locus (*lpf1*) is located in an O157-specific island (OI-141) of approximately 5.9 kb, and inserted in the *yhjX-yhjW* intergenic region (in relation to the *E. coli* K-12 chromosome). The *lpf1* operon contains six genes (*lpfABCC'DE*) sim-ilar in sequence and gene order to the *Sal-monella lpfABCDE* genes (101). Expression of the STEC O157:H7 *lpf1* operon in a non-fimbriated *E. coli* ORN172 strain was reported to increase adherence to HeLa and MDCK cells, and peritrichous short fimbriae were observed (101). It was further demonstrated that *stx*-positive and *stx*-negative STEC O157:H7 mutated in the *lpfA1* gene (encoding the major fimbrial subunit) exhibited a re-duction in adherence to epithelial cells and displayed a diffuse adherence pattern (87, 101). A recent study showed that STEC O157:H7 adhered more abundantly to surfaces coated with fibronectin, laminin, and collagen IV and that a reduced binding of the bacteria to these extracellular matrix proteins was observed for an *lpf* mutant (*lpf1*) strain (102). This study demonstrated that Lpf1 and ex-tracellular matrix proteins interact, and their interaction may contribute to STEC O157:H7 colonization of the gastrointestinal tract. The second *lpf* operon, the *lpf2* locus, is approxi-mately 6.8 kb and located in OI-154; it is inserted in the *glmS-pstS* intergenic region

(98). The *lpf2* locus contains five genes (*lpfABCDD'*) but lacks an *lpfE* homolog, and instead, the *lpfD* gene is duplicated in O157 strains. The in vitro adherence phenotype conferred by this locus is not well understood. When a nonfimbriated *E. coli* K-12 strain was used to express the *lpf2* locus, a less adherent phenotype to Caco-2 intestinal cells was ob-served (98). However, a previous study using a random transposon mutagenesis showed that an insertion in the *lpfD2* gene caused increased bacterial adherence to HeLa cells (103). Further, disruption of the gene encoding the major fimbrial subunit, the *lpfA2* gene, resulted in a reduction in initial adherence to Caco-2 cells, although adherence to HeLa cells was unaffected (98). Finally, expression of the *lpf2* gene in a nonfimbriated *E. coli* strain resulted in the appearance of thin, fibrilla-like structures on the bacterial surface that were structurally different from those observed for a strain expressing the cloned *lpf1* genes (98).

Homologs of *lpf* genes have also been iden-tified for non-O157 STEC strains, and their role in adherence has also been explored (Fig. 4). It was found that a Tn*5phoA* mutant of the LEE-negative STEC O113:H21 strain exhibited re-duced adherence to Chinese hamster ovary-K1 (CHO-K1) cells, and further analysis mapped the mutation to a gene with homology to the *lpfD2* gene (104). Sequencing analysis demon-strated that the STEC O113:H21 strain pos-sesses an *lpf* operon (also referred as *lpf*O113) containing four genes (*lpfABCD* genes) found at the same chromosomal location as the STEC O157:H7 *lpf2* locus (OI-154) (104). Inactivation of the *lpfA*O113 gene in STEC O113:H21 resulted in a significant reduction in micro-colony formation on CHO-K1 cells, and when the *lpf*O113 genes were introduced into the nonfimbriated *E. coli* K-12 strain ORN103, the bacteria adhered in a localized pattern, as op-posed to a diffuse adherence, indicating that the Lpf2 fimbria homologs may promote interbacterial interactions (104).

Due to the relatively subtle effects of *lpf* mutations on adherence in vitro, coupled with the divergent findings from in vivo or organ culture experiments, the precise role of Lpf in *E. coli* O157:H7 adherence remains somewhat unclear. Therefore, the role of the *E. coli* O157:H7 *lpf* loci was further tested in an infant rabbit model, which mimics the diarrhea and gut pathology, including the histopathological A/E lesions, seen in patients with STEC infection. By performing competition experiments between *E. coli* O157:H7 and an isogenic *lpf1 lpf2* double mutant, it was found that the mutant was outcompeted in the ileum, cecum, and midcolon of rabbits, confirming that Lpf contributes to intestinal colonization (105). Unexpectedly in this study, it was observed that the *lpf1 lpf2* double mutant showed an increased adherence to colonic epithelial cells in vitro, and transmission electron microscopy revealed curli-like structures on the surface of this mutant. Interestingly, deletion of *csgA* per se did not appear to affect intestinal colonization. Therefore, in addition to conclusively demonstrating that Lpf contributes to *E. coli* O157:H7 intestinal colonization, the authors indicated that the regulatory mechanisms controlling expression of Lpf and curli are interconnected (105).

E. COLI O104:H4 OUTBREAK STRAIN AS EAEC

As mentioned before, the O104:H4 outbreak strain represents a rare combination of EAEC characteristics and Stx2 expression; only sporadic cases of disease associated with Stx-producing EAEC have previously been described in Germany (106, 107), France (108, 109), the United Kingdom (110), the Republic of Georgia (13), and Japan (111). EAEC was first described in 1987, based on the characteristic adherence phenotype with cultured HEp-2 cells (stacked-brick appearance on epithelial cells) (Fig. 2). This biological test still remains the gold standard of diagnosis, but it does not distinguish between pathogenic and nonpathogenic strains. Subsequently, a number of virulence factors have been described for EAEC strains and include adhesins

(e.g., AAF/I to III), heat-stable enterotoxin, transporters, and other secreted proteins (e.g., the serine protease autotransporter Pic, which has mucinase activity), as well as multiple factors contributing to EAEC-induced inflammation. However, none of these factors are found in all EAEC isolates, and no single factor has ever been implicated in EAEC virulence. Thereby, EAEC isolates are genetically a heterogeneous group of *E. coli* strains (112). EAEC strains were first associated with persistent diarrhea in infants from developing countries; since then they have increasingly been linked as a cause of acute and persistent diarrhea in young infants and children in developing and industrialized countries, in individuals infected with human immunodeficiency virus, as a cause of acute diarrhea in travelers from industrialized regions, and with foodborne outbreaks. A major effect of EAEC infection is on the malnourished children in developing countries (112). Thus, EAEC without Stx causes diarrhea in persons who reside in developed countries, such as the United Kingdom (113, 114), Switzerland (115), and Japan (116), although EAEC is more commonly associated with acute and persistent (>14 days) diarrhea of infants, children, human immunodeficiency virus-positive persons, and travelers to developing countries (112, 117). EAEC isolates are a highly heterogeneous group of bacteria (118, 119), and each isolate carries a particular subset of EAEC-associated virulence genes; no single virulence factor is consistently associated with EAEC pathogenesis. The addition of *stx2* to the EAEC virulence gene repertoire, however, has led to a pathogen that has the capacity to cause disease on a large scale with a potentially deadly outcome due to the possibility of the development of HUS (Fig. 4).

AggR and AggR-Regulated Genes (AAF/I, Dispersin, Dispersin Transporter)

AggR is a member of the AraC/XylS family of bacterial transcriptional activators (10, 120), exhibiting the greatest levels of amino acid identity with the CfaD (68%), Rns (66%), and CsvR (62%) regulators of enterotoxigenic *E. coli* (10). Proteins belonging to this family are defined by a conserved 99-amino-acid C terminus domain and regulate diverse cellular functions including metabolism, stress response, and the synthesis of virulence factors. Multiple epidemiologic studies suggest that strains expressing AggR are more likely to cause diarrheal disease than those without it, proposing the term "typical EAEC" to describe strains harboring the AggR regulon (121, 122). A number of AggR-regulated genes have been described previously in archetype EAEC strain 042. The genes encoding AAF were the first found to be regulated by AggR (10), followed by *aap* (encoding the dispersin surface protein) (123) and the Aat secretion system, which is required for transport of dispersin to the bacterial surface (124). AggR also activates expression of the Aai type VI secretion system (T6SS) in strain 042 (125), though the role of Aai in EAEC virulence remains unknown.

Electron microscopy studies demonstrated different fimbriae in several EAEC strains (126–128). The best characterized are AAF I, II, and III, which are responsible for the expression of aggregative adherence (15, 17, 129). AAFs are encoded in high-molecular-weight plasmids (designated pAA), and their biogenesis employs the chaperone-usher secretion pathway (11, 12, 15). These fimbriae are members of the Dr superfamily of adhesins that includes strains of UPEC and diarrheagenic *E. coli* with a genetic organization consisting of chaperone, usher, and pilin subunits (13, 14). Expression of AAF genes requires AggR (10); this protein regulates the genes involved in fimbrial biogenesis for both AAF/I and AAF/II. AAF/I are flexible, bundleforming fimbriae, and the genes responsible for their biogenesis are located in two unlinked regions of the plasmid: region 1, containing the genes that encode the pilin, chaperonin, and usher, and region 2 (separated by 9 kb), containing the regulator-encoding gene (designated *aggR*) (10, 15, 16).

aggR is located in an open reading frame of 794 bp that encodes a protein with a predicted molecular size of 29.4 kDa. The cloned *aggR* gene is sufficient to complement a region 1 clone to confer AAF/I expression, while an *aggR* mutant is negative for AAF/I expression.

Genomic studies of the pAA plasmid revealed a small open reading frame just upstream of the *aggR* gene in most EAEC strains. This gene was found to encode a 10-kDa secreted protein that was recognized by the sera of volunteers fed strain 042. A null mutant of this gene revealed a unique hyperaggregative phenotype, and scanning electron microscopy of the bacterial strains showed collapse of the AAF onto the bacterial cell surface. The protein product of this gene was therefore termed antiaggregation protein (Aap), nicknamed dispersin (123). Dispersin is secreted to the surface of EAEC strains and binds noncovalently to LPS of the outer membrane. Data suggest that the mechanism of dispersin's effect may be mediated predominantly through its ability to neutralize the strong negative charge of the LPS, so that the AAFs, which carry a strong positive charge, are free to splay out from the surface and bind distant sites (130). Interestingly, the secretion of dispersin is dependent on the presence of an ABC transporter complex also encoded on plasmid pAA (124). The genes encoding this transporter (the Aat complex) correspond to the site of the previously cryptic aggregative adherence probe. As both dispersin and the Aat complex are under the control of AggR, the latter protein is emerging as the central regulator of virulence functions in EAEC.

The AggR regulon is not restricted to the pAA plasmid. It has been found that in addition to the plasmid-encoded genes, AggR regulates a chromosomal operon inserted on a PAI at the *pheU* locus (125). AggR activates the expression of chromosomal genes, including 25 contiguous genes (*aaiA–Y*), which are localized to a 117-kb PAI inserted at *pheU*. Many of these genes have homologs in other gramnegative bacteria and were recently proposed to constitute T6SS. AaiC was identified as a secreted protein that has no apparent homologs within GenBank. EAEC strains carrying in-frame deletions of *aaiB*, *aaiG*, *aaiO*, or *aaiP* synthesized AaiC, but AaiC secretion was abolished. Cloning of *aai* genes into *E. coli* HB101 suggested that *aaiA–P* are sufficient for AaiC secretion.

Recent studies in the Nataro laboratory have revealed that AggR positively regulates its own expression in a complex fashion. AggR binds directly and specifically to two sites flanking the *aggR* promoter. Additionally, *aggR* promoter was found to be positively regulated by the DNA-binding protein FIS and negatively regulated by the global regulator H-NS. EAEC present in the mouse intestine possessed relatively high levels of *fis* promoter and *aggR* promoter activity and a low level of *hns* promoter when compared with in vitro experiments. The data provide significant insights into the regulation cascade leading to *aggR* expression in the mammalian intestine during EAEC infection (131). A recent study showed that there are at least 44 AggR-regulated genes in the genome of EAEC strain 042. Twenty-five of the 44 genes were previously known to be so regulated, identified as part of the Aai T6SS (16 genes), the dispersin secretion system (5 genes), and the AAF/II fimbrial biogenesis system (4 genes). Sixteen of the 44 genes are predicted to encode hypothetical proteins, and only 5 of these genes showed homology to other genes encoding known bacterial proteins, suggesting new virulence-related functions (132).

AggR has also a role in the inflammatory response against EAEC. Jiang et al. (121) reported that IL-8 levels were higher in feces of patients infected with *aggR*- or *aafA*-containing strains compared with those infected with strains negative for these factors. Recently, it was also shown that EAEC strains harboring *aggR*, *aggA,* and *aap* were more likely to cause IL-8 induction of >4,100 pg/ml from nonpolarized HCT-8 intestinal epithelial cells than EAEC negative for those genes (133). In fact, polarized T84 intestinal cells were found to release IL-8 even when infected with strain 042 mutated in the major flagellar subunit FliC. IL-8 release from polarized T84 cells was found to require the AggR activator and the AAF fimbriae, and IL-8 release was significantly less when cells were infected with mutants in the minor fimbrial subunit AafB (134).

SPATEs (Pic, SepA, SigA)

SPATEs and other autotransporters use a type V secretion system for export to the extracellular space (135, 136). The autotransporters contain all the information necessary for passage through the inner membrane and the outer membrane. To mediate its own secretion, an autotransporter contains three functional domains: an N-terminal signal sequence, an extracellular passenger domain, and a C-terminal β-barrel domain. The signal sequence initiates Sec-dependent transport across the inner membrane and is proteolytically removed in the periplasmic space. The C-terminal domain forms a β-barrel pore in the outer membrane, which facilitates the delivery of the passenger domain to the extracellular space. The passenger domains of some autotransporters remain anchored to the extracellular face of the outer membrane, but SPATEs are released from the bacterial cell by proteolytic nicking of a site between the β-barrel pore and the passenger domain. The mature, secreted SPATEs are 104- to 110-kDa toxins that contain a typical N-terminal serine protease catalytic domain followed by a highly conserved β-helix motif, which is present in nearly all autotransporters (135, 137). Although the general process of SPATE secretion is understood, the details of many events in SPATE biogenesis (the chaperone function in the periplasm, the mechanism of β-barrel insertion into the outer membrane, the translocation pathway across the outer membrane, the proteolytic release of the mature protein from the outer membrane, etc.) remain unresolved (135).

It has been proposed that SPATEs can be divided phylogenetically into two distinct classes, designated 1 and 2 (138). Class 1 SPATEs are cytotoxic in vitro and induce mucosal damage on intestinal explants (Fig. 4). Although the actions of class 1 SPATEs are not fully understood, several have been shown to enter eukaryotic cells and to cleave cytoskeletal proteins (21, 139, 140) while the class 2 SPATEs induce mucus release, cleave mucin, and confer a subtle competitive advantage in mucosal colonization (20, 141, 142) (Fig. 4). A study to determine the prevalence of SPATEs in EAEC was performed by seeking 10 genes encoding serine protease autotransporter toxins in a collection of clinical EAEC isolates. Eighty-six percent of EAEC strains harbored genes encoding one or more class I cytotoxic SPATE proteins (Pet, Sat, EspP, or SigA). Two class II noncytotoxic SPATE genes were found among EAEC strains: *pic* and *sepA*, each originally described in *S. flexneri* 2a. Using a multiplex PCR for five SPATE genes (*pet*, *sat*, *sigA*, *pic*, and *sepA*), the authors found that most of the *Shigella* sp. isolates also harbored more than one SPATE, whereas members of most other *E. coli* pathotypes rarely harbored a cytotoxic SPATE gene. SPATEs may be relevant to the pathogenesis of both EAEC and *Shigella* spp. (143).

Pic is localized in the EAEC chromosome (118, 141), and it was found to be identical to a protein termed Shmu (*Shigella* mucinase), which is encoded on the *Shigellashe* PAI (144). Through functional analysis of Pic, it was found that its proteolytic site is involved in Pic mucinase activity, serum resistance, and hemagglutination. Phenotypes identified for Pic suggest that it is involved in the early stages of the pathogenesis and most probably promotes the intestinal colonization (141, 144). Pic binds mucin, and this binding was blocked in competition assays using monosaccharide constituents of the oligosaccharide side chains of mucin. Moreover, Pic mucinolytic activity decreased when sialic acid was removed from mucin. Thus, Pic is a mucinase with lectin-like activity that can be related to its reported hemagglutinin activity (19). Recently, it was shown that Pic induces hypersecretion of

mucus, which was accompanied by an increase in the number of mucus-containing goblet cells (Fig. 4). This finding is in accord with one of the hallmarks of EAEC: the formation of biofilm, which comprises a mucous layer with immersed bacteria in the intestines of patients. Interestingly, an isogenic *pic* mutant (EAEC Δ*pic*) is unable to cause this mucus hypersecretion. Furthermore, purified Pic was also able to induce intestinal mucus hypersecretion. Thus, Pic mucinase is responsible for one of the pathophysiologic features of the diarrhea mediated by EAEC (20). It has also been shown that Pic protease promotes intestinal colonization and growth in the presence of mucin, suggesting a novel metabolic role for the Pic mucinase in EAEC colonization. Interestingly, it has been found that Pic targets a broad range of human leukocyte adhesion proteins. Substrate specificity is restricted to glycoproteins rich in O-linked glycans, including CD43, CD44, CD45, CD93, CD162 (P-selectin glycoprotein ligand 1), and the surface-attached chemokine fractalkine, all implicated in leukocyte trafficking, migration, and inflammation. Additionally, exposure of human leukocytes to purified Pic results in polymorphonuclear cell activation, but impaired chemotaxis and transmigration; Pic-treated T cells undergo programmed cell death (145).

Besides Pic, another SPATE from class II found in *E. coli* O104:H4 is SepA. In 1995, Benjelloun-Touimi et al. (22) described a Tsh homolog designated SepA (for *Shigella* extracellular protein), which is the major extracellular protein of *S. flexneri* (22). In 1994, the first SPATE was described as a temperature-sensitive hemagglutinin (Tsh) in avian-pathogenic *E. coli*, which causes disseminated infections in birds (146). Investigation of the proteolytic activity of SepA by using a wide range of synthetic peptides found that SepA hydrolyzed several of these substrates and that the activity was inhibited by the serine protease inhibitor phenylmethylulfonyl fluoride (147). Several SepA-hydrolyzed peptides were described as specific substrates for cathepsin G, a serine protease produced by polymorphonuclear leukocytes that was proposed to play a role in inflammation. However, unlike cathepsin G, SepA degraded neither fibronectin nor angiotensin I and had no effect on the aggregation of human platelets. The presence of *sepA* on the virulence plasmid, as well as the recognition of SepA by the sera of monkeys infected with *Shigella* sp., suggested that SepA might be involved in *Shigella* pathogenicity (22). However, construction and phenotypic characterization of a *sepA* mutant suggested that SepA is required neither for entry into cultured cells nor for intracellular dissemination. Nevertheless, the *sepA* mutant demonstrated a reduced ability to induce both mucosal atrophy and tissue inflammation in the rabbit ligated ileal loop model, indicating that SepA may play a role in tissue invasion, although this hypothesis remains to be elucidated (22).

On the other hand, a SPATE of class I has been reported in O104:H4, SigA. An initial report showed that the *she* PAI, which contains the *pic* (*she*) gene, also contains a gene encoding a second immunoglobulin A protease-like homolog, *sigA*, lying 3.6 kb downstream and in an inverted orientation with respect to *pic* (144). Functional analysis showed that SigA is a secreted temperature-regulated serine protease capable of degrading casein. Experiments similar to those used with Pet (another SPATE of EAEC) revealed that SigA is cytopathic for HEp-2 cells, suggesting that it may be a cell-altering toxin with a role in the pathogenesis of *Shigella* infections. Indeed, it was found that SigA binds specifically to HEp-2 cells and degrades recombinant human αII spectrin (α-fodrin) in vitro and also cleaves intracellular fodrin in situ, causing its redistribution within cells, suggesting that the cytotoxic and enterotoxic effects mediated by SigA are likely associated with the degradation of epithelial fodrin (21) (Fig. 4). Furthermore, SigA was at least partly responsible for the ability of *S. flexneri* to stimulate fluid accumulation in ligated rabbit ileal loops (148). In the case of Pet, it is a cytoskeleton-altering toxin (136), because it induces contraction of the cytoskeleton, loss of actin stress fibers, and release of focal contacts in HEp-2 and HT29/C1 cell monolayers, followed by complete cell rounding and detachment. Interestingly, Pet cytotoxicity and enterotoxicity depend on Pet serine protease activity (149). It has been shown that Pet enters the eukaryotic cell and that trafficking through the vesicular system is required for the induction of cytopathic effects. Thus, after clathrin-mediated endocytosis, Pet undergoes a retrograde trafficking to the endoplasmic reticulum to be translocated into the cytosol (150, 151). Finally, an intracellular target, α-fodrin (αII spectrin), has been found for Pet. Pet binds and cleaves epithelial fodrin (between M1198 and V1199) in vitro and in vivo, causing fodrin redistribution within the cells, to form intracellular aggregates as membrane blebs (139).

Shigella Enterotoxin 1 (Set1)

ShET1 toxin is a subunit toxin encoded by *setA* and *setB*, which are thought to form an oligomeric toxin consisting of a single 20-kDa SetA protein associated with a pentamer of five 7-kDa B subunits (SetB) (18). ShET1 appears to induce intestinal secretion via cyclic AMP and cyclic GMP; however, the precise mechanism of action and detailed biochemistry remain inconclusive. Unusually, the *setAB* genes are encoded within the *pic* gene but on the complementary strand and thus have the same prevalence characteristics and disease associations as *pic* (141). However, the role of the ShET1 in EAEC pathogenesis has not been studied, even though it can work on adenylyl cyclase (Fig. 4). In the case of the *S. flexneri* 2a strain, culture filtrates cause significant fluid accumulation in rabbit ileal loops, when the bacteria are grown in iron-depleted medium. Also, testing in Ussing chambers showed a greater rise in potential difference and short circuit current with such filtrates compared with the medium control. Ultrafiltration and gel exclusion size fractionation of M4243 filtrate revealed that the activity was in the fraction of approximately 60 kDa. It is thought that ShET1 is elaborated in vivo, since it elicits an immune response and may be important in the pathogenesis of diarrheal illness due to *S. flexneri* (152).

CONCLUSIONS

The genotypes, phenotypes, and phylogeny of the outbreak isolates demonstrate that the *E. coli* O104:H4 outbreak strain is a clone that combines virulence potentials of two different pathogens: STEC and EAEC. Thus, EAEC of serotype O104:H4 is by itself an emerging serovariant that has acquired an original set of virulence factors. All shared virulence profiles combining typical STEC (*stx2, iha, lpf*O26, *lpf*O113) and EAEC (*aggA, aggR, set1, pic, aap*) loci and expressed phenotypes that define STEC and EAEC, including production of Stx2 and aggregative adherence to epithelial cells (Fig. 4). Additionally, isolates displayed a high antibiotic resistance, specifically those related to *β*-lactamase.

EAEC strains of serotype O104:H4 contain a large set of virulence-associated genes regulated by the AggR transcription factor (Fig. 1). These include the pAA plasmid genes encoding the AAF, which anchor the bacterium to the intestinal mucosa and induce inflammation (Fig. 2), and a protein-coat secretion system (Aat), including its secreted protein, dispersin. A switch of the virulence plasmid (pAA) together with the type of the AAF could be an additional explanation for the higher virulence of this outbreak strain. Other adhesion factors (Iha, Lfp) could help this outbreak strain as a synergistic or initial adhesion factor to guarantee an efficient adhesion and perhaps Stx delivery. Additionally, EAEC strains of serotype O104:H4 produce a variable number of SPATEs, implicated in mucosal damage and colonization. It allows speculation that the combined activity of these SPATEs, together with other EAEC virulence factors, accounts for the increased uptake of Stx into the systemic circulation, resulting in the high rates of HUS (Fig. 4).

Although more studies are needed to explain the increased virulence, *E. coli* O104:H4 shows that mixed virulence profiles in enteric pathogens introduced into susceptible populations can cause serious outbreaks and have terrible consequences for infected people.

ACKNOWLEDGMENTS

The author was supported by a Conacyt grant (128490). I thank Paul S. Ugalde for the artistic work in Fig. 1 and 4 and Lucia Chavez-Dueñas for organizing the reference database. I declare no conflicts of interest with regard to the manuscript.

CITATION

Navarro-Garcia F. 2014. *Escherichia coli* O104:H4 pathogenesis: an enteroaggregative *E. coli*/Shiga toxin-producing *E. coli* explosive cocktail of high virulence. Microbiol Spectrum 2(4):EHEC-0008-2013.

REFERENCES

1. **Frank C, Werber D, Cramer JP, Askar M, Faber M, an der Heiden M, Bernard H, Fruth A, Prager R, Spode A, Wadl M, Zoufaly A, Jordan S, Kemper MJ, Follin P, Muller L, King LA, Rosner B, Buchholz U, Stark K, Krause G.** 2011. Epidemic profile of Shiga-toxin-producing *Escherichia coli* O104:H4 outbreak in Germany. *N Engl J Med* **365:**1771–1780.

2. **Gault G, Weill FX, Mariani-Kurkdjian P, Jourdan-da Silva N, King L, Aldabe B, Charron M, Ong N, Castor C, Mace M, Bingen E, Noel H, Vaillant V, Bone A, Vendrely B, Delmas Y, Combe C, Bercion R, d'Andigne E, Desjardin M, de Valk H, Rolland P.** 2011. Outbreak of haemolytic uraemic syndrome and bloody diarrhoea due to *Escherichia coli* O104:H4, south-west France, June 2011. *Euro Surveill* **16.**

3. **Bielaszewska M, Mellmann A, Zhang W, Kock R, Fruth A, Bauwens A, Peters G, Karch H.** 2011. Characterisation of the *Escherichia coli* strain associated with an outbreak of haemolytic uraemic syndrome in Germany, 2011: a microbiological study. *Lancet Infect Dis* **11:**671–676.

4. **Mellmann A, Harmsen D, Cummings CA, Zentz EB, Leopold SR, Rico A, Prior K, Szczepanowski R, Ji Y, Zhang W, McLaughlin SF, Henkhaus JK, Leopold B, Bielaszewska M, Prager R, Brzoska PM, Moore RL, Guenther S, Rothberg JM, Karch H.** 2011. Prospective genomic characterization of the German enterohemorrhagic *Escherichia* coli O104:H4 outbreak by rapid next generation sequencing technology. *PLoS One* **6:**e22751.

5. **Rasko DA, Webster DR, Sahl JW, Bashir A, Boisen N, Scheutz F, Paxinos EE, Sebra R, Chin CS, Iliopoulos D, Klammer A, Peluso P, Lee L, Kislyuk AO, Bullard J, Kasarskis A, Wang S, Eid J, Rank D, Redman JC, Steyert SR, Frimodt-Moller J, Struve C, Petersen AM, Krogfelt KA, Nataro JP, Schadt EE, Waldor MK.** 2011. Origins of the *E. coli* strain causing an outbreak of hemolytic-uremic syndrome in Germany. *N Engl J Med* **365:**709–717.

6. **Wieler LH, Semmler T, Eichhorn I, Antao EM, Kinnemann B, Geue L, Karch H, Guenther S, Bethe A.** 2011. No evidence of the Shiga toxin-producing *E. coli* O104:H4 outbreak strain or enteroaggregative *E. coli* (EAEC) found in cattle faeces in northern Germany, the hotspot of the 2011 HUS outbreak area. *Gut Pathog* **3:**17.

7. **Auvray F, Dilasser F, Bibbal D, Kerouredan M, Oswald E, Brugere H.** 2012. French cattle is not a reservoir of the highly virulent enteroaggregative Shiga toxin-producing *Escherichia coli* of serotype O104:H4. *Vet Microbiol* **158:**443–445.

8. **Monecke S, Mariani-Kurkdjian P, Bingen E, Weill FX, Baliere C, Slickers P, Ehricht R.** 2011. Presence of enterohemorrhagic *Escherichia coli* ST678/O104:H4 in France prior to 2011. *Appl Environ Microbiol* **77:**8784–8786.

9. **Huang DB, Mohanty A, DuPont HL, Okhuysen PC, Chiang T.** 2006. A review of an emerging enteric pathogen: enteroaggregative *Escherichia coli*. *J Med Microbiol* **55:**1303–1311.

10. **Nataro JP, Yikang D, Yingkang D, Walker K.** 1994. AggR, a transcriptional activator of aggregative adherence fimbria I expression in enteroaggregative *Escherichia coli*. *J Bacteriol* **176:**4691–4699.

11. **Elias WP Jr, Czeczulin JR, Henderson IR, Trabulsi LR, Nataro JP.** 1999. Organization of biogenesis genes for aggregative adherence fimbria II defines a virulence gene cluster in enteroaggregative *Escherichia coli*. *J Bacteriol* **181:**1779–1785.

12. **Bernier C, Gounon P, Le Bouguenec C.** 2002. Identification of an aggregative adhesion fimbria (AAF) type III-encoding operon in enteroaggregative *Escherichia coli* as a sensitive probe for detecting the AAF-encoding operon family. *Infect Immun* **70:**4302–4311.

13. **Servin AL.** 2005. Pathogenesis of Afa/Dr diffusely adhering *Escherichia coli*. *Clin Microbiol Rev* **18:**264–292.

14. **Boisen N, Struve C, Scheutz F, Krogfelt KA, Nataro JP.** 2008. New adhesin of enteroaggregative *Escherichia coli* related to the Afa/Dr/AAF family. *Infect Immun* **76:**3281–3292.

15. **Nataro JP, Deng Y, Maneval DR, German AL, Martin WC, Levine MM.** 1992. Aggregative adherence fimbriae I of enteroaggregative *Escherichia coli* mediate adherence to HEp-2 cells and hemagglutination of human erythrocytes. *Infect Immun* **60:**2297–2304.

16. **Savarino SJ, Fox P, Deng Y, Nataro JP.** 1994. Identification and characterization of a gene cluster mediating enteroaggregative *Escherichia coli* aggregative adherence fimbria I biogenesis. *J Bacteriol* **176:**4949–4957.

17. **Behrens M, Sheikh J, Nataro JP.** 2002. Regulation of the overlapping pic/set locus in *Shigella flexneri* and enteroaggregative *Escherichia coli*. *Infect Immun* **70:**2915–2925.

18. **Fasano A, Noriega FR, Liao FM, Wang W, Levine MM.** 1997. Effect of shigella enterotoxin 1 (ShET1) on rabbit intestine in vitro and in vivo. *Gut* **40:**505–511.

19. **Gutierrez-Jimenez J, Arciniega I, Navarro-Garcia F.** 2008. The serine protease motif of Pic mediates a dose-dependent mucolytic activity after binding to sugar constituents of the mucin substrate. *Microb Pathog* **45:**115–123.

20. **Navarro-Garcia F, Gutierrez-Jimenez J, Garcia-Tovar C, Castro LA, Salazar-Gonzalez H, Cordova V.** 2010. Pic, an autotransporter protein secreted by different pathogens in the *Enterobacteriaceae* family, is a potent mucus secretagogue. *Infect Immun* **78:**4101–4109.

21. **Al-Hasani K, Navarro-Garcia F, Huerta J, Sakellaris H, Adler B.** 2009. The immunogenic SigA enterotoxin of *Shigella flexneri* 2a binds to HEp-2 cells and induces fodrin redistribution in intoxicated epithelial cells. *PLoS One* **4:**e8223.

22. **Benjelloun-Touimi Z, Sansonetti PJ, Parsot C.** 1995. SepA, the major extracellular protein of *Shigella flexneri*: autonomous secretion and involvement in tissue invasion. *Mol Microbiol* **17:**123–135.

23. **Paton AW, Paton JC.** 1998. Detection and characterization of Shiga toxigenic *Escherichia coli* by using multiplex PCR assays for stx1, stx2, eaeA, enterohemorrhagic *E. coli* hlyA, rfbO111, and rfbO157. *J Clin Microbiol* **36:**598–602.

24. **Scheutz F, Teel LD, Beutin L, Pierard D, Buvens G, Karch H, Mellmann A, Caprioli A, Tozzoli R, Morabito S, Strockbine NA, Melton-Celsa AR, Sanchez M, Persson S, O'Brien AD.** 2005. Multicenter evaluation of a sequence-based

protocol for subtyping Shiga toxins and standardizing Stx nomenclature. *J Clin Microbiol* **50**:2951–263.

25. **Friedrich AW, Bielaszewska M, Zhang WL, Pulz M, Kuczius T, Ammon A, Karch H.** 2002. *Escherichia coli* harboring Shiga toxin 2 gene variants: frequency and association with clinical symptoms. *J Infect Dis* **185**:74–84.

26. **Bielaszewska M, Friedrich AW, Aldick T, Schurk-Bulgrin R, Karch H.** 2006. Shiga toxin activatable by intestinal mucus in *Escherichia coli* isolated from humans: predictor for a severe clinical outcome. *Clin Infect Dis* **43**:1160–1167.

27. **Mey AR, Wyckoff EE, Oglesby AG, Rab E, Taylor RK, Payne SM.** 2002. Identification of the *Vibrio cholerae* enterobactin receptors VctA and IrgA: IrgA is not required for virulence. *Infect Immun* **70**:3419–3426.

28. **Tarr PI, Bilge SS, Vary JC Jr, Jelacic S, Habeeb RL, Ward TR, Baylor MR, Besser TE.** 2000. Iha: a novel *Escherichia coli* O157:H7 adherence-conferring molecule encoded on a recently acquired chromosomal island of conserved structure. *Infect Immun* **68**:1400–1407.

29. **Leveille S, Caza M, Johnson JR, Clabots C, Sabri M, Dozois CM.** 2006. Iha from an *Escherichia coli* urinary tract infection outbreak clonal group A strain is expressed in vivo in the mouse urinary tract and functions as a catecholate siderophore receptor. *Infect Immun* **74**:3427–3436.

30. **Pierard D, De Greve H, Haesebrouck F, Mainil J.** 2012. O157:H7 and O104:H4 Vero/Shiga toxin-producing *Escherichia coli* outbreaks: respective role of cattle and humans. *Vet Res* **43**:13.

31. **Gyles CL.** 2007. Shiga toxin-producing *Escherichia coli*: an overview. *J Anim Sci* **85**:E45–E62.

32. **Gould LH, Demma L, Jones TF, Hurd S, Vugia DJ, Smith K, Shiferaw B, Segler S, Palmer A, Zansky S, Griffin PM.** 2009. Hemolytic uremic syndrome and death in persons with *Escherichia coli* O157:H7 infection, foodborne diseases active surveillance network sites, 2000–2006. *Clin Infect Dis* **49**:1480–1485.

33. **Kaper JB, Nataro JP, Mobley HL.** 2004. Pathogenic *Escherichia coli*. *Nat Rev Microbiol* **2**:123–140.

34. **Knutton S, Lloyd DR, McNeish AS.** 1987. Adhesion of enteropathogenic *Escherichia coli* to human intestinal enterocytes and cultured human intestinal mucosa. *Infect Immun* **55**:69–77.

35. **Elliott SJ, Wainwright LA, McDaniel TK, Jarvis KG, Deng YK, Lai LC, McNamara BP, Donnenberg MS, Kaper JB.** 1998. The complete sequence of the locus of enterocyte effacement (LEE) from enteropathogenic *Escherichia coli* E2348/69. *Mol Microbiol* **28**:1–4.

36. **Elliott SJ, Sperandio V, Giron JA, Shin S, Mellies JL, Wainwright L, Hutcheson SW, McDaniel TK, Kaper JB.** 2000. The locus of enterocyte effacement (LEE)-encoded regulator controls expression of both LEE- and non-LEE-encoded virulence factors in enteropathogenic and enterohemorrhagic *Escherichia coli*. *Infect Immun* **68**:6115–6126.

37. **Deng W, Li Y, Hardwidge PR, Frey EA, Pfuetzner RA, Lee S, Gruenheid S, Strynakda NC, Puente JL, Finlay BB.** 2005. Regulation of type III secretion hierarchy of translocators and effectors in attaching and effacing bacterial pathogens. *Infect Immun* **73**:2135–2146.

38. **Jerse AE, Yu J, Tall BD, Kaper JB.** 1990. A genetic locus of enteropathogenic *Escherichia coli* necessary for the production of attaching and effacing lesions on tissue culture cells. *Proc Natl Acad Sci USA* **87**:7839–7843.

39. **Abe A, de Grado M, Pfuetzner RA, Sanchez-Sanmartin C, Devinney R, Puente JL, Strynadka NC, Finlay BB.** 1999. Enteropathogenic *Escherichia coli* translocated intimin receptor, Tir, requires a specific chaperone for stable secretion. *Mol Microbiol* **33**:1162–1175.

40. **Elliott SJ, Hutcheson SW, Dubois MS, Mellies JL, Wainwright LA, Batchelor M, Frankel G, Knutton S, Kaper JB.** 1999. Identification of CesT, a chaperone for the type III secretion of Tir in enteropathogenic *Escherichia coli*. *Mol Microbiol* **33**:1176–1189.

41. **Kenny B, DeVinney R, Stein M, Reinscheid DJ, Frey EA, Finlay BB.** 1997. Enteropathogenic *E. coli* (EPEC) transfers its receptor for intimate adherence into mammalian cells. *Cell* **91**:511–520.

42. **Navarro-Garcia F, Serapio-Palacios A, Ugalde-Silva P, Tapia-Pastrana G, Chavez-Duenas L.** 2013. Actin cytoskeleton manipulation by effector proteins secreted by diarrheagenic *Escherichia coli* pathotypes. *Biomed Res Int* **2013**:374395.

43. **Karmali MA.** 2003. The medical significance of Shiga toxin-producing *Escherichia coli* infections. An overview. *Methods Mol Med* **73**:1–7.

44. **Gal-Mor O, Finlay BB.** 2006. Pathogenicity islands: a molecular toolbox for bacterial virulence. *Cell Microbiol* **8**:1707–1719.

45. **Coombes BK, Gilmour MW, Goodman CD.** 2011. The evolution of virulence in non-O157 Shiga toxin-producing *Escherichia coli*. *Front Microbiol* **2**:90.

46. **Ju W, Shen J, Toro M, Zhao S, Meng J.** 2013. Distribution of pathogenicity islands OI-122, OI-43/48, OI-57 and a high-pathogenicity island (in Shiga toxin-producing *Escherichia coli*. *Appl Environ Microbiol* **79**:3406–3412.

47. **Schmidt H, Zhang WL, Hemmrich U, Jelacic S, Brunder W, Tarr PI, Dobrindt U, Hacker J,**

Karch H. 2001. Identification and characterization of a novel genomic island integrated at *sel C* in locus of enterocyte effacement-negative, Shiga toxin-producing *Escherichia coli*. *Infect Immun* **69**:6863–6873.

48. Newton HJ, Sloan J, Bulach DM, Seemann T, Allison CC, Tauschek M, Robins-Browne RM, Paton JC, Whittam TS, Paton AW, Hartland EL. 2009. Shiga toxin-producing *Escherichia coli* strains negative for locus of enterocyte effacement. *Emerg Infect Dis* **15**:372–380.

49. Xu X, McAteer SP, Tree JJ, Shaw DJ, Wolfson EB, Beatson SA, Roe AJ, Allison LJ, Chase-Topping ME, Mahajan A, Tozzoli R, Woolhouse ME, Morabito S, Gally DL. 2012. Lysogeny with Shiga toxin 2-encoding bacteriophages represses type III secretion in enterohemorrhagic *Escherichia coli*. *PLoS Pathog* **8**:e1002672.

50. Imamovic L, Jofre J, Schmidt H, Serra-Moreno R, Muniesa M. 2009. Phage-mediated Shiga toxin 2 gene transfer in food and water. *Appl Environ Microbiol* **75**:1764–1768.

51. Strockbine NA, Jackson MP, Sung LM, Holmes RK, O'Brien AD. 1988. Cloning and sequencing of the genes for Shiga toxin from *Shigella dysenteriae* type 1. *J Bacteriol* **170**:1116–1122.

52. Tesh VL. Induction of apoptosis by Shiga toxins. *Future Microbiol* **5**:431–453.

53. Tam P, Mahfoud R, Nutikka A, Khine AA, Binnington B, Paroutis P, Lingwood C. 2008. Differential intracellular transport and binding of verotoxin 1 and verotoxin 2 to globotriaosylceramide-containing lipid assemblies. *J Cell Physiol* **216**:750–763.

54. Chark D, Nutikka A, Trusevych N, Kuzmina J, Lingwood C. 2004. Differential carbohydrate epitope recognition of globotriaosyl ceramide by verotoxins and a monoclonal antibody. *Eur J Biochem* **271**:405–417.

55. Rutjes NW, Binnington BA, Smith CR, Maloney MD, Lingwood CA. 2002. Differential tissue targeting and pathogenesis of verotoxins 1 and 2 in the mouse animal model. *Kidney Int* **62**:832–845.

56. Lingwood CA. 1996. Role of verotoxin receptors in pathogenesis. *Trends Microbiol* **4**:147–153.

57. Okuda T, Tokuda N, Numata S, Ito M, Ohta M, Kawamura K, Wiels J, Urano T, Tajima O, Furukawa K. 2006. Targeted disruption of Gb3/CD77 synthase gene resulted in the complete deletion of globo-series glycosphingolipids and loss of sensitivity to verotoxins. *J Biol Chem* **281**:10230–10235.

58. Bast DJ, Banerjee L, Clark C, Read RJ, Brunton JL. 1999. The identification of three biologically relevant globotriaosyl ceramide receptor binding sites on the Verotoxin 1 B subunit. *Mol Microbiol* **32**:953–960.

59. Schweppe CH, Bielaszewska M, Pohlentz G, Friedrich AW, Buntemeyer H, Schmidt MA, Kim KS, Peter-Katalinic J, Karch H, Muthing J. 2008. Glycosphingolipids in vascular endothelial cells: relationship of heterogeneity in Gb3Cer/CD77 receptor expression with differential Shiga toxin 1 cytotoxicity. *Glycoconj J* **25**:291–304.

60. Romer W, Berland L, Chambon V, Gaus K, Windschiegl B, Tenza D, Aly MR, Fraisier V, Florent JC, Perrais D, Lamaze C, Raposo G, Steinem C, Sens P, Bassereau P, Johannes L. 2007. Shiga toxin induces tubular membrane invaginations for its uptake into cells. *Nature* **450**:670–675.

61. Falguieres T, Mallard F, Baron C, Hanau D, Lingwood C, Goud B, Salamero J, Johannes L. 2001. Targeting of Shiga toxin B-subunit to retrograde transport route in association with detergent-resistant membranes. *Mol Biol Cell* **12**:2453–2468.

62. Spilsberg B, Llorente A, Sandvig K. 2007. Polyunsaturated fatty acids regulate Shiga toxin transport. *Biochem Biophys Res Commun* **364**:283–288.

63. Mahfoud R, Manis A, Binnington B, Ackerley C, Lingwood CA. 2010. A major fraction of glycosphingolipids in model and cellular cholesterol-containing membranes is undetectable by their binding proteins. *J Biol Chem* **285**:36049–36059.

64. Lingwood CA, Binnington B, Manis A, Branch DR. 2010. Globotriaosyl ceramide receptor function – where membrane structure and pathology intersect. *FEBS Lett* **584**:1879–1886.

65. Boyd B, Lingwood C. 1989. Verotoxin receptor glycolipid in human renal tissue. *Nephron* **51**:207–210.

66. Lingwood CA. 1994. Verotoxin-binding in human renal sections. *Nephron* **66**:21–28.

67. Pudymaitis A, Armstrong G, Lingwood CA. 1991. Verotoxin-resistant cell clones are deficient in the glycolipid globotriosylceramide: differential basis of phenotype. *Arch Biochem Biophys* **286**:448–452.

68. Sandvig K, Garred O, Prydz K, Kozlov JV, Hansen SH, van Deurs B. 1992. Retrograde transport of endocytosed Shiga toxin to the endoplasmic reticulum. *Nature* **358**:510–512.

69. Sandvig K, Bergan J, Dyve AB, Skotland T, Torgersen ML. 2010. Endocytosis and retrograde transport of *Shiga toxin*. *Toxicon* **56**:1181–1185.

70. McCluskey AJ, Poon GM, Bolewska-Pedyczak E, Srikumar T, Jeram SM, Raught B, Gariepy J. 2008. The catalytic subunit of Shiga-like toxin 1 interacts with ribosomal stalk proteins and is inhibited by their conserved C-terminal domain. *J Mol Biol* **378**:375–386.

71. **Endo Y, Tsurugi K, Yutsudo T, Takeda Y, Ogasawara T, Igarashi K.** 1988. Site of action of a Vero toxin (VT2) from *Escherichia coli* O157:H7 and of Shiga toxin on eukaryotic ribosomes. RNA N-glycosidase activity of the toxins. *Eur J Biochem* **171**:4–50.

72. **Obrig TG, Moran TP, Brown JE.** 1987. The mode of action of Shiga toxin on peptide elongation of eukaryotic protein synthesis. *Biochem J* **244**:287–294.

73. **Obrig TG, Del Vecchio PJ, Brown JE, Moran TP, Rowland BM, Judge TK, Rothman SW.** 1988. Direct cytotoxic action of Shiga toxin on human vascular endothelial cells. *Infect Immun* **56**:2373–2378.

74. **Jandhyala DM, Ahluwalia A, Obrig T, Thorpe CM.** 2008. ZAK: a MAP3Kinase that transduces Shiga toxin- and ricin-induced proinflammatory cytokine expression. *Cell Microbiol* **10**:1468–1477.

75. **Iordanov MS, Pribnow D, Magun JL, Dinh TH, Pearson JA, Chen SL, Magun BE.** 1997. Ribotoxic stress response: activation of the stress-activated protein kinase JNK1 by inhibitors of the peptidyl transferase reaction and by sequence-specific RNA damage to the alpha-sarcin/ricin loop in the 28S rRNA. *Mol Cell Biol* **17**:3373–3381.

76. **Walchli S, Aasheim HC, Skanland SS, Spilsberg B, Torgersen ML, Rosendal KR, Sandvig K.** 2009. Characterization of clathrin and Syk interaction upon Shiga toxin binding. *Cell Signal* **21**:1161–1168.

77. **Nakao H, Takeda T.** 2000. *Escherichia coli* Shiga toxin. *J Nat Toxins* **9**:299–313.

78. **Orth D, Grif K, Khan AB, Naim A, Dierich MP, Wurzner R.** 2007. The Shiga toxin genotype rather than the amount of Shiga toxin or the cytotoxicity of Shiga toxin in vitro correlates with the appearance of the hemolytic uremic syndrome. *Diagn Microbiol Infect Dis* **59**:235–242.

79. **Scheiring J, Andreoli SP, Zimmerhackl LB.** 2008. Treatment and outcome of Shiga-toxin-associated hemolytic uremic syndrome (HUS). *Pediatr Nephrol* **23**:1749–1760.

80. **Allison HE.** 2007. Stx-phages: drivers and mediators of the evolution of STEC and STEC-like pathogens. *Future Microbiol* **2**:165–174.

81. **Herold S, Karch H, Schmidt H.** 2004. Shiga toxin-encoding bacteriophages—genomes in motion. *Int J Med Microbiol* **294**:115–121.

82. **Fuller CA, Pellino CA, Flagler MJ, Strasser JE, Weiss AA.** Shiga toxin subtypes display dramatic differences in potency. *Infect Immun* **79**:1329–1337.

83. **de Sablet T, Bertin Y, Vareille M, Girardeau JP, Garrivier A, Gobert AP, Martin C.** 2008. Differential expression of stx2 variants in Shiga toxin-producing *Escherichia coli* belonging to seropathotypes A and C. *Microbiology* **154**:176–186.

84. **Zhang X, McDaniel AD, Wolf LE, Keusch GT, Waldor MK, Acheson DW.** 2000. Quinolone antibiotics induce Shiga toxin-encoding bacteriophages, toxin production, and death in mice. *J Infect Dis* **181**:664–670.

85. **Wong CS, Jelacic S, Habeeb RL, Watkins SL, Tarr PI.** 2000. The risk of the hemolytic-uremic syndrome after antibiotic treatment of *Escherichia coli* O157:H7 infections. *N Engl J Med* **342**:1930–1936.

86. **Johnson JR, Russo TA, Tarr PI, Carlino U, Bilge SS, Vary JC Jr, Stell AL.** 2000. Molecular epidemiological and phylogenetic associations of two novel putative virulence genes, *iha* and *iroN* (E. coli), among *Escherichia coli* isolates from patients with urosepsis. *Infect Immun* **68**:3040–3047.

87. **Toma C, Martinez Espinosa E, Song T, Miliwebsky E, Chinen I, Iyoda S, Iwanaga M, Rivas M.** 2004. Distribution of putative adhesins in different seropathotypes of Shiga toxin-producing *Escherichia coli*. *J Clin Microbiol* **42**:4937–4946.

88. **Johnson JR, Jelacic S, Schoening LM, Clabots C, Shaikh N, Mobley HL, Tarr PI.** 2005. The IrgA homologue adhesin Iha is an *Escherichia coli* virulence factor in murine urinary tract infection. *Infect Immun* **73**:965–971.

89. **Postle K, Kadner RJ.** 2003. Touch and go: tying TonB to transport. *Mol Microbiol* **49**:869–882.

90. **Rashid RA, Tarr PI, Moseley SL.** 2006. Expression of the *Escherichia coli* IrgA homolog adhesin is regulated by the ferric uptake regulation protein. *Microb Pathog* **41**:207–217.

91. **Litwin CM, Calderwood SB.** 1993. Role of iron in regulation of virulence genes. *Clin Microbiol Rev* **6**:137–149.

92. **Touati D.** 2000. Iron and oxidative stress in bacteria. *Arch Biochem Biophys* **373**:1–6.

93. **Herold S, Paton JC, Srimanote P, Paton AW.** 2009. Differential effects of short-chain fatty acids and iron on expression of iha in Shiga-toxigenic *Escherichia coli*. *Microbiology* **155**:3554–3563.

94. **Ewers C, Li G, Wilking H, Kiessling S, Alt K, Antao EM, Laturnus C, Diehl I, Glodde S, Homeier T, Bohnke U, Steinruck H, Philipp HC, aWieler LH.** 2007. Avian pathogenic, uropathogenic, and newborn meningitis-causing *Escherichia coli*: how closely related are they? *Int J Med Microbiol* **297**:163–176.

95. **Ons E, Bleyen N, Tuntufye HN, Vandemaele F, Goddeeris BM.** 2007. High prevalence iron receptor genes of avian pathogenic *Escherichia coli*. *Avian Pathol* **36**:411–414.

96. **Rodriguez-Siek KE, Giddings CW, Doetkott C, Johnson TJ, Fakhr MK, Nolan LK.** 2005. Comparison of *Escherichia coli* isolates implicated in human urinary tract infection and avian colibacillosis. *Microbiology* **151**:2097–2110.

97. **Baumler AJ, Heffron F.** 1995. Identification and sequence analysis of *lpfABCDE*, a putative fimbrial operon of *Salmonella typhimurium*. *J Bacteriol* **177**:2087–2097.

98. **Torres AG, Kanack KJ, Tutt CB, Popov V, Kaper JB.** 2004. Characterization of the second long polar (LP) fimbriae of *Escherichia coli* O157:H7 and distribution of LP fimbriae in other pathogenic *E. coli* strains. *FEMS Microbiol Lett* **238**:333–344.

99. **Jordan DM, Cornick N, Torres AG, Dean-Nystrom EA, Kaper JB, Moon HW.** 2004. Long polar fimbriae contribute to colonization by *Escherichia coli* O157:H7 in vivo. *Infect Immun* **72**:6168–6171.

100. **Torres AG, Milflores-Flores L, Garcia-Gallegos JG, Patel SD, Best A, La Ragione RM, Martinez-Laguna Y, Woodward MJ.** 2007. Environmental regulation and colonization attributes of the long polar fimbriae (LPF) of *Escherichia coli* O157:H7. *Int J Med Microbiol* **297**:177–185.

101. **Torres AG, Giron JA, Perna NT, Burland V, Blattner FR, Avelino-Flores F, Kaper JB.** 2002. Identification and characterization of *lpfABCC'DE*, a fimbrial operon of enterohemorrhagic *Escherichia coli* O157:H7. *Infect Immun* **70**:5416–5427.

102. **Farfan MJ, Cantero L, Vidal R, Botkin DJ, Torres AG.** 2011. Long polar fimbriae of enterohemorrhagic *Escherichia coli* O157:H7 bind to extracellular matrix proteins. *Infect Immun* **79**:3744–3750.

103. **Torres AG, Kaper JB.** 2003. Multiple elements controlling adherence of enterohemorrhagic *Escherichia coli* O157:H7 to HeLa cells. *Infect Immun* **71**:4985–4995.

104. **Doughty S, Sloan J, Bennett-Wood V, Robertson M, Robins-Browne RM, Hartland EL.** 2002. Identification of a novel fimbrial gene cluster related to long polar fimbriae in locus of enterocyte effacement-negative strains of enterohemorrhagic *Escherichia coli*. *Infect Immun* **70**:6761–6769.

105. **Lloyd SJ, Ritchie JM, Torres AG.** 2012. Fimbriation and curliation in *Escherichia coli* O157:H7: a paradigm of intestinal and environmental colonization. *Gut Microbes* **3**:272–276.

106. **Bockemuhl J, Aleksic S, Karch H.** 1992. Serological and biochemical properties of Shiga-like toxin (verocytotoxin)-producing strains of *Escherichia coli*, other than O-group 157, from patients in Germany. *Zentralbl Bakteriol* **276**:189–195.

107. **Mellmann A, Lu S, Karch H, Xu JG, Harmsen D, Schmidt MA, Bielaszewska M.** 2008. Recycling of Shiga toxin 2 genes in sorbitol-fermenting enterohemorrhagic *Escherichia coli* O157:NM. *Appl Environ Microbiol* **74**:67–72.

108. **Boudailliez B, Berquin P, Mariani-Kurkdjian P, Ilef D, Cuvelier B, Capek I, Tribout B, Bingen E, Piussan C.** 1997. Possible person-to-person transmission of *Escherichia coli* O111–associated hemolytic uremic syndrome. *Pediatr Nephrol* **11**:36–39.

109. **Morabito S, Karch H, Mariani-Kurkdjian P, Schmidt H, Minelli F, Bingen E, Caprioli A.** 1998. Enteroaggregative, Shiga toxin-producing *Escherichia coli* O111:H2 associated with an outbreak of hemolytic-uremic syndrome. *J Clin Microbiol* **36**:840–842.

110. **Willshaw GA, Scotland SM, Smith HR, Rowe B.** 1992. Properties of Vero cytotoxin-producing *Escherichia coli* of human origin of O serogroups other than O157. *J Infect Dis* **166**:797–802.

111. **Iyoda S, Terajima J, Wada A, Izumiya H, Tamura K, Watanabe H.** 2000. Molecular epidemiology of enterohemorrhagic *Escherichia coli*. *Nihon Saikingaku Zasshi* **55**:29–36.

112. **Estrada-Garcia T, Navarro-Garcia F.** 2012. Enteroaggregative *Escherichia coli* pathotype: a genetically heterogeneous emerging foodborne enteropathogen. *FEMS Immunol Med Microbiol* **66**:281–298.

113. **Smith HR, Cheasty T, Rowe B.** 1997. Enteroaggregative *Escherichia coli* and outbreaks of gastroenteritis in UK. *Lancet* **350**:814–815.

114. **Tompkins DS, Hudson MJ, Smith HR, Eglin RP, Wheeler JG, Brett MM, Owen RJ, Brazier JS, Cumberland P, King V, Cook PE.** 1999. A study of infectious intestinal disease in England: microbiological findings in cases and controls. *Commun Dis Public Health* **2**:108–113.

115. **Pabst WL, Altwegg M, Kind C, Mirjanic S, Hardegger D, Nadal D.** 2003. Prevalence of enteroaggregative *Escherichia coli* among children with and without diarrhea in Switzerland. *J Clin Microbiol* **41**:2289–2293.

116. **Itoh Y, Nagano I, Kunishima M, Ezaki T.** 1997. Laboratory investigation of enteroaggregative *Escherichia coli* O untypeable:H10 associated with a massive outbreak of gastrointestinal illness. *J Clin Microbiol* **35**:2546–2550.

117. **Harrington SM, Dudley EG, Nataro JP.** 2006. Pathogenesis of enteroaggregative *Escherichia coli* infection. *FEMS Microbiol Lett* **254**:12–18.

118. **Czeczulin JR, Whittam TS, Henderson IR, Navarro-Garcia F, Nataro JP.** 1999. Phylogenetic analysis of enteroaggregative and diffusely adherent *Escherichia coli*. *Infect Immun* **67**:2692–2699.

119. Okeke IN, Wallace-Gadsden F, Simons HR, Matthews N, Labar AS, Hwang J, Wain J. 2010. Multi-locus sequence typing of entero-aggregative *Escherichia coli* isolates from Nigerian children uncovers multiple lineages. *PLoS One* **5:** e14093.

120. Gallegos MT, Michan C, Ramos JL. 1993. The XylS/AraC family of regulators. *Nucleic Acids Res* **21:**807–810.

121. Jiang ZD, Greenberg D, Nataro JP, Steffen R, DuPont HL. 2002. Rate of occurrence and pathogenic effect of enteroaggregative *Escherichia coli* virulence factors in international travelers. *J Clin Microbiol* **40:**4185–4190.

122. Huang DB, Mohamed JA, Nataro JP, DuPont HL, Jiang ZD, Okhuysen PC. 2007. Virulence characteristics and the molecular epidemiology of enteroaggregative *Escherichia coli* isolates from travellers to developing countries. *J Med Microbiol* **56:**1386–1392.

123. Sheikh J, Czeczulin JR, Harrington S, Hicks S, Henderson IR, Le Bouguenec C, Gounon P, Phillips A, Nataro JP. 2002. A novel dispersin protein in enteroaggregative *Escherichia coli*. *J Clin Invest* **110:**1329–1337.

124. Nishi J, Sheikh J, Mizuguchi K, Luisi B, Burland V, Boutin A, Rose DJ, Blattner FR, Nataro JP. 2003. The export of coat protein from entero-aggregative *Escherichia coli* by a specific ATP-binding cassette transporter system. *J Biol Chem* **278:**45680–45689.

125. Dudley EG, Thomson NR, Parkhill J, Morin NP, Nataro JP. 2006. Proteomic and microarray characterization of the AggR regulon identifies a pheU pathogenicity island in enteroaggregative *Escherichia coli*. *Mol Microbiol* **61:**1267–1282.

126. Knutton S, Shaw RK, Bhan MK, Smith HR, McConnell MM, Cheasty T, Williams PH, Baldwin TJ. 1992. Ability of enteroaggregative *Escherichia coli* strains to adhere in vitro to human intestinal mucosa. *Infect Immun* **60:**2083–2091.

127. Suzart S, Guth BE, Pedroso MZ, Okafor UM, Gomes TA. 2001. Diversity of surface structures and virulence genetic markers among entero-aggregative *Escherichia coli* (EAEC) strains with and without the EAEC DNA probe sequence. *FEMS Microbiol Lett* **201:**163–168.

128. Vial PA, Robins-Browne R, Lior H, Prado V, Kaper JB, Nataro JP, Maneval D, Elsayed A, Levine MM. 1988. Characterization of entero-adherent-aggregative *Escherichia coli*, a putative agent of diarrheal disease. *J Infect Dis* **158:**70–79.

129. Czeczulin JR, Balepur S, Hicks S, Phillips A, Hall R, Kothary MH, Navarro-Garcia F, Nataro JP. 1997. Aggregative adherence fimbria II, a second fimbrial antigen mediating aggregative adherence in enteroaggregative *Escherichia coli*. *Infect Immun* **65:**4135–4145.

130. Velarde JJ, Varney KM, Inman KG, Farfan M, Dudley E, Fletcher J, Weber DJ, Nataro JP. 2007. Solution structure of the novel dispersin protein of enteroaggregative *Escherichia coli*. *Mol Microbiol* **66:**1123–1135.

131. Rossiter AE, Browning DF, Leyton DL, Johnson MD, Godfrey RE, Wardius CA, Desvaux M, Cunningham AF, Ruiz-Perez F, Nataro JP, Busby SJ, and Henderson IR. 2011. Transcription of the plasmid-encoded toxin gene from enteroaggregative *Escherichia coli* is regulated by a novel co-activation mechanism involving CRP and Fis. *Mol Microbiol* **81:**179–191.

132. Morin N, Santiago AE, Ernst RK, Guillot SJ, Nataro JP. 2013. Characterization of the AggR regulon in enteroaggregative *Escherichia coli*. *Infect Immun* **81:**122–132.

133. Huang DB, DuPont HL, Jiang ZD, Carlin L, Okhuysen PC. 2004. Interleukin-8 response in an intestinal HCT-8 cell line infected with enteroaggregative and enterotoxigenic *Escherichia coli*. *Clin Diagn Lab Immunol* **11:**548–551.

134. Harrington SM, Strauman MC, Abe CM, Nataro JP. 2005. Aggregative adherence fimbriae contribute to the inflammatory response of epithelial cells infected with enteroaggregative *Escherichia coli*. *Cell Microbiol* **7:**1565–1578.

135. Henderson IR, Navarro-Garcia F, Desvaux M, Fernandez RC, Ala'Aldeen D. 2004. Type V protein secretion pathway: the autotransporter story. *Microbiol Mol Biol Rev* **68:**692–744.

136. Navarro-Garcia F, Elias WP. 2011. Autotransporters and virulence of enteroaggregative *E. coli*. *Gut Microbes* **2:**13–24.

137. Dautin N, Bernstein HD. 2007. Protein secretion in gram-negative bacteria via the autotransporter pathway. *Annu Rev Microbiol* **61:**89–112.

138. Dutta S, Lalitha PV, Ware LA, Barbosa A, Moch JK, Vassell MA, Fileta BB, Kitov S, Kolodny N, Heppner DG, Haynes JD, Lanar DE. 2002. Purification, characterization, and im-munogenicity of the refolded ectodomain of the *Plasmodium falciparum* apical membrane antigen 1 expressed in *Escherichia coli*. *Infect Immun* **70:**3101–3110.

139. Canizalez-Roman A, Navarro-Garcia F. 2003. Fodrin CaM-binding domain cleavage by Pet from enteroaggregative *Escherichia coli* leads to actin cytoskeletal disruption. *Mol Microbiol* **48:**947–958.

140. Navarro-Garcia F, Canizalez-Roman A, Luna J, Sears C, Nataro JP. 2001. Plasmid-encoded toxin of enteroaggregative *Escherichia coli* is internalized by epithelial cells. *Infect Immun* **69:**1053–1060.

141. **Henderson IR, Czeczulin J, Eslava C, Noriega F, Nataro JP.** 1999. Characterization of *pic*, a secreted protease of *Shigella flexneri* and enteroaggregative *Escherichia coli*. *Infect Immun* **67:** 5587–5596.

142. **Harrington SM, Sheikh J, Henderson IR, Ruiz-Perez F, Cohen PS, Nataro JP.** 2009. The Pic protease of enteroaggregative *Escherichia coli* promotes intestinal colonization and growth in the presence of mucin. *Infect Immun* **77:**2465–2473.

143. **Boisen N, Ruiz-Perez F, Scheutz F, Krogfelt KA, Nataro JP.** 2009. Short report: high prevalence of serine protease autotransporter cytotoxins among strains of enteroaggregative *Escherichia coli*. *Am J Trop Med Hyg* **80:**294–301.

144. **Rajakumar K, Sasakawa C, Adler B.** 1997. Use of a novel approach, termed island probing, identifies the *Shigella flexneri* she pathogenicity island which encodes a homolog of the immunoglobulin A protease-like family of proteins. *Infect Immun* **65:**4606–4614.

145. **Ruiz-Perez F, Wahid R, Faherty CS, Kolappaswamy K, Rodriguez L, Santiago A, Murphy E, Cross A, Sztein MB, Nataro JP.** 2011. Serine protease autotransporters from *Shigella flexneri* and pathogenic *Escherichia coli* target a broad range of leukocyte glycoproteins. *Proc Natl Acad Sci USA* **108:**12881–12886.

146. **Provence DL, Curtiss R 3rd.** 1994. Isolation and characterization of a gene involved in hemagglutination by an avian pathogenic *Escherichia coli* strain. *Infect Immun* **62:**1369–1380.

147. **Benjelloun-Touimi Z, Si Tahar M, Montecucco C, Sansonetti PJ, Parsot C.** 1998. SepA, the 110 kDa protein secreted by *Shigella flexneri*: two-domain structure and proteolytic activity. *Microbiology* **144(Pt 7):**1815–1822.

148. **Al-Hasani K, Henderson IR, Sakellaris H, Rajakumar K, Grant T, Nataro JP, Robins-Browne R, Adler B.** 2000. The sigA gene which is borne on the she pathogenicity island of *Shigella flexneri* 2a encodes an exported cytopathic protease involved in intestinal fluid accumulation. *Infect Immun* **68:**2457–2463.

149. **Navarro-Garcia F, Sears C, Eslava C, Cravioto A, Nataro JP.** 1999. Cytoskeletal effects induced by pet, the serine protease enterotoxin of enteroaggregative *Escherichia coli*. *Infect Immun* **67:** 2184–2192.

150. **Navarro-Garcia F, Canizalez-Roman A, Burlingame KE, Teter K, Vidal JE.** 2007. Pet, a non-AB toxin, is transported and translocated into epithelial cells by a retrograde trafficking pathway. *Infect Immun* **75:**2101–2109.

151. **Navarro-Garcia F, Canizalez-Roman A, Vidal JE, Salazar MI.** 2007. Intoxication of epithelial cells by plasmid-encoded toxin requires clathrin-mediated endocytosis. *Microbiology* **153:**2828–2838.

152. **Fasano A, Noriega FR, Maneval DR Jr, Chanasongcram S, Russell R, Guandalini S, Levine MM.** 1995. *Shigella* enterotoxin 1: an enterotoxin of *Shigella flexneri* 2a active in rabbit small intestine in vivo and in vitro. *J Clin Invest* **95:**2853–2861.

THE WAY FORWARD

The Way Forward

27

VANESSA SPERANDIO[1]

RESEARCH INTO THE MOLECULAR MECHANISMS OF EHEC VIRULENCE

We have observed a vertical leap in our understanding of EHEC's virulence. In the past edition of this book, the locus of enterocyte effacement (LEE) and its encoded type 3 secretion system (T3SS) had been recently discovered (1, 2). However, few effectors were known at the time, with Tir (3) and intimin (4) dominating research on the molecular mechanisms involved in the formation of attaching and effacing (AE) lesions. Structural insights into the T3SS came later, with the description of the EscF needle (5) and the EspA filament (6) forming the unique translocon of the EHEC and EPEC T3SSs. The number of effectors quickly expanded from the six LEE-encoded effectors, to the first hints that effectors encoded outside of the LEE existed (7), to the large expansion of their repertoire (8). Next-generation sequencing of many EHEC genomes also highlighted the fact that different strains of EHEC and enteropathogenic *E. coli* (EPEC) carry different combinations of these effectors (9). Vigorous research was initially devoted to understanding the mechanism through which EHEC engaged the actin cytoskeleton to form AE lesions. These studies involved Tir and intimin interactions, but also extensive studies on the EspFu/TccP effectors (3, 10–16). More recently, studies of non-AE-related effectors and their role in

[1]Department of Microbiology and Department of Biochemistry, The University of Texas Southwestern Medical Center, Dallas, TX 75390.

Enterohemorrhagic Escherichia coli *and Other Shiga Toxin-Producing* E. coli
Edited by Vanessa Sperandio and Carolyn J. Hovde
© 2015 American Society for Microbiology, Washington, DC
doi:10.1128/9781555818791.ch27

more discrete actin rearrangements, as well as in modulation of the host immune response, have taken the front seat (17–24). Looking forward, we need to understand how different combinations of T3SS effectors affect the virulence potential of EHEC strains. We are also starting to study the hierarchy of secretion of these effectors (25, 26) and how they work in concert. Knowledge of which effectors are acting within a mammalian cell at any given time, and how their functions amplify or antagonize their phenotypes, is the next frontier in understanding the role of these proteins in the bacterial/host interplay. Another still unresolved issue is how the T3SS is regulated to shift from secreting the translocon proteins (EspA, EspB, and EspD) to secreting *bona fide* effectors within epithelial cells. There is also the question of how the EspA filament is disassembled during the infection process to allow the close contact between the bacteria and the host. Finally, the big question that remains open is, how does EHEC cause diarrhea in the human intestine, and which are the main players in this disease process?

An ongoing puzzle that remains largely unanswered is how EHEC initially adheres to the host enterocytes. Is this adherence mediated by the T3SS *per se*, and if so, what is the receptor? Would this implicate the EspA filament and maybe EspB and EspD? Conversely, could this adherence be mediated by other fimbrial adhesins? EHEC does encode for several adhesins in its genome; certain adhesins such as the long polar fimbriae (Lpf) have been implicated in tissue tropism, and other structures have been shown to contribute to cattle carriage of this pathogen (27). However, a clear picture of the contribution of these adhesive structures to EHEC intestinal adherence and their role in intestinal tropism is still missing.

Although we still have many mysteries to solve in the intestinal phase of EHEC infection, we are also searching for many answers regarding Shiga toxin. The mechanism of action of the A subunit is known, but independently, Shiga toxin in combination with lipopolysaccharide has profound immune modulation properties that we don't completely understand (28). There is also not enough information on the central nervous system action of Shiga toxin, where research is still in its infancy although this is a key aspect of EHEC-mediated disease (29). We still have to deal with the mystery of why some patients develop Stx-mediated hemolytic uremic syndrome (HUS) and others don't, let alone explaining the age distribution usually associated with HUS. Could this be dependent on the host's Gb3 affinity to Shiga toxin? Is age distribution dependent on changes in the gastrointestinal (GI) microbiota? It is documented that usually children under 5 years of age are more susceptible to HUS. As it turns out, one's GI microbiota does not become stable until the age of 5 (30, 31). The influence of microbiota, and potentially probiotics, on Shiga toxin-mediated disease has again been at center stage. Fukuda et al. (32) reported that strains of *Bifidobacterium* (largely employed as probiotics) with enhanced ability to produce acetate can inhibit Shiga toxin translocation from the intestinal lumen to the blood, having a protective effect against EHEC infections. Hence, these studies beg the question, do differences in the microbiota composition change the host's susceptibility to EHEC infection? Which role does dysbiosis play in susceptibility to GI pathogens?

EHEC/MICROBIOTA/HOST RELATIONSHIPS

EHEC colonizes the human colon, where it encounters trillions of bacterial species that comprise the microbiota. EHEC "learned" to exploit signal and nutritional cues provided by this microbiota to time regulation of its virulence traits. EHEC senses the signal AI-3 produced by the microbiota, the levels of the carbon sources glucose and fucose, and the nitrogen source ethanolamine to fine-tune expression of the LEE genes (33–36). Cues from the microbiota also seem to affect the production of Shiga toxin (37) and, as mentioned above (32), the host's susceptibility to this toxin. Given the profound association that humans have with their GI microbiota (38) and the interplay between the microbial GI flora and enteric GI pathogens, several questions will populate future research. Is the microbiota protective against pathogens, or can it be exploited by pathogens to gauge their niche within the intestine? Does EHEC adapt itself metabolically to subvert competition for nutrients with the microbiota? Do differences in microbiota affect host susceptibility to EHEC infection? Indeed, there is a report suggesting that microbiota transplants from susceptible mice to *Citrobacter rodentium* (a murine surrogate model for EHEC infections) to resistant mice, and vice versa, changed host susceptibility to infection, suggesting the very exciting possibility that manipulations of the GI microbiota composition could have a protective effect against EHEC infections (39). Moreover, recent research suggested that virulence traits such as the T3SS are actually competition tools employed by pathogens to compete for colonization niches within the GI tract and that they are only required in the presence of the competing microbial flora (40). The Kamada study goes as far as suggesting an Achilles heel to GI infections, in which *C. rodentium* promotes overgrowth of *Gammaproteobacteria* during infection, which ultimately compete with sugar sources, eliminating the pathogen from the GI tract and functioning as a natural "probiotics treatment."

Besides sensing cues from the microbiota to colonize the human GI tract, EHEC also adapted to sense acyl-homoserine lactone (AHL) signals produced by the rumen microbiota to modulate gene expression toward successful colonization of its main reservoir, ruminants (41).

In addition to "reading" chemical cues provided by the GI microbiota, EHEC also senses chemical signals from the host, specifically the stress hormones epinephrine and norepinephrine. The sensing of these host

signals and the microbiota AI-3 converge at the biochemical level to the receptor QseC, which controls a complex virulence program within the EHEC cell (33, 42–44). However, there are many other human hormones present in the GI tract that can be exploited by EHEC as excellent cues towards successful host colonization. Consequently, one has to ask, which other human hormones can EHEC respond to? How will responses to different human hormones be integrated? Do medications that alter human signaling processes affect susceptibility to EHEC disease?

TREATMENT

The main worry surrounding EHEC infections is the development of HUS, which is caused by Shiga toxin (45). Because Shiga toxins are encoded by lysogenic phages and their expression is coupled to the phage entering its lytic cycle (46, 47), the use of antibiotics to treat EHEC infections is controversial, due to their potential induction of toxin expression. In fact, several studies demonstrated an increased risk for the development of HUS following antibiotic treatment (48, 49). However, some antibiotics do not seem to increase Shiga toxin release or increase the risk of HUS development (50). Nevertheless, treatment of EHEC infections with antibiotics is controversial, and the usual recommendation is to perform supportive treatment by giving isotonic saline to patients to avoid dehydration (51).

Translation of basic knowledge of EHEC pathogenesis allowed the investigation of new strategies to treat EHEC infections. Many potential new therapeutics have been developed, the majority of them designed to prevent toxin binding to its Gb3 receptor or to limit Gb3 receptor generation. There are also therapies that inhibit toxin trafficking, processing, or activity (reviewed in chapter 22 of this volume). Other therapeutics prevent Shiga toxin and LEE expression, or kill EHEC without promoting Shiga toxin release (reviewed in chapter 22). However, although many new therapies are being developed and are currently under different stages of testing (ranging from efficacy *in vitro*, to animal models, to clinical trials), no unified new therapeutic to treat EHEC infections is currently used in clinics. Consequently, the full development of novel therapies to treat EHEC infections and either prevent or ameliorate HUS is still one of the biggest challenges in the field.

RESERVOIRS

Although many recent outbreaks have been associated with produce, the main reservoir for EHEC is ruminants. Much effort has been put forth to control infection of meat products by EHEC within abattoirs. Also, many vaccines have been developed to control EHEC colonization of cattle (reviewed in chapters 26 to 31). Complete eradication of EHEC from its environmental reservoir is quite a challenge! However, it is foreseeable that controlling its presence in its environmental reservoir will lead to fewer outbreaks.

STEC is part of the normal intestinal microbiological flora of ruminants. Its extent and diversity are staggering, with more than 400 STEC serotypes identified as transiently colonizing cattle. The 1994 ruling by the U.S. Department of Agriculture-Food Safety and Inspection Service of *E. coli* O157:H7 as an adulterant in raw ground beef, and the subsequent 2011 designations of O26, O45, O103, O111, O121, and O145 as adulterants in certain fresh beef products, are more than challenging for the industry and will have economic consequences. The adulterant status of these STEC strains does not seem fully scientifically based, since serotype does not equal pathogen and other more prevalent and more virulent bacteria, such as *Salmonella*, have not been designated as adulterants in fresh product. To meet this challenge, improved accurate, rapid methodologies to test fresh products (meat and vegetables) for EHEC will be required.

Nonetheless, investigations of STEC in animal reservoirs, in the farm environment, and in foods all have great potential to improve human health. Several of the remaining big issues in this area of research include understanding the following:

- the role of Shiga toxin in the bovine gastrointestinal tract and in the environment
- the carriage of EHEC by cattle and the impact of the competing microbiota, immune response, and molecular mechanisms of microbial attachment to mucosa, hide, water troughs, etc.
- the survival of EHEC in the farm environment and farm management practices that impact environmental reservoirs (water, feed, birds, etc.)
- seasonal prevalence (summer for *E. coli* O157:H7, but spring and fall for the other six EHEC)
- effective preharvest and processing interventions

EPIDEMIOLOGY

EHEC outbreaks occur throughout the globe. The serotype most prevalent in the U.S. is usually O157:H7; however, there is a steep increase in outbreaks caused by non-O157 serotypes (reviewed in chapters 11, 16, and 23 of this volume). Additionally, the rapid evolution of *E. coli* is responsible for the quick rise of more virulent or different types of STEC. The O157 spinach outbreak strain had two copies of Shiga toxin, leading to a higher percentage of HUS cases than usual in this outbreak. The German O104:H4 STEC strains are more closely related to enteraggregative *E. coli* (EAEC) than to EHEC and evolved from EAEC acquiring a Stx prophage (chapters 32 and 33). Epidemiological surveys generally include serotyping, detection of Shiga toxin by various methods (enzyme-linked immunosorbent assays, PCR, etc.), multiplex PCRs including defined sets of virulence factors, and restriction fragment length polymorphisms

(RFLPs). The speed and accessibility of next-generation sequencing are revolutionizing and increasing the speed with which outbreak strains are identified. A striking example is the German outbreak where, through international collaboration among sequencing and epidemiology laboratories and real-time public genomic data release, the outbreak strain was quickly identified. Although next-generation sequencing-based epidemiology is powerful, several challenges exist to its worldwide implementation on EHEC epidemiological surveys, especially in developing nations, where the cost of this technology is still a significant hurdle. Moreover, a real bottleneck is the bioinformatics power needed to process these large genomic datasets.

CONCLUDING REMARKS

As a community, EHEC researchers, public health officials, medical doctors, veterinarians, and food safety experts are living in exciting and revolutionary times. Our understanding of EHEC virulence and evolution has dramatically increased, leading to designs of new potential therapies, vaccines, and detection methods. Globalization and the speed of information sharing through the press and social networks have led to faster identification and control of outbreaks.

REFERENCES

1. **McDaniel TK, Jarvis KG, Donnenberg MS, Kaper JB.** 1995. A genetic locus of enterocyte effacement conserved among diverse enterobacterial pathogens. *Proc Natl Acad Sci USA* **92:**1664–1668.

2. **Jarvis KG, Giron JA, Jerse AE, McDaniel TK, Donnenberg MS, Kaper JB.** 1995. Enteropathogenic *Escherichia coli* contains a putative type III secretion system necessary for the export of proteins involved in attaching and effacing lesion formation. *Proc Natl Acad Sci USA* **92:**7996–8000.

3. **Kenny B, DeVinney R, Stein M, Reinscheid DJ, Frey EA, Finlay BB.** 1997. Enteropathogenic *E. coli* (EPEC) transfers its receptor for intimate adherence into mammalian cells. *Cell* **91:**511–520.

4. **Jerse AE, Yu J, Tall BD, Kaper JB.** 1990. A genetic locus of enteropathogenic *Escherichia coli* necessary for the production of attaching and effacing lesions on tissue culture cells. *Proc Natl Acad Sci USA* **87:**7839–7843.

5. **Sekiya K, Ohishi M, Ogino T, Tamano K, Sasakawa C, Abe A.** 2001. Supermolecular structure of the enteropathogenic *Escherichia coli* type III secretion system and its direct interaction with the EspA-sheath-like structure. *Proc Natl Acad Sci USA* **98:**11638–11643.

6. **Knutton S, Rosenshine I, Pallen MJ, Nisan I, Neves BC, Bain C, Wolff C, Dougan G, Frankel G.** 1998. A novel EspA-associated surface organelle of enteropathogenic *Escherichia coli* involved in protein translocation into epithelial cells. *EMBO J* **17:**2166–2176.

7. **Deng W, Puente JL, Gruenheid S, Li Y, Vallance BA, Vazquez A, Barba J, Ibarra JA, O'Donnell P, Metalnikov P, Ashman K, Lee S, Goode D, Pawson T, Finlay BB.** 2004. Dissecting virulence: systematic and functional analyses of a pathogenicity island. *Proc Natl Acad Sci USA* **101:**3597–3602.

8. **Tobe T, Beatson SA, Taniguchi H, Abe H, Bailey CM, Fivian A, Younis R, Matthews S, Marches O, Frankel G, Hayashi T, Pallen MJ.** 2006. An extensive repertoire of type III secretion effectors in *Escherichia coli* O157 and the role of lambdoid phages in their dissemination. *Proc Natl Acad Sci USA* **103:**14941–14946.

9. **Hazen TH, Sahl JW, Fraser CM, Donnenberg MS, Scheutz F, Rasko DA.** 2013. Refining the pathovar paradigm via phylogenomics of the attaching and effacing *Escherichia coli*. *Proc Natl Acad Sci USA* **110:**12810–12815.

10. **de Grado M, Abe A, Gauthier A, Steele-Mortimer O, DeVinney R, Finlay BB.** 1999. Identification of the intimin-binding domain of Tir of enteropathogenic *Escherichia coli*. *Cell Microbiol* **1:**7–17.

11. **DeVinney R, Stein M, Reinscheid D, Abe A, Ruschkowski S, Finlay BB.** 1999. Enterohemorrhagic *Escherichia coli* O157:H7 produces Tir, which is translocated to the host cell membrane but is not tyrosine phosphorylated. *Infect Immun* **67:**2389–2398.

12. **Luo Y, Frey EA, Pfuetzner RA, Creagh AL, Knoechel DG, Haynes CA, Finlay BB, Strynadka NC.** 2000. Crystal structure of enteropathogenic *Escherichia coli* intimin-receptor complex. *Nature* **405:**1073–1077.

13. **Hartland EL, Batchelor M, Delahay RM, Hale C, Matthews S, Dougan G, Knutton S, Connerton I, Frankel G.** 1999. Binding of intimin from enteropathogenic *Escherichia coli* to Tir and to host cells. *Mol Microbiol* **32:**151–158.

14. **Garmendia J, Phillips AD, Carlier MF, Chong Y, Schuller S, Marches O, Dahan S, Oswald E, Shaw RK, Knutton S, Frankel G.** 2004. TccP is an enterohaemorrhagic *Escherichia coli* O157:H7 type III effector protein that couples Tir to the actin-cytoskeleton. *Cell Microbiol* **6:**1167–1183.

15. **Liu H, Magoun L, Luperchio S, Schauer DB, Leong JM.** 1999. The Tir-binding region of enterohaemorrhagic *Escherichia coli* intimin is sufficient to trigger actin condensation after bacterial-induced host cell signalling. *Mol Microbiol* **34:**67–81.

16. **Campellone KG, Robbins D, Leong JM.** 2004. EspFU is a translocated EHEC effector that interacts with Tir and N-WASP and promotes Nck-independent actin assembly. *Dev Cell* **7:**217–228.

17. **Alto NM, Scott JD.** 2004. The role of A-Kinase anchoring proteins in cAMP-mediated signal transduction pathways. *Cell Biochem Biophys* **40:**201–208.

18. **Orchard RC, Kittisopikul M, Altschuler SJ, Wu LF, Suel GM, Alto NM.** 2012. Identification of F-actin as the dynamic hub in a microbial-induced GTPase polarity circuit. *Cell* **148:**803–815.

19. **Wong AR, Clements A, Raymond B, Crepin VF, Frankel G.** 2012. The interplay between the *Escherichia coli* Rho guanine nucleotide exchange factor effectors and the mammalian RhoGEF inhibitor EspH. *MBio* **3:**e00250-11. doi:10.1128/mBio.00250-11.

20. **Hemrajani C, Berger CN, Robinson KS, Marches O, Mousnier A, Frankel G.** 2010. NleH effectors interact with Bax inhibitor-1 to block apoptosis during enteropathogenic *Escherichia coli* infection. *Proc Natl Acad Sci USA* **107:**3129–3134.

21. **Pearson JS, Riedmaier P, Marches O, Frankel G, Hartland EL.** 2011. A type III effector protease NleC from enteropathogenic *Escherichia coli* targets NF-kappaB for degradation. *Mol Microbiol* **80:**219–230.

22. **Hartland EL, Leong JM.** 2013. Enteropathogenic and enterohemorrhagic *E. coli*: ecology, pathogenesis, and evolution. *Front Cell Infect Microbiol* **3:**15.

23. **Gao X, Wang X, Pham TH, Feuerbacher LA, Lubos ML, Huang M, Olsen R, Mushegian A, Slawson C, Hardwidge PR.** 2013. NleB, a bacterial effector with glycosyltransferase activity, targets GAPDH function to inhibit NF-kappaB activation. *Cell Host Microbe* **13:**87–99.

24. **Pham TH, Gao X, Tsai K, Olsen R, Wan F, Hardwidge PR.** 2012. Functional differences and interactions between the *Escherichia coli* type III secretion system effectors NleH1 and NleH2. *Infect Immun* **80:**2133–2140.

25. **Berger CN, Crepin VF, Baruch K, Mousnier A, Rosenshine I, Frankel G.** 2012. EspZ of enteropathogenic and enterohemorrhagic *Escherichia coli* regulates type III secretion system protein translocation. *MBio* **3:**e00317-12. doi:10.1128/mBio.00317-12.

26. **Mills E, Baruch K, Aviv G, Nitzan M, Rosenshine I.** 2013. Dynamics of the type III secretion system activity of enteropathogenic *Escherichia coli*. *MBio* **4:**e00303-13. doi:10.1128/mBio.00303-13.

27. **Farfan MJ, Torres AG.** 2012. Molecular mechanisms that mediate colonization of Shiga toxin-producing *Escherichia coli* strains. *Infect Immun* **80:**903–913.

28. **Obrig TG, Karpman D.** 2012. Shiga toxin pathogenesis: kidney complications and renal failure. *Curr Top Microbiol Immunol* **357:**105–136.

29. **Obata F, Tohyama K, Bonev AD, Kolling GL, Keepers TR, Gross LK, Nelson MT, Sato S, Obrig TG.** 2008. Shiga toxin 2 affects the central nervous system through receptor globotriaosylceramide localized to neurons. *J Infect Dis* **198:**1398–1406.

30. **Murray CS, Tannock GW, Simon MA, Harmsen HJ, Welling GW, Custovic A, Woodcock A.** 2005. Fecal microbiota in sensitized wheezy and non-sensitized non-wheezy children: a nested case-control study. *Clin Exp Allergy* **35:**741–745.

31. **Tannock GW.** 2005. New perceptions of the gut microbiota: implications for future research. *Gastroenterol Clin North Am* **34:**361–382, vii.

32. **Fukuda S, Toh H, Hase K, Oshima K, Nakanishi Y, Yoshimura K, Tobe T, Clarke JM, Topping DL, Suzuki T, Taylor TD, Itoh K, Kikuchi J, Morita H, Hattori M, Ohno H.** 2011. Bifidobacteria can protect from enteropathogenic infection through production of acetate. *Nature* **469:**543–547.

33. **Sperandio V, Torres AG, Jarvis B, Nataro JP, Kaper JB.** 2003. Bacteria-host communication: the language of hormones. *Proc Natl Acad Sci USA* **100:**8951–8956.

34. **Pacheco AR, Curtis MM, Ritchie JM, Munera D, Waldor MK, Moreira CG, Sperandio V.** 2012. Fucose sensing regulates bacterial intestinal colonization. *Nature* **492:**113–117.

35. **Njoroge JW, Nguyen Y, Curtis MM, Moreira CG, Sperandio V.** 2012. Virulence meets metabolism: Cra and KdpE gene regulation in enterohemorrhagic *Escherichia coli*. *MBio* **3:**e00280-12. doi:10.1128/mBio.00280-12.

36. **Kendall MM, Gruber CC, Parker CT, Sperandio V.** 2012. Ethanolamine controls expression of genes encoding components involved in interkingdom signaling and virulence in enterohemorrhagic *Escherichia coli* O157:H7. *MBio* **3:**e00050-12. doi:10.1128/mBio.00050-12.

37. **de Sablet T, Chassard C, Bernalier-Donadille A, Vareille M, Gobert AP, Martin C.** 2009. Human microbiota-secreted factors inhibit shiga toxin synthesis by enterohemorrhagic *Escherichia coli* O157:H7. *Infect Immun* **77:**783–790.

38. **Gordon JI, Klaenhammer TR.** 2011. A rendez-vous with our microbes. *Proc Natl Acad Sci USA* **108(Suppl 1):**4513–4515.

39. **Willing BP, Vacharaksa A, Croxen M, Thanachayanont T, Finlay BB.** 2011. Altering host resistance to infections through microbial transplantation. *PLoS One* **6:**e26988. doi:10.1371/journal.pone.0026988

40. **Kamada N, Kim YG, Sham HP, Vallance BA, Puente JL, Martens EC, Nunez G.** 2012. Regulated virulence controls the ability of a pathogen to compete with the gut microbiota. *Science* **336:**1325–1329.

41. **Hughes DT, Terekhova DA, Liou L, Hovde CJ, Sahl JW, Patankar AV, Gonzalez JE, Edrington TS, Rasko DA, Sperandio V.** 2010. Chemical sensing in mammalian host-bacterial commensal associations. *Proc Natl Acad Sci USA* **107:**9831–9836.

42. **Clarke MB, Hughes DT, Zhu C, Boedeker EC, Sperandio V.** 2006. The QseC sensor kinase: a bacterial adrenergic receptor. *Proc Natl Acad Sci USA* **103:**10420–10425.

43. **Rasko DA, Moreira CG, Li DR, Reading NC, Ritchie JM, Waldor MK, Williams N, Taussig R, Wei S, Roth M, Hughes DT, Huntley JF, Fina MW, Falck JR, Sperandio V.** 2008. Targeting QseC signaling and virulence for antibiotic development. *Science* **321:**1078–1080.

44. **Hughes DT, Clarke MB, Yamamoto K, Rasko DA, Sperandio V.** 2009. The QseC adrenergic signaling cascade in enterohemorrhagic *E. coli* (EHEC). *PLoS Pathog* **5:**e1000553. doi:10.1371/journal.ppat.1000553.

45. **Karmali MA, Petric M, Lim C, Fleming PC, Arbus GS, Lior H.** 1985. The association between idiopathic hemolytic uremic syndrome and infection by verotoxin-producing *Escherichia coli*. *J Infect Dis* **151:**775–782.

46. **Neely MN, Friedman DI.** 1998. Functional and genetic analysis of regulatory regions of coliphage H-19B: location of shiga-like toxin and lysis genes suggest a role for phage functions in toxin release. *Mol Microbiol* **28:**1255–1267.

47. **Wagner PL, Livny J, Neely MN, Acheson DW, Friedman DI, Waldor MK.** 2002. Bacteriophage control of Shiga toxin 1 production and release by *Escherichia coli*. *Mol Microbiol* **44:**957–970.

48. **Slutsker L, Ries AA, Maloney K, Wells JG, Greene KD, Griffin PM.** 1998. A nationwide case-control study of *Escherichia coli* O157:H7 infection in the United States. *J Infect Dis* **177:**962–966.

49. **Wong CS, Jelacic S, Habeeb RL, Watkins SL, Tarr PI.** 2000. The risk of the hemolytic-uremic syndrome after antibiotic treatment of *Escherichia coli* O157:H7 infections. *N Engl J Med* **342:**1930–1936.

50. **Panos GZ, Betsi GI, Falagas ME.** 2006. Systematic review: are antibiotics detrimental or beneficial for the treatment of patients with *Escherichia coli* O157:H7 infection? *Aliment Pharmacol Ther* **24:**731–742.

51. **Holtz LR, Neill MA, Tarr PI.** 2009. Acute bloody diarrhea: a medical emergency for patients of all ages. *Gastroenterology* **136:**1887–1898.

Index

A1 and A2 fragments, 513
AAF fimbriae, 518
aaiC gene, 30
Aat protein-coat secretion system, 507
AB5 structure, 510
Abattoir
    postharvest interventions in, 437–456
    preharvest interventions in, 421–436
Abl protein, 107
Acetic acid, for beef processing, 442, 449–450
N-Acetyl-ᴅ-glucosamine, 113
Acid tolerance, 186, 382
    EHEC, 176–177
    STEC, 366
Acid washes, for beef processing, 442–445, 449–451
Actin, activation of, 106–107, 115
Acuity pyramid, 302
Acute respiratory distress syndrome, in HUS, 309
Acyl-homoserine-lactone autoinducers, 182,
    405–406, 412–413
Adenosine A2a receptor protein, 77
adfO gene, 22
Adherence, to epithelium, 5–6
Adhesins, 131–155
AdiA protein, in acid tolerance, 176–177
adk gene, 21
Aerobic plate counts, 445–446
aggR gene and AggR protein, 30, 506, 517–519
Aggregative adherence fimbriae, 8, 506–507
AHLs (acyl-homoserine-lactone autoinducers), 182,
    405–406, 412–413
AI-2 and AI-3 quorum sensing molecules, 177–178
AI autoinducers, 404–406
Alpacas, STEC in, 212
Ammonium hydroxide, for beef processing, 442
Amyloid component B, for STEC, 344, 349–350
Amyloid protein, 80
Anemia, hemolytic, 6–7
Animal models
    for antibiotic use, 343
    for EHEC, 157–174
    for STEC, 220–221
Animal reservoirs
    classification of, 20–21
    for *E. coli* O157:H7, 437–441, 458–463

research on, 536
    for STEC, 211–230, 264, 278, 458–463
Animal waste
    organism survival in, 427
    VTEC in, 463–465
Antelopes, STEC in, 212
Antiaggregation protein, 518
Antibiotics
    EHEC infections and, 403–404
    for HUS, controversy over, 303–306
    for isolating organisms, 272
    for meat processing, 441–445, 467–468
    for STEC infections, controversy over, 342–343
    for typing, 274
Antibodies
    EHEC, 384
    lipopolysaccharide, 479–481
Anti-inflammatory action, of short-chain fatty acids,
    404
Antisilencers, 180
Anuria, in HUS, 303, 306
AP transcription factors, 323
Apoptosis, 85, 116, 383, 385, 388, 390, 513
Aquaporin 4, 85
arcA gene, 21
Arg protein, 107
Arginine dependent system, for acid tolerance,
    176–177
argW gene, 25–26
aroE gene, 21
Arp2/3 complex, 107, 115–116
Arrhythmogenic media, 272
aspC gene, 21
Asymptomatic carriers, 299
    of EHEC, 205
    ruminants as, 214
    of STEC, 221
Attaching and effacing (A/E) lesions, 5–6, 97–130,
    132–134, 179–182, 254, 326–327, 412–413, 509
Autoinducers, 182, 412–413
Autotransporters, 142–144, 517

Baboons, as disease models, 166–167
Bacterial adrenergic sensor kinases, 100
Bacteriophage lysotyping, 274

Bacteriophages, 25–27, 109
    for beef processing, 442, 446, 451
    STEC in, 249
Bacteroides thetaiotaomicron, 408, 412
Band 0.03, 386
Barriers, to colonization, 382
Basal ganglia lesions, in HUS, 308–309
Bax inhibitor-1, 116
BBG *Escherichia coli*, 26–27
Bcl-2 proteins, 513
Beef, *see also* Cattle
    ground, outbreaks in, 7–8, 56, 256, 301, 372,
        440–441
    hides and carcasses of, *see* Hides and carcasses
    outbreaks due to, 203–205
    processing of
    postharvest interventions for, 437–456
    preharvest interventions for, 421–436
    STEC in, 438–441, 466–468
Behavioral factors, STEC susceptibility and, 367
bfpA gene, 56
"Big five" genes
    for EHEC testing, 269
    for STEC testing, 268
"Big seven" genes, for EHEC testing, 269
"Big seven group," for STEC testing, 268
Big Six adulterants, 509
"Big six" genes, for produce testing, 238
Biological reservoir, 425
Biotyping, 274
Birds, STEC in, 212, 218, 429
Bison, STEC in, 212, 216
Bovine animals, *see also* Cattle
    STEC in, 212–214
Brain, Stx effects on, 81–91, 390
Brefeldin A, for STEC, 351
Butyrate, 179
    anti-inflammatory role of, 404, 411
    in EHEC colonization, 382

C-9, for STEC, 344, 347–348
Caenorhabditis elegans, 428
Cah protein, 144
Calcium imaging, 90
Calcium-binding antigen 43, 144
Canines
    as disease models, 166–167, 220–221
    STEC in, 212, 220, 429
Capillary defects, in central nervous system lesions,
    83, 85
Carbohydrate metabolism, disruption of, 404,
    408–411
Carbohydrate recognition, 178–179, 187
Carcasses, *see* Hides and carcasses
Cardiovascular disorders, in HUS, 309
Carriage, *see also* Animal reservoirs
    in asymptomatic humans, 205, 214, 299

Carvacrol, for beef processing, 450–451
Caspase-3 activation assay, 85, 90
Cat(s), STEC in, 212, 220, 429
Catabolite repressor/activator protein, 178–179, 184,
    408
Catecholamines, 382
Cathelicidin, 382
Cattle, *see also* Beef
    as disease models, 220–221
    EHEC in, 56, 203–205, 412–413
    E. coli O157:H7 in, 421–456
    STEC in, 212–214, 245, 264, 362–367, 426–427,
        460–462
    vaccines for, 423–425, 487–501
CDC42, 115
CD40L, 386
Cell binding, of Stxs, 41–43
C-EnterNet, 362–363
Centers for Disease Control and Prevention,
    247–248, 362
Central nervous system, Stx effects on, 81–91, 161,
    390
Ces proteins, 102
Cetylpyridinium chloride, for beef processing, 445,
    447
CfaD protein, 518
CG *Escherichia coli*, 26–27
Chaperone-usher secretion pathway, 8, 507
Cheese, STEC contamination of, 466
Chemical dehairing, 446–447
Chemoattractants, 77
Chemokine(s), 77–79, 383–384, 388–390
Chemokine ligand 1, 79
Chickens
    as disease models, 166–167, 220–221
    STEC in, 212, 218–219
Chitosan acetate, for beef processing, 451
Chlorine compounds, for beef processing, 442, 445,
    447, 450
Chlorofoam, for beef processing, 447
Chloroquine, for STEC, 345, 351
Cholesterol, in lipid rafts, 41–42
Cif protein, 109, 116
Citric acid, for beef processing, 442, 450
Citrobacter rodentium animal models, 157, 164–166,
    324–326
Classification, of *Escherichia coli*, 5–6, 17–18, 20–21
Clathrin, in cell binding, 42
Cleaning, of cattle hides and carcasses, 441–452
clpX gene, 21
Coagulation cascade, 388
Coliforms, in produce, 232–233
Colitis, hemorrhagic, *see* Hemorrhagic colitis
Collagen, 138
Colonization
    EHEC, 382–384, 412–413
    STEC, 438

Colony hybridization, 272
Colony immunoblot test, for Stxs, 266–267
Coma, in HUS, 308–309
Commission for European Normalization, 268–269
Complement, activation of, 390–391, 513
Complement receptor, 114
Composting, of animal waste, 464
Contamination
    with *E. coli* O157:H7, 422–423, 425–429
    interventions for, 441–453
    of milk, 213
    prevalence of, 440–441
    sources of, 438–440
    with STEC, 211–230
    vaccination preventing, 493
    of water, 213, 218, 232, 464–465
COPII coat protein, 184
Costs, of STEC infections, 364–365
Cra protein, 178–179, 184, 408
CRISPR sequences, 276
Critical Control Point Program, 421, 452–453
Crl protein, 140
Cross-contamination, 440, 494
Cross-immunity, 384
CsgA protein (curli), 66, 139–140
CsvR protein, 518
Cultural factors, STEC susceptibility and, 367
Culture, 201, 271–273
Cultured substances, for beef processing, 443
Curli, 66, 139–140, 183
cyaA gene, 21
Cysteine methyltransferase, 328
Cytokeratins, 116
Cytoskeleton, modification of, 115–116
Cytotoxicity assays, for Stxs, 265–266

Dairy products, contamination of, 465–466
Dalsy, for STEC, 344, 348
Decontamination, of cattle, 422
Deer, STEC in, 212, 215–216, 429
Defensins, 382
Dehairing, chemical, 446–447
Department of Agriculture, Microbiological Data Program, 233–234
Diabetes mellitus, in HUS, 309
Dialysis, for HUS, 306
Diarrhea
    in HUS, 310–311
    pathogenesis of, 116–117
Diffusely adherent *Escherichia coli*, 17
Diploscapter, 428
Direct screening, for, STEC, 271–272
Dispersin, 507, 518
DksA protein, 181
DNA analysis, for Stxs, 267–270
dnaG gene, 21

Dogs
    as disease models, 166–167, 220–221
    STEC in, 212, 220, 429
Donkeys, STEC in, 216–217
Dr superfamily, 8, 507
Drinking water, contamination of, in farms, 213
DsrA protein, 186
Ducks, STEC in, 212, 218
Dysbiosis, EHEC infections and, 403–417

eae gene, 6, 23, 30, 46, 56, 104, 132, 237, 238, 277–278
EAEC (enteroaggregative *Escherichia coli*), 17
    genomics of, 63–65
    outbreaks, 505–529
    pathogenesis of, 505–529
EC55989, 508
ECP (*Escherichia coli* common pilus) protein, 140–141, 178, 183
Eculizumab
    for HUS, 310
    for STEC, 345, 352
EDL933 isolate, of *Escherichia coli*, 57, 61–62, 99–100, 102, 113, 183, 185
Eeyarestatin, for STEC, 345, 351
efa gene and Efa protein, 29, 183
efa1 gene and Efa1 protein, 100, 117–118
Effector molecules, injection of, 184–185
Eha proteins, 142–143, 183, 240
EHEC (enterohemorrhagic *Escherichia coli*), 17
    adhesins of, 131–155
    animal models for, 157–174
    colonization by, 382–384
    definitions of, 300
    with EAEC characteristics, 506
    emerging strains of, 269
    genomics of, 55–71
    host response to, 381–402
    inflammatory response in, 321–339
    locus of enterocyte effacement of, 97–130
    microbiota interactions with, 403–417
    nomenclature of, 56
    outbreaks of, 19, 199–209
    pathogenesis of, 381–402
    phylogeny, 22
    research topics for, 533–539
    testing for, 237
    treatment of, 341–358
    vaccines for, 477–485
    virulence gene regulation by, 175–195
EHEC factor for adherence 1, 117–118
ehxA gene and EhxA protein, 186, 238
EibG protein, 145
EivF protein, 182
Elands, STEC in, 212
Electrophysiologic studies, 90
ELF protein, 141
Elk, STEC in, 216

Endothelial cells, 83, 85, 90
Enrichment, for STEC testing, 235–241, 270–271
Enteroaggregative *Escherichia coli* (EAEC), *see*
    EAEC
Enterohemolysin, 20, 186, 239–240, 273
Enterohemorrhagic *Escherichia coli* (EHEC), *see*
    EHEC
Enteroinvasive *Escherichia coli*, 17
Enteropathogenic *Escherichia coli* (EPEC), 17, 20, 56
Enterotoxigenic *Escherichia coli* (ETEC), 17, 234
Enterotoxins, 234, 521–522
Environment
    E. coli O157:H7 survival in, 427–429
    STEC in, 264
    water contamination in, 213, 218, 232
Enzyme immunoassays, 301–302
Enzyme-linked immunosorbent assays, for Stxs,
    266–267
EPEC (enteropathogenic *Escherichia coli*), 17
    nomenclature of, 56
    pathologic effects of, 20
Epidemiology, *see also* Outbreaks
    E. coli O104:H4, 8–96–7
    E. coli O157:H7, 6–7, 55–71, 421–422, 437
    research on, 536–537
    STEC, 299–300, 361–380
    in animals, 211–230
    detection methods for, 263–295
Epinephrine/norepinephrine signaling, 187, 406
Epithelial cells, 99, 322
Equine animals, STEC in, 212, 216–217
Erk protein, 107, 323
esc gene family, 101–102
Escherichia coli, as indicator, for produce quality,
    232–233
Escherichia coli common pilus protein, 140–141, 178,
    183
Escherichia coli immunoglobulin-binding protein,
    145
Escherichia coli laminin-binding fimbriae, 141
Escherichia coli O26, 248–249
Escherichia coli O103:H25, 462
Escherichia coli O104:H4, 253, 255–257, 373
    history of, 4, 8–10
    pathogenesis of, 505–529
Escherichia coli O113:H21, 441
Escherichia coli O127:H6, 99, 104–106
Escherichia coli O157:H7
    animal models for, 157–174
    animal reservoirs for, 211–230
    in cattle, 437–456
    cattle preharvest interventions for, 421–436
    clinical perspectives of, 299–310; *see also*
        Hemolytic-uremic syndrome;
        Hemorrhagic colitis
    epidemiology of, 6–7, 55–71, 421–422, 437
    evolution of, 301

genomics of, 55–71
genotyping of, 430
history of, 4–8, 55–57
in Japan, 199–209
locus of enterocyte effacement, 104–117
outbreaks of, 18–20, 56, 231–244, 437
pathogenic effects of, 65–66, 384–385
in produce, 231–244
reservoirs for, 57, 458–463
risk factors for, 361–380
typing of, 57–60
vaccines for, 423–425, 477–485, 487–501
Esp protein family, 99–118, 134, 143, 181, 184–185,
    333, 408–409, 413, 423–424, 491
ETEC (enterotoxigenic *Escherichia coli*), 17, 234
Ethanolamine, 179, 186, 187
EtrA protein, 182
European Centre for Disease Prevention and
    Control, 246–247, 254–255, 363
European Food Safety Authority, 249, 255
European Food Safety on Biological Hazards
    (BIOHAZ) panel, 19–20
European Surveillance System, 19–20
European Union, surveillance in, 245–259
European Union Reference Laboratories, 251–253
*European Union Summary Report on Trends and
    Sources of Zoonoses, Zoonotic Agents and
    Food-Borne Outbreaks*, 249–250
Evolution
    of *E. coli* O157:H7, 65
    of STEC, 371
Exo2, for STEC, 345, 351
External quality assessment, 250–253

F9 fimbriae, 141
F9 protein, 183
Fabry's disease, 41
facD gene, 21
Factor H, 391
Farms, STEC contamination in, 211–230
Fatty acids, short-chain, 404, 410–411, 515–516
FCγ receptor, 114
FDA Bacteriological Analytical Manual, 239
Fecal coliforms, in produce, 232–233
Fecal pat prevalence, 422–423
Feces
    interventions for reducing contamination by,
        441–453
    organism survival in, 463–465
    STEC in, 211–230, 421–429, 438–440
Fenugreek sprouts, 205, 373, 505
Feral pigs, STEC in, 212, 217–218, 429, 462–463
Ferrets
    as disease models, 166–167
    STEC in, 212
Fiber, dietary, microbiota interactions with, 411
Fibrin deposition, 78, 80

Fibronectin, 138
Filaments, in Type III secretion system, 102–103
fimA gene, 142
Fimbriae, 136–145, 183, 518
Fimbrial locus, for vaccine development, 425
Fish, STEC in, 212, 219
Flagella, 144–145, 177–178, 186
Flagellin, 411, 413
    in inflammatory response, 323–324
    vaccine based on, 425, 491
flg genes, 411
flh genes, 177–178
Fli proteins, 177–178
flicC gene, 144–145, 276
Flies, STEC in, 220, 428–429
Fluid management, for HUS, 303
Fluorescein actin stain, 6, 98–99
Fodrin, 521
Follicle-associated epithelium, 105, 136–137, 139, 382
Food and Waterborne Diseases and Zoonoses, 246–248, 250, 254–255
Food safety
    cattle preharvest interventions for, 421–436
    fresh produce testing for, 231–244
    risk analysis in, 361–380
    surveillance for, 245–259
FoodNet, 248, 250–251, 362, 364–365
Fresh produce, *see* Produce
Fruit, *see* Produce
Fucose, 179, 409–410
fumC gene, 21
Fur protein, 515–516
Furin inhibitors, for STEC, 345, 351–352
FusK protein, 179
Fyn protein, 107

Gad proteins, in acid tolerance, 176–177
Galactose catabolism, 187
1,4-Galactosyltransferase, 90
Gammaproteobacteria, 410
Gastrointestinal disorders, in HUS, 309
Gb3 polymers, 344, 348, 511–512
Gb3 receptor, 385–386
Geese, STEC in, 212, 218
Gel electrophoresis, 267
Gene(s), 21–23; *see also specific genes*
    Stx A subunit, 40
    Stxs, 5
Gene array analysis, of gene response, 79–80
Gene probe hybridization, 267–268
Genome sequence-based high-resolution genotyping, 59–60
Genome sequences, Q genes, 27–28
Genomic(s)
    EHEC, 55–71
    E. coli O157:H7, 55–71

Genomic islands, 28–29
Genotyping, *of Escherichia coli* O157, 430
Glial cells, 90
Globotriaosylceramide, 38, 40, 75, 90, 158, 411
Gluconeogenesis, 178–179, 186
Glucose metabolism, disruption of, 404, 408–411
Glucose-repressed system, for acid tolerance, 176–177
Glutamate decarboxylase acid resistance system, 412
Glutamate-dependent system, for acid tolerance, 176–177
Glutamic acid, as active site, 39, 40
Glycosphingolipid (G3), 41–43, 45
Glycosyltranferases, 330–331
gnd gene, 276
Goats, STEC in, 212, 214–215, 462
Goblet cells, 520
Golgi matrix protein, 116
GrlRA regulators, 181
Ground beef, outbreaks of, 7–8, 56, 256, 301, 372, 440–441
grpE gene, 21
Guanine nucleotide exchange factors, 115
gyrB gene, 21

H antigen, for typing, 275–276
ha gene and Iha protein, 145, 183, 382, 508, 510, 515–516
Hamburger outbreaks, 7–8, 56, 256, 301, 372, 440–441
Hazard Analysis and Critical Control Points, 452–453
HCP protein, 141–142, 324
Health care costs, of STEC infections, 364–365
Heat-labile enterotoxin, 234
Heat-stable enterotoxin, 234
Hematologic complications, in HUS, 308
Hemodialysis, for HUS, 306
Hemolysis, in HUS, 308
Hemolytic anemia, in HUS, 6–7
Hemolytic-uremic syndrome
    acuity pyramid of, 302
    animal models for, 157–174
    clinical perspectives of, 299–319
    definitions of, 300
    diagnosis of, 300–302
    history of, 6–7, 300–301
    outbreaks of, 23–25, 29–30, 45, 203
    pathogenesis of, 45–46, 75–95, 381–402, 505–529
    prognosis for, 302, 306
    research on, 533–534
    risk assessment for, 278–282
    surveillance for, 246–249, 254, 256
    toxins of, 185–186
Hemorrhagic colitis
    animal models for, 157–174

Hemorrhagic colitis (*continued*)
history of, 7–8
outbreaks of, 18
pathogenesis of, 508
risk assessment for, 278–282
Hemorrhagic *Escherichia coli* pilus, 141–142
Herd immunity, 493
Hfq protein, 181
Hides and carcasses
chemical dehairing of, 446–447
cleaning of, 441–452
contamination of, 422–423, 428, 438–453
interventions for reducing contamination in, 441–453, 467–468
Histidine sensors, 406
Histopathology, 83, 85, 159–160
History, 4–9
of *E. coli* O104:H4, 4–10
of *E. coli* O157:H7, 55–57
of hemolytic-uremic syndrome, 6–7
H-NS protein, 133, 140, 176, 180
Holotoxin, 75, 513
Horizontal gene transfer, 65
Horses, STEC in, 212, 216–217, 429
Host(s)
response to EHEC, 381–402
of STEC, 211–230, 429
Host factors
in animal studies, 167–168
in STEC susceptibility, 366–367
Houseflies, STEC in, 220, 428–429
Hp90 protein, 104, 134
HUS, *see* Hemolytic-uremic syndrome
HuSAP, for STEC, 344, 349–350
Hybridoma cell lines, for Stxs, 266–267
Hydrogen peroxide, for dehairing, 446
Hyperkalemia, in HUS, 306–307
Hypertension, in HUS, 307–308
Hyperuricemia, in HUS, 307
Hypoalbuminemia, in HUS, 307
Hypobromous acid, for beef processing, 443, 447
Hypocalcemia, in HUS, 303, 306–307
Hypochlorous acid, for beef processing, 443
Hypocomplementemia, 390–391
Hypokalemia, in HUS, 306–307
Hyponatremia, in HUS, 306–307

icd genes, 21
IκB kinase, 323
Imatinib, for STEC, 345, 352
Immunity, innate, modulation of, 113–118
Immunoblot test, 266–267, 270, 272
Immunodetection assays, 90
Immunological assays, for Stxs, 266–267
Immunomagnetic separation assays, 239, 272
Incubation time, 271

Infectious Disease Surveillance Center (Japan), 200–201
Infectious Diseases Control Law (Japan), 200
Inflammatory bowel syndrome, 404
Inflammatory response, in EHEC infections, 321–339
Inhibitory proteins, 113
Innate immunity, modulation of, 113–118
Insects, STEC in, 220
Insulin receptor substrate p53 (IRSp53), 107–108, 134
Insulin receptor tyrosine kinase substrate (IRTKS), 107–108, 134
Integrative elements, 328
Integrins, 106, 134–136
Interferon(s), in inflammatory response, 325
Interferon-gamma-inducible protein-10 (IP-10), 79–80
Interleukin(s), 322–327, 383–384, 388–389
International Escherichia and Klebsiella Centre, 275
International Organization for Standardization, technical specification, 249, 252–253
Intimin, 6, 46, 104–106, 133–136, 237, 459, 509
types of, 28–29
vaccine based on, 491
IrgA protein, 508
Iron transport, vaccines blocking, 424
IRSp53 (insulin receptor substrate p53), 107–108, 134
IRTKS (insulin receptor tyrosine kinase substrate), 107–108, 134
IS printing system, 201
Isolation, of STEC, 271–273

Jack in the Box *Escherichia coli* O157:H7 outbreak, 421–422
Japan, EHEC outbreaks in, 199–209
JNK, 323

Kdp proteins, 178–179, 182, 184, 407–408
Keratinocyte-derived chemokine, 79
Kidney failure, *see also* Hemolytic-uremic syndrome
in HUS, 6–7, 309–310
pathogenesis of, 388–390
Knife trimming, of contaminated spots, for beef processing, 448–449

Laboratories
reference, 235, 250–253
for surveillance, 246–253
Lactic acid, for beef processing, 443–445, 449–450, 451
Lactoferrin, for beef processing, 443
Lairage, contamination during, 439
Laminin, 138
Lauramide arginine ethyl ester preparation, for beef processing, 443

LED209 small molecule, for STEC, 343–346
Ler protein, 133, 180–182, 410
Lettuce
  coliforms in, 233
  STEC in, 235
Leucine-responsive protein, 410
Leukocytosis, 388–389
lifA gene, 29, 100, 117–118
Lineage-specific polymorphism typing assay,
  59–60
Lipid rafts, 41, 512
Lipocalin-2, 80
Lipopolysaccharide, 77–81, 383–387, 478–482
Livestock, *see* Cattle; Goats; Sheep
Llamas, STEC in, 212
Locus of enterocyte effacement, 6, 20, 22–23, 28–29,
  45–46, 97–130, 132–145, 179–184, 265, 366, 509
Long-polar fimbriae, 136–139, 516–517
Loop-mediated isothermal amplification, 267
lpf genes and Lpf proteins, 136–139, 183, 425,
  516–517
lrp gene and Lrp protein, 382, 410
Lux proteins, 405–406
lysP gene, 21

Macrophage inflammatory proteins, 45, 77–78, 324,
  386
Macropinocytosis, 42, 382
Maintenance host population, 429
Manganese, for STEC, 345, 351
Mannose-binding lectin pathway, 391
Map protein, 109–113, 115–117, 184
Mat protein, 140–141
mdh gene, 21
Meat, from cattle, *see* Beef
Meat Inspection Act of 1906, 421
Mesangiolysis, 514
Metalloproteinases, 329–330
Methyltransferase, 113
Mice
  as disease models, 164–166, 220–221, 324–326
  STEC in, 212, 219
Microarray analysis, 79–80, 276
Microbiological Data Program, 233–235
Microbiology Laboratory Guidebook, 441–442
Microbiota, EHEC interactions with, 403–417,
  534–535
Milk, STEC contamination of, 213, 218, 465–466
mirA gene, 25
Mitochondrial-associated protein (Map protein),
  109–113, 115–117, 184
Mitogen-activated protein kinases, 322–323, 383,
  513
Mitomycin C, for enrichment, 270
MlrA protein, 140
MMA-tet, for STEC, 344, 349
Molecular methods

  research on, 533–534
  for typing, 274–278
Molecular risk assessment, 265
Molecular syringe, 179–180
Monitoring, for public health, 245–259
Monkeys, as disease models, 167
Monoclonal antibodies, for STEC, 345, 350–351
Monocyte(s), in inflammatory response, 385–387
Monocyte chemotactic protein 1, 77–78, 324
Monosialotetrahexosylganglioside, 41
Most probable number method, 233
Motility, 177–178
mtlD gene, 21
Muc1 protein, 144
Mucins, 333, 408–409, 520
Multilocus sequence typing, 21–22, 234, 274–275,
  277
Multilocus variable number tandem repeat analysis,
  58–59, 277
Multiple hurdle treatment, for beef processing,
  451–452
Multiple-locus variable-number tandem-repeat
  analysis, 201
Multiplex PCR assays, 268
  for ETEC, 234
  for produce, 238
  for STEC, 267
Murine interleukin-8 mimic, 79
Murine species, *see* Mice
mutS gene, 21
MyD88, 326, 383–384

Na(+)/H(+) exchanger regulatory factor 2
  (NHERF2), 116–117
National Epidemiological Surveillance of Infectious
  Diseases (Japan), 200
National Health Surveillance System (Argentina),
  363
National reference laboratories, 250 251
Nationally Notifiable Diseases Surveillance System,
  362
Nck protein, 107
NEDD 8 protein, 109
Nematodes
  as disease models, 166–167
  E. coli O157:H7 in, 428
Neural Wiskott-Aldrich syndrome protein, 107–109,
  115
Neurologic disorders, in HUS, 308–309
Neutrophils
  in inflammatory response, 322, 324, 333, 384–385,
  387
  migration of, 78–79
NHERF2 [(Na)/H(+) exchanger regulatory factor
  2], 116–117
Nisin, for beef processing, 443–444
Nitrobenzyl-thioinosine, for STEC, 345, 351

nle genes and Nle proteins, 29, 112, 116, 184–185, 277–279, 282, 328–332
NM strain, typing of, 276
NOD proteins, 333
Nomenclature, of EHEC, 56
Nonselective culture media, 272
Novobiocin, for enrichment, 270
NRIR sequences, 185
Nuclear factor-kappa B, 113, 322–323, 326–328, 331–333, 383
Nucleolin, 106, 135–136
Nucleotide polymorphism-derived genotyping, 21–22
Nucleotide sequencing, 277–278
Nutrient deprivation, 181
Nutrient signaling, 407–411
N-WASP (neural Wiskott-Aldrich syndrome protein), 107–109, 115

O antigen, 275–276, 478
Octamer-based genome scanning typing assay, 59
OI-122 pathognicity island, 29
O-island regulators, 182
Oligoanuria, in HUS, 303, 306
Open reading frames, 282
orf gene family, 101
Organic acids, for beef processing, 449–451
Organic produce, vs. conventional produce, 233
OSP vaccine, 478–482
OspZ protein, 328
Oubain, for STEC, 345, 352
Outbreaks, 18–20
    EAEC, 459, 505–529
    EHEC, 158–160
    E. coli O157:H7, 18–20, 56–57, 231–244, 437
    hemolytic-uremic syndrome, 23–25, 29–30, 45
    hemorrhagic colitis, 18
    produce-related, 231–232
    sequencing studies of, 62–65
    STEC, 361–380
    surveillance and, 245–259
Outer membrane protein A, 145, 183
Ozone, for beef processing, 444–445, 450

p38, 323
pagC gene, 29
Pancreatitis, in HUS, 309
Paralysis, 81, 83, 85
Pasteurization, steam, for beef processing, 448–449
patC gene, 25
Pathogen Reduction and Hazard Analysis, 421
Pathogen-associated molecular patterns, 383–384
Pathogenesis
    of *E. coli* O104:H4, 505–529
    of *E. coli* O157:H7, 65–66, 384–385
    of EHEC, 381–402

    of hemolytic-uremic syndrome, 45–46, 75–95, 381–382, 505–529
    of hemorrhagic colitis, 508
Pathogenicity islands, 365–366, 509
Pathotypes, 17–23, 29–31, 264, 274–278
pch genes and Pch proteins, 181, 182, 410
PCR testing
    for ground beef, 441
    for produce, 236–241
    for Stxs, 266–270, 276, 278
Penn State University, *Escherichia coli* reference laboratory in, 235
PerC protein, 182
Periharvest interventions, for *Escherichia coli* O157:H7, 437–456
Peripheral nervous system, Stx effects on, 90
Peritoneal dialysis, for HUS, 306
Peroxyacetic acid preparations, for beef processing, 444, 450–451
Pet protein, 521
Pets, STEC in, 220, 429
Petting zoos, STEC in, 214
Phage lysotyping, 274, 277
Phagocytosis, inhibition of, 114
phe genes, 100
Phenotyping, 275–276
Phosphoric acid, for beef processing, 447, 450
Phylogenetic analysis-derived genotyping, 21–22
Pic protein, 507–508, 517, 519–521
Pigeons, STEC in, 212, 218
Pigs
    as disease models, 160–161
    STEC in, 212, 217, 429, 462–463
Pilin, 507
Plasmapheresis, for HUS, 310
Plasmid profiles, for typing, 274
Platelets, in inflammatory response, 386–390
Polyclonal antibodies, for STEC, 344–345, 350–351
Porcine edema disease, 217, 278, 462–463
Postharvest interventions, 437–456, 467
Poultry, STEC in, 212, 218–219, 429
PpdD protein, 141–142
ppGpp protein, 181
Predicted protein D, 141–142
Preharvest interventions, 467
    for *E. coli* O157:H7, 421–436
    vaccination, 423–425, 487–501
prfC gene, 25
Primates, as disease models, 220–221
Probiotics, 344, 349, 411
Produce, *see also specific items, e.g.,* Spinach
    EHEC in, 203–205
    ETEC in, 234
    organism attachment to, 65–66
    Shiga-toxin producing organisms in, 231–244
    STEC in, 231–244
Prophages, 109, 185–186, 514

Protozoa, *Escherichia coli* O157:H7 in, 428
Public health
    cattle vaccination and, 487–501
    genome sequencing and, 66
    Japanese outbreak, 199–209
    microbiologic surveillance for, 245–259
    taxonomy and, 29–31
Pulsed-field gel electrophoresis, 21, 58, 201,
    276–277, 364, 439
PulseNet, 201, 248, 364
purD gene, 21
Pure Food and Drug Act, 421
Pyocin, for STEC, 343, 344

Q antiterminator protein, 27–28
Qse proteins, 133, 167–168, 177–178, 182, 184, 186,
    187, 406–407
Quorum sensing, 182, 382, 404–411

Rab protein, 116
Rabbits
    as disease models, 161–164, 220–221
    STEC in, 212, 219–220
Rac proteins, 115, 333
Raccoons, STEC in, 212, 220
RANTES, 77, 385, 389
Rats
    as disease models, 167, 220–221
    STEC in, 212, 219
Ready-to-eat produce, *see* Produce
recA gene and RecA protein, 21, 185–186
Recto-anal junction, 213–215, 412
Reference laboratories, 235, 250–253
Regulations, for STEC monitoring, 255–257
Rel proteins, 181, 322–323
Research, topics for, 533–539
Reservoirs, *see also* Animal reservoirs
    for *E. coli* O157:H7, 57, 422–429
    research on, 536
    for STEC, 264
Restriction fragment length polymorphism, 267, 277
Retinal hemorrhage, in HUS, 309
Retro-2, for STEC, 345, 351
Retrograde trafficking, 43, 512–513
rfbE gene, 276
Rho proteins, 115
Ribotoxic stress response, 44, 513
Risk assessment, 254–255
    for cattle contamination, 425
    for STEC, 235–241, 265, 278–282
Risk factors, for STEC, 361–380
RNaseE, 181
Rns protein, 518
Rodents
    as disease models, 220–221
    STEC in, 212, 219
ror gene family, 101–102

Rotifers, *Escherichia coli* O157:H7 in, 428
Rpo proteins, 140, 176
Ruminants, *see also* Cattle; Goats; Sheep
    EHEC in, 412–413
    STEC in, 212, 214–215, 249–250

Saa protein, 143–144, 237–238, 240
Sab protein, 143–144
Safety, food, *see* Food safety
Sakai isolate, of *Escherichia coli*, 57, 61–62, 185
Sakai, Japan, EHEC outbreaks in, 199–209, 372
Salads, ready-to-eat, 231–244
Salicylidene acylhydrazides, 102
Sanitizers, for beef processing, 441–445
scbB gene, 25
SdiA protein, 177, 182, 405–406, 413
Seasonal variation, in cattle contamination,
    422–423, 424, 426–430, 439
Seizures, 83
sel genes, 100
Selective culture media, 272–273
sen gene, 29
sep gene family and Sep proteins, 101–102, 109, 181,
    507–508, 519–521
Septrin, 521
Serine protease autotransporters of
    Enterobacteriaceae (SPATEs), 8, 507–508,
    519–521
Serologic tests, for Stxs, 266–267
Serotypes, 17–18, 29–31
Serotyping, 274
serU gene, 25
Serum bactericidal assay, 480
Sfp (sorbitol-fermenting fimbriae protein), 141
Shedding, of bacteria, 213–214, 424, 440, 488–489
Sheep, STEC in, 212, 214–215, 429, 438, 462
Shellac coating, for beef processing, 446
Shellfish, STEC in, 212, 219
ShET1 toxin, 508, 521–522
SHIBAM medium, 240
Shiga toxin-producing *Escherichia coli* (STEC), 17;
    see also Stx(s)
SHP-1 protein, 332
Siderophore receptor and porin cattle vaccine,
    423–425, 491–492
SigA protein, 507–508, 519–521
Single nucleotide polymorphism typing, 60, 276
Slaughter, of cattle
    postharvest interventions for, 437–456
    preharvest interventions of, 421–436
Slide agglutination tests, 275
Small molecules, for STEC, 343, 344–346
SNX9 protein, 115
Sodium decyl sulfate, for beef processing, 451
Sodium hydroxide, for beef processing, 447
Sodium sulfide preparation, for dehairing, 446
Sodium-D-glucose transporter, 117

Solid agar plates, 272
SopE protein, 115, 333
Sorbitol-fermenting *Escherichia coli*, 237, 301
Sorbitol-fermenting fimbriae protein, 141
Sorbitol-MacConkey agar, 301–302
SPATEs (serine protease autotransporters of
     Enterobacteriaceae), 8, 507–508, 519–521
Spinach
   coliforms in, 233
   outbreaks related to, 56–57, 65–66, 235, 236, 366,
     372
   STEC in, 235
Spinal cord, Stx effects on, 81–91
SpoT protein, 181
Sprouts, outbreaks from, 8–9, 256, 372–373, 505
SRP proteins, 423–425, 491–492
ssrA gene, 25
Stacked-brick adherence pattern, 506–507, 517
STARFISH, for STEC, 344, 348
Starlings, STEC in, 212, 218, 429
stcE gene and StcE protein, 22, 183, 184, 385
Steam treatment, for beef processing, 448–449
STEC (Shiga toxin-producing *Escherichia coli*), *see
     also specific organisms, e.g., Escherichia coli*
     O157:H7
   animal reservoirs for, 211–230
   classification of, 19
   clinical perspectives of, 299–319; *see also*
        Hemolytic-uremic syndrome; Hemorrhagic
        colitis
   definitions of, 300
   detection of, 263–295, 441–442
   disease caused by, 23–25
   genomic islands of, 28–29
   isolation of, 271–273
   nonhuman sources of, 263–295
   outbreaks, 19, 361–380, 465
   pathologic effects of, 18
   phylogeny, 21–22
   in produce, 231–244
   risk analysis for, 235–241, 278–282
   risk factors for, 361–380
   serotypes of, 29–31, 362
   subtypes of, 264–265
   surveillance for, 245–259
   testing for, 235–241
   treatment of, 341–358
   types of, 264, 274–278
Stress response, ribotoxic, 44
Stress, STEC susceptibility and, 366–367
Stringent response, 181
Stroke, in HUS, 308
Stromal cell-derived pathway, 389–390
Structures, of Stxs, 38–41
Stx(s), 37–53; *see also specific subtypes*
   actions of, 43–44
   animal models for, 157–174

   cell binding of, 41–43
   detection of, 265–271
   discovery of, 2, 37
   diseases associated with, 38, 45–46
   food containing, 234–235
   genetic analysis of, 40–41
   history of, 4–9
   overview of, 37–38
   pathogenic effects of, 75–95, 381–402
   pathophysiology of, 508
   in prophages, 185–186
   retrograde trafficking of, 43
   structure of, 39–41
   subtypes of, 265
   typing of, 38–39
   vaccines based on, 481
Stx1
   bacteriophages associated with, 25–27
   discovery of, 37
   nomenclature of, 37–38
   overview of, 37–38
   pathologic effects of, 18
   properties of, 23–25
   variants of, 24–25
Stx1a
   action of, 44–45
   genetic analysis of, 40–41
   nomenclature of, 37–38
   structure of, 39–41
   versus Stx2a, 44–45
Stx2, 24
   nomenclature of, 37–38
   pathogenic effects of, 18, 510–515
   properties of, 23–25
   variants of, 24–25
Stx2a, 23–24
   action of, 44–45
   disease associated with, 45
   nomenclature of, 37–38
   versus Stx1a, 44–45
   subtypes of, 39
stx2b gene, 25
stx2c gene, 24
Stx2d, 23–24, 39
Stx2e, 25
Stxb, 24
Stxc, 24
subAB gene, 236–238, 240
Subendothelial matrix proteins, 387
Subtilase cytotoxin, 237–238
Subtypes
   STEC, 274–278
   toxins, 23–25
Sugar, for beef processing, 443
SUPER TWIG, for STEC, 344, 348
Super-shedding animals, 213–214, 367, 424, 460, 489
Surveillance

EHEC infections, 200–202
for public health, 245–259
STEC infections, 362–364
Swine, *see* Pigs
SYNSORB, for STEC, 344, 348

TAB proteins, 113, 328
TaqMan minor groove primer, 268
Taxonomy, 17–21, 29–31
Tccp protein, 107–108, 184
Tellurite media, 272–273
Temperature-sensitive hemagglutinin, 520
ter genes, 273
Terminal deoxynucleotidyltransferase-mediated
    dUTP-biotin nick and labeling (TUNEL), 85, 90
Terminology, 17–18
Tetrahymena, 428
Thrombocytopenia, 6–7, 386
Thrombosis
    pathogenesis of, 387–388
    renal, 77–79
Tir protein, 28–29, 133–136, 184, 332, 509
    in intimin binding, 105–109, 113
    vaccine based on, 491
TIRAP protein, 384
Tissue factor, 387, 390
Tissue tropism, 135–136, 139
Toll-like receptors, 323, 326, 383–384, 386–387
torS/T gene, 25
Total microbial populations, in produce, 232
ToxB protein, 117–118
Toxins, 23–25; *see also specific toxins*
TRAF2 (tumor necrosis factor receptor-associated
    factor 2), 113, 328
TRAM protein, 384
Transfusions, for HUS, 308
Translocons, 101–104
Transmission
    of EHEC, 204–205
    of STEC, 264
Treatment
    of EHEC infections, 341–358
    of HUS, 299–319
    research on, 533–534
    of STEC infections, 341–358
TRIF protein, 384
Trisodium phosphate, for beef processing, 447
Tropism, extension of, 105
Tubulin, 116
Tumor necrosis factor, 324–325, 330, 390
Tumor necrosis factor alpha, 85, 90, 383
Tumor necrosis factor receptor-associated factor 2
    (TRAF2), 113
TUNEL (terminal deoxynucleotidyltransferase-
    mediated dUTP-biotin nick and labeling), 85,
    90
Turkeys, STEC in, 212

TVP, for STEC, 344, 349
Type I fimbriae, 142
Type III secretion system, 28–29, 101–102, 109–113,
    132, 178–184, 490–491
    cattle vaccines based in, 423–425
    proinflammatory activity of, 332–333
Type V secretion system, 507
Type VI secretion system, 518
Typing, of *Escherichia coli* O157:H7, 57–60

Ubiquitin, 109
UhpB protein, 409
uidA gene, 21
Urtoxazumab, for STEC, 345, 350

Vaccines
    cattle, 423–425, 487–501
    for humans, 477–485
    intimin, 136
Vacuuming, with steam, for beef processing,
    448–449
Vascular cell adhesion molecule-1, 79
Vegetables, *see* Produce
Verocytotoxigenic *Escherichia coli* reverse passive
    latex agglutination assay, for Stxs, 266–267
Verocytotoxin-producing *Escherichia coli*, *see*
    VTEC
Verotoxins, *see also* Stx(s)
    terminology of, 4
Veterinary reference laboratories, 251–253
Virulence factors, 366
    of EHEC, 97–130, 381–402
    microbiota interactions and, 407–411
    for typing, 56
Virulence gene regulation, 175–195
Volume expansion, for HUS, 303
Volunteer studies, of EHEC, 159
von Willebrand factor, 387
VTEC (verocytotoxin-producing *Escherichia coli*),
    17; *see also* EAEC; STEC
    outbreaks of, 20
    pathologic effects of, 18
    public health approach to, 457–476
    reservoirs for, 457–476
vtx1 gene, 25
vtx2 gene, 25
vtx2a gene, 24–25
vtx2c gene, 27

Washed blood agar medium, 239–240
Washing methods, for cattle hide, 441, 445–446.448
Water
    contamination of, 232, 464–465
    in environment, 213, 218
    as reservoir, 427–428
Water buffalo, STEC in, 212, 215
Water fowl, STEC in, 212, 218

Whole-genome mapping, 59
Whole-genome sequence typing, 60
Wildlife, STEC in, 429
Wiskott-Aldrich syndrome protein, 107–109, 115
wrbA gene, 25–26
WxxxE proteins, 115, 333
wyz gene, 276
wzx gene, 276

Yaks, STEC in, 212

Ycb proteins, 183
yciD gene, 25
Yeast-2-hybrid analysis, 101
yecE gene, 25
yehV gene, 25–25
yjbM gene, 25
ynfH gene, 25

Z4322 gene, 29, 117–118
Zinc, for STEC, 344, 346

About the Editors

Vanessa Sperandio is a Professor in the Departments of Microbiology and Biochemistry at UT Southwestern. Dr. Sperandio's research investigates chemical, stress, and nutritional signaling at the interface among the mammalian host, beneficial microbiota, and invading pathogens. Her laboratory research focuses on how bacterial cells sense several mammalian hormones leading to rewiring and reprogramming of bacterial transcription towards host and niche adaptation. Dr. Sperandio is chair-elect of Division D of the ASM and was elected a fellow of the American Academy of Microbiology in 2013. She is the recipient of the 2015 ASM Eli-Lilly and Company-Elanco research award and a winner of the GSK Discovery Fast Track challenge program 2014. She was a Pew Latin American Fellow in Biological Sciences, an Ellison Medical Foundation New Scholar in Global Infectious Diseases, a Burroughs Wellcome Fund Investigator in Pathogenesis of Infectious Diseases, and a Kavli Frontiers of Science Fellow from the National Academy of Science since 2007. She currently serves on the editorial boards of *mBio*, *Infection and Immunity*, *Journal of Bacteriology*, and *Gut Pathogens*.

Carolyn J. Hovde is a University Distinguished Professor who has served as the Idaho NIH INBRE Director since 2006. Dr. Hovde's laboratory studies *E. coli* O157:H7 with a primary focus on understanding the relationship between this human pathogen and its silent reservoir, healthy cattle. Dr. Hovde has authored more than 100 scientific publications and holds one patent. She has been the recipient of numerous honors and awards including election as a Fellow to the American Association for the Advancement of Science and winner of the ASM Carski Foundation Distinguished Undergraduate Teaching Award. She currently serves as the President of the National Association of IDeA Principal Investigators.